Highlights in Theoretical Chemistry

Vol. 2
Series Editors: Ch.J. Cramer • D.G. Truhlar

For further volumes:
http://www.springer.com/series/11166

Nino Russo • Nadia Rega • Carlo Adamo
Volume Editors

Vincenzo Barone

A Festschrift from Theoretical Chemistry Accounts

With contributions from

Carlo Adamo • Giuliano Alagona • Carlos Alemán
Massimiliano Aschi • Alessandro Bagno • Lotfi Belkhiri
Malgorzata Biczysko • Abdou Boucekkine • Philippe Carbonniere
Maurizio Casarin • Mauro Causà • Luigi Cavallo
Ugo Cosentino • Orlando Crescenzi • Francesco Ferrante
Luca De Gioia • Henry K. Hall Jr. • Denis Jacquemin
Frédéric Labat • Jean-Paul Malrieu • Andrea di Matteo
Maria Cristina Menziani • Giuseppe Milano • Christophe Morell
Andrea Peluso • Antonino Polimeno • Christian Silvio Pomelli
Claude Pouchan • Cristina Puzzarini • Gopalan Rajaraman
Maria J. Ramos • Nadia Rega • Nino Russo • Fabio Della Sala
H. Bernhard Schlegel • Bruna Clara De Simone • Mauro Stener
Jacopo Tomasi • Federico Totti

Springer

Volume Editors
Nino Russo
Dipartimento di Chimica
Università della Calabria
Arcavacata di Rende
Italy

Nadia Rega
Dipartimento di Chimica
Università degli Studi di Napoli Federico II
Naples
Italy

Carlo Adamo
Lab. d'Electrochimie
Ecole Nat. Sup. de Chimie de Paris
Paris
France

Originally Published in Theor Chem Acc, Volume 131 (2012)

ISSN 2194-8666 ISSN 2194-8674 (electronic)
ISBN 978-3-642-34461-9 ISBN 978-3-642-34462-6 (eBook)
DOI 10.1007/978-3-642-34462-6
Springer Heidelberg New York Dordrecht London

Library of Congress Control Number: 2013934835

Printed on acid-free paper

Springer is part of Springer Science+Business Media (www.springer.com)

Contents

Theor Chem Acc (2012) 131:1226
DOI 10.1007/s00214-012-1226-7

PREFACE

Preface to the special collection in honour of Vincenzo Barone

Nino Russo · Nadia Rega · Carlo Adamo

Published online: 5 June 2012

The 60th birthday of Vincenzo Barone

Vincenzo Barone is Full Professor in Theoretical and Computational Chemistry at the Scuola Normale Superiore of Pisa and President of the Italian Chemical Society (SCI).

Vincenzo Barone was born in Ancona (Italy) on November 1952. After his high school education, Enzo (as his friends call him) studied at the Department of Chemistry of the University of Naples, and in 1976, he received his degree (Laurea) in Chemistry discussing the thesis under the supervision of Professor Giuseppe Del Re, one of the pioneers of theoretical chemistry in Italy. He then moved as post-doctoral fellow to the University of Grenoble (France) where he worked under the supervision of

Professor Robert Subra. We can date back to this period his interest in the chemistry of organic radicals, a field in which over the years he has produced major and widely recognized contributions.

From 1978 to 1981, Enzo was Assistant Professor in General Chemistry at University of Calabria (Italy), where he contributed to the creation of the Laboratory of Theoretical and Computational Chemistry (currently led by Professor Nino Russo). During this time, he was visiting scientist in well-known international Laboratories in France (Ecole Normale Supérieure, Paris; Université Joseph Fourier, Grenoble), Canada (Université de Montreal) and Germany (University of Erlangen-Nürnberg).

After a period as Associated Professor at the University of Naples, he become Full Professor of Theoretical Chemistry in 1994. In Naples, Enzo created the Theoretical and Computational chemistry group in which many young researchers found a fertile substrate for learning and working in exciting research fields. In few years, the group reached a remarkable reputation in the international

Published as part of the special collection of articles celebrating the 60th birthday of Professor Vincenzo Barone.

N. Russo
Dipartimento di Chimica, Università della Calabria,
via Pietro Bucci, 87036 Arcavacata di Rende, Italy
e-mail: nrusso@unical.it

N. Rega
Dipartimento di chimica, Università Degli Studi di Napoli
Federico II, Complesso Universitario Monte S. Angelo
via Cintia, 80126 Naples, Italy
e-mail: nadia.rega@unina.it

C. Adamo (✉)
Laboratoire d'Electrochimie, Chimie des Interfaces
et Modélisation pour l'Energie, Ecole Nationale
Supérieure de Chimie de Paris, 11,
rue Pierre et Marie Curie, 75231 Paris, France
e-mail: carlo-adamo@chimie-paristech.fr

C. Adamo
Institute Universitaire de France,
103, Boulevard Saint Michel, 75005 Paris, France

scientific community. After this stimulating experience, he moved to Pisa in 2008, firstly as Director of the Institute for chemical and physical processes (IPCF, a Research Institute of the Italian National Research Council) and later as Full Professor in Theoretical and Computational Chemistry at Scuola Normale Superiore, one of the most prestigious universities in Italy.

Also in these two institutions, he attracted countless guest scientists, graduates, post-graduates and post-docs from all over the world. To all the co-workers, Enzo always ensured a congenial atmosphere, allowing the team to work with enthusiasm and efficacy. Many of his former students and post-doc fellows are now professors and/or leading scientists in universities and research centres in Italy and around the world.

In the last years, Enzo has promoted a number of scientific initiatives (e.g. the M^3-Village Networks, the Idea Virtual Laboratory and the CODECS COST Action) that have strongly contributed to the development of theoretical and computational chemistry in Italy and in Europe. Furthermore, he has played a seminal role in the creation of the Computational Chemistry Division of the Italian Chemical Society.

His scientific profile is really impressive: he is author of more than 500 papers, which have received over 25,000 citations, with an h-index of 64; in 2009, he was included among the ISI highly cited researchers for the Chemistry category.

Enzo was also invited to hundreds of lectures in national and foreign institutions or congresses, and he is member of the editorial board of prestigious journals (e.g. *Phys. Chem. Chem. Phys.*, *Theor. Chem. Acc.*, *J. Comput. Chem.*) of the scientific council of many scientific institutions, fellow of the International Academy of Quantum Molecular Science and of the Royal Society of Chemistry, and chairman of different national and international scientific events. Furthermore, he is a co-author of Gaussian, one of the most popular quantum chemistry codes.

Recently, he has received the prestigious Sacconi Medal of the Italian Chemical Society for his contribution to the development of theoretical and computational chemistry.

Of equal importance is the educational activity of Enzo: until now, he has been supervisor of about 20 master and more than 15 PhD students and has directed over 20 post-docs in modern theoretical chemistry research.

Enzo has developed his research activities in many fields of theoretical and computational chemistry, with the overall goal to establish a theoretical microscopic approach for the study of structural, dynamic, spectroscopic and reactive properties of materials, biomolecules and more in general soft matter in complex environments. His contributions to the accurate description of solvent effects and nuclear motions represent a logical element of this strategy.

Of particular interest is Enzo's conception of theoretical and computational chemistry as an "experimental" tool that can be used not only to achieve chemical accuracy in computed physico-chemical properties, but also to unravel and understand phenomena at the atomistic level. This has led Enzo to adopt in his research a strongly multidisciplinary approach, which is focused in the first place on theoretical developments and on multiscale descriptions, but also addresses the potentialities of computer architectures and of grid implementations.

More in details, the areas in which Enzo has provided remarkable contributions include the following:

- Chemical bond and molecular reactivity. In this context, we only mention the work on the absorption of atoms and molecules on solid surfaces, the chemical properties and reactivity of radical species, conformational analysis and photo-electron spectroscopy of biphenyl-like systems;
- Density functional theory (DFT). Already in 1992, Enzo was convinced of the potential of density functional theory to deal with chemical problems. His contributions include the development and validation of new exchange–correlation functionals (such as mPW1PW and PBE0), the validation of DFT for open-shell systems and, more recently, the implementation and parameterization of tight-binding DFT and of its time-dependent extension;
- Solvation theory with special emphasis on the polarizable continuum model (PCM). The approach has been developed and implemented for the evaluation of solvation free energies, including solvent effects on molecular structures in both ground and excited electronic states. In the last years, he has also been involved in the development of a mixed discrete-continuum model especially useful for computational spectroscopy applications. Enzo's work in this field has contributed to make of PCM the most widely used tool to study molecular systems in condensed phases at quantum mechanical level;
- Computational spectroscopy. Enzo's studies on harmonic and anharmonic treatments of molecular vibrations paved the route to unprecedented results in the field. In the last years, his scientific interests have focused on theoretical developments instrumental to the accurate simulation of spectral line-shapes for large molecular systems in condensed phases; as an example of the potential of the approach, one can cite the computation of entire EPR spectra of different molecular systems, in excellent agreement with the experimental counterparts.

This short extract of the Enzo's scientific contributions is inevitably reductive; the interested reader will easily find in the current literature his fundamental contributions to the advancement of Theoretical and Computational Chemistry. We are therefore confident that in the next decades, Enzo Barone will continue to develop and apply new theoretical approaches and will thus keep contributing to a better knowledge of Chemistry.

Finally, we would like to underline the genuine personal character of Enzo, his ability to be a friend for his students, co-workers and colleagues, his helpful attitude and his broad cultural interests. For a long time, Enzo was a teetotaller, but over the last 20 years, he has become able to appreciate a glass of good wine. For this reason, we will accompany our wishes for his 60th birthday by opening a bottle of the best Italian wine in order to toast with all the numerous friends he has in the world. *Forza Enzo, sono solo i primi sessant'anni!* (Go Enzo, these are just the first 60 years!)

We are grateful to the many authors who contributed to this special issue of TCA dedicated to the 60th birthday of Enzo Barone and to the numerous referees who have dedicated their valuable time to evaluate the manuscripts. We are also grateful to Christopher Cramer for his enthusiastic acceptation of our proposal for this special issue and to the TCA staff for its help all along the editorial work.

Theor Chem Acc (2012) 131:1126
DOI 10.1007/s00214-012-1126-x

REGULAR ARTICLE

Ab initio classical trajectory calculations of 1,3-cyclobutanedione radical cation dissociation

Jia Zhou · H. Bernhard Schlegel

Received: 1 March 2011 / Accepted: 11 August 2011 / Published online: 14 February 2012

Abstract The dissociation of 1,3-cyclobutanedione radical cation was studied by ab initio direct classical trajectory calculations at the BH&HLYP/6-31G(d) level of theory. A microcanonical ensemble using normal mode sampling was constructed by distributing 10 kcal/mol of excess energy above the transition state for the tautomerization of the keto-enol to the diketo form. A total of 210 trajectories were run starting from this transition state, yielding chemically activated 1, 3-cyclobutanedione radical cation. The majority of the trajectories resulted in $CH_2CO^{+\cdot}$ + CH_2CO, with the activated CC bond breaking nearly twice as often as the spectator CC bond. The non-statistical behavior is observed because the rate of energy redistribution within the molecule is comparable to or slower than the dissociation rates. In addition to the expected products, dissociation to $CH_2COCH_2^{+\cdot}$ + CO and formation of a proton-transferred product $HCCO^{\cdot}$ + CH_3CO^+ were also seen in some of the trajectories.

Keywords Cyclobutanedione · Ab initio MD · Branching ratio · Radical cation

1 Introduction

A chemically activated species can dissociate in a non-statistical manner if the rate for dissociation is faster than the rate for intramolecular energy redistribution. Acetone radical cation is an archetypal example of this process and has been studied experimentally and theoretically over the past 35 years [1–12]. Isomerization from the enol form to the keto form activates the newly formed methyl group that dissociates preferentially. Energy also flows to the other methyl group resulting in its dissociation at a slower rate and with a different energy distribution. The observed branching ratio is ca 1.5:1 in favor of the newly formed methyl group [1–9, 11, 12]. Similar to acetone radical cation, our ab initio MD study of pentane-2,4-dione radical cation [13] showed a competition between thermodynamics and non-statistical behavior in C–C bond dissociation. The activated terminal C–C bond had a lower dissociation probability than the weaker neighboring C–C bond. The two interior C–C bonds are equal in strength, but the one closer to the activated bond dissociates ca 20 times more frequently. Based on the above studies, it would be interesting to examine other chemically activated systems that would allow the flow of energy through several bonds of equal strength, thereby avoiding any thermodynamic bias. 1,3-Cyclobutanedione radical cation could be a good candidate. Upon isomerization to the diketo form, the energy from the activated C–C bond can flow sequentially to three other C–C bonds, potentially resulting in greater variety of non-statistical behavior than acetone radical cation. Experimentally, this reaction might be difficult to study because side reactions may compete with isomerization. However, in a theoretical study, the starting conditions can be chosen to avoid possible side reactions and focus on the competition between intramolecular energy flow and unimolecular dissociation (Scheme 1).

2 Method

Similar to our previous studies on acetone radical cation [11, 12] and pentanedione radical cation [13], we have used

Dedicated to Professor Vincenzo Barone and published as part of the special collection of articles celebrating his 60th birthday.

J. Zhou · H. B. Schlegel (✉)
Department of Chemistry, Wayne State University, Detroit, MI 48202, USA
e-mail: hbs@chem.wayne.edu

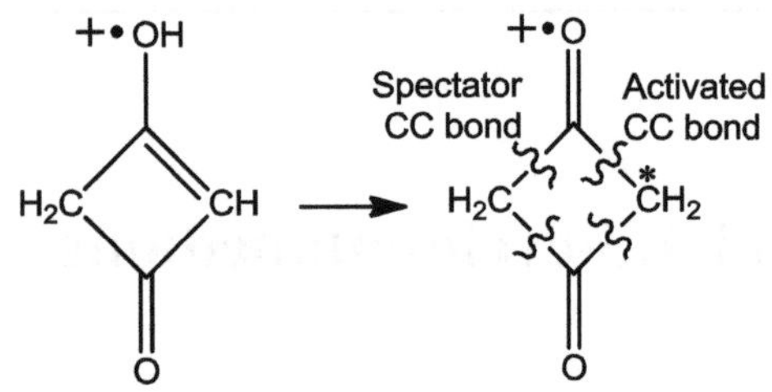

Scheme 1 Dissociation of 1,3-cyclobutanedione radical cation

ab initio classical trajectory calculations to study the non-statistical dissociation of 1,3-cyclobutanedione radical cation. The Gaussian suite of programs [14] was used for the ab initio electronic structure and molecular dynamics calculations. The geometries of the minima and transition states were optimized by Hartree–Fock (HF), hybrid density functional theory (B3LYP and BH&HLYP) [15–17] and second-order Møller–Plesset perturbation theory (MP2) [18]. Higher accuracy energy differences were calculated by the CBS-QB3 [19] and G4 [20] methods. The G4 calculations have a mean absolute deviation of 0.80 kcal/mol for 270 enthalpies of formation in the G3/05 test set and are used as a reference standard.

Ab initio classical trajectories were computed with BH&HLYP/6-31G(d) since this level of theory was in good agreement with the more accurate but more expensive methods. A Hessian-based predictor–corrector method [21, 22] was used to integrate the classical trajectories using a step size of 0.25 $amu^{1/2}$ bohr. The energy was conserved to better than 1×10^{-5} hartree, and the angular momentum was conserved to 1×10^{-8} ħ. Trajectories were initiated at the transition state for the keto-enol to diketo tautomerization. A microcannonical ensemble of initial states was constructed using normal mode sampling [23, 24]. As in our previous studies on acetone and pentanedione radical cations, a total energy of 10 kcal/mol above the zero point energy was added to the transition state. This initial energy was distributed among the 23 vibrational modes and translation along the transition vector toward the product. The total angular momentum was set to zero (corresponding to a rotationally cold distribution), and the phases of the vibrational modes were chosen randomly. The initial conditions are similar to those used previously for acetone radical cation [11, 12] and pentanedione radical cation [13]. A total of 210 trajectories were integrated starting from the transition state. Trajectories were terminated after 400 fs or when the centers of mass of the fragments were 8 bohr apart and the gradient between the fragments was less than 1×10^{-5} hartree/bohr.

3 Structures and energetics

The optimized geometries of the diketo and keto-enol forms of cyclobutanedione radical cation, various intermediates,

Fig. 1 Structures and selected geometric parameters of stationary ▶ points on the cyclobutanedione radical cation potential energy surface optimized at the BH&HLYP/6-31G(d) and G4 (B3LYP/6-31G(2df,p)) levels of theory (*top* and *bottom rows*, respectively). For structure **10**, QCISD/6-311G(d) value is listed. Bond distances are in angstroms. Relative energies in kcal/mol (given in parentheses) were calculated at the G4 level of theory

transition states and products are shown in Fig. 1 for a number of levels of theory. The relative energies of these structures at the G4 level of theory are summarized in Fig. 2 and in Table 1. In its diketo form, 1,3-cyclobutanedione radical cation, **1**, has D_{2h} symmetry with a planar heavy-atom skeleton. The keto-enol form, **3**, also has a planar heavy-atom skeleton and lies 4.1 kcal/mol higher in energy than **1**. The keto-enol to diketo tautomerization of **3** via **TS2** to **1** produces a chemically activated cyclobutanedione radical cation. Breaking one CC bond has a barrier of only 6.6 kcal/mol and produces $CH_2C(O)CH_2CO^{+\cdot}$, **5**. Because of the release of ring strain, **5** is 14.6 kcal/mol more stable than **1**. Two CC dissociation paths lead from **5** to products with similar energies. One path yields $CH_2C(O)CH_2^{+\cdot}$ + CO, **6**, and requires 33.4 kcal/mol. The other path produces $CH_2CO^{+\cdot}$ + CH_2CO, **9**, via transition state **TS7** and intermediate **8** and has a barrier of 30.6 kcal/mol. The central CC bond in complex **8** is an odd electron bond and is very long (ca 2.4 Å). Nevertheless, it has a dissociation energy of 20.7 kcal/mol. Similar to our previous study on pentanedione radical cation [13], a proton transfer can occur between the products. This process happens via complex **10**, transition state **TS11** and complex **12**. The product CH_3CO^+ + $HCCO^\cdot$, **13**, is only 2 kcal/mol higher than **9**.

The relative energies of selected stationary points on the potential energy surface have been computed at a number of levels of theory and are compared in Table 1. Energies are calculated relative to $CH_2CO^{+\cdot}$ + CH_2CO, **9**, and the G4 relative energies are taken as reference values in the comparisons. As expected, the CBS-QB3 relative energies agree well with the G4 values, with mean average deviation (MAD) of 1.0 kcal/mol. Since the DFT methods were not able to locate structure **10**, QCISD geometries were used instead. The Hartree–Fock calculations greatly underestimate the stability of **1**, **TS2** and **8** relative to **9** and fail to locate structure **12**. The MP2 performs similar to B3LYP at predicting the relative energies (MAD of 3.7 and 3.6 kcal/mol, respectively) except that MP2 fails to locate the **TS4** structure. The BH&HLYP functional yields a lower MAD (1.9 kcal/mol), **TS2** is 3.5 kcal/mol too high, and **3** is 6.4 kcal/mol too stable, but the differences for the remaining structures are all less than 3 kcal/mol. Since the BH&HLYP calculations are in significantly better agreement with the G4 values than the HF, MP2 and B3LYP calculations, they have been used for the molecular dynamics simulations.

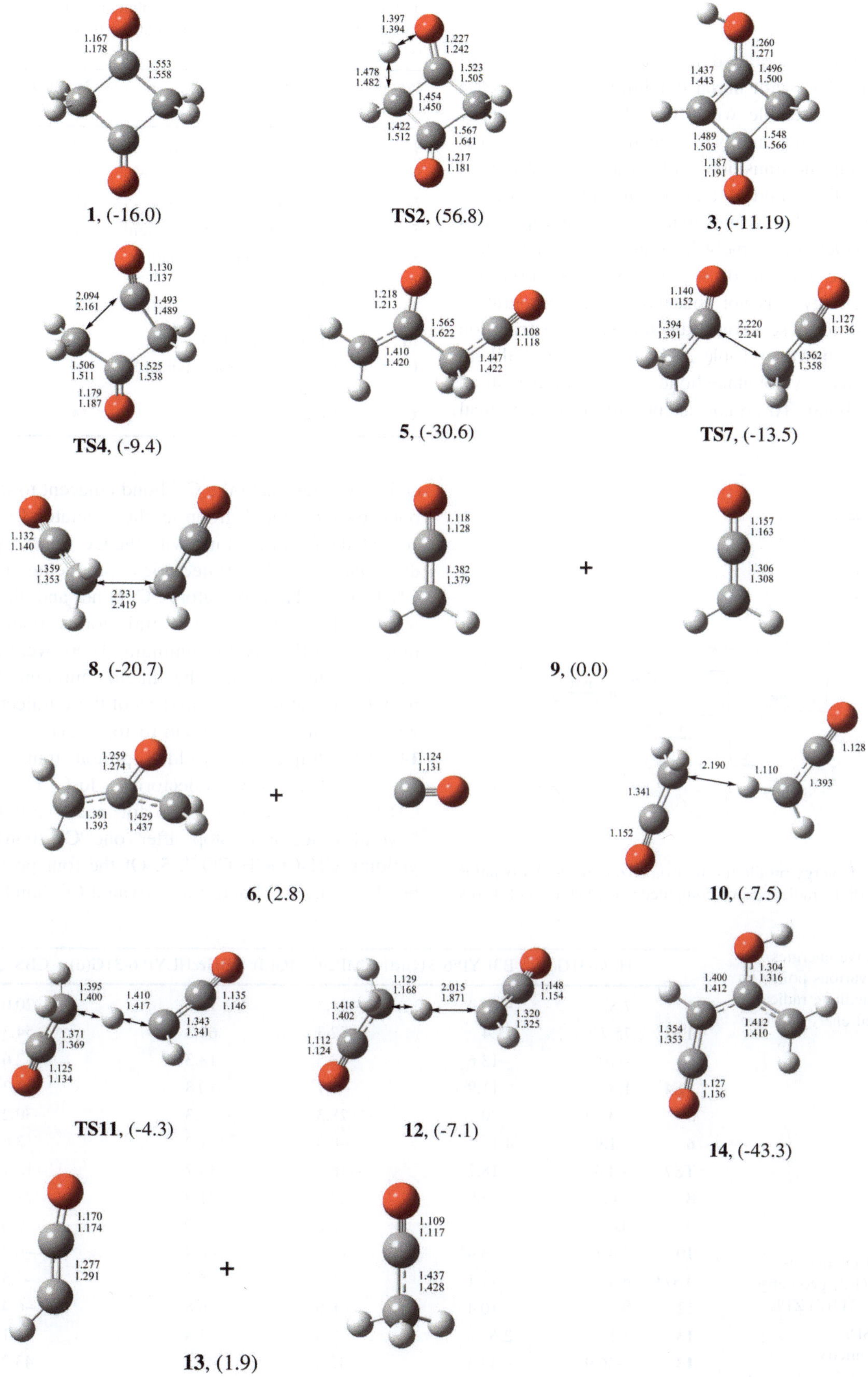
1, (-16.0)
TS2, (56.8)
3, (-11.19)
TS4, (-9.4)
5, (-30.6)
TS7, (-13.5)
8, (-20.7)
+
9, (0.0)
+
6, (2.8)
10, (-7.5)
TS11, (-4.3)
12, (-7.1)
14, (-43.3)
+
13, (1.9)

4 Dynamics

The trajectories were started at the keto-enol to diketo transition state **TS2** with initial conditions chosen from a microcanonical ensemble with 10 kcal/mol extra energy above the zero point energy of the transition state, and translation along the transition vector was directed toward the products. This modest amount of initial energy is somewhat arbitrary but assures that most of the trajectories reach the products in a timely fashion. Of the 210 trajectories that were integrated, seven had to be discarded because the energy was not conserved or the integration failed. Excluding these seven trajectories, the distribution of products is shown in Table 2. The keto-enol to diketo reaction produces cyclobutanedione radical cation with an activated CC bond. To be consistent with acetone radical cation, we designate the CC bond adjacent to the activated bond by a carbonyl group as the spectator CC bond. The activated CC bond, along with the CC bond opposite to it, dissociated in 93 trajectories producing $CH_2CO^{+\cdot}$ + CH_2CO, **9**. The spectator CC bond and the CC bond opposite it dissociated in 50 trajectories. Four trajectories remained in the diketo minimum. There were 20 trajectories that did not reach the diketo minimum but crossed back to the keto-enol isomer; 15 of these trajectories ended with breaking of a CC bond to form $CH_2C(OH)CHCO^{+\cdot}$, **14**. Two trajectories yielded proton transfer products, **13**. Another two trajectories lost CO producing $CH_2COCH_2^{+\cdot}$ + CO, **6** (one for each carbonyl group). Several trajectories stop after one CC bond breaking, yielding $CH_2COCH_2CO^{+\cdot}$, **5**. Of the four possibilities for breaking one CC bond, the activated CC bond broke most

Table 2 Branching ratios for cyclobutanedione radical cation dissociation from molecular dynamics at the BH&HLYP/6-31G(d) level of theory

Product structure	Description	Number of trajectories	Percentages
9	Active bond dissociation	93	45.8
9	Spectator bond dissociation	50	24.6
5	Active bond dissociation	17	8.4
5	Spectator bond dissociation	7	3.4
5	Opposite bond dissociation	8	4.0
6	CO elimination	2	1.0
13	Proton transfer	2	1.0
14	$CH_2C(OH)CHCO^{+\cdot}$	15	7.5
1	Cyclobutanedione radical cation	4	2.0
3	Enol isomer	5	2.5

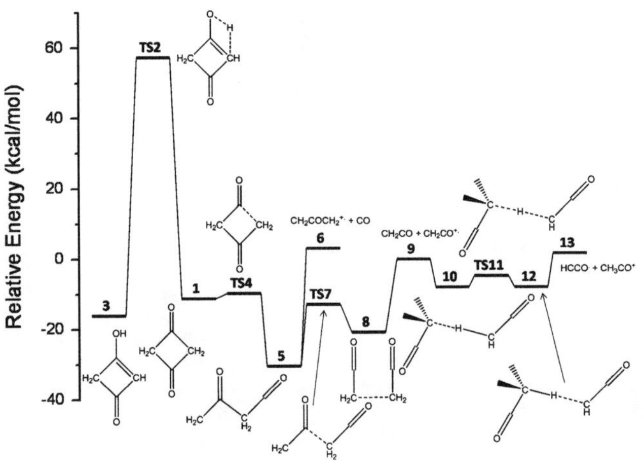

Fig. 2 Potential energy profile for the isomerization and dissociation of cyclobutanedione radical cation computed at the G4 level of theory

Table 1 Relative energies (kcal/mol) of various points on the cyclobutanedione radical cation potential energy surface

	HF/6-31G(d)	B3LYP/6-31G(d)	MP2/6-31G(d)	BH&HLYP/6-31G(d)	CBS-QB3	G4
1	7.8	−24.4	−21.3	−18.6	−20.0	−16.0
TS2	75.7	51.4	72.3	60.3	54.3	56.8
3	−10	−13.6	−3.5	−18.3	−12.6	−11.9
TS4	1.8	−11.9	−9.7[a]	−10.8	−8.3	−9.4
5	−31.9	−30	−28.3	−33.3	−30.2	−30.6
6	−1.9	4.1	−0.4	1.2	3.6	2.8
TS7	−4.4	−18.7	−18.2	−14.7	−12.1	−13.5
8	−6.0	−27.9	−25.9	−21.9	−21.5	−20.7
9	0.0	0.0	0.0	0.0	0.0	0.0
10	−4.1	−13.4[a]	−8.6	−8.4[a]	−6.5[b]	−7.5[b]
TS11	6.5	−11.1	−6.5	−5.2	−3.5	−4.3
12	[c]	−10.4	−8.6	−6.8	−6.4	−7.1
13	5.1	2.6	2.8	3.8	2.1	1.9
14	−36.9	−44.3	−42.2	−45.6	−43.2	−43.3
MAD	8.4	3.6	3.7	1.9	1.0	

[a] Single-point energy at QCISD/6-311G(d) geometry plus QCISD/6-311(d) ZPE

[b] Using QCISD/6-311G(d) geometry

[c] Stationary point could not be located

frequently (17 trajectories), followed by the spectator CC bond (seven trajectories) and the other two bonds (four trajectories each).

In acetone radical cation, the two CC bonds have equal bond energies and are expected to dissociate at equal rates. The observed branching ratio of ca 1.5–1 thus indicates significant non-statistical behavior. In pentanedione, the active versus spectator bond dissociation ratio is complicated by differences in the bond strengths. Nevertheless, comparison with RRKM (Rice–Ramsperger–Kassel–Marcus) rate calculations indicates significant non-statistical behavior. The situation is simpler for cyclobutanedione radical cation, since all of the CC bonds have the same dissociation energy. There are two ways to produce $CH_2CO^{+\cdot} + CH_2CO$, **9**, depending on whether the active CC bond or spectator CC bond breaks. Statistically, they would be expected to have the same dissociate rate, but the molecular dynamics calculations yield a ratio of $93{:}50 = 1.86$ for active versus spectator CC bond dissociation. This is a greater deviation from statistical behavior than for acetone radical cation with the same initial conditions. As in acetone radical cation, it is the result of competition between the rate of dissociation and the rate of intramolecular energy flow. In solution, collisions with the solvent may help redistribute the energy and thus reduce the non-statistical behavior. Thus, the non-statistical behavior should be greatest in the gas phase.

Acknowledgments This work was supported by a grant from the National Science Foundation (CHE0910858). Computer time made available by Computing Grid of Wayne State University is gratefully acknowledged.

References

1. Depke G, Lifshitz C, Schwarz H, Tzidony E (1981) Angew Chem Int Ed 20:792
2. McAdoo DJ, Hudson CE (1984) Int J Mass Spectrom Ion Processes 59:77
3. McAdoo DJ, McLafferty FW, Smith JS (1970) J Am Chem Soc 92:6343
4. Turecek F, Hanus V (1984) Organic Mass Spectrom 19:631
5. Heinrich N, Louage F, Lifshitz C, Schwarz H (1988) J Am Chem Soc 110:8183
6. Lifshitz C, Tzidony E (1981) Int J Mass Spectrom Ion Physics 39:181
7. Lifshitz C (1983) J Phys Chem 87:2304
8. Lifshitz C, Tzidony E, Terwilliger DT, Hudson CE (1980) Adv Mass Spectrom 8A:859
9. Osterheld TH, Brauman JI (1993) J Am Chem Soc 115:10311
10. Nummela JA, Carpenter BK (2002) J Am Chem Soc 124:8512
11. Anand S, Schlegel HB (2004) Phys Chem Chem Phys 6:5166
12. Zhou J, Schlegel HB (2008) J Phys Chem A 112:13121
13. Zhou J, Schlegel HB (2009) J Phys Chem A 113:1453
14. Frisch MJ, Trucks GW, Schlegel HB, Scuseria GE, Robb MA, Cheeseman JR, Montgomery JA Jr, Vreven T, Kudin KN, Burant JC, Millam JM, Iyengar SS, Tomasi J, Barone V, Mennucci B, Cossi M, Scalmani G, Rega N, Petersson GA, Nakatsuji H, Hada M, Ehara M, Toyota K, Fukuda R, Hasegawa J, Ishida M, Nakajima T, Honda Y, Kitao O, Nakai H, Klene M, Li X, Knox JE, Hratchian HP, Cross JB, Bakken V, Adamo C, Jaramillo J, Gomperts R, Stratmann RE, Yazyev O, Austin AJ, Cammi R, Pomelli C, Ochterski JW, Ayala PY, Morokuma K, Voth GA, Salvador P, Dannenberg JJ, Zakrzewski VG, Dapprich S, Daniels AD, Strain MC, Farkas O, Malick DK, Rabuck AD, Raghavachari K, Foresman JB, Ortiz JV, Cui Q, Baboul AG, Clifford S, Cioslowski J, Stefanov BB, Liu G, Liashenko A, Piskorz P, Komaromi I, Martin RL, Fox DJ, Keith T, Al-Laham MA, Peng CY, Nanayakkara A, Challacombe M, Gill PMW, Johnson B, Chen W, Wong MW, Gonzalez C, Pople JA (2007) Gaussian DV, revision F02. Gaussian Inc, Wallingford, CT, USA
15. Becke AD (1993) J Chem Phys 98:1372
16. Becke AD (1993) J Chem Phys 98:5648
17. Lee C, Yang W, Parr RD (1988) Phys Rev B 37:785
18. Moller C, Plesset MS (1934) Phys Rev 46:618
19. Montgomery JA, Ochterski JW, Peterson GA (1994) J Chem Phys 101:5900
20. Larry AC, Paul CR, Krishnan R (2007) J Chem Phys 126:084108
21. Bakken V, Millam JM, Schlegel HB (1999) J Chem Phys 111:8773
22. Millam JM, Bakken V, Chen W, Hase WL, Schlegel HB (1999) J Chem Phys 111:3800
23. Hase WL (1998) In: Schleyer PvR, Allinger NL, Clark T, Gasteiger J, Kollman PA, Schaefer III HF, Schreiner PR (eds) Encyclopedia of computational chemistry. Wiley, Chichester, p 402
24. Peslherbe GH, Wang H, Hase WL (1999) Adv Chem Phys 105:171

Theor Chem Acc (2012) 131:1130
DOI 10.1007/s00214-012-1130-1

REGULAR ARTICLE

New insights into oxidation properties and band structure of fluorescein dyes from ab initio calculations

Francesco Buonocore · Andrea di Matteo

Received: 16 June 2011 / Accepted: 6 August 2011 / Published online: 14 February 2012

Abstract During recent years, several publications have investigated the electrical bistability of spin cast films of halogenated fluorescein dyes. In the present contribution, we simulate the excited states of single fluorescein dyes with time-dependent density functional theory (TD-DFT) and we analyzed the band structure of the corresponding molecular crystals with DFT. More precisely, the molecules examined are fluorescein, erythrosine B, and rose bengal. We consider the molecular crystals of fluorescein in salt and lactone forms as well as erythrosine B. Rose bengal showed high quantum yield of the triplet state and high electronic affinity. Therefore, the rose bengal has very strong oxidation properties and it is able to form electrically bistable thin oxide layer. The poor crystal order and small bandwidths of fluorescein in salt form and erythrosine B indicated high resistivity for both crystals.

Keywords Fluoresceins · Electrical bistability · DFT · TD-DFT

Dedicated to Professor Vincenzo Barone and published as part of the special collection of articles celebrating his 60th birthday.

F. Buonocore · A. d. Matteo (✉)
STMicroelectronics Srl, via Remo de Feo 1,
Arzano, Naples, Italy
e-mail: andrea.di-matteo@st.com

F. Buonocore
e-mail: francesco.buonocore@st.com

F. Buonocore · A. d. Matteo
IMAST Scarl P.le E. Fermi, 1 Localita' Granatello,
80055 Portici, Naples, Italy

1 Introduction

In recent years, several publications [1–6] have been reported investigating the electrical bistability of spin cast films of halogenated fluorescein dyes (FDs). Indeed, the switch from a low conduction state to a high conduction state has been reported in the current–voltage electrical characteristic when a voltage threshold $+|V_{th}|$ is reached. The switch is reversible in the sense that the electrical characteristic commutates back to the original state of conduction for a voltage $-|V_{th}|$. In the experiments, the organic film is spin cast on ITO or ZnO substrate and aluminum top electrode is evaporated. In particular, an ON/OFF ratio as high as 350,000 has been reported for electrical current of the rose bengal (RB) films, attributed to the very low OFF current, also compared to the OFF current of the fluorescein films indeed with a quite low ON/OFF ratio. Those promising developments suggest it might be possible to demonstrate and to build up high-density memory devices based on electrically bistable fluorescein films.

Until present days, several mechanisms have been proposed to explain and understand the electrical bistability of FDs, and in our opinion, the debate is still open.

The proposed mechanisms can be resumed as follows: (a) the electrical external field (bias), induces a redox equilibrium between two states: the two redox states inside bandgap have different alignments with respect to the electrodes work function and the electrical response is due to different injection or extraction probability of hole/electron observed for the two states; (b) the electrical external field (bias), induces a charge transfer of electrons from one of the electrodes to the FD molecules increasing the conjugation, by increasing the coplanarity between xanthene and phenyl ring moieties (conformational change); (c) the organic material acts as an insulating layer in which a conductive path can be generated by one-dimensional "filament" connecting some conductive elements by quantum tunneling: the conductive elements can be either metal ions diffused from electrodes or oxygen vacancies; (d) the mechanism of switching is an interface effect between organic layer and top electrode (aluminum or other metal) where the resistive switching ratio depends on the nature of the oxide and layer thickness.

In the models (a) and (b), the resistive switching is an *intrinsic* effect due to the properties of the organic material, while for mechanisms (c) and (d), the resistive switching is an *extrinsic* effect due either to the electro-formed filament or to the oxide at the metal electrode-organic layer interface.

When the resistive switch was evidenced the first time [1], it was attributed to the change of conjugation (mechanism b). Moreover, Fazzi et al. [7] made accurate calculations and estimated the level alignment at the organic/inorganic interfaces with ITO and aluminum of the neutral and charged molecules (mechanism a). Recently, some papers [5, 6] suggested that the mechanism of switching is completely extrinsic. Jakobsson [5] evidenced the existence of switchable hot spots due to the existence of highly conducting filaments (mechanism c). Kärthauser [6] supported the hypothesis of a conduction switch generated by the thin aluminum oxide layer at the interface (mechanism d). Indeed, it has been observed that binary oxide layers can be electrically bistable [8–17]. Moreover, as it will be discussed later, the RB has an efficient electron transfer with oxygen to give the superoxide anion, O_2^{*-}. It is worthy underline that the four proposed mechanisms (a), (b), (c), and (d) originating the electrical bistability of FD film/electrodes system are not only alternatives to each other, but they could also coexist in the inorganic–organic structure.

The models (a) and (b) are very simple and they can be used to explain the electrical response of a single molecule addressed by two electrodes. However, we stress that the behavior of molecular systems is correlated with both the bulk and hetero junctions interface effects, which are completely neglected in the single molecule models. Therefore, the investigation of the molecular crystal details and the chemical processes at the organic/electrode interface are necessary.

In this paper, more insights into the role of the FD molecules in the generation of the oxide at the interface are given, which is probably responsible for the resistive switching in the extrinsic mechanism (d). Indeed, FDs possess excited triplet states of appropriate energy to be efficient sensitizers of singlet oxygen that could oxidize electrodes at a faster rate than the reaction involving triplet oxygen. FDs exhibited high adsorption coefficient in the green area of the visible spectrum (480–550 nm) and high quantum yield of the triplet state with long triplet states lifetimes. Therefore, xanthene dyes possess the required properties to be photosensitizer of singlet oxygen with high yields. The mechanism of photosensitization of singlet oxygen is described later. Increasing the atomic mass of halogen substituents increases also the yield of intersystem crossing to the triplet state of the dye; for this reason, RB is a more efficient photosensitizer. Moreover, we observe that there is a lack of papers appeared in literature concerning the electronic structure of FD crystals and the correlation with electrical conduction properties. Our investigations originated from the well-known aggregation phenomena in xanthene dyes. The charge transport properties in organic crystals depend on the packing of the molecules chains and on the order in the solid state as well as on the density of impurities and structural defects.

The purposes of the present paper are as follows: (1) to gain more insights into how the FD molecular layer could promote the formation of the metal oxide at the interface between molecular film and top electrode where the electrical switching is supposed to occur; (2) to analyze the different molecular packing and the band structures of the FD molecular crystals in order to gain information on the mobility in the crystal by comparing the bandwidths of the bands related to the charge transport.

The molecules under examination are fluorescein (FL) in quinone form, FL in lactone form, erythrosine B (ERY), and rose bengal (RB). The molecular structures are sketched in Fig. 1. The molecular crystals considered are FL in salt form, FL in lactone form and ERY.

We calculate the electronic properties of the single molecules within the time-dependent density functional theory (TD-DFT) approach in order to understand the chemical processes that could be implicated in the formation of a metal oxide layer at the top electrode. The TD-DFT has been demonstrated to be successful for modeling dye spectral properties and it can yield results with high accuracy at relatively low computational cost in

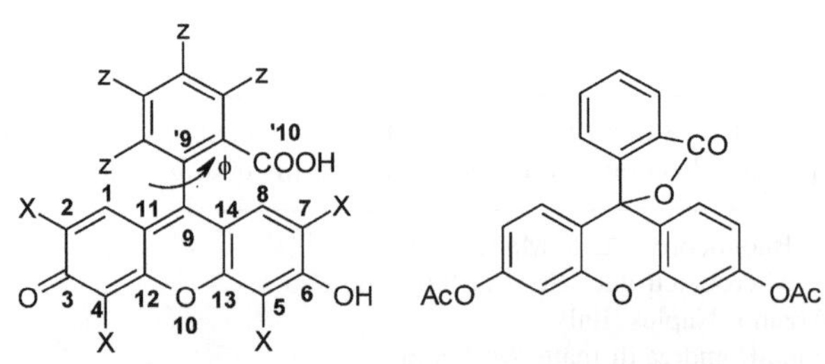

	X	Z
Rose Bengal	I	Cl
Erythrosine	I	H
Fluorescein	H	H

Fig. 1 Molecular structures of the investigated fluorescein dye molecules. The lactone form of the fluorescein is also shown

comparison with highly accurate methods such as multireference configuration interaction (MRC) or complete active space with second-order perturbation theory (CASPT2), which are applicable only to small molecules up to 20 atoms [18, 19]. Moreover, Barone et al. [20] have developed a theoretical multilevel approach for studying the photophysical properties of organic dyes in solution. In particular, they have used accurate force field parameters to simulate MD trajectories. Fluorescence has been calculated at the TD-DFT level on configurations sampled from MD trajectories in order to include time-dependent solvent effects in the computed emission spectrum; this multilevel approach yielded good agreement with respect to experiments. In our case, the electrical bistability has been observed in spin cast films of FDs in which could be relevant the solid state description of electronic properties of molecules.

For the modeling of the electronic properties of the organic crystals, we use the density functional theory (DFT). Although DFT approximation cannot account for van der Waals forces, which are responsible for the cohesion of the molecular solids, this theory provides reliable predictions in covalent and metallic systems. In particular, the use of experimental crystal geometry can guarantee an accurate electronic description. Therefore, we calculated the electronic structure of some fluorescein forms at the GGA level, importing the geometrical structures from crystallographic experiments [21–23] without performing DFT optimization of the geometries.

1.1 Computational details

The crystal DFT calculations in the following section were carried out with the CASTEP code [24] within the generalized gradient approximation (GGA) and with the expression for the exchange and correlation energy introduced by Perdew, Burke, and Ernzerhof (PBE) [25]. The ultrasoft pseudopotentials [26] for atoms and waves up to an energy cutoff of 340 eV were used. A $2 \times 2 \times 3$ Monkhorst and Pack mesh of k-points has been used to sample the Brillouin zone.

The single molecule calculations were performed using Gaussian03 program package [27]. The geometrical parameters of the molecules have been fully optimized at the PBE0 [28] and PBE levels using the LANL2DZ pseudopotentials [29] and corresponding basis set for Cl and I atoms, and 6–31G(d) basis set for all the other atoms. A larger basis set has been used for TDDFT computations, taking the 6–31 + G(d,p) basis set for the lighter atoms. The PBE0 has shown to overcome many of the problems encountered with standard functional, providing accurate excitation energies.

2 Results and discussion

2.1 Crystals

The symmetry system of the unit cell of the FL salt is monoclinic and the edges are a = 13.64 Å, b = 12.60 Å, c = 13.79 Å. The crystallographic angles are $\alpha = 90.000$, $\beta = 97.875$, and $\gamma = 90.000$. There are four molecules per unit cell. The symmetry of the FL in lactone form is triclinic, and the edges of the unit cell are a = 11.04 Å, b = 11.73 Å, and c = 7.37 Å. The crystallographic angles are $\alpha = 95.759$, $\beta = 97.429$, and $\gamma = 94.830$. There are two molecules per unit cell. For simplicity, Na ions were removed from the crystal structure of the salt and the molecules were hydrogenated to restore the neutrality of the cell, thus ionic bands could be ignored. Images of the unit cells of the FL crystals are reported in Fig. 2.

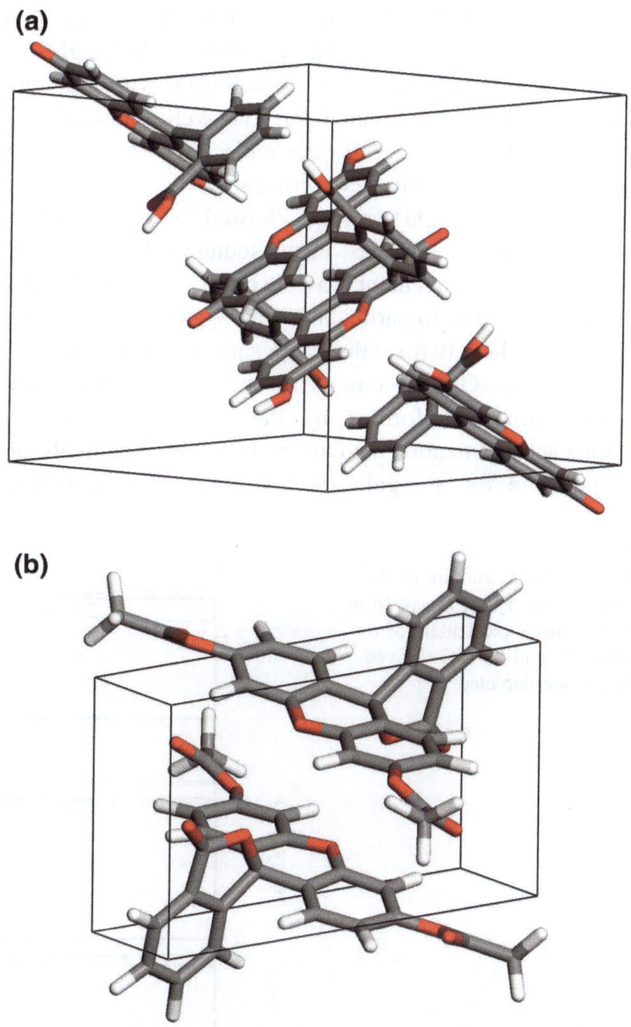

Fig. 2 Unit cell of (**a**) the fluorescein in salt form (from reference 22) and **b** the fluorescein in lactone form (from reference 21) molecular crystals. In the salt, sodium ions have been removed and the molecules hydrogenated (see text for more details)

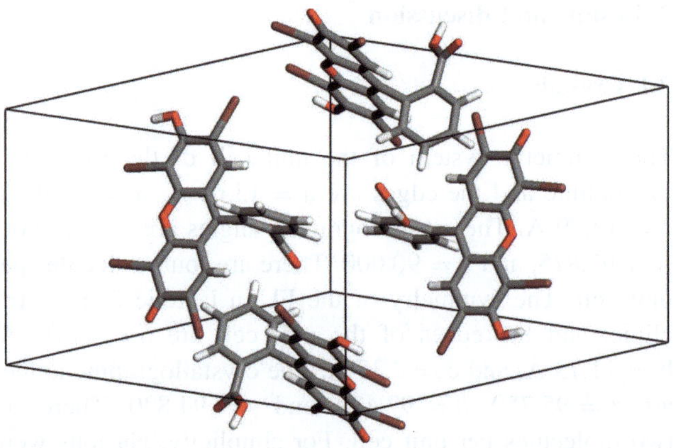

Fig. 3 Unit cell of the erythrosine B molecular crystal (from reference 23). Ethanol and ions have been removed and the molecules hydrogenated (see text for more details)

The symmetry system of the unit cell of the ERY B is monoclinic and the edges are a = 16.13 Å, b = 12.94 Å, and c = 16.37 Å (Fig. 3). The crystallographic angles are $\alpha = 90.000$, $\beta = 134.737$, and $\gamma = 90.000$ with four molecules per unit cell. Ethanol and ions were removed from the crystal structure, while the molecules were hydrogenated to guarantee the electroneutrality.

The three single crystals showed a very different arrangement of molecules; in the sodium salt form of FL, an irregular arrangement was observed along (010) crystallographic axis. In particular, a pair aggregation form was found, in which two xanthene rings are alternated with two phenyl rings. On the contrary, in the lactone form, aggregation structures formed by the xanthene rings in a coplanar organization and in co-facial arrangements are responsible for a regular π–π stacking. This particular structure organization is expected to have a strong impact on the transport properties and to bring significant difference with respect to sodium salt form. Finally, in the ERY B, an irregular arrangement along (010) direction was observed as in FL salt. However, in this case, xanthene and phenyl rings were alternated with partial overlap of π orbitals.

In Figs. 4 and 5, the band structures of the FL crystals in salt and lactone form are plotted and only the energy window between −1.0 and 3.5 eV near the top of the valence band, where the zero of the energy is set, is shown. The band structures of the crystals exhibited small dispersions, which reflected the weak chemical intermolecular interactions. The bandwidth of the HOMO (LUMO)-derived bands were 0.01 (0.04) eV for the salt and 0.03 (0.12) eV for the lactone form. The band gap was $E_g = 1.8$ eV in salt form and $E_g = 3.1$ eV in lactone form. It is well known that GGA functionals underestimate bandgap; nonetheless, we can compare calculated energies for different crystals. In the salt, there are four molecules, and thus each of the HOMO and LUMO bands are formed by 4 bands, with twofold degeneration removed only along the Z–G, G–Y, and B–D paths of the Brillouin zone as plotted in the inset of Fig. 4, where the details of the HOMO and LUMO bands are shown. In the lactone form with two molecules per unit cell, two HOMO- and two LUMO-derived bands are present that are non-degenerate due to the symmetry of the unit cell.

It is interesting to investigate the crystal arrangement and band structure of the ERY crystal (Fig. 6), which is characterized by a molecular symmetry different from that of FL salt. There are four molecules in the unit cell and four HOMO (LUMO)-derived bands with bandwidth

Fig. 4 Band structure of the fluorescein crystal in salt form. In the insets particulars of the HOMO- and LUMO-derived bands are depicted

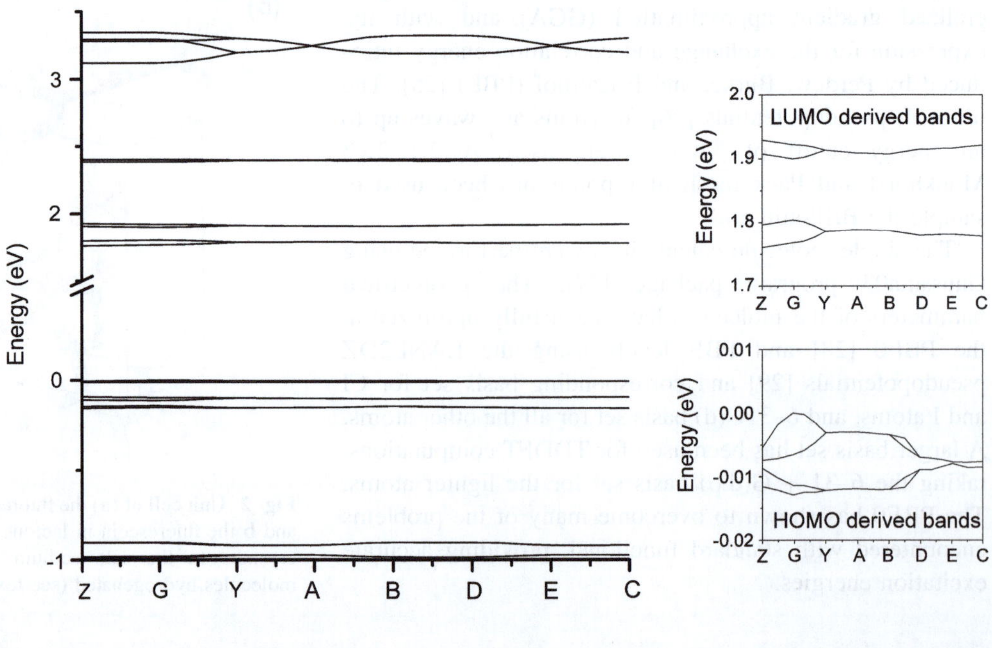

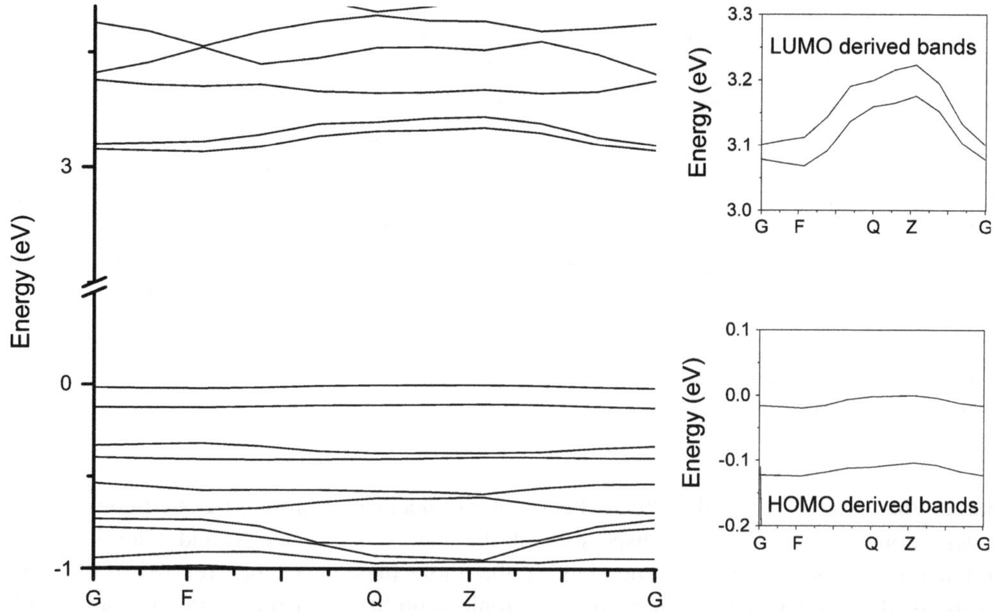

Fig. 5 Band structure of the fluorescein crystal in lactone form. In the insets particulars of the HOMO- and LUMO-derived bands are depicted

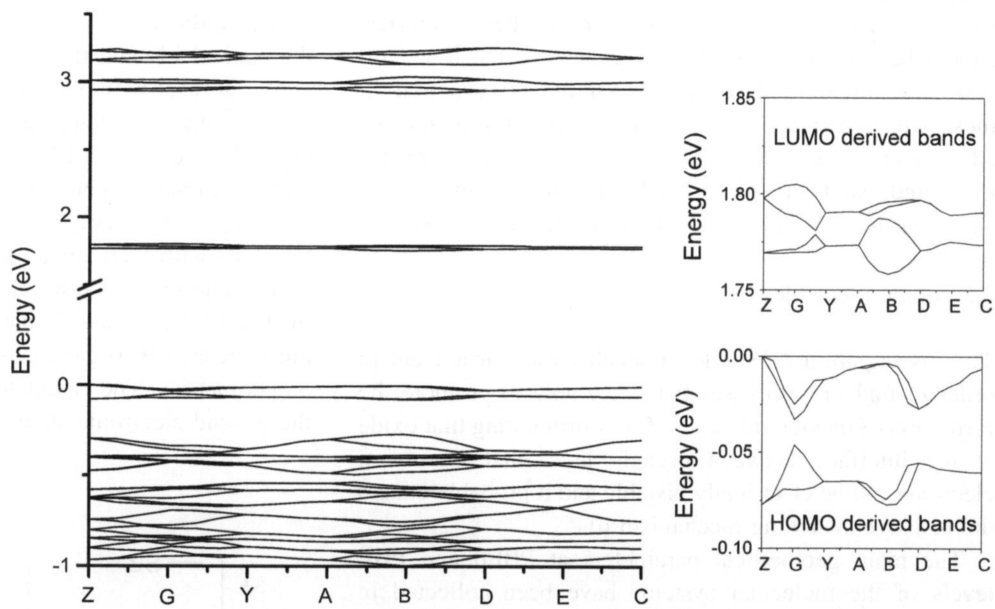

Fig. 6 Band structure of the erythrosine B crystal. In the insets particulars of the HOMO- and LUMO-derived bands are depicted

0.02 (0.03) eV, while the bandgap is $E_g = 1.8$ eV. Therefore, near the energy gap, the band structure is similar to that of the FL salt, but in this case, we have twofold degeneration along Y–A, D–E, and E–C Brillouin zone paths.

We have also calculated the optical absorptions of crystals and single molecules in periodic cells based on the dielectric function expressed in terms of the matrix elements for electronic interband transitions [30]. The DFT single molecule results were obtained by placing a molecule in a large periodic cell in such a way that intermolecular interactions were negligible. We have introduced a 0.5 eV scissor operator [31] to reproduce the absorption spectra of FL and ERY [32]. The results are reported in Table 2.

The simple morphological investigation of the crystal structures can give some insights into the transport properties of the crystals. Indeed, the FL in lactone form stacks in the crystal keeping the molecular planes parallel and allowing a coherent overlap of the π-orbitals. This is not the case for salt and ERY B, where molecular planes change direction in the same unit cell. Therefore, the lactone form should demonstrate improved mobility because of the better molecular order. Also, the band structures gave information on the transport properties of the crystals. It was shown that the bandwidth of the lactone form is

Table 1 Main geometrical parameter (Å and degrees)

	Acid fluorescein			Lactone form			Erythrosine			Rose bengal	
	Exp[a]	Vacuum		Exp[b]	Vacuum		Exp[c]	Vacuum		Vacuum	
		PBE0	PBE		PBE0	PBE		PBE0	PBE	PBE0	PBE
C9–C′9	1.499	1.496	1.495	1.501	1.522	1.520	1.469	1.501	1.494	1.501	1.494
C9–O′10				1.511	1.477	1.498					
$C_{13}O_{10}C_{12}$	120.6	121.4	120.8	120.9	119.2	118.5	121.8	121.5	121.0	121.8	121.2
$C_{14}C_9C_{11}$	119.2	119.6	119.4	117.5	111.3	111.4	119.5	119.7	119.4	120.3	119.9
ϕ	79.21	93.68	105.4				80.95	93.3	96.4	89.88	99.9

The values have been computed at the PBE0 and PBE levels

[a] Ref [22], [b]ref [21], [c]ref [23]

larger than that of the salt, due to better intermolecular interactions. In fact, a large dispersion of the bands is related to a good electronic coupling of the molecules forming the crystal that enhances the electronic mobility within the structure. Very flat bands are a sign of poor interactions, and this was the feature found for FL salt and ERY B, so that the mobility offered by these materials should be low. This can be at the base of the low OFF current reported in the molecular films, in particular in those with halogens. The lack of the crystal structure for RB hindered the investigation of the molecular arrangement and band structure of RB crystal. However, it is expected that this crystal is similar to the ERY B crystal.

2.2 Single molecules

The investigation on single molecules was carried out in order to find the processes that were able to promote the formation of metal oxide layer. It is worth noting that oxide at the interface between organic layer and aluminum electrode can be electrically bistable and it probably causes the resistive switching mechanism d).

The main geometrical parameters at PBE0 and PBE levels of the molecular systems have been collected in Table 1 together with the available X-ray data [21–23]. The atom labeling of FL is reported in Fig. 1. The same labeling was applied to all the molecules of the family. There is an overall satisfactory agreement between the theoretical and the available experimental data. As a general remark, we can state that the systems have similar geometrical rearrangements. It is noteworthy to underline that the torsion angles at PBE0 level are closer to experimental data than corresponding values at PBE level. This is induced by the particular tendency of the PBE approach, compared to that of PBE0, to stabilize conjugated systems.

FDs are photosensitizers capable of adsorbing light in the visible spectrum and using that energy to excite oxygen to its singlet state [33]. Excitation is generally obtained through a one-photon transition between the singlet ground state S_0 and a singlet excited state S_n of the FD. Then, the S_n state relaxes to the lowest excited singlet state S_1. The triplet state T_1 of the FD is generated by intersystem crossing. The lifetime of the T_1 state is longer than that of the S_1 state, so that it can react in one of the following way: (I) free radicals are produced by electron transfer between the excited FD and the molecular substrate. The free radicals can react with oxygen to form active oxygen species, such as singlet oxygen and the superoxide radical anion O_2^{*-}; (II) the excited FD collides with triplet oxygen generating singlet oxygen via an energy transfer process. Both processes can promote the formation of oxide at the hetero-interface with metal electrodes.

The energy gap between the singlet ground state S_0 and the lowest triplet state T_1 has been analyzed by carrying out unrestricted PBE0 calculations both at the corresponding S_0 geometries. The calculated energy gaps with respect to the ground electronic state are reported in Fig. 7.

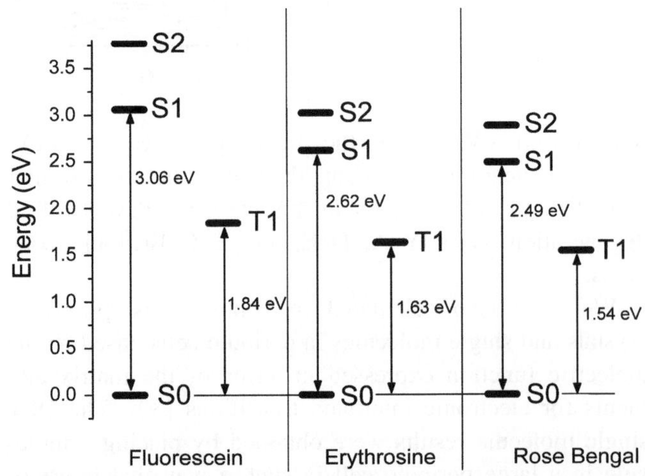

Fig. 7 Excited energy level for singlet and triplet state of fluorescein, erythrosine, and rose bengal

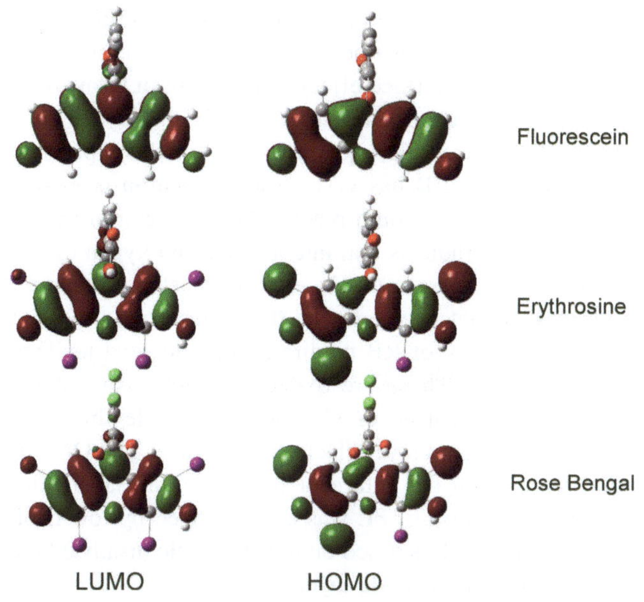

Fig. 8 LUMO and HOMO molecular orbitals of the investigated molecules

In all of the considered systems, the triplet state corresponds to an excitation from the HOMO to the LUMO. In going from the FL to the RB, the S_0/T_1 vertical gap decreases from 1.84 to 1.54 eV. This behavior can be rationalized in terms of the HOMO shape depicted in Fig. 8; in the ERY and RB with heavy halogen mass; this orbital has a significant antibonding C–I. The antibonding character decreases upon excitation of one electron, as it can be deduced from LUMO shapes and then the stability of the molecule increases, decreasing T_1 energy. It can be said that halogens increase the polarizability of the molecule, playing a key role in the systematic decrease in energy gap going from FL to RB, by increasing the atomic weight of the substituent halogens.

The UV spectra have been evaluated using TD-DFT approach, and the results are collected in Table 2.

For the absorption peak calculated with the plane wave basis set, the scissor operator was applied to correct the error of the band gap intrinsic in the GGA-PBE calculation within DFT. In the PBE0 level of calculation within TD-DFT, although no scissor operator was introduced, a quite strong discrepancy between experimental and theoretical values was found in the case of FL. Nevertheless, it is also clear that, as discussed by Neckers [32], the formation of aggregates modified the absorption spectrum, decreasing the quantum yield of singlet oxygen. The lactone form showed a band at high energy due to broken of xantene ring conjugation. The inter-ring cyclization of the lactone changed the hybridization on C9 from sp^2 to sp^3 and shifted the absorption band to low wavelength.

The calculated energy difference between triplet T_1 and the lowest excited singlet state S_1 is reported in Table 3. It is well known that intersystem crossing from S_1/T_1 can be evaluated by Fermi's Golden Rule taking into account the vibronic coupling between the two states. Indeed, the energy gap S_1/T_1 can be considered like a rough evaluation of the quantum yield of the triplet state. Low S_1/T_1 gaps correspond to high triplet yield. Preliminary calculations on the FL in lactone form have shown that the triplet state had a very high energy; therefore, the triplet yield was not enough appreciable to include this molecule in the present discussion.

Among the molecules considered, the RB is the molecule with the strongest oxidation properties in its triplet state. In fact, as evidenced before, the triplet quantum yield was very high and the corresponding singlet oxygen yield was $\phi_\Delta = 0.76$, with intense absorption bands in the green area of the visible spectrum [34]. Therefore, the reduction in the molecule and the consequent formation of oxide layer are highly probable.

The electron affinities (EA) were evaluated as the energy released when an electron was added to a neutral molecule and computed as the energy difference between the neutral form and the anion: $EA = E(M) - E(M^-)$.

The ionization potential (IP) was evaluated as the amount of energy required to remove an electron from a molecule and computed as the energy difference between the cation and the neutral molecule: $IP = E(M^+) - E(M)$. The EA and IP shown in the Fig. 9 have been evaluated taking into account the zero-point energy corrections.

Table 2 Computed TDDFT vertical excitation energies in nm units at PBE0 level for the lowest singlet state

	Acid fluorescein			Lactone form		Erythrosine			Bengal rose	
	TDDFT	DFT	Exp	TDDFT	DFT	TDDFT	DFT	Exp	TDDFT	Exp
S0-S1	404	478 (507)	490	231	253 (260)	472	531 (554)	526	497	549
S0-S2	389					410			430	
S0-S3	328					334			345	

The bands are evaluated using 6–31 + G** basis set at the LANL2DZ geometries. DFT optical absorption peaks in nm units for electronic interband transitions are also reported for crystals; in parenthesis are shown the values for isolated molecules. Experimental data are from Ref. [32]

Table 3 Computed energy gap (eV) between S1 and T1 states

	Energy gap (eV) S1-T1
Fluorescein	1.22
Erythrosine	0.99
Rose bengal	0.95

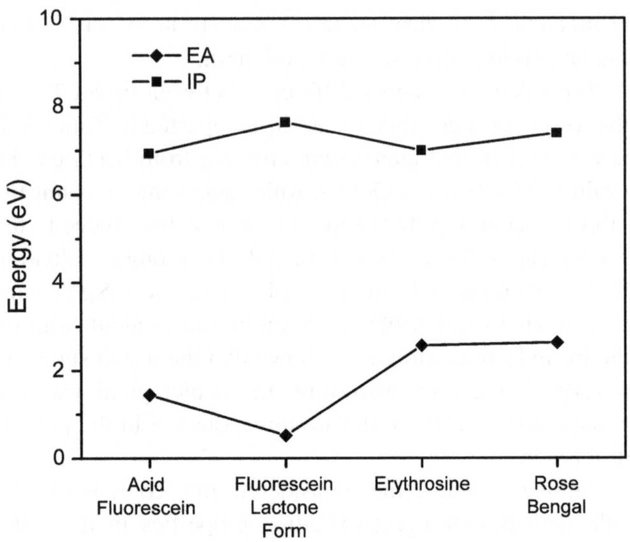

Fig. 9 Electronic affinities and ionization potentials of the molecules

Although the oxidation and reduction potentials were evaluated in solution, a systematic investigation of IP and EA provided an idea about the redox behavior of our systems. A systematic regular behavior can be observed for all the systems with the exception of FL in lactone form. In particular, an increase of the polarizability and of the number of halogen atoms enhances the EA and IP. The high electronic affinity observed for RB supported the hypothesis that the molecule show high tendency to reduce itself oxidizing the electrode.

3 Conclusions

FL in lactone form crystal stacks regularly keeping parallel planes and allowing a coherent overlap of the π-orbitals. Instead, the crystals of the FL salt and ERY B revealed a structure with poor order and weak intermolecular interactions. The irregular spatial arrangement was reflected in the band structures with small bandwidths. Very flat bands indicated poor intermolecular interactions and low charge mobility as well as high resistivity. However, it was found that ERY B and FL salt have similar bandwidths, so that FDs including xanthene rings with halogen substituent atoms have resistivity comparable to that of FL salt without halogen atoms. Therefore, the improved ON/OFF ratio of FDs with halogen substituent atoms would be explained in terms of hetero-interface effects rather than in terms of the internal arrangement of the molecules in the crystal.

The RB showed high quantum yield of the triplet state and, as a result, RB has very strong oxidation properties. Indeed, triplet state could react either by electron transfer producing free radicals that interacted with oxygen to form active oxygen species or via energy transfer process that generates singlet oxygen. On the other hand, the high electronic affinity of RB confirmed the elevated tendency to reduce itself. Those are oxidation mechanisms able to support the formation of electrically bistable thin oxide layer at the organic layer/electrode interface. The ON/OFF ratio depends on the nature of the oxide and layer thickness. Lactone form of FL was characterized by low triplet yield that made films based on this molecule disadvantaged in the formation of oxide layers.

It was found that RB is the molecule more prone to promote the formation of an electrically bistable aluminum oxide layer at the electrode. Further calculations are required to include vibronic contributions, which are essentials to achieve a full picture of the photosensitization mechanisms.

Acknowledgments This study was supported by TRIPODE project DM 20160 of the MIUR, Italy.

References

1. Bandyopadhyay A, Pal AJ (2003) J Phys Chem B 107:2531
2. Bandyopadhyay A, Pal AJ (2004) Appl Phys Lett 84:999
3. Bandyopadhyay A, Pal AJ (2003) Adv Mater 15:1949
4. Bandyopadhyay A, Pal AJ (2005) J Phys Chem B 109:6084
5. Jakobsson FLE, Crispin X, Cölle M, Büchel M, de Leeuw DM, Berggren M (2007) Org Electron 8:559
6. Kärthauser S, Lüssem B, Weides M (2006) J Appl Phys 100: 94504
7. Fazzi D, Castiglioni C, Negri F (2010) Phys Chem Chem Phys 12:1600
8. Schmidt PE, Davila C, Esqueda P, Callarotti R (1975) J Solid State Chem 12:293
9. Ansari AA, Qadeer A (1985) J Phys D 18:911
10. Shirodkar VS, Chourey AG (1988) J Mater Sci Lett 7:1235
11. Gloos K, Koppinen PJ, Pekola JP (2003) J Phys Condens Matter 15:1733
12. Seo S, Lee MJ, Seo DH, Jeoung EJ, Suh DS, Joung YS, Yoo IK, Hwang IR, Kim SH, Byun IS, Kim JS, Choi JS, Park BH (2004) Appl Phys Lett 85:5655
13. Cölle M, Büchel M, de Leeuw DM (2006) Org Electron 7:305
14. Yang JJ, Pickett MD, Li X, Ohlberg DAA, Stewart DR, Williams RS (2008) Nat Nanotechnol 3:429
15. Yoshida TK, Noshiro H, Sugiyama Y (2007) Appl Phys Lett 91:223510
16. Yang MK, Park JW, Ko TK, Lee JK (2009) Appl Phys Lett 95: 042105
17. Sun B, Liu L, Xu N, Gao B, Wang Y, Han D, Liu X, Han R, Kang J (2009) Jpn J Appl Phys 48:04C061

18. Jacquemine D, Perpete EA, Ciofini I, Adamo C (2009) Acc Chem Res 42:326
19. Drew A, Head-Gordon M (2005) Chem Rev 105:4009
20. Barone V, Bloino J, Monti S, Pedone A, Prampolini G (2011) Phys Chem Chem Phys 13:2160
21. Knudsen KD, Pattison P, Fitch AN, Cernik RJ (1998) Angew Chem Int Ed 37:2340
22. Yamaguchi K, Tamura Z, Maeda M (1997) Acta Cryst C 53:284
23. Cody V (1987) Acta Cryst C4(3):705
24. Clark SJ, Segall MD, Pickard CJ, Hasnip PJ, Probert MJ, Refson K, Payne MC (2005) Z Kristallogr 220:567
25. Perdew JP, Burke K, Ernzerhof M (1996) Phys Rev Lett 77:3865
26. Vanderbilt D (1990) Phys Rev B 41:7892
27. Gaussian 03 Revision C.02 (2004) Frisch MJ, Trucks GW, Schlegel HB, Scuseria GE, Robb MA, Cheeseman JR, Montgomery Jr. JA, Vreven T, Kudin KN, Burant JC, Millam JM, Iyengar SS, Tomasi J, Barone V, Mennucci B, Cossi M, Scalmani G, Rega N, Petersson GA, Nakatsuji H, Hada M, Ehara M, Toyota K, Fukuda R, Hasegawa J, Ishida M, Nakajima T, Honda Y, Kitao O, Nakai H, Klene M, Li X, Knox JE, Hratchian HP, Cross JB, Bakken V, Adamo C, Jaramillo J, Gomperts R, Stratmann RE, Yazyev O, Austin AJ, Cammi R, Pomelli C, Ochterski JW, Ayala PY, Morokuma K, Voth GA, Salvador P, Dannenberg JJ, Zakrzewski VG, Dapprich S, Daniels AD, Strain MC, Farkas O, Malick DK, Rabuck AD, Raghavachari K, Foresman JB, Ortiz JV, Cui Q, Baboul AG, Clifford S, Cioslowski J, Stefanov BB, Liu G, Liashenko A, Piskorz P, Komaromi I, Martin RL, Fox DJ, Keith T, Al-Laham MA, Peng CY, Nanayakkara A, Challacombe M, Gill PMW, Johnson B, Chen W, Wong MW, Gonzalez C, Pople JA, Gaussian Inc, Wallingford CT
28. Adamo C, Barone V (1999) J Chem Phys 110:6158
29. Hay PJ, Wadt WR (1985) J Chem Phys 82:284
30. Read AJ, Needs RJ (1991) Phys Rev B 44:13071
31. Godby RW (1992) Top Appl Phys 69:88
32. Neckers DC (1989) J Photochem Photobiol A Chem 47:1
33. DeRosa MC, Crutchley RJ (2002) Coord Chem Rev 233–234:351
34. Redmond RW, Gamlin JN (1999) Photochem Photobiol 70:391

Theor Chem Acc (2012) 131:1133
DOI 10.1007/s00214-012-1133-y

REGULAR ARTICLE

Thermodynamic and stereochemical aspects of the polymerizability of glycolide and lactide

Carlos Alemán · Oscar Bertran · K. N. Houk · Anne Buyle Padías · Henry K. Hall Jr.

Received: 15 June 2011 / Accepted: 25 August 2011 / Published online: 16 February 2012

Abstract The ring-opening polymerizations of the dilactones glycolide and the *S,S*- and *S,R*-stereoisomers of lactide were studied using quantum mechanical methods. The ring strain and the conformational distribution of these cyclic monomers and of the polymers were calculated, and the effect of the medium on the polymerization was predicted, for both bulk and solution. The polymerizability of the three monomers in the gas phase, that is, nonpolar medium, is much greater than that of δ-valerolactone or 1,4-dioxan-2-one. This difference vanishes in the polar medium chloroform, which is attributed to the fact that, while all of these monomers possess polar cis-lactone bonds, the three dilactones possess small dipole moments. The data are combined to give polymerization enthalpy and free energy values. The four stereoregular lactide polymers did not differ significantly in energy. Accordingly, the ability to synthesize any one of these rests on catalyst specificity ("polymer chain-end control"). Although introduction of sterically demanding methyl groups into glycolide is expected to favor coiled conformations and decrease polymerizability, this was not found to be the case. Good agreement of calculated values with experimental data from the literature was achieved.

Keywords Lactides · Polyesters · Quantum mechanics · Ring-opening polymerization · Stereochemistry · Steric effects

Dedicated to Professor Vincenzo Barone and published as part of the special collection of articles celebrating his 60th birthday.

C. Alemán
Departament d'Enginyeria Química, E.T.S. d'Enginyers Industrials de Barcelona, Universitat Politècnica de Catalunya, Diagonal 647, 08028 Barcelona, Spain

C. Alemán (✉)
Center for Research in Nano-Engineering, Universitat Politècnica de Catalunya, Campus Sud, Edifici C', C/Pasqual i Vila s/n, 08028 Barcelona, Spain
e-mail: carlos.aleman@upc.edu

O. Bertran
Departament d'Enginyeria Química, EUETII, Universitat Politècnica de Catalunya, Pça Rei 15, 08700 Igualada, Spain

K. N. Houk
Department of Chemistry and Biochemistry, University of California, Los Angeles, CA 90095-1569, USA

A. B. Padías · H. K. Hall Jr. (✉)
Department of Chemistry and Biochemistry, The University of Arizona, Tucson, AZ 85721, USA
e-mail: hkh@u.arizona.edu

1 Introduction

Polymers are essential materials for modern civilization. Although at present most of the materials in use originate from petroleum, the search for polymers from renewable resources is assuming great urgency. Among these, polylactic acid (PLA) is a shining example of what can be accomplished in this direction [1]. Lactic acid is a renewable resource derived from the starch of either corn or sugar beets, which is fermented to form glucose and consequently converted to lactic acid. Although the polymer can be synthesized by polycondensation from lactic acid, the preferred route is by ring-opening polymerization of its cyclic dimer, lactide, 3,6-dimethyl-1,4-dioxane-2,5-dione. A flowering of research activity has accomplished the lactide polymerization to a variety of stereoisomeric polymers displaying properties ranging from amorphous glasses to highly crystalline materials. These polymers

Scheme 1 Polymerizations and model reactions for glycolide and lactide

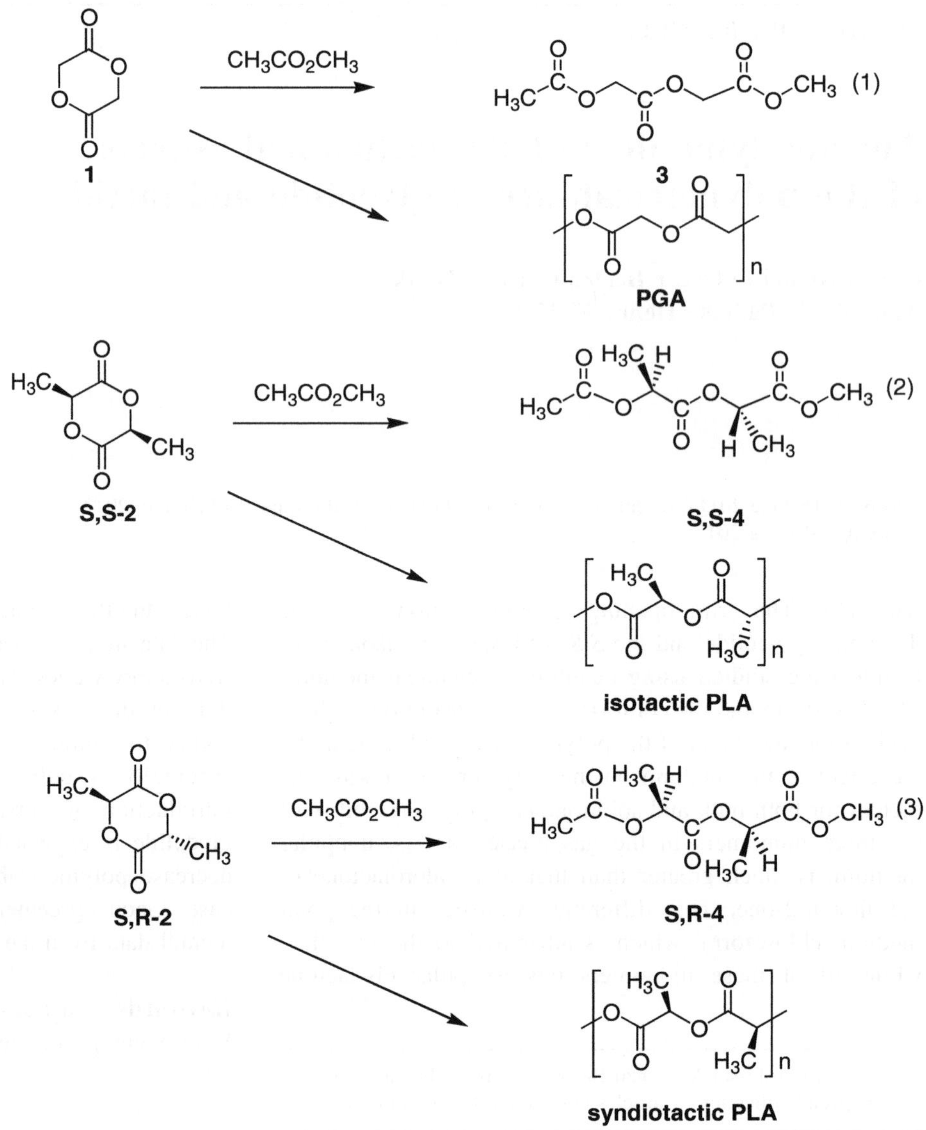

compete as single-use packaging materials, films, and other commodity uses. However, they find their most outstanding use in biomedical devices such as sutures and implants; these devices slowly hydrolyze back to lactic acid and reenter the Krebs cycle. More recently, these polymers have been employed as scaffolds for tissue engineering and in drug delivery systems. Excellent reviews of the many chemical and practical aspects of lactone polymerization have been published [2–13].

The parent lactone lacking substituents, glycolide, or 1,4-dioxane-2,5-dione, **1**, yields only a single polyglycolide (PGA) isomer (Scheme 1). Lactide on the other hand exists as different stereoisomers: the *meso*-lactide, ***S,R-2***, and the two chiral forms, ***S,S-2*** or ***R,R-2***. Polymerization of pure ***S,S-2*** or ***R,R-2*** leads to isotactic PLA in which all the chiral carbons have the same absolute configuration throughout the chain. Syndiotactic PLA, which shows alternate configurations of the sequential stereocenters, is derived from stereoselective polymerization of ***S,R-2***. Finally, heterotactic PLA is obtained by alternating linkages formed by ***R,R*** and ***S,S*** stereocenters. Copolymerization of glycolide with either of the lactides gives further novel materials for biomedical uses.

Recently, we studied the ring-opening polymerization of monolactones by quantum mechanical methods. The ring strain and the conformational distribution of these cyclic monomers and of the polymers were calculated [14–16]. Consistent with extensive studies of carbonyl-containing molecules [17–19], a high preference for *gauche*-coiled conformations was found to strongly influence the outcomes. Finally, the effect of the medium on the polymerization was determined, both in bulk and in solution. These data were combined to give polymerization enthalpy and free energy changes in good agreement with literature experimental values [14–16].

Our current work applies these methods to the ring-opening polymerization of the dilactones glycolide (**1**) and lactide, the ***S*,*S*-2** and ***S*,*R*-2** stereoisomers being considered for the latter. Beyond calculations analogous to those carried out previously, we calculate the stereochemistry of the propagations leading to the described structures.

2 Methods

Quantum mechanical calculations were performed with Gaussian 03 [20]. All geometry optimizations were performed using the MP2 [21] method combined the 6-31G(d) basis set [22]. Frequency analyses were carried out to verify the nature of all the stationary points and to calculate the zero-point vibrational energies (ZPVE) and both thermal and entropic corrections at 298 K. All thermodynamic corrections were calculated using the standard expressions for an ideal gas in the canonical ensemble. In addition, the electronic energies of all the minima were re-evaluated using single-point calculations at the MP2/6-311++G(d,p) level [23]. Accordingly, the best estimate to the free energy in the gas phase (ΔG^{gp}) was obtained by adding the statistical corrections calculated at the MP2/6-31G(d) level to the latter electronic energies.

To obtain an estimation of the environmental effects, single-point calculations were conducted on the gas-phase optimized structures using a SCRF model. Specifically, the polarizable continuum model (PCM) developed by Tomasi et al. [24–26] was used to simulate chloroform ($\varepsilon = 4.9$) as solvent. The PCM method represents the polarization of the liquid by a charge density appearing on the surface of the cavity created in the solvent. This cavity is built using a molecular shape algorithm. PCM calculations were performed in the framework of the HF/6-311++G(d,p) level using the standard protocol. Within this context, it should be recalled that previous studies indicated that the free energy of solvation (ΔG_{sol}) obtained using solute geometry relaxations in solution and single-point calculations on the optimized geometries in gas phase are almost identical [27–29]. The conformational free energies in chloroform solution (ΔG^{chl}) were estimated using the classical thermodynamics scheme, that is, adding ΔG_{sol} to ΔG^{gp}.

The buildup procedure [30–32] was applied to determine the conformational profile of the model compounds of the polymers. This procedure is based on the assumption that each fragment of a given molecule is conformationally independent of the other fragments in the molecule. Thus, model compounds were divided into fragments. The minimum energy conformations of these small fragments were characterized using a systematic conformational search procedure and joined to form the overall conformations that were taken as starting points for geometry optimizations. In the systematic conformational search procedure, each flexible dihedral angle was expected to have three minima, the number of minima that may be anticipated for the potential energy hypersurfaces (PEHs) of a compound with n flexible dihedral angle being 3^n.

3 Results and discussion

3.1 Conformational analyses of linear monomeric oligomers

The linear diesters **3** and **4** are the repeating units of the PGA and PLA (Scheme 1). However, the calculations were started with the even simpler monomeric units **5** and ***S*-6**, from which the more complex structures can be built up, as shown in Scheme 2.

The minimum energy conformations of **5** and ***S*-6** were determined using a systematic search procedure. Thus, nine starting structures were generated for each compound by combining the three minima of each flexible dihedral angle (χ_1 and χ_2 in Scheme 2), which converged into 5 (one unique and two twofold degenerate) and 7 minima, respectively, after geometry optimization at the MP2/6-31G(d) level. To ascertain the effect of the medium, calculations were carried out for both the gas phase and for chloroform solution.

Table 1 lists the dihedral angles $\{\chi_i\}$, the ΔG^{gp} (gas phase), and the ΔG^{chl} (chloroform solution) of the minimum energy conformations obtained for **5** and ***S*-6**. As can be seen, the two conformations of lower energy are very similar for the two compounds. Thus, the global minimum (**5a** and ***S*-6a**) adopts a partially folded conformation with one dihedral angle χ_1 arranged in *gauche*$^-$ ($\sim -70°$) and the other one extended ($\chi_2 \sim 170°$). In contrast, both χ_1 and χ_2 are folded in the second minimum (**5b** and ***S*-6b**); this conformation is destabilized with respect to the former by <0.7 and 0.5 kcal/mol in the gas phase and chloroform solution, respectively. The fully extended conformations, in which χ_1 and χ_2 show a *trans* conformation, are disfavored with respect to the global minimum by 1.4 (**5c**) and 2.2 kcal/mol (***S*-6f**) in the gas phase. However, the bulk solvent stabilizes these conformers by 0.9 and 0.7 kcal/mol, respectively.

3.2 Conformational analyses of linear dimeric oligomers

Turning to the linear dimers **3** and ***S*,*S*-4**, 25 and 49 starting structures, respectively, were constructed, using the

Scheme 2 Schematics for the calculations

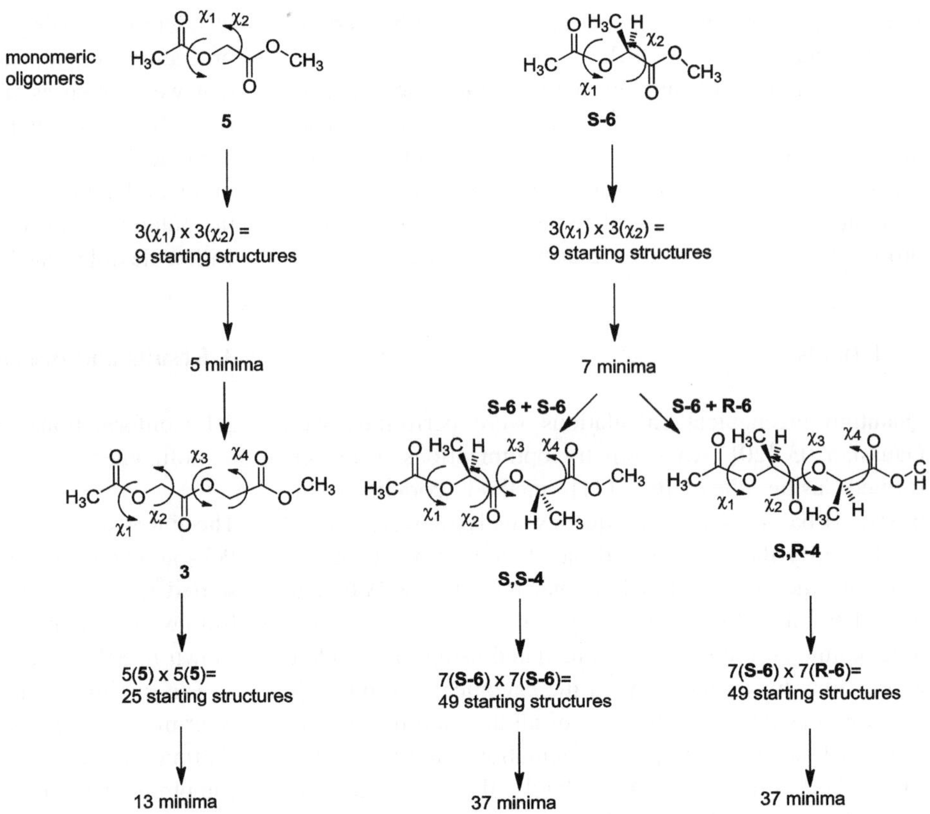

Table 1 Relevant dihedral angles and thermodynamical parameters (in kcal/mol) calculated for the minimum energy conformations of monomeric linear compounds **5** and ***S*-6**

5					*S*-6				
#	χ_1	χ_2	ΔG^{gp}	ΔG^{chl}	#	χ_1	χ_2	ΔG^{gp}	ΔG^{chl}
5a	−71.4	175.5	0.0[a]	0.0[b]	***S*-6a**	−67.0	166.1	0.0[c]	0.0[d]
5b	−66.6	−17.6	0.7	0.5	***S*-6b**	−63.0	−23.9	0.6	0.2
5c	180.0	−179.9	1.4	0.5	***S*-6c**	52.0	35.2	2.4	1.6
					***S*-6d**	57.0	−158.2	2.8	2.2
					***S*-6e**	−145.7	50.8	1.8	2.0
					***S*-6f**	−151.1	177.8	2.2	1.5
					***S*-6g**	−155.5	−47.4	2.4	2.3

Geometry optimizations and frequency calculations were performed at the MP2/6-31G(d) level. Electronic energies and solvation energies were derived from single-point calculations at the MP2/6-311++G(d,p) and PCM-HF/6-311++G(d,p) levels, respectively

ΔG^{gp} and ΔG^{chl} correspond to the relative free energies in the gas phase and chloroform solution, respectively

[a] $G = -495.014230$ a.u.

[b] $\Delta G_{sol} = -0.31$ kcal/mol

[c] $G = -534.189801$ a.u.

[d] $\Delta G_{sol} = 1.35$ kcal/mol

following buildup procedure: 5(minima of **5**) × 5(minima of **5**) and 7(minima of ***S*-6**) × 7(minima of ***S*-6**). The 49 starting structures of ***S,R*-4** were generated using the same process: 7(minima of ***S*-6**) × 7(minima of ***R*-6**), whereby the minima for ***R*-6** are obtained by changing the sign of the dihedral angles of the ***S*-6** minima. Scheme 2 summarizes the procedure used to determine the conformational preferences of these linear model compounds of PGA and PLA.

Application of the buildup procedure led to 13 twofold degenerated minimum energy conformations for **3**. However, only 5 and 8 such minima offer ΔG^{gp} and/or ΔG^{chl} values lower than 1.5 kcal/mol, respectively. Table 2 lists

Table 2 Relevant dihedral angles and thermodynamical parameters (in kcal/mol) calculated for the significant minimum energy conformations of dimeric linear compounds **3**, ***S*,*S*-4** and ***S*,*R*-4**

#	χ_1	χ_2	χ_3	χ_4	ΔG^{gp}	ΔG^{chl}
3						
3a	69.8	−170.8	71.9	−176.8	0.0	0.0
3b	−70.8	174.9	63.6	15.1	1.2	1.3
3c	78.2	170.1	−72.3	175.2	1.4	0.9
3d	−65.1	−21.1	72.4	−175.5	1.4	1.1
3e	70.5	−172.9	63.3	15.9	1.5	1.4
3f	71.0	−174.8	173.3	174.0	2.3	1.1
3g	173.9	167.8	−71.3	177.6	2.3	1.3
3h	66.6	17.6	−71.4	−175.5	2.6	1.5
***S*,*S*-4**						
***S*,*S*-4a**	−65.1	160.4	−67.8	168.8	0.0	0.3
***S*,*S*-4b**	−65.4	161.8	−63.7	−21.6	0.6	0.0
***S*,*S*-4c**	−66.5	−32.7	−64.0	−24.6	1.3	1.0
***S*,*S*-4d**	−65.5	−33.8	−69.9	158.3	1.5	1.0
***S*,*S*-4e**	−60.4	−27.1	54.5	33.9	2.6	1.2
***S*,*S*-4f**	−60.9	−26.0	−144.8	32.4	2.7	1.4
***S*,*R*-4**						
***S*,*R*-4a**	−62.0	−25.6	68.7	−168.0	0.0	0.4
***S*,*R*-4b**	−68.3	170.4	60.5	22.0	0.3	0.8
***S*,*R*-4c**	−61.9	−25.5	65.0	21.7	0.6	0.0
***S*,*R*-4d**	−73.2	−179.7	67.6	−163.2	1.0	0.7
***S*,*R*-4e**	−156.2	−41.3	63.3	22.1	1.5	1.0
***S*,*R*-4f**	−64.2	159.9	−52.5	−33.8	1.6	1.2

Geometry optimizations and frequency calculations were performed at the MP2/6-31G(d) level. Electronic energies and solvation energies were derived from single-point calculations at the MP2/6-311++G(d,p) and PCM-HF/6-311++G(d,p) levels, respectively

the dihedral angles {χ_i with $i = 1$–4}, ΔG^{gp} and ΔG^{chl} for these highly populated structures, whose populations add up to 92% when calculated using conformational statistic weights. The lowest energy minimum **3a** in the gas phase (population = 66.5 %) is also the most favored conformation in chloroform solution (population = 47.5%). In this conformation, the two repeating units adopt an identical arrangement with {χ_1, χ_3} and {χ_2, χ_4} showing *gauche* and *trans* conformations, respectively. A detailed analysis of the other minima indicates that the population of the coiled conformations, that is, those in which three or more dihedral angles are arranged in *gauche*$^+$ or *gauche*$^-$, is 24% and 26% in the gas phase and chloroform solution, respectively. The population of the extended and semi-extended conformations, in which three or more dihedral angles adopt a *trans* arrangement, increases from 3% (gas phase) to 17% when the bulk solvent is considered. The fully extended conformation is the least favored minimum in the gas phase ($\Delta G^{gp} = 3.7$ kcal/mol), although its stability increases in solution ($\Delta G^{chl} = 1.6$ kcal/mol).

Table 2 includes the minima characterized for ***S*,*S*-4** and ***S*,*R*-4** with ΔG^{gp} and/or ΔG^{chl} lower than 1.5 kcal/mol. Only four and five structures have significant populations in the gas phase (6 in chloroform solution). The buildup procedure for these stereoisomers provided 37 and 39 minima, respectively, distributed in a $\Delta G^{gp}/\Delta G^{chl}$ interval of 7.3/5.7 and 7.1/6.6 kcal/mol, which is significantly larger than the values for **3** (3.7/2.5 kcal/mol). The lowest energy minimum in the gas phase ***S*,*S*-4a** is similar to **3a** and consists of two repeating units with the same conformation, in which {χ_1, χ_3} and {χ_2, χ_4} are arranged in *gauche*$^-$ and *trans*, respectively. In contrast, the equivalent conformation ***S*,*R*-4d** is disfavored by 1.0 kcal/mol, while the lowest energy minimum has a different dihedral angle χ_2, which is arranged in *gauche*$^-$. Interestingly, the minimum ***S*,*R*-4a** corresponds to ***S*,*S*-4d**, which is destabilized by 1.5 kcal/mol with respect to ***S*,*S*-4a**. The predicted population for the coiled conformations of ***S*,*S*-4** and ***S*,*R*-4** in the gas phase (chloroform solution) is 36% (66%) and 84% (82%), respectively. Consequently, the population of

Table 3 Selected geometrical parameters around the ester bond for the lowest minimum energy conformations calculated for **1**, ***S*,*S*-2**, and ***S*,*R*-2**

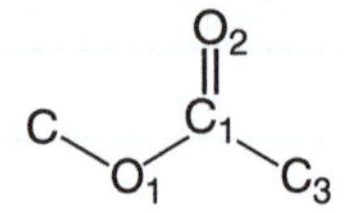

# (in deg.)	1[a]	*S*,*S*-2[a]	*S*,*R*-2	γ-BL[b]	δ-VL[b]	ε-CL[c]	MA (*s-trans*)	MA (*s-cis*)
$O_1–C_1–C_3$	114.7	114.7	116.3/116.7	108.9	118.6	118.7	110.5	110.5
$O_1–C_1=O_2$	121.4	121.0	120.5/121.2	122.4	118.7	122.9	123.4	118.5
$O_2=C_1–C_3$	123.9	124.3	123.1/122.1	128.7	122.4	118.3	126.0	126.0
$C–O_1–C_1–C_3$	9.5	7.3	8.1/12.9	2.4	−20.7	−2.8	180.0	0.0
$C–O_1–C_1=O_2$	−170.2	−172.8	−171.6/−167.1	−177.3	164.8	179.2	0.0	180.0

Parameters for the *s-trans* and *s-cis* conformers of methyl acetate (MA) as well as for γ-butyrolactone (γ-BL), δ-valerolactone (δ-VL) and ε-caprolactone (ε-CL) are also included for comparison

[a] The geometric parameters of the two ester bonds are identical

[b] From reference [14]

[c] From reference [15]

the extended and semi-extended conformations for ***S*,*S*-4** and ***S*,*R*-4** is extremely low, that is, lower than 2% independent of the environment. Thus, the replacement of one hydrogen atom in the repeat unit of PGA by the more sterically demanding methyl group results in drastic changes compared to **3** and enhances the tendency in PLA chains to adopt coiled conformations. Moreover, this tendency seems to be higher in syndiotactic PLA than in isotactic PLA.

3.3 Conformational analysis of cyclic dimers; strain and stability of the dilactones

Table 3 lists the most relevant geometric parameters around the ester group for the minimum energy conformations of **1**, ***S*,*S*-2** and ***S*,*R*-2**, as well as results obtained for optimized s-*cis* and s-*trans* forms of methyl acetate (MA) and other lactones previously reported, *ithat is*, γ-butyrolactone (γ-BL), δ-valerolactone (δ-VL), and ε-caprolactone (ε-CL) [14–16]. The bond and dihedral angles calculated for **1**, ***S*,*S*-2** and ***S*,*R*-2** are different from the values of s-*trans* MA, considered to be the strain-free reference. The strain deformation predicted for these three dilactones is approximately halfway between that found for γ-BL (small ring) and those obtained for δ-VL and ε-CL (larger rings).

Comparison of the free energies calculated for ***S*,*S*-2** and ***S*,*R*-2** indicates that the former is more stable by 1.2 and 2.2 kcal/mol in the gas phase and chloroform solution, respectively. These calculated values are in agreement with crystallographic data, which show that these dilactones consistently adopt a twisted boat form. In the ***S*,*S***-isomer, the two methyl groups adopt cis-diequatorial positions, while the ***S*,*R***-isomer adopts an almost planar twisted boat form with one methyl in equatorial position, while the other one has to be axial [33, 34].

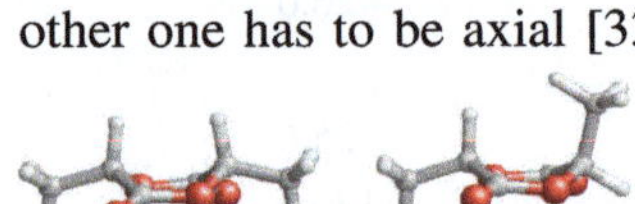

***S*,*S*-2** ***S*,*R*-2**

3.4 Ring-opening reactions

Reactions (1), (2), and (3) of the lactones with methyl acetate (Scheme 1) are simple models for the ring-opening polymerizations of **1**, ***S*,*S*-2**, and ***S*,*R*-2**, respectively. Specifically, reaction (2) models the polymerization of ***S*,*S*-2** to isotactic PLA, while reaction (3) models the polymerization of ***S*,*R*-2** to syndiotactic PLA. The enthalpies and Gibbs free energies in the gas phase (ΔH^{gp}_{rxn} and ΔG^{gp}_{rxn}) and in chloroform solution (ΔH^{chl}_{rxn} and ΔG^{chl}_{rxn}) of the ring-opening reaction for these cyclic compounds were calculated using the minimum energy structures provided by the buildup procedure for **3**, ***S*,*S*-4** and ***S*,*R*-4**. Table 4 lists the ΔH^{gp}_{rxn}, ΔG^{gp}_{rxn}, ΔH^{chl}_{rxn}, and ΔG^{chl}_{rxn} values calculated for reactions (1), (2), and (3).

The ΔG^{gp}_{rxn} values obtained for reactions (1), (2), and (3) are −7.4, −5.1, and −11.2 kcal/mol, respectively, indicating that all three are exergonic processes. These values demonstrate that the polymerizability of the ***S*,*S*-2** dilactone is predicted to be significantly lower than that of its stereoisomer ***S*,*R*-2**, in part because the former is more stable. The intrinsic polymerizability of **1** is predicted to be intermediate between those of the two lactides, while both δ-valerolactone and 1,4-dioxan-2-one display much lower polymerizabilities.

These results show that the incorporation of the methyl substituents does not produce drastic changes from the glycolide dimer **3**, even though the replacement of one

Table 4 Thermochemistry[a] of reactions of methyl acetate with cyclic dimers **1**, ***S,S*-2**, and ***S,R*-2** (reactions 1, 2, and 3, respectively; Scheme 1) in the gas phase and chloroform solution. Thermochemical parameters for six-membered mono-lactones δ-valerolactone (**δ-VL**, reaction 4)[b] and 1,4-dioxan-2-one (**Dox**, reaction 5)[c] have been included in the table for comparison

		ΔH_{rxn}^{gp}	ΔG_{rxn}^{gp}	ΔH_{rxn}^{chl}	ΔG_{rxn}^{chl}
1	Reaction (1)	−14.6	−7.4	−7.5	−0.5
***S,S*-2**	Reaction (2)	−16.1	−5.1	−3.8	−0.8
***S,R*-2**	Reaction (3)	−24.0	−11.2	−11.0	0.3
δ-VL	Reaction (4)	−16.2	−1.9	−12.6	−2.7
Dox	Reaction (5)	−13.2	−0.5	−8.5	−0.8

[a] In kcal/mol ΔH_{rxn}^{gp} and ΔG_{rxn}^{gp} are the enthalpy and free energy in the gas phase, respectively. ΔH_{rxn}^{chl} and ΔG_{rxn}^{chl} are the enthalpy and free energy in chloroform solution, respectively

[b] Calculated in reference [14] considering the following reaction as 4:

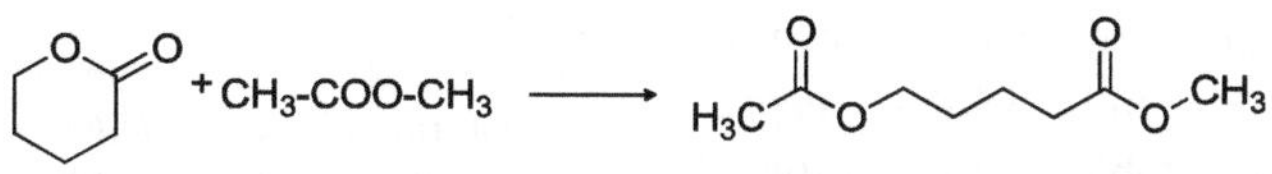

δ-VL

[c] Calculated in reference [15] considering the following reaction as 5:

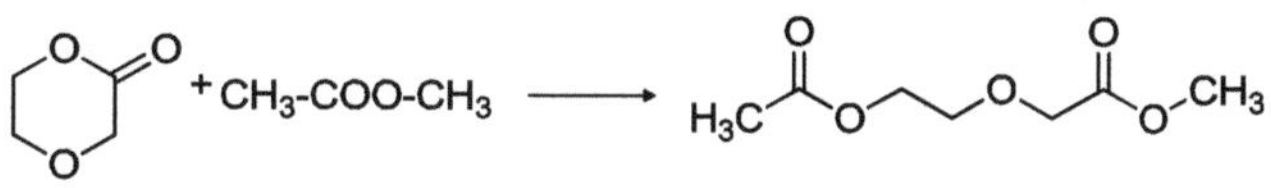

Dox

hydrogen atom in the repeating unit of PGA by the more sterically demanding methyl group enhances the tendency in PLA chains to adopt coiled conformations. This tendency is higher in isotactic PLA, (–SS–)$_n$, than in syndiotactic PLA, (–SR–)$_n$.

The ΔG_{rxn}^{chl} values for reactions (1)–(3) in the more polar medium chloroform are calculated considering statistic weights of the minimum energy conformations found for the linear compounds in chloroform, and they are −0.5, −0.8, and +0.3 kcal/mol, respectively. These values reveal that the solvent plays a very important role in the thermodynamics of the ring-opening reaction of the three dilactones, and the reactions become much less exergonic in a polar solution than in the gas phase. Previously, we have noted the importance of solvation in ring-opening polymerizations: A highly polar lactone (*E*-isomer) is transformed into a linear, much less polar, polymer (*Z*-isomer) [14–16]. Solute–solvent interactions are stronger for the dilactones than for the linear compounds; the ΔG_{sol} values being about 7–9 kcal/mol more favorable for the former than for the latter ones. This is in contrast to our previous studies of δ-valerolactone and 1,4-dioxan-2-one; the ΔG values for these polymerizations are basically the same in the nonpolar gas phase or the polar chloroform solution. The reason for this divergent behavior is that solvation of glycolide and lactide is different from that of monolactones; all of these monomers are highly polar, but glycolide and lactide present stronger interactions with the solvent because of their two ester groups.

The difference between reactions (2) and (3) arises because the interaction with the solvent is more favorable for ***S,S*-2** ($\Delta G_{sol} = -6.3$ kcal/mol) than for ***S,R*-2** ($\Delta G_{sol} = -5.4$ kcal/mol), possibly because it is more planar and therefore the carbonyl groups are more accessible. The slightly endergonic character calculated for reaction (3) is probably a consequence of the limitations of the SCRF model used in this work; continuum SCRF models not only neglect the configurational sampling of the solvent molecules, but also omit the local anisotropies produced by the solvent molecules of the first solvation shell. The influence of these effects, especially of the latter one, increases with the molecular size and with the number of polar groups [14, 15].

3.5 Stereoisomeric lactide polymers

As described above, reactions (2) and (3) in Scheme 1 are models for the formation of isotactic and syndiotactic PLA, respectively. However, modeling the formation of heterotactic polymer cannot be carried out from the dimers, but requires examination of the appropriate tetramers. The computational difficulties increase rapidly as the number of degrees of freedom in the molecules becomes higher, so we have used an approximation method. Specifically, the buildup procedure was applied considering only the energy minima for ***S,S*-4** and ***S,R*-4** with $\Delta G^{gp} \leq 1.5$ kcal/mol, that is, ***S,S*-4a**–***S,S*-4d** and ***S,R*-4a**–***S,R*-4e**, respectively, in Table 2. Accordingly, a total of 4 × 4 = 16 starting structures were constructed and subsequently optimized at the MP2/6-31G(d) level for the –SSS– and –SRR– tetramers, while 5 × 5 = 25 starting structures were considered for the –SRSR– and –SRRS– sequences. The nature of the minimum state of the resulting conformations was not confirmed through frequency analyses, even though single-point calculations were carried out at the PCM-HF/6-311++G(d,p) (chloroform) level. Therefore, this biased search procedure was used to estimate only the ΔH values in both the gas phase and chloroform solution, but not the ΔG values.

The gas phase (empty symbols) and chloroform (filled symbols) energies of the conformations obtained for the four tetramers are shown in Figure 1 relative to the most stable arrangement of the –SRSR– and –SSSS–, respectively. In the gas phase, the most stable conformation of each of the four tetramers is within a relative 1.0 kcal/mol energy interval; this interval increases to 2.7 kcal/mol in

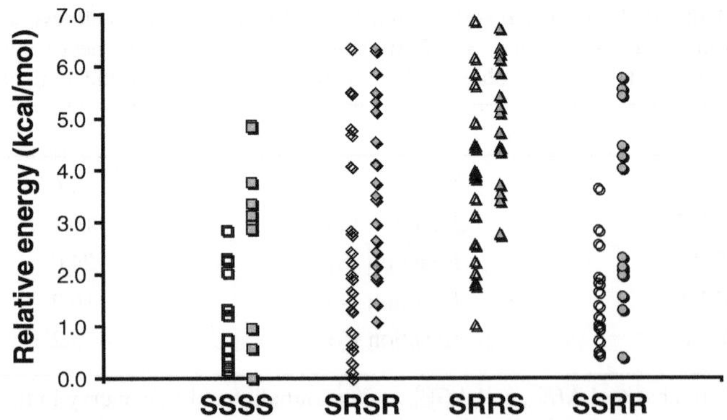

Fig. 1 Relative energies of the conformations optimized at the MP2/6-31G(d) level for the –SSSS–, –SRSR–, –SRRS– and –SSRR– tetramers. Relative energies in the gas phase (*empty symbols*) and chloroform solution (*filled symbols*) are relative to the most stable conformation of the –SRSR– and –SSSS– sequence, respectively. Solvation energies were derived from single-point calculations at the PCM-HF/6-311++G(d,p) level

Table 5 Comparison of the Enthalpies of Reaction (in kcal/mol) for the Dimer and Tetramer Formations in the Gas Phase and in Chloroform

Reaction	ΔH^{gp}_{rxn}	ΔH^{chl}_{rxn}
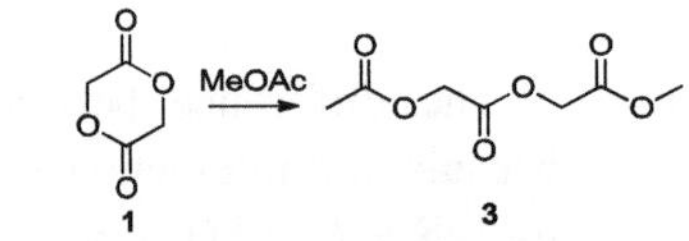	−14.6	−7.5
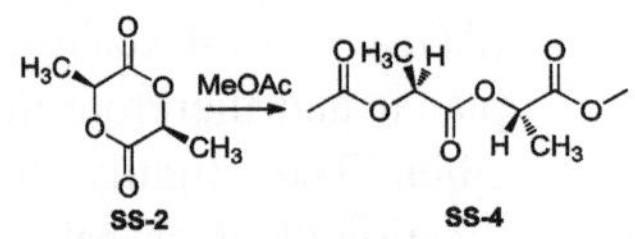	−16.1	−3.8
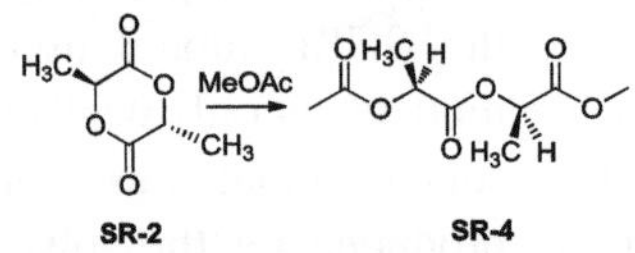	−24.0	−11.0
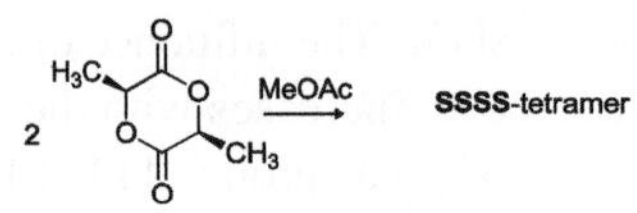	−37.3	−23.7
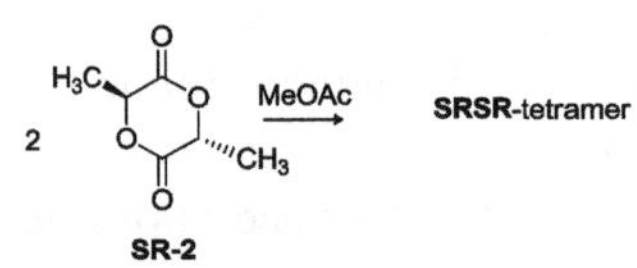	−39.0	−25.0
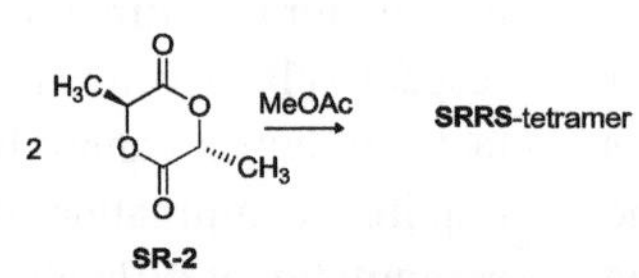	−37.9	−23.5
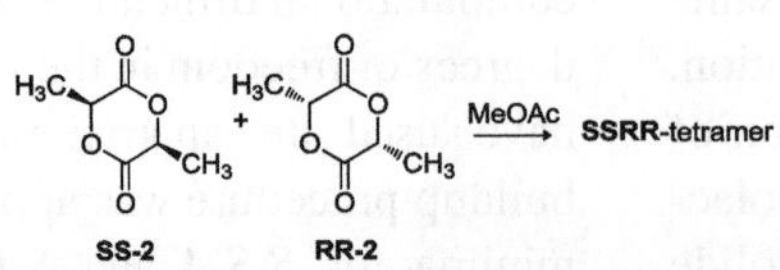	−38.7	−25.6

chloroform solution. Application of conformational statistic weights using Boltzmann probabilities to these sets of conformations indicates that the –SRSR– and –SSSS– tetramers are isoenergetic in the gas phase, while the –SSRR– and –SRRS– are destabilized by only 0.3 and 1.1 kcal/mol. In chloroform solution, the –SSRR–, –SRSR–, and –SRRS– tetramers are destabilized by 0.3, 1.4, and 2.8 kcal/mol, respectively, compared to the –SSSS– one. The similarity among the energies of the more stable structures identified for the different sequences should be attributed to the attractive van der Waals interactions produced by the side methyl groups, which are satisfactorily reproduced at the MP2 level [35–38]. These interactions are maximized by small rearrangements, facilitated by the intrinsic conformational flexibility of the compounds under study.

These small conformational energy differences produced very comparable ΔH^{gp}_{rxn} and ΔH^{chl}_{rxn} values for the four sequences, as reflected in Table 5 for the corresponding formation reactions. Thus, ΔH^{gp}_{rxn} and ΔH^{chl}_{rxn} range from -37.3 to -39.0 and from -23.5 to -25.6 kcal/mol, respectively, that is, the ΔH interval between the different monomer sequences is 1.7 and 2.1 kcal/mol in the gas phase and chloroform, respectively. Comparison of these reaction enthalpies with those obtained for ***S*,*S*-2** and ***S*,*R*-2**, also included in Table 5 indicates that the remarkable thermodynamic stereoselectivity found for the dimers is considerably reduced when the polymerization propagates to tetramers, and presumably beyond. This is not unexpected in view of the distance between one chiral center and the next in the growing polymer chain. Accordingly, the ability to prepare the various stereo–isomeric polymers depends on the exquisite tailoring of the sterically crowded chiral polymerization catalysts ("polymer chain-end control").

3.6 Comparison of calculated polymerization enthalpies with experimental values

The calculated polymerization enthalpies for glycolide and lactide (Table 4) can be compared with values from the literature. It should be noted that the calculations were performed in chloroform solution ($\varepsilon = 4.9$) because the dielectric constant of cyclic dilactones decreases very rapidly once the polymerization reaction starts, the final value for PLA being $\varepsilon = 3.1$ [39]. Agreement of these values with one another and with our calculated values is excellent. The experimental value for ΔH_p for polymerization for glycolide in bulk is -6.3 kcal/mol [40], -5.5 kcal/mol for the S,S-lactide [4, 41], and -5.7 kcal/mol for the R,S-lactide [42], with the calculated values being -7.5, -3.8, and -11.0 kcal/mol, respectively.

Lactone polymerization involves converting an *E* ester isomer to the *Z* isomer. The latter is stabilized by 5–8 kcal/mol due to antiparallel dipole alignment, p-orbital on oxygen overlap with the carbonyl group, and lack of steric nonbonded interactions, favoring polymer over monomer. Glycolide and lactides, having two *E* ester groups, might be even more destabilized by ~10–16 kcal/mole, and the ΔG^{gp}_{rxn} values clearly favor the ring-opening of dilactones with respect to monolactones. However, their antiparallel alignment is a stabilizing factor [2]; so the polymerization enthalpy is less than might be expected for a molecule containing two *E* ester isomers.

Acknowledgments Computer resources were generously provided by the Centre de Supercomputació de Catalunya (CESCA). Financial support from the MICINN and FEDER (MAT2009-09138) and Generalitat de Catalunya (research group 2009 SGR 925 and XRQTC) is gratefully acknowledged. Support for the research of C.A. was received through the prize "ICREA Academia" for excellence in research funded by the Generalitat de Catalunya.

References

1. Duda A, Penczek S (2001) Mechanisms of aliphatic polyester formation. In: Steintuechel A, Doi Y (eds) Biopolymers, vol 3b. Polyesters II—properties and chemical synthesis, Ch 12. Wiley-VCH, Weinheim
2. Dechy-Cabaret O, Martin-Vaca B, Bourissou D (2004) Chem Rev 104:6147–6176
3. Zhong Z, Dijkstra PJ, Feijen J (2004) J Biomat Sci Polym Ed 15:929–946
4. Duda A, Kowalski A, Libiszowski J, Penczek S (2005) Macromol Symp 224:71–84
5. Cirugomane A, Thomas CM, Carpentier J-F (2007) Pure Appl Chem 79:2013–2030
6. Kamber NE, Joong W, Waymouth RM, Pratt RC, Lohmeijer BGG, Hedrick JL (2007) Chem Rev 107:5813–5840
7. Bourrisou D, Moebs-Sanchez S, Martin-Vaca B (2007) Comptes Rendus Chimie 10:775–794
8. Platel RH, Hodgson LM, Williams CK (2008) Polym Rev 48:11–63
9. Chisholm MH, Zhou Z (2008) Stereoselective polymerization of lactide. In: Bough LS, Canich JAM (eds) Stereoselective polymerizations with single-site catalysts. CRC Press, Boca Raton, pp 645–660
10. Stanford MJ, Dove A (2010) Chem Soc Rev 39:486–494
11. Chisholm MH (2010) Pure Appl Chem 82:1647–1662
12. Wheaton C, Hayes P, Ireland B (2009) Dalton Trans 4832–4846
13. Houk KN, Jabbari A, Hall HK Jr, Aleman C (2008) J Org Chem 1573:2674–2678
14. Aleman C, Betran O, Casanovas J, Houk KN, Hall HK Jr (2009) J Org Chem 74:6237–6244
15. Alemán C, Casanovas J, Zanuy D, Hall HK Jr (2005) J Org Chem 70:2950–2956
16. Alemán C, Casanovas J, Hall HK Jr (2005) J Org Chem 70:7731–7736
17. Navarro E, Alemán C, Puiggalí J (1995) J Am Chem Soc 117:7307–7310
18. Alemán C, Navarro E, Puiggali J (1995) J Org Chem 60:6135–6140
19. Alemán C, Navarro E, Puiggalí J (1996) J Phys Chem 100:16131–16136
20. Frisch MJ, Trucks GW, Schlegel HB, Scuseria GE, Robb MA, Cheeseman JR, Montgomery JA, Vreven T Jr, Kudin KN, Burant

JC, Millam JM, Iyengar SS, Tomasi J, Barone V, Mennucci B, Cossi M, Scalmani G, Rega N, Petersson GA, Nakatsuji H, Hada M, Ehara M, Toyota K, Fukuda R, Hasegawa J, Ishida M, Nakajima T, Honda Y, Kitao O, Nakai H, Klene M, Li X, Knox JE, Hratchian HP, Cross JB, Adamo C, Jaramillo J, Gomperts R, Stratmann RE, Yazyev O, Austin AJ, Cammi R, Pomelli C, Ochterski JW, Ayala PY, Morokuma K, Voth GA, Salvador P, Dannenberg JJ, Zakrzewski VG, Dapprich S, Daniels AD, Strain M, Farkas O, Malick DK, Rabuck AD, Raghavachari K, Foresman JB, Ortiz JV, Cui Q, Baboul AG, Clifford S, Cioslowski J, Stefanov BB, Liu G, Liashenko A, Piskorz P, Komaromi I, Martin RL, Fox DJ, Keith T, Al-Laham MA, Peng CY, Nanayakkara A, Challacombe M, Gill PMW, Johnson B, Chen W, Wong MW, Gonzalez C, Pople JA (2003) Gaussian 03, revision B.02. Gaussian Inc., Pittsburgh
21. Møller C, Plesset MS (1934) Phys Rev 46:618–622
22. Hariharan PC, Pople JA (1973) Theor Chim Acta 23:213–222
23. Frisch MJ, Pople JA, Binkley JS (1984) J Chem Phys 80:3265–3269
24. Miertus S, Scrocco E, Tomasi J (1981) Chem Phys 55:117–129
25. Miertus S, Tomasi J (1982) Chem Phys 65:239–245
26. Tomasi J, Mennucci B, Cammi R (2005) Chem Rev 105:2999–3094
27. Hawkins GD, Cramer CJ, Truhlar DG (1998) J Phys Chem B 102:3257–3271
28. Jang YH, Goddard WA III, Noyes KT, Sowers LC, Hwang S, Chung DS (2003) J Phys Chem B 107:344–357
29. Iribarren JI, Casanovas J, Zanuy D, Alemán C (2004) Chem Phys 302:77–83
30. Gibson KD, Scheraga HA (1987) J Comput Chem 8:826–834
31. Vasquez M, Scheraga HA (1988) J Biomol Struct Dyn 5:705–755
32. Vasquez M, Scheraga HA (1988) J Biomol Struct Dyn 5:757–784
33. Chisholm MH, Eilerts NW, Huffman JC, Iyer SS, Pacold M, Phomphrai K (2000) J Am Chem Soc 122:11845–11854
34. Van Hummel GJ, Harkema S (1982) Acta Cryst B38:1679–1681
35. Li AHT, Chao S (2006) J Chem Phys 125:094312
36. Johnson ER, DiLabio GA (2006) Chem Phys Lett 419:333–339
37. Fomine S, Tlenkopatchev M, Martinez S, Fomina L (2002) J Phys Chem A 106:3941–3946
38. Rodríguez-Ropero F, Casanovas J, Alemán C (2008) J Comput Chem 29:69–78
39. Katsuyoshi S, Shigetaka K (2005) IEEE J Trans Fundam Mater 125:204–208
40. Chujo K, Kobayashi H, Suzuki J, Tokuhara S, Tanake M (1967) Makromol Chem 100:262–266
41. Duda A, Penczek S (1990) Macromolecules 23:1636–1639
42. Wang Y, Hillmyer MA (2000) Macromolecules 33:7395–7403

Theor Chem Acc (2012) 131:1135
DOI 10.1007/s00214-012-1135-9

REGULAR ARTICLE

A theoretical study on CH_2N_2 isomers: structure and energetics

Cristina Puzzarini · Alberto Gambi

Received: 29 June 2011 / Accepted: 25 August 2011 / Published online: 18 February 2012

Abstract Five CH_2N_2 isomers, namely cyanamide, carbodiimide, diazomethane, isocyanamide and nitrilimine, have been investigated at a high level of accuracy. The singles and doubles coupled-cluster method including a perturbational correction for connected triple excitations, CCSD(T), in conjunction with correlation-consistent basis sets ranging in size from triple to quintuple zeta have been employed. Extrapolation to the complete basis set limit has been used with treatments of core-valence correlation effects in order to accurately predict structures, relative energies as well as N–H and C–H bond dissociation energies. The latter required to also investigate the HNNC radical with the same methodology used for CH_2N_2 isomers, while HCNN and HNCN data are available in the literature by the same authors (Puzzarini and Gambi in J Chem Phys 122:064316, 2005). For all the species studied, harmonic vibrational frequencies have also been evaluated at the CCSD(T) level in order to obtain zero-point corrections to total energies.

Keywords CH_2N_2 non-cyclic isomers · Molecular structure · Energetics

Dedicated to Professor Vincenzo Barone and published as part of the special collection of articles celebrating his 60th birthday.

C. Puzzarini (✉)
Dipartimento di Chimica "G. Ciamician",
Università di Bologna, Via Selmi 2, 40126 Bologna, Italy
e-mail: cristina.puzzarini@unibo.it

A. Gambi
Dipartimento di Chimica, Fisica e Ambiente,
Università di Udine, Via Cotonificio 108, 33100 Udine, Italy

1 Introduction

Cyanamide (H_2NCN) and its tautomer carbodiimide (HNCNH), diazomethane (H_2CNN) and isocyanamide (H_2NNC) with their tautomer nitrilimine (HNNCH), together with the three-membered ring diazirine (c-H_2CN_2), are the six experimentally known CH_2N_2 isomers and the most stable species on the singlet potential energy surface of the CH_2N_2 system. These are important molecules entering several chemical processes. Cyanamide is of considerable interest because in addition to its industrial [1] and agricultural uses is also a key molecule in prebiotic [2] synthesis and is one of the species that have been detected in the interstellar medium [3]. Carbodiimides are important reactants for peptides and nucleotide reactions [4], whereas diazomethane provided chemists with a versatile synthetic reagent, especially in the methylation of acids and alcohols [5]. Due to the easy generation of carbenes by thermolysis or photolysis, diazirine has extensively been used for preparations and mechanistic studies [6]. Relevant to the present investigation is that its molecular structures and energetics have been recently investigated at high level of theory by the authors themselves [7]. Isocyanamide is subject to fast rearrangement to cyanamide, but despite its unstable nature, valence ionization spectra [8] and rotation-inversion spectra in the millimeter-wave region have been observed [9]. Because of its character as reactive intermediate, there is very little direct experimental evidence on the structure of nitrilimine [10, 11], the first matrix-spectroscopic identification of HCNNH being reported only 15 years ago [12].

The HCN_2 radicals, which can be obtained by the removal of a hydrogen atom from the CH_2N_2 isomers, are highly reactive transient species that play an important role in various fields such as atmospheric processes, especially in relation to combustion products. Relevant to these reactions are the diazomethyl (HCNN) and cyanomidyl (HNCN) radicals, whose energetics and structural properties have been recently investigated at high level of theory [13]. The diazomethyl radical can be obtained from the C–H bond cleavage of diazomethane or from the N–H cleavage of nitrilimine. As far as the cyanomidyl radical is concerned, it can be generated by N–H bond cleavage of cyanamide or carbodiimide. Finally, from the N–H bond breaking in isocyanamide or the C–H cleavage of nitrilimine, the HNNC (isocyanomidyl) radical can be obtained. Although considered feasible [12], the latter has never been experimentally detected. On the other hand, its presence in the electronic spectra of cyanogen radical reactions has been suggested by Basco et al. [14].

In this work, the five isomeric non-cyclic species, namely cyanamide, carbodiimide, diazomethane, isocyanamide and nitrilimine, have been accurately investigated by means of state-of-the-art computational techniques. Calculations were also extended to the HNNC radical in order to determine all possible N–H and C–H dissociation energies.

The manuscript is organized as follows. In the next section, the methodology used is explained together with the corresponding quantum-chemical details. Thereafter, the results are reported and discussed with emphasis on molecular structures of the species investigated (Sect. 3.1) and the corresponding energetics (Sect. 3.2).

2 Computational details

Throughout the present study, the coupled-cluster method with single and double excitations with a perturbative treatment of connected triples [CCSD(T)] has been used [16, 17]. This method has been found suitable for describing the systems under investigation, as confirmed by the values of the coupled-cluster T_1 diagnostic [18, 19] (i.e., 0.01–0.03, where the largest value applies to the open-shell case). Test computations, carried out at the multi-configuration self-consistent field (MCSCF) [20, 21] level, showed that all the species considered are well described by a single-reference wave function. Correlation-consistent basis sets, the valence cc-pVnZ (n = T,Q,5) [22] sets, have been used in conjunction with the CCSD(T) method in the frozen-core approximation. All the computations have been carried out with the MOLPRO suite of programs [15].

For all the species considered, geometry optimizations have been performed using numerical gradients, as implemented in MOLPRO [15, 23, 24]. The step sizes used were 0.0005 Å for bond lengths and 0.1 degrees for bond angles. As for convergence criteria, both the maximum component of the gradient and the maximum component of the step have been constrained to be less than 1.0×10^{-6} a.u.

Since a hierarchical sequence of bases has been considered, the systematic trend of the optimized geometrical parameters can be exploited to estimate the complete basis set (CBS) limit. As demonstrated by the recent study from one of the author [25], it can be assumed that the convergence behavior of the structural parameters mimics that of the energy. The consolidated $1/n^3$ extrapolation form [26] has been used to describe the convergence of the correlation contribution, and $n = 4$ (which stands for cc-pVQZ) and $n = 5$ (for cc-pV5Z) have been employed. The extrapolated correlation contributions have then been added to the corresponding Hartree–Fock self-consistent field (SCF) geometrical parameters optimized employing the cc-pV6Z basis, as this level of theory is expected to provide the CBS SCF limit. The extrapolated structures obtained in such a way will be denoted as r(CBS) in the reminder of the text. As concerns total energies, an approach similar to that employed for structural parameters has been followed to derive the CBS limit. In detail, the correlation contribution has been extrapolated to the complete basis limit by means of the n^{-3} formula [26, 27]:

$$\Delta E^{\text{corr}}(n) = \Delta E^{\text{corr}}_{\infty} + A'n^{-3}, \tag{1}$$

and added to the SCF CBS limit, evaluated as follows [28, 29]

$$E^{\text{SCF}}(n) = E^{\text{SCF}}_{\infty} + B'\exp(-C'n). \tag{2}$$

To take into account the effects of core-valence (CV) correlation, geometry optimizations have been carried out also including all electrons in the correlation treatment. The weighted core-valence correlation-consistent cc-pwCVQZ basis set [30] has been used in this step. In regards to the geometrical parameters, making use of the additivity approximation, the core-correlation corrections have been added to the extrapolated structures r(CBS) as:

$$\begin{aligned}(\text{CBS} + \text{CV}) = r(\text{CBS}) + r(\text{wCVQZ, all}) \\ - r(\text{wCVQZ, valence}),\end{aligned} \tag{3}$$

where r(wCVQZ, all) and r(wCVQZ, valence) are the geometries optimized at the CCSD(T)/cc-pwCVQZ level correlating all and only valence electrons, respectively. Analogously, for total energies, the CV corrections have been calculated as

$$\Delta E_{\text{CV}} = E_{\text{core+val}} - E_{\text{val}}, \tag{4}$$

where $E_{\text{core+val}}$ is the CCSD(T) total energy obtained by correlating all electrons, and E_{val} is the CCSD(T) total energy obtained in the frozen-core approximation, both in the cc-pwCVQZ basis set.

In a similar way, the effects of diffuse functions (aug) in the basis set have been considered for both geometry and energy:

$$\Delta r_{\text{aug}} = r(\text{CCSD(T)/aug-cc-pVTZ}) - r(\text{CCSD(T)/cc-pVTZ}) \quad (5)$$

and

$$\Delta E_{\text{aug}} = E(\text{CCSD(T)/aug-cc-pVTZ}) - E(\text{CCSD(T)/cc-pVTZ}). \quad (6)$$

These corrections have then been added to the corresponding CBS+CV values in order to obtain the best estimated equilibrium structures and energies.

Isomerization from cyanamide to carbodiimide and from diazomethane to nitrilimine has also been investigated. Geometry optimizations of the connecting transition states have been carried out at the CCSD(T) method in conjunction with triple- and quadruple-zeta basis sets. As mentioned for the investigation of minimum structures, test computations have been carried out to check the suitability of the coupled-cluster method in describing these transition states. Furthermore, to have all the information required for studying the C–H and N–H dissociation energies, the tetratomic HNNC radical species has also been studied at the coupled-cluster level employing the same methodology mentioned above. Being HNNC an open-shell species, the variant denoted R/UCCSD(T) [31–33] has been employed. This is based on restricted open-shell Hartree–Fock (ROHF) orbitals but spin unrestricted in the solution of the CCSD equations. For the other tetratomic species involved, we refer to our previous investigation reported in Ref. [13].

Finally, to obtain the zero-point vibrational (ZPV) corrections to total energy within the harmonic approximation,

$$E_{\text{ZPV}}^{\text{harm}} = \frac{1}{2}\sum_i d_i \omega_i, \quad (7)$$

harmonic frequencies for all the species investigated have been computed at the CCSD(T)/cc-pVQZ level of theory in the frozen-core approximation. The harmonic force fields have been evaluated in a cartesian coordinate representation at the corresponding optimized geometry, and as in the case of energy gradients, the hessian matrix was calculated numerically [23, 24] by finite differences. Following the

Table 1 Equilibrium geometries and energies of nitrilimine, HCNNH (^1A), computed at the coupled-cluster level of theory employing different basis sets

	N–H (Å)	N–N (Å)	C–N (Å)	C–H (Å)	∠HNN (deg.)	∠NNC (deg.)	∠NCH (deg.)	T_1^a (∠HNNC) (deg.)	T_2^a (∠NNCH) (deg.)	Energy (Hartree)
MP2/6-311+G(2df,2p)[b]	1.020	1.256	1.181	1.067	108.6	170.3	144.48	−137.2	227.4	
QCISD/6-311G**[b]	1.022	1.244	1.204	1.081	107.1	169.1	128.9	−130.6	219.8	
B3LYP/6-311+G(d,p)[c]	1.023	1.242	1.189	1.076	109.5	169.8	133.0	−137.2	227.3	
CASSCF/3-21G(d)[d]	1.016	1.233	1.205	1.073	109.2	166.8	127.2	−131.4	219.9	
B3LYP/6-311++G(3df,2p)[e]	1.0216	1.2379	1.1858	1.0744	109.86	169.75	132.48	−135.52	224.81	
aug-cc-pVTZ	1.0226	1.2506	1.2001	1.0772	108.15	168.50	130.08	−131.85	220.83	−148.4707934
cc-pVTZ	1.0219	1.2477	1.2019	1.0771	107.95	168.66	129.09	−130.84	220.01	−148.4587066
cc-pVQZ	1.0203	1.2457	1.1969	1.0758	108.25	168.85	130.10	−131.12	220.23	−148.5027860
cc-pV5Z	1.0202	1.2457	1.1956	1.0754	108.37	168.88	130.49	−131.32	220.38	−148.5165868
cc-pwCVQZ(f.c.)	1.0204	1.2454	1.1965	1.0758	108.25	168.83	130.09	−131.22	220.32	−148.5061322
cc-pwCVQZ(all)	1.0190	1.2431	1.1930	1.0738	108.45	169.10	130.89	−131.97	221.17	−148.6733044
CBS[f]	1.0202	1.2457	1.1952	1.0752	108.42	168.89	130.65	−131.45	220.48	−148.5290394
CBS+CV[g]	1.0188	1.2434	1.1917	1.0732	108.62	169.16	131.45	−132.20	221.33	−148.6962117
CBS+CV+aug[h]	1.0195	1.2463	1.1899	1.0733	108.82	169.00	132.44	−133.21	222.15	−148.7082985

Extrapolated geometries and energies to valence correlation limit (CBS) and best estimates values (CBS+CV and CBS+CV+aug) are also reported

'f.c.' and 'all' mean CCSD(T) calculations correlating only valence and all electrons, respectively

[a] Dihedral angles

[b] Ref. [42]

[c] Ref. [11]

[d] Ref. [43]

[e] Ref. [41]

[f] CBS geometry and energy. See text

[g] CV correction added to the corresponding CBS limit. See text

[h] Best estimated geometry and energy. See text

standard notation, in Eq. 7, d_i and ω_i are the degeneracy and the harmonic frequency of the ith vibrational normal mode, respectively.

3 Results and discussion

3.1 Molecular structure

The equilibrium geometries and energies of the five CH_2N_2 isomers investigated in the present study as obtained at the CCSD(T) level employing different basis sets are summarized in Tables 1, 2, 3, 4 and 5. For all the molecules, the extrapolation to the CBS limit has been carried out as explained in the methodology section, and the results are also listed in Tables 1, 2, 3, 4 and 5. The subsequent inclusion of the core-correlation and diffuse-function corrections provided our best estimated values (denoted CBS+CV+aug).

From the results of Tables 1, 2, 3, 4 and 5, it is first observed that the valence correlation limit is nearly reached at the CCSD(T)/cc-pV5Z level; in fact, the discrepancies between r(V5Z) and r(CBS) are in all cases smaller than 0.0005 Å for bonds and in general smaller than 0.2 degrees for angles. The largest deviations are presented by dihedral angles; for instance, for the denoted T_2 angle of cyanamide, a discrepancy as large as 0.57 degrees is observed (see Table 4). As far as core-valence corrections are concerned, they are small but not negligible, that is, they are of the order of 0.001–0.004 Å for bond lengths and 0.1–0.3 degrees for angles. Once again, the largest effect is noted for the T_2 angle of cyanamide, for which a correction of 0.6 degrees is observed. With a few exception, the effects of the inclusion of diffuse functions in the basis set are small,

Table 2 Equilibrium geometries and energies of diazomethane, CH_2NN ($^1A_1$), computed at the coupled-cluster level of theory employing different basis sets

	C–H (Å)	C–N (Å)	N–N (Å)	∠HCN (deg.)	∠CNN (deg.)	T[a] (∠HNNN) (deg.)	Energy (Hartree)
B3LYP/6-311+G(d,p)[b]	1.078	1.295	1.137	118.0	180.0	180.0	
CASPT3/cc-pVTZ[c]	1.071	1.303	1.128	117.2	180.0	180.0	
CCSD(T)/aug-cc-pVTZ[d]	1.0760	1.3030	1.1434	117.35	180.0	180.0	
MRCI/cc-pVTZ[e]	1.068	1.295	1.125	117.4	180.0	180.0	−148.417234
B3LYP/6-311++G(3df,2p)[f]	1.0758	1.2910	1.1323	117.98	180.0	180.0	
aug-cc-pVTZ	1.0761	1.3029	1.1434	117.37	180.0	180.0	−148.5093657
cc-pVTZ	1.0751	1.3027	1.1426	117.39	180.0	180.0	−148.4987697
cc-pVQZ	1.0746	1.2996	1.1395	117.36	180.0	180.0	−148.5418702
cc-pV5Z	1.0745	1.2991	1.1389	117.34	180.0	180.0	−148.5552657
cc-pwCVQZ(f.c.)	1.0745	1.2993	1.1392	117.36	180.0	180.0	−148.5452495
cc-pwCVQZ(all)	1.0732	1.2965	1.1370	117.42	180.0	180.0	−148.7124624
CBS[g]	1.0745	1.2990	1.1388	117.34	180.0	180.0	−148.5674199
CBS+CV[h]	1.0732	1.2962	1.1366	117.40	180.0	180.0	−148.7346327
CBS+CV+aug[i]	1.0742	1.2964	1.1374	117.38	180.0	180.0	−148.7452287
Experiment[j] r_s	1.077	1.300	1.139	116.95	180.0	180.0	
Experiment[m] r_0/r_s	1.075	1.300	1.140	117.0	180.0	180.0	

Extrapolated geometries and energies to valence correlation limit (CBS) and best estimates values (CBS+CV and CBS+CV+aug) are also reported

'f.c.' and 'all' mean CCSD(T) calculations correlating only valence and all electrons, respectively

[a] Dihedral angle

[b] Ref. [11]

[c] Ref. [44]

[d] Ref. [45]

[e] Ref. [46]

[f] Ref. [41]

[g] CBS geometry and energy. See text

[h] CV correction added to the corresponding CBS limit. See text

[i] Best estimated geometry and energy. See text

[j] Refs. [50, 51]

[m] Refs. [52, 53]. For more details, see original papers

Table 3 Equilibrium geometries and energies of carbodiimide, HNCNH (^1A), computed at the coupled-cluster level of theory employing different basis sets

	N–H (Å)	C–N (Å)	∠HNC (deg.)	∠NCN (deg.)	T[a] (deg.)	Energy (Hartree)
CCSD(T)/cc-pVQZ+CV(cc-pCVTZ)[b]	1.0098	1.2242	117.84	170.62	44.59	
MP2/cc-pVQZ[c]	1.008	1.225	119.0	170.0	46.5	
CCSD/6-31G(d,p)[c]	1.102	1.233	116.9	170.5	46.8	
B3LYP/6-311++G(3df,2p)[d]	1.0105	1.2185	120.34	170.83	44.6	
aug-cc-pVTZ	1.0127	1.2302	117.38	170.25	44.53	−148.5564791
cc-pVTZ	1.0119	1.2300	116.70	170.57	44.62	−148.5438347
cc-pVQZ	1.0104	1.2264	117.50	170.53	44.59	−148.5888910
cc-pV5Z	1.0102	1.2257	117.78	170.50	44.59	−148.6028899
cc-pwCVQZ(f.c.)	1.0104	1.2261	117.51	170.54	44.59	−148.5922566
cc-pwCVQZ(all)	1.0090	1.2232	117.90	170.65	44.60	−148.7596429
CBS[e]	1.0102	1.2255	117.88	170.48	44.59	−148.6155649
CBS+CV[f]	1.0088	1.2226	118.27	170.59	44.60	−148.7829513
CBS+CV+aug[g]	1.0096	1.2228	118.95	170.27	44.51	−148.7955957
Experiment r_s^h	1.0074	1.2242	118.63	170.63	44.50	

Extrapolated geometries and energies to valence correlation limit (CBS) and best estimates values (CBS+CV and CBS+CV+aug) are also reported

'f.c.' and 'all' mean CCSD(T) calculations correlating only valence and all electrons, respectively

[a] The two dihedral angles $\angle H_1NCN$ and $\angle NCNH_2$ are defined as T + 180.0 degrees and T − 180.0 degrees, respectively

[b] Ref. [47]

[c] Ref. [48]

[d] Ref. [41]

[e] CBS geometry and energy. See text

[f] CV correction added to the corresponding CBS limit. See text

[g] Best estimated geometry and energy. See text

[h] Ref. [54]

the corrections being smaller than 0.001 Å for bonds and smaller than 0.5 degrees. From a comparison among the five isomers, we note that, while the N–H and C–H distances vary a little from one species to another, the N–N and C–N bond lengths may change significantly depending on the corresponding bond order. Before proceeding with commenting on the comparison of our results with previous theoretical investigations and available experimental data, a few comments on the expected accuracy of our molecular structures are warranted. On the basis of the magnitude of the various contributions and the literature on this topic (see for example Refs. [13, 34–40]), the accuracy of CBS+CV and CBS+CV+aug type geometries is expected to be of the order of 0.001–0.002 Å for distances and of about 0.05–0.1 degrees for angles. These can be considered as conservative estimates, which also account for the approximations made and for neglecting higher-order correlation effects.

From Tables 1, 2, 3, 4 and 5, it is evident that all isomers considered have been previously studied at different levels of theory [11, 41–48], but we report the first systematic investigation of their molecular structure. As in most cases the method and/or the basis set used are inferior to ours, a detailed comparison of these results to ours is not too meaningful. We only note an overall agreement. As concerns experiment, data are available only for diazomethane, carbodiimide and cyanamide [49–52, 54]. In all cases, we note a good agreement, and in particular, the recent, thorough investigation of the molecular structure of cyanamide, which provided the corresponding equilibrium geometry [49], allows us to confirm the accuracy estimate discussed above.

The results for the molecular structure of the HNNC radical, investigated following the methodology presented in the previous section, are given in Table 6. Concerning the convergence to the CBS limit, the magnitude of the CV corrections and the accuracy of the final CBS+CV geometry, we note that the conclusions drawn above for the five CH_2N_2 isomers also apply to this molecule, even if they are not straightforwardly applicable to open-shell species as for the latter the situation is more involved from a theoretical point of view.

As mentioned in the methodology section, the transition states along the isomerization path from cyanamide to carbodiimide and from diazomethane to nitrilimine have been optimized at the CCSD(T)/cc-pVTZ and CCSD(T)/

Table 4 Equilibrium geometries and energies of cyanamide, NH_2CN ($^1A'$), computed at the coupled-cluster level of theory employing different basis sets

	N–H (Å)	N–C (Å)	C–N (Å)	∠HNC (deg.)	∠NCN (deg.)	T_1^a (∠HNCH) (deg.)	T_2^a (∠NCNH) (deg.)	Energy (Hartree)
MP2/cc-pVQZ[b]	1.007	1.347	1.169	113.4	176.8			
B3LYP/6-311++G(3df,2p)[c]	1.0093	1.3392	1.1545	115.9	177.32	111.83	−111.83	
aug-cc-pVTZ	1.0109	1.3565	1.1658	113.05	176.81	128.26	−116.10	−148.5625089
cc-pVTZ	1.0101	1.3578	1.1655	112.47	176.79	126.55	−116.99	−148.5509322
cc-pVQZ	1.0086	1.3531	1.1621	113.07	176.86	128.33	−116.24	−148.5950219
cc-pV5Z	1.0084	1.3520	1.1615	113.28	176.93	129.04	−115.51	−148.6087072
cc-pwCVQZ(f.c.)	1.0086	1.3528	1.1618	113.09	176.86	128.40	−115.81	−148.5984090
cc-pwCVQZ(all)	1.0073	1.3493	1.1593	113.39	176.96	129.34	−115.26	−148.7658267
CBS[d]	1.0083	1.3517	1.1613	113.35	176.98	129.33	−114.94	−148.6211350
CBS+CV[e]	1.0070	1.3482	1.1588	113.65	177.08	130.27	−114.39	−148.7885527
CBS+CV+aug[f]	1.0078	1.3469	1.1591	113.23	177.10	132.08	−113.50	−148.8001294
Experiment r_e, I (semi-exp)[g]	1.0063(1)	1.3475(4)	1.1594(4)	112.69(2)	[176.96]			
Experiment r_e, II (semi-exp)[g]	1.0059(1)	1.3470(2)	1.1594(2)	112.74(2)	178.22(17)			
Experiment r_m^{1L} [g]	1.0065(15)	1.3450(8)	1.1620(8)	112.62(14)	[180.0]			

Extrapolated geometries and energies to valence correlation limit (CBS) and best estimates values (CBS+CV and CBS+CV+aug) are also reported 'f.c.' and 'all' mean CCSD(T) calculations correlating only valence and all electrons, respectively

[a] Dihedral angle

[b] Ref. [48]

[c] Ref. [41]

[d] CBS geometry and energy. See text

[e] CV correction added to the corresponding CBS limit. See text

[f] Best estimated geometry and energy. See text

[g] Ref. [49]. Angle between the N–C bond and the ∠(HNH) bisector: 42.64(32) [r_m^{1L}], 43.35(2) [r_e (semi-exp),I] and 42.78(9) [r_e (semi-exp),II] degrees

cc-pVQZ levels of theory. Harmonic frequency calculations confirmed the transition state character of the optimized geometries. The corresponding results are collected in Table 7. We note that in both cases, the transition state is a cyclic, non-planar form. For the first isomerization, the N–C–N frame is close to linearity and one H connects the two nitrogens. For the second one, the C–N–N frame is strongly bent, forming an angle close to 90 degrees. According to Ref. [55], the accuracy of molecular structures optimized at the CCSD(T)/cc-pVQZ is expected to be of the order of 0.002–0.004 Å for distances and 0.1–0.3 degrees for angles. From the energy values reported in Tables 1, 2, 3, and 4, the barriers to isomerization have been derived. We note that in both cases, the barrier is high, and therefore, we expect that the isomerization hardly happens.

3.2 Energetics

On the basis of the total energies summarized in Tables 1, 2, 3, 4 and 5, the relative stability of the five isomers can be discussed. By including the results for diazirine [7] (the extrapolation to the CBS limit has been repeated following the methodology introduced above), the relative energies of the most stable CH_2N_2 isomers have been obtained. These are reported in Table 8 with respect to cyanamide, which is the lowest-lying isomer, and schematized in Fig. 1. First of all, the stability order, given as energetic order, is:

$$\text{cyanamide} < \text{carbodiimide} < \text{diazomethane} < \text{diazirine} < \text{isocyanamide} < \text{nitrilimine},$$

where cyanamide is only ~3 kcal/mol stabler than carbodiimide, but as discussed above, the two isomers are separated by a barrier as high as ~80 kcal/mol. The other isomers are definitely less stable, with diazomethane ~32 kcal/mol higher in energy, diazirine ~43 kcal/mol, isocyanamide ~45 kcal/mol and nitrilimine ~56 kcal/mol. There is thus a small energy difference between diazirine and isocyanamide, but there is no direct path connecting these two isomers. From Table 8, we note that the convergence to the CBS limit is smooth, and the CCSD(T)/cc-pV5Z results are already converged; in fact, the differences between the latter and the CBS values are smaller than 0.2 kcal/mol. By enlarging the basis set, there is not an unique trend: the energy difference increases for

Table 5 Equilibrium geometries and energies of isocyanamide, NH_2NC ($^1A'$), computed at the coupled-cluster level of theory employing different basis sets

	N–H (Å)	N–N (Å)	N–C (Å)	∠HNN (deg.)	∠NNC (deg.)	T_1^a (∠*HNNH*) (deg.)	T_2^a (∠CNNH) (deg.)	Energy (Hartree)
B3LYP/6-311+G(d,p)[b]	1.016	1.356	1.173	109.3	174.1	120.0		
B3LYP/6-311++G(3df,2p)[c]	1.0146	1.3547	1.1676	109.29	174.23	120.05	−120.05	
aug-cc-pVTZ	1.0159	1.3700	1.1788	107.41	174.22	116.28	−122.28	−148.4917144
cc-pVTZ	1.0148	1.3692	1.1789	107.29	174.33	115.85	−123.47	−148.4800091
cc-pVQZ	1.0133	1.3648	1.1751	107.62	174.40	116.65	−122.24	−148.5240210
cc-pV5Z	1.0132	1.3638	1.1744	107.75	174.42	116.95	−121.84	−148.5374871
cc-pwCVQZ(f.c.)	1.0134	1.3643	1.1746	107.64	174.34	116.67	−122.19	−148.5273196
cc-pwCVQZ(all)	1.0120	1.3607	1.1719	107.87	174.55	117.19	−119.27	−148.6941112
CBS[d]	1.0132	1.3635	1.1740	107.80	174.43	117.13	−121.70	−148.5497324
CBS+CV[e]	1.0118	1.3599	1.1713	108.03	174.64	117.65	−118.78	−148.7165009
CBS+CV+aug[f]	1.0129	1.3607	1.1712	108.15	174.53	118.08	−117.59	−148.7282062

Extrapolated geometries and energies to valence correlation limit (CBS) and best estimates values (CBS+CV and CBS+CV+aug) are also reported

'f.c.' and 'all' mean CCSD(T) calculations correlating only valence and all electrons, respectively

[a] Dihedral angle

[b] Ref. [11]

[c] Ref. [41]

[d] CBS geometry and energy. See text

[e] CV correction added to the corresponding CBS limit. See text

[f] Best estimated geometry and energy. See text

Table 6 Equilibrium geometries and energies of isocyanomidyl radical, HNNC ($^2A''$), computed at the coupled-cluster level of theory employing different basis sets

	H–N (Å)	N–N (Å)	N–C (Å)	∠HNN (deg.)	∠NNC (deg.)	Energy (Hartree)
MP2/cc-pVTZ(all)[a]	1.021	1.298	1.180	105.6	173.9	
B3LYP/6-311++G(3df,2p)[b]	1.0281	1.2870	1.1879	107.37	171.44	
CCSD(T)/aug-cc-pVTZ[c]	1.024	1.297	1.191	106.1	171.6	
aug-cc-pVTZ	1.0284	1.3037	1.1961	105.86	171.72	−147.8364668
cc-pVTZ	1.0276	1.3035	1.1959	105.66	172.22	−147.8262512
cc-pVQZ	1.0260	1.2990	1.1924	105.97	172.14	−147.8674980
cc-pV5Z	1.0258	1.2981	1.1918	106.07	172.11	−147.8803260
cc-pwCVQZ(f.c.)	1.0261	1.2987	1.1920	105.97	172.11	−147.8708520
cc-pwCVQZ(all)	1.0248	1.2957	1.1893	106.14	172.24	−148.0370570
CBS[d]	1.0257	1.2978	1.1916	106.11	172.10	−147.8920230
CBS+CV[e]	1.0244	1.2948	1.1889	106.28	172.23	−148.0582280
CBS+CV+aug[f]	1.0252	1.2950	1.1891	106.48	171.73	−148.0684436

Extrapolated geometries and energies to valence correlation limit (CBS) and best estimates values (CBS+CV and CBS+CV+aug) are also reported

'f.c.' and 'all' mean (R/U)CCSD(T) calculations correlating only valence and all electrons, respectively

[a] Ref. [56]

[b] Ref. [41]

[c] Ref. [57]

[d] CBS geometry and energy. See text

[e] CV correction added to the corresponding CBS limit. See text

[f] Best estimated geometry and energy. See text

Table 7 CCSD(T) optimized geometries and energies for the carbodiimide–cyanamide transition state (TS1) and the nitrilimine–diazomethane transition state (TS2)

	H_1–N_2 (Å)	N_2–C_3 (Å)	C_3–N_4 (Å)	N_4–H_5 (Å)	$\angle H_1N_2C_3$ (deg.)	$\angle N_2C_3N_4$ (deg.)	$\angle C_3N_4H_5$ (deg.)	T_1^a (deg.)	T_2^a (deg.)	Energy[b] (cyanamide) (kcal/mol)	Energy[b] (carbodiimide) (kcal/mol)
TS1											
cc-pVTZ	1.7722	1.2186	1.3417	1.0215	40.35	188.09	108.54	75.43	186.37	83.8	79.3
cc-pVQZ	1.7418	1.2130	1.3337	1.0196	41.71	187.93	109.24	99.33	187.93	83.7 (79.1)	79.8 (76.2)

	H_1–C_2 (Å)	C_2–N_3 (Å)	N_3–N_4 (Å)	N_4–H_5 (Å)	$\angle H_1C_2N_3$ (deg.)	$\angle C_2N_3N_4$ (deg.)	$\angle N_3N_4H_5$ (deg.)	T_1^a (deg.)	T_2^a (deg.)	Energy[b] (diazomethane) (deg.)	Energy[b] (nitrilimine) (deg.)
TS2											
cc-pVTZ	1.1086	1.4321	1.2551	1.1543	103.16	88.87	104.10	−98.56	−7.47	81.5	56.4
cc-pVQZ	1.1072	1.4271	1.2513	1.1582	103.36	88.79	103.87	−98.64	−7.56	82.2 (79.1)	57.7 (54.5)

[a] For TS1: $T_1 = \angle H_1N_2C_3N_4$ and $T_2 = \angle N_2C_3N_4H_5$. For TS2: $T_1 = \angle H_1C_2N_3N_4$ and $T_2 = \angle C_2N_3N_4H_5$

[b] Relative energy with respect to the isomer given below. In parentheses, the ZPV corrected value is given

Table 8 Relative energy (kcal/mol) of CH_2N_2 isomers with respect to cyanamide at the coupled-cluster level of theory employing different basis sets

	Carbodiimide	Diazomethane	Diazirine	Isocyanamide	Nitrilimine
B3LYP/6-311++G(3df,2p)[a]	−1.59	28.37	45.07	44.39	51.78
cc-pVQZ	3.85	33.35	42.38	44.55	57.88
cc-pV5Z	3.65	33.54	42.46	44.69	57.81
CBS	3.50	33.71	42.53	44.81	57.79
CBS+CV	3.51	33.84	43.22	45.21	57.94
CBS+CV+ZPV[b]	2.60	32.14	42.67	44.95	56.39
CBS+CV+aug	2.84	34.45	–	45.13	57.62
CBS+CV+aug+ZPV[b]	1.94	32.75	–	44.87	56.07

CBS, CBS+CV and CBS+CV+aug values are also reported

[a] ZPV correction included, Ref. [41]

[b] ZPV corrections at the CCSD(T)/cc-pVQZ level within the harmonic approximation

diazomethane, diazirine and isocyanamide, while it decreases for carbodiimide and nitrilimine. CV corrections are in all cases positive, but for their extent, a general conclusion cannot be drawn; in fact, they vary from being negligible, 0.01 kcal/mol, in the case of carbodiimide to being relevant, 0.7 kcal/mol, for diazirine. In all cases, inclusion of diffuse functions in the basis set leads to corrections smaller than 1 kcal/mol: they range from about 0.3 kcal/mol for nitrilimine to about 0.9 kcal/mol for isocyanamide. While they are positive for diazomethane and isocyanamide, they are negative for carbodiimide and nitrilimine. To the best of our knowledge, the comparison to previous theoretical calculations is limited to those at the B3LYP/6-311++G(3df,2p) level of Ref. [41]. Differences of some kcal/mol are observed, and in the latter work, a wrong stability order was obtained with carbodiimide predicted stabler than cyanamide and isocyanamide stabler than diazirine. In view of the high-level calculations carried out in the present investigation, we consider our results more reliable.

From the total energies reported in Tables 1, 2, 3, 4, 5, and 6 and those available in Ref. [13], the C–H and N–H equilibrium dissociation energies corresponding to the following reactions can be evaluated:

$$H_2NCN \rightarrow HNCN + H, \quad (8)$$

$$HNCNH \rightarrow HNCN + H, \quad (9)$$

$$H_2CNN \rightarrow HCNN + H, \quad (10)$$

$$HCNNH \rightarrow HCNN + H, \quad (11)$$

$$HCNNH \rightarrow HNNC + H, \quad (12)$$

$$H_2NNC \rightarrow HNNC + H. \quad (13)$$

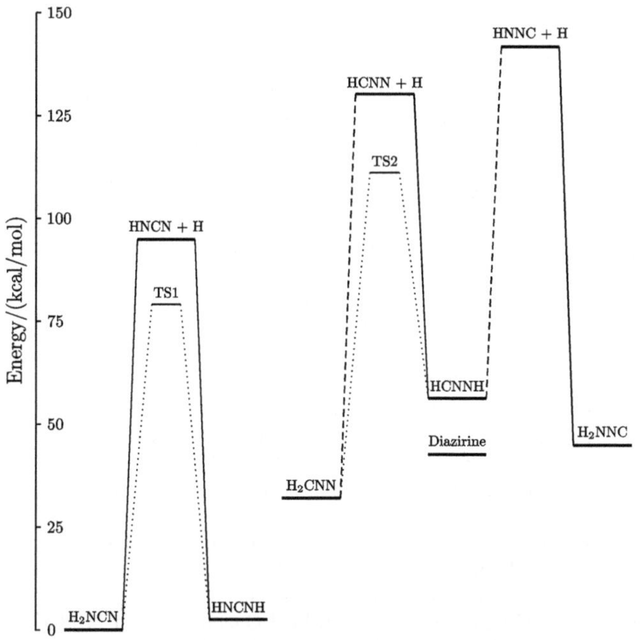

Fig. 1 Relative energy levels for the CH_2N_2 isomers with respect to cyanamide. Transition states TS1 and TS2 are also reported *(dotted lines)*. The dissociation limit involving HCNN, HNCN and HNNC radicals is indicated: C–H bond cleavage in dashed line and N–H bond breaking in continuous line. All data taken from Tables 7, 8, and 9

The results obtained are collected in Table 9. Here, we mention that for HCNN and HNCN, the required extrapolations of the total energies reported in Ref. [13] have been carried out following the methodology presented above. The total energies extrapolated to the CBS limit have been employed in order to determine the complete basis set limit of the dissociation energies. First of all, we note that by enlarging the basis set, the dissociation energy increases. As noted for relative energies, the convergence is smooth and essentially reached at the CCSD(T)/cc-pV5Z level, the differences with respect to the CBS limit being smaller than 0.4 kcal/mol. Subsequently, by adding the core-valence corrections to the CBS limit, best estimates of the equilibrium dissociation energies have been derived. CV corrections are positive and of the order of 0.4–0.6 kcal/mol. For the two reactions involving the dissociation to HNNC+H, the effect of additional diffuse functions in the basis set can be also estimated. The corresponding corrections are positive and on the order of 1 kcal/mol. On the basis of the above discussion and of the literature on this topic (see, for example, Refs. [37–39, 55]), our best estimated dissociation energies are expected to fulfill the chemical accuracy, that is, they are expected to have an accuracy better than 1 kcal/mol. The addition of the appropriate zero-point differences evaluated from the harmonic frequencies (Eq. 7) provides the D_0 dissociation energies, which are also reported in Table 9. As expected, ZPV corrections lower the dissociation energies, and they are important corrections to be taken into account for this property as they amount to 7–9 kcal/mol. As well known, the use of the harmonic approximation overestimates the ZPV corrections; in the present case, the relevant term is the anharmonic contribution connected to C–H or N–H bonds which can be estimated on the order of −0.3 to −0.5 kcal/mol. Concerning the literature, to the best of our knowledge, there are no experimental data to compare with, while the theoretical values are limited to those at the B3LYP/6-311++G(3df,2p) level deducible from the energies obtained in Ref. [41]. In view of the limited accuracy of the B3LYP/6-311++G(3df,2p) level pointed out when discussing relative energies (see Table 8), the corresponding results are not collected in Table 9.

4 Concluding remarks

The structural and energetics of cyanamide, carbodiimide, diazomethane, isocyanamide and nitrilimine as well as of the HNNC radical have been accurately investigated using the CCSD(T) method with systematic treatments of basis

Table 9 C–H and N–H dissociation energies (kcal/mol) for the five CH_2N_2 isomers considered

	cc-pVQZ	cc-pV5Z	CBS	CBS+CV	CBS+CV+ZPV[a]	CBS+CV+aug	CBS+CV+aug+ZPV[a]
C–H dissociation							
$H_2CNN \rightarrow HCNN + H$	105.04	105.34	105.46	105.93	98.18		
$HCNNH \rightarrow HNNC + H$	84.93	85.51	85.98	86.59	78.62	87.87	79.90
N–H dissociation							
$H_2NCN \rightarrow HNCN + H$	103.09	103.51	103.79	104.16	94.96		
$H_2NNC \rightarrow HNNC + H$	98.25	98.63	98.96	99.32	90.05	100.36	91.10
$HCNNH \rightarrow HCNN + H$	80.52	81.07	81.38	81.82	73.93		
$HNCNH \rightarrow HNCN + H$	99.24	99.86	100.29	100.65	92.36		

[a] ZPV corrections at the CCSD(T)/cc-pVQZ level within the harmonic approximation

set truncation errors, and core-valence correlation and diffuse-function effects. Where available the present results are in excellent agreement with experiment. As far as molecular structures are concerned, the accuracy obtained in the present study is estimated to be 0.001–0.002 Å for bond lengths and 0.05–0.1 degrees for angles. The relative and dissociation energies derived are estimated to fulfill the so-called chemical accuracy, that is, they are expected to be affected by uncertainties of the order of 1 kcal/mol.

Acknowledgments This work has been supported by University of Bologna (RFO funds). Authors also thank the CINECA supercomputing center for a grant of computer time on the IBM SP5 and SP6 (HP10C7LE76) machines.

References

1. Maier-Greiner UH, Klaus BA, Estermeier LM, Hartmann GR (1991) Angew Chem Int Ed Engl 30:1314
2. Turner BE, Kislayakov AG, List HS, Kaifu N (1975) Astrophys J 201:149
3. Matthews HE, Sears T (1986) Astrophys J 300:766
4. Williams A, Ibrahim IT (1981) Chem Rev 81:589
5. Streitwieser A Jr, Heathcock CH (1985) Introduction to organic chemistry, 3rd edn. Macmillan Publishing Company, New York
6. Liu, MTH (eds) (1987) The chemistry of diazirines, vol I and II. CRC Press, Boca Raton
7. Puzzarini C, Gambi A (2004) Chem Phys 306:131
8. Cambi R, von Niessen W, Schirmer J (1984) Chem Phys 86:389
9. Schäfer E, Winnewisser M (1982) Ber Bunsenges Chem Phys 86:780
10. Bertrand G, Wentrup C (1994) Angew Chem Int Ed Engl 33:527
11. Goldberg N, Fiedler A, Schwarz H (1994) Helv Chim Acta 77:2354
12. Maier G, Eckwerk J, Bothur A, Reisenauer HP, Schmidt C (1996) Liebigs Ann 1041
13. Puzzarini C, Gambi A (2005) J Chem Phys 122:064316
14. Basco N, Yee KK (1968) Chem Commun (Lond) 150
15. Werner H-J, Knowles PJ, Lindh R, Manby FR, Schütz M, Celani P, Korona T, Rauhut G, Amos RD, Bernhardsson A, Berning A, Cooper DL, Deegan MJO, Dobbyn AJ, Eckert F, Hampel C, Hetzer G, Lloyd AW, McNicholas SJ, Meyer W, Mura ME, Nicklass A, Palmieri A, Pitzer R, Schumann U, Stoll H, Stone A J, Tarroni R, Thorsteinsson T (2006) MOLPRO, version 2006.1, a package of ab initio programs
16. Raghavachari K, Trucks GW, Pople JA, Head-Gordon M (1989) Chem Phys Lett 156:479
17. Hampel C, Peterson KA, Werner HJ (1992) Chem Phys Lett 190:1
18. Lee TJ, Taylor PR (1989) Int J Quantum Chem Symp 23:199
19. Lee TJ, Scuseria GE (1995) In: Langhoff SR (ed) Quantum mechanical electronic structure calculations with chemical accuracy. Kluwer, Dordrecht, p 47
20. Werner H-J, Knowles PJ (1985) J Chem Phys 82:5053
21. Knowles PJ, Werner H-J (1985) Chem Phys Lett 115:259
22. Dunning TH Jr (1989) J Chem Phys 90:1007
23. Eckert F, Pulay P, Werner H-J (1997) J Comp Chem 18:1473
24. El Azhary A, Rauhut G, Pulay P, Werner H-J (1998) J Chem Phys 108:5185
25. Puzzarini C (2009) J Phys Chem 113A:14530
26. Helgaker T, Klopper W, Koch H, Noga J (1997) J Chem Phys 106:9639
27. Halkier A, Helgaker T, Jørgensen P, Klopper W, Koch H, Olsen J, Wilson AK (1998) Chem Phys Lett 286:243
28. Feller D (1992) J Chem Phys 96:6104
29. Feller D (1993) J Chem Phys 98:7059
30. Peterson KA, Dunning TH Jr (2002) J Chem Phys 117:10548
31. Scuseria GE (1991) Chem Phys Lett 176:27
32. Watts JD, Gauss J, Bartlett RJ (1993) J Chem Phys 98:8718
33. Knowles PJ, Hampel C, Werner H-J (1994) J Chem Phys 99:5219
34. Puzzarini C (2005) J Chem Phys 123:024313
35. Puzzarini C, Cazzoli G, Gambi A, Gauss J (2006) J Chem Phys 125:054307
36. Puzzarini C, Cazzoli G, Baldacci A, Baldan A, Michauk C, Gauss J (2007) J Chem Phys 127:164302
37. Puzzarini C (2008) Chem Phys 346:45
38. Puzzarini C, Barone V (2010) Int J Quantum Chem 110:637
39. Puzzarini C (2010) Int J Quantum Chem 110:2483
40. Puzzarini C, Barone V (2011) Phys Chem Chem Phys 13:7158
41. Du B, Zhang W (2006) Int J Quantum Chem 106:1827
42. Wong MW, Wentrup C (1993) J Am Chem Soc 115:7743
43. Kuhler K, Palmer RT, Wittkamp BL, Hoffmann MR (1996) J Mol Struct (THEOCHEM) 360:41
44. Papakondylis A, Mavridis A (1999) J Phys Chem A 103:1255
45. Dixon DA, de Jong WA, Peterson KA, McMahon TB (2005) J Phys Chem A 109:4073
46. Kerkines ISK, P. Carsky P, A. Mavridis A (2005) J Phys Chem A 109:10148
47. Koput J, Jabs W, Winnewisser M (1998) Chem Phys Lett 295:462
48. Tordini F, Bencini A, Bruschi M, De Gioia L, Zampella G, Fantucci P (2003) J Phys Chem A 107:1188
49. Kraśnicki A, Kisiel Z, Jabs W, Winnewisser BP, Winnewisser M (2011) J Mol Spectrosc 267:144
50. Cox AP, Thomas LF, Sheridan J (1958) Nature (Lond) 181:1000
51. Sheridan J (1962) Adv Mol Spectrosc Proc IVth Int Meet Mol Spectrosc 1:139
52. Harmony MD, Laurie VW, Kuczkowski RL, Schwendeman RH, Ramsay DA, Lovas FJ, Lafferty WJ, Maki AG (1979) J Phys Chem Ref Data 8:619
53. Moore CB (1963) J Chem Phys 39:1884
54. Jabs W, Winnewisser M, Belov SP, Lewen F, Maiwald F, Winnewisser G (1999) Molec Phys 97:213
55. Helgaker T, Jørgensen P, Olsen J (2000) Electronic-structure theory. Wiley, Chichester
56. Kurosaki Y, Takayanagi T (1999) J Phys Chem A 103:9323
57. Berman MR, Tsuchiya T, Gregušová A, Perera SA, Bartlett RJ (2007) J Phys Chem A 111:6894

Theor Chem Acc (2012) 131:1140
DOI 10.1007/s00214-012-1140-z

REGULAR ARTICLE

Computational ^{19}F NMR. 1. General features

Giacomo Saielli · Riccardo Bini · Alessandro Bagno

Please note an erratum was published subsequent to publication of the original article. For reasons of consistency with the electronic record, the article and the corresponding erratum are reproduced here as published in the journal.

Received: 14 July 2011 / Accepted: 2 September 2011 / Published online: 15 February 2012

Abstract Fluorine-19 NMR chemical shifts have been calculated for a wide variety of fluorine-containing inorganic and organic molecules by relativistic DFT methods. The agreement with experimental values, spanning the whole range from ClF to FOOF, is satisfactory but somewhat less accurate than for comparable light nuclei. ^{19}F shifts in uranium chlorofluorides have been analyzed in detail, and the poor agreement with experiment is partly rationalized.

Keywords DFT calculations · NMR spectroscopy · Structure elucidation · Fluorine

1 Introduction

Fluorine is an outstanding chemical element, in that it forms compounds with practically all other elements, including most noble gases, and—despite being scarce in living systems—a wide variety of organic molecules. Indeed, many fluorinated molecules and polymers are of paramount industrial and economic importance.

The structural elucidation of fluorinated molecules generally takes advantage of NMR spectroscopy through the fluorine-19 nucleus ($I = ½$, 100% natural abundance), to the extent that ^{19}F belongs to the so-called "common" nuclei. In fact, ^{19}F NMR exhibits a number of favorable factors for experimental studies, like a high magnetogyric ratio (close to that of ^{1}H), narrow spectral lines, and wide chemical shift range. It then comes as no surprise that the amount of published ^{19}F NMR data is impressive. Such data (which include chemical shifts and coupling constants) span organic and inorganic species and have been collected in several reviews and books [1–3].

Many ^{19}F spectra exhibit several resonances that may not always be assigned in a straightforward manner. Moreover, unusual fluorine compounds may give rise to spectra that are difficult to interpret. In such cases, it would be desirable to deploy modeling tools to enable such assignments independently. Finally, the widespread use of fluorinated drugs (of which fluorouracil is the best known example) offers opportunities for ^{19}F NMR to investigate their metabolic paths [4–7]. Lately, incorporation of fluorine in unnatural amino acids has been proposed as a method for spectral simplification [8].

The known chemical shift range spanned by general fluorine compounds is ca. 1,300 ppm, but much narrower (ca. 300 ppm) in organofluorine compounds. The high sensitivity of such shifts to even modest changes in electronic structure has been widely exploited for the investigation of substituent effects. Thus, a large database of ^{19}F shifts in variously substituted fluorobenzene (but also bicyclo[2.2.2]octane etc.) derivatives is also available [3, 9].

Couplings involving ^{19}F are noteworthy for extending over more bonds than is typical for other nuclei; ^{4}J and ^{5}J are very common and of substantial magnitude. Indeed, several such long-range couplings have been ascribed to a through-space overlap of non-bonding orbitals, thereby

Dedicated to Professor Vincenzo Barone and published as part of the special collection of articles celebrating his 60th birthday.

G. Saielli
Istituto per la Tecnologia delle Membrane del CNR, Unità di Padova, Via Marzolo 1, 35131 Padua, Italy

R. Bini · A. Bagno (✉)
Dipartimento di Scienze Chimiche, Università di Padova, Via Marzolo 1, 35131 Padua, Italy
e-mail: alessandro.bagno@unipd.it

giving rise to "through-space couplings", which have been known for a long time [10–13].

Recent advancements in density functional theory (DFT) have allowed to model the NMR properties of a wide variety of nuclei in virtually all kinds of molecules and complexes [14]. It is in fact possible to predict the required molecular properties (nuclear shieldings and couplings) to an accuracy that enables one to use such predictions for structure determination. This has been demonstrated for "routine" nuclei such as ^{1}H or ^{13}C [15–18] as well as for unusual heteronuclei as ^{99}Ru, ^{119}Sn, ^{183}W, ^{195}Pt, or ^{199}Hg (to mention just those for which fairly systematic studies have been carried out) [18–25].

Nevertheless, and despite the obvious advantages outlined above, ^{19}F NMR has not been extensively studied by computation. Some early computational work (aimed at benchmarking and validation) included F_2 as test molecule, which proved to be challenging [26–28]. Interest in modeling ^{19}F shifts has steadily continued, and several papers have appeared in the last decade aimed at general structural issues, solution chemistry and physical organic chemistry, [29–46] solid-state issues (especially the prediction of chemical shift tensors), [47–64] *J*-couplings [65–76], and paramagnetic shifts [77, 78]. However, little if any attempt at a systematic investigation has been presented [41].

Reasons for this nucleus lagging behind others might also include the well-known sensitivity of ^{19}F chemical shifts to environmental effects such as solvation, which places some uncertainty on the experimental data (indeed, spreads over tens of ppm are not uncommon in published data sources).

In this preliminary broad-range investigation, we have probed into the feasibility and accuracy of DFT predictions for the widest possible variety of molecular fluorine compounds (several inorganic fluorides are not molecular, but are rather complicated oligomeric or polymeric systems). Several simple organofluorine derivatives were also included, with no attempt at modeling the specific features of fluorinated functional groups. In order to accommodate the wide range of structures for which data are available, which include heavy-element fluorides, a relativistic method has been consistently adopted (see Sect. 2).

Particular attention was devoted to the ^{19}F chemical shifts of uranium(VI) chlorofluorides, UF_nCl_{6-n} [79]. For such species, it proved impossible to model their ^{19}F shifts with a wide choice of density functionals and effective core potential basis sets, and the origin of such a failure was amply discussed [80–83]. However, all such calculations were not calibrated against a well-behaved data set, whereas such a comparison is expected to highlight possible shortcomings of these particular compounds with respect to fluorides at large. Drawing on such detailed work, we ran a corresponding set of calculations in the context of a broader investigation, but with a wider choice of theoretical levels than for most other fluorine compounds.

2 Computational details

Most calculations were carried out with ADF [84, 85] and associated routines for NMR shieldings [27, 86, 87]. In most cases, we have adopted GGA (BP86 [88, 89] and BLYP [88, 90–92]) and the hybrid PBE0 [93, 94] functionals with the zero-order regular approximation (ZORA); a triple-zeta, twice-polarized Slater basis set (TZ2P) was employed for all atoms. Geometry optimizations were performed at the ZORA scalar (ZSC) level, whereas NMR calculations at the ZORA spin–orbit (ZSO) level. For U(VI) compounds, we also employed the SAOP model functional [95] together with the same relativistic treatment and basis set; the latter calculations were run at the PBE0 geometry. Calculated chemical shifts are referenced to $CFCl_3$ ($\sigma_{ref} = 133$, 120, 182 and 196 ppm with BP86, BLYP, PBE0 and SAOP, respectively) and are the sum of the diamagnetic, paramagnetic, and spin–orbit terms ($\sigma = \sigma_d + \sigma_p + \sigma_{SO}$); thus, $\delta_{calc} = \sigma_{ref} - \sigma$. The effect of dispersive interactions on the geometry was evaluated by running optimizations with Grimme's DFT-D3 correction term [96, 97] to the hybrid B3LYP [98] functional. For this protocol, shieldings have been calculated at the optimized geometries with the hybrid PBE0 functional. Some additional calculations were performed with the dispersion including M06 functional [99]. These calculations employed Gaussian 09 [100] with the 6-311G(d,p) basis for geometry optimization and cc-pVTZ for NMR shieldings, without relativistic corrections. For such results, the reference shielding is 166 ppm. Finally, in selected instances, MP2 ab initio calculations were run, using the cc-pVQZ both during optimization and for shielding calculations ($\sigma_{ref} = 208$ ppm). The results are evaluated from the statistical error of the least-squares regression $\delta_{calc} = b\delta_{exp} + a$ and the mean absolute error ($MAE = \Sigma_i|\delta_i^{calc} - \delta_i^{exp}|/n$).

3 Results

For this study, we restricted our scope to a variety of fluoro compounds spanning the whole known chemical shift range, from the most shielded (ClF, $\delta = -448$) to the most deshielded (FOOF, $\delta = 825$ ppm). Compounds featuring bonds to B, C, N, O, P, Si, S, Cl, Se, Ge, As, Xe, W, and U forming well-defined molecular species were included. The complete list of compounds is presented in Table 1 together with literature sources of experimental shifts and the corresponding calculated values. Calculated shifts are plotted against experimental values in Figs. 1 and 2.

Table 1 Experimental and calculated ^{19}F chemical shifts

Species	δ_{exp}	Refs.	BP86	BLYP	PBE0	PBE0-D3	M06
$CFCl_3$	0.0	[1]	0.0	0.0	0.0	0.0	0.0
$CFHCl_2$	−80.8	[2]	−97.3	−98.3	−87.4	−90.2	79.3
CF_2Cl_2	−6.8	[1]	−17.0	−19.5	−9.2	−10.9	−12.2
CF_2HCl	−71.8	[2]	−94.1	−97.5	−78.4	−82.7	−81.0
CF_4	−61.7	[2]	−88.1	−93.2	−67.0	−71.0	−76.3
CHF_3	−78.3	[2]	−105.5	−110.5	−84.6	−89.9	−90.6
CF_3Cl	−28.6	[1]	−46.8	−50.8	−31.8	−35.3	16.2
CF_3Br	−17.9	[2]	−32.7	−36.2	−19.5	−22.6	−28.7
CH_2F_2	−143.4	[2]	−175.8	−180.7	−151.7	−158.2	−155.2
CH_3F	−271.9	[2]	−322.2	−330.1	−288.8	−295.9	−285.3
CH≡CF	−210.0	[1]	−258.6	−272.2	−226.3	−233.9	−220.1
CF≡CF	−261.3	[3]	−319.0	−330.9	−283.2	−290.1	−278.9
C_6F_6	−162.9	[1]	−189.6	−199.4	−169.0	−175.2	−168.5
CF_3OH	−54.5	[1]	−82.9	−87.5	−62.4	−66.4	−71.6
CH_3OF	120.3	[1]	121.5	142.7	58.5	76.5	68.4
C**F**$_3$OF	−72.3	[1]	−96.5	−100.5	−78.4	−82.0	−85.1
CF_3O**F**	147.1	[1]	150.3	184.0	78.2	109.6	84.2
$(CF_3)_3N$	−58.6	[1]	−79.6	−85.1	−62.5	−66.4	−70.5
F_2C=O	−23.0	[1]	−37.9	−45.0	−21.1	−24.9	−28.2
F_2C=S	40.5	[1]	39.2	33.5	51.3	48.8	52.7
$(CF_3)_2C$=O	−84.6	[1]	−102.2	−106.9	−83.7	−88.3	−91.7
FCN	−156.0	[1]	−203.9	−216.6	−180.7	−186.6	−174.0
CF_3CN	−56.3	[1]	−78.7	−83.4	−63.1	−67.2	−72.8
F_2	422.9	[1]	410.1	461.8	321.7	370.6	369.7
FOH	21.0	[1]	−0.1	23.7	−40.0	−21.4	−29.8
OF_2	250.0	[1]	250.4	290.9	184.3	215.2	177.3
FOOF	825.0	[1]	829.8	944.0	423.8	548.4	603.5
FOOF (exptl geometry)	825.0	[1]	752.0	755.0	762.6		858.6
$FOClO_3$	219.4	[1]	205.7	237.9	145.4	168.2	134.0
FOSF_5	188.0	[1]	200.2	232.4	126.3	127.0	156.6
NH_2F	−155.6	[1]	−221.3	−275.8	−220.0	−215.8	−214.8
NHF_2	−11.2	[1]	−6.8	−104.9	−27.3	−17.3	−39.9
NF_3	146.9	[1]	170.9	195.8	124.5	142.4	109.3
$NFCl_2$	129.0	[1]	156.9	184.4	92.6	112.9	95.1
NF_2=NF_2	60.4	[1]	58.9	79.9	27.0	44.0	19.7
(*E*)-FN=NF	94.9	[1]	163.8	125.9	127.4	148.2	135.3
F_3NO	367.0	[1]	429.1	491.3	345.3	399.9	320.1
BF_3	−126.8	[1]	−164.0	−176.9	−132.1	−141.1	−132.0
BBr_2F	−0.9	[1]	−11.3	−21.5	11.6	4.1	12.2
CH_3BF_2	−68.8	[1]	−98.1	−111.4	−68.6	−77.7	−7.7
$(CH_3)_2BF$	−19.3	[1]	−36.2	−49.4	−12.6	−22.7	−22.4
SF_6	56.5	[1]	99.1	112.6	78.9	90.0	88.5
PF_3	−34.0	[1]	−48.2	−56.2	−28.0	−34.2	−43.1
POF_3	−88.9	[1]	−109.6	−120.4	−87.1	−92.5	−89.0
ClF	−419.4	[1]	−497.9	−496.6	−484.8	−480.3	−482.9
ClF_3 (1F)	7.7	[1]	−12.7	53.9	19.4	35.2	68.5
ClF_3 (2F)	118.1	[1]	266.7	279.6	156.1	180.6	194.8
ClF_5 (1F)	428.8	[1]	556.7	631.3	462.2	515.3	598.3

Table 1 continued

Species	δ_{exp}	Refs.	BP86	BLYP	PBE0	PBE0-D3	M06
ClF_5 (4F)	259.8	[1]	354.3	407.4	276.3	314.5	335.0
SiH_3F	−217.0	[1]	−273.8	−286.2	−231.6	−242.5	−229.2
$(CH_3)_3SiF$	−159.0	[1]	−211.0	−223.5	−175.9	−185.7	−175.0
SiF_4	−163.3	[1]	−200.8	−215.4	−167.1	−176.1	−167.1
SiF_5^-	−136.0	[1]	−189.5	−205.7	−149.7	−163.8	−148.4
GeF_4	−172.0	[1]	−168.7	−173.9	−164.7	−168.7	−150.8
SeF_6	55.0	[1]	186.9	17.0	106.6	128.0	120.8
SO_2F_2	33.5	[1]	45.0	48.9	41.7	146.4	53.8
WF_6	166.0	[1]	161.2	159.1	168.2		
AsF_5 (av.)	−66.0	[1]	−12.9	−6.8	−2.9		
AsF_6^-	−69.5	[1]	−36.6	−32.4	−53.7		
XeF_2	−183.4	[101]	−123.3	−136.9	−177.9		
XeF_4	−17.8	[101]	22.4	39.5	−8.3		
XeF_6	126.4	[101]	160.1	182.8	112.8		
PF_6^-	−70.8	[1]	−96.2	−105.1	−76.7		
UF_6	764.0	[79]	831.4	836.0	771.1		
UF_5Cl, A_4X (A_4)	781.5	[79]	811.9	812.7	772.8		
UF_5Cl, A_4X (X)	762.0	[79]	781.6	779.7	767.7		
cis-UF_4Cl_2, A_2X_2 (A_2)	785.8	[79]	799.5	796.3	774.9		
cis-UF_4Cl_2, A_2X_2 (X_2)	760.0	[79]	765.5	760.5	768.0		
trans-UF_4Cl_2 (A_4)	755.5	[79]	800.3	796.6	781.6		
cis-UF_2Cl_4 (A_2)	781.0	[79]	762.5	739.7	777.7		
trans-UF_2Cl_4 (A_2)	778.9	[79]	784.5	778.9	788.9		
$UFCl_5$	774.3	[79]	743.6	734.5	786.7		
mer-UF_3Cl_3, A_2X (A_2)	782.6	[79]	790.7	786.1	781.8		
mer-UF_3Cl_3, A_2X (X)	753.0	[79]	755.5	747.9	775.4		
fac-UF_3Cl_3	786.4	[79]	754.3	747.9	770.1		

All shifts in ppm, referenced to $CFCl_3$. More data on U(VI) complexes are presented separately in Table 4

3.1 General features

Inspection of Fig. 1 shows that the BP86 functional performs fairly well in predicting ^{19}F spectra, with a MAE of 33 ppm. General trends are correctly reproduced, albeit with some notable outliers (SeF_6 and ClF_5 have an absolute deviation >100 ppm). Nevertheless, the correlation line has an almost unit slope through ca. 1,300 ppm, including the extreme values for ClF and FOOF. The latter compound is known to be difficult to model computationally; for this reason, chemical shifts reported in Figs. 1 and 2 were obtained for the experimental structure (except for the protocol labeled PBE0-D3) and will be discussed separately. Notably, with this functional, the shift of elemental fluorine has a deviation of only 13 ppm from experiment, that is, much less than the MAE. Uranium chlorofluorides lie in the deshielded region, as expected; however, no trend is apparent in the relatively small range encompassed by the data. Analogous calculations with the BLYP functional perform rather similarly: the correlation coefficient is practically unchanged, and the MAE (43 ppm) is slightly higher. Uranium chlorofluorides still do not exhibit a recognizable trend.

The hybrid functional PBE0 provided the best performance, with even less scatter and an improved correlation coefficient and MAE (21 ppm). There are also some notable differences: (a) surprisingly, F_2 is poorly modeled (deviation of 101 ppm, much higher than the MAE). (b) Whereas in the previous two cases, the calculated values for UF_nCl_{6-n} are quite scattered, with PBE0, they are compressed into a 20-ppm range, and are therefore remarkably constant (still without a recognizable trend, though). The results concerning uranium compounds will be discussed in detail separately.

In relative terms, a MAE of 20–40 ppm is just 2–3% of the total shift range (1,300 ppm), that is, slightly higher than the known uncertainty of computed ^{13}C (ca. 1 ppm, or 0.5%) and ^{1}H shifts (0.2 ppm, or 2%) [18]. As a preliminary conclusion, it seems that commonly used functionals lead to a moderately good performance, which should enable one to draw meaningful inferences

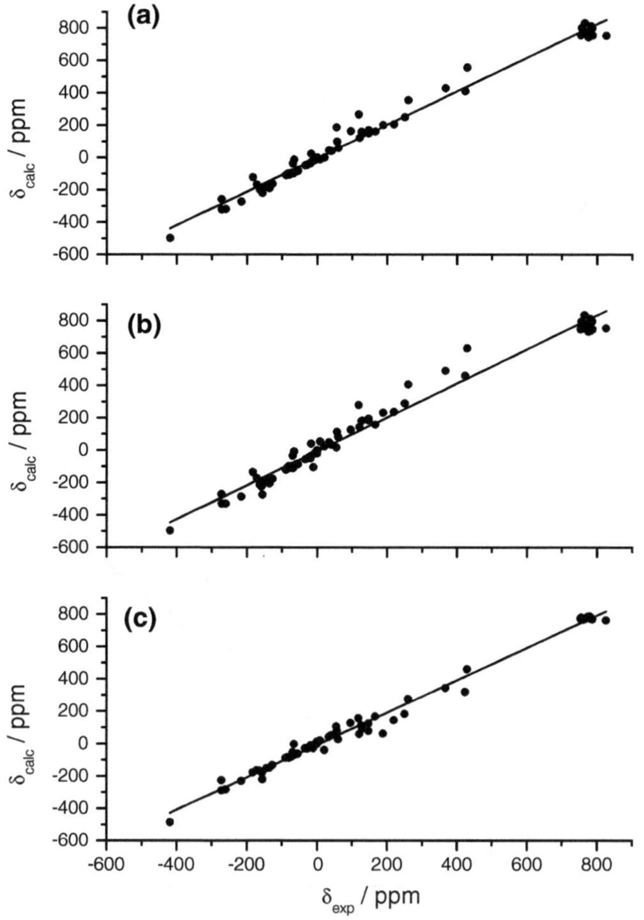

Fig. 1 Correlation of experimental with calculated ^{19}F chemical shifts. The correlation spans the whole known range from ClF ($\delta = -448$ ppm) to FOOF ($\delta = 825$ ppm). Calculations at the ZSO/TZ2P level. **a** BP86, **b** BLYP, **c** PBE0. See text for the data points of FOOF. The parameters of the linear regression $\delta_{\rm calc} = b\delta_{\rm exp} + a$ are as follows: BP86: $b = 1.039$, $a = -6.0$ ppm, $r^2 = 0.985$. BLYP: $b = 1.054$, $a = -8.5$ ppm, $r^2 = 0.977$. PBE0: $b = 1.004$, $a = -9.1$ ppm, $r^2 = 0.992$

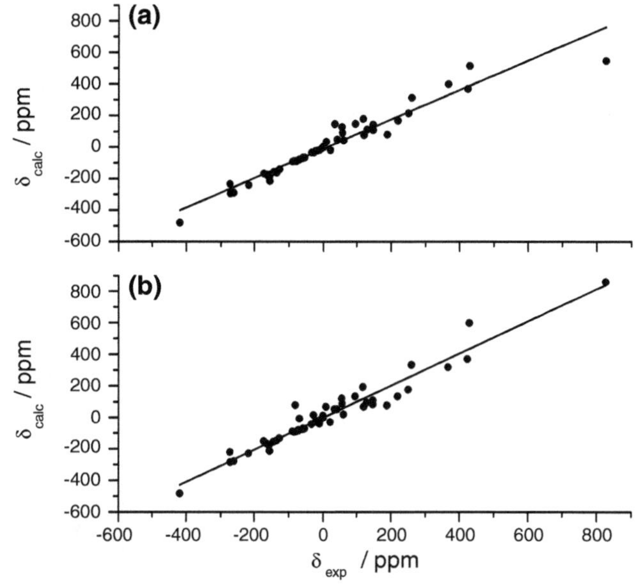

Fig. 2 Correlation of experimental with calculated ^{19}F chemical shifts with density functionals including dispersion (same data as in Fig. 1). **a** PBE0-D3, **b** M06 (see text). The parameters of the linear fits $\delta_{\rm calc} = b\delta_{\rm exp} + a$ are: PBE0-D3: $b = 0.947$, $a = -9.5$ ppm, $r^2 = 0.937$. M06: $b = 1.032$, $a = -2.4$ ppm, $r^2 = 0.947$

concerning at least the range of ^{19}F shifts for unknown compounds. In practice, this implies that the predictive capability is somewhat limited.

3.2 Including dispersion effects

One of the reasons of the relatively poor performance of the protocols tested might be the lack of a proper treatment of dispersive interactions, which may be important for fluorine owing to its larger polarizability than, for example, hydrogen or carbon (also owing to its non-bonding electron pairs). Dispersive interactions might affect molecular properties such as shieldings directly, or by affecting the molecular geometries. Thus, we have firstly evaluated the effect of including dispersive interactions in the geometry optimization by including the D3 correction term introduced by Grimme [96, 97] (in the ADF implementation, this correction is not included in the shielding calculation). In these additional tests, we selected only the subset of molecules without heavy atoms. For these tests, we have chosen the widely used hybrid functional B3LYP for the optimization (i.e., ZSC-B3LYP-D3/TZ2P), while the shieldings were calculated at the ZSO-PBE0/TZ2P level. In the case of FOOF (dealt with separately below), we have employed the optimized geometry, in order to test the effect of dispersive interactions on the geometry itself, even though all other calculations employed the experimental geometry. As we can see by comparing the correlation in Fig. 2a with those of Fig. 1, there is not any noticeable improvement in the results.

As a further test, we also considered the recent M06 functional, which has been reported to be capable to treat weak dispersive interactions to a good level of approximation [99]. (For FOOF the experimental geometry was again used.) However, the correlations reported in Fig. 2b are even poorer than those obtained with the protocols of Fig. 1.

For a small subset of compounds, we finally resorted to the very expensive MP2 ab initio method. Such results (Table 2; Fig. 3) do not show a significant improvement compared to DFT methods, given the larger MAE of 57 ppm (although including the difficult case of FOOF) and the slope of the correlation which is lower than unity (0.87). Moreover, it must be remarked that the demands on computer time and resources would prohibit an extensive use of this method in all but the smallest molecules.

Table 2 Experimental and calculated ^{19}F chemical shifts at the MP2 level

Species	δ_{exp}	MP2
$CFCl_3$	0	0
ClF	−419.4	−479.5
F_2	422.9	332.8
FOH	21	−3.1
FOF	250	196.8
FOOF	825	630.5
ClF_3 (1F)	7.7	−22.5
ClF_3 (2F)	118.1	118.8

See Table 1 for references. MP2/cc-pVQZ//MP2/cc-pVQZ calculations

Table 3 Experimental and calculated geometrical parameters of FOOF

	r_{OO} (Å)	r_{OF} (Å)	α_{OOF} (°)	φ_{FOOF} (°)
Exptl.[a]	1.216	1.586	109.2	88.1
ZSC-BP86/TZ2P	1.204	1.603	111.3	89.3
ZSC-BLYP/TZ2P	1.468	1.398	98.3	90.3
ZSC-PBE0/TZ2P	1.233	1.483	108.8	87.5
ZSC-B3LYP-D3/TZ2P	1.229	1.529	109.4	87.8
M06/cc-pVTZ	1.211	1.524	108.9	86.9
MP2/cc-pVQZ	1.200	1.547	98.3	87.7

[a] Ref. [104]

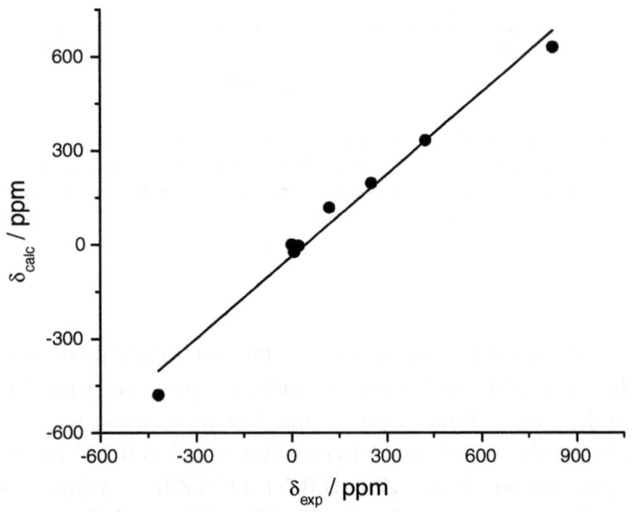

Fig. 3 Correlation of experimental with calculated ^{19}F chemical shifts at the MP2 level. The parameters of the linear fits $\delta_{calc} = b\delta_{exp} + a$ are as follows: $b = 0.872$, $a = -36.8$ ppm, $r^2 = 0.981$

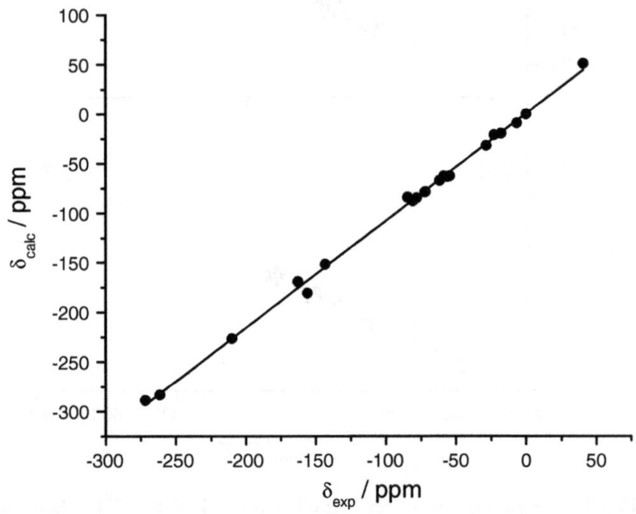

Fig. 4 Correlation of experimental with calculated (PBE0 level) ^{19}F chemical shifts for organofluorine compounds. The parameters of the linear fits $\delta_{calc} = b\delta_{exp} + a$ are follows: $b = 1.081$, $a = 0.7$ ppm, $r^2 = 0.998$

3.3 The case of FOOF

The geometry and electronic structure of FOOF is a known difficult case for computational chemistry [102, 103]. Here, we have computed ^{19}F shifts for geometries optimized at the respective levels, as well as with the experimental geometry; [104] (Table 3).

Table 3 shows that the geometries predicted by BP86 and the hybrid functionals PBE0, M06, and B3LYP are not far from the experimental structure, while BLYP is in large error. Nevertheless, the dependence of ^{19}F shielding on the structural parameters is so strong that even small deviations produce large errors in the calculated chemical shifts. With BP86 and BLYP (likely accidentally), the result is better with the optimized geometry; with PBE0, the optimized geometry yields a very poor shift, whereas the one obtained from the experimental geometry essentially fits the other data (Table 1). It then seems that a reasonable agreement with experiment can only be attained using the experimental geometry, above and beyond the difficulties shown above for fluorine compounds. It must also be noted, however, that the reported experimental value of 825 ppm [1] derives from an extrapolation to infinite dilution and −83 °C of the shift (865 ppm) determined for the neat liquid at −128 °C [105]. Given the approximations involved and the high sensitivity to experimental conditions, it seems that a reliable experimental $\delta(^{19}F)$ value for FOOF is hardly available.

3.4 Organic molecules

Even though the scope of this paper is to highlight the main broad features of ^{19}F NMR modeling, it is instructive to have a closer look at the subset of organic molecules (here intended as limited to carbon-bound fluorine atoms). Such data are presented in Fig. 4 (PBE0 results only).

The MAE is ca. 8 ppm, that is, 3% of the 320-ppm range; as before, inclusion of dispersion effects does not lead to major improvements. Hence, in relative terms, the accuracy is not higher than for the general case. However, the scatter is much smaller, which suggests that the predictive capability may be substantially higher. This is not too surprising in view of the overall greater similarity of bonding, although bonds between F and most common C hybridizations (sp^3, sp^2, *sp,* and carbonyl) are included. A detailed investigation of organofluorine compounds will be the subject of a forthcoming paper.

3.5 Uranium chlorofluorides

Treatment of UF_6 with Me_3SiCl or UCl_6 at −75 °C was reported to afford a mixture of mixed U(VI) chlorofluorides UF_nCl_{6-n} ($n = 1$–5) [79]. ^{19}F NMR spectra display several signals of widely varying intensity, which were assigned to all possible ligand and stereochemical combinations (UF_6, UF_5Cl, *cis*-UF_4Cl_2, *trans*-UF_4Cl_2, *mer*-UF_3Cl_3, *fac*-UF_3Cl_3, *cis*-UF_2Cl_4, *trans*-UF_2Cl_4, and $UFCl_5$). Some of these complexes are expected to exhibit well-defined spin systems, giving rise to distinctive coupling patterns between ^{19}F nuclei (A_6, A_5X, A_4X_2, A_4, A_2X, A_3, A_2, A_2, A, respectively); indeed, signal pairs matching these expectations were found, leading to positive identification of (apart from the known spectrum of UF_6) UF_5Cl, *cis*-UF_4Cl_2, and *mer*-UF_3Cl_3. However, some (*trans*-UF_4Cl_2, *fac*-UF_3Cl_3, *cis*-UF_2Cl_4, *trans*-UF_2Cl_4, and $UFCl_5$) have all equivalent ^{19}F nuclei and therefore give rise to singlets which could not be assigned directly, but rather on the basis of assumptions regarding the relative amount of reactants. It must also be noted that the range of ^{19}F shifts spanned by the data is only ca. 50 ppm, that is, comparable with the MAE values obtained so far.

Schreckenbach [80–82] and Marsden [83] analyzed computationally in great detail the thermochemistry and spectroscopic properties of the above complexes with a wide choice of density functionals and basis sets. The first study [80–82] employed all-electron Slater basis sets in association with Pauli and ZORA relativistic methods and compared with non-relativistic calculations with relativistically corrected effective core potential (RECP) basis sets for uranium. The second one [83] examined a variety of RECP methods with an emphasis on the effect of core size. In all cases, the computed NMR ^{19}F shifts showed a large scattering, and even a qualitative match with the experimental values could hardly be attained.

The RECP approach is questionable whenever nuclear properties are sought, owing to their incorrect asymptotic behavior in the vicinity of the nucleus [80–82]. Nevertheless, it has shown predictive power if (as is usually the case) lighter atoms are modeled by an all-electron basis set; [83, 106, 107] results are predictably better if the core is kept small [82]. Some predictions can be made even for the heavy atom itself, for example, regarding the broad features of ^{183}W NMR of several tungsten complexes, albeit with large systematic errors [108].

Given the good performance of the ZORA-DFT framework in the prediction of NMR shifts for many heavy-atom nuclei such as ^{183}W, [19–25] we reckoned that the approach could be extended to heavier atoms as well. Given the expected difficulty of this problem, we calculated the shifts with the SAOP model functional in addition to BP86 and PBE0, since SAOP was reported to provide improved ^{183}W shieldings at a modest computational cost [109].

These results are reported in Table 4 and Fig. 5. As could be anticipated from Fig. 1, the range of calculated shifts is essentially correct at ca. 760 ppm with BP86 and PBE0, but much lower (530 ppm) with SAOP. Nevertheless, the correlation with experiment is weak; evidently none of the theoretical levels adopted here is sufficiently accurate to make quantitative predictions of such shifts. Some differences are apparent—shifts computed with BP86 and SAOP show much more variability (over a

Table 4 Experimental and calculated ^{19}F chemical shifts of U(VI) chlorofluorides

Species	δ_{exp}	Spin system	BP86	PBE0	SAOP
UF_6	764.0	A_6	831.7	771.5	603.0
UF_5Cl	781.5	A_4X (A_4)	812.2	773.2	571.5
	762.0	A_4X (X)	781.9	768.1	555.3
cis-UF_4Cl_2	785.8	A_2X_2 (A_2)	799.8	775.3	555.5
	760.0	A_2X_2 (X_2)	765.8	768.4	536.9
trans-UF_4Cl_2	755.5	A_4	800.6	782.0	560.4
mer-UF_3Cl_3	782.6	A_2X (A_2)	791.0	782.2	548.4
	753.0	A_2X (X)	755.8	775.8	527.0
fac-UF_3Cl_3	786.4	A_3	754.6	770.5	526.0
cis-UF_2Cl_4	781.0	A_2	762.8	778.1	520.4
trans-UF_2Cl_4	746.1	A_2	784.8	789.3	543.8
$UFCl_5$	774.3	A	743.9	787.1	516.8

Same theoretical levels as in Table 1, except SAOP (single-point calculation at the PBE0 geometry)

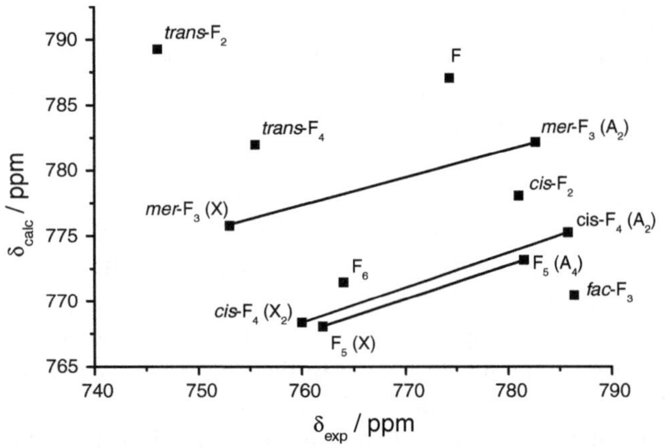

Fig. 5 Correlation of experimental with calculated ^{19}F chemical shifts of uranium chlorofluorides (PBE0 level; with other methods the outcome is similar). *Lines* connect pairs of non-equivalent spins in A_mX_n spin systems

100-ppm range) than those with PBE0; in the latter case, all shifts are compressed into a mere 21-ppm range.

Even though ^{235}U shieldings are not experimentally accessible owing to its very unfavorable NMR properties, we examined its pattern of variation across the range of complexes, in order to highlight possible computational shortcomings. Following Refs. [80–82], we confirmed that $\sigma(^{235}U)$ values (PBE0) follow a regular and well-defined trend, the shielding increasing linearly (by ca. 2,000 ppm per F) with the number of fluorine atoms, for example, from −15,000 (UCl_6) to −456 ppm (UF_6); complexes with the same number of fluorine atoms differ by ca. 300 ppm. Overall, the results indicate no evident flaw in the modeling of their electronic structure, nor call for substantial differences with the lighter congeners.

A major general reason for this rather poor performance lies of course in the fact that the small variability is comparable with the MAE of the calculations. Nevertheless, some features are noteworthy: in all cases where two ^{19}F signals are present and can be assigned with confidence from the splitting patterns, that is, UF_5Cl, *cis*-UF_4Cl_2, and *mer*-UF_3Cl_3, relative trends within each molecule are correct at all theoretical levels. Thus, for example, the X signal of the UF_5Cl spin system is correctly predicted to be more shielded than A_4, etc. (see Fig. 5), although computed differences are smaller than they should be.

We can then divide the UF_nCl_{6-n} series into a "safe" domain (comprising UF_5Cl, *cis*-UF_4Cl_2, *mer*-UF_3Cl_3, and UF_6); for these, assignments are essentially unquestionable, and computed results are fairly well behaved. The remaining ones (*trans*-UF_4Cl_2, *fac*-UF_3Cl_3, *cis*-UF_2Cl_4, *trans*-UF_2Cl_4, and $UFCl_5$) have been assigned only indirectly. (For example, we recall that the signal attributed to *cis*-UF_2Cl_4 was very weak).

4 Discussion

4.1 General issues

At the present stage, it is rather difficult to disentangle the errors produced by geometry optimization converging to an incorrect structure from those coming directly from the calculations of the NMR property. The case of FOOF we discussed in some detail shows that a correct geometry is indeed essential; shielding changes of the order of 10^2 ppm are observed for structures differing in the F–O distance by 10^{-2} Å. However, inclusion of dispersive effects in the geometry alone (B3LYP-D3 optimizations) or in the geometry and shielding (M06 calculations) has not shown significant improvement over standard functionals.

Apart from "external" sources of error, such as solvent effects, two major issues may adversely affect the general quality of computed chemical shifts, that is, deficiencies in the density functional and (for heavy-atom fluorides) the method for dealing with relativistic effects.

In our study, we have emphasized the applicability of a few density functionals to a wide range of fluorinated molecules, rather than focusing on a small test set of molecules with a wide choice of methods. We have preferred this avenue because of the extremely wide variability of composition and structure that fluorinated molecules exhibit, thereby highlighting the scope of any given method. While of course such a scope awaits for further studies, emerging features are (a) hybrid functionals (PBE0 and M06) perform better than GGA ones (BP86, BLYP); (b) inclusion of dispersion leads to negligible improvement; and (c) a common ab initio method such as MP2 also does not perform better (in this case, the scope and applicability would be limited by the high computational cost, though).

Similarly, only ZORA has been used here as relativistic treatment. Whereas it is by no means the only one available, ZORA has the advantage of being the most widely employed one (thanks also to the availability of implementations that include both scalar effects and spin–orbit coupling on NMR parameters), thereby allowing for a comparison with existing data. It should be remarked, however, that other methods have been implemented (see e.g., [110] for a comparison of such methods for ^{199}Hg). Probably the only instance where relativistic effects should be seriously called into play concerns uranium chlorofluorides; indeed, the rather poor performance of all methods considered here (or in the literature at large) might be due in part to the need of improving their treatment, which may be challenging for actinides.

A further issue that may deserve some attention concerns the sizable isotope effect on ^{19}F shifts in certain compounds, notably those containing chlorine atoms in a

F–C–Cl or F–C–C–Cl arrangement (quite common in chlorofluorocarbons). The reference compound ($CFCl_3$) is known to exhibit a three-line signal due to isotope shifts between isotopomers differing in $^{35}Cl/^{37}Cl$ content, so that criticisms have been leveled at its suitability as a reference signal. Computational predictions of such isotope shifts [40] are of the order of 20 ppm for ^{19}F in $CFCl_3$; however, experimental measurements in several chlorofluorocarbons have shown that such shifts give rise to easily detectable splittings, but amount to no more than 5–7 ppb when F and Cl are separated by 2 bonds [111]. This disagreement is certainly worth further investigations, because the scope of these vibrational corrections is not yet known.

A significant improvement in the performance is observed when the correlation is limited to a similar class of molecules or bond types, such as in the organic molecules subset. This is encouraging, since in practical applications of the computational tool, one will have to distinguish between analogous systems, and the predictive power in such instances seems to offer a considerably better prospect. Moreover, the huge body of data concerning organofluorine molecules will certainly provide further testing ground for these concepts. The case of uranium chlorofluorides confirms these observations, since relative chemical shifts within the same molecule are fairly correctly reproduced.

A further point of concern is the rather wide range of experimental values that are often reported for each molecule, which points to strong external effects influencing the observed shift. This is not unreasonable, in view of the polarizability of fluorine and its exposure to solvent in most molecules; on the other hand, the vast majority of data presented here pertain to weakly polar molecules, whose data were collected in non-polar solvents. In such cases, continuum solvent models are known to provide extremely small effects, whereas calculations on the fluoride anion solvated by various molecules show some 70-ppm variability in $\delta(F^-)$ [112]. While the case of F^- may be an anomalously high value for solvent effects, this issue is clearly an open one. DFT calculations with explicit solvent molecules are notoriously prone to artifacts; [113] in this respect, an avenue for progress might be the development of continuum models of perfluorinated solvents, often the solvents of choice for such molecules.

4.2 Uranium chlorofluorides

The observations collected in this work indicate that within a given structure, the computational levels that we have presented have sufficient predictive power. What are, then, the reasons for our general inability to model these NMR spectra? As mentioned before, part of the problem lies in the general accuracy that can be attained, which borders with the variability of the data. On the other hand, we have shown that dispersion effects do not lead to any improvement and can hardly be called into question here. Probably there is room for methodological improvement in the quality of density functional and relativistic treatment for actinides. Indeed, the knowledge base of applications of ZORA to actinides is very limited.

A specific chemical feature of uranium complexes may, however, be that the thermochemistry and dynamics of the reactions leading to their formation are not known. Thus, for example, ligand exchange among the species may lead to apparent chemical shifts that cannot be compared with calculated ones. All in all, this set of fluorine compounds does not seem to present outstanding intrinsic peculiarities if compared to other fluorinated compounds, but is probably much more complex than it looks, and would require more detailed studies before a comparison with calculated NMR spectra is attempted.

5 Conclusions

We have investigated the performance of some DFT and ab initio protocols for the prediction of fluorine chemical shifts using a broad selection of compounds and X–F bonds, encompassing the entire range of resonances of ^{19}F. We generally observe a satisfactory agreement, with statistical parameters such as the r^2 correlation coefficient or the MAE only slightly worse than those obtained from DFT protocols applied to the calculation of chemical shifts of nuclei such as ^1H or ^{13}C. Nevertheless, the overall accuracy values remain significantly lower and may compromise applicability to structural problems. Various factors might contribute, both at the experimental and at the computational level (density functional and relativistic methods, dispersion, isotope effects, and solvent effects), but have been scarcely studied in the literature and indicate ways of further development.

Acknowledgments This work was financially supported by the University of Padova (postdoctoral fellowship to R.B.). Computational resources were provided by the Laboratorio Interdipartimentale di Chimica Computazionale (LICC) at the Department of Chemistry of the University of Padova.

References

1. Berger S, Braun S, Kalinowski H-O (1994) NMR Spektroskopie von Nichtmetallen, vol 4, ^{19}F-NMR-Spektroskopie. Thieme, Stuttgart
2. Everett TS (1995) Nuclear magnetic resonance spectroscopy of organofluorine compounds. In: Hudlicky M, Pavlath AE (eds) Chemistry of organic fluorine compounds II. American Chemical Society, Washington, DC, pp 1037–1086
3. Dolbier WR Jr (2009) Guide to fluorine NMR for organic chemists. Wiley, Hoboken

4. Bachert P (1998) Prog NMR Spectrosc 33:1
5. Lutz NW, Hull WE (1999) NMR Biomed 12:237
6. Malet-Martino M, Gilard V, Desmoulin F, Martino R (2006) Chim Clin Acta 366:61
7. Martino R, Gilard V, Desmoulin F, Malet-Martino M (2006) Chemotherapy 52:215
8. Cellitti SE, Jones DH, Lagpacan L, Hao X, Zhang Q, Hu H, Brittain SM, Brinker A, Caldwell J, Bursulaya B, Spraggon G, Brock A, Ryu Y, Uno T, Schultz PG, Geierstanger BH (2008) J Am Chem Soc 130:9268
9. Hansch C, Leo A, Taft RW (1991) Chem Rev 91:165
10. Jaime-Figueroa S, Kurz LJ, Liu Y, Cruz R (2000) Spectrochim Acta 56A:1167
11. Adcock W, Bullpitt M, Kitching W (1977) J Org Chem 42:2411
12. Gribble GW, Keavy DJ, Olson ER, Rae ID, Staffa A, Herr TH, Ferraro MB, Contreras RH (1991) Magn Reson Chem 29:422
13. Mallory FB, Mallory CW, Butler KE, Lewis MB, Xia AQ, Luzik ED Jr, Fredenburgh LE, Ramanjulu MM, Van QN, Francl MM, Freed DA, Wray CC, Hann C, Nerz-Stormes M, Carroll PJ, Chirlian LE (2000) J Am Chem Soc 122:4108
14. Kaupp M, Bühl M, Malkin VG (eds) (2004) Calculation of NMR and EPR parameters. Wiley-VCH, Weinheim
15. Palermo G, Riccio R, Bifulco G (2010) J Org Chem 75:1982
16. Smith SG, Goodman JM (2010) J Am Chem Soc 132:12946
17. Saielli G, Nicolaou KC, Ortiz A, Zhang H, Bagno A (2011) J Am Chem Soc 133:6072
18. Bagno A, Saielli G (2007) Theor Chem Acc 117:603 (and references therein)
19. Autschbach J, Zheng S (2009) Annu Rep NMR Spectrosc 67:1
20. Casella G, Ferrante F, Saielli G (2010) Org Biomol Chem 8:2711
21. Bagno A, Bini R (2010) Angew Chem Int Ed 49:1083
22. Bühl M, Peters D, Herges R (2009) Dalton Trans 6037
23. Waller MP, Geethalakshmi KR, Bühl M (2008) J Phys Chem B 112:5813
24. Chen F, Ma GB, Bernard GM, Cavell RG, McDonald R, Ferguson MJ, Wasylishen RE (2010) J Am Chem Soc 132:5479
25. Justino LLG, Ramos ML, Kaupp M, Burrows HD, Fiolhais C, Gil VMS (2009) Dalton Trans 9735
26. Malkin VG, Malkina OL, Casida M, Salahub DR (1994) J Am Chem Soc 116:5898
27. Schreckenbach G, Ziegler T (1995) J Phys Chem 99:606
28. Lee AM, Handy NC, Colwell SM (1995) J Chem Phys 103:10095
29. Shaghaghi H, Ebrahimi H, Tafazzoli M, Jalali-Heravi M (2010) J Fluor Chem 131:47
30. Abbandonato G, Catalano D, Marini A (2010) Langmuir 26:16762
31. Chen Y, Zhao L, Xu CM, Liu ZC, Frenking G (2009) J Mol Struct (Theochem) 905:40
32. Adcock W (2009) J Phys Org Chem 22:1065
33. Shainyan BA, Ushakov IA, Tolstikova LL, Koch A, Kleinpeter E (2008) Tetrahedron 64:5208
34. Finze M, Reiss GJ (2008) Eur J Inorg Chem 2321
35. Nishida M, Fukaya H, Hayakawa Y, Ono T, Fujii K, Uekusa H (2006) Helv Chim Acta 89:2671
36. Prakash GKS, Mathew T, Marinez ER, Esteves PM, Rasul G, Olah GA (2006) J Org Chem 71:3952
37. Okazaki T, Laali KK (2005) J Org Chem 70:9139
38. Forgeron MAM, Wasylishen RE, Penner GH (2004) J Phys Chem A 108:4751
39. Pavone M, Barone V, Ciofini I, Adamo C (2004) J Chem Phys 120:9167
40. Dransfeld A (2004) Chem Phys 298:47
41. Fukaya H, Ono T (2004) J Comput Chem 25:51
42. Adcock W, Peralta JE, Contreras RH (2003) Magn Reson Chem 41:503
43. Kvicala J, Czernek J, Bohm S, Paleta O (2002) J Fluor Chem 116:121
44. DerHovanessian A, Rablen PR, Jain A (2000) J Phys Chem A 104:6056
45. Tanuma T, Irisawa J, Ohnishi K (2000) J Fluor Chem 102:205
46. Kaupp M, Malkina OL, Malkin VG (1999) J Comput Chem 20:1304
47. Elavarasi SB, Dorai K (2010) Chem Phys Lett 489:248
48. Weber J, Schmedt auf der Günne J (2010) Phys Chem Chem Phys 12:583
49. Griffin JM, Yates JR, Berry AJ, Wimperis S, Ashbrook SE (2010) J Am Chem Soc 132:15651
50. Carnevale D, Wormald P, Ameduri B, Tayouo R, Ashbrook SE (2009) Macromolecules 42:5652
51. Zheng AM, Liu SB, Deng F (2009) J Phys Chem C 113:15018
52. Moon C, Brunklaus G, Sebastiani D, Rudzevich Y, Bohmer V, Spiess HW (2009) Phys Chem Chem Phys 11:9241
53. Sternberg U, Klipfel M, Grage SL, Witter R, Ulrich AS (2009) Phys Chem Chem Phys 11:7048
54. Vogt FG, Katrincic LM, Long ST, Mueller RL, Carlton RA, Sun YT, Johnson MN, Copley RCB, Light ME (2008) J Pharm Sci 97:4756
55. Zhao X, Devries JS, McDonald R, Sykes BD (2007) J Magn Reson 187:88
56. Tatsuno H, Aimi K, Ando S (2007) Magn Reson Chem 45:401
57. Robbins AJ, Ng WTK, Jochym D, Keal TW, Clark SJ, Tozer DJ, Hodgkinson P (2007) Phys Chem Chem Phys 9:2389
58. Pulido A, Sastre G, Corma A (2006) ChemPhysChem 7:1092
59. Buznik VM, Gabuda SP, Kozlova SG, Ignatieva LN (2005) J Struct Chem 46:87
60. Body M, Silly G, Legein C, Buzare JY (2005) J Phys Chem B 109:10270
61. Yates JR, Dobbins SE, Pickard CJ, Mauri F, Ghi PY, Harris RK (2005) Phys Chem Chem Phys 7:1402
62. Cai SH, Yu XY, Chen Z, Wan HL (2003) Magn Reson Chem 41:902
63. Solis D, Ferraro MB, Facelli JC (2002) J Mol Struct 602:159
64. Sanders LK, Oldfield E (2001) J Phys Chem A 105:8098
65. Vilcachagua JD, Ducati LC, Rittner R, Contreras RH, Tormena CF (2011) J Phys Chem A 115:1272
66. Ebrahimi A, Habibi-Khorassani M, Doosti M (2010) Chem Phys Lett 491:11
67. Jaszunski M, Vaara J (2009) Phys Chem Chem Phys 11:4136
68. Kruck M, Munoz MP, Bishop HL, Frost CG, Chapman CJ, Kociok-Kohn G, Butts CP, Lloyd-Jones GC (2008) Chem Eur J 14:7808
69. Czerski I, Kamienska-Trela K, Kozminski W, Ratajczyk T, Szymanski S, Wojcik J (2007) Magn Reson Chem 45:1040
70. Hennig M, Munzarová ML, Bermel W, Scott LG, Sklenár V, Williamson JR (2006) J Am Chem Soc 128:5851
71. Contreras RH, Esteban AL, Diez E, Head NJ, Della EW (2006) Mol Phys 104:485
72. Kapicka L, Dastych D, Richterova V, Alberti M, Kubacek P (2005) Magn Reson Chem 43:294
73. Feindel KW, Wasylishen RE (2004) Magn Reson Chem 42:S158
74. Barone V, Provasi PF, Peralta JE, Snyder JP, Sauer SPA, Contreras RH (2003) J Phys Chem A 107:4748
75. Lantto P, Vaara J, Helgaker T (2002) J Chem Phys 117:5998
76. Benedict H, Shenderovich IG, Malkina OL, Malkin VG, Denisov GS, Golubev NS, Limbach HH (2000) J Am Chem Soc 122:1979
77. Cavallotti C, Metrangolo P, Meyer F, Recupero F, Resnati G (2008) J Phys Chem A 112:9911
78. Mao JH, Zhang Y, Oldfield E (2002) J Am Chem Soc 124:13911
79. Downs AJ, Gardner CJ (1984) J Chem Soc Dalton Trans 2127

80. Schreckenbach G, Wolf SW, Ziegler T (2000) J Phys Chem A 104:8244
81. Schreckenbach G (2002) Inorg Chem 41:6560
82. Schreckenbach G (2005) Int J Quantum Chem 101:372
83. Iché-Tarrat N, Marsden CJ (2008) J Phys Chem A 112:7632
84. te Velde G, Bickelhaupt FM, van Gisbergen SJA, Fonseca Guerra C, Baerends EJ, Snijders JG, Ziegler T (2001) J Comput Chem 22:931
85. Fonseca Guerra C, Snijders JG, te Velde G, Baerends EJ (1998) Theor Chem Acc 99:391. ADF2009.01, SCM, Theoretical Chemistry, Vrije Universiteit, Amsterdam, The Netherlands. http://www.scm.com
86. Wolff SK, Ziegler T, van Lenthe E, Baerends EJ (1999) J Chem Phys 110:7689
87. Krykunov M, Ziegler T, van Lenthe E (2009) J Phys Chem A 113:11495
88. Becke AD (1988) Phys Rev A 38:3098
89. Perdew JP, Wang Y (1986) Phys Rev B 33:8822
90. Lee C, Yang W, Parr RG (1988) Phys Rev B 37:785
91. Johnson BG, Gill PMW, Pople JA (1993) J Chem Phys 98:5612
92. Russo TV, Martin RL, Hay PJ (1994) J Chem Phys 101:7729
93. Ernzerhof M, Scuseria G (1999) J Chem Phys 110:5029
94. Adamo C, Barone V (1999) J Chem Phys 110:6158
95. Gritsenko OV, Schipper PRT, Baerends EJ (1999) Chem Phys Lett 302:199
96. Grimme S (2006) J Comput Chem 27:1787
97. Grimme S, Anthony J, Ehrlich S, Krieg H (2010) J Chem Phys 132:154104
98. Stephens PJ, Devlin FJ, Chabalowski CF, Frisch MJ (1994) J Phys Chem 98:11623
99. Zhao Y, Truhlar DG (2008) Theor Chem Acc 120:215
100. Frisch MJ, Trucks GW, Schlegel HB, Scuseria GE, Robb MA, Cheeseman JR, Scalmani G, Barone V, Mennucci B, Petersson GA, Nakatsuji H, Caricato M, Li X, Hratchian HP, Izmaylov AF, Bloino J, Zheng G, Sonnenberg JL, Hada M, Ehara M, Toyota K, Fukuda R, Hasegawa J, Ishida M, Nakajima T, Honda Y, Kitao O, Nakai H, Vreven T, Montgomery JrJA, Peralta JE, Ogliaro F, Bearpark M, Heyd JJ, Brothers E, Kudin KN, Staroverov VN, Kobayashi R, Normand J, Raghavachari K, Rendell A, Burant JC, Iyengar SS, Tomasi J,Cossi M, Rega N, Millam NJ, Klene M, Knox JE, Cross JB, Bakken V, Adamo C, Jaramillo J, Gomperts R, Stratmann RE, Yazyev O, Austin AJ, Cammi R, Pomelli C, Ochterski JW, Martin RL, Morokuma K, Zakrzewski VG, Voth GA, Salvador P, Dannenberg JJ, Dapprich S, Daniels AD, Farkas Ö, Foresman JB,Ortiz JV, Cioslowski J, Fox DJ (2009) Gaussian 09, Revision B01. Gaussian, Inc., Wallingford
101. Gerken M, Hazendonk P, Nieboer J, Schrobilgen GJ (2004) J Fluor Chem 125:1163
102. Kraka E, He Y, Cremer D (2001) J Phys Chem A 105:3269
103. Menconi G, Wilson PJ, Tozer DJ (2001) J Chem Phys 114:3958
104. Hedberg L, Hedberg K, Eller PG, Ryan RR (1988) Inorg Chem 27:232
105. Nikitin IV, Rosolovskii V Ya (1971) Russ Chem Rev 40:889
106. Bühl M, Gaemers S, Elsevier CJ (2000) Chem Eur J 6:3272
107. Straka M, Kaupp M (2005) Chem Phys 311:45
108. Bagno A, Bonchio M (2000) Chem Phys Lett 317:123
109. Vankova N, Heine T, Kortz U (2009) Eur J Inorg Chem 5102
110. Arcisauskaite V, Melo JI, Hemmingsen L, Sauer SPA (2011) J Chem Phys 135:044306
111. Tordeux M, Wakselman C, Jarjayes O, Beguin CG (2001) Magn Reson Chem 39:301 (and references therein)
112. Gerken M, Boatz JA, Kornath A, Haiges R, Schneider S, Schroer T, Christe KO (2002) J Fluor Chem 116:49
113. Bagno A, Rastrelli F, Saielli G (2005) Prog NMR Spectrosc 47:41

Theor Chem Acc (2012) 131:1283
DOI 10.1007/s00214-012-1283-y

ERRATUM

Erratum to: Computational ^{19}F NMR. 1. General features

Giacomo Saielli · Riccardo Bini · Alessandro Bagno

Published online: 13 November 2012

Erratum to: Theor Chem Acc (2012) 131:1140
DOI 10.1007/s00214-012-1140-z

Owing to an erroneous interchange of literature experimental data in the original publication, the interpretation of the computed results for uranium chlorofluorides should be partly modified as follows.

The ^{19}F spectra reported by Downs (ref. [17]) for UF_nCl_{6-n} could, in some cases, be assigned with confidence owing to their splitting patterns: this holds for UF_5Cl (A_4X), *cis*-UF_4Cl_2 (A_2X_2) and *mer*-UF_3Cl_3 (A_2X). (However, the assignment for *cis*-UF_4Cl_2 was only tentative)

However, we erroneously swapped the entries for each spin pair in the above compounds, which led to a seemingly better agreement with the calculations. In the paper we had suggested that, while the general performance of the calculations was unsatisfactory, individual trends in ^{19}F chemical shifts were correct for each of the above molecules. However, after reordering the data this appears not to be the case; the correlation is poor even for the above compounds. The affected items are Tables 1 and 4 (pairs of experimental shifts for UF_5Cl, *cis*-UF_4Cl_2 and *mer*-UF_3Cl_3 should be exchanged), and Fig. 5. For brevity, the revised situation is depicted in Fig. 1 (which replaces Fig. 5).

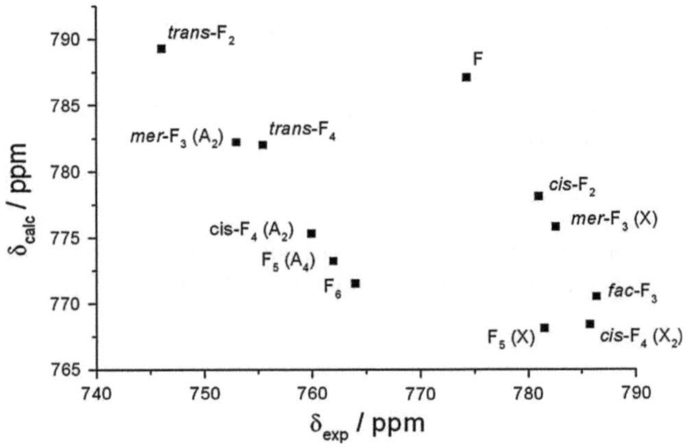

Fig. 5 Correlation of experimental with calculated ^{19}F chemical shifts of uranium chlorofluorides

General trends (Fig. 1) are unchanged since the interval spanned by UF_nCl_{6-n} is just ca. 40 ppm, over a full range of 1,300 ppm. The overall conclusions are, therefore, not affected.

Acknowledgments We thank Dr. Georg Schreckenbach for alerting us of this mistake.

The online version of the original article can be found under doi:10.1007/s00214-012-1140-z.

G. Saielli
Istituto per la Tecnologia delle Membrane del CNR, Unità di Padova, Via Marzolo 1, 35131 Padua, Italy

R. Bini · A. Bagno (✉)
Dipartimento di Scienze Chimiche, Università di Padova, Via Marzolo 1, 35131 Padua, Italy
e-mail: alessandro.bagno@unipd.it

Theor Chem Acc (2012) 131:1141
DOI 10.1007/s00214-012-1141-y

REGULAR ARTICLE

The nature of the C–As bonds in arsaalkynes: an atoms in molecules and electron localization function study

**Tiziana Marino · Maria C. Michelini ·
Nino Russo · Emilia Sicilia · Marirosa Toscano**

Received: 1 July 2011 / Accepted: 2 September 2011 / Published online: 22 February 2012

Abstract The nature of the C–As bonds present in a series of six neutral and anionic arsaalkynes was investigated. The optimized geometries, vibrational frequencies, and wavefunctions of all the studied species were computed using density functional theory (B3LYP/6-311++G(2d,2p)). To understand the bonding characteristics, Atoms in Molecules theory and topological analysis of the electron localization function were used. For comparison, natural bond orbital analysis was also performed. The results suggest that in five of the six studied compounds, the C–As bonds are essentially triple bonds, with similar characteristics to the C–P analogues in phosphaalkynes.

Keywords Arsaalkynes · Phosphaalkynes · Atoms in molecules · Electron localization function · Density functional theory

1 Introduction

The chemistry of multiple bond compounds between group 14 elements and heavier pnictogen atoms has been the subject of substantial study [1–9]. These species are interesting from the synthetic point of view, because of the possibility of obtaining new ligands in coordination chemistry or good precursors in material chemistry. Moreover, formation of multiple bonds by third- and forth-row elements is intriguing from a theoretical point of view. New findings continue to challenge widely accepted rules of bonding. As an example, for a long time, the so-called double bond rule, which establishes that thermally stable multiple bond compounds can be formed only by elements of the second period, was not brought into question. However, after the discovery of the first phosphaalkyne, HCP, by Gier in 1961 [10] and the synthesis in 1981 of the first stable compound involving a heavier element (*t*BuCP) [11], this rule was dismissed and the detailed study of the chemistry and properties of phosphaalkynes became a rich area of research. In spite of the analogies between isoelectronic low-coordinated phosphorus compounds and arsenic compounds, the chemistry of arsaalkynes was not similarly investigated owing to the inherent instability of arsenic compounds, and their inaccessibility by synthetic protocols conventionally used in the preparation of phosphaalkynes. The number of known arsaalkynes is consequently much lower.

The first stable arsaalkyne, whose synthesis and structural characterization was reported in 1986, was Mes*-C$\equiv$As (Mes* = 2,4,6-tri-*tert*-butylphenyl) [12]. The presence of the bulky Mes* group is mainly responsible for the thermal stability of this compound. Further stabilization effects could come from the interaction of the C$\equiv$As triple bond with the aryl π system. Subsequent studies showed that when the less bulky tert-butyl group is employed, the resulting arsaalkyne has only a transient existence [13]. Treatment of a platinum complex with Mes*C$\equiv$As resulted in the synthesis of an arsaalkyne-transition metal complex $Pt(PPh_3)_2\{\eta^2{\equiv}AsCMes^*\}$ that was structurally

---

Dedicated to Professor Vincenzo Barone and published as part of the special collection of articles celebrating his 60th birthday.

Electronic supplementary material The online version of this article (doi:10.1007/s00214-012-1141-y) contains supplementary material, which is available to authorized users.

T. Marino · M. C. Michelini · N. Russo (✉) · E. Sicilia · M. Toscano
Dipartimento di Chimica, Università della Calabria, 87030 Arcavacata di Rende, Italy
e-mail: nrusso@unical.it

characterized by X-ray spectroscopy [14] together with the same arsaalkyne Mes*C≡As ligand. More recently, an unexpectedly stable C≡As anionic compound, $[(CF_3)_3BCAs]^-$, was synthesized and characterized together with its lighter congener $[(CF_3)_3BCP]^-$ [15]. Thermally unstable Me–C≡As decomposes well below room temperature [16]; the parent methylidynearsine HC≡As, the third known unsubstitued heteroalkyne, was only very recently synthesized and characterized by mass spectrometry and UV-photoelectron spectroscopy [17].

Theoretical studies of some of the above-mentioned compounds containing a C≡As triple bond have been previously reported. The first theoretical investigation of methylidynearsine HC≡As dates back to 1984 and deals with the theoretical estimation of the gas-phase basicity of molecules having a C–As multiple bond [18]. Geometrical structures of Me–C≡As, and its N and P analogues, were optimized at MP2 level using the LANL1DZ basis set [16]. Mó and collaborators reported the results of a combined theoretical and experimental study of gas-phase acidity of H–C≡As and Me–C≡As and phosphorus-containing analogues [19]. More recently, several theoretical approaches were employed to evaluate the methylidynearsine ionization potential [17]. Electronic and structural properties of neutral, cationic, and anionic methylidynearsine were examined at QCISD, MP2, and B3LYP levels with the 6-311++G** basis set [20]. In a recent study, the vibrational spectrum of HC≡As was calculated using DFT, including anharmonic corrections [21].

We present here a theoretical investigation aimed to gain insight into the nature of C–As bonds in a series of neutral and anionic arsaalkynes. Different bonding analysis methodologies were used, and the obtained bond descriptions are compared. Comparison is made of the present results with previously reported studies [22] of the C–P bonds of phosphaalkynes.

2 Computational methods

The geometries of neutral and anionic species were fully optimized, and all stationary points were characterized by harmonic frequency analysis. The B3LYP hybrid functional [23, 24] was employed with the 6-311++*G*(2d,2p) basis sets, as implemented in Gaussian03 package [25]. In the case of Pt, the LanL2DZ effective core potential [26] was used with its split valence basis sets. The wavefunctions of all optimized structures were obtained at the same level of theory.

Two different topological methodologies were used: the Atoms in Molecules (AIM) theory and analysis of the electron localization function (ELF). The AIM theory is based on the topological analysis of the electron density, $\rho(r)$ [27]. Bonds are defined by a bond path, which is necessary for chemical bonding regardless of its nature. The chemical nature of the bonding may be characterized and classified according to the properties of the bond critical points (BCPs). Topologically, a BCP corresponds to a point in the real space where the gradient of the density, $\nabla\rho_{BCP}$, is zero and where the curvature of ρ_{BCP}, expressed through the three eigenvalues of the diagonalized Hessian of ρ_{BCP}, is positive for an eigenvector linking two atomic centers (λ_3) and negative for the two others (λ_1, λ_2) perpendicular to it. A chemical bond thus results from the competition of the parallel expansion of ρ, which separates charges in their respective atomic basins (λ_3) and the perpendicular contraction of ρ toward a bond path (λ_1, λ_2). The dominant effect is measured by the Laplacian of ρ_{BCP}, $\nabla^2\rho_{BCP}$. Values of ρ_{BCP} greater than 0.2 are typical of covalent bonds, and $\nabla^2\rho_{BCP}$ is generally less than zero for such interactions, reflecting the concentration of electron density along the bond path linking the bonded atoms. Thus, when $\nabla^2\rho_{BCP} < 0$, charge is concentrated at the critical point, while when $\nabla^2\rho_{BCP} > 0$, charge is locally depleted. In strongly polar bonding, there is a significant accumulation of electron density between the nuclei, as in all shared interactions, but the Laplacian can be of either sign.

Additional information of the chemical bond can be obtained through energetic considerations related to the interplay of changes in the potential and kinetic energies. This can be done by applying the local expression of the virial theorem to the critical point, $1/2\nabla^2\rho_{BCP} = 2G_{BCP} + V_{BCP}$, where G_{BCP} and V_{BCP} are the kinetic and potential energy densities at the BCP, respectively [28]. Cremer and Kraka demonstrated that the sign of the total energy, $H_{BPC} = G_{BCP} + V_{BCP}$, is an index of the amount of covalency in the chemical interaction [28, 29]. H_{BCP} is negative for interactions with significant sharing of electrons.

Two other bond properties are usually analyzed in the framework of AIM analysis when the interest is focused on the multiplicity of the bonds, that is, the delocalization index and the ellipticity. The delocalization index, δ(A,B), accounts for the electrons delocalized or shared between the atomic basins of atoms A and B. For bonded atoms, δ(A,B) yields a measure of the bond order between them [30, 31]. A simple relationship between the delocalization index and the formal bond order (number of Lewis bonded pairs) has been found for molecular bonds with equally shared pairs. For polar bonds, there is no longer such a simple relationship. Instead, it has been reported that the delocalization index tends to decrease with the increased electronegativity difference of the atoms involved in the bond [32]. The ellipticity, ε, measures the extent to which density is preferentially accumulated in a given plane containing the bond path [33–35]. The ellipticity is defined as $\varepsilon = \lambda_1/\lambda_2 - 1$, where $|\lambda_1| \geq |\lambda_2|$. If $\lambda_1 = \lambda_2$, then $\varepsilon = 0$ and the bond is cylindrically symmetric. Values of ε greater than zero indicate partial π-character in a bond or electronic distortion away from σ-symmetry along the path. Double bonds are usually characterized by significant ellipticity values, whereas for triple bonds the bonding regains its cylindrical symmetry resulting in values of ellipticity close to zero [35]. It has been shown that the ellipticity can vary along the bond paths with heteropolar bonds [36]. As a consequence, its value at the BCP could be misleading.

In the present work, the BCPs were analyzed in terms of electron density ρ_{BCP}, its Laplacian $\nabla^2\rho_{BCP}$, and the total energy density at the BCP, H_{BPC}. The delocalization index, δ(C,As), and the ellipticity of the C–As bonds are also reported. AIM calculations were performed using AIMAll [37].

The topological description of the chemical bond proposed by Silvi and Savin [38] is based on the gradient field analysis of the ELF of Becke and Edgecombe [39]. ELF is a scalar function whose gradient field enables to carry out a partition of the molecular space into basins of attractors closely related to electron pair domains of the VSEPR model of molecular geometry [40]. The ELF exhibits maxima at the most probable positions of localized electron pairs, and each maximum is surrounded by a basin in which there is an increased probability of finding an electron pair. These basins correspond to the qualitative electron pair domains of the VSEPR model and have the same geometry as the VSEPR domains [40]. The valence shell of a molecule consists of two types of basin: polysynaptic basins (generally disynaptic), which belong to two atomic valence shells and the monosynaptic ones, which belong to only one valence shell, and which qualitatively correspond to nonbonding valence density. The valence basins are labeled by *V* followed by a list of the atomic symbols of the centers of the valence shells, that is, *V*(A) and *V*(A,B) for a monosynaptic and a disynaptic basin [41]. The electronic population of a synaptic basin, $\bar{N}$, is obtained as the integral of the one-electron density over the basin. The variance of the basin population, $\sigma^2(\bar{N})$, represents the quantum–mechanical uncertainty of the basin population and is a result of the delocalization of electrons [42]. It is also written as the sum of the contributions of all other basins and is used as a normalization factor to represent the cross-contribution between two specific basins. The relative fluctuation of a basin can be expressed as $\lambda = \sigma^2(\bar{N})/\bar{N}$, namely, the variance of the basin normalized for the electronic population [42, 43]. This quantity is very useful to describe the degree of electron delocalization. ELF analysis has provided very useful pieces of information on the bonding for a wide range of chemical cases [44]. ELF analysis was performed using the TopMoD package [45, 46], and the results were visualized with the help of the Molekel 4.3 software [47].

With the aim of comparison, the bonding features of all the studied molecules and the net atomic charges were also analyzed by means of the natural bond orbital (NBO) and the natural population analysis (NPA) [48, 49].

3 Results and discussion

The optimized geometries of the studied molecules are shown in Fig. 1. The computed frequencies of the C–As stretching modes and the carbon and arsenic NPA charges are collected in Table 1.

The dearth of experimental results for compounds containing C–As triple bonds precludes extensive comparison with the computed geometrical and vibrational properties, as was done in our previous study of analogous systems containing a C–P triple bond [22]. The exceptions are $[(CF_3)_3BCAs]^-$ and Mes*CAs; for which geometrical parameters and vibrational frequencies were determined [14, 15]. For all other examined arsaalkynes, no experimental vibrational frequency is available. The present results can, however, be compared with previous theoretical computations for HCAs, CAs^-, and CH_3CAs species [16–21].

The simplest studied species are **(1)** and **(2)** in Fig. 1, in which the C≡As bond lengths are equal to 1.716 and 1.650 Å, respectively. These values are in good agreement with the QCISD/6-311+*G(df,p)* C≡As bond lengths of 1.724 and 1.658 Å in **(1)** and **(2)**, respectively [19]. Predictably, as a consequence of deprotonation, the charged methylidynearsine species **(1)** has an elongated bond.

In contrast to the unsubstituted compound **(2)**, the ethylidynearsine CH_3CAs derivative **(3)** of Fig. 1 has been synthetized [16]. For this species, the C≡As bond distance computed here (1.660 Å) agrees well with the value of 1.658 Å obtained at QCISD/6-311+*G(df,p)* level [19]. As in the case of analogues with phosphorous, the C≡As bond lengthens from HCAs to CH_3CAs.

Our computed B3LYP vibrational frequencies for the C≡As stretching mode are very similar to previously reported computed values. In particular, for CAs^-, our value of 998 cm^{-1} is very close to that of 983 cm^{-1} previously reported [19]. Analogously, the B3LYP frequencies of 1109 and 1461 cm^{-1} for **(2)** and **(3)** agree well with literature values of 1097 and 1482 cm^{-1}, respectively [19].

The only experimental study of the *t*BuCAs species **(4)** employed it in the synthetic pathway to arsaalkynes, but only as transient species [13]. Our computed geometry of **(4)** is very similar to that obtained for the analogous compound of phosphorous [22]. As expected due to the different sizes of the triply bonded atoms, the C≡As

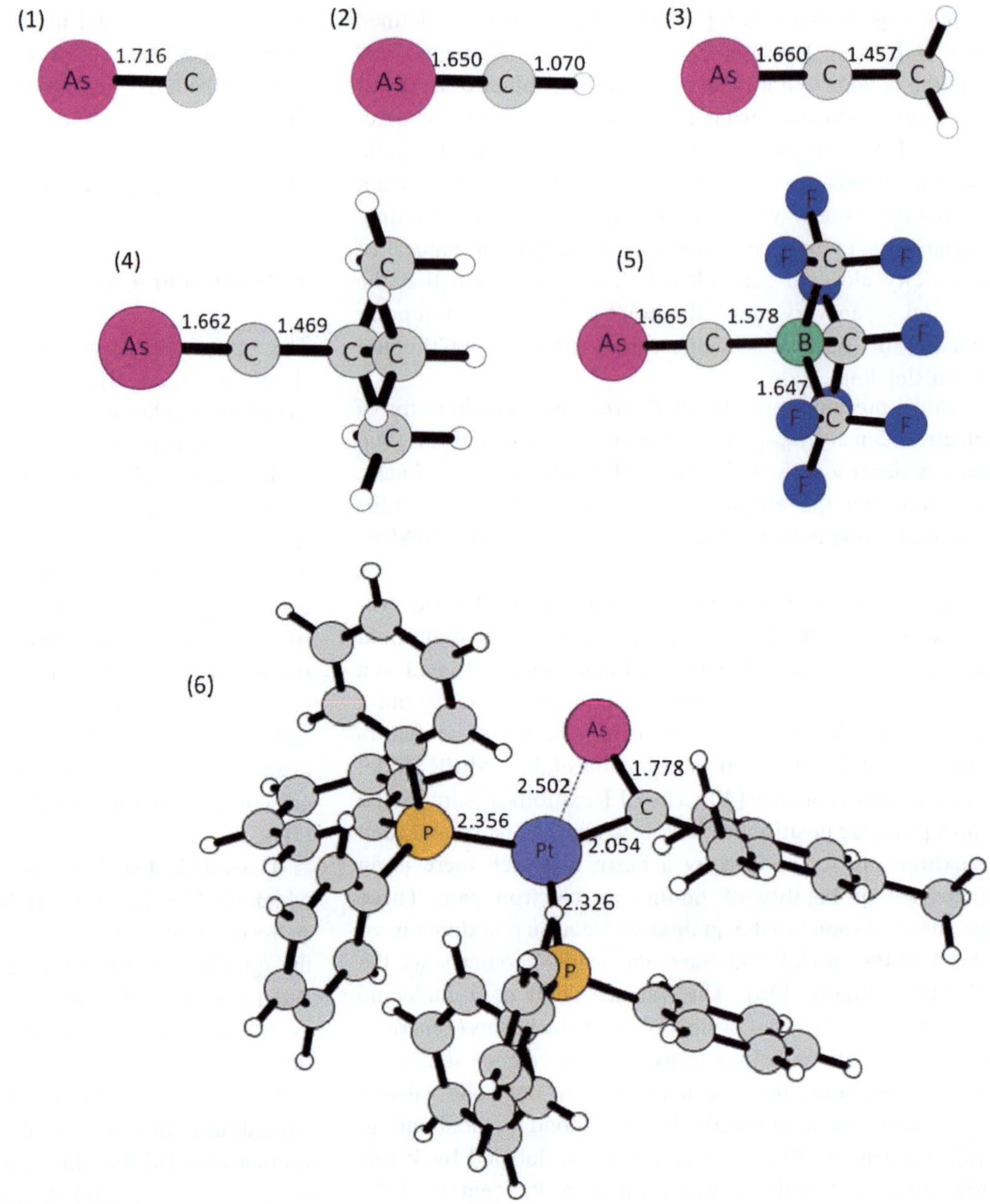

Fig. 1 Selected geometrical parameters of the optimized structures of: (**1**) CAs^-; (**2**) HCAs; (**3**) CH_3CAs; (**4**) *t*BuCAs; (**5**) $[(CF_3)_3BCAs)]^-$; and (**6**) $Pt(PPh_3)_2\{\eta^2\text{-}AsC(C_6H_2Me_3)\}$

Table 1 B3LYP harmonic vibrational frequencies (cm^{-1}) for the C≡As bond stretching and natural atomic charges (|e|) on carbon (qC) and arsenic (qAs) atoms

Species	Frequency	Q_C	Q_{As}
CAs^-	998 (983[a])	−0.84	−0.16
HCAs	1109 (1097[a])	−0.77	0.56
CH_3CAs	1461 (1482[a])	−0.53	0.51
*t*BuCAs	1451	−0.53	0.53
$[(CF_3)_3BCAs)]^-$	1350 (1325[b])	−0.67	0.42
$Pt(PPh_3)_2\{\eta^2\text{-}AsC(C_6H_2Me_3)\}$	–	−0.55	0.43

Available frequency values arising from previous calculations are also reported (in parentheses)

[a] Taken from ref. [19]

[b] Taken from ref. [15]

distance (1.662 Å) is longer than the C≡P distance (1.546 Å). Consequently, the vibrational stretching frequency of the C≡As bond in *t*BuCAs (1451 cm^{-1}) is lower than that of the computed C≡P bond stretch in *t*BuCP (1573 cm^{-1} [22]).

The optimized structure of anionic $[(CF_3)_3BCAs]^-$ ((**5**) in Fig. 1) has much in common with the geometry of species (**4**). Species (**5**) and the neutral arsaalkyne-transition metal complex (**6**) are examples of kinetically stable anionic arsaalkynes [14, 15]. The computed C–As bond length of 1.665 Å is in good agreement with the measured value of 1.675 Å and is longer (by 0.1 Å) than the computed C–P distance in the corresponding phosphaethynyl compound [22]. The computed B–C–As angle (179.9°) is in accord with the measured value of 177.5°. Finally, the computed stretching frequency (1350 cm^{-1}) agrees very well with the experimental value of 1325 cm^{-1} [14].

The arsaalkyne-transition metal complex (**6**) (Fig. 1) is a simplified model of the Mes*CAs system, which was the first reported stable arsaalkyne compound [12]. The simplified model involves the replacement of tert-butyl with methyl groups. The optimized molecular structure of (**6**) shows that the arsaalkyne bond coordinates the platinum center in a side-on (η^2) fashion. The coordination environment around the metal approximates a square planar structure like that observed in the X-ray crystal structure of Mes*CAs [14]. The computed As≡C bond distance (1.778 Å) and C–C≡As angle (40.0°) are in very good agreement with the experimental values (1.786 Å and 39.5°) [14].

A comparison of the optimized C–As bond lengths of compounds (**1**) to (**5**) (Fig. 1) with the C–P bond distances of analogous compounds [22] indicates a systematic elongation of the bonds between 0.114 and 0.118 Å, when substituting As by P atoms. A slightly larger difference (0.121 Å) is found between the computed C–As distance in (**6**) and the C–P bond length of a similar compound reported in [22].

3.1 AIM bonding analysis

The properties of the electron density at the C–As BCPs of compounds (**1**) to (**6**) are summarized in Table 2. All values of ρ_{BCP} suggest significant concentration of electrons at the bond critical point (c.a. 0.20 a.u.). These values are only slightly lower than the corresponding values for the C–P bonds in analogue species [22]. In all cases, the bond critical points are in regions of positive values of $\nabla^2\rho_{BCP}$ (Table 2).

The computed C–As bond lengths increase in going from compound (**2**) to (**6**) by almost 0.13 Å. The optimized bond distance for CAs^- is 1.716 Å, placing this molecule between (**5**) and (**6**), according to its C–As bond length. Therefore, the concentration of charge in the bonding region correlates fairly well with the bond length, being lowest for compound (**6**) ($d_{C\text{–}As}$ = 1.778 Å) and highest for HCAs ($d_{C\text{–}As}$ = 1.650 Å). Compound (**5**) deviates slightly from this trend. In fact, the charge density increases slightly (+0.005 au) with respect to compound (**4**), despite the small increase in the C–As bond length (+0.003 Å).

The total electronic energy density at the C–As BCP of all the studied species is negative, which indicates that there is significant sharing of electrons despite the positive values of the Laplacian of the density at the same point. There is a good correlation between the amount of charge density and the stabilization energy at the BCP. H_{BCP} is more negative for compound (**2**), which was found to have the highest value of charge density at the BCP, while the lowest absolute value was found for compound (6), which has the lowest ρ_{BCP} (see Table 2).

The delocalization index, δ(C,As), ranges between 2.62 and 2.90 for compounds (**1**) to (**5**) and drops to 1.78 in compound (**6**). The ellipticity of the C–As bonds of compounds (**1**) to (**5**) is very close to zero ($\varepsilon < 10^{-4}$), whereas for compound (**6**) that value rises to 0.05.

It is known that the electron density at the BCP generally shows a good correlation with the bond length and the bond order. However, the actual value of the electron density varies depending on the characteristics of the particular bond that is being analyzed. The main goal of this work has been to give insight into the characteristics of the C–As bonds in a series of compounds presenting formal triple bonds and analyze similarities and differences with similar bonds formed by the lighter members of the group. With this aim, it is useful to perform a comparative analysis of the AIM properties of the bond of interest in a series of compounds. In particular, a set of homologue species containing C–As, C–P, and C–N bonds with different bond orders were chosen as a benchmark. The molecules considered were hydrogen cyanide (HC≡N), methanimine (H_2C=NH), methylamine (H_3C–NH_2), and the corresponding phosphorus and arsenic homologues. The optimized geometries of these molecules are shown in Fig. 2, whereas the computed AIM properties are summarized in Table 3.

An analysis of the results presented in Table 3 shows that in the case of C–N bonds the electron density at the BCP increases by a factor of 1.53 in going from single to double bonds and by a factor of 1.25 in going from double to triple bonds. The same factors drop to 1.29 and 1.15 for C–P bonds and to 1.36 and 1.18 for C–As bonds. Therefore, the reinforcement of the bonds due to the formation of double and triple bonds is notably lower in the case of C–P and C–As bonds. A comparison of the C–N bonds of HC≡N, H_2C=NH, and H_3C–NH_2 with its C–P and C–As homologues shows that the charge density at the C–N

Table 2 AIM topological properties of the C–As bonds of compounds (**1**) to (**6**)

	ρ_{BCP}^{a}	$\nabla^2\rho_{BCP}^{b}$	H_{BCP}^{c}	$\delta(C,As)^{d}$	Q_C, Q_{As}^{e}
CAs^{-f} (**1**)	0.199	0.135	−0.155	2.90	−0.92, −0.08
HCAs (**2**)	0.208	0.385	−0.159	2.76	−0.76, 0.65
CH_3CAs (**3**)	0.203	0.387	−0.152	2.64	−0.72, 0.55
*t*BuCAs (**4**)	0.202	0.379	−0.152	2.62	−0.74, 0.55
$[(CF_3)_3BAsC]^-$ (**5**)	0.207	0.331	−0.159	2.72	−1.15, 0.42
$Pt(PPh_3)_2\{\eta^2\text{-}AsC(C_6H_2Me_3)\}$ (**6**)	0.170	0.129	−0.121	1.78	−0.73, 0.31

[a] Electron density at the BCP in a.u.

[b] Laplacian of the electron density at the BCP in a.u.

[c] Total electronic energy density at the C–As BCP

[d] The electron delocalization index, **δ**(C,As), accounts for the electrons shared between the C and As atoms

[e] AIM atomic charges for the C and As atoms

[f] With the aim of comparison, the corresponding values for CN^- are: $\rho_{BCP} = 0.481$, $\nabla^2\rho_{BCP} = -0.770$, $\delta(C,N) = 2.40$, $H_{BCP} = -0.907$, Q_C, $Q_{As} = 0.49, -1.49$. For CP^- are: $\rho_{BCP} = 0.213$, $\nabla^2\rho_{BCP} = 0.320$, $\delta(C,As) = 2.80$, $H_{BCP} = -0.219$, Q_C, $Q_{As} = -1.37, 0.37$

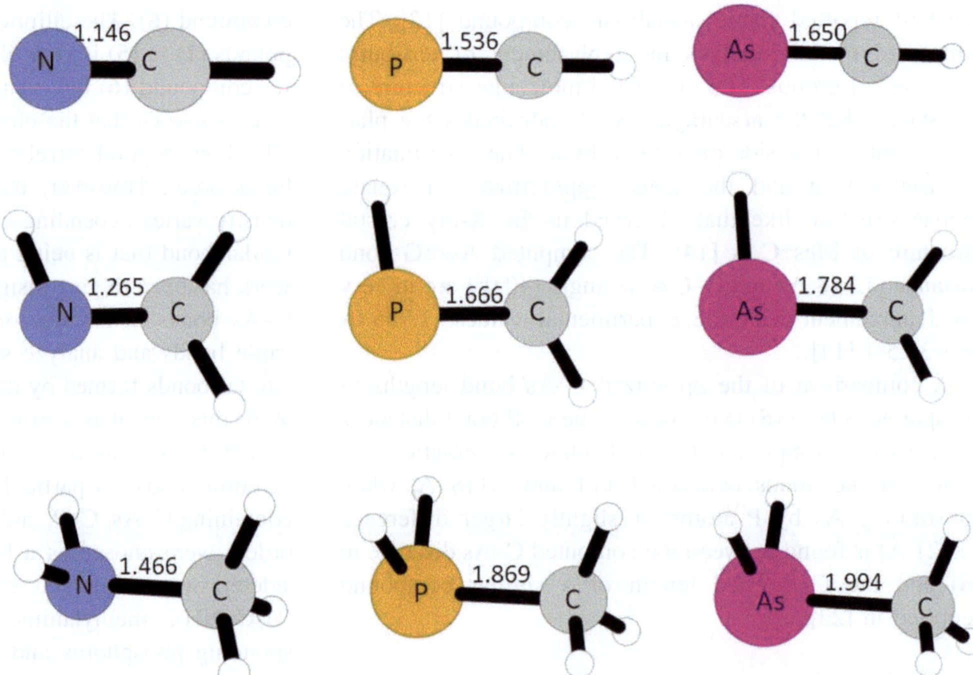

Fig. 2 Optimized C–N, C–P, and C–As bond lengths of: HCN, H_2CNH, and H_3CNH_2 (*first column*); HCP, H_2CPH, and H_3CPH_2 (*second column*) and HCAs, H_2CAsH, and H_3CAsH_2 (*third column*)

BCPs is higher than the same quantity at the C–P bond by a factor of between 1.74 and 2.25 (depending on the order of the bond) and by a factor of between 2.02 and 2.38 for the C–As bond. The difference increases for higher bond orders. The Laplacian of the density at the BCP is negative in all the C–N studied molecules, as well as for the C–P and C–As single bonds of methylphosphine and methylarsine. In contrast, $\nabla^2\rho_{BCP}$ is positive in the case of the double bonds of H_2CPH and H_2CAsH; and the triple bonds of HCP and HCAs. H_{BCP} is negative in all the benchmark molecules, and the amount of this energy correlates fairly well with the charge density at the BCP. As expected, the bond ellipticity reaches a maximum in double bonds, being zero in the case of HCN and its homologues, and close to zero for the single bonded species (less than 0.065). The delocalization index is very close to one (0.93–1.04) for the H_3CXH_2 molecules, between 1.70 and 1.85 for H_2CXH and between 2.56 and 2.76 for HCX (X = N, P, As). The delocalization indices for single, double, and triple bonds for various molecules are reported in literature [50–52]. There is agreement for δ(A,B) to usually be lower than the corresponding bond order. For instance, in C–C bonds, this index takes values of 0.95, 1.89, and 2.85 for single, double, and triple bonds [52]. As previously mentioned, a

Table 3 AIM topological properties of the C–N bonds of HCN, H_2CNH, and H_3CNH_2; C–P bonds of HCP, H_2CPH, and H_3CPH_2; and C–As bonds of HCAs, H_2CAsH, and H_3CAsH_2

	Formal C–N bond order	ρ^{a}_{BCP}	$\nabla^2\rho^{b}_{BCP}$	H^{c}_{BCP}	$\delta(C,N)^{d}$	ε^{e}	Q_C, Q_N^{f}
HCN	3	0.496	−0.325	−0.933	2.60	0.000	0.88, −1.05
H_2CNH	2	0.398	−1.182	−0.649	1.80	0.219	0.67, −1.00
H_3CNH_2	1	0.260	−0.626	−0.252	1.04	0.038	0.36, −0.91
	Formal C–P bond order	ρ^{a}_{BCP}	$\nabla^2\rho^{b}_{BCP}$	H^{d}_{BCP}	$\delta(C,P)^{c}$	ε	Q_C, Q^{e}_P
HCP	3	0.220	0.634	−0.215	2.56	0.000	−1.17, 1.06
H_2CPH	2	0.192	0.075	−0.195	1.70	0.384	−0.80, 1.19
H_3CPH_2	1	0.149	−0.274	−0.115	0.93	0.061	−0.33, 1.24
	Formal C–As bond order	ρ^{a}_{BCP}	$\nabla^2\rho^{b}_{BCP}$	H^{d}_{BCP}	$\delta(C,As)^{c}$	ε	Q_C, Q^{e}_{As}
HCAs	3	0.208	0.385	−0.159	2.76	0.000[g]	−0.76, 0.65
H_2CAsH	2	0.176	0.050	−0.130	1.85	0.263	−0.50, 0.72
H_3CAsH_2	1	0.129	−0.075	−0.073	0.96	0.027	−0.24, 0.81

[a] Electron density at the BCP in a.u.

[b] Laplacian of the electron density at the BCP in a.u.

[c] Total electronic energy density at the BCPs

[d] The electron delocalization index, δ(C,X), accounts for the electrons shared between atoms (X = N, P, and As)

[e] The bond ellipticity

[f] AIM atomic charges for the C, N, P, and As atoms

[g] The actual value is 2.2 10^{-11}

larger departure is expected when the bond pairs are not equally shared between the bonded atoms.

From all the above, some preliminary conclusions that reveal the main electronic characteristics of the C–As bonds in the studied species can be drawn. The C–As bonds of compounds (**1**) to (**6**) show significant concentration of charge density at the BCPs (c.a 0.2 au) together with positive values of $\nabla^2\rho_{BCP}$ and negative values of H_{BCP}. Therefore, despite the charge density is locally depleted in the bonding region, the condition $H_{BCP} < 0$ holds for all the studied bonds, which permits to qualitatively classify those bonds as shared in nature. Despite the known dependency of the delocalization index with the electronegative difference of the bonded atoms, the present results seem to indicate that this index gives a reliable indication of the bond orders for the case under study. According to this index, compounds (**1**) to (**5**) are characterized by the presence of a triple C–As bond, whereas the value of this parameter in compound (**6**) is closer to those characteristic of double bonds (Table 3). The values of ellipticity at the BCPs confirm that compounds (**1**) to (**5**) are cylindrically symmetric. Even when there is an important increase in the ellipticity at the C–As BCP of compound (**6**), the computed value is still out of the range of double bonded C–As bonds (Table 3). Therefore, no definitive conclusions can be drawn on the basis of this index.

In summary, the topological analysis of the charge density indicates that the C–As bonds of compounds (**1**) to (**5**) can be described as triple bonds, weakened by an important depletion of the charge density in the bond region. The C–As bond of compound (**6**) shows some differences with those of compound (**5**), that is, the characteristics of that bond are closer to those of typical C–As double bonds.

3.2 ELF bonding analysis

The main information concerning the C–As bonds of compounds (**1**) to (**6**) obtained by the topological analysis of the ELF function is presented in Table 4. The basin populations of the disynaptic and monosynaptic basins involving the As atoms, *V*(C,As) and *V*(As), are reported together with the electronic population of the second disynaptic basin involving the C atom engaged in the C–As bond, *V*(C,R). The basin variances, $\sigma^2(\bar{N})$, and the relative fluctuations, $\lambda(\bar{N})$, are included in the same table. Figures 3 and 4 show the ELF isosurfaces for compounds (**1**) to (**5**) and (**6**), respectively.

ELF analysis was also performed on the benchmark molecules, namely, HC≡N, H_2C=NH, H_3C–NH_2, and the phosphorus and arsenic analogues. The obtained results are reported in Table 5, whereas the ELF isosurfaces are shown in Fig. 5.

Table 4 ELF topological properties of the C–As bonds of compounds (**1**) to (**6**)

	V(C,As), $\sigma^2(\bar{N})$, $\lambda(\bar{N})$	V(C,R), $\sigma^2(\bar{N})$, $\lambda(\bar{N})$	V(As), $\sigma^2(\bar{N})$, $\lambda(\bar{N})$
CAs^{-a} (**1**)	2.82, 1.47, 0.52	2.72, 1.01, 0.37[b]	4.72, 2.10, 0.44
HCAs (**2**)	4.20, 1.72, 0.41	2.31, 0.71, 0.31 (R=H)	3.82, 1.86, 0.49
CH_3CAs (**3**)	3 × 1.46, 3 × 0.94, 0.64	2.14, 1.07, 0.50 (R=C)	3.89, 1.88, 0.48
*t*BuCAs (**4**)	3 × 1.44, 3 × 0.93, 0.65	2.20, 1.10, 0.50 (R=C)	3.94, 1.90, 0.48
$[(CF_3)_3BCAs)]^-$ (**5**)	3 × 1.33, 3 × 0.89, 0.67	2.37, 1.11, 0.47 (R=B)	4.04, 1.92, 0.48
$Pt(PPh_3)_2\{\eta^2\text{-}AsC(C_6H_2Me_3)\}$ (**6**)	2.91, 1.56, 0.54	1.64, 1.08, 0.66 (R=Pt)	3.79, 1.96, 0.52
		2.25, 1.11, 0.49 (R=C)	

The basin populations, $\bar{N}$, the variance of the basin populations, $\sigma^2(\bar{N})$, and the relative fluctuation, $\lambda(\bar{N})$, are given in electrons

[a] With the aim of comparison, the corresponding values for CN^- are: V(C,N), $\sigma^2(\bar{N})$, $\lambda(\bar{N})$ = 3.39, 1.45, 0.43 and V(N), $\sigma^2(\bar{N})$, $\lambda(\bar{N})$ = 3.54, 1.27, 0.36. For CP^- : V(C,P), $\sigma^2(\bar{N})$, $\lambda(\bar{N})$ = 2.91, 1.39, 0.48 and V(P), $\sigma^2(\bar{N})$, $\lambda(\bar{N})$ = 4.22, 1.55, 0.37

[b] This value corresponds to the population and the variance of the monosynaptic V(C) basin

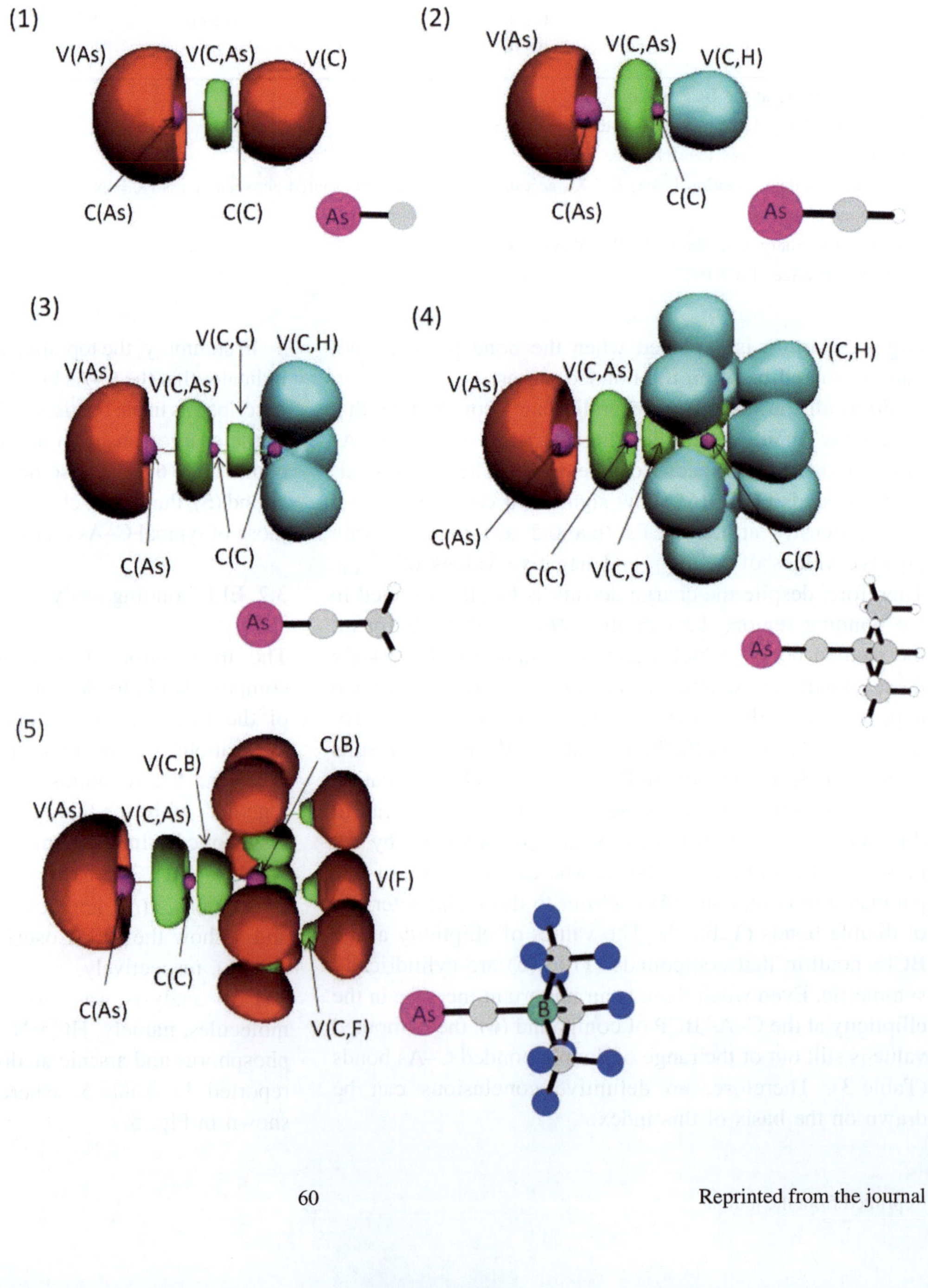

Fig. 3 ELF isosurfaces ($\eta = 0.75$) for optimized structures of: (**1**) CAs^-; (**2**) HCAs; (**3**) CH_3CAs; (**4**) *t*BuCAs; and (**5**) $[(CF_3)_3BCAs)]^-$. Core basins are represented in *magenta*, valence monosynaptic in *red*, protonated disynaptic in *light blue*, and valence disynaptic in *green*. To help the visualization, *bold* and *sticks* representations of the optimized structures are shown in the *insets*

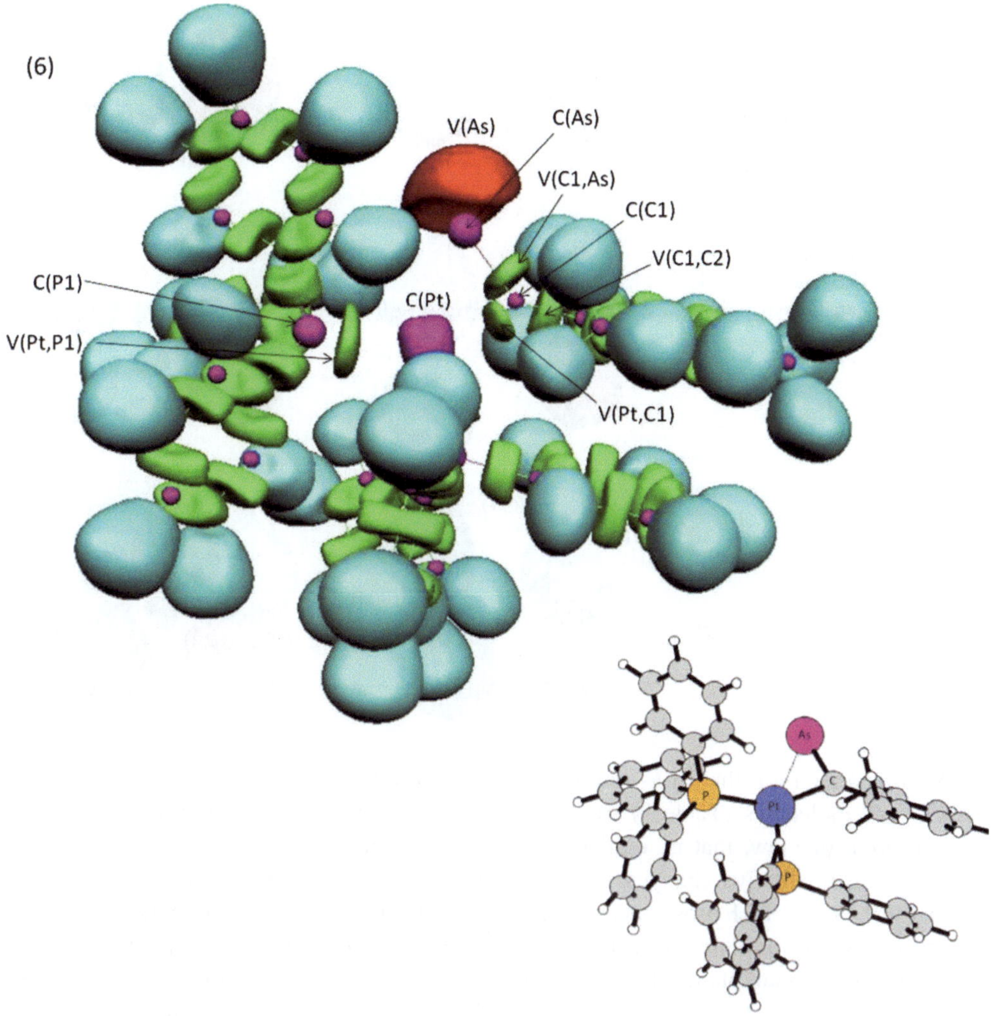

Fig. 4 ELF isosurfaces ($\eta = 0.75$) for the optimized structure of compound (**6**), $Pt(PPh_3)_2\{\eta^2\text{-}AsC(C_6H_2Me_3)\}$. Core basins are represented in *magenta*, valence monosynaptic in *red*, protonated disynaptic in *light blue*, and valence disynaptic in *green*. To help the visualization, a *bold* and *sticks* representation of the optimized structure are shown in the *insets*

Table 5 ELF topological properties of the C–N bonds of HCN, H_2CNH, and H_3CNH_2; C–P bonds of HCP, H_2CPH, and H_3CPH_2; and C–As bonds of HCAs, H_2CAsH, and H_3CAsH_2

	$V(C,N)$, $\sigma^2(\bar{N})$, $\lambda(\bar{N})$	$V(N,H)$, $\sigma^2(\bar{N})$, $\lambda(\bar{N})$	$V(N)$, $\sigma^2(\bar{N})$, $\lambda(\bar{N})$
HCN	4.16, 1.53, 0.37	2.32, 0.68, 0.29[a]	3.32, 1.20, 0.36
H_2CNH	2.93, 1.38, 0.47	1.95, 0.75, 0.38	2.67, 1.08, 0.40
H_3CNH_2	1.65, 0.95, 0.58	1.95, 0.77, 0.39	2.18, 0.99, 0.45
	$V(C,P)$, $\sigma^2(\bar{N})$, $\lambda(\bar{N})$	$V(P,H)$, $\sigma^2(\bar{N})$, $\lambda(\bar{N})$	$V(P)$, $\sigma^2(\bar{N})$, $\lambda(\bar{N})$
HCP	4.08, 1.57, 0.38	2.31, 0.70, 0.30[a]	3.45, 1.34, 0.39
H_2CPH	3.01, 1.38, 0.46	2.03, 0.61, 0.30	2.61, 1.05, 0.40
H_3CPH_2	1.80, 0.98, 0.54	1.99, 0.62, 0.31	2.12, 0.85, 0.40
	$V(C,As)$, $\sigma^2(\bar{N})$, $\lambda(\bar{N})$	$V(As,H)$, $\sigma^2(\bar{N})$, $\lambda(\bar{N})$	$V(As)$, $\sigma^2(\bar{N})$, $\lambda(\bar{N})$
HCAs	4.20, 1.72, 0.41	2.31, 0.71, 0.31[a]	3.82, 1.86, 0.49
H_2CAsH	2 × 1.54, 2 × 0.96, 0.62	2.07, 0.71, 0.34	2.96, 1.46, 0.49
H_3CAsH_2	1.76, 1.03, 0.59	2.03, 0.71, 0.35	2.47, 1.22, 0.49

The basin populations, $\bar{N}$, the variance of the basin populations, $\sigma^2(\bar{N})$, and the relative fluctuation, $\lambda(\bar{N})$, are given in electrons

[a] This value corresponds to the population and the variance of the $V(C,H)$ basin

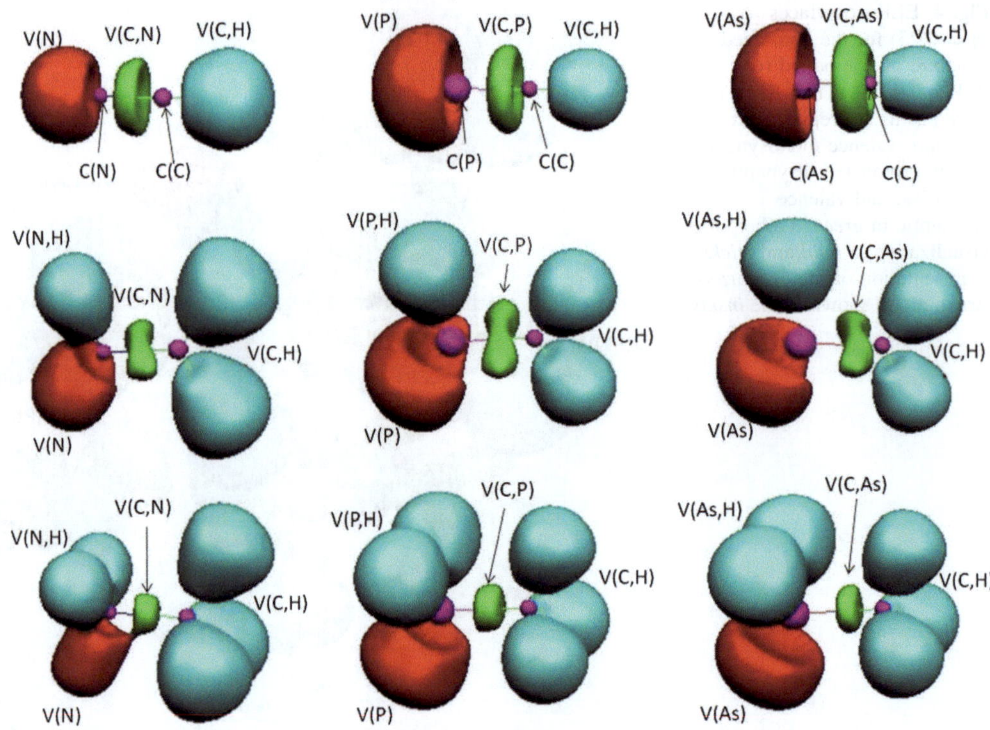

Fig. 5 ELF isosurfaces ($\eta = 0.8$ for *first* and *second column*, 0.75 for *third column*) for optimized structures of: HCN, H_2CNH, and H_3CNH_2 (*first column*); HCP, H_2CPH, and H_3CPH_2 (*second column*) and HCAs, H_2CAsH, and H_3CAsH_2 (*third column*). Core basins are represented in *magenta*, valence monosynaptic in *red*, protonated disynaptic in *light blue*, and valence disynaptic in *green*

A comparison of the ELF results performed on the benchmark molecules (Table 5) shows that from a topological point of view, that is, number and type of basins, there are no qualitative changes when substituting the N atom by P or As. ELF isosurfaces reveal that disynaptic basins resembling prolate spheroids elongated above and below the molecular plane (depicted in green in Fig. 5), which are characteristic of double bonds, for H_2C=NH and its phosphorus and arsenic analogues, whereas torus-like disynaptic basins of cylindrical symmetry typical of triple bonds are present in HC≡N and its homologues. As can be seen in Fig. 5, the disynaptic valence basins are always located closer to the most electronegative atom. The ELF descriptions of HCN, H_2C=NH, and H_2C=PH obtained here agree with the same type of study performed by Chesnut [53].

There is a good correlation between the calculated mean values of the basin populations and the formal bond order. The disynaptic valence electron populations are between 1.65 and 1.80 e^- for formal single bonds, between 2.93 and 3.08 e^- for double, and between 4.08 and 4.20 e^- for triple bonds. There is an important increase in the electron delocalization in going from N to As, as evidenced by the rise in the values of the relative fluctuation of the disynaptic and monosynaptic basins involving those atoms (Table 5). The electron population of the monosynaptic valence basins increases with the increase in the formal bond order, and the highest values involve As atoms. As a consequence of the high monosynaptic electron population, the disynaptic basins representing the triple bonds have an electron population notably lower than the expected six electrons. Previous theoretical studies have shown that this is usually the case in molecules containing triple bonds in the presence of lone pairs, and the effect has been called lone-pair bond weakening effect (LPBWE) [54]. The bonding population has been shown to be usually significantly lower than the expected six electrons, at the expense of increased lone-pair populations [22, 54, 55].

The electron population of the disynaptic *V*(C,As) basins of compounds (**1**) to (**6**) ranges from 2.82 to 4.38 electrons. In all cases, the main atomic contribution to these basins comes from the C atom (between 71 and 77% of the total population). As evidenced in Table 3, three *V*(C,As) basins have been located for CH_3CAs, *t*BuCAs, and $[(CF_3)_3BCAs]^-$, each of them with a basin population of between 1.33 and 1.46 e^-. It is known that generally there is no direct correspondence between the number of valence disynaptic attractors representing the multiple bond in topological analysis of ELF and the multiplicity of the bond. Previous studies have concluded that the multiplicity of disynaptic basins can be better explained by symmetry considerations [44]. As for the benchmark molecules, the monosynaptic *V*(As) basins have electron populations that are much higher than the expected value for a formal single lone pair (2 electrons). Those values range between 3.79 and 4.72 e^-. Both valence basins, *V*(As) and *V*(C,As), are characterized for having very high relative fluctuations (0.41–0.67), which indicates a great degree of electron delocalization.

A comparison of Figs. 3 and 5 puts in evidence the similarity between the torus-like disynaptic *V*(C,As) basins of compounds (**1**) to (**5**), and the ones present in HC≡N and its homologues. The *V*(C,As) basin of compound (6) departs from the cylindrical symmetry that characterizes the same basin in compounds (**1**) to (**5**) (see Figure S1, Supporting Information for a better view of this basin).

The first studied compound, CAs^-, has special characteristics due to the presence of lone pairs on both bonded atoms. In fact, the disynaptic basin population is the lowest of all the studied compounds, due to the extensive delocalization of the bonding electrons with both monosynaptic valence basins, *V*(C) and *V*(As), which have electronic population notably higher than the expected 2 electrons (Table 4). A comparison of the characteristic of the bonds of CN^-, CP^-, and CAs^- shows an important drop in the electron population of the disynaptic basin involving the two atoms forming the bond, from the 3.39 e^- of CN^- to the 2.82 e^- of CAs^-. This drop is accompanied by an increase in the population of the monosynaptic basins, from the 3.54 e^- of *V*(N) to the 4.72 e^- of *V*(As). As observed in the benchmark molecules, there is also an important increase in the relative fluctuation of the basins, which indicates an increase in electron delocalization between the disynaptic and monosynaptic basins in going from CN^- to CAs^-.

The topological representation obtained from ELF analysis is usually interpreted in terms of superposition of mesomeric structures [56, 57]. Therefore, the valence population can be considered as the result of the weighted contribution of different mesomeric structures of that molecule. Within this interpretation, the trend observed in the disynaptic basin population in going from CN^- to CAs^- indicates that the weight of the mesomeric structure representing a triple bond between the two atoms is less important in the case of CAs^-. ELF analysis, therefore, supports the findings obtained by AIM, according to which there is an important drop of the BCP charge density in going from CN^- to CAs^- (see footnote of Table 2).

There is an important increase in the disynaptic *V*(C,As) basin population in going from CAs^- to HCAs, that is, from 2.82 to 4.20 e^-. As previously mentioned, the depleted *V*(C,As) population is due to the high electron delocalization with the lone pairs of both, C and As atoms.

Compounds (**2**) to (**5**) show very close values of electron population of the *V*(C,As) valence basin, and the small changes observed can be rationalized considering the characteristics of the group bonded to the C atom engaged in the C–As bond. Going from HCAs to CH_3CAs, there is a small increase in the *V*(C,As) electron population, that is, from 4.20 to 4.38 (3 × 1.46) e^-, which confirms the greater electron-releasing character of the CH_3 group compared with the inductive effect of the H atom. There are not evident changes in the *V*(C,As) population in going from CH_3CAs to *t*BuCAs, despite the expected higher inductive effect of the *tert*-butyl group; but there is an important decrease of basin population in compound (**5**). The strong inductive effect of the electronegative CF_3 groups provokes a depopulation of the *V*(C,As) basin. In $[(CF_3)_3BCAs]^-$, the boron atom in engaged in four covalent bonds with four different carbon atoms, as evidenced by the presence of four disynaptic *V*(C,B) valence basins with electron populations of between 2.20 and 2.37 e^-. The contribution of boron to these valence populations is of only 13%. Therefore, the C atom engaged in the C–As bond contributes importantly to the formation of the B–C bond, and this reduces the availability of charge for the formation of the C–As bond. This is evidenced by the lower contribution of electron charge from that atom to the *V*(C,As) valence basin (74% in contrast to the 77% of compounds (2), (3), and (4)).

Compound (6) has a much lower *V*(C,As) electronic population, 2.91 e^-, with almost 74% of that population coming from the C atom. The torus-like symmetry typical of triple bonds is somewhat lost in the *V*(C,As) valence basin of this compound (see Figure S1, Supporting Information). The C atom engaged in the C–As bond participates also in other two valence basins: one with a second C atom and another with the Pt atom, *V*(C1,Pt), as shown in Fig. 4. Pt is involved in the formation of three valence basins, the previously mentioned *V*(C1,Pt), and two valence basins *V*(Pt,P1) and *V*(Pt,P2) containing a valence population of c.a. 2.22 e^-. In all cases, the contribution of the Pt atom to the total electronic population of the disynaptic basins is very low, that is, 20%.

The bonding description of complex (**6**) obtained by ELF is supported by AIM analysis. In fact, three BCPs involving Pt atom were localized, a Pt–C BCP with a charge density of 0.128 a.u. and two Pt–P BCPs with charge densities of around 0.100 au. In all cases, the Laplacian is positive (between 0.16 and 0.21 a.u.), and the total energy density H_{BCP} is negative.

3.3 NBO analysis

In general, NBO analysis confirms the triple bond nature between C and As atoms in all studied species except (**6**). As an example of a bond analysis, the σ bond and the two π bonds characterizing the *t*BuCAs species derive from the overlap between the hybrid sp and pure p orbitals of the C and As atoms, in accordance with the following composition: $\sigma(CAs) = 0.81(sp)_C + 0.58(sp^{3.33})_{As}$ and $\pi(CAs) = 0.74(p)_C + 0.67(p)_{As}$. In the other structures, the compositions are nearly the same. Similar results have been obtained in the analogues involving the phosphorus atom.

In the case of the complex (**6**), the bonding analysis obtained by NBO is different since the C≡As bond is described as a double bond whose σ- and π-components arise from the orbital overlaps $0.80(sp^{1.68})_C + 0.59(sp^{5.06})_{As}$ and $0.72(p)_C + 0.68(p)_{As}$, respectively. This result confirms the experimental observation according to which the value of the C–As bond length (1.786 Å) is closer to that predicted for an As=C bond in arsaalkenes. The analogous compound of phosphorus shows, instead, a C–P triple bond [22].

4 Conclusions

In this work, we have attempted to give insight into the nature of the C–As bonds of a series of compounds by means of a comparative analysis of the bonding characteristics, which were analyzed using different methodologies (AIM, ELF, NBO).

The results of the present study can be summarized as follows:

1. According to the topological analysis of the charge density, the C–As bonds of all the studied compounds show a significant concentration of charge at the BCP. All of them are characterized by positive values of the Laplacian of the charge density and negative values of total energy density at the BCP, which indicates that the nature of the bonds can be characterized as shared. An analysis of diverse AIM properties indicates that the C–As bonds of compounds (**1**) to (**5**) can be described as triple bonds weakened by an important depletion of the charge density in the bond region. The studied AIM topological properties seem to indicate that the C–As bond of compound (**6**) has characteristics closer to those of a double bond.
2. The results of ELF analysis support the conclusions drawn from AIM analysis. From a topological point of view, the valence disynaptic basins, which represent that C–As bonds of compounds (**1**) to (**5**), have torus-like shapes, typical of triple bonds. The valence basin electron populations are much lower than the expected value of six electrons due to an important electron delocalization between the disynaptic *V*(C,As) basins and the monosynaptic valence basins representing the As lone pairs. This type of depleted valence basins has been often found in molecules containing triple bonds in the presence of lone pairs. The disynaptic valence basin representing the C–As bond of complex (**6**) has a more distorted shape, and the valence population is markedly lower than that of compounds (**1**) to (**5**).
3. AIM and ELF topological analyses performed on a series of homologue molecules containing single, double, and triple C–N, C–S, and C–As bonds have permitted evaluation of the bond properties changes that take place when substituting nitrogen atom by its heavier congeners. AIM analysis shows an important decrease of the BCP charge density in going from N to As, especially in the case of triple bonds. The Laplacian of the charge density increases from the large negative values of the C–N bonds to positive values in the case of multiple C–P and C–As bonds. ELF analysis shows that the substitution of N by P or As does not change qualitatively the topological properties of the studied molecules. The main effect of that substitution is an increase in the electron delocalization, which is highest for molecules containing C≡As triple bonds.
4. NBO analysis agrees well with the indications arising from the topological methodologies confirming that C and As atoms are linked together by a triple bond in all the studied species, with the only exception of the compound (**6**).

Acknowledgments We gratefully acknowledge the Dipartimento di Chimica, Università della Calabria for financial aid.

References

1. Barrau J, Esoudié J, Satgé J (1990) Chem Rev 90:283
2. Grev RS (1991) Adv Organomet Chem 33:125
3. Esoudié J, Couret C, Ranaivonjatovo H, Satgé J (1994) Coord Chem Rev 130:427
4. Barrau J, Rima G (1998) Coord Chem Rev 178:593
5. Esoudie' J, Couret C, Ranaivonjatovo H (1998) Coord Chem Rev 178:565
6. Tokitoh N, Matsumoto T, Okazaki R (1999) Bull Chem Soc Jpn 72:1665
7. Power PP (1999) Chem Rev 99:3463
8. Leigh WJ (1999) Pure Appl Chem 71:453
9. Tokitoh N (1999) Pure Appl Chem 71:495
10. Gier TE (1961) J Am Chem Soc 83:1769
11. Becker G, Gresser G, Uhl W (1981) Z Naturforsch B 36:16
12. Märkl G, Sejpka H (1986) Angew Chem Int Ed 25:264
13. Hitchcock PB, Johnson JA, Nixon J F (1993) Angew Chem Int Ed 32:103
14. Hitchcock PB, Johnson JA, Nixon JF J Chem Soc Chem Commun 2061
15. Finze M, Bernhardt E, Willner H, Lehmann CW (2004) Angew Chem Int Ed 43:4160
16. Guillemin JC, Lassalle L, Dréan P, Wlodarczak G, Demaison J (1994) J Am Chem Soc 116:8930
17. Guillemin JC, Chrostowska A, Dargelos A, Nguyen TXM, Graciaa A, Guenot P (2008) Chem Commun 4204
18. Lohr LL, Scheiner AC (1984) THEOCHEM 18:195
19. Mó O, Yanez M, Guillemin JC, Riague EH, Gal JF, Maria PC, Poliart CD (2002) Chem Eur J 8:4919
20. Nazari F, Ansari N (2010) J Mol Model 16:1075
21. Marchal R, Bégué D, Chrostowska A, Pouchan C (2010) Chem Phys Lett 493:24
22. Lucas MF, Michelini MC, Russo N, Sicilia E (2008) J Chem Theory Comput 4:397
23. Becke AD (1993) J Chem Phys 98:5648
24. Stephens PJ, Devlin FJ, Chabalowski CF, Frisch MJ (1994) J Phys Chem 98:11623

25. Frisch MJ, Trucks GW, Schlegel HB, Scuseria GE, Robb MA, Cheeseman JR, Montgomery JA, Jr, Vreven T, Kudin KN, Burant JC, Millam JM, Iyengar SS, Tomasi J, Barone V, Mennucci B, Cossi M, Scalmani G, Rega N, Petersson GA, Nakatsuji H, Hada M, Ehara M, Toyota K, Fukuda R, Hasegawa J, Ishida M, Nakajima T, Honda Y, Kitao O, Nakai H, Klene M, Li X, Knox JE, Hratchian HP, Cross JB, Bakken V, Adamo C, Jaramillo J, Gomperts R, Stratmann RE, Yazyev O, Austin AJ, Cammi R, Pomelli C, Ochterski JW, Ayala PY, Morokuma K, Voth GA, Salvador P, Dannenberg JJ, Zakrzewski VG, Dapprich S, Daniels AD, Strain MC, Farkas O, Malick DK, Rabuck AD, Raghavachari K, Foresman JB, Ortiz JV, Cui Q, Baboul AG, Clifford S, Cioslowski J, Stefanov BB, Liu G, Liashenko A, Piskorz P, Komaromi I, Martin RL, Fox DJ, Keith T, Al-Laham MA, PengCY,Nanayakkara A, ChallacombeM,Gill PMW, Johnson B, Chen W, Wong MW, Gonzalez C, Pople JA (2004) Gaussian Inc., Wallingford
26. Wadt WR, Hay PJ (1985) J Chem Phys 82:284
27. Bader RFW (1990) Atoms in molecules. A quantum theory. Oxford University Press, Oxford
28. Cremer D, Kraka E (1984) Croat Chem Acta 57:1259
29. Cremer D, Kraka E (1984) Angew Chem Int Ed Engl 23:627
30. Fradera X, Austen MA, Bader RFW (1999) J Phys Chem A 103:304
31. Fradera X, Poater J, Simon S, Duran M, Solà M (2002) Theor Chem Acc 108:214
32. Poater J, Duran M, Solà M, Silvi B (2005) Chem Rev 105:3911
33. Cremer D, Kraka E, Slee TS, Bader RFW, Lau CDH, Nguyen-Dang TT, McDougall PJ (1983) J Am Chem Soc 105:5069
34. Bader RFW, Slee TS, Cremer D, Kraka E (1983) J Am Chem Soc 105:5061
35. Schrer W, Sirsch P, Shorokhov DM, Tafipolsky M, McGrady GS, Gullo E (2003) Chem Eur J 9:6057
36. Cheesman JR, Carroll MT, Bader RFW (1988) Chem Phys Lett 143:450
37. Keith TA (2011) AIMAll (Version 11.05.16) (aim.tkgristmill.com)
38. Silvi B, Savin A (1994) Nature 371:683
39. Becke AD, Edgecombe KE (1990) J Chem Phys 92:5397
40. Gillespie RJ, Robinson EA (2007) J Comput Chem 28:87
41. Savin A, Silvi B, Colonna N (1996) Can J Chem 74:1088
42. Noury S, Colonna N, Savin A, Silvi B (1998) J Molec Struct 450:59
43. Silvi B (2004) Phys Chem Chem Phys 6:256
44. Silvi B, Fourré I, Alikhani ME (2005) Monatshefte für Chemie 136:855
45. Noury S, Krokidis X, Fuster F, Silvi B (1997) TopMoD package, Universitè Pierre et Marie Curie
46. Noury S, Krokidis X, Fuster F, Silvi B (1999) Comput Chem 23:597
47. Flükiger P, Lüthi H-P, Portmann S, Weber J (2000) MOLEKEL 4.0. Swiss National Supercomputing Centre CSCS, Manno
48. Reed AE, Weinhold FJ (1985) Chem Phys 83:1736–1740
49. Reed AE, Curtiss LA, Weinhold F (1988) Chem Rev 88:899–926
50. Chesnut DB (2003) J Phys Chem A 107:4307
51. Chesnut DB (2001) Chem Phys 271:9
52. Sánchez-González A, Martínez-Garcia H, Melchor S, Dobado JA (2004) J Phys Chem A 108:9188
53. Chesnut DB (2000) Heterat Chem 11:341
54. Shaik S, Danovich D, Silvi B, Lauvergnat DL, Hiberty PC (2005) Chem Eur J 11:6358
55. Matito E, Silvi B, Duran M, Solà M (2006) J Chem Phys 125:024301
56. Lepetit C, Silvi B, Chauvin R (2003) J Phys Chem 107:464
57. Pilme J, Silve B, Alikhani ME (2005) J Phys Chem A 109:10028

Theor Chem Acc (2012) 131:1142
DOI 10.1007/s00214-012-1142-x

REGULAR ARTICLE

The fate of branched and linear isomers in the rhodium-catalyzed hydroformylation of 3,4,4-trimethylpent-1-ene

Giuliano Alagona · Caterina Ghio

Received: 13 July 2011 / Accepted: 2 September 2011 / Published online: 18 February 2012

Abstract In nonreversible hydroformylations, the computational evaluation of regio- and stereoselectivities from the relative energy barriers of the transition states (TS) for the alkyl-Rh intermediate formation step is possible, provided all low energy conformers are considered. In contrast, in reversible hydroformylations, also the subsequent reaction steps need to be taken into account to shed some light on mechanistic details. Thus, an extensive comparison of branched (B) and linear (L) reaction pathways for the Rh-catalyzed hydroformylation of 3,4,4-trimethylpent-1-ene (a bulky chiral substrate), going from a number of reactant complexes to products, has been carried out to rationalize the experimental result that pointed to reaction reversibility, although the value of the regioselectivity ratio (B:L = 15:85), based on alkyl-Rh TS free energies, computed under the hypothesis of nonreversibility, was in satisfactory agreement with the experimental one (5:95). A density functional theory approach at the B3P86/6-31G* level coupled to effective core potentials for Rh in the LanL2DZ valence basis set has been employed. By comparing the activation free energies involved in the various steps for the different reactant adducts, interestingly a similar behavior along all the linear pathways is found: the alkyl-Rh formation TS presents the highest barrier; thus, the reaction is nonreversible for all the linear isomers that invariably proceed to yield the linear aldehyde. Conversely, the behavior is quite different along the branched pathways. While some branched isomers eventually produce the corresponding aldehydes, two of the others follow distinct competing pathways, because β-hydride elimination occurs (a) to the terminal olefin-Rh complex (the starting material) that reacts again with the original regioselectivity, increasing the linear fraction, when the CO addition and insertion TS are higher than the alkyl-Rh TS and (b) to the internal olefin-Rh complex when the CO addition and insertion TS are higher than the relevant β-hydride elimination TS, but not than the alkyl-Rh TS. The solvent effect on the reversible profile, evaluated either in the supermolecule approach by adding a benzene molecule to the calculations or in the IEF-PCM framework ($\varepsilon = 2.247$), does not bring about any substantial change in the profile, leaving unaltered the conclusions reached.

Keywords Density functional calculations · B3P86/6-31G* · ECP/LanL2DZ · Reaction mechanisms · Hydroformylation reaction · *Oxo* process · β-Elimination

Dedicated to Professor Vincenzo Barone and published as part of the special collection of articles celebrating his 60th birthday.

Electronic supplementary material The online version of this article (doi:10.1007/s00214-012-1142-x) contains supplementary material, which is available to authorized users.

G. Alagona (✉) · C. Ghio
Molecular Modeling Lab, IPCF-CNR, Pisa UOS,
Via Moruzzi 1, 56124 Pisa, Italy
e-mail: G.Alagona@ipcf.cnr.it

C. Ghio
e-mail: C.Ghio@ipcf.cnr.it

1 Introduction

The alkene hydroformylation or *oxo* process, though developed in 1938 [1, 2], is still one of the most important catalytic reactions to produce aldehydes [3, 4], bearing a very versatile chemical function, easily converted into the corresponding alcohols employed as detergent, in the fragrance industry, and for several other applications of high economic value. The hydroformylation is a typical homogeneous process catalyzed by transition metal complexes, that is, low valent cobalt or mainly rhodium catalysts [5–7], unmodified or modified with phosphine ligands [7–22]. Phosphine-modified rhodium catalysts are in general active at high temperature only ($T \geq 100$ °C), whereas unmodified Rh catalysts work well even at room temperature (rt) or below $T \approx 60$ °C [23]. Under mild hydroformylation conditions, $Rh_4(CO)_{12}$, the unmodified catalytic precursor, is transformed into H–Rh$(CO)_3$, the Rh tricarbonyl hydride, which is the catalytic active species in the hydroformylation reaction [24], whose schematic representation is shown in Scheme 1. Depending on the catalyst orientation with respect to the double bond, either branched (B) or linear (L) intermediates are produced, eventually leading to B (with a new stereocenter) or L aldehydes. Because of the economic value of the reaction products, it is thus important to control the reaction regioselectivity (B:L).

The computational evaluation [25] (and even prediction) of the reaction regioselectivity, feasible provided the reaction is nonreversible, can be obtained from the relative stabilities of the rhodium-alkyl intermediate transition states (TS) [26–37] with the following expression:

$$\mathrm{B}:\mathrm{L} = k_{\mathrm{B}}:k_{\mathrm{L}} = \sum e^{-\Delta G^{\neq}_{\mathrm{B}}/RT}:\sum e^{-\Delta G^{\neq}_{\mathrm{L}}/RT} \approx \sum e^{-\Delta E^{\neq}_{\mathrm{B}}/RT}:\sum e^{-\Delta E^{\neq}_{\mathrm{L}}/RT} \quad (1)$$

where k is the reaction rate, $\Delta G^{\neq}$ is the TS free energy, and $\Delta E^{\neq}$ is the TS internal energy. Zero-point energies and thermal corrections were computed in the rigid rotor-harmonic oscillator approximation to obtain free energies [38]. The internal energy is only an approximation to the free energy valid when the entropic contributions are not large. In the case of very flexible substrates, such as 1-hexene, the internal energy can provide better results than the free energy because of hindered rotations considered as real vibrations [28].

When the R substituent on the terminal olefin is a chiral group such as $-CH(CH_3)C(CH_3)_3$ (the $-C(CH_3)_3$ group is sometimes named tBu for short in what follows, Fig. 1), the branched aldehyde appears in two diastereomeric forms, b and b', displayed in Scheme 2. It is worth noting that $b + b' = \mathrm{B}$, giving the total fraction of branched aldehydes.

The diastereoselectivity (b:b') for nonreversible reactions can be evaluated with a formula analogous to that used to calculate the regioselectivity ratio:

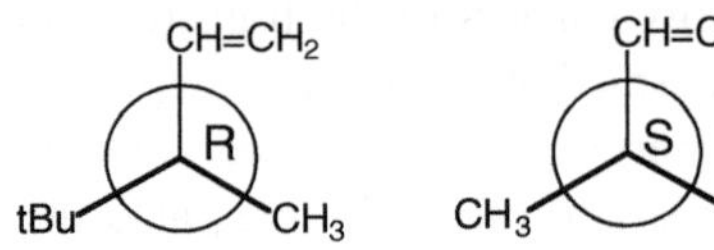

Fig. 1 Schematic representation of the $-CH(CH_3)C(CH_3)_3$ group chirality in the terminal olefin

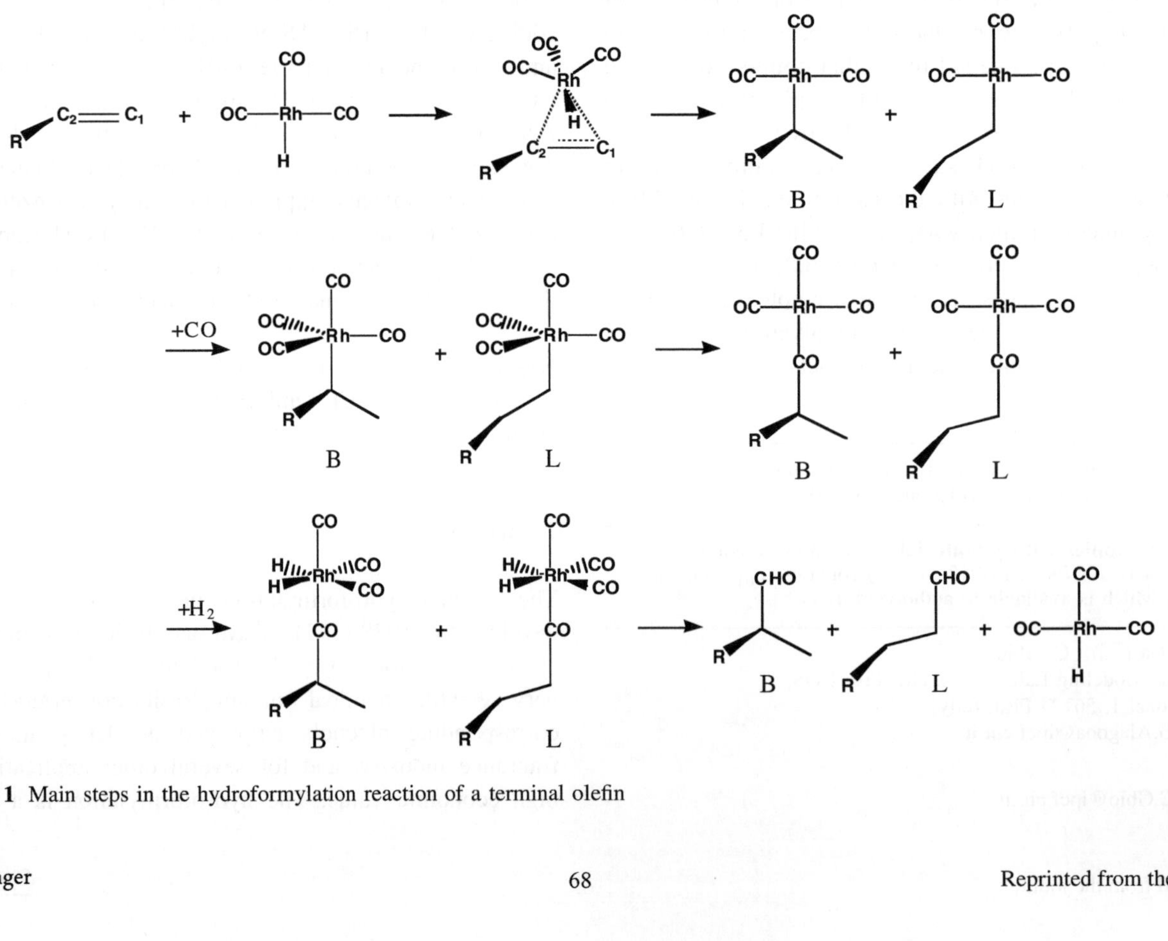

Scheme 1 Main steps in the hydroformylation reaction of a terminal olefin

Scheme 2 The hydroformylation reaction for a chiral substrate

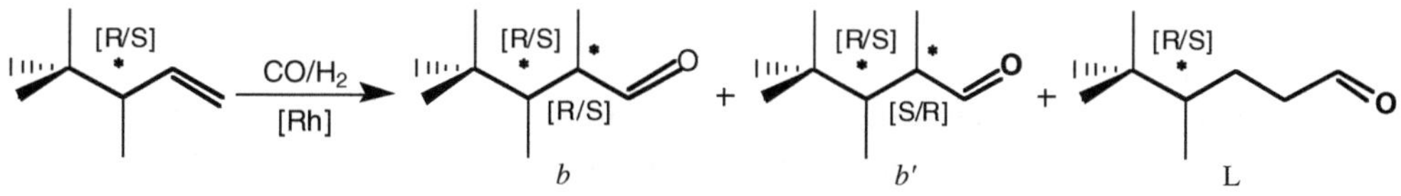

$$b : b' = k_b : k_{b'} = \sum e^{-\Delta G_b^{\neq}/RT} : \sum e^{-\Delta G_{b'}^{\neq}/RT} \approx \sum e^{-\Delta E_b^{\neq}/RT} : \sum e^{-\Delta E_{b'}^{\neq}/RT} \quad (2)$$

The main interest of 3,4,4-trimethylpent-1-ene, the substrate displayed in Scheme 2, is due to the fact that deuterioformylation experiments put forward reaction reversibility at room temperature [37], albeit well-assessed computational methods, applied under the hypothesis of nonreversibility, satisfactorily reproduced the experimental regioselectivity (B:L = 5:95 vs a computed value of 15:85). The diastereoselectivity, being related just to 5% of the reaction products, is much less important, because even experimental values can be uncertain and prone to errors.

It was thus necessary to understand the origin of the reaction reversibility, examining all the TS and intermediates that cannot be experimentally observed for the various reaction steps.

From the computational viewpoint, the presence of rotatable bonds and two chiral forms remarkably increases the number of isomers to be considered in the conformational search. Moreover, the attack of the catalyst can occur onto both faces of the double bond.

2 Computational details

All the calculations have been carried out with the Gaussian 03 system of programs [39] in the density functional theory (DFT) framework, making use of a hybrid method, B3P86 [40, 41], with the 6-31G* basis set [42] specifying 5D (pure d functions) for using a uniform description of the basis set in the absence of Rh. Effective core potentials for Rh[1] in the LanL2DZ valence basis set [43–45] have been employed. We resorted to the B3P86[2] functionals because they were successfully used in most of our previous studies on regio- and diastereoselectivity of the Rh-catalyzed hydroformylation of several olefins [28–37] and compared well with MP2 results [28], despite lacking dispersion. Regrettably, for systems of this size, MP2 calculations with larger basis sets[3] at the free energy level are far beyond what our current computer facility can afford.

Zero-point energy (ZPE) and thermal corrections to compute the Gibbs free energy at $T = 298.15$ K and 1 atm for the stationary points located along the pathways made use of the rigid rotator-harmonic oscillator approximation [38]:

$$(G(298) = E(0) + \text{ZPE} + \Delta H(0-298) - T\Delta S(0-298)) \quad (3)$$

The stationary points were characterized via vibrational frequency analysis: minima, that is, intermediates (Int) or isolated species, have all frequencies real, while transition states (TS) have one imaginary frequency. Of course, all normal modes associated with imaginary frequencies corresponded to the expected TS; thus, it was not necessary to carry out intrinsic reaction coordinate calculations.

The solvent (benzene) effect was considered either using the supermolecule approach, that is, a single molecule of benzene was added to the system, or the polarizable continuum model in the integral equation formalism framework (IEF-PCM) [49–56]. In the latter case, the benzene dielectric constant ($\varepsilon = 2.247$) was used for the solvent that surrounds the solute embedded in a cavity of interlocking spheres with radii determined via the united-atom topological model. The total free energy in solution, G_{tot}, is given by the sum of electrostatic and dispersion-repulsion-cavitation free energy terms, that is, $G_{tot} = G_{el} + G_{drc}$. Thermal averages are included only for the solvent, while other free energy contributions are missing [56]. Thus, free energy corrections are to be added to stationary points in solution, computing ZPE and thermal corrections in solution on solute geometries optimized in that phase:

$$G_{tot}(298) = G_{tot}(0) + [\text{ZPE} + \Delta H(0-298) - T\Delta S(0-298)]_{sol} \quad (4)$$

Molecular structures have been drawn with Molden [57] that was utilized also to visualize the normal modes.

3 Structures and energies

Before discussing the mechanistic details, it is advisable to describe the species involved in the various steps of the hydroformylation reaction.

[1] Notice that to simplify the notation, only B3P86/6-31G* is reported throughout.

[2] Following the suggestion by Jonas and Thiel [46], who systematically investigated the performance of a variety of basis sets and methods (HF, DFT, and MP2) in transition metal carbonyls but Rh ones, we initially adopted BP86. Tests carried out in Ref. [28] on hydroformylations with Rh carbonyls supported the use of B3 instead.

[3] MP2 is often unreliable with transition metal systems [47, 48], if coupled to a basis set with limited virtual space, as is the case of 6-31G*.

3.1 Reactant complexes

To explore the reaction mechanism, it is necessary to locate all the structures of the reactant complexes. For most of the terminal olefins considered in Ref. [28], there was a single conformation of the reactant. The complex with H–Rh(CO)$_3$, the catalyst square-planar active species, was placed on top of the double bond (first step structure in Scheme 1) with the roughly aligned H–Rh–CO moiety perpendicular to it ($\theta \approx \pm 90°$, with $\theta = C_1C_2RhH$). When $\theta \approx 90°$, the H at Rh points toward the observer (or forward, **f**) as the R substituent, while when $\theta \approx -90°$ it points away from the observer (or backward, **b**). Obviously, to produce the B or L intermediates, eventually leading to the B or L aldehydes, the H–Rh–CO moiety has to rotate by about 90° in either sense to reach $\theta \approx 180$ or 0°, respectively. Interested readers can find in Ref. [28] the relevant potential energy surface computed with a smaller basis set for a typical case.

For 3,4,4-trimethylpent-1-ene (**TMP$_1$** for short), in principle there are three reactant complexes for each diastereomer and for each arrangement ($\theta \approx \pm 90°$) of the catalyst. Due to the threefold symmetry of the chiral substituent defined via the δ dihedral angle (with $\delta = H_3C_3C_2H_2$), when the Hs at the chiral centers are in antiperiplanar ($\delta \approx 180°$) or $\pm$ synclinal ($\delta \approx \pm 60°$) conformations, the corresponding structures are named **0** (**a**), **1** (**+s**), and **2** (**−s**), respectively. The reactant lowest energy adducts (named Reac_$\delta\theta$) are reported in Table 1. The double bond face of attack needs not be considered because the structure obtained attacking from above the molecular plane depicted in Scheme 2 (i.e. (2*S*,3*S*)-Reac) is the mirror image of that obtained attacking from below ((2*R*,3*R*)-Reac) and so on.

Table 1 B3P86/6-31G* relative energies and free energies (kcal/mol) of the reactant complexes with some geometrical parameters (degrees)

	System	ΔE	$H_3C_3C_2C_1$	$H_3C_3C_2H_2$	ΔG (298)
2*S*,3*R*	Reac_0b	0[a]	28.0	−175.4	0[b]
	Reac_0f	0.73	28.6	−175.6	0.60
	Reac_1b	4.34	−91.2	61.0	5.26
	Reac_1f	4.73	−86.4	63.8	4.93
2*S*,3*S*	Reac_2f	1.35	132.0	−69.2	0.91
	Reac_2b	1.82	125.0	−75.2	1.54
	Reac_0b	3.72	41.9	−169.6	4.38
	Reac_0f	5.38	34.9	−178.0	5.96
	Reac_1f	7.11	−113.2	43.4	7.92
	Reac_1b	7.56	−120.6	39.4	8.05

Hydrogens are named after the C they belong to

[a] Ref. energy = −766.946541 E_h

[b] Ref. free energy at 298.15 K and 1 atm = −766.738577 E_h

The Can–Ingold–Prelog priority rules assign the chirality based on the weight of the substituents at the chiral center, but along the reaction pathway, some of those substituents change, and this fact would produce different chiralities for the same chiral substituent that can be confusing. Since from the experimental point of view, the regio- and diastereoselectivities are determined on the aldehydes, in our previous investigations we invariably referred to the aldehyde chirality also for the Rh-alkyl TS and intermediates. We are going to follow an analogous approach here, keeping the chiral substituent unaltered along each reaction path. When the starting olefin is 3*R*, the branched aldehyde turns out to be 2*R*,3*R* and 2*S*,3*R* when the catalyst attacks the double bond from below and from above.

For the 2*S*,3*R* diastereomer no Reac_2 structure was located despite several constrained optimizations had been performed for that purpose.

The flexible rotation about θ, considered for 2*S*,3*R*-Reac_0, is displayed in Fig. 2. In the $\theta \approx 270°$ region, at $\theta = -87.5°$, the most stable reactant complex is located, 2*S*,3*R*-Reac_0b, displayed in Fig. 3 with atom number clues. As stated in the short name definition, the H on Rh points backward, on the opposite side of the substituent, while in 2*S*,3*R*-Reac_0f at $\theta = 86.3°$, it is on the same side. This arrangement is just slightly less favorable (by 0.7 kcal/mol) than 2*S*,3*R*-Reac_0b. This is not however a general rule as can be derived from a perusal of Table 1. Due to the steric hindrance of the –CH(CH$_3$)C(CH$_3$)$_3$ substituent, no stable reactant complex is found in the $\theta \approx 180°$ region. Conversely, the reactant complex in the $\theta \approx 0°$ region ($\theta = -18.13°$) is higher in energy than 2*S*,3*R*-Reac_0b by nearly 4.4 kcal/mol.

The potential energy for the $H_3C_3C_2C_1$ flexible torsion was investigated for the lowest energy "backward" reactant complexes of **TMP$_1$**, and the obtained profiles for 2*S*,3*R*-Reac_0b (solid line) and 2*S*,3*S*-Reac_2b (dotted

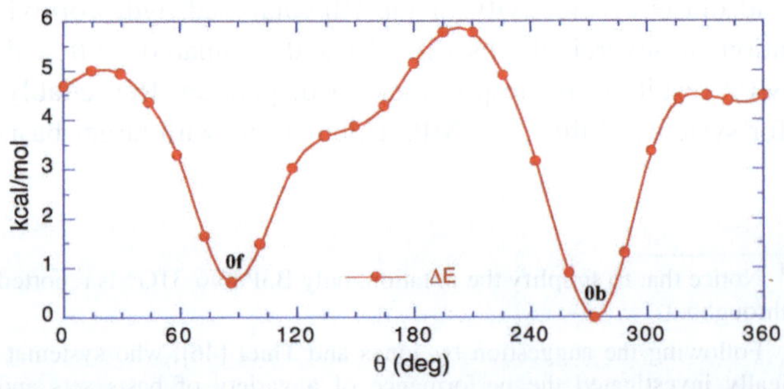

Fig. 2 B3P86/6-31G* potential energy profile (kcal/mol) for the rotation about $\theta = C_1C_2RhH$ of the catalyst in the 2*S*,3*R*-Reac_0 reactant complex

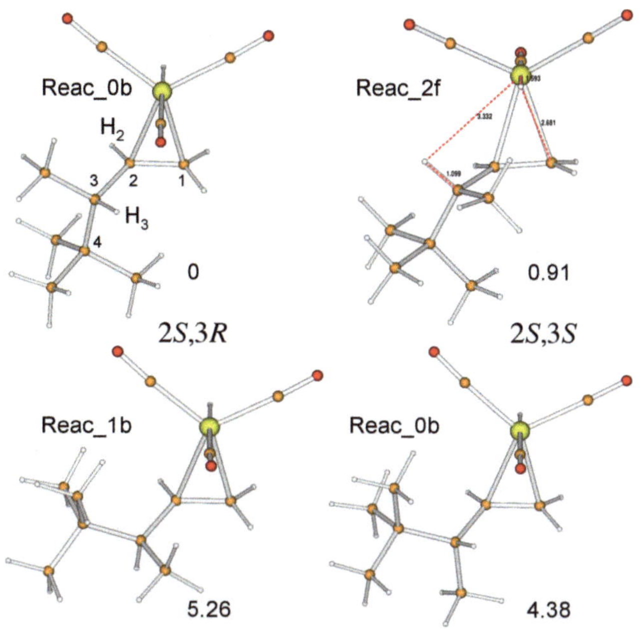

Fig. 3 B3P86/6-31G* structures and relative free energies (kcal/mol) of four reactant complexes

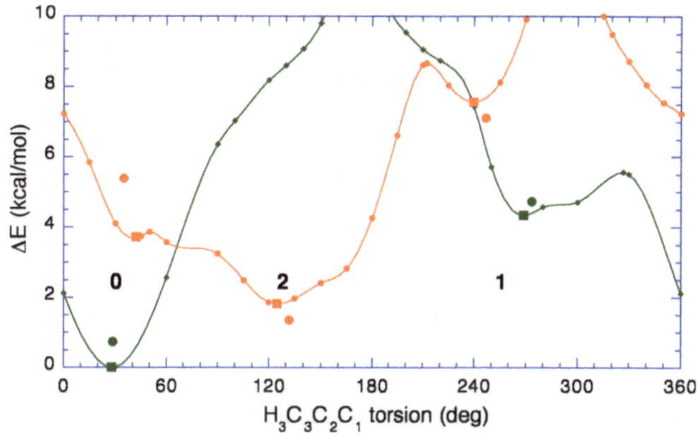

Fig. 4 B3P86/6-31G* $H_3C_3C_2C_1$ flexible potential energy scan for (2*S*,3*R*)- and (2*S*,3*S*)-Reac_δb in *solid* and *dotted lines*, respectively. The locations of the various reactant complex local minima are also shown (*squares* = backward orientation of H at Rh, *circles* = forward orientation of H at Rh)

line) are displayed in Fig. 4. In the regions about 180° for 2*S*,3*R*-Reac_0b and about 300° for 2*S*,3*S*-Reac_2b, there is a remarkable increase in the potential energy because tBu arrives in close contact with the catalyst carbonyl groups. To avoid bumping into them, the bulky group rotates about the C_3C_4 bond, and actually every attempt to locate transition states in those regions produced C_3C_4 rotational TS.

In any case, the various 2*S*,3*R* rotamers should easily interconvert into each other as well as the 2*S*,3*S* ones. Due to the Curtin–Hammett principle [58–60], their relative stability is not important since the TS along the branched or linear reaction paths are significantly higher, as shown in the next Section, than the barriers between them ($\sim$5.5 kcal/mol for 2*S*,3*R*; $\sim$2 or $\sim$7 kcal/mol for 2*S*,3*S*). The locations of the other species (**f**) are also shown in Fig. 4, because there is an obvious correlation between $H_3C_3C_2C_1$ and $H_3C_3C_2H_2$ (Table 1). Of course, the Reac_δb structures (squares) lie exactly on the profiles, while the Reac_δf ones (circles) are located nearby. Interestingly enough, in the region corresponding to **2**, there is only a small shoulder in the (2*S*,3*R*)-Reac potential energy profile: the reason why a minimum in that region was not found is due to the short separation between the tBu group and the catalyst. Conversely, for (2*S*,3*S*)-Reac, all three conformers turn out to be stable.

3.2 Alkyl-Rh transition state structures

As put forward in the introduction, under the hypothesis of nonreversibility, the reaction selectivity can be determined from the relative energies or free energies of the Rh-alkyl formation TS, provided all the low-lying TS structures are located [22, 29, 30]. At the TS, the system takes a square-planar pyramidal (SP) arrangement. As basal moieties, the pyramid features the H bonded to Rh and the CO group opposite to it along one of the diagonals, while along the other there are one of the former double bond C atoms and one of the CO in the plane roughly perpendicular to H–Rh–CO. The remaining CO takes the apical position,[4] in general pointing in the opposite direction with respect to the substituent in linear TS (i.e., $\alpha \approx -90°$, with $\alpha = C_1C_2RhC_{ap}$) and in the same direction in branched ones. A careful TS conformational search is thus necessary, considering also the possible swap of the apical group position.

A number of branched and linear TS are reported in Table 2 and some of them are displayed in Figs. 5, 6, 7, with **0**, **1**, and **2** corresponding to the value of the $H_3C_3C_2H_2$ dihedral angle, as for the reactant complexes (see Fig. 4).

Along the branched path, a single case of apical CO swap was found for the least stable TS*b*′1 diastereomer that produced an even less stable and distorted SP arrangement (TS*b*′1ap). Both structures are displayed in Fig. 5.

The branched TS to obtain the alkyl-Rh$(CO)_3$ intermediates feature a remarkable similarity in the SP arrangement of the catalyst onto the former double bond (displayed in Fig. 6), as can be inferred from the RhH, HC_2, and RhC_{ap} separations reported in Table 2. The different stability of the various structures depends on the catalyst-substituent interaction, especially on the position taken by the bulky group (tBu).

[4] Occasionally, the CO groups in the plane roughly perpendicular to H–Rh–CO can swap their positions, thus inverting the pyramid orientation.

Table 2 B3P86/6-31G* relative energies and free energies (kcal/mol) of branched and linear alkyl-Rh transition states with respect to the most stable one at each level and some geometrical parameters (Å and degrees)

System[a]	ΔE	RhH	HC_2	RhC_{ap}	$C_1C_2RhC_{ap}$	IF[b]	ΔG
TS*b*0	0[c]	1.648	1.608	1.983	88.6	689.i	1.08
TS*b*1	3.47	1.645	1.622	1.981	86.7	702.i	5.22
TS*b*′2	0.74	1.653	1.606	1.985	86.2	720.i	1.89
TS*b*′0	3.59	1.641	1.642	1.988	82.5	697.i	4.72
TS*b*′1	6.92	1.654	1.567	1.979	87.9	721.i	8.73
TS*b*′1ap	7.84	1.659	1.574	1.953	−121.4	714.i	10.11
TS*l*0	0.12	1.647	1.677	1.998	−104.2	682.i	0[d]
TS*l*1	1.79	1.649	1.686	1.988	−93.4	645.i	2.61
TS*l*2	6.80	1.655	1.615	1.997	−88.9	684.i	7.81
TS*l*′2	0.33	1.650	1.644	1.998	−97.8	679.i	1.10
TS*l*′0	2.20	1.638	1.666	1.976	92.7	651.i	3.23
TS*l*′1	6.65	1.646	1.758	2.017	−95.7	554.i	8.18

[a] The prime corresponds to the *3S* chirality in the aldehyde

[b] Imaginary frequency (cm^{-1})

[c] Ref. energy = −766.927591 E_h, energy barrier = 11.891 kcal/mol

[d] Ref. free energy at 298.15 K and 1 atm = −766.722959 E_h, free energy barrier = 9.801 kcal/mol

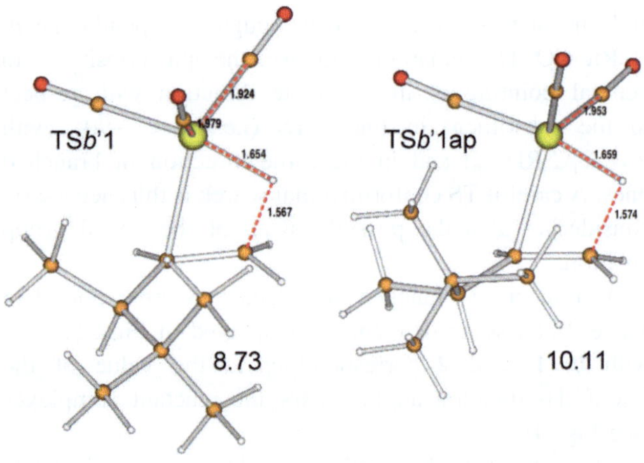

Fig. 5 TS structures and relative free energies (kcal/mol) for TS*b*′1 with swapped apical CO

The four most stable linear TS structures are displayed in Fig. 7. Since in the linear TS the catalyst is mostly located farther apart from the substituent due to the CO apical group orientation, the RhC_{ap} separation is on the average larger than in branched TS. Only TS*l*′0, the least favorable among the four, has a positive value of the $C_1C_2RhC_{ap}$ dihedral angle (Table 2).

Despite several attempts to obtain other linear TS constraining either the substituent or the apical CO group in particular orientations, the optimizations to the TS always produced already located TS structures once constraints were relieved, because of the tBu steric hindrance. Of course, this occurred for the branched TS as well; but for

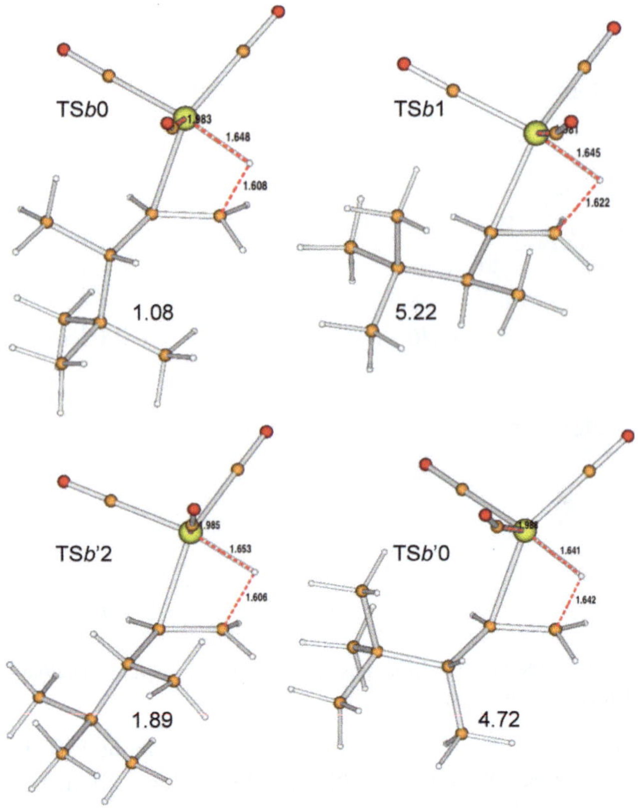

Fig. 6 B3P86/6-31G* structures and relative free energies (kcal/mol) of four branched TS

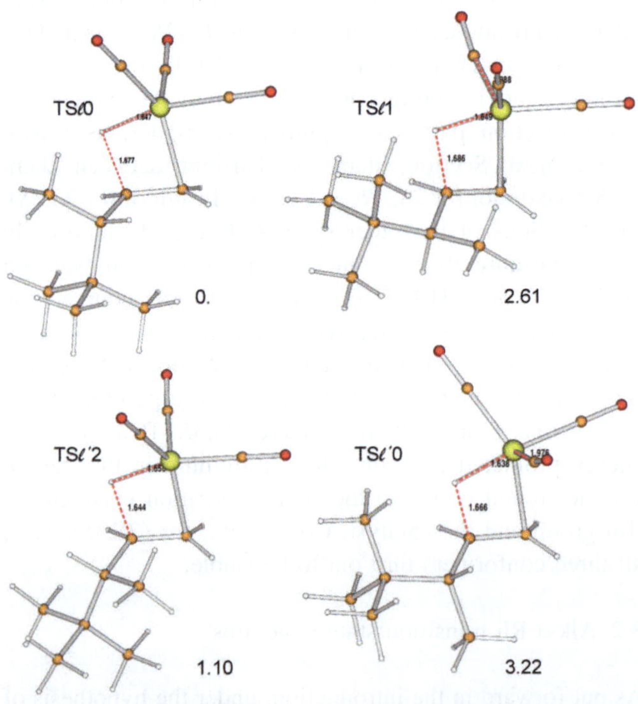

Fig. 7 B3P86/6-31G* structures and relative free energies of four linear TS with respect to the most stable one (kcal/mol)

Table 3 B3P86/6-31G* relative energies and free energies (kcal/mol) of branched and linear alkyl-Rh tricarbonyl intermediates with respect to 2*S*,3*R*-Reac_0b and some geometrical parameters (Å and degrees)

System[a]	ΔE^b	RhH_n	RhH_β	$H_3C_3C_2C_1$	$H_3C_3C_2H_2$	ΔG^c
Tri*b*1	−6.21	2.277	4.054	−53.6	61.7	−3.34
Tri*b*0	−3.24	2.488	3.379	39.0	161.8	−1.39
Tri*b*2	−2.66	2.440	2.898	174.9	−68.0	−0.14
Tri*b*′2	−5.81	3.392	2.374	141.0	−96.2	−3.89
Tri*b*′0	−5.73	2.162	3.222	66.5	−175.3	−3.43
Tri*b*′1	−2.80	2.271	4.045	−32.9	80.3	0.11
System[a]	ΔE^b	RhH_n	RhH	$H_3C_3C_2C_1$	$H_3C_3C_2H_2$	ΔG^c
Tri*l*0r	−9.76	3.030	3.097	−41.8	80.2	−8.80
Tri*l*1	−9.45	2.310	3.218	−35.5	84.4	−8.05
Tri*l*2	−5.46	2.283	3.225	−165.5	−47.4	−2.61
Tri*l*′0r	−9.77	3.150	2.990	44.5	167.3	−8.37
Tri*l*′0	−9.48	2.714	3.254	48.5	170.1	−8.05
Tri*l*′2r	−7.74	3.152	2.917	−179.8	−54.7	−5.83
Tri*l*′2	−6.85	2.349	3.251	−179.6	−56.4	−4.58
Tri*l*′1	−4.72	2.250	3.192	−43.8	74.0	−2.72

[a] The prime corresponds to the *3S* chirality in the reactant complexes

[b] Ref. energy = −766.946541 E_h

[c] Ref. free energy at 298.15 K and 1 atm = −766.738577 E_h

the linear TS the favorable position of Rh placed on the terminal C_1 atom, instead than on the more crowded C_2, might suggest that other TS structures exist. Therefore, it was necessary to absolutely exclude this possibility to computationally evaluate with confidence the regioselectivity under the hypothesis of nonreversibility.

From the values reported in Table 2, discarding those larger than 4 kcal/mol that do not affect the result, the regioselectivity ratios (B:L) turn out to be 47:53 and 14.7:85.3, respectively, when based on potential energies or free energies. Thus, for this substrate, the potential energy seemingly is not a satisfactory approximation to the free energy, since the experimental regioselectivity is 5:95.[5]

Although the agreement between theory and experiment might be in principle acceptable considering the free energy-based regioselectivity ratio, experimental results indicate that this particular reaction is reversible. It was however difficult to understand the origin of that behavior, even though deuterioformylation experiments had been carried out for that purpose [37]. This is the reason why all steps of the reaction mechanism are being computationally examined.

Besides all TS for the alkyl-Rh formation step that need to be computed to get reliable values of the reaction selectivity [29, 30], in our mechanistic investigation of the subsequent steps all the distinct reactant complexes are to be taken into account, due to the low barriers separating them in Fig. 4.

3.3 Alkyl-Rh tricarbonyl intermediates

The first branched and linear intermediates along the hydroformylation reaction mechanism are the alkyl-$Rh(CO)_3$ ones (Scheme 1), where the metal forms a four-coordinate alkyl-tricarbonyl complex, hereafter named Tri for short. The relative energies and free energies of some stable structures are reported in Table 3 together with a number of geometrical parameters. RhH_n is the separation between Rh and the closest hydrogen (H_n) belonging to the methyl/tBu groups for the branched isomers or even located at C_2 or C_3 for the linear ones. This distance can give a hint about possible agostic interactions preventing a facile approach to Rh of the fourth CO group whose addition produces alkyl-$Rh(CO)_4$, named Tetra for short, the next reaction intermediate. The RhH_β distance (reported only for the branched intermediates) is the separation between Rh and the H in β position (the H at C_3, also named H_3). A short value of RhH_β suggests that β-elimination to give the internal olefin is possible, as depicted in Scheme 3.

Conversely, the distance between Rh and one of the H at the adjacent methyl group (C_1) is not reported for Tri*bn*, although β-elimination to give the starting olefin might occur as well. In such a case however the TS structure and energy cannot change with respect to that already determined for the forward reaction. The $H_3C_3C_2C_1$ and $H_3C_3C_2H_2$ dihedral angles are reported to indicate which species is the likely ancestor of each intermediate. RhH, reported only for the linear intermediates to distinguish the newly attached H from H_2, is the distance between Rh and its H just transferred to C_2.

[5] Interestingly enough, the potential- and free-energy-based diastereoselectivity ratios (*b*:*b*′) are conversely very similar to each other, that is, 78:22 (ΔE) and 79:21 (ΔG); the difference between potential- and free-energy-based regioselectivity ratios thus depends on the linear regioisomers. Using a formula analogous to Eq. 2 and replacing *b* and *b*′ with the computationally available *l* and *l*′ populations, the two *l*:*l*′ potential- and free-energy-based diastereoselectivity ratios actually turn out to be 59:41 and 86.3:13.7, respectively.

Scheme 3 β-Elimination to produce the internal olefin complex with H–Rh(CO)$_3$

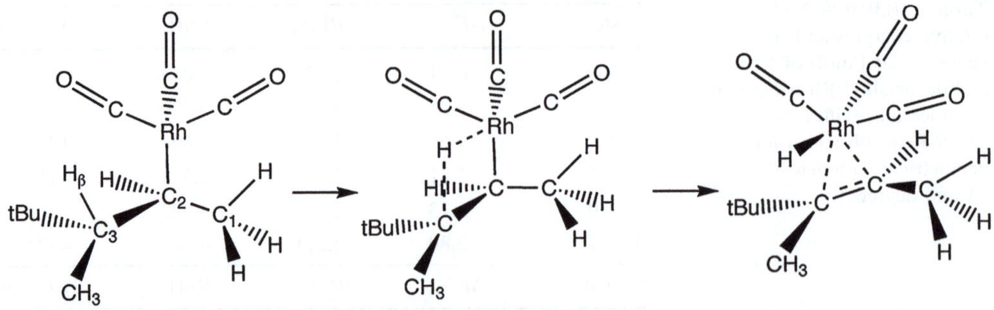

Trib1 (displayed in Fig. 8 together with the other most favorable branched tricarbonyl intermediates) is stabilized by the agostic interaction with one of the tBu H. The RhH_β separation conversely is very large and prevents β-elimination to give the internal olefin. Although Trib1 has the most favorable potential energy, it ranks only third when considering the free energy: Trib′2 and Trib′0 turn out to be indeed somewhat more favorable. The RhH_β separation is close enough to suggest the possibility of β-elimination to yield the internal olefin primarily in Trib′2 where agostic interactions are feeble. In Trib′0, the free coordination site of Rh is engaged in an agostic interaction with an H of the tBu group.

Interestingly enough, a Trib2 intermediate (Table 3), less stable than Trib1 and Trib0, is located at this stage.

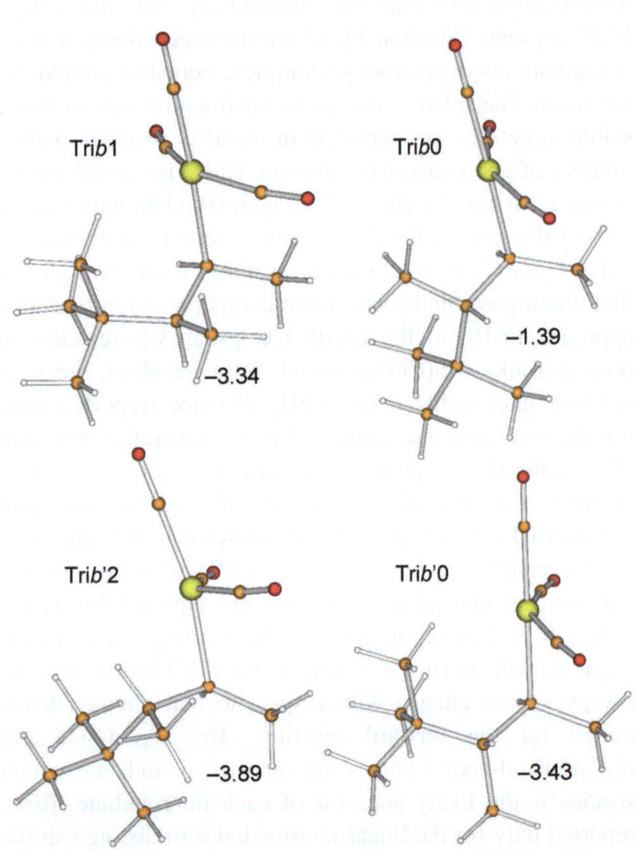

Fig. 8 B3P86/6-31G* structures and relative free energies (kcal/mol) of four branched tricarbonyl intermediates

Following backwards the reaction pathway, TSb0 was reached, indicating that two channels are present on the potential energy surface down from TSb0. Thus Reac_0 is the common ancestor of Trib0 and Trib2.

In analogy to what done for the reactant complexes, the $H_3C_3C_2C_1$ flexible potential energy scan for one (2*S*,3*S*) of the tricarbonyl intermediates has been investigated to determine the height of the barriers dividing the various rotamers. The potential energy profile is displayed in Fig. 9 with the free energies of the stationary points superimposed. The free energy barriers turn out to be slightly higher than the potential energy ones, because the local minima free energies differ less that the TS ones from the underlying profile, but they should still allow the interconversion among Trib′1, Trib′0, and Trib′2.

From Table 3, it appears evident that the branched tricarbonyl intermediates are on the average less stable than the linear ones (four of them are displayed in Fig. 10). Moreover, the linear structures can easily rotate about the C_1C_2 axis taking a zig-zag planar arrangement with Rh in a transoid position with respect to C_3, thus reducing the steric hindrance between the catalyst and the alkyl substituent. In order to distinguish those structures an "r" has been appended to their names: Tril′0r is shown in the upper left corner of Fig. 10 in comparison with Tril′0.

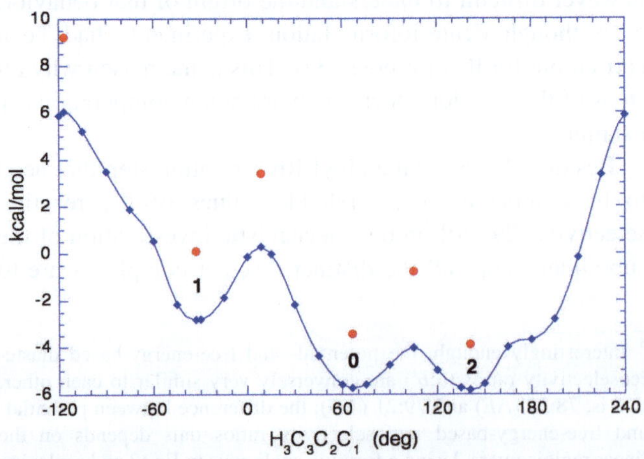

Fig. 9 B3P86/6-31G* $H_3C_3C_2C_1$ flexible potential energy scan for (2*S*,3*S*)-tricarbonyl intermediates (Trib′ for short). The locations of the free energy stationary points are also shown (*solid circles*)

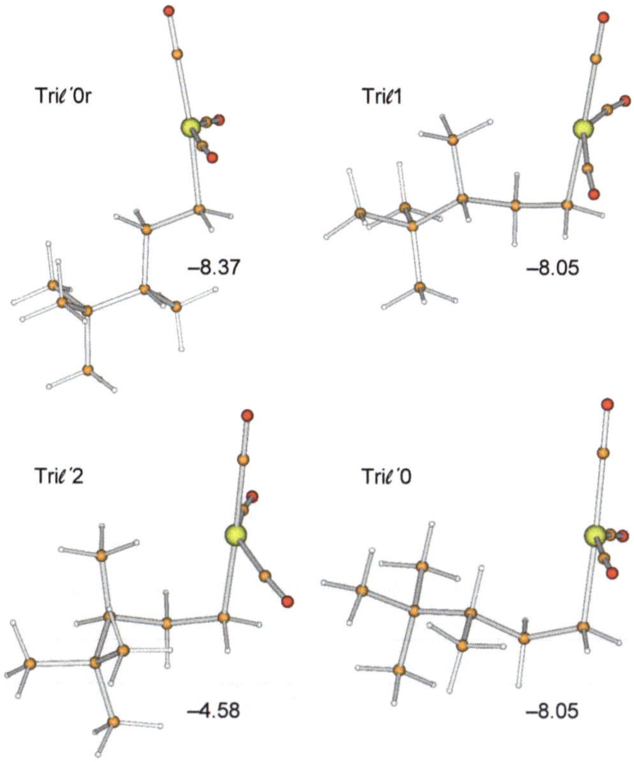

Fig. 10 B3P86/6-31G* structures and relative free energies (kcal/mol) of four linear tricarbonyl intermediates

3.4 Fourth CO approaching paths and alkyl-Rh tetracarbonyl intermediates

The issue of agostic interactions has been stressed in the discussion above because the attack of the fourth CO to alkyl-Rh tricarbonyl intermediates to give the alkyl-Rh tetracarbonyl ones (five-coordinate complexes) can present a barrier if all Rh coordination sites are already saturated, leading to a roughly trigonal–bipyramidal (TBP) arrangement with two CO groups in equatorial position and one CO in axial position as the alkyl moiety. Because of this fact, two approaching paths of the fourth CO to Rh have been examined [in the plane passing through Rh and its axial ligands and containing the bisector of its equatorial CO ligands] from two almost opposite directions, one (named "in") from inside the equatorial CO ligands and the other (named "out") from outside the equatorial CO ligands, that is, along the agostic ligand or the empty Rh valence.

The behavior of the branched (Tri*b*) and linear (Tri*l*) intermediates is remarkably different. All approaching paths of CO to Tri*l* intermediates spontaneously produce the tetracarbonyl intermediate without any barrier. A very small barrier was located only in the Tri*l*′0 "out" path, displayed in Fig. 11a as compared to the relevant "in" path along which the reaction obviously proceeds. By contrast, all approaching paths of CO to any Tri*b* (see, for instance, the case of Tri*b*1, shown in Fig. 11b) are to surmount a barrier, sharply enhanced when considering the free energy, reported in Table 4. At this point, due to change in the number of species considered (Tri + CO → Tetra),[6] it is necessary to take into account the free energy. For the linear stationary points after the alkyl-Rh TS, there are two Hs at C_2. The $H_3C_3C_2H_2$ and $H_3C_3C_2H_n$ dihedral angles (the latter with the newly added H) are compared only in Table 4. Then, track of the original value of $H_3C_3C_2H_2$ is kept in the linear names.

Relative energies and free energies with respect to the most stable reactant complex plus a noninteracting CO for a number of branched TS for the CO addition and of their relevant tetracarbonyl intermediates are reported in Table 4 together with few geometrical parameters.

In general, the same tetracarbonyl intermediate was located regardless of the approaching path followed by the fourth CO group. The only exception was found in the case of Tetra*b*1, where the significant difference between the intermediates obtained from the "in" and "out" approaching paths is due to the diverse torsion (by about 50°) in the aliphatic part and to the corresponding slight rotation of the $Rh(CO)_4$ group. The superimposition of the two structures is displayed in Fig. S1 of the electronic supplementary information (ESI).

Few branched TS are displayed in Fig. 12, while in the bottom part, in addition to two TS structures, an alkyl-Rh tetracarbonyl intermediate Tetra*b*′2, originating from TS-Tri*b*′2_in (and "out"), is shown at the right hand side. In general, a TBP arrangement at Rh is found for the branched intermediates, with a quite limited reorientation in the alkyl substituent. Thus, additional Tetra*b* structures are not displayed because they can easily be inferred from those of the corresponding TS.

For the CO addition to linear tricarbonyl intermediates, only a TS structure (reported in Table 4) was located along the "out" approaching path, as shown in Fig. 11. It is however to be noted that no barrier was found along the "in" approaching path that is thus viable and clearly preferred to give the relevant linear tetracarbonyl intermediate. Due to the high flexibility of the linear C backbone, as remarked above, most of the favorable linear conformers (some of them are reported in Table 4 and displayed in Fig. 13) feature C_3 in anti position with respect to Rh. This arrangement, not attainable for branched regioisomers, allows a larger separation between the R substituent and the catalyst, thus relieving the steric hindrance.

[6] Thus two species react to give just one new compound.

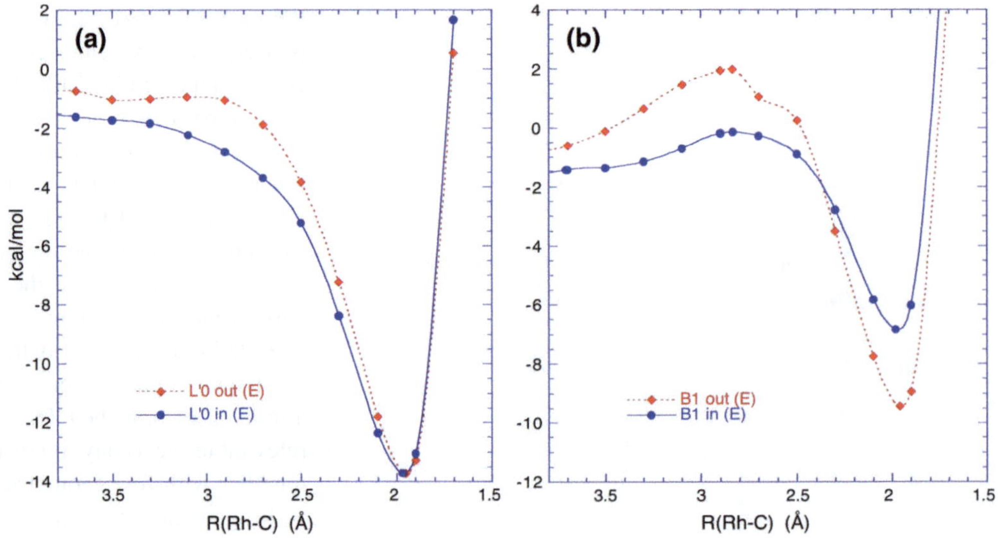

Fig. 11 B3P86/6-31G* potential energy profiles (kcal/mol) along the "in" (*solid circles*) and "out" (*diamonds*) approaching paths of the fourth CO to: **a** Tri*l*′0 (with respect to Tri*l*′0 + CO) and **b** Tri*b*1 (with respect to Tri*b*0 + CO)

Table 4 B3P86/6-31G* relative energies and free energies (kcal/mol) of TS for the CO addition to branched Rh tricarbonyl intermediates with respect to Reac_0b + CO

System[a]	ΔE[b]	Rh...CO	$H_3C_3C_2H_2$	IF[c]	ΔG[d]
TS-Tri*b*1_in	−6.35	2.836	62.4	65.i	5.36
TS-Tri*b*1_out	−4.22	2.838	80.0	72.i	7.37
TS-Tri*b*0_in	−5.63	3.299	−178.9	31.i	4.59
TS-Tri*b*0_out	−4.40	2.828	179.3	69.i	7.96
TS-Tri*b*2_in	−2.62	2.992	−69.4	58.i	8.62
TS-Tri*b*′1_out	−3.39	2.930	80.4	60.i	8.27
TS-Tri*b*′2_out	−6.57	3.338	−94.8	21.i	3.34
TS-Tri*b*′2_in	−6.01	3.046	−92.8	54.i	4.58
TS-Tri*b*′0_out	−2.83	2.632	−178.7	100.i	9.85
TS-Tri*l*′0_out	−10.41	3.105	172.4	33.i	−0.37

System[a]	ΔE	Rh...CO	$H_3C_3C_2H_n$	ΔG
Tetra*b*1_out	−15.65	1.960	128.8	−0.54
Tetra*b*1_in	−13.05	1.982	75.3	2.26
Tetra*b*0	−13.21	1.996	179.9	1.18
Tetra*b*2_in	−12.63	1.984	−55.5	2.61
Tetra*b*′2	−18.61	1.965	−87.8	−3.56
Tetra*b*′1_out	−11.43	1.951	74.3	3.12
Tetra*b*′0_out	−9.36	1.949	−165.1	4.39

System[a]	ΔE	$H_3C_3C_2H_2$	$H_3C_3C_2H_n$	ΔG
Tetra*l*0r	−25.24	−161.4	83.1	−12.43
Tetra*l*1r	−23.08	54.9	−57.9	−9.56
Tetra*l*′2r	−25.31	−83.8	160.7	−11.27
Tetra*l*′0	−23.19	169.8	−75.8	−8.95

Branched and linear Rh tetracarbonyl intermediates are also reported with some geometrical parameters (Å and degrees)

All reference values are reported in Table S1 of the ESI

[a] The prime corresponds to the *3S* chirality in the reactant complexes

[b] Ref. energy (Reac_0b + CO) = −880.493185 E_h

[c] Imaginary frequency (cm^{-1})

[d] Ref. free energy (Reac_0b + CO) at 298.15 K and 1 atm = −880.299291 E_h

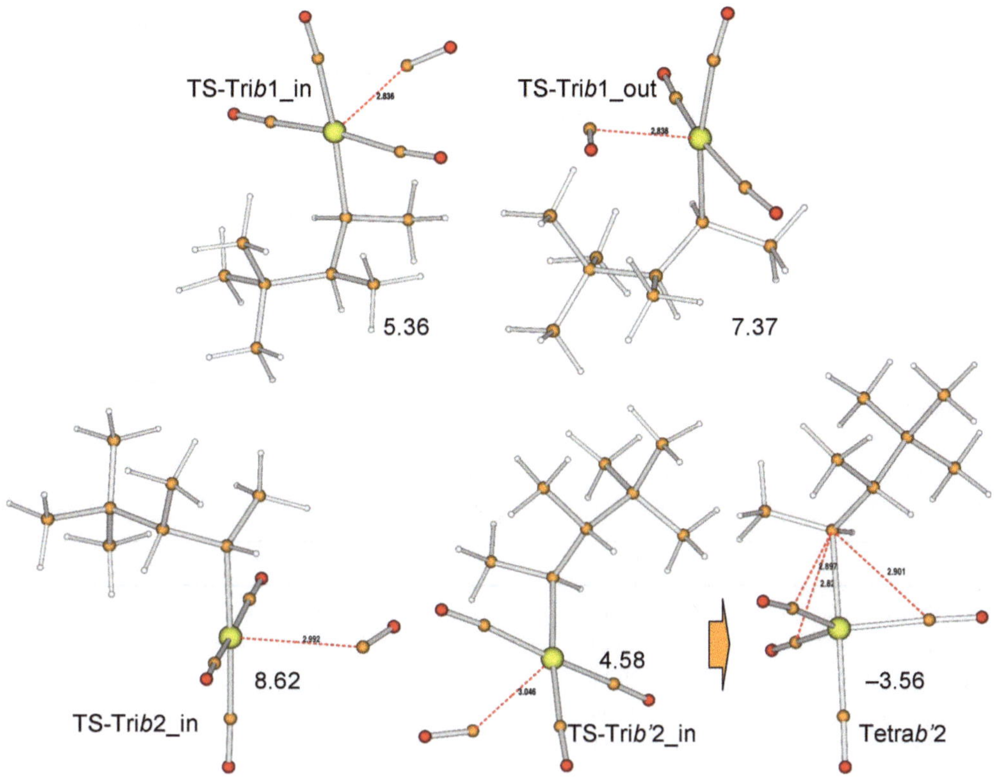

Fig. 12 B3P86/6-31G* structures and relative free energies (kcal/mol) of four branched TS for the CO addition to tricarbonyl species and a branched tetracarbonyl intermediate (*lower right corner*)

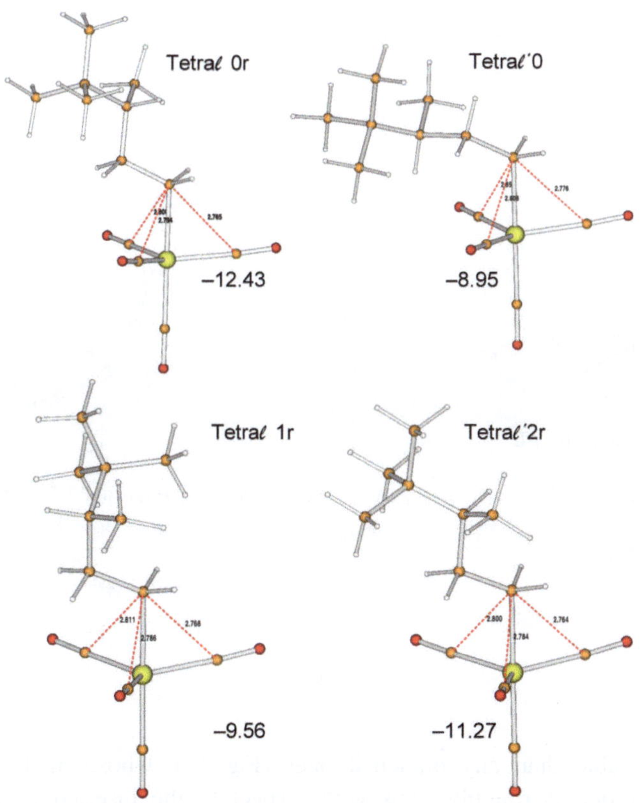

Fig. 13 B3P86/6-31G* structures and relative free energies (kcal/mol) for some linear tetracarbonyl intermediates

3.5 CO migratory insertion TS and Rh-acyl intermediates

The next step in the reaction is the migratory insertion of one of the CO groups into the alkyl ligand to give an Rh-acyl complex. Since the three equatorial CO groups are in principle equivalent, the starting structures for TS optimizations have been model-built using in turn one of them.[7] Only the most favorable optimized structure for each conformer is however reported in Table 5. All branched TS are significantly high in relative free energy (8–13 kcal/mol) with respect to Reac_0b + CO, whereas the linear ones are very favorable: their relative free energies are just 1–4 kcal/mol. Interested readers can find some structures of the CO insertion TS for branched and linear isomers in Figs. S2 and S3, respectively, of the ESI.

The TS barrier for the CO insertion step higher than that for the alkyl-Rh intermediate formation along few branched pathways is responsible for the reaction reversibility with return to the initial reactant complex, as discussed later on in the proper section. The subsequent reaction steps have been thoroughly examined only for conformers

[7] For branched structures, out of the two with a suitable orientation.

Table 5 B3P86/6-31G* relative energies and free energies (kcal/mol) of the CO migratory insertion TS for branched and linear Rh tetracarbonyl intermediates with respect to Reac_0b + CO

System[a]	ΔE	Rh–C_2	C–C_2	$H_3C_3C_2H_2$	IF[b]	ΔG
TS-insCO*b*0	−6.10	2.468	2.177	−174.3	209.i	8.16
TS-insCO*b*1	−4.41	2.470	2.252	77.3	179.i	9.48
TS-insCO*b*2	−1.69	2.448	2.181	−49.7	212.i	13.24
TS-insCO*b*′0	−4.19	2.463	2.209	−169.0	190.i	9.91
TS-insCO*b*′1	−2.41	2.456	2.271	73.9	173.i	11.70
TS-insCO*b*′2	−6.84	2.444	2.140	−93.0	227.i	7.41

System[a]	ΔE	Rh–C_1	C–C_1	Rh–C	IF[b]	ΔG
TS-insCO*l* 0	−11.03	2.383	2.060	1.884	265.i	1.36
TS-insCO*l* 1	−9.14	2.385	2.051	1.883	265.i	4.41
TS-insCO*l*′0	−11.52	2.391	2.070	1.883	262.i	2.14
TS-insCO*l*′2	−11.16	2.380	2.044	1.883	277.i	1.20

System[a]	ΔE	$H_3C_3C_2H_2$	$H_3C_3C_2C_1$	ΔG
AcylRh-*b*1	−30.79	73.9	−167.4	−15.91
AcylRh-*b*2	−30.48	−72.5	35.1	−14.86
AcylRh-*b*0	−29.11	−172.1	−55.3	−14.74
AcylRh-*b*′1	−27.59	77.1	−167.5	−12.65
AcylRh-*b*′2	−32.11	−86.2	26.0	−16.70
AcylRh-*l* 0	−33.27	−164.6	−42.6	−19.08
AcylRh-*l* 1	−32.28	56.4	−178.8	−17.47
AcylRh-*l*′0	−34.09	164.5	43.3	−19.74
AcylRh-*l*′2	−33.60	−77.8	43.3	−19.88

Branched and linear Rh-acyl intermediates are also reported

Ref. energies and free energies in Table 4; geometrical parameters in Å and degrees

[a] The prime corresponds to the *3S* chirality in the reactant complexes

[b] Imaginary frequency (cm^{-1})

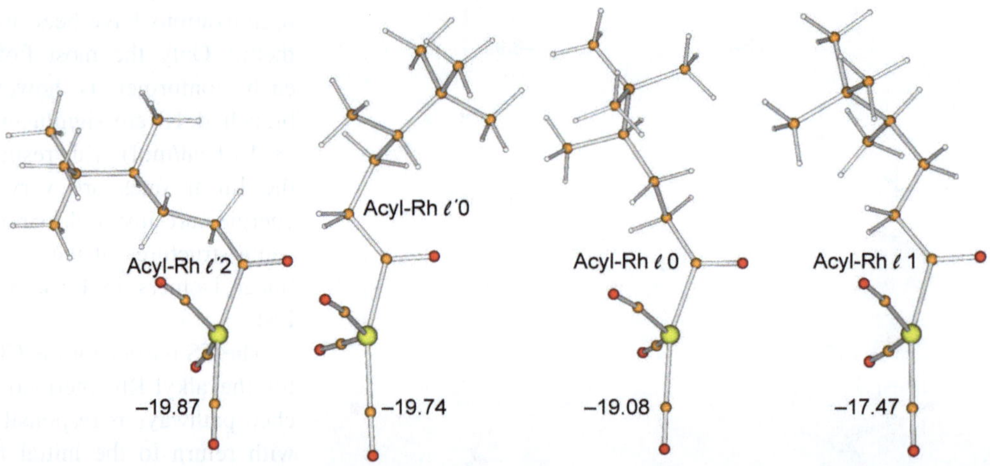

Fig. 14 B3P86/6-31G* structures and relative free energies (kcal/mol) for some acyl-Rh linear intermediates

not already bound to return again to the starting material, that is, for those without TS higher than the alkyl-Rh one.

Due to the flexibility of their aliphatic backbone, linear acyl-Rh intermediates (Fig. 14) are generally much more stable than any branched ones (Fig. 15), whose methyl group is roughly trans with respect to the inserted CO, although the stability of the intermediates is not as important as the barrier heights.

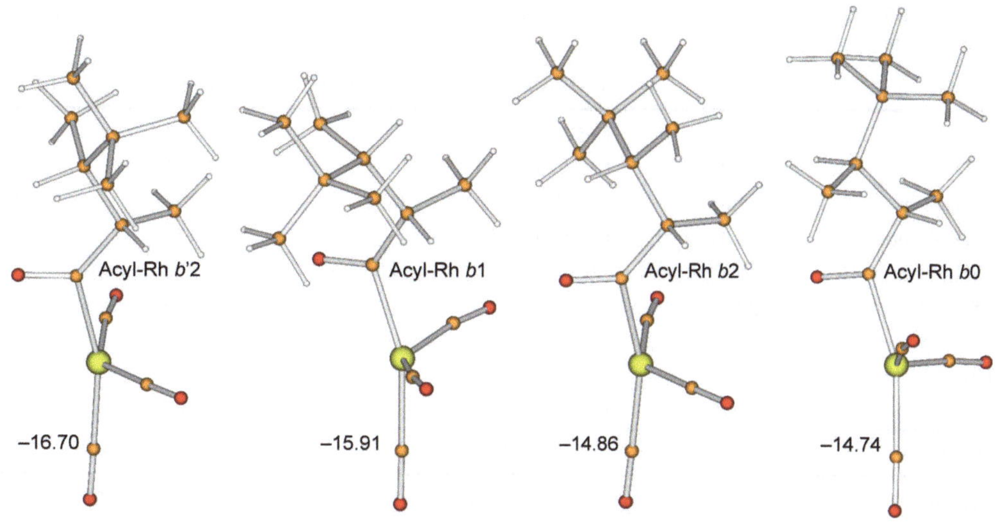

Fig. 15 B3P86/6-31G* structures and relative free energies (kcal/mol) for some acyl-Rh branched intermediates

3.6 H_2 oxidative addition

For the approaching paths of H_2 to Rh, a strategy similar to that used to investigate the attack of the fourth CO was followed, that is, "in" and "out" paths with respect to the bisector of the CO ligands to Rh have been examined, since the four-coordinated acyl-Rh complex is not planar and none of them can be excluded in advance. The two possible reaction modes in the H_2 oxidative addition with the H–H bond axis either perpendicular or parallel to the acyl–Rh bond [61] have been examined. The results are reported in Table 6 for TS and the relevant intermediates. The parallel arrangement (indicated appending a "p" to the structure name in tables and figures) has been obtained only by relaxing the structure arrived at after imposing that constraint and is remarkably less favorable than the perpendicular ones, as can be inferred from a perusal of Table 6.

As for the other intermediates, that is, alkyl-Rh(CO)$_3$ + CO, the complexes with H_2 at large separation have been determined first. In most cases, both Hs are located in the same plane of the basal CO groups (structures and relative energies are reported in Figs. S4 and S5 of the ESI for complexes with H_2 of linear and branched Rh-acyls, respectively).

At the TS for H_2 addition, both Hs are located in the basal CO plane (H–H distance of about 1.15–1.16 Å). Branched and linear TS are shown in Figs. S6 and S7, respectively, of the ESI. When the Hs are perpendicular to the basal CO plane (in other words, when they were initially parallel to the acyl-Rh bond), their separation increases to about 1.26 Å. This occurs however only after a constrained optimization, for the "in" approach of H_2 to the branched (3*S*)-Rh-acyl. The comparison between the "out" and "in" TS structures for H_2 addition to the branched (3*S*)-Rh-acyl system can be found in Fig. S8 of the ESI.[8] The six-coordinate branched and linear intermediates after H_2 addition are displayed in Figs. S9 and S10, respectively, of the ESI. They do not remarkably differ from the relevant TS apart from the decreased RhH distances and the H···H separation that becomes slightly greater than 2 Å.

3.7 Aldehyde reductive elimination

In the aldehyde reductive elimination, one of the H coordinated to Rh moves onto the inserted carbonyl C atom, yielding the product aldehyde, while the other remains bonded to Rh, which is released as H–Rh(CO)$_3$ in the vicinity of the aldehyde. The TS have been computed considering either H moving to give the aldehyde. In Table 7, however, only the most favorable TS for branched and linear alkyls are reported, together with a few geometrical parameters. The $H_3C_3C_2H_2$ dihedral angle is reported only for the branched isomers, because for the linear ones it is devoid of significance. Interestingly enough, in the case of TS for branched isomers, the H…CO separation is somewhat larger than for linear TS. A few low free energy structures are shown in Figs. S11–S13 of the ESI for branched and linear TS isomers. In Fig. S13, the linear aldehyde complexes with H–Rh(CO)$_3$, derived from the TS structures shown, are also displayed. In Fig. S14, three branched aldehyde complexes with H–Rh(CO)$_3$ are displayed. The linear aldehyde complexes with H–Rh(CO)$_3$ are invariably more stable than the branched ones (Table 7).

[8] The intermediate corresponding to "TSH_2b'2 in p" is "H_2Int-*b*'2 in p" (reported in Table 6 and displayed in Fig. S8) that evolves to the branched aldehyde through "TS–H_2b'2 in p" (ΔG=15.13 kcal/mol), that is, with a much higher barrier than when the hydrogens are in the basal CO plane.

Table 6 B3P86/6-31G* relative energies and free energies (kcal/mol) of the H_2 oxidative addition TS for branched and linear Rh-acyl intermediates

System[a]	ΔE^{b}	Rh–H/Rh–H	C–H/C–H	H–H	$H_3C_3C_2H_2$	IF[c]	ΔG^{d}
TSH_2b0	−16.61	1.657/1.658		1.167	−169.7	1008.i	7.54
TSH_2b1 in	−17.98	1.657/1.661		1.161	68.9	1002.i	5.52
TSH_2b2 in	−14.82	1.647/1.666		1.153	−72.0	967.i	9.51
TS$H_2b'2$ out	−18.91	1.654/1.664		1.157	−81.8	977.i	5.09
TS$H_2b'2$ in p	−10.18	1.636/1.686	3.48/3.88	1.262	−86.9	998.i	14.03
TSH_2l 0	−22.19	1.658		1.168		1000.i	0.67
TSH_2l 1	−19.95	1.657/1.659		1.168		999.i	3.36
TS$H_2l'2$	−22.19	1.658		1.167		1003.i	0.31
TS$H_2l'0$	−21.75	1.658		1.169		1001.i	1.53
System[a]	ΔE^{b}	Rh–H	C–H/C–H	H–H	$H_3C_3C_2H_2$		ΔG^{d}
+H_2Int-$b1$	−26.94	1.596/1.601		2.043	67.7		−0.90
+H_2Int-$b2$	−24.28	1.593/1.595		2.065	−71.7		2.63
+H_2Int-$b0$	−25.39	1.599/1.597		2.041	−170.8		1.21
+H_2Int-$b'2$ out	−27.66	1.604/1.589		2.031	−81.5		−1.10
+H_2Int-$b'2$ in	−27.39	1.591/1.602		2.041	−88.8		−0.83
+H_2Int-$b'2$ in p	−19.69	1.594/1.654	2.76	2.180			7.94
+H_2Int-$l'0$	−30.45	1.593/1.602		2.038			−4.62
+H_2Int-$l'2$[e]	−30.92	1.593/1.603		2.037			−5.50
+H_2Int-l 0[e]	−30.92	1.593/1.603		2.037			−5.51
+H_2Int-l 1	−28.62	1.593/1.603		2.037			−3.01

Branched and linear H_2 oxidative addition intermediates are also reported

[a] The prime corresponds to the *3S* chirality in the reactant complexes; geometrical parameters in Å and degrees

[b] Ref. energy (Reac_0b + CO + H_2) = −881.704361 E_h

[c] Imaginary frequency (cm^{-1})

[d] Ref. free energy (Reac_0b + CO + H_2) at 298.15 K and 1 atm = −881.511827 E_h

[e] Optical antipodes = mirror images

4 Reaction mechanisms

The reaction profiles collecting all the aforementioned elementary steps and their transition states (olefin coordination, olefin insertion, CO addition and subsequent insertion, H_2 oxidative addition and aldehyde reductive elimination, as proposed by Wilkinson et al. [62, 63]) have been drawn using the relative free energies of the various TS and intermediates for each individual conformer, that is, maintaining the alkyl arrangements as similar as possible, consistent with the data reported in the tables.

4.1 Linear reaction profiles

The linear profiles, displayed in Fig. 16, are very similar to each other and feature the alkyl-Rh TS much higher than the other subsequent TS, thus indicating that the linear isomers easily proceed to aldehyde. Both alkyl-Rh tricarbonyl and tetracarbonyl intermediates have been displayed, although in the presence of CO the linear tetracarbonyl intermediates form directly, unlike the branched ones.

4.2 Branched reaction profiles

Also, some of the branched regioisomers, namely *b*0 (profiles in Fig. 17), *b*′2 (profiles in Fig. 18), *b*1 and *b*′1 (profiles in Fig. 19), may proceed to aldehyde although branched pathways are not as favorable as linear ones.

Obviously, the TS to return back to the initial reactant complex (involving the terminal olefin) is exactly the same as that leading to the alkyl-Rh tricarbonyl intermediate, while β-elimination to the internal olefin complex, in principle possible when the H at C_3 is in a favorable position to bind to Rh, has also been computed and superimposed onto the profile. The β-elimination (to the internal olefin) TS is even higher in free energy than the alkyl-Rh TS for the *b*0 regioisomer (Fig. 17), whereas it is

Table 7 B3P86/6-31G* relative energies and free energies (kcal/mol) of the H_2 reductive elimination TS for branched and linear intermediates

System[a]	ΔE[b]	Rh–H/Rh–H	H–H/H–C	$H_3C_3C_2H_2$	IF[c]	ΔG[d]
TS–H_2b0	−16.25	1.589/1.672	2.165/1.545	−171.7	710.i	9.66
TS–H_2b1	−17.85	1.589/1.670	2.173/1.563	61.5	735.i	7.60
TS–H_2b2	−15.20	1.585/1.670	2.205/1.581	−76.9	737.i	10.56
TS–$H_2b'2$ out	−19.07	1.675/1.589	2.166/1.557	−82.8	705.i	6.83
TS–$H_2l'0$	−20.75	1.672/1.591	2.177/1.528		719.i	4.21
TS–$H_2l'2$	−21.12	1.672/1.591	2.174/1.526		720.i	2.76
TS–H_2l 0	−21.14	1.672/1.591	2.174/1.525		718.i	3.31
TS–H_2l 1	−18.88	1.672/1.591	2.176/1.527		716.i	6.02

System[a]	ΔE[b]	$H_3C_3C_2C_1$	OCC_2H_2	$H_3C_3C_2H_2$	ΔG[d]
Rh-Ald-$b1$	−40.29	−26.9	−119.4	92.9	−17.20
Rh-Ald-$b2$	−39.55	173.9	−106.4	−67.6	−15.67
Rh-Ald-$b0$	−37.65	61.4	−15.8	−173.6	−14.27

System[a]	ΔE[b]	$H_3C_3C_2C_1$	OCC_1C_2	ΔG[d]
Rh-Ald-$l'0$	−41.53	41.55	−4.56	−19.18
Rh-Ald-$l'2$	−41.81	42.36	−0.43	−19.95

Few branched and linear aldehyde complexes are also reported

[a] The prime corresponds to the *3S* chirality in the reactant complexes; geometrical parameters in Å and degrees

[b] Ref. energy + CO + H_2 = −881.704361 E_h

[c] Imaginary frequency (cm^{-1})

[d] Ref. free energy + CO + H_2 at 298.15 K and 1 atm = −881.511827 E_h

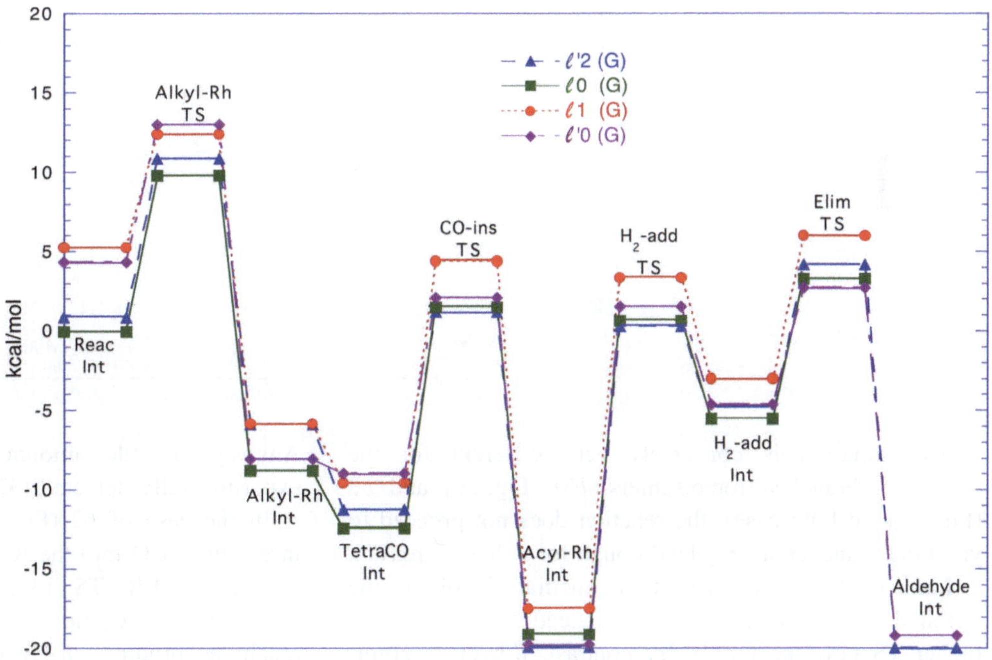

Fig. 16 B3P86/6-31G* relative free energy profiles (kcal/mol) for the linear regioisomers

lower than the alkyl-Rh TS for $b'2$ (Fig. 18). In this latter case, it is anyway higher than all the subsequent TS; thus, also $b'2$ yields the branched aldehyde.

Both $b1$ (upper left corner of Fig. 8) and $b'1$ cannot undergo β-elimination to give the internal olefin intermediate as such, because their H in β is on the opposite side with respect to the catalyst. The $b'1$ profile (Fig. 19) is less favorable than that of $b1$. Interestingly enough, the $b1$ barrier is only 9.8 kcal/mol, although the $b1$ reactant complex is somewhat high in free energy.

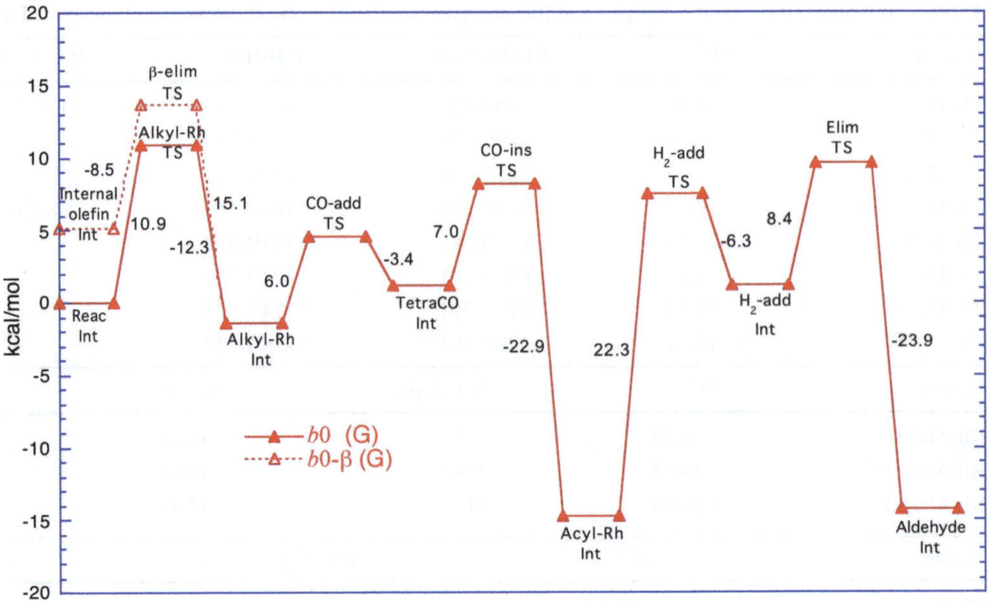

Fig. 17 B3P86/6-31G* relative free energy profiles (kcal/mol) for the *b*0 regioisomer

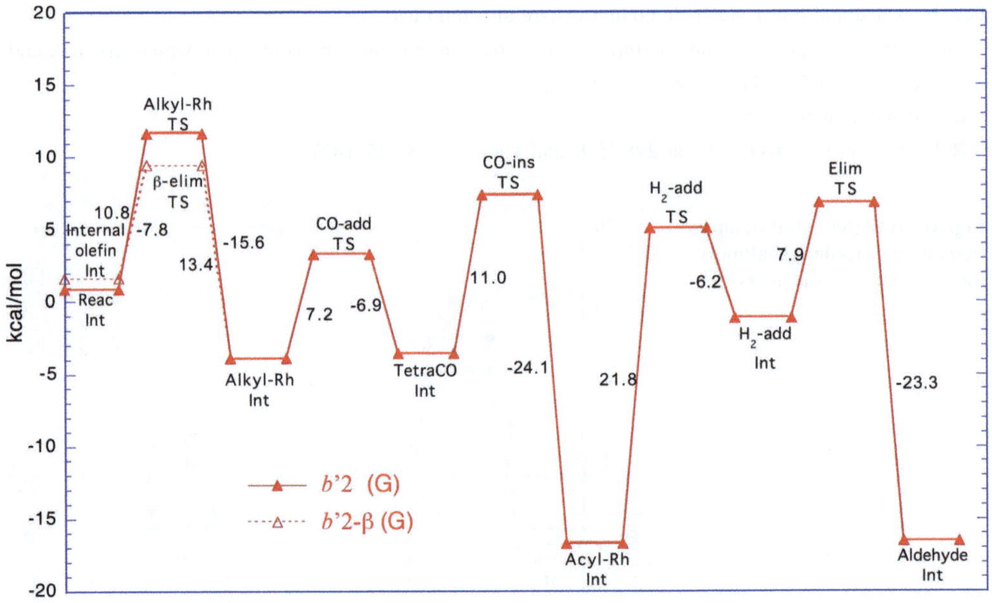

Fig. 18 B3P86/6-31G* relative free energy profiles (kcal/mol) for the *b*′2 regioisomer

The situation is conversely very different for the remaining branched regioisomers, *b*′0 (Fig. 20) and *b*2 (Fig. 21). In both cases, the reaction does not proceed to yield the branched aldehyde although for a diverse reason.

For *b*′0, the initial reactant intermediate involving the terminal olefin cannot be obtained indeed due to the much higher TS (14.5 kcal/mol). By contrast, *β*-hydride elimination to give the internal olefin intermediate might occur, because the corresponding TS is lower than the CO addition and CO insertion ones (9.5 vs 9.8 and 9.9 kcal/mol, respectively). However, due to the small energy difference among the computed TS, it is difficult in this case to predict with confidence exclusively the internal olefin formation. Anyway, a little amount of internal olefin has been experimentally detected [37].

In the case of *b*2 (Fig. 21), the TS for the migratory insertion of CO into the Rh–C bond turns out to be higher than the alkyl-Rh TS (13.2 vs 10.9 kcal/mol, respectively). Therefore, the reaction cannot proceed further toward the aldehyde product. *β*-Hydride elimination to give the internal olefin intermediate cannot occur either, because its TS (13.8 kcal/mol) is higher than the CO insertion TS. Consequently, the terminal olefin complex forms and goes through the linear and branched pathways again with the same regioselectivity as before, thus enriching the linear fraction.

Fig. 19 B3P86/6-31G* relative free energy profiles (kcal/mol) for the $b1$ and $b'1$ regioisomers

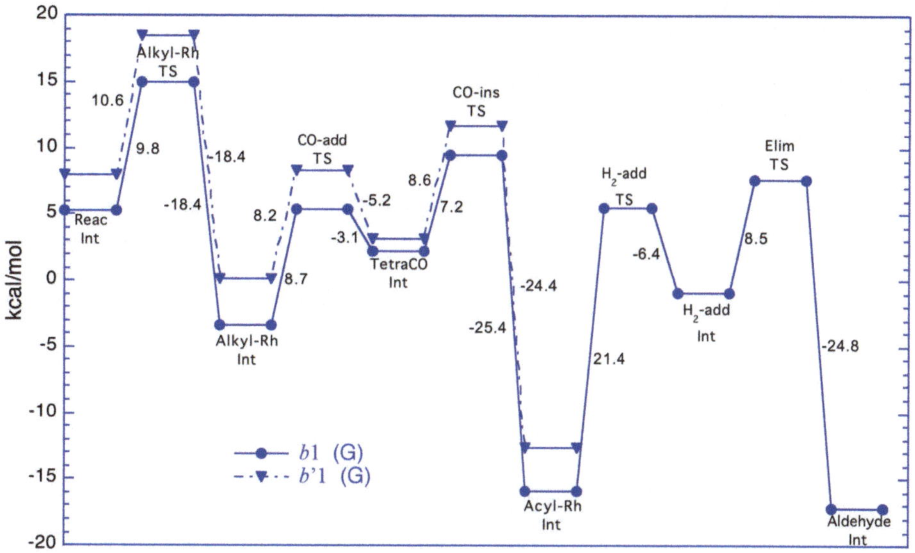

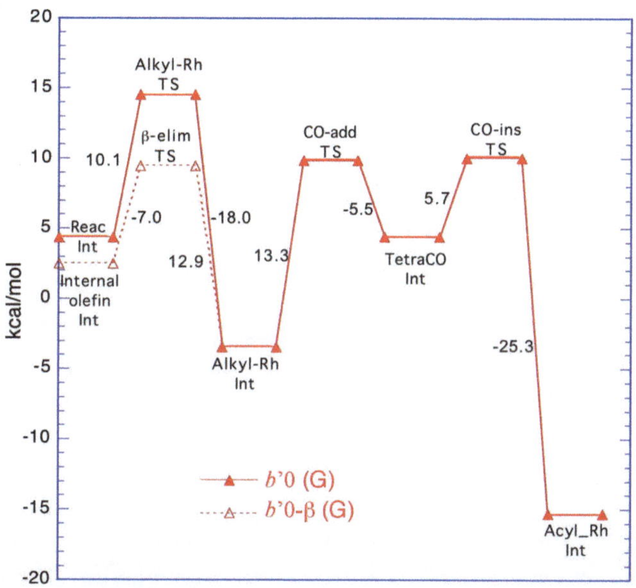

Fig. 20 B3P86/6-31G* relative free energy profiles (kcal/mol) for the $b'0$ regioisomer

4.3 Solvent effects

Since the influence of the environment might affect the profile, a tentative evaluation of its contribution has been performed with the supermolecule approach. In a previous study by other authors [61], where the full cycle of olefin (ethene) hydroformylation was considered with a phosphine-modified Rh catalyst, another substrate molecule was added to mimic the solvent coordinating effect to the metal. Due to the size and the conformational flexibility of the substrate here considered, a single molecule of the solvent used in the experiments (benzene) was added to the computational setting for $b2$[9] in suitable starting positions. New geometry optimizations have been carried out starting from the optimized structures in vacuo. ZPE and thermal corrections have been added to obtain the free energy of each stationary point for the supermolecule. For each species, the lowest free energy stationary points have been also drawn in Fig. 21 (solid and empty circles) for comparison. From the resulting profiles, it is evident that the presence of benzene does not affect the alkyl-Rh TS or the CO insertion TS, which maintain unaltered their free energy difference. The main effect is resented by the TS for the fourth CO approach to Rh that becomes 10.4 kcal/mol (it was 8.6 kcal/mol in vacuo) because the coordinating solvent hampers its path. The TS obtained including benzene are displayed in Fig. 22, while all stationary points along that pathway are shown in Fig. S15 of the ESI.

The β-hydride elimination TS to give the internal olefin intermediate is only marginally influenced by the solvent, decreasing by 0.7 kcal/mol, while a larger effect is visible on the internal olefin itself. The relevant structures including benzene are displayed in Fig. 23.

Utilizing benzene ($\varepsilon = 2.247$) in the continuum solvent framework (IEF-PCM on radii determined from the united-atom topological model) and allowing the solute stationary points to relax, a negligible effect is found on solute geometries and free energies ($G_{tot}(298)$). This is not unexpected because the benzene dielectric constant is very close to vacuum, and thus the reaction field produced should be small. From the superimposition of the profiles

[9] The most critical case because of TS relative values, involving β-elimination leading either to internal or initial olefin.

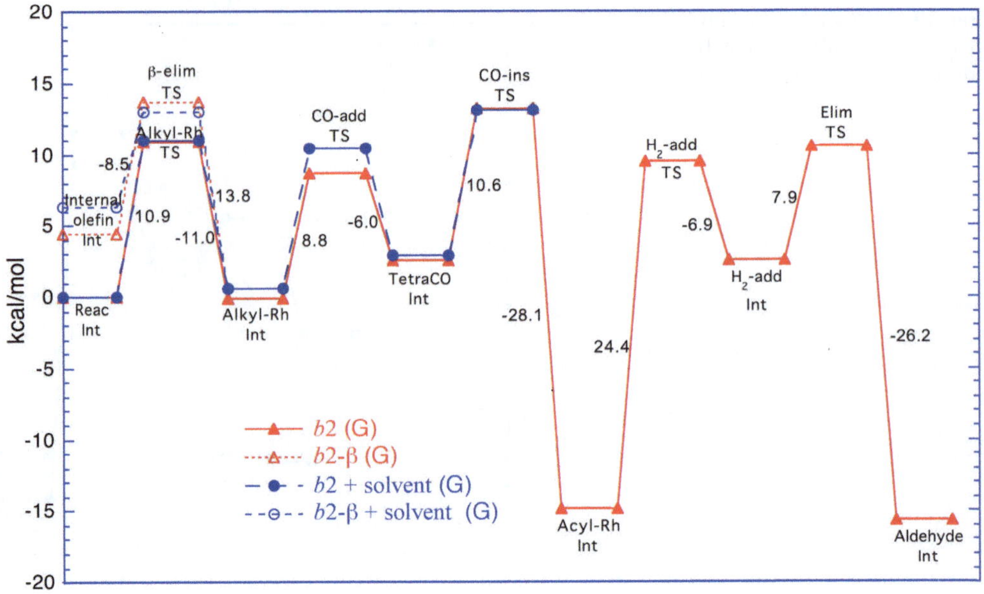

Fig. 21 B3P86/6-31G* relative free energy profiles (kcal/mol) for the *b*2 regioisomer in vacuo and in the presence of a solvent molecule (benzene). Values related to the in vacuo profiles

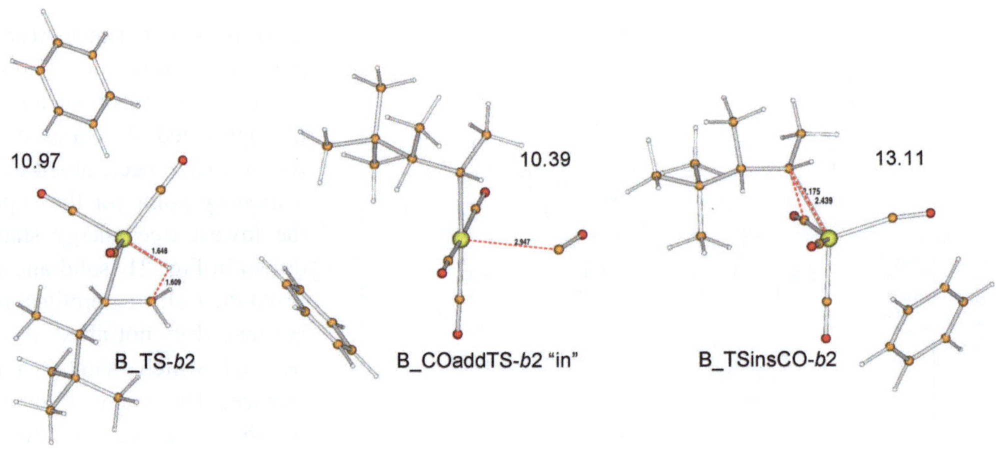

Fig. 22 B3P86/6-31G* TS structures and relative free energies (kcal/mol) for the *b*2 regioisomer in the presence of a solvent molecule (benzene), indicated by the prefix B_

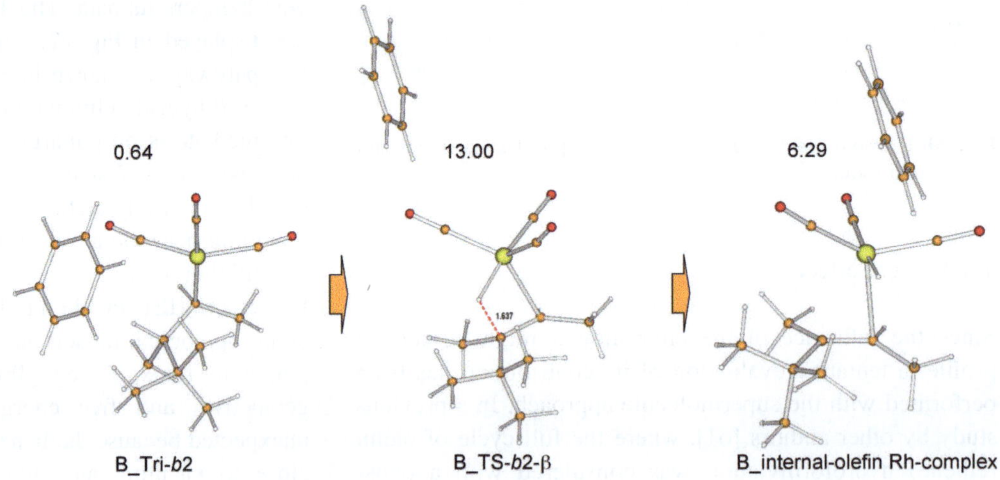

Fig. 23 B3P86/6-31G* structures and relative free energies (kcal/mol) for the *β*-elimination stationary points including a solvent molecule (benzene) along the *b*2-*β* branched pathway leading to the internal olefin-Rh complex, reported in Fig. 21 (*empty circles*)

along the branched pathway (Fig. S16 of the ESI), no difference is detected for the Rh-alkyl intermediate and the *β*-elimination process to give the internal olefin intermediate, while an analogous stabilization is observed for the alkyl-Rh and CO insertion TS. Opposite, yet limited, effects are found on the CO addition TS (whose barrier increases) and on the tetracarbonyl intermediate. Similar negligible effects have been obtained with PCM on metal catalyzed reactions by other authors as well using gas phase-optimized geometries either in benzene [64] or in diethyl ether solution [65] with the B3LYP hybrid functionals.

5 Conclusions

The branched to linear (B:L) regioselectivity ratio in the hydroformylation of 3,4,4-trimethylpent-1-ene (**TMP$_1$**) carried out with an unmodified Rh catalyst and computed under the hypothesis of nonreversibility from the relative populations of branched and linear alkyl-Rh transition states (TS) at the B3P86/6-31G* level, utilizing effective core potentials for Rh in the LanL2DZ valence basis set, turned out to be 47:53 and 15:85 at the potential and free energy levels, respectively. Curiosity was elicited by the potential energy results, obtained much earlier than the free energy ones, in our experimental colleagues spurring them to carry out deuterioformylation experiments to confirm the remarkable preference in favor of linear aldehydes (5:95). Those experiments put forward a reversible behavior for the branched intermediates, that is, β-hydride elimination. Thus, although the agreement between experiment and theory at the free energy level was rather satisfactory, the full cycle of olefin hydroformylation for **TMP$_1$**, a chiral substrate, has been computationally investigated to figure out the origin of the β-hydride elimination observed. Geometry optimizations to locate a huge number of stationary points (TS and intermediates) for branched and linear conformers involved in the catalytic cycle have been performed for two configurations, allowing the Re and Si faces of attack to be considered (as a matter of fact, the experiment was carried out on the racemic mixture). In addition, harmonic vibrational frequency calculations to evaluate zero-point and thermal corrections have been computed.

The free energy profiles along the linear pathways are fairly similar, maintaining the mutual positions of the various regioisomers almost unaltered, and indicate that all the 85% of the linear fraction formed at the alkyl-Rh TS invariably produces at room temperature the linear aldehyde. Interestingly, for linear regioisomers in the presence of CO, tetracarbonyl intermediates form without any activation barrier.

The free energy profiles along the branched pathways have been plotted separately for each regioisomer to permit to single out their features. Considering the $b0$ regioisomer (**0** is the most stable reactant complex), the alkyl-Rh TS and the aldehyde elimination TS are very close in free energy and β-hydride elimination might occur, while that to give the internal olefin can be excluded, unlike for the $b'0$ regioisomer that brings about the internal olefin. Both $b1$ and $b'1$ regioisomers proceed to aldehyde, although **1** is the least stable reactant complex. It is anyway worth to be considered because of the Curtin–Hammett principle. The $b'2$ regioisomer whose population is low (high alkyl-Rh TS) also proceeds to give the branched aldehyde. Conversely, the much more populated $b2$ regioisomer returns to the initial olefin complex that proceeds to aldehyde with the same regioselectivity as before (15:85), thus further increasing the linear fraction.

The solvent effect, included via either the supermolecule approach, by adding a single molecule of benzene, or the benzene reaction field in the polarizable continuum model framework (IEF-PCM), does not substantially modify the picture of the reaction obtained in vacuo.

A final comment is necessary. This investigation indicates that regrettably it is almost impossible to predict selectivities based on alkyl-Rh transition state relative energies or free energies alone, because reaction reversibility may occur even at room temperature (where usually nonreversibility is assumed). This implies additional calculations along the reaction pathways to verify whether higher TS are to be surmounted by some regio- or diastereoisomers.

References

1. Roelen O (1938) (Ruhrchemie AG). D. B. P. 849 458
2. Roelen O (1953) Chem Zentr 927–929
3. Cornils B, Herrmann WA (1996) Applied homogeneous catalysis with organometallic compounds, vol 1. Wiley, New York, pp 3–25
4. Papadogianakis G, Sheldon RA (1996) New J Chem 20:175–185
5. Süss-Fink G, Meister G (1993) Adv Organomet Chem 35:41–134
6. Versluis L, Ziegler T, Fan L (1990) Inorg Chem 29:4530–4536
7. Wender I, Pino P (1977) Organic syntheses via metal carbonyls, vol 2. Wiley, New York
8. Falbe J (1980) New nyntheses with carbon monoxide. Springer, Berlin
9. Pignolet LH (ed) (1983) Homogeneous catalysis with metal phosphine complexes. Plenum, New York
10. Casey CP, Petrovich LM (1995) J Am Chem Soc 117:6007–6014
11. Casey CP, Paulsen EL, Bettenmueller EW, Proft BR, Matter BA, Powell DR (1999) J Am Chem Soc 119:11817–11825
12. Van Leeuwen PWNM, Kamer PCJ, Reek JNH (1999) Pure Appl Chem 71:1443–1452
13. Nozaki K, Sakai N, Nanno T, Higashijima T, Mano S, Horiuchi T, Takaya H (1997) J Am Chem Soc 119:4413–4423 and refs quoted therein
14. Herrmann WA, Kohlpaintner CW, Herdtwck E, Kiprof P (1991) Inorg Chem 30:4271–4275
15. Herrmann WA, Schmid R, Kohlpaintner CW, Priermeier T (1995) Organometallics 14:1961–1968
16. Decker SA, Cundari TR (2001) Organometallics 20:2827–2841
17. Decker SA, Cundari TR (2001) J Organomet Chem 635:132–141
18. Decker SA, Cundari TR (2002) New J Chem 26:129–135

19. Rocha WR, De Almeida WB (2000) Int J Quantum Chem 78:42–51
20. Dias RP, Prates MSL Jr, De Almeida WB, Rocha WR (2011) Int J Quantum Chem 111:1280–1292
21. Carbó JJ, Maseras F, Bo C, van Leeuwen PW (2001) J Am Chem Soc 123:7630–7637
22. Landis CR, Uddin J (2002) J Chem Soc, Dalton Trans 729–742
23. Beller M, Cornils B, Frohning CD, Kohlpaintner CW (1995) J Mol Cat A Chem 104:17–85
24. Pidun U, Frenking G (1998) Chem Eur J 4:522–540
25. Torrent M, Solà M, Frenking G (2000) Chem Rev 100:439–493 and refs quoted therein
26. Consiglio G, Pino P (1982) Top Curr Chem 105:77–123
27. Gleich D, Schmid R, Herrmann WA (1998) Organometallics 17:4828–4834
28. Alagona G, Ghio C, Lazzaroni R, Settambolo R (2001) Organometallics 20:5394–5404
29. Alagona G, Ghio C, Lazzaroni R, Settambolo R (2004) Inorg Chim Acta 357:2980–2988
30. Alagona G, Ghio C (2005) J Organomet Chem 690:2339–2350
31. Settambolo R, Rocchiccioli S, Lazzaroni R, Alagona G (2006) Lett Org Chem 3:10–12
32. Alagona G, Ghio C, Rocchiccioli S (2007) J Mol Model 13:823–837
33. Lazzaroni R, Settambolo R, Marchetti M, Paganelli S, Alagona G, Ghio C (2008) Inorg Chim Acta 362:1641–1644
34. Ghio C, Lazzaroni R, Alagona G, (2009) Eur J Inorg Chem 98–103
35. Lazzaroni R, Settambolo R, Alagona G, Ghio C (2010) Coord Chem Rev 254:696–706. Erratum in: (2011) Coord Chem Rev 255:3031
36. Alagona G, Lazzaroni R, Ghio C (2011) J Mol Model 17:2275–2284. Erratum. doi:10.1007/s00894-012-1373-8
37. Lazzaroni R, Settambolo R, Alagona G, Ghio C (2012) J Mol Cat A Chem. doi:10.1016/j.molcata.2011.12.021
38. McQuarrie DA (2000) Statistical mechanics. University Science Book, Sausalito
39. Frisch MJ, Trucks GW, Schlegel HB, Scuseria GE, Robb MA, Cheeseman JR, Montgomery JA Jr, Vreven T, Kudin KN, Burant JC, Millam JM, Iyengar SS, Tomasi J, Barone V, Mennucci B, Cossi M, Scalmani G, Rega N, Petersson GA, Nakatsuji H, Hada M, Ehara M, Toyota K, Fukuda R, Hasegawa J, Ishida M, Nakajima T, Honda Y, Kitao O, Nakai H, Klene M, Li X, Knox JE, Hratchian HP, Cross JB, Bakken V, Adamo C, Jaramillo J, Gomperts R, Stratmann RE, Yazyev O, Austin AJ, Cammi R, Pomelli C, Ochterski JW, Ayala PY, Morokuma K, Voth GA, Salvador P, Dannenberg JJ, Zakrzewski VG, Dapprich S, Daniels AD, Strain MC, Farkas O, Malick DK, Rabuck AD, Raghavachari K, Foresman JB, Ortiz JV, Cui Q, Baboul AG, Clifford S, Cioslowski J, Stefanov BB, Liu G, Liashenko A, Piskorz P, Komaromi I, Martin RL, Fox DJ, Keith T, Al-Laham MA, Peng CY, Nanayakkara A, Challacombe M, Gill PMW, Johnson B, Chen W, Wong MW, Gonzalez C, Pople JA (2004) Gaussian 03, Revision C.02, Gaussian Inc., Wallingford
40. Becke AD (1993) J Chem Phys 98:5648–5652
41. Perdew JP (1986) Phys Rev B 33:8822–8824
42. Hehre WJ, Radom L, Schleyer PVR, Pople JA (1986) Ab initio molecular orbital theory. Wiley, New York
43. Dunning TH Jr, Hay PJ (1976) In: Schaefer HF III (ed) Modern theoretical chemistry. Plenum, New York, pp 1–28
44. Hay PJ, Wadt WR (1985) J Chem Phys 82:270–283
45. Hay PJ, Wadt WR (1985) J Chem Phys 82:299–310
46. Jonas V, Thiel W (1995) J Chem Phys 102:8474–8484
47. Goh SR, Marynick DS (2002) Organometallics 21:2262–2267
48. Frenking G, Antes I, Boehme M, Dapprich S, Ehlers AW, Jonas V, Nehaus A, Otto M, Stegmann R, Veldjamp A, Vyboishchikov SF (1996) In: Lipkowitz KB, Boyd DB (eds) Reviews in computational chemistry, vol 8. VCH, New York, pp 63–144
49. Tomasi J, Persico M (1994) Chem Rev 94:2027–2094
50. Barone V, Cossi M, Tomasi J (1997) J Chem Phys 107:3210–3221
51. Cancès E, Mennucci B, Tomasi J (1997) J Chem Phys 107:3032–3041
52. Cancès E, Mennucci B (1998) J Chem Phys 109:249–259
53. Cancès E, Mennucci B, Tomasi J (1998) J Chem Phys 109:260–266
54. Amovilli C, Barone V, Cammi R, Cancès E, Cossi M, Mennucci B, Pomelli CS, Tomasi J (1998) Adv Quantum Chem 32:227–261
55. Tomasi J, Mennucci B, Cammi R (2005) Chem Rev 105:2999–3094
56. Tomasi J (2007) In: Mennucci B, Cammi R (eds) Continuum solvation models in chemical physics: from theory to applications. Wiley, New York, pp 1–28
57. Schaftenaar G, Noordik JH (2000) J Comput Aided Mol Design 14:123–134
58. Curtin DY (1954) Rec Chem Prog 15:111–128
59. Hammett LP (1970) Physical organic chemistry, 2nd edn. McGraw-Hill, New York, pp 119–120 Chapter 5
60. Seeman JI (1983) Chem Rev 83:84–134
61. Matsubara T, Koga N, Ding Y, Musaev DG, Morokuma K (1997) Organometallics 16:1065–1078
62. Yagupsky G, Brown CK, Wilkinson G (1969) J Chem Soc Chem Comm 1244–1245
63. Yagupsky G, Brown CK, Wilkinson G (1970) J Chem Soc A 1392–1401
64. Tang D, Qin S, Su Z, Hu C (2007) Organometallics 26:33–47
65. Nowroozi-Isfahani T, Musaev DG, McDonald FE, Morokuma K (2005) Organometallics 24:2921–2929

Theor Chem Acc (2012) 131:1147
DOI 10.1007/s00214-012-1147-5

REGULAR ARTICLE

First-principles simulations of the ^{27}Al and ^{17}O solid-state NMR spectra of the $CaAl_2Si_3O_{10}$ glass

Alfonso Pedone · Elisa Gambuzzi · Gianluca Malavasi · Maria Cristina Menziani

Received: 23 June 2011 / Accepted: 2 September 2011 / Published online: 15 February 2012

Abstract The local and medium-range structure of the $20CaO{\cdot}20Al_2O_3{\cdot}60SiO_2$ glass generated by classical molecular dynamics simulations has been compared to NMR experiments by computing the ^{27}Al and ^{17}O NMR parameters and NMR spectra from first-principles simulations. The calculation of the NMR parameters (chemical shielding and quadrupolar parameters), which are then used to simulate solid-state MAS and 3QMAS NMR spectra, is achieved by the gauge including projector augmented-wave and the projector augmented-wave methods on the DFT-PBE relaxed structure. The NMR spectra calculated with the present approach are found to be in excellent agreement with the experimental data, providing an unambiguous view of the local and medium-range structure of aluminosilicate glasses.

Keywords First-principles simulations · NMR spectra · Silicate glass · 3D structure

1 Introduction

In the recent decades, silica-based glasses have emerged as fundamental materials for new technological applications regarding data processing and transmission (optical fiber amplifiers, semiconductor devices), nuclear waste confinement and biomaterials (bioactive glasses).

To rationalize and predict the macroscopic thermodynamic and transport properties of such materials, a microscopic understanding of their structure and dynamics is required [1, 2].

Solid-state NMR spectroscopy has proven to be a very useful and versatile technique for studying the structure and the dynamics of non-crystalline materials such as oxide glasses and melts since long time [1, 3]. The effectiveness of NMR spectroscopy relies on the strong correlation between NMR observables (chemical shifts, quadrupolar parameters, couplings between nuclides) and the short- and medium-range structure of spin active nuclei constituting the glass [1].

The directional character of many NMR interactions coupled with the rigidity of structural units in non-crystalline solids results in heavily broad lineshapes, which can be envisioned as being composed of a series of narrow lines, each coming from a different crystallite orientation with respect to the external magnetic field. For this reason, more effort over the last 40 years has been devoted to the development of ways to average the anisotropic broadening in an artificial way, either by manipulating the spins or by manipulating the sample.

Initial experiments were focused on magic angle spinning [4–7] (MAS), which has been extensively applied to cancel out the effect of dipolar and chemical shift anisotropies of spin ½ nuclei (such as ^{29}Si) in silica-based glasses. This technique fails to remove the second-order quadrupolar effects present for nuclei with half-integer nuclear spin (so-called half-integer quadrupolar nuclei, here ^{17}O ($I = 5/2$), ^{43}Ca ($I = 7/2$), ^{27}Al ($I = 5/2$)); therefore, new techniques such as dynamic angle spinning (DAS), double rotation (DOR), or multiquantum magic angle spinning (MQMAS) have been developed [8, 9].

Dedicated to Professor Vincenzo Barone and published as part of the special collection of articles celebrating his 60th birthday.

A. Pedone · E. Gambuzzi · G. Malavasi · M. C. Menziani (✉)
Dipartimento di Chimica, Università degli Studi di Modena e Reggio Emilia, Via Campi 183, 41125 Modena, Italy
e-mail: menziani@unimo.it

Although experiments of this kind allow a high resolution in crystalline materials, the disorder and variety of structural units (different coordination numbers and Qn distributions, where n is the number of bridging oxygens (BO) around a network former species T = Si/Al) present in glasses make the interpretation difficult.

Traditionally, the interpretation of spectra of amorphous materials of unknown structure has been generally based on empirical correlations derived from the study of crystalline materials with known structure. However, there is still a need for much more work of this type, to establish and refine correlations between NMR observables and structural parameters.

Theoretical calculations of NMR parameters and structural correlations have been very helpful in interpreting spectra of glasses [10–12]. Since NMR interactions are essentially dependent upon local properties and are dominated by bonding in the first few coordination spheres, most of the theoretical investigations were performed on small molecular clusters using atomic orbital basis sets [13, 14]. The accuracy of this approach is limited by the choice of parameters such as the size of the cluster, the termination of the dangling bonds produced on cutting the cluster from the infinite solid (for covalent solids), or the treatment of the electrostatic interactions. Moreover, in the special case of amorphous materials, the cluster approach does not account for the correlations between structural factors that exist in solids and disorder in glasses.

These problems have been overcome with the introduction of effective methods specially devised to calculate NMR parameters in extended systems described within periodic boundary conditions, plane wave basis sets and density functional theory [15, 16].

The chemical shifts and the electric field gradients can be obtained from first principle with the gauge including projector augmented-wave (GIPAW) and the projector augmented-wave (PAW) methods, respectively [15, 16]. These methods are able to deal with large systems (up to hundreds of atoms), and exhibit an outstanding accuracy, as demonstrated recently for various crystalline and amorphous crystalline solids [12, 16–19].

Recently, Pedone et al. [10, 11] have reproduced, for the first time, the MAS and 3QMAS NMR solid-state spectra of spin active nuclei (^{29}Si, ^{31}P, ^{43}Ca, ^{17}O and ^{23}Na) for the $CaSiO_3$ and the $46.1SiO_2{\cdot}24.4Na_2O{\cdot}26.9CaO{\cdot}2.6P_2O_5$ (45S5) glasses using a combined classical molecular dynamics simulations for generating the glass structure and periodic DFT calculations to obtain NMR parameters. The latter were introduced in spin-effective Hamiltonian encoded into a home-made package named fpNMR to simulate the spectra [10].

This integrated computational approach proved to be very useful to interpret the experimental spectra of the aforementioned nuclei and to gain new insights into the atomistic structure of very complex oxide glasses such as calcium-silicate and phospho-silicate glasses containing sodium and calcium ions.

The aim of the present work is to describe and apply the integrated computational procedure developed in the recent years to the simulation of the ^{27}Al, ^{17}O MAS and MQMAS solid-state spectra of a calcium aluminum–silicate (CAS) glass of composition $20CaO{\cdot}20Al_2O_3{\cdot}60SiO_2$, in order to obtain an unambiguous view of its local and medium-range structure around aluminum and oxygen atoms.

2 Computational details

2.1 Glass generation

Two three-dimensional structures of a calcium aluminosilicate glass with composition $20CaO{\cdot}20Al_2O_3{\cdot}60SiO_2$ were generated by means of classical molecular dynamics simulations. The shell model approach [20, 21] is used in the potential to include polarization effects, by taking into account the large polarizability of oxygen ions. In this model, the total charge Z of the oxygen ions is split between a core (of charge Z + Y) and a massless shell (of charge –Y), which are coupled by a harmonic spring. Besides the damped harmonic interaction with the corresponding core, the oxygen shells interact with each other and with Si, Al and Ca cations through a short-range Buckingham term, whereas coulombic forces act between all species which bear full formal charges. Three-body screened harmonic potentials are used to control the intra-tetrahedral O–Si–O and O–Al–O angles during dynamics. The expression for the complete potential model is therefore

$$U_{ij}(r_{ij}) = \frac{q_i q_j}{r_{ij}} + A_{ij} e^{-\left(\frac{r_{ij}}{\rho_{ij}}\right)} - \frac{C_{ij}}{r_{ij}^6} \qquad (1)$$

$$U_{ijk}(\theta_{ijk}) = \frac{1}{2} k_b (\theta_{ijk} - \theta_0)^2 \exp\left\{ -\left(\frac{r_{ij}}{\rho_{ijk}} + \frac{r_{jk}}{\rho_{ijk}} \right) \right\} \qquad (2)$$

$$U_{\text{core-shell}}(r_{\text{core-shell}}) = \frac{1}{2} k_s (r_{\text{core-shell}})^2 \qquad (3)$$

The complete potential model and the parameters [22] used are reported in Table 1. The leap-frog algorithm encoded in the DL_POLY package [23] has been used to integrate the equation of motions with a time step of 0.2 fs small enough to control the high-frequency motion of the core–shell spring during MD simulations [24].

The initial configurations were generated by placing randomly 250 atoms in a cubic box of 14.98 Å accordingly to the experimental density of 2.624 g/cm^3. The systems

Table 1 Shell model interatomic potential: analytic functions and parameters [22]

	A (eV)	ρ (Å)	C (eV Å^6)
Buckingham $Ae^{-r/\rho} - C/r^6$			
O_s–O_s	22,764.30	0.1490	27.88
Si–O_s	1,283.91	0.32052	10.661580
Ca–O_s	2,152.3566	0.309227	0.09944
Al–O_s	1,460.3000	0.29912	0.0000000
	k_b (eV rad^{-2})	θ_0 (deg)	ρ (Å)
Three-body potential $\frac{1}{2}k_b(\theta-\theta_0)^2\exp(-[r_{12}/\rho + r_{13}/\rho])$			
O–Si–O	100.0	109.47	1.0
O–Al–O	100.0	109.47	1.0
		k_s (eV Å^{-2})	Y (e)
Core–shell potential $1/2k_s r^2$			
O_c–O_s		74.92	−2.8482

were heated and hold at 3,200 K for 100 ps in the NVT ensemble ensuring a suitable melting of the samples. The liquids were then cooled to 300 K at a nominal cooling rate of −5 K/ps. The resulting glass structures were subjected to a final NVT trajectory of 200 ps [2].

2.2 NMR parameters calculation

NMR calculations were carried out with the CASTEP [25] density functional theory (DFT) code using the GIPAW [15] algorithm, which allows the reconstruction of the all electrons wave function in the presence of a magnetic field. The generalized gradient approximation (GGA) PBE [26] functional was employed, and the core–valence interactions were described by ultrasoft pseudopotentials generated on the fly [27]. For ^{17}O, the 2s and 2p orbitals were considered as valence states with a core radius of 1.3 Å; for ^{29}Si and ^{27}Al, a core radius of 1.8 Å was used with 3s and 3p valence orbitals, while for ^{43}Ca, a core radius of 2.0 Å was used with 3s, 3p and 4s valence states.

For the PAW and GIPAW calculations, we used two projectors in each s and p angular momentum channel for O, and in the s, p and d channel for Si, Al and Ca.

It is well known that in the PBE approximation, the energy of Ca 3d orbital is too low and the hybridization with O 2p orbitals is overestimated [17]. As a consequence, the ^{17}O chemical shifts computed with PBE are affected by very large errors (up to 124 ppm) for O sites close to Ca atoms. To overcome this problem, a modified Ca pseudopotential with shifted 3d orbitals of 3.2 eV has been used as proposed by Profeta et al. [17].

Before computing the NMR parameters, constant pressure geometry optimizations of the classical generated models were performed at the Γ point [28]. Wave functions were expanded in plane waves with the kinetic energy cutoff of 700 eV, this has been demonstrated to be long enough to reach energy and NMR chemical shift converged values [10, 29].

The shielding tensor $(\overleftrightarrow{\sigma})$ is defined as the proportionality factor between the induced and the externally applied magnetic fields at the positions of the nuclei:

$$\overleftrightarrow{\sigma}(R) = \vec{B}_{\text{ind}}(R)/\vec{B}, \tag{4}$$

where $\vec{B}_{\text{ind}}(R)$ is the inhomogeneous magnetic field induced by the external field $\vec{B}$ and it can be calculated according to the Biot-Savart law:

$$\vec{B}_{\text{ind}}(\vec{r}) = \frac{\mu_0}{4\pi}\int d^3\vec{r}'\,\frac{\vec{r}'-\vec{r}}{\left|\vec{r}'-\vec{r}\right|^3}\times\vec{j}\left(\vec{r}'\right), \tag{5}$$

where the induced current can be obtained from the first-order perturbation theory [15].

Then, the isotropic chemical shift, δ_{iso}, can be obtained from the isotropic chemical shielding, σ_{iso}:

$$\sigma_{\text{iso}} = {}^1\!/_3 Tr\{\overleftrightarrow{\sigma}\} \tag{6}$$

$$\delta_{\text{iso}} = -(\sigma_{\text{iso}} - \sigma_{\text{ref}}), \tag{7}$$

where σ_{ref} is a reference isotropic shielding. From the principal components of the symmetric part of the chemical shift tensor, the magnitude, δ_{CSA}, and asymmetry, η_{CSA}, of the chemical shift anisotropy (CSA) can be calculated:

$$\delta_{\text{CSA}} = \delta_{ZZ} - \delta_{\text{iso}} \tag{8}$$

$$\eta_{\text{CSA}} = \frac{(\delta_{YY} - \delta_{XX})}{(\delta_{ZZ} - \delta_{\text{iso}})}, \tag{9}$$

where δ_{ii} are the principal components of the symmetric chemical shift tensor such that $|\delta_{ZZ} - \delta_{iso}| \geq |\delta_{YY} - \delta_{iso}| \geq |\delta_{XX} - \delta_{iso}|$.

Nuclei with a spin >½ have quadrupolar parameters related to the traceless electric field gradient tensor $\vec{V}(\vec{r})$:

$$V_{\alpha\beta}(\vec{r}) = \frac{\partial E_\alpha(\vec{r})}{\partial r_\beta} - \frac{1}{3}\delta_{\alpha\beta}\sum_\gamma \frac{\partial E_\gamma(\vec{r})}{\partial r_\gamma}, \qquad (10)$$

where α, β and γ denote the Cartesian coordinates x, y, and z, and $E_\alpha(\vec{r})$ is the local electric field at the position $(\vec{r})$. If the eigenvalues of the EFG tensor are labeled V_{xx}, V_{yy}, V_{zz} so that $|V_{zz}| > |V_{yy}| > |V_{xx}|$, then one can define the quadrupolar coupling constant as

$$C_Q = \frac{eQV_{zz}^{PAF}}{h} \qquad (11)$$

And the asymmetry parameter as:

$$\eta_Q = \frac{V_{xx} - V_{yy}}{V_{zz}} \qquad (12)$$

In this work, the δ_{iso} of ^{27}Al, ^{17}O and ^{43}Ca were evaluated using the shielding references (σ_{ref}) of 555.4, 260.5 and 1,134.1 ppm obtained by α-corundum, α-cristobalite and CaO, respectively, and the experimentally determined quadrupolar moment, eQ, of 146.6, 25.58 and 40.9 mB was used for ^{27}Al, ^{17}O and ^{43}Ca atoms [30].

The simulation of the solid-state MAS and 3QMAS NMR spectra from the CASTEP outputs have been carried out by means of the software fpNMR [10, 31]. For each atomic site i, the NMR spectrum of interest, say $I_i(\vec{v})$, is calculated and then co-added to yield the final NMR spectra $I(\vec{v}) = \sum_i I_i(\vec{v})$. The software allows the summation to be restricted to a given chemical speciation of the considered atom and to merge the spectra of several structural models to increase the statistical data.

Finally, it is worth noting that the aforementioned approach is a common procedure in ab initio calculations of glasses and enables only short-range relaxations at the ab initio level, whereas the medium-range structure, such as the connectivity of the glass-forming sites, is "frozen" to the initial configuration obtained by classical MD.

However, several papers have demonstrated that the present approach is adequate to study the local structure, such as the coordination environment of ions in bioactive glasses [32], and important vibrational features [33–35] and NMR parameters [11, 36].

3 Results and discussion

The structure of alkali or alkaline earth aluminosilicate glasses has since long been understood in terms of a three-dimensional connected structure of $[SiO_4]^{4-}$ and $[AlO_4]^{5-}$ tetrahedra linked at the corners by bridging oxygen (BO) atoms. In this picture, a fully connected network structure can be maintained if the charge deficit created by the substitution of Si^{4+} with Al^{3+} is compensated by 'charge balancing' cations such as Na^+ or Ca^{2+} located close to aluminum.

According to this model, the glass reaches maximum polymerization for a total compensation of charges by the cations, for instance, for calcium the system results fully polymerized at a ratio $[CaO]/[Al_2O_3] = 1$.

Any excess of Ca^{2+} will acquire non-bridging oxygens (NBO) between themselves and a tetrahedral cation (Al, Si), thus interrupting the network connectivity.

However, recent experiments have shown that the compensated continuous random network model is not entirely correct for some aluminosilicate compositions. In particular, both precise measurements of the viscosity of Na- and Ca-aluminosilicate liquids [37] and high-resolution ^{17}O MQMAS experiments [38] indicate that the network in aluminosilicate glasses with a tectosilicate composition (that is, where the ratio $[CaO]/[Al_2O_3]$ is equal to one) is not fully polymerized and that NBO exist in the system. At the same time, to preserve the total number of Al–O and Si–O bonds, the presence of NBO is compensated by oxygen triclusters, also called three bridging oxygens (TBO). That is, oxygen bounded to three Al and/or Si atoms which are fourfold-coordinated.

To elucidate better the local environment around aluminum ions and corroborate recent experimental evidences on the oxygen speciation, the theoretical ^{27}Al and ^{17}O MAS and 3Q-MAS spectra of the structural models generated by classical molecular dynamics simulations will be analyzed in details in the following paragraphs.

3.1 ^{27}Al NMR spectroscopy

Figure 1a shows the ^{27}Al MAS NMR spectrum of the CAS glass obtained as the summation of the NMR spectra of the two structural models containing 250 atoms each. Then, the merged spectrum has been normalized to be comparable to the experimental one which has been digitalized from Ref. [39] and reported in the bottom panel of the figure. Both the experimental and the simulated spectra were collected/simulated at a magnetic field of 11.7 T and at a spinning frequency of 14 kHz.

The simulated ^{27}Al MAS spectrum is in fairly agreement with the experimental one, which shows a chemical shift of about 50 ppm confirming that all the aluminum atoms are found in tetrahedral configuration.

Among the 60 Al atoms present in the two structural models studied only one is fivefold coordinated with $\delta_{iso} = 10.7$ ppm, $\delta_{CSA} = -15.6$ ppm, $\eta_{CSA} = 0.45$,

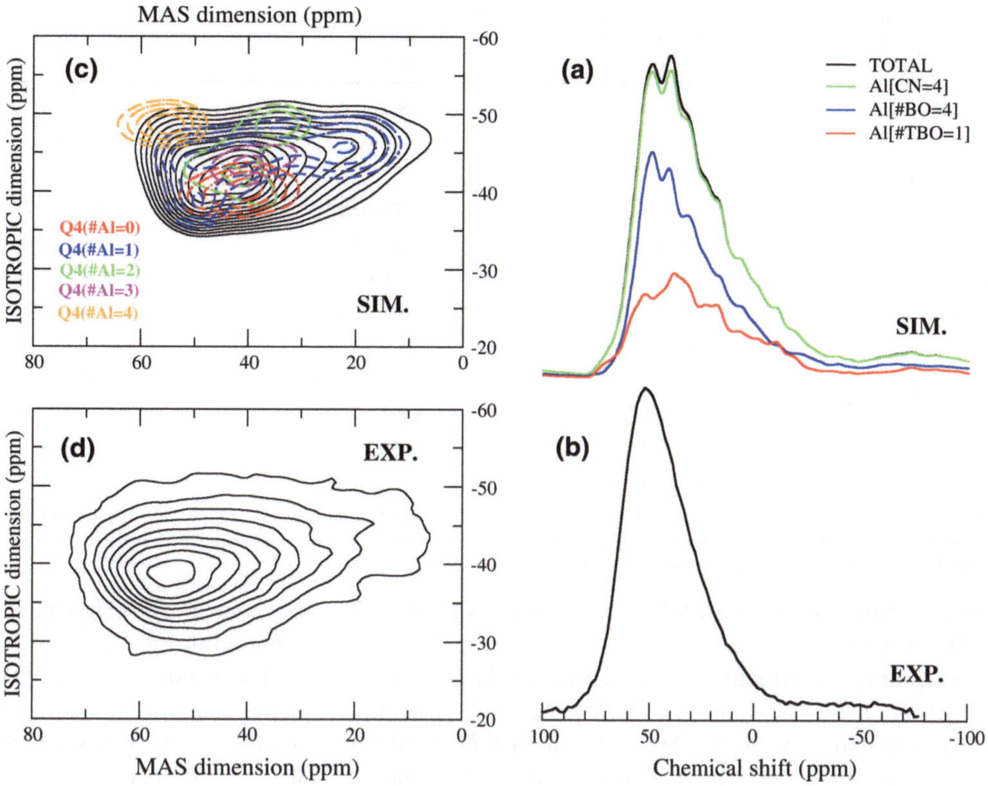

Fig. 1 **a** Simulated and **b** experimental ^{27}Al MAS spectra of the CAS glass. The *green line* represents the spectrum of fourfold-coordinated Al ions without NBO connected with them; the *blue line* represents the spectrum of Al ions connected to 4 BO, while the *red line* represents the spectrum of Al ions connected to 1 TBO and 3 BO. **c** Simulated and **d** experimental ^{27}Al 3Q-MAS spectra of the CAS glass. The signals generated by Al ions with 0, 1, 2, 3 and 4 Al ions in the second coordination sphere are also reported in the theoretical spectrum

$C_Q = 6.5$ MHz and $\eta_Q = 0.51$ and one is a Q^3 species with $\delta_{iso} = 62.9$ ppm. Therefore, penta-coordinated Al atoms, if present in the experimental structure of the glass, should be detectable as a shoulder in the tail of the spectrum. Conversely, the signals coming from Q^3 species could not be easily extracted from the experimental spectra since they range in the same region as for Q^4 species.

Hence, the structure of the glass is governed by aluminum atoms surrounded by 4 oxygens of the BO or TBO type, the spectrum of which is represented as the green line in Fig. 1a.

Despite the good agreement between the simulated and the experimental spectra, some inconsistencies are present. In particular, the latter is narrower than the former probably because of a distorted tetrahedral arrangement of the aluminum atoms in the modeled structure. In fact, although both the mean O–Si–O and O–Al–O bond angles are located at the same value of 109°, the standard deviation is greater for Al (9.9°) than for Si (4.7°). These findings are in nice agreement with previous Car-Parrinello MD simulations carried out by Benoit et al. [40] on aluminosilicate glasses with a tectosilicate composition.

The second noticeable inconsistency in the simulated ^{27}Al MAS spectrum regards the presence of a second peak at around 40 ppm, which is imputable to Al atoms connected to 1 TBO (reported as red line signal in Fig. 1a) besides the signal of Al atoms connected to only 4 BO atoms which is located at around 50 ppm.

Indeed, in the models obtained 40% of the Al atoms have at least 1 TBO in the first coordination sphere, a value that is probably too high. The presence of TBO in the first coordination sphere leads to a non-ideal tetrahedral arrangement because of the reduction of the O–Al–O angles and thus to a greater standard deviation for this angle and a broader ^{27}Al MAS NMR spectrum.

The high amount of TBO found in our models is in agreement with Car-Parrinello MD simulations [40], suggesting that their presence might be a consequence of the small system sizes and the very fast quenching rate that we are forced to use in our computational investigations [2, 41] but inaccuracies due to the empirical force field employed cannot be excluded.

The mean structural features (bond lengths and bond angles) and NMR parameters (isotropic chemical shift, chemical shift anisotropy and quadrupolar coupling constants) of Al ions with 1 or 2 TBO in the first coordination sphere are reported in Table 2. Interestingly, while the isotropic chemical shift does not seem to be affected by the presence of TBO since it lies at around 60–61 ppm, a value close to the experimental value of 66 ppm reported by Angeli et al. [39], the quadrupolar coupling constant (C_Q) increases with the amount of TBO in the first coordination

Table 2 Structural features and computed NMR parameters of aluminum in the CAS glass

	Al	Al (Q^4)	Al (#TBO = 1)	Al (#TBO = 2)
# of atoms	60	58	22	2
Al–O (Å)	1.77 (0.07)	1.77 (0.02)	1.77 (0.04)	1.80
Al–NBO	1.75	–	–	–
Al–BO	1.76 (0.03)	1.76 (0.02)	1.76 (0.04)	1.74
Al–TBO	1.83 (0.05)	1.82 (0.05)	1.82 (0.05)	1.86
O–Al–O (°)	109.2 (7.8)	109.3 (9.0)	109.1 (11.1)	109.4
Al–O–T (°)	132.3 (16.5)	132.4 (16.3)	128.8 (17.9)	124.6
δ_{iso} (ppm)	61.1 (10.3)	61.9 (8.0)	60.4 (13.2)	61.0
δ_{CSA} (ppm)	−8.8 (29.2)	−9.1 (29.6)	−7.4 (30.3)	−19.1
η_{CSA}	0.62 (0.22)	0.63 (0.22)	0.57 (0.24)	0.59
C_Q (MHz)	9.49 (2.8)	9.58 (2.8)	9.86 (3.0)	11.86
η_Q	0.64 (0.23)	0.64 (0.23)	0.61 (0.20)	0.80
P_Q (MHz)	10.14 (9.9)	10.24 (9.9)	10.48 (3.0)	13.0

Standard deviations are reported in parenthesis

sphere from 9.58 to 11.86 MHz for Al atoms with 0 to 2 TBO, respectively.

Regarding the structural features connected to the short-range order of Al ions, the DFT-optimized structural models provide a mean Al–O bond length of 1.77 Å, which is in very good agreement with experimental distances measured in aluminosilicate glasses and crystalline analogous of different compositions [42–46].

The Al–O bond increases with the number of TBO connected to the central Al atom, while the mean Al–O–T angle decreases from 132.4° to 124.6° for Al atoms surrounded by 0 and 2 TBO, respectively.

As expected, the Al–NBO bond distance (1.75 Å) is shorter than the Al–BO one (1.76 Å) which in turn is shorter than the Al–TBO bond distance (1.83 Å).

Further precious information on the medium-range structure and the Al/Si ratio in aluminosilicate glasses can be gained by analyzing the ^{27}Al 3QMAS spectrum as reported in Fig. 1c.

The simulated ^{27}Al 3QMAS spectrum is shown in the upper panel of the figure together with the contributions of tetrahedral Al ions surrounded by $n = 0,\ldots,4$ Al atoms in the second sphere. These units are denoted with the following symbolism [Q4(#Al = n)], while the mean structural and NMR parameters of the different speciation of Al atoms are reported in Table 3.

The detailed analysis of the second coordination sphere of Al ions shows that the medium-range order of the studied glass is governed by Q^4(4Si) (27%), Q^4(3Si,1Al) (40%) and Q^4(2Si,2Al) (22%) units which are depicted in Fig. 2.

Table 3 shows that the isotropic chemical shift increases with the amount of Al ions in the second coordination sphere from 58.7 ppm (for Q^4(4Si)) to 72.1 ppm (for Q^4(4Al)) as well as the quadrupolar coupling constant C_Q

Table 3 Mean structural features and computed NMR parameters of the Q^4 species of aluminum depending on the network former present in the second coordination sphere

	Q^4(4Si)	Q^4(3Si,1Al)	Q^4(2Si,2Al)	Q^4(1Si,3Al)	Q^4(4Al)
# of atoms	16	24	13	4	3
Al–O (Å)	1.77 (0.01)	1.77 (0.02)	1.78 (0.05)	1.78 (0.03)	1.76 (0.02)
Al–BO	1.76 (0.01)	1.76 (0.02)	1.76 (0.05)	1.77 (0.02)	1.75 (0.01)
Al–TBO	1.91	1.85 (0.06)	1.81 (0.08)	1.82 (0.07)	1.78 (0.02)
O–Al–O (°)	109.4 (8.3)	109.2 (9.5)	108.9 (12.8)	109.0 (8.8)	109.4 (7.5)
Al–O–T (°)	136.3 (14.7)	134.3 (17.8)	129.9 (14.3)	122.6 (15.9)	124.6 (11.3)
δ_{iso} (ppm)	58.7 (5.8)	61.0 (8.9)	60.7 (16.6)	64.4 (1.5)	72.1 (6.6)
δ_{CSA} (ppm)	−2.5 (24.0)	−9.8 (32.0)	−13.5 (31.0)	−22.0 (28.0)	3.5 (24.7)
η_{CSA}	0.67 (0.14)	0.65 (0.22)	0.51 (0.14)	0.56 (0.16)	0.74 (0.24)
C_Q (MHz)	8.61 (2.6)	10.05 (2.9)	10.17 (3.4)	8.71 (1.6)	7.7 (1.4)
η_Q	0.62 (0.12)	0.63 (0.23)	0.60 (0.18)	0.76 (0.19)	0.77 (0.14)
P_Q (MHz)	9.18 (2.6)	10.73 (3.0)	10.80 (3.5)	9.55 (3.5)	8.46 (1.4)

Standard deviations are reported in parenthesis

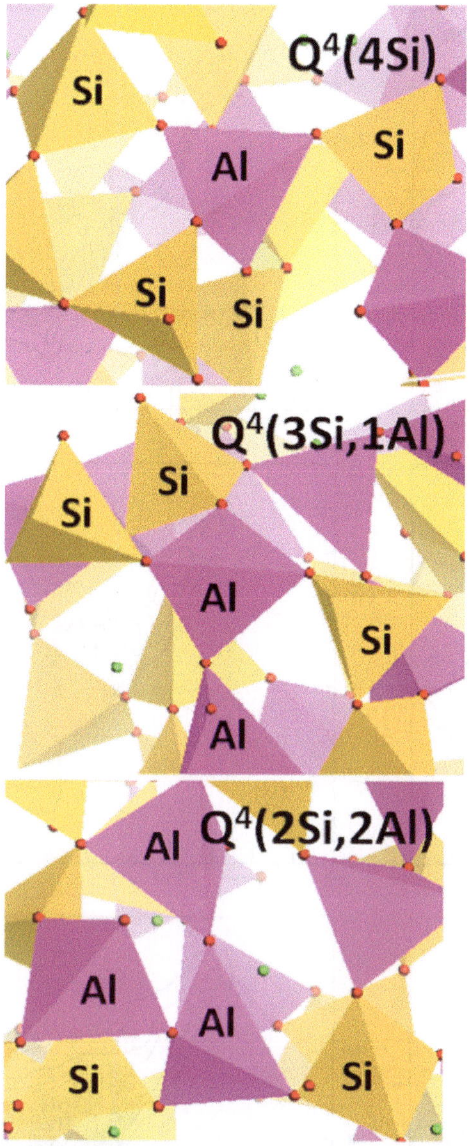

Fig. 2 Local environment (medium-range order) of each AlO_4 tetrahedron found in the CAS glass. *Violet tetrahedral* represents AlO_4 units, while the *yellow* ones represent SiO_4 units

which increases from 8.6 MHz (for Q^4(4Si)) to 10.2 MHz (for Q^4(2Si,2Al).

The presence of Al ions in the surrounding of AlO_4 units gives rise to a distortion in the tetrahedra as highlighted by the anisotropic chemical shielding (δ_{CSA}) whose absolute value increases from 2.5 ppm for Q^4(4Si) to 22 ppm Q^4(1Si,3Al) and by the standard deviation of the O–Al–O bond angle which increases from 8.8° to 12.5° for Q^4(4Si) and Q^4(2Si,2Al), respectively.

Another important structural parameter that is usually found to affect the ^{27}Al isotropic chemical shift is the average Al–O–T bond angle, where T is the tetrahedral atom Si or Al [47]. In Fig. 3, the ^{27}Al δ_{iso} is plotted against the mean

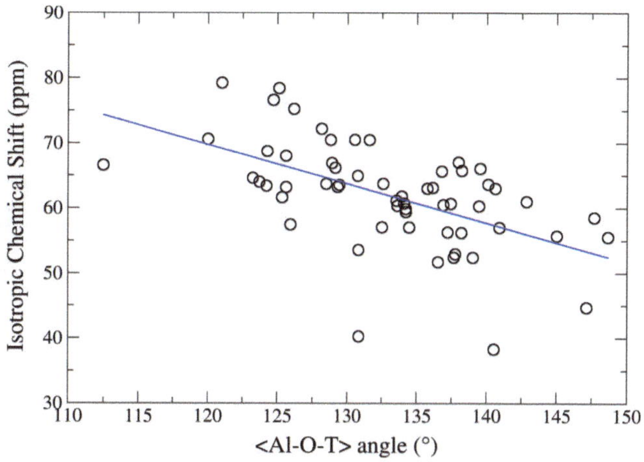

Fig. 3 Simulated ^{27}Al isotropic chemical shift as a function of the mean <Al–O–T> angle

<Al–O–T> angle for the CAS glass studied. Linear regression analysis of the data as $\delta_{iso} = a \cdot \langle Al - O - T \rangle + b$ points yields a slope of −0.60 and intercept of 142.27° with a poor correlation coefficient ($R^2 = 0.303$).

As it has been underlined in a recent paper by some of us [10], the presence of Ca ions in the glass structure yields worst correlations with respect to that obtained when the modifier is an alkaline ion such as Na or Li for which the linear fit yields R^2 values in the range 0.7–0.9 [36]. However, notwithstanding the poor correlation coefficient, it can be noticed that by increasing Al–O–T bond angle, the ^{27}Al chemical shift decreases, thus confirming the results of previous experimental and theoretical studies regarding the relationship between δ_{iso} and Al–O–T bond angle [48, 49].

The rationale behind the correlation between the shielding properties of Al atoms and the values of the Al–O–T angle has been unveiled by Liu et al. [47] in their ab initio NMR shielding calculations, natural bonding orbital analysis and natural chemical shielding analysis performed on model clusters with different Al–O–T angles. Based on Hartree–Fock calculations, they demonstrated that the opening of the Al–O–T bond angle leads to a slight increase in the population of the Al–O and Si–O bond orbital electrons and a dramatic change in the bond orbital shapes and hybridization (with an increase in the s character and a decrease in bond bending). Moreover, the movement of one lone pair on the oxygen toward Al or Si contributes to the shielding of both Al and Si.

3.2 ^{17}O NMR spectroscopy

Oxygen-17 NMR is an ideal method for the direct investigation of whether NBO and TBO are indeed present in aluminum silicate glasses in which the [CaO]/[Al_2O_3] ratio is equal to one. In a pioneering work, Stebbins and Xue

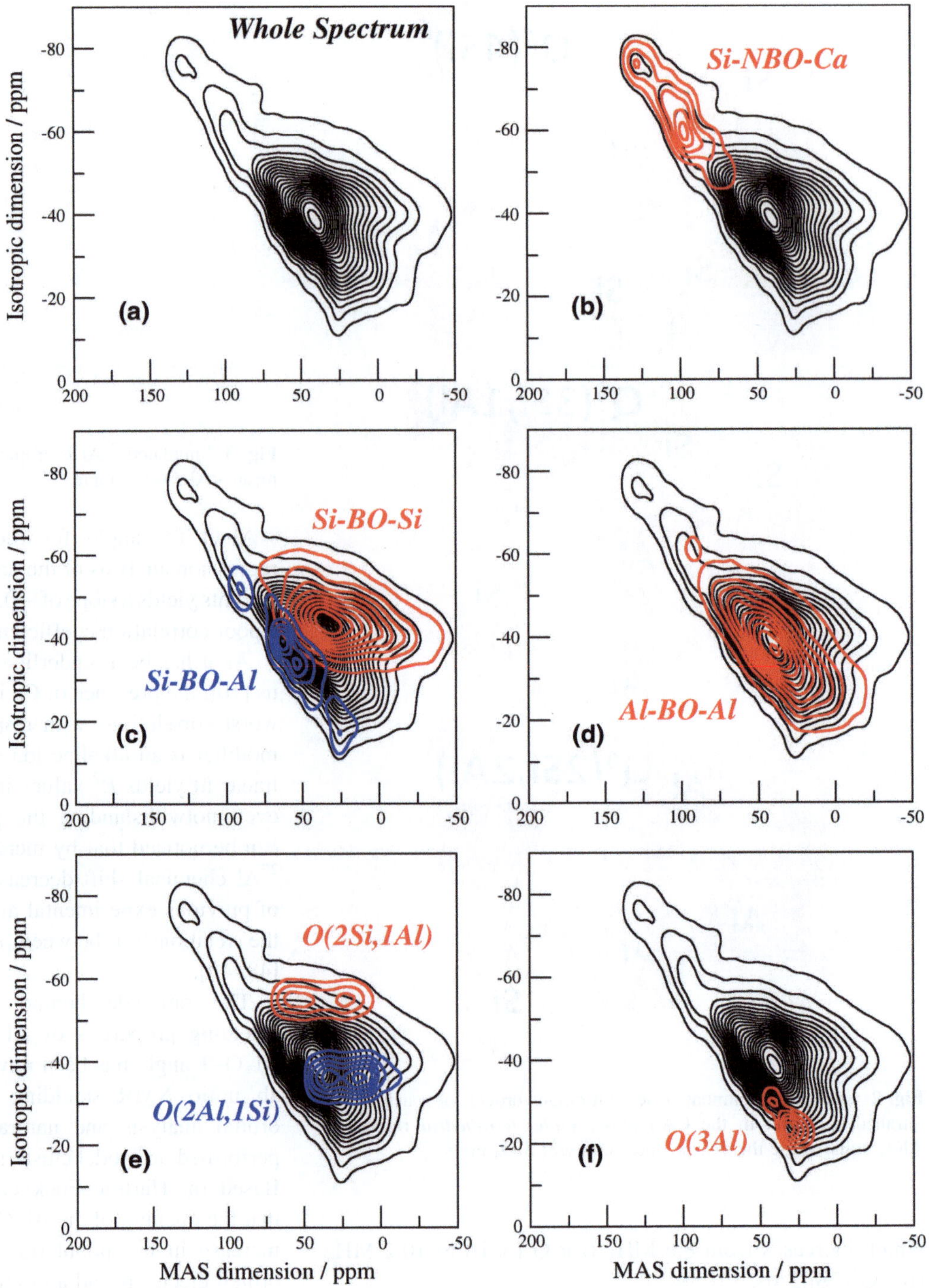

Fig. 4 **a** Simulated oxygen-17 3QMAS NMR spectrum at 11.75 T of the CAS glass. The signals of the different oxygen speciation are also superimposed in the panels from **b** to **f**

[38] found that the 3QMAS spectrum of the $CaAl_2Si_2O_8$ was composed essentially by four peaks denoted as A, B, C and D. Peak A was associated with NBO atoms whose isotropic chemical shift (δ_{iso}) and quadrupolar coupling constant (C_Q) were calculated to be 113 ± 2 ppm and 2.9 ± 0.2 MHz. The peak B, assigned to Si–O–Al bonds, was the main one with $\delta_{iso} = 61$ ppm and $C_Q = 3.5$ MHz, while peak C (50 ppm, 5.5 MHz) was associated with Si–O–Si bonds. Finally, they detected a small peak, called D, with $\delta_{iso} = 20$ ppm and $C_Q = 2.3$ MHz, which was supposed to be the signal arising from OAl_3 or OAl_2Si triclusters, in which Al remains in four-coordination.

To help in the interpretation of the experimental spectrum, the simulated ^{17}O 3QMAS NMR spectrum is reported in Fig. 4, together with the spectra generated for the different oxygen speciation present in our models (see panels from b to f).

It is worth to highlight here that albeit the glass studied in this paper has a different composition with respect to that investigated by Stebbins et al. [38] both are

Table 4 Computed NMR parameters and population of oxygen speciation

	δ_{iso} (ppm)	C_Q (MHz)	η_Q	Population (%)
OSi_2	61.0 (10.9)	5.18 (0.54)	0.41 (0.26)	28.5
OAl_2	62.6 (17.7)	2.26 (0.43)	0.62 (0.25)	6.4
OSiAl	58.2 (15.6)	3.80 (0.46)	0.53 (0.27)	55.8
OSi	110.9 (16.1)	2.85 (0.45)	0.44 (0.25)	5.4
OAl	159.5	1.18	0.45	0.3
OSi_2Al	77.4 (1.1)	5.53 (0.19)	0.32 (0.19)	0.6
$OSiAl_2$	50.1 (2.8)	4.66 (0.32)	0.49 (0.27)	1.1
OAl_3	41.9 (11.9)	3.15 (0.71)	0.38 (0.21)	1.9

Standard deviations are reported in parenthesis

tectosilicate glasses with a unitary $[CaO]/[Al_2O_3]$ ratio and thus the same type of structural units (NBO and TBO) is present in the structure. Although the site populations might be slightly different leading to differences in the shape of the NMR spectrum, the peak positions of these important structural features are only slightly affected by the glass composition and the comparison of the ^{17}O NMR spectra between the two glasses is justified.

The computed mean NMR parameters (δ_{iso}, C_Q, η_Q) and the population of the T-NBO, T–BO–T and TBO structural units are reported in Table 4. Hopefully, these values might be used to guide the fitting of the experimental ^{17}O 3QMAS spectrum for resolving and quantifying better the different structural units present in the glass in future works.

It is worth noticing that both the simulated spectrum and the NMR parameters of the different species are in good agreement with those reported by Stebbins and Xue [38].

In particular, the structural model obtained in this work contains 5.4% of NBO connected to Si atoms with $\delta_{iso} = 111$ ppm and $C_Q = 2.85$ MHz and 0.3% of NBO bounded to Al whose NMR parameters (159 ppm, 1.2 MHz) are very different with respect to the ones of NBO–Si units. Therefore, if present in the experimental structure, they should be easily detectable by both ^{17}O MAS and 3QMAS NMR experiments; however, so far, no experimental evidence has been found regarding the presence of these structural units in glasses with a tectosilicate composition as that studied in this work.

Among the bridging oxygens, 55.4% of them form Si–O–Al units (58.2 ppm, 3.8 MHz), 28.5% are present as Si–O–Si units (61 ppm, 5.2 MHz) and 6.4% as Al–O–Al units (62.6 ppm, 2.3 MHz). These units have similar isotropic chemical shifts but different C_Q values, which make the Si–O–Al and Si–O–Si units quite distinguishable in the bi-dimensional spectrum. Instead, the Al–O–Al peak that accompanies Al/Si disorder in the glass is hidden by the Si–O–Al and Si–O–Si peaks (see Fig. 4c, d).

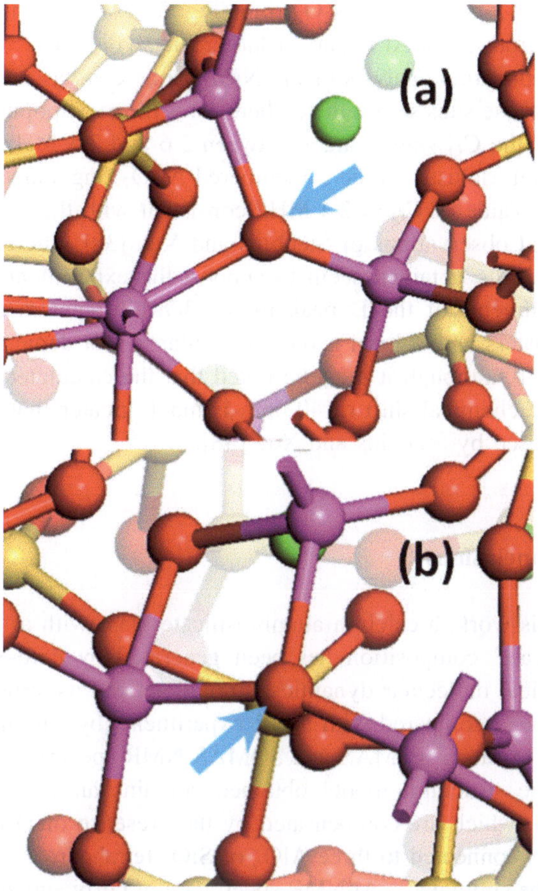

Fig. 5 Local geometry of OAl_3 sites found in the simulated glass. The oxygen tricluster is highlighted by a *light blue arrow*. **a** Al atoms (*violet spheres*) contribute to the formation of large rings. **b** Al atoms form a 2-membered Al_2O_2 ring

As already reported in previous Car-Parrinello molecular dynamics simulations of CAS glasses with similar compositions [40], also in our structural models, the excess number of NBO is compensated by the existence of OAl_2Si (1.1%), $OAlSi_2$ (0.6%) and OAl_3 (1.9%) tricluster oxygens whose contributions to the 3QMAS spectrum are shown in Fig. 4e, f.

Figure 4e and Table 4 clearly reveal that the NMR parameters of the OAl_2Si (50.1 ppm, 4.66 MHz) and $OAlSi_2$ (77.4 ppm, 5.53 MHz) are in the same range covered by the Si–O–Si and Si–O–Al sites, and since these triclusters are present in the glass in very small quantities, it is very unlikely that 3QMAS NMR could detect them. Conversely, the NMR parameters of the OAl_3 triclusters are close to the edge of the main Si–O–Al peak and their position (see Fig. 4f) is in nice agreement with the peak D observed by Stebbins et al. [38].

The five OAl_3 sites found in the simulated glass structure are present in two structural motifs shown in Fig. 5; in particular, in one of these OAl_3 sites, two of the Al atoms form a 2-membered Al_2O_2 ring (Fig. 5b), whereas in the other four sites, Al is part of larger rings (Fig. 5a).

A punctual analysis of the NMR parameters associated with these sites shows that, whereas for the most populated motif the C_Q value ranges between 2.6 and 3.5 MHz, the oxygen tricluster in the 2-membered Al_2O_2 ring shows the small value of $C_Q = 2.4$ MHz consistent with the experimental observations of Stebbins and Xue [38]. Therefore, the results obtained seem to support the experimentalist' assignments of the D peak in the 3QMAS spectra to O triclusters with a local geometry similar to that depicted in Fig. 5b, although it must be noted that the calculated isotropic chemical shift of 40 ppm is much greater than that proposed by Stebbins and Xue [38].

4 Conclusions

In this work, a calcium aluminosilicate glass with a 'tectosilicate' composition has been generated by core–shell classical molecular dynamics simulations and its structure has been compared with NMR experiments by simulating the ^{27}Al and ^{17}O MAS and 3QMAS NMR spectra.

The structural model obtained contains an excess of NBO, which are compensated by the presence of O triclusters connected to three AlO_4 or SiO_4 tetrahedra.

The simulated ^{27}Al NMR spectra are in fairly agreement with the experimental ones and they confirm that all the aluminum atoms are found in tetrahedral configuration. However, it seems that the structural model obtained in this paper overestimates the amount of Al atoms (40%) connected to at least 1 TBO in their first coordination sphere. The presence of TBO leads to a non-ideal tetrahedral geometry around Al and increases the value of the quadrupolar coupling constant C_Q. As a consequence, the MAS NMR spectra result broader with the appearance of a double peak not detected experimentally.

The detailed analysis of the second coordination sphere of Al ions shows that the medium-range order of the studied glass is governed by Q^4(4Si) (27%), Q^4(3Si,1Al) (40%) and Q^4(2Si2Al) (22%) species. Both the isotropic chemical shift and C_Q increase with the amount of Al ions in the second coordination sphere.

The analysis of the oxygen site populations and of the ^{17}O NMR spectra support the experimentalist' assignments of the peaks in the 3QMAS spectra, reported by Stebbins.

In particular, the structural model obtained in this work contains 5.4% of NBO connected to Si atoms, 55.4% of the oxygens form Si–O–Al units, 28.5% are present as Si–O–Si units, 6.4% are present as Al–O–Al units and 3.6% as O triclusters. Among the O triclusters, only those in the 2-membered Al_2O_2 rings shows the small value of $C_Q = 2.4$ MHz consistent with the experimental observations of Stebbins and Xue.

Acknowledgments The authors thank the Italian Ministry of University and Research for funding (Project COFIN2008, prot. 2008J9RNB3 "Integrazione Temporale per l'Evoluzione Molecolare"). A. P. thanks Dr. Thibault Charpentier of the CEA, IRAMIS, Service Interdisciplinaire sur les Systémes Moléculaires et Matériaux (Paris, France) for fruitful discussions.

References

1. Massiot D, Fayon F, Montouillout V, Pellerin N, Hiet J, Roiland C, Florian P, Coutures J-P, Cormier L, Neuville DR (2008) J Non-Cryst Solids 354(2–9):249–254
2. Pedone A (2009) J Phys Chem C 113:20773–20784
3. Stebbins JF (1995) In: Stebbins JF, McMillan PF, Dingwell DB (eds) Reviews in mineralogy, vol 32. MSA, Washington DC, p 191
4. Dupree E, Pettifer RF (1984) Nature (London) 308:523
5. Linati L, Lusvardi G, Malavasi G, Menabue L, Menziani MC, Mustarelli P, Pedone A, Segre U (2008) J Non-Cryst Solids 354:84–89
6. Maekawa H, Maekawa T, Kawamura K, Yokokawa T (1991) J Non-Cryst Solids 127:53
7. Voigt U, Lammert H, Eckert H, Heuer A (2005) Phys Rev B 72:64207
8. Frydman L, Harwod JS (1995) J Am Chem Soc 117:5367–5368
9. Medek A, Harwood S, Frydman L (1995) J Am Chem Soc 117:12779–12787
10. Pedone A, Charpentier T, Menziani MC (2010) Phys Chem Chem Phys 12:5064–6066
11. Pedone A, Charpentier T, Malavasi G, Menziani MC (2010) Chem Mater 22:5644–5652
12. Charpentier T, Ispas S, Profeta M, Mauri F, Pickard CJ (2004) J Phys Chem B 108:4147–4161
13. Pedone A, Pavone M, Menziani MC, Barone V (2008) J Chem Theory Comput 4:2130–2140
14. Tossell JA, Horbach J (2005) J Phys Chem B 109:1794–1797
15. Pickard CJ, Mauri F (2001) Phys Rev B 63:245101
16. Profeta M, Mauri F, Pickard CJ (2003) J Am Chem Soc 125:541–548
17. Profeta M, Benoit M, Mauri F, Pickard CJ (2004) J Am Chem Soc 126:12628–12635
18. Tielens F, Gervais C, Lambert JF, Mauri F, Costa D (2008) Chem Mater 20:3336–3344
19. Ferlat G, Charpentier T, Seitsonen AP, Takada A, Lazzeri M, Cormier L, Calas G, Mauri F (2008) Phys Rev Lett 101:065504
20. Dick G, Overhauser AW (1958) Phys Rev 112:90
21. Pedone A, Corno M, Civalleri B, Malavasi G, Menziani MC, Segre U, Ugliengo P (2007) J Mater Chem 17:2061–2068
22. Tilocca A, de Leeuw NH, Cormack AN (2006) Phys Rev B 73:104209
23. Smith W, Forester TR (1996) J Mol Graph 14:136
24. Pedone A, Malavasi G, Menziani MC (2009) J Phys Chem C 113:15723–15730
25. Segall MD, Lindan PJD, Probert MJ, Pickard CJ, Hasnip PJ, Clarck SJ, Payne MC (2002) J Phys: Condens Matter 14:2717
26. Perdew JP, Burke K, Ernzerhof M (1996) Phys Rev Lett 77:3865
27. Yates JR, Pickard CJ, Mauri F (2007) Phys Rev B 76:024401
28. Monkhorst H, Pack JD (1976) Phys Rev B 13:5188
29. Choi M, Matsunaga K, Oba F, Tanaka I (2009) J Phys Chem C 113:3869–3873

30. Pyykkò P (2001) Mol Phys 99:1617
31. Charpentier T, Kroll P, Mauri F (2009) J Phys Chem C 113:7917–7929
32. Malavasi G, Menziani MC, Pedone A, Civalleri B, Corno M, Ugliengo P (2007) Theor Chem Acc 117:933–942
33. Tilocca A (2007) Phys Rev B 76:224202
34. Corno M, Pedone A, Dovesi R, Ugliengo P (2008) Chem Mater 20:56105621
35. Corno M, Pedone A (2009) Chem Phys Lett 476:218–222
36. Angeli F, Villain O, Schuller S, Ispas S, Charpentier T (2011) Geochim Cosmochim Acta 75:2453–2469
37. Toplis MJ, Dingwell DB (1996) Trans Am Geophys Un 77:F848
38. Stebbins JF, Xu Z (1997) Nature 390:60–62
39. Angeli F, Gaillard M, Jollivet P, Charpentier T (2007) Chem Phys Lett 440:324–328
40. Benoit M, Profeta M, Mauri F, Pickard CJ, Tuckerman ME (2005) J Phys Chem B 109:6052–6060
41. Pota M, Pedone A, Malavasi G, Durante C, Cocchi M, Menziani MC (2010) Comput Mater Sci 47:739
42. Brown GE, Gibbs GV, Ribbe PH (1969) Am Miner 54:1044
43. Taylor M, Brown GE (1979) Geochim Cosmochim Acta 43:61
44. Himmel B, Weigelt J, Gerber T, Nofz M (1991) J Non-Cryst Solids 136:27
45. Petrov V, Gerber T, Himmel B (1998) Phys Rev B 58:11982
46. Petrov V, Billinger SJL, Shastri SD, Himmel B (2000) Phys Rev Lett 85:3436
47. Liu Y, Nekvasil H, Tossell JA (2005) J Phys Chem A 109:3060–3066
48. Farnan I, Grandinetti PJ, Baltisberger JH, Stebbins JF, Werner U, Eastman MA, Pines A (1992) Nature 358:31
49. Tossell JA (2001) Rev Miner 42:435

Theor Chem Acc (2012) 131:1150
DOI 10.1007/s00214-012-1150-x

REGULAR ARTICLE

On the kinetics and thermodynamics of S–X (X = H, CH_3, SCH_3, $COCH_3$, and CN) cleavage in the formation of self-assembled monolayers of alkylthiols on Au(111)

Madhavan Jaccob · Gopalan Rajaraman · Federico Totti

Received: 14 July 2011 / Accepted: 13 September 2011 / Published online: 18 February 2012

Abstract Abounding potential technological applications is one of the many reasons why adsorption of aliphatic thiols on gold surface is a subject of intense research by many research groups. Understanding and exploring the nature of adsorbed species, the site of adsorption and the nature of interaction between adsorbed species and the gold surface using experimental and theoretical investigations is an active area of pursuit. However, despite a large number of investigations to understand the atomistic structures of thiols on Au(111), some of the basic issues are still unaddressed. For instance, there is still no clear information about the mechanism of adsorption of alkylthiol on gold surface. Furthermore, the reactivity and mechanism of adsorption of alkylthiol is likely to differ when gold adatoms and/or vacancies in the gold layers are considered. In this work, we have tackled these issues by computing the stationary states involved in the thiols adsorption in order to shed light on the kinetics aspects of adsorption process. In this respect, we have considered a variety of thiols into consideration such as methylthiol, dimethylsulfide, dimethyldisulfide, thioacetates, and thiocyanates. We have also considered the cleavage mechanism in the clean and the reconstructed surface scenario and the structure, energetics and spin densities have been computed using electronic structure calculations. For all the studied cases, an homolytic cleavage of CH_3S–X (X = H, CH_3, SCH_3, CN, and $COCH_3$) bond has been found to occur upon adsorption on the gold surface.

Keywords DFT · Thiols · Kinetics and the mechanism of adsorption · Gold surface

Dedicated to Professor Vincenzo Barone and published as part of the special collection of articles celebrating his 60th birthday.

Electronic supplementary material The online version of this article (doi:10.1007/s00214-012-1150-x) contains supplementary material, which is available to authorized users.

M. Jaccob · G. Rajaraman (✉)
Department of Chemistry,
Indian Institute of Technology Bombay,
Powai, Mumbai 400076, India
e-mail: rajaraman@chem.iitb.ac.in

F. Totti (✉)
Dipartimento di Chimica, Polo Scientifico,
INSTM Università degli Studi di Firenze,
via della Lastruccia 3, 50019 Sesto fiorentino, Italy
e-mail: federico.totti@unifi.it

1 Introduction

In recent years, tremendous efforts have been put forward to develop the next generation electronic devices based on the organic/inorganic moieties with high conductivity and optical properties layered on gold surface as self-assembled monolayers (SAMs) [1–9]. Adsorption of aliphatic thiols on gold surface is a subject of interest pursuing their potential applications in nanoproduction, spintronics, biological sensoring, magnetic memory banks, etc. [10–15]. These potential applications of SAMs of thiols on gold surface motivate many researchers to explore the nature of adsorbed species, site of adsorption and the nature of interaction between adsorbed species and the gold surface using experimental and theoretical investigations [36–54]. Earlier experimental and theoretical report suggested that the SAMs of methylthiol preferred to adopt a ($\sqrt{3} \times \sqrt{3}$)R30° structure and may manifest c(2 × 4) superstructure upon varying the chain length; the S head group seems to predominately attack the fcc (face-centered cubic) site of Au(111) surface [16–21]. However, the adsorption site is still a matter of great controversy as many possibilities have been lately suggested, including a surface reconstruction and presence of defects on the Au(111) surface leading to the presence of Au adatoms and/or vacancies in the surface monolayers [23–32].

For a long time, a combined experimental and DFT methods unequivocally suggested that adsorption on the bridge site is on top of gold atoms slightly shifted to the fcc hollow if defects are not taken into consideration [33–35]. However, in recent years this perception has changed as surmounting experimental and theoretical evidences suggest that a sandwich complex of two thiols and one gold adatom is the most preferred structure for a low concentration coverage as it is evident from the low temperature STMs and DFT calculations [36, 37, 42, 43]. Undoubtedly, among a large varieties of thiols reported, studies on methyl thiols (MeSH) are the most common [22–54]. Despite a large number of investigations to understand the atomistic structures of thiols on Au(111), some of the basic issues are still unaddressed. One of the outstanding issues in this respect is the cleavage of S–H bonds of methylthiol (MeSH) upon adsorption. Based on electrochemical data, Schlenoff et al. [55] reported that the adsorption of alkylthiols on the gold surface occurs exothermically (reaction energy is -5.5 kcal/mol) with evolution of H_2. This was further supported by XPS, Fourier transform infrared (FTIR) spectroscopy, Fourier transform mass spectrometry, electrochemistry, Raman spectroscopy, and theoretical methods [55–68]. More conclusive evidences have been emerged recently on the release of H_2 by the S–H bond cleavage on various 4-nitrophenylthiols during adsorption on gold surface [69]. A very recent DFT study evaluating the thermodynamics of S–H cleavage reveals that the S–H bond preferentially cleaves with formation of H_2 rather than an Au–H species [70]. On the other hand, there are reports contradicting the cleavage of S–H bond, especially at low coverage regime. Such studies, in fact, reveal that the S–H bond is intact on the clean gold surface [63, 71, 72]. Although theoretical studies in general support the homolytic cleavage of S–H bond [73], the conclusion from calculations was derived solely from thermodynamic stability data without addressing the kinetic barrier involved in the S–H cleavage. Since a prohibitive barrier to reach the transition state will naturally eliminate the possibility of the S–H cleavage, the evaluation of the kinetic barrier height for the S–H cleavage is mandatory to prove/disprove the S–H cleavage in SAMs. Here, we aim to address this issue through quantum chemical calculations on model complexes of SAMs of thiols.

Additionally, we have also considered other variants of thiols routinely used in experiments to probe the S–X (X = H, CH_3, SCH_3, CN, and $COCH_3$) bond cleavage in order to compare them with the S–H case. In fact, among the presented cases above, the S–X bond cleavage is still controversial in some cases (X = Me) [74–77] while it has been unequivocally proved to be cleaved in others (X = SCH_3, $COCH_3$ and CN) [78–92]. In the case of dialkylsulfides (CH_3SCH_3), a combined spectroscopic data suggest that there is no cleavage of S–C bond and sulfur interacts with the metal surface through a dative bond during the formation of SAM [74–77]. SAMs formed from dialkylsulfides are in general found to be less stable than those obtained from thiols and disulfides [93–104]. In contrast to this, Porter et al. [103] reported that S–C bond cleavage occurs in organosulfides, resulting the SAMs that are exactly similar to those of disulfides and thiols. On the other hand, Schlenoff et al. [55] reported that based on the coverage measurements, S–C bond cleavage is minimal. At the same time, both experimental and theoretical studies on other alkyl thiols, such as dimethyldisulphide (DMDS), reveal that the S–S bonds cleaves homolytically leaving a MeS· radical to be adsorbed firmly on the surface. The kinetics of such cleavage has been experimentally measured for DMDS indicating activation energy of 65 kJ/mol [105, 106].

In order to overcome the oxidizing ability of alkylethiol under the conditions of air, light, and moisture [86, 107–110], alternative precursors based on the organic thiocyanates [78–84] and alkylthioacetates [85, 86, 88–90, 92, 93, 111, 112] of gold thiolate assemblies have been synthesized. During the chemisorption of alkylthiocyanates on gold surface, Au–S bond is formed through removal of cyano group in the form of $[Au(CN)_2]^-$ via surface-mediated reduction of the thiocyanate [80, 85]. In contrast to the thiocyanates, chemisorption of the thioacetates on the Au(111) surface is believed to be occurring through a polar transition state associated with a markedly slower rate than that observed for the alkylthiols [86, 90].

In this framework, we have recently explored [73] the thermodynamic aspects of methyl and various substituted thiol molecules using periodic mixed Gaussians-plane waves DFT calculations [73–114]. Our calculations suggest that homolytic cleavage is energetically more favorable and reconstructed Au(111) surface models are energetically favored over clean unreconstructed Au(111) surfaces. Computations using TPSS functional [73, 113] clearly indicate that an energetic gain of -42.7 kcal/mol is obtained when chemisorption of $CH_3S^{\cdot}$ occurs on Au(111) surface. Moreover, the dissociative mechanism occurs with an energy of -7.2 kcal/mol when the formation of H_2 molecule during the adsorption process is taken into account. Despite the availability of many experimental and theoretical reports on the thermodynamic aspects of various alkylthiols on reconstructed and clean Au(111) surface, there is no clear information about the mechanism of adsorption of alkylthiol on gold surface. Furthermore, the reactivity and mechanism of adsorption of alkylthiol is likely to differ when gold adatoms and/or vacancies in the gold layers are considered. Here, we are planning to tackle these issues by computing the transition states of the S–X bond cleavage to shed light on the kinetics aspects of adsorption process. In this respect, we have considered a variety of thiols into consideration such as methylthiol, dimethylsulfide, DMDS, thioacetates, and thiocyanates. We have considered the cleavage mechanism in both clean and reconstructed surface scenario and discussed the structure, energetics, and spin densities using electronic structure calculations. In addition to kinetics, the formation energies for the most stable radical species have also been computed and discussed.

2 Computational details

All the calculations were performed using Jaguar suit of packages [115] with a hybrid B3LYP functional [116, 117] with a LACVP basis set [118–120]. LACVP basis set comprises the LANL2DZ [121–123] basis set for gold and 6–31G basis for other atoms. Considering computational cost in mind in calculating transition states, a single gold layer model of appropriate size varying from 16 to 20 gold atoms was chosen to mimic the monolayer arrangements. Two optimization procedures were attempted: at the first approximation, we have relaxed all the atoms (full optimization acronym as Fopt) and the corresponding transition states have single imaginary frequency that corresponds to a desired reaction coordinate on the potential energy surface. A slight distortion of Au(111) surface is detected in this procedure, and this is much probably due to the lack of more layers on our model systems. Therefore, we have decided to freeze the first layer of gold atoms as expected in SAM arrangement (partial optimization acronym as Popt). The transition state computed using this procedure, however, had multiple imaginary frequencies due to frozen atoms in the coordinates. The transition states in these cases were characterized by first largest negative frequency, and their nature was graphically verified by using Molden software [124]. Additionally, to mimic the surface scenario, we have decided to add point charges to the explicit gold atoms in order to have a three-layer model with total 192 gold atoms (in total, 192 gold point charges have been added to the existing 16 gold atom models). For the computation of the point charges, a periodic calculation using CP2K program [125] was performed. CP2K is a hybrid basis set formalism known as Gaussian and Plane Wave Method (GPW where the Kohn–Sham orbitals are expanded in terms of contracted Gaussian type orbitals (GTO), while an auxiliary plane wave basis set is used to expand the electronic charge density. A double-ζ GTH basis set and their relativistic norm-conserving pseudo potentials (Goedecker, Teter, and Hutter) [126–128] were used in addition to a plane wave basis set with an energy cutoff of 350 Ry. The Mulliken charges computed by this procedure were then taken to pursue computation of transition states with Jaguar suite of programs. Free energy corrections were computed at 298.15 K.

In order to have another and more reliable check of the reliability of the energetics computed using one-layer model, we have compared the reaction energies of the archetypal species MeSH and MeSMe in clean and reconstructed scenario computed using periodic DFT calculations performed with a unit cell containing three layers of gold atoms [75]. The compared results are reported in Table S1. More computational details about these calculations are discussed in Ref. [75]. The computed reaction energies with both approaches give a consistent trend and a pretty constant deviation providing confidence on our computed energetics using the single-layer model.

3 Results and discussion

3.1 Studies on methylthiol

The chemisorption of methylthiol on gold surface has been modeled by a single layer of Au(111) with 16 gold atoms. Table 1 lists the activation energy of S–H cleavage and reaction energy of the adsorption of methylthiol on gold surface obtained using various theoretical approach listed above. The energy values (barrier height, $\Delta G^{\#}$, and reaction energy, ΔG_r) were computed using the following equations unless otherwise mentioned,

$$\Delta G^{\#} = G^{\#}(\mathrm{RS}\cdots X - \{\mathrm{Au_{ad}}\}\mathrm{Au}(111)) - G(\mathrm{RSX} - \{\mathrm{Au_{ad}}\}\mathrm{Au}(111))$$

$$\Delta G_r = \left[G(\mathrm{RS}^{\cdot} - \{\mathrm{Au_{ad}}\}\mathrm{Au}(111)) + G\left(\frac{1}{2}X_2\right)\right] - G(\mathrm{RSX} - \{\mathrm{Au_{ad}}\}\mathrm{Au}(111))$$

where $\{\mathrm{Au_{ad}}\}$ indicates the presence of one gold adatom and X = H, CH_3, SCH_3, CN, and $COCH_3$.

Selected structural parameters of the optimized structures reported in Fig. 1 are presented in Table 2. The barrier height, $\Delta G^{\#}$, for homolytic cleavage of methylthiol on gold surface is found to be 45.5 kcal/mol, and the free energy of the reaction, ΔG_r, is found to be endothermic by 14.7 kcal/mol. Since all atoms are relaxed, the layer is no longer rigid resulting in a slight deviation from the planar Au(111) surface expected otherwise.

This implicates the shortcoming of the model employed. Being unable to increase the thickness of the gold slab and

Table 1 Activation and reaction energies of the dissociation of MeSH on the clean and reconstructed surface

	Clean surface			Reconstructed surface
	Fopt	Popt	Popt-dimeric	Popt
ΔG[#]	45.5	72.6[a]	20.7	55.4
ΔG_r	14.7	10.1	−14.9	0.2

[a] Upon partial charge, the barrier reduces to 42.4 kcal/mol. *Fopt* Full optimization, *Popt* partial optimization, *Popt-dimeric* partial optimization used for the dimeric case

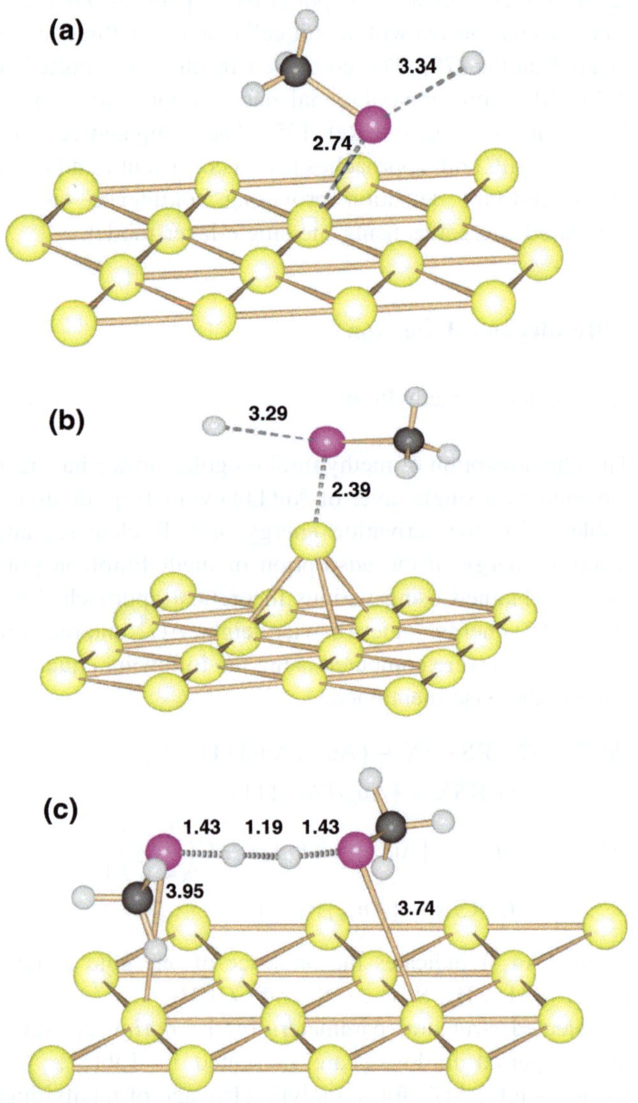

Fig. 1 Computed transition states of MeSH on clean and reconstructed surface (S = pink, H = white, C = gray, Au = yellow)

not expecting a sever surface modification due to strong Au–Au bond, in order to avoid any artificial surface deformation, further calculations were performed by freezing the layered gold atoms.

In this partial optimization of gold surface case, the activation barrier has increased from 45.5 to 72.6 kcal/mol and the reaction energy is computed to be endothermic by 10.1 kcal/mol. The reactant MeSH is adsorbed on the surface with rather a weak Au–S interaction as the sulfur-gold bond lengths are in the range of ~3.5 Å. The sulfur atom sits almost in the middle of an Au triangle in the layer (See Fig. 1) (FCC or HCP hole in SAMs) with all three Au–S bond lengths almost equal in magnitude with a tilt angle ζ of 103.7°. In the transition state, as the S–H bond elongates in a homolytic fashion, sulfur gains significant spin density leading to a stronger Au–S interaction (Fig. 2). This is evident from the shorter Au–S bond lengths detected in the transition state structure with 2.74, 2.87, and 3.08 Å as the nearest Au–S interaction. The tilt angle with respect to the methyl group is found to 121.1°, which is larger than that observed in the reactant. It is important to note here that the tilt angle ζ can be seen as an index of the electron delocalization of the adsorbate on the Au(111) [129]. The S–H bond then cleaves leading to the generation of MeS˙, which adsorbed firmly on the surface with a much shorter Au–S bond length.

The sulfur atom as product moves rather to a bridge position with the two Au–S bond lengths of 2.815 and 2.900 Å. The homolytic cleavage of the S–H bond is also evident from the abundant spin density of opposite sign detected in the transition state for the sulfur and hydrogen atoms (Fig. 2). Further, calculations performed using the point charges model reduces the barrier to 42.4 kcal/mol compared with 72.6 kcal/mol obtained without the addition of the point charges. The reduction in barrier height is therefore expected as this introduces an electrostatic stabilization of the transition state. Such a large reduction in the barrier height by the addition of point charges reveals that a more realistic model employing several layers of gold atoms or point charges is expected to reduce the barrier height significantly and yield more realistic barrier heights for the reaction to occur at normal conditions. The point charge calculations have been performed only on one model complex, but we believe the relative reduction in the barrier heights to remain approximately the same among different thiols computed here (see below) as also suggested by comparing the reaction energies with more complex models (see Table S1). The S–H bond length is found to be 1.38 Å in the reactant while it elongates in the transition state up to 3.75 Å.

The reactivity and adsorption phenomena is expected to differ when Au adatoms and/or vacancies are present in the top layer of the Au(111). It is evidenced from the reported theoretical and experimental results that the presence of surface gold adatom leads to stabilization of different species in SAMs compared to what is known from the unreconstructed surface scenario. A large number of recent

Table 2 Selected structural parameters of adsorption and dissociation of methyl thiol on clean and reconstructed surface along with spin densities and Mulliken charges (parenthesis) on S and H atoms

	Spin densities		Structural parameters		
	S	H	S–H	Au–S	Au–S–C
Popt					
R	0.0 (0.13)	0.0 (0.07)	1.37	3.54	103.7
TS	0.26 (0.29)	−0.97 (0.02)	3.34	2.74	121.1
$Popt_{dimeric}$					
R	0.0	0.0	1.37	3.34	99.3 (107.3)[a]
TS	0.0 (0.16)	0.0 (−0.07)	1.43	3.74	87.5 (77.6)[a]
$Popt_{Add}$					
R	0.003 (0.21)	0.0 (0.14)	1.38	2.51	99.9
TS	−0.23 (0.23)	0.95 (0.03)	3.28	2.38	102.7
$Popt_{partial\ charge}$					
R	0.0 (0.16)	0.0 (0.09)	1.38	3.20	118.6
TS	−0.01 (0.40)	0.0 (0.21)	3.75	2.51	98.8

[a] Au–S_2–C bond angle, *Fopt* Full optimization, *Popt* partial optimization, *Popt-dimeric* partial optimization used for dimeric case

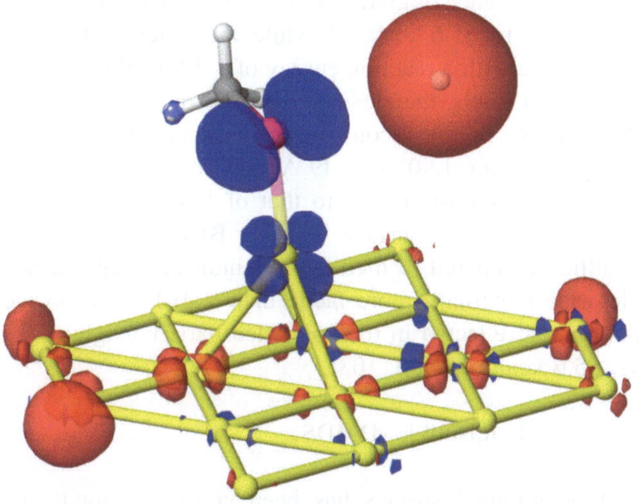

Fig. 2 DFT computed spin density plot of adatom transition state of methyl thiol species on reconstructed surface (S = pink, H = white, C = gray, Au = yellow)

literature supports a reconstructed surface as the most likely model for the SAMs. Therefore, presence of Au adatoms has been taken into consideration for the S–H cleavage. Since we are interested in kinetics of various alkanethiol chemisorption on gold surface, the formation of adatom/vacancies has not been taken into account as this requires much larger models for the computation. The formation of add atom/vacancies has been computed using more realistic three-layer model using periodic DFT methods, and the results are summarized elsewhere [75] (e.g., adatom formation is found to be exothermic by 3.3 kcal/mol). The optimized geometry of the transition states of the methyl thiols on the reconstructed surface is shown in Fig. 1b. Computation yields a barrier height of 55.4 kcal/mol for the S–H cleavage, and the reaction is thermo neutral. Presence of Au adatom substantially lowers the barrier by 17.2 kcal/mol, and the excessive endothermicity observed is brought down to thermonetural conditions. The MeSH in the reconstructed surface prefers a quasi-atop configuration with a tilt angle of 99.9° with Au–S bond length of 2.52 Å. However, the Au–S bond strengthens in the transition state as it gets reduced to 2.39 Å with a tilt angle of 102.8°. The S–H bond elongates to 3.28 Å compared to 1.38 Å observed at the reactant, instead. Moreover, the sulfur prefers a perfect atop configuration. A lower energy barrier in the presence of Au adatom reveals that the defects present in the surface likely to speed up the cleavage process and aid firm chemisorption and ordered SAMs. It has to be noted here that the barrier height might get reduced further upon the addition of point charges as in the unreconstructed scenario discussed earlier.

Although the reactions computed so far are endothermic, it is worth to mention that ΔG_r values have been computed from the adsorbed methyl thiol species. Since the adsorption of thiols on Au(111) is found to be exothermic in nature, the reaction energy from the Au(111) and free reactant is likely to be exothermic. To check this argument, we have computed the adsorption energy of MeSH on the more likely experimental scenario, that is, the reconstructed case. The reaction is found to be exothermic by −9.8 kcal/mol. and this brings down the total reaction energy into exothermic region by −9.6 kcal/mol (see Fig. 3; note here that the adatom formation that was estimated to be exothermic by −3.3 kcal/mol from our periodic DFT study [73] has not been taken into account in this energetics equation).

The transition state modeling performed so far releases hydrogen atom, which upon recombination with hydrogen produces H_2 molecule as it is also evident from the experiments. In an ordered $(\sqrt{3} \times \sqrt{3})R30°$ SAM arrangement, since the two methylthiols are close to each

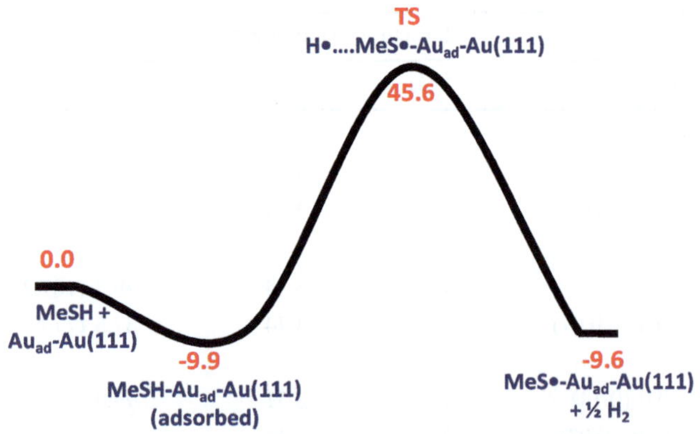

Fig. 3 Computed PES for the reaction of MeSH on Au(111)

other, it is likely that the cleavage of the two methylthiols S–H bonds and formation of H_2 can occur simultaneously through a concerted transition state. This is especially true for high-coverage regime where the methylthiols arrange with one another with a short intermolecular distance. An attempt to compute such transition state was successful and yields a very low barrier height of 20.7 kcal/mol. Consequently, now the reaction energy falls in the exothermic regime by −14.9 kcal/mol. The optimized structure of the transition state is shown in Fig. 1c. At the transition state, two MeSH groups are in *trans* configuration [130] with a S–S distance of 4.05 Å. The S–H bonds are elongated to 1.43 Å while the H···H interaction is at 1.20 Å. Although this co-operative transition state is likely to be the realistic model for the S–H cleavage, since a very close sulfur–sulfur interaction is mandatory for this transition state to occur, it is very likely that this happens in high-coverage regime. On the other hand, in low coverage regime, the single molecule TS presented before could be the most likely.

These results clearly illustrate that the dissociation of the S–H bond is homolytic in nature without a prohibitive kinetic barrier and is also likely to be aided by the defects on the surface. A co-operative mechanism is instead expected when short intermolecular interactions, that is, high coverage, are present.

3.2 Dimethylsulfide (DMS)

As a natural extension to methylthiols, we have considered dimethylsulfide adsorption on Au(111). There are several experimental reports on the dimethylsulfide adsorption on Au(111) [100–104]. In this case, a homolytic S–C cleavage is expected as the resulting radical species is likely to adsorb firmly on the surface. This assumption is supported by experimental bond dissociation energy (BDE) of the S–C bond in DMS, which is much lower than the S–H bond in the methylthiols. Additionally, there is also convincing spectroscopic evidence where S–C cleavage has been observed for some alkylthiols on Ag surface [131–137]. Considering these facts, the $CH_3S\text{-}CH_3$ cleavage transition state has been modeled and computed. The activation and reaction energies are given in Table 3, while selected structural parameters of the optimized structure are given in Table 4. The optimized geometry of adsorption of DMS on clean and reconstructed surface is presented in Fig. 4. The barrier for S–C bond cleavage after the adsorption of DMS on Au(111) is found to be 69.6 kcal/mol on clean surface, and the reaction is found to be endothermic by 3.9 kcal/mol. The S–C bond length is found to 1.89 Å in the reactant while it elongates in the transition state up to 3.55 Å (Fig. 4a). A very long S–C bond in transition state indicates a product-like transition state. Spin densities of −0.22 on the sulfur atom and −1.12 and 0.01 on the two carbon atoms are observed. Such scenario clearly shows that the dissociation of S–C bond occurs homolytically.

In the presence of gold adatoms, the barrier gets reduced from 69.6 to 55.6 kcal/mol while the reaction becomes exothermic with a reaction energy of −3.6 kcal/mol. As in the methyl thiols, the cleavage is aided in the presence of Au adatom. In the reconstructed surface, the S–C bond elongates from 1.90 to 4.49 Å and the computed spin densities is almost similar to that of S–C bond cleavage occurs in the clean surface. A lower BDE for the dimethylsulfide compared to methylthiols indicates that the S–C cleavage is perhaps more facile than S–H bond cleavage, and this is reflected in the computed barrier heights (72.6 vs. 55.6 kcal/mol) [55, 103].

3.3 Dimethyldisulfide (DMDS)

A few thiolated species has been chosen to model the kinetics where the cleavage has been clearly envisaged with both experimental and theoretical studies. On this account, dimethyldisulfide takes the prime place as there are numerous experimental and theoretical studies on its adsorption on Au(111). The transition state geometry of dissociation of DMDS on clean and reconstructed surface is presented in Fig. 5. Activation and reaction energies and

Table 3 Activation and reaction energies of the dissociation of various alkylthiol and their alternatives on clean and reconstructed surface

	Partial optimization			
	Clean surface		Constructed surface	
	$\Delta G^{\#}$	ΔG_r	$\Delta G^{\#}$	ΔG_r
CH_3–S–CH_3	69.6	3.9	55.6	−3.6
CH_3–S–S–CH_3	17.47	−4.2	23.7	−2.4
CH_3–S–OAc	–	–	13.9	7.2
CH_3–S–CN	46.7	–	6.4	–

Table 4 Selected structural parameters of adsorption and dissociation of DMS, thioacetate, thiocyanate on clean and reconstructed Au(111) surface along with spin densities on S and H atoms

	Spin densities		Structural parameters		
	S	C	S–C	Au–S	Au–S–C
DMS					
Popt					
R	0.33	−0.66	1.89	3.17	107.4
TS	−0.22	1.12	3.55	2.71	127.7
$Popt_{Add}$					
R	0.39	−0.63	1.90	2.50	100.2
TS	−0.22	1.13	4.49	2.39	71.1
Thioacetate					
$Popt_{Add}$					
R	0.39	−0.63	1.99	2.77	96.9
TS	0.0	0.0	2.13	2.78	91.1
Thiocyanate					
Popt					
R	0.0	0.0	1.75	3.52	91.9
TS	−0.44	0.22	2.43	2.84	115.9
$Popt_{Add}$					
R	0.0 (0.58)	0.0 (−0.07)	1.76	2.56	99.5
TS	0.44 (0.59)	0.0 (−0.02)	4.42Å	2.42	104.0

structural parameters of adsorption of DMDS on clean and reconstructed gold surface are given in Tables 3 and 5. The barrier for dissociation of DMDS on the Au(111) after the adsorption is found to 17.5 kcal/mol, and the reaction is exothermic by −4.2 kcal/mol. Combined low-energy electron diffraction, Auger electron spectroscopy, X-ray photoelectron spectroscopy, and line of sight mass spectrometry have been used to study the adsorption and desorption of DMDS on Au(111). According to these works, the S–S bond clearly cleaves on the surface and the estimated barrier height is to be in the range of 14.6–15.3 kcal/mol [105].

In the dissociation transition state, S–S bond length is 4.28 Å while during adsorption S–S bond length is 2.27 Å (Fig. 3b). In the dissociation transition state, DMDS molecule shows significant spin densities on the two S atoms ensuring that dissociation of DMDS occur homolytically. The computed structure and barrier heights are in good agreement with earlier experimental and theoretical observations [105].

3.4 Thiocyanates

Experimental studies reveal that thiocyanates chemisorb on gold surface after the cleavage of S–CN bond and, moreover, high-quality SAM is formed after the occurrence of surface-mediated reduction of thiocyanates on Au(111) surface [82]. In order to find out the kind of cleavage and whether surface-mediated reduction or homolytic cleavage occurs during the dissociation of S–C bond of thioacetate, we have modeled the transition state for dissociation of S–C(CN) bond of thioacetates on the clean and reconstructed Au(111) surface (see Tables 3, 4; Fig. 6). The energy barrier for dissociation of thiocyanates on the clean surface is 46.7 kcal/mol. In the dissociation transition state, Au–S and S–C bond distance is found to be 2.84 and 4.32 Å. During adsorption, Au–S and S–C bond distance is found to be 3.52 and 1.75 Å. In the dissociation transition state, thiocyanates has −0.44 and 0.22 on the sulfur and carbon atoms. This clearly shows the homolytic cleavage occurs during the dissociation of thiocyanates on the clean Au(111) surface. The barrier for the dissociation of thiocyanate on the reconstructed surface is very low with just 6.4 kcal/mol and reveals that thiocyanates among other studies have the lowest kinetic requirement for the cleavage to occur and this explains more regular ordered SAMs formed in general with thiocyanates [82–84].

3.5 Thioacetates

Thioacetates are routinely used in the preparation of SAMs as they serve as best protecting groups from unwarranted reactions to occur. Moreover, it has been shown that experimentally the thioacetates undergo spontaneous deprotection upon adsorption, leading to radical thiols to be adsorbed on the surface [91, 92]. This approach has been

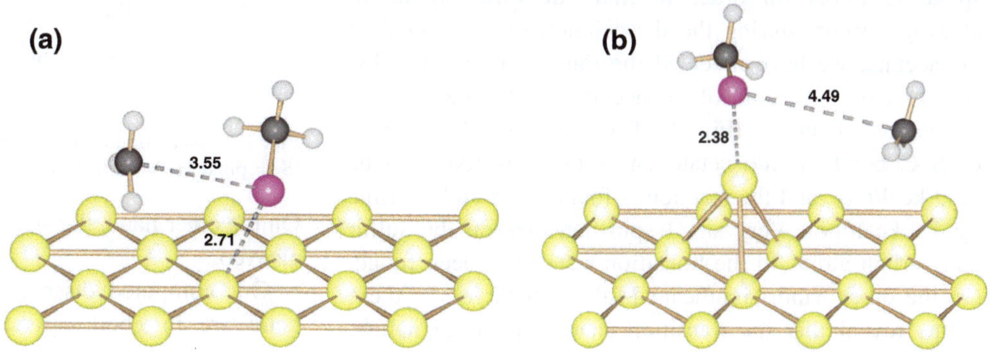

Fig. 4 Dissociation transition of DMS on clean and reconstructed surface (S = pink, H = white, C = gray, Au = yellow)

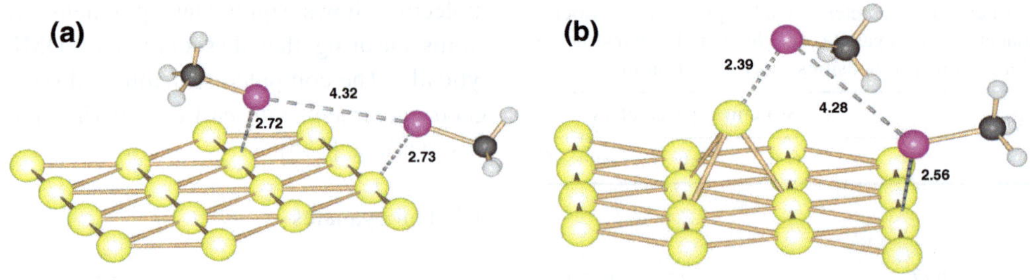

Fig. 5 Dissociation transition of DMDS on clean and reconstructed surface (S = pink, H = white, C = gray, Au = yellow)

Table 5 Activation and reaction energies of the dissociation of DMDS on clean and reconstructed Au(111) surface

	Spin densities		Structural parameters					
	S	S	S–S	Au–S	Au–S_1–S_2	Au–S_2–S_1	Au–S_1–C	Au–S_2–C
Popt								
R	0.0	0.0	2.25	3.34	103.0	70.7	153.9	104.5
TS	0.02	0.15	4.32	2.73	85.9	92.1	83.3	101.9
$Popt_{Add}$								
R	0.0	0.0	2.27	2.50	104.5		99.5	
TS	0.49	0.20	4.28	2.39	84.7	101.4	99.4	90.9

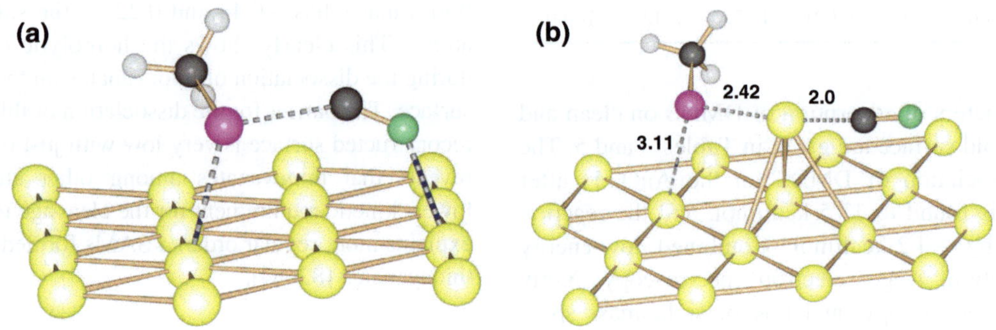

Fig. 6 Dissociation transition of thiocyanate on the clean and reconstructed surface (S = pink, H = white, C = gray, Au = yellow, N = green)

successively used for the synthesis of SAMs of different nature [92]. An earlier experimental study clearly suggests that dissociation of long chain alkylethioactetates after the adsorption occurs through a polar transition states, and the long chain alkylthioacetates adsorb slower than the corresponding thiols. In order to find out which mode of cleavage occurs during the dissociation of S–C bond of thioacetate, we have modeled the transition state for dissociation of S–C bond of thioacetates on the most likely reconstructed surface (Fig. 7). The barrier for dissociation of S–C bond of thioacetate on Au(111) is found to be 13.9 kcal/mol, and the reaction is found to be endothermic by 7.2 kcal/mol. Very small spin densities on the sulfur and carbon atoms in the transition state have been found. On the other hand, significant Mulliken charges on these atoms reveal that the transition state is polar in nature.

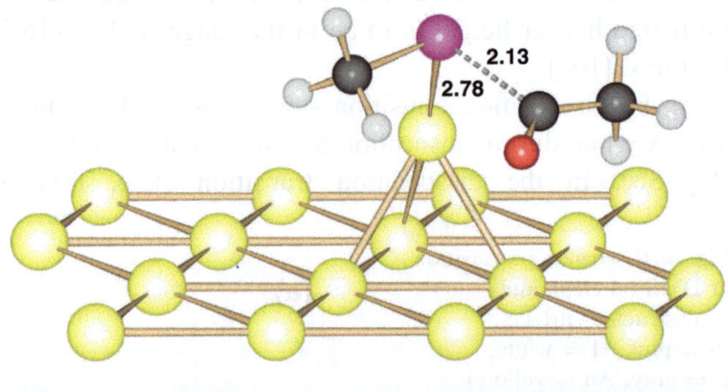

Fig. 7 Dissociation transition of thioacetate on reconstructed surface (S = pink, H = white, C = gray, Au = yellow, Red = Oxygen)

On the other hand, the stabilized products are homolytically cleaved.

Our computed dissociation transitions state and energetic are in good agreement with earlier experimental observations [88, 91, 92]. Our calculations seem therefore to suggest a homolytic cleavage assisted by a polarized transition state.

4 Conclusions

The present study aims to probe the kinetic and thermodynamic aspects of the S–X (X = Me, S, Acetyl, CN) cleavage of various alkylthiols during the adsorption on clean and reconstructed Au(111) surface using DFT methods. The transition state for each species has been modeled on a single layer of gold atoms, and the barrier height and the reaction energies were then computed. The following points emerge from our calculations, and some of these conclusion are also supported by independent experiments,:

1. The cleavage of S–H bond in MeSH occurs homolytically on clean and reconstructed Au surface. The homolytic cleavage of S–H bond is found to have lower barrier height on reconstructed surface, that is, in the presence of gold adatoms. This result reveals that the adatom is supposed to aid the cleavage of S–H bonds resulting in stable and ordered SAMs. Additionally, our calculations show that the cleaved hydrogen atoms can recombine and leave as H_2, as also experimentally detected.
2. In this framework, we have shown the intermolecular distances play an important role in the S–H cleavage kinetics. Short inter-thiols distances favor a concerted transition state with a lower activation energy than the one found for an isolated thiol molecule on gold. This leads us to conclude that, at least in high-coverage regime, a cooperative cleavage of the S–H bond with a spontaneous release of the H_2 is expected.
3. Our calculations suggest that the S–C bond in the DMS is also likely to cleave homolytically. Such suggestion is supported by a lower activation energy than the one found for the S–H cleavage of the methylthiol.
4. The computed activation free energy for the S–S bond cleavage of the DMDS is found to be in agreement with the experimentally reported values. The computed barrier height reveals that the S–S cleavage rates are likely to be faster than the S–H and S–C cleavage of the MeSH and DMS, respectively.
5. Among the various substituted alkylthiols, the following order is found for the cleavage kinetics: CH_3–S–CN > CH_3–S–OAc > CH_3–S–S–CH_3 > CH_3–S–CH_3 > CH_3–S–H. This clearly illustrates the role of electronegativity on the homolytic cleavage of alkylthiols on reconstructed gold surface.

Acknowledgments GR would like to acknowledge the financial support from the Government of India through Department of Science and Technology (SR/S1/IC-41/2010) and generous computational resources from Indian Institute of Technology Bombay. MJ thanks Indian Institute of Technology Bombay for financial support through Institute Post Doctoral Fellowship. FT acknowledges FP7-STREP: MOLSPINQIP–ICT-211284 'Molecular Spin Clusters for Quantum Information Processing', ERC-2010-AdG—no 267746 MolNanoMaS—'Molecular Nanomagnets at Surfaces: Novel Phenomena for Spin-based Technologies', and PRIN 2008—FZK5AC, 'Strutture molecolari e nanocristalline con funzionalità magnetiche, foto-magnetiche e foto-emettitrici e loro organizzazione su superfici, in film polimerici o in sol–gel'.

References

1. Ulman A (1996) Chem Rev 96:1533
2. Bumm LA, Arnold JJ, Cygan MT, Dunbar TD, Burgin TP, Jones L, Allara DL, Tour JM, Weiss PS (1996) Science 271:1705
3. Poirier GE (1997) Chem Rev 97:1117
4. Reed MA, Zhou C, Muller CJ, Burgin TP, Tour JM (1997) Science 278:252
5. Ulman A (2001) Acc Chem Res 34:855
6. Nitzan A, Ratner MA (2003) Science 300:1384
7. Love JC, Estroff LA, Kriebel JK, Nuzzo RG, Whitesides GM (2005) Chem Rev 105:1103
8. Heimel G, Romaner L, Zojer E, Bredas JL (2008) Acc Chem Res 41:721
9. Chen W, Wee ATS (2009) J Electron Spectros Relat Phenomena 172:54
10. Kumar A, Abbott NL, Biebuyck HA, Enoc K, Whitesides GM (1995) Acc Chem Res 28:219
11. Houston JE, Kim IH (2002) Acc Chem Res 35:547
12. Woodruff DP (2006) Phys Chem Chem Phys 10:7211
13. Vericat C, Vela ME, Benitez GA, Martin Gago JA, Torrelles X, Salvarezza RC (2006) J Phys Condens Matter 18:867
14. Dubois LH, Nuzzo RG (1992) Ann Rev Phys Chem 43:437
15. Kramer S, Fuierer RR, Gorman CB (2003) Chem Rev 103:4367
16. Chidsey CED, Liu G-Y, Rowntree YP, Scoles G (1989) J Chem Phys 91:4421
17. Alves CA, Smith EL, Porter MD (1992) J Am Chem Soc 114:1222
18. Poirier GE, Tarlov MJ, Rushneier HE (1994) Langmuir 10:3383
19. Fenter P, Eisenberger P, Liang KS (1993) Phys Rev Lett 70:2447
20. Camillone N, Chidsey CED, Liu G-Y, Scoles G (1993) J Phys Chem 98:3503
21. Poirier GE, Tarlov MJ (1994) Langmuir 10:2859
22. Poitier GE, Pylant ED (1996) Science 272:1145
23. Sellers H, Ulman A, Shnidman Y, Eilers JE (1993) J Am Chem Soc 115:9389
24. Gronbeck H, Curioni A, Andreoni W (2000) J Am Chem Soc 122:3839
25. Yourdshahyan Y, Zhang HK, Rappe AM (2001) Phys Rev B 63:81405
26. Tachibana M, Yoshizawa K, Ogawa A, Fujimoto H, Hofmann R (2002) J Phys Chem B 106:12727
27. Hayashi T, Morikawa Y, Nozoye H (2001) J Chem Phys 114:7615
28. Vargas C, Giannozzi P, Selloni A, Scoles G (2001) J Phys Chem B 105:9509
29. Gottschalck J, Hammer B (2002) J Chem Phys 116:784

30. Molina ML, Hammer B (2002) Chem Phys Lett 360:264
31. Morikawa Y, Liew CC, Nozoye H (2002) Surf Sci 514:389
32. Roper MG, Skegg MP, Fisher CJ, Lee JJ, Dhank VR, Woodruff DP, Jones RG (2004) Chem Phys Lett 389:87
33. Kondoh H, Iwasaki M, Shimada T, Amemiya K, Yokoyama T, Ohta T, Shimomura M, Kono S (2003) Phys Rev Lett 90:066102
34. Yu M, Bovet N, Satterley CJ, Bengió S, Lovelock KRJ, Milligan PK, Jones RG, Woodruff DP, Dhanak V (2006) Phys Rev Lett 97:166102
35. Maksymovych P, Sorescu DC, Dougherty D, Yates TY Jr (2005) J Phys Chem B 109:22463
36. Maksymovych P, Sorescu DC, Yates TY Jr (2006) Phys Rev Lett 97:146103
37. Grönbeck H, Häkkinen H (2007) J Phys Chem B 111:3325
38. Mazzarello R, Cossaro A, Verdini A, Rousseau R, Casalis L, Danisman MF, Floreano L, Scandolo S, Morgante A, Scoles G (2007) Phys Rev Lett 98:016102
39. Nogoya A, Morikawa Y (2007) J Chem Phys 19:365245
40. Wang J-G, Selloni A (2007) J Phys Chem C 111:12149
41. Cossaro A, Mazzarello R, Rousseau R, Casalis L, Verdini A, Kohlmeyer A, Floreano L, Scandolo S, Morgante A, Klein ML, Scoles G (2008) Science 321:943
42. Maksymovych P, Yates JT (2008) J Am Chem Soc 130:7518
43. Carro P, Salvarezza R, Torres D, Illas F (2008) J Phys Chem C 112:19121
44. Kautz NA, Kandel SA (2008) J Am Chem Soc 130:6908
45. Voznyy O, Dubowski JJ (2009) Langmuir 25:7353
46. Groenbeck H, Haekkinen H, Whetten RL (2008) J Phys Chem C 112:15940
47. Jiang D (2009) Chem Phys Lett 477:90
48. Franke A, Pehlke E (2009) Phys Rev B 79:235441
49. Jiang D, Dai S (2009) J Phys Chem C 113:3763
50. Wang J, Selloni A (2009) J Phys Chem C 113:3763
51. Chaudhuri A, Lerotholi TJ, Jackson DC, Woodruff DP, Jones RG (2009) Phys Rev B 79:195439
52. Chaudhuri A, Lerotholi TJ, Jackson DC, Woodruff DP, Dhanak V (2009) Phy Rev Lett 102:126101
53. Torres E, Blumenau AT, Biedermann PU (2009) Phy Rev B: Condens Matter 79:075440
54. Voznyy O, Dubowski JJ, Yates JT, Maksymovych P (2009) J Am Chem Soc 113:12989
55. Schlenoff JB, Li M, Ly H (1995) J Am Chem Soc 117:12528
56. Walczak MW, Chung C, Stole SM, Widrig CA, Porter MD (1991) J Am Chem Soc 113:2370
57. Brust M, Walker M, Bethell D, Schiffrin DJ, Whyman R (1994) J Chem Soc, Chem Commun 801
58. Thomas RC, Sun L, Crooks M (1991) Langmuir 7:620
59. Chailapakul O, Sun L, Xu C, Crooks M (1993) J Am Chem Soc 115:12459
60. Porter MD, Bright TB, Allara DL, Chidsey CED (1987) J Am Chem Soc 109:3559
61. Nuzzo RG, Fusco FA, Allara DL (1987) J Am Chem Soc 109:2358
62. Bain CD, Biebuyck HA, Whitesides GM (1989) Langmuir 5:723
63. Nuzzo RG, Zegarski BR, Dubois LH (1987) J Am Chem Soc 109:733
64. Nuzzo RG, Dubois LH, Allara DL (1990) J Am Chem Soc 112:558
65. Li Y, Huang J, McIver RT Jr, Hemminger JC (1992) J Am Chem Soc 114:2428
66. Widrig CA, Chung C, Porter MD (1991) J Electroanal Chem 310:335
67. Bryant MA, Pemberton JE (1991) J Am Chem Soc 113:3630
68. Bryant MA, Pemberton JE (1991) J Am Chem Soc 113:8284
69. Kankate L, Turchanin A, Gölzhäauser A (2009) Langmuir 25:10435
70. Tielens F, Santos E (2010) J Phys Chem C 114:9444
71. RzezinickaII, Lee JS, Maksymovych P, Yates JT (2005) J Phys Chem B 109:15992
72. Hasan M, Bethell D, Brust M (2002) J Am Chem Soc 124:1132
73. Rajaraman G, Caneschi A, Gatteschi D, Totti F (2011) Phys Chem Chem Phys 13:3886
74. Beulen MWJ, Bugler J, Lammerink B, Geurts FAJ, Biemond EMEF, Van Leerdam KGC, Van Veggel FCJM, Engbersen JFJ, Reinhoudt DN (1998) Langmuir 14:6424
75. Friggeri A, Schoenherr H, van Manen H-J, Huisman B-H, Vancso GJ, Huskens J, van Veggel FCJM, Reinhoudt DN (2000) Langmuir 16:7757
76. Faull JD, Gupta VK (2002) Langmuir 18:6584
77. Liebau M, Janssen HM, Inoue K, Shinkai S, Huskens J, Sijbesma RP, Meijer EW, Reinhoudt DN (2002) Langmuir 18:674
78. Yamada T, Sekine R, Sawaguchi T (2000) J Chem Phys 113:1217
79. Houmam A, Hamed EM, Hapiot P, Motto JM, Schwan AL (2003) J Am Chem Soc 125:12676
80. Houmam A, Hamed EM, Still IWJ (2003) J Am Chem Soc 125:7258
81. Kordis J, Gingerich KA, Seyse RJ (1974) J Chem Phys 61:5114
82. Coi Y, Jeong Y, Chung H, Tto E, Hara M, Noh J (2008) Langmuir 24:91
83. Dietz O, Rayon VM, Frenking G (2003) Inorg Chem 42:4977
84. Ciszek JW, Stewart MP, Tour JM (2004) J Am Chem Soc 126:13172
85. Singh A, Dahanayaka DH, Biswas A, Bumm L, Haltermanm RL (2010) Langmuir 26:13221
86. Tour JM, Jones LII, Pearson DL, Lamba JJS, Burgin TP, Whitesides GM, Allara DL, Parikh AN, Atre SV (1995) J Am Chem Soc 117:9529
87. Cai L, Yao Y, Yang J, Price DW, Tour JM (2002) Chem Mater 14:2905
88. Stapleton JJ, Harder P, Daniel TA, Reinard MD, Yao Y, Price DW, Tour JM, Allara DL (2003) Langmuir 19:8245
89. Park T, Kang H, Choi I, Chung H, Ito E, Hara M, Noh J (2009) Bull Korean Chem Soc 30:441
90. Kang Y, Won D, Kim S, Seo K, Choi H, Lee G, Noh Z, Lee T, Lee C (2004) Mater Sci Eng C 24:43
91. Béthencourt MI, Srisombat L-O, Chinwangso P, Lee TR (2009) Langmuir 25:1265
92. Rodriguez-Douton MJ, Mannini M, Armelao L, Barra A-L, Tancini E, Sessoli R, Cornia A (2011) Chem Commun 47:1467
93. Park T, Kang H, Kim Y, Lee S, Noh J (2011) Bull Korean Chem Soc 32:39
94. Noh J, Murase T, Nakajima K, Lee H, Hara M (2000) J Phys Chem B 104:7411
95. Jung C, Dannenberger O, Xu Y, Buck M, Grunze M (1998) Langmuir 14:1103
96. Lee H, He Z, Hussey CL, Mattern DL (1998) Chem Mater 10:4148
97. Leavy MC, Bhattacharyya S, Cleland WE Jr, Hussey CL (1999) Langmuir 15:6582
98. Takiguchi H, Sato K, Ishida T, Abe K, Yase K, Tamada K (2000) Langmuir 16:1703
99. Noh J, Kato HS, Kawai M, Hara M (2002) J Phys Chem B 106:13268
100. Beulen MWJ, Huisman B-H, van der Heijden PA, van Veggel FCJM, Simons MG, Biemond EMEF, de Lange PJ, Reinhoudt DN (1996) Langmuir 12:6170
101. Noh J, Nakamura F, Kim J, Lee H, Hara M (2002) Mol Cryst Liq Cryst 377:165
102. Schoenherr H, Vancso GJ, Huisman B-H, Van Veggel FCJM, Reinhoudt DN (1999) Langmuir 15:5541
103. Zhong C-J, Porter MD (1994) J Am Chem Soc 116:11616

104. Zhong C-J, Brush RC, Anderegg J, Porter MD (1999) Langmuir 15:518
105. Roper MG, Jones RG (2008) Phys Chem Chem Phys 10:1336
106. Cometto FP, Macagno VA, Paredes-Olivera P, Patrito EM, Ascolani H, Zampieri G (2010) J Phys Chem C 114:10183
107. Khayankarn O, Pearson RA, Verghese N, Shafi A (2005) J Adhes 81:941
108. Béthencourt MI, Barriet D, Frangi NM, Lee TR (2005) J Adhes 81:1031
109. Colorado R Jr, Lee TR (2003) Langmuir 19:3288
110. Zhang X, Chabal YJ, Christman SB, Chaban EE, Garfunkel E (2001) J Vac Sci Technol A 19:1725
111. Cai L, Yao Y, Yang J, Price DW, Tour JM (2002) Chem Mater 14:2905
112. Béthencourt MI, Srisombat L-o, Chinwangso P, Lee TR (2009) Langmuir 25:1265
113. Bencini A, Rajaraman G, Totti F, Tusa M (2009) Superlattices Microstruct 46:4
114. Rajaraman G, Caneschi A, Gatteschi D, Totti F (2010) J Mat Chem 20:10747
115. Jaguar 7.0, Schrödinger, Portland, OR 97204
116. Becke AD (1993) J Chem Phys 98:5648
117. Stephens PJ, Devlin FJ, Chabalowski CS, Frisch MJ (1994) J Phys Chem 98:11623
118. Hay PJ, Wadt WR (1985) J Chem Phys 82:270
119. Hay PJ, Wadt WR (1985) J Chem Phys 82:284
120. Hay PJ, Wadt WR (1985) J Chem Phys 82:299
121. Hariharan PC, Pople JA (1973) Theor Chim Acta 28:213
122. Francl MM, Petro WJ, Hehre WJ, Binkley JS, Gordon MS, DeFrees DJ, Pople JA (1982) J Chem Phys 77:3654
123. Rassolov V, Pople JA, Ratner M, Windus TL (1998) J Chem Phys 109:1223
124. Schaftenaar G, Noordik JH (2000) Molden: a pre- and post-processing program for molecular and electronic structures. J Comput Aided Mol Design 14:1
125. Mundy CJ, Mohamed F, Schiffman F, Tabacchi G, Forbert H, Kuo W, Hutter J, Krack M, Iannuzzi M, McGrath M, Guidon M, Kuehne TD, Laino T, VandeVondele J, Weber V (2000) CP2K software package, http://cp2k.berlios.de
126. VandeVondele J, Krack M, Mohamed F, Parrinello M, Chassaing T, Hutter J (2005) Comput Phys Commun 167:103
127. Lippert G, Hutter J, Parrinello M (1997) Mol Phys 92:477
128. Lippert G, Hutter J, Parrinello M (1999) Theor Chem Acc 103:124
129. Kong B, Kim Y, Choi SI (2008) Bull Korean Chem Soc 29:1843
130. Cometto FP, Macagno VA, Paredes-Olivera P, Patrito EM, Ascolani H, Zampieri, G (2010) J Phys Chem C 114:10183
131. Kim KL, Lee SJ, Kim K (2004) J Phys Chem B 108:9216
132. Baadji N, Piacenza M, Tugsuz T, Della SF, Maruccio G, Sanvito S (2009) Nat Mater 8:813
133. Zulczewski G, Sanvito S, Coey M (2009) Nat Mater 8:693
134. Zhang R, Ma G, Li R, Qian Z, Shen Z, Zhao X, Hou S, Sanvito S (2009) J Phys Condens Matter 21:335301
135. Giusti A, Charron G, Mazerat S, Compain J-D, Mialane P, Dolbecq A, Riviere E, Wernsdorfer W, Ngo B, Keita B, Nadjo L, Filoramo A, Bourgoin J-P, Mallah T (2009) Angew Chem Int Ed 48:4949
136. Wernsdorfer W (2009) Nat Nanotechnol 4:145
137. Bogani L, Wernsdorfer W (2008) Nat Mater 7:179

Theor Chem Acc (2012) 131:1151
DOI 10.1007/s00214-012-1151-9

REGULAR ARTICLE

A relativistic DFT study of magnetic exchange coupling in ketimide bimetallic uranium(IV) complexes

Samir Meskaldji · Abdellah Zaiter · Lotfi Belkhiri · Abdou Boucekkine

Received: 23 July 2011 / Accepted: 13 September 2011 / Published online: 15 February 2012

Abstract Magnetic exchange couplings in bis(ketimide) binuclear U^{IV}/U^{IV} complexes $[Cp'_2UCl]_2(\mu\text{-ketimide})$ diuranium(IV) and $[(C_5H_5)_2(Cl)An]_2(\mu\text{-ketimide})$ ($Cp' = C_5Me_4Et$; ketimide = $N{=}CMe\text{-}(C_6H_4)\text{-}MeC{=}N$) have been investigated computationally using relativistic density functional theory (DFT) combined with the broken symmetry (BS) approach. Using the B3LYP hybrid functional, the BS ground state of these U^{IV}/U^{IV} $5f^2$–$5f^2$ complexes has been found of lower energy than the high spin (HS) quintet state, indicating an antiferromagnetic character (estimated coupling constant $|J| < 5\ cm^{-1}$) which has not yet been evidenced unambiguously experimentally. On the contrary, the BP86 GGA functional overestimates greatly the antiferromagnetic character of the complexes ($|J| > 100\ cm^{-1}$). As recently reported for para-bis(imido) $[(C_5H_5)_3U]_2$ (μ-imido) uranium(V) complex, spin polarization is mainly responsible for the antiferromagnetic coupling through the π-network orbital pathway within the bis(ketimide) bridge. Furthermore, spin polarization is exalted by the combined roles of the 5*f* metal orbitals and of the π-conjugated ketimide bridging ligand which permit electronic communication between the two uranium atoms albeit separated by a distance of the order of 10 Å. The MO analysis clarifies which MOs contribute to the antiferromagnetic coupling in the binuclear complexes under consideration and brings to light the 5*f* orbitals driving contribution.

Dedicated to Professor Vincenzo Barone and published as part of the special collection of articles celebrating his 60th birthday.

S. Meskaldji · A. Zaiter · L. Belkhiri (✉)
URCHEMS, Université Mentouri de Constantine,
25017 Constantine, Algeria
e-mail: lotfi.belkhiri@umc.edu.dz

S. Meskaldji
ENSET, 21000 Skikda, Algeria

A. Boucekkine (✉)
Laboratoire Sciences chimiques de Rennes,
UMR CNRS 6226, Université de Rennes 1,
Campus de Beaulieu, 35042 Rennes Cedex, France
e-mail: abdou.boucekkine@univ-rennes1.fr

Keywords Ketimide biuranium(IV) complexes · Magnetic exchange coupling · ZORA/B3LYP · Broken symmetry

1 Introduction

Since the first report of the single-molecular magnetic compound with the Cu_2Gd core in 1985 [1], the chemistry of this new class of fascinating transition d-metals and lanthanide containing materials continue to attract much attention at both experimental [1–14] and theoretical [15–46] levels. Molecules containing a finite number of interacting spin centres (e.g. paramagnetic ions) connected through a bridging ligand provide ideal opportunities to study basic concepts of magnetism, for example, superexchange interactions as early expressed by Kahn [47]. They constitute model compounds for the development of single-molecule magnets (SMM) and multifunctional materials [4–6].

From a theoretical point of view, quantifying magnetic coupling can be achieved by exchange coupling constant *J* calculations [20, 29, 32]. In the framework of density functional theory (DFT), the *J* constant is generally extracted from high spin (HS) and broken symmetry (BS) state energy computations [23–27, 48–55]. This technique introduced initially by Noodlemann et al. [56–58] was successfully used in combination with the hybrid B3LYP functional, in the case of molecular radicals [15–19] and polynuclear d-transition and mixed 4*f*–*d* complexes [16, 20–46]. During last decade, much progress has been obtained in elucidation of magneto-structural correlations [32–36, 43, 46]. In the case of actinide compounds, it must be noted the renewed interest for some bi- and trimetallic species that could exhibit magnetic properties [59–62]. Indeed, recent studies have shown that they can also potentially be exploited in producing SMMs [60]. Furthermore, such systems are of fundamental interest owing to the involvement of the actinide 5*f* electrons in magnetic exchange interactions to be compared to the 4*f* orbitals of lanthanides [59, 60]. Although an important number of examples of these actinide-containing molecules featuring clear signature of magnetic exchange coupling have been reported [63–89], their theoretical study remains a challenge for quantum chemistry methods as stressed by Rinehart et al. [62].

A successful strategy in promoting interactions between actinide metallic ions has been the use of covalently linked bridging ligands. Indeed, magnetic coupling is attributed to indirect exchange interactions between spin centres connected by bridging functionalized ligands [60–63, 77, 78]. Thus, the use of an aromatic spacer bearing two functionalized metallic centres and combined with the radial extension of 5*f* orbitals should facilitate significant metal–metal electronic and magnetic interactions. For example, the imido linkage (imido = phenylenediimido ligand) was the first used in the diuranium(V) para-imido $[(MeC_5H_4)_3U]_2(\mu\text{-}1,4\text{-}N_2C_6H_4)$ complex which exhibits a significant antiferromagnetic $5f^1$–$5f^1$ coupling [63]. Since a number of fascinating binuclear uranium(III), (IV) and (V) systems have been synthesized and their magnetic properties brought to light by experimental susceptibility measurements [64–89]. These systems include the binuclear amide complex $[U(\eta^8\text{-}C_8H_8)]_2[\mu\text{-}\eta^4{:}\eta^4\text{-}HN(CH_2)_3N(CH_2)_2N(CH_2)_3NH]$ [79], arene bridging diuranium(III) complexes, such as $[(Mes(^tBu)N)_2U]_2(\mu\text{-}\eta^6{:}\eta^6\text{-}C_7H_8)$ and $[(Cp^*)_2U]_2(\mu\text{-}\eta^6{:}\eta^6\text{-}C_6H_6)$ systems [85], cylooctatetraene $(\mu\text{-}\eta^8{:}\eta^8\text{-}C_8H_8)U_2[NC(^tBu)MesN]_6$ (Mes = 2,4,6-$C_6H_2Me_3$) [84], polyuranyl UO_2^+ complexes [87], trimetallic actinide complex $[Cp^*_2An\{N{=}C(Bz)(tpy\text{-}An'Cp^*_2)\}_2]$ exhibiting mixed valency (An=Th^{IV}, An′=U^{IV}) [88] and Pyrazolate-bridged $[U(Me_2Pz)_4]_2$ (Me_2Pz = 3,5-dimethylpyrazolate) [71] diuranium(IV) complexes, are also likely to exhibit interesting magnetic properties.

Furthermore, recent reports in synthetic efforts toward paramagnetic actinide-containing assemblies indicate that ketimide (1,4-phenylenediketimide) ligands could lead to diverse and interesting magnetic behaviours [74–77]. Consequently, simple bis(ketimide) $[(C_5Me_4Et)_2(Cl)An]_2(\mu\text{-}\{N{=}CMe\text{-}(C_6H_4)\text{-}MeC{=}N\})$ binuclear An^{IV}/An^{IV} (Th, U) complexes [75] were synthesized; in the uranium case, metal–metal electronic communication could occur between the two $5f^2$–$5f^2$ spin centres through the aromatic spacer ligand.

More recently, a para- and meta-diethynylbenzene (DEB) bridging ligand have been used in binuclear isovalent U(IV)/U(IV) complexes with $[(NN'_3)_2U_2(DEB)]$ formula [78]. Experimental and theoretical DFT investigations of their magnetic properties have shown that both isomers exhibit a rather weak ferromagnetic and antiferromagnetic character for meta- and para-DEB, respectively [78]. A new bis(imido) diuranium(V) complex $[U(N^tBu)_2(I)(^tBu_2bpy)]_2$ exhibiting an moderate antiferromagnetic coupling between two metallic $5f^1$–$5f^1$ spin active centres has been synthesized and theoretically investigated [77]. For this system, B3LYP calculations predicted correctly that the BS state is lower in energy than the triplet HS state, providing an exchange coupling constant J (-12 cm^{-1}) in agreement with the experimental fitting of susceptibility measurements.

As aforementioned, exploiting and modelling the magnetic properties of actinide complexes still require a fundamental understanding of how to promote and modulate electronic delocalization and magnetic exchange interactions between spin centres [59–89]. In addition to this fundamental interest, the actinide coordination chemistry features a rich interplay of metal-bonding interactions, substitution pattern and the type of π network bridging ligand effects, whose understanding is crucial to the development of actinide magneto-structural chemistry and to settle longstanding questions about the role of 5*f* orbitals in metal–metal electronic and magnetic interactions through aromatic spacers [62].

In this work, we are particularly interested in the $[Cp'_2UCl]_2(\mu\text{-ketimide})$ diuranium(IV) complexes (Cp′ = C_5Me_4Et; ketimide = 1,4-phenylenediketimide) [75, 77] for which no systematic theoretical study has been carried out up to now, to our knowledge. This interesting system contains two active $5f^2$ electrons per centre connected by the bis(ketimide) aromatic π-spacer (Fig. 1). Experimentally, an antiferromagnetic character is expected for the complex in question, but was not completely brought to light by susceptibility measurements, while as judged by voltammetry, ‘the complex displays appreciable electronic communication between the metal centres through the

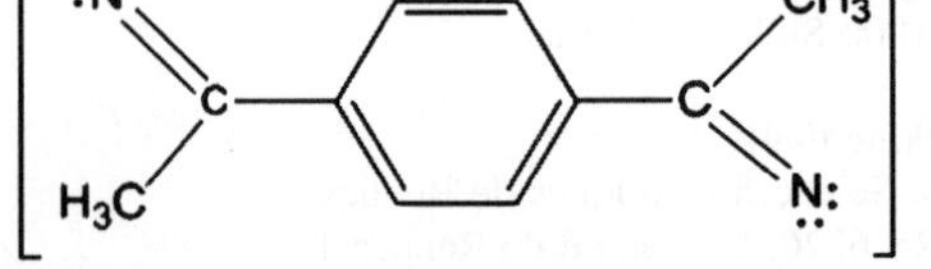

Fig. 1 The dianionic bis(ketimide) ligand

π-system of the dianionic diketimide bridging ligand' [75]. We have recently carried out [90] relativistic DFT investigations on bis(imido) diuranium ($5f^1$–$5f^1$) complexes in the framework of the zero order regular approximation (ZORA) predicting ferromagnetic and antiferromagnetic coupling, respectively, for the meta- and para-bis(imido) isomers, in agreement with experimental data [60, 63]. These results encouraged us to apply the same methodology to study the magnetic exchange interactions for the above mentioned diuranium(IV) ketimide system. Our theoretical study is likely to help the experimentalists in their investigations.

We shall first consider a model system henceforward called U_2ketimide where the Cp′ ligand of the complex has been replaced by an unsubstituted Cp ring and then the actual complex. The driving role of the aromatic spacer, that is, the bis(ketimide) bridging ligand and that of the uranium(IV) 5*f* orbitals, will be particularly investigated. As suggested by the authors of ref. 20b, we shall also consider its hypothetical congener ThUketimide complex that has not yet been isolated experimentally, in order to study the effect of replacing one paramagnetic U(IV) $5f^2$ centre by a diamagnetic Th(IV) $5f^0$ one.

2 Description of the model and computational details

The magnetic interaction between two atomic spins is usually described by the phenomenological Hamiltonian of Heisenberg, Dirac and Van Vleck (HDvV) [28–31, 91, 92] given by $\hat{H} = -2\,J\hat{S}_1 \cdot \hat{S}_2$, where $\hat{S}_1$ and $\hat{S}_2$ are the respective local spin angular momentum operators. A positive sign of the coupling constant *J* indicates a ferromagnetic interaction (parallel alignment of spins), whereas the negative sign indicates an antiferromagnetic interaction (anti parallel alignment of spins). It must be noted that to estimate the coupling constant *J*, one has to calculate energy differences that are often smaller than ~0.5 kcal/mol (few tens cm^{-1}) [24, 29]. Accurate quantum mechanical evaluation of the coupling constant requires the use of multideterminantal post-Hartree–Fock calculations [24, 93]. However, such ab initio methodology is computationally challenging for great size systems and its use is reduced to small molecules or simplified models [93]. An alternative is provided by the DFT/BS approach [14] which has been proved to be one of the most efficient tools to investigate the magnetic properties of transition metal complexes [16, 20–46, 94–97]. In numerous cases, good agreement has been achieved with the experimental data using the BS recipe in conjunction with the B3LYP functional [98, 99]. In some cases, a mixing of Hartree–Fock exchange between 33% and 50% was found to lead to better results than the 20% used in B3LYP ([49] and ref. therein).

All computational results reported in this work were performed using the Amsterdam Density Functional (ADF2010.02 release) program [100]. Relativistic corrections have being introduced via the zero order regular approximation (ZORA) [101–104] accounting for scalar relativistic effects. Spin–orbit corrections to the energy difference between the high spin (HS) and BS states have not been considered. These corrections are of atomic nature [105–112], that is, not very sensitive to the environment of the metal ion for a given oxidation state. In the present study, the metal electron configuration is $5f^2$ for each U(IV) ion, whatever the HS or BS state, so that neglecting spin–orbit corrections seems to be a valid assumption. Recent DFT calculations of magnetic exchange coupling for biuranium(IV) and (V) species [77, 78, 90] did not also include spin–orbit corrections.

The GGA Becke–Perdew (BP86) [113, 114] and B3LYP hybrid functionals have been used. For all elements, Triple-ζ Slater-type valence orbitals (STO) augmented by one set of polarization functions were used taken from the ADF/ZORA/TZP database. The frozen-core approximation where the core density is obtained from four-component Dirac–Slater calculations has been applied for all atoms in the case of BP86 calculations: 1*s* core electrons were frozen, respectively, for carbon C[1*s*], nitrogen N[1*s*] and Cl[2*p*]. The actinide U[5*d*] valence space of the heavy elements includes the 5*f*/6*s*/6*p*/6*d*/7*s*/7*p* shells (14 valence electrons). The B3LYP computations are all electron ones.

Our previous work [105–109] and several recent studies [68, 69, 116–120] have shown that ZORA/BP86/TZP computations reproduce experimental geometries and ground state properties of f–element compounds with a satisfying accuracy.

It is noteworthy that several stages were needed to obtain converged results observing an aufbau electron configuration with the electronic structure $5f^2$–$5f^2$ for the U_2ketimide binuclear complex under consideration. The HS state geometry optimization at the ZORA/BP86/TZP level has been carried out first; the smearing procedure implemented in ADF, permitting a fractional occupation of the 5*f* orbitals by the four unpaired electrons ($5f^2$–$5f^2$), was turned on. Then, the electron smearing was progressively reduced until reaching the required 5*f* electrons distribution. Finally, the BS state energy (E_{BS}) has been obtained performing a single point calculation using the MOs of the HS structure as starting guess and changing the spin on the second uranium atom.

Molecular structure drawings, spin densities and molecular orbital plots were generated using the ADF-GUI auxiliary program [100].

3 Results and discussion

3.1 Description of the molecular geometries

The ketimide (1,4-phenylenediketimide) dianionic ligand (Fig. 1) has shown through recent spectroscopic and electronic structure reports [74–77, 115, 119, 120], a great ability to promote ligand–metal charge transfer. It proved also to be an efficient bridging ligand for high-nuclearity molecular frameworks [88]. In our case, the greater radial extension of the 5*f* valence orbitals of uranium can potentially provide increased overlap with the bridging ligand, as a result of the U(5*f*)–N π-bonding interaction that involves the nitrogen lone pair, thereby enhancing the communication between the bridged metal centres [77, 90].

As aforementioned, our targeted systems are the bimetallic model $[(Cp)_2UCl]_2(\mu$-ketimide) called U_2ketimide complex, $[(C_5Me_4Et)_2UCl]_2(\mu$-ketimide) the X-ray characterized [75] one and the hypothetical ThUketimide species. In order to compare the homobimetallic U(IV)/U(IV) U_2ketimide and the mixed Th(IV)/U(IV) ThUketimide complexes, their molecular geometries were fully optimized in C_s symmetry at the ZORA/BP86/TZP level. The optimized geometries have been obtained considering the HS triplet $5f^0$–$5f^2$ ($S = 1$) and quintet $5f^2$–$5f^2$ ($S = 2$) spin states for the ThUketimide and U_2ketimide, respectively. The obtained molecular structures of the model ThUketimide and U_2ketimide complexes are depicted on Fig. 2.

In Table 1 are given relevant optimized bond distances and angles of the ThUketimide and U_2ketimide model complexes, respectively, in their triplet and quintet HS state in the C_s symmetry. Available X-ray structural parameters for the experimental $[(C_5Me_4Et)_2UCl]_2(\mu$-ketimide) and $[(C_5Me_5)_2ThCl]_2(\mu$-ketimide) complexes [75] are also given in this table. The given geometrical parameters are defined on Fig. 2.

As it can be seen in Table 1, there is a good overall agreement between the optimized geometry of the U/U species and the available X-ray structure, reminding that the latter bears more crowded C_5Me_4Et ligands [75] explaining partly why the U–N computed distance is underestimated. Furthermore, comparing the two Th/U and U/U optimized structures, it can be seen that differences in metal–ligand bond lengths are partly due to Th(IV) vs. U(IV) ionic radii variations, the radius of Th(IV) being 0.05 Å larger than the U(IV) one [121] but also to the bonds covalent character which is higher for uranium.

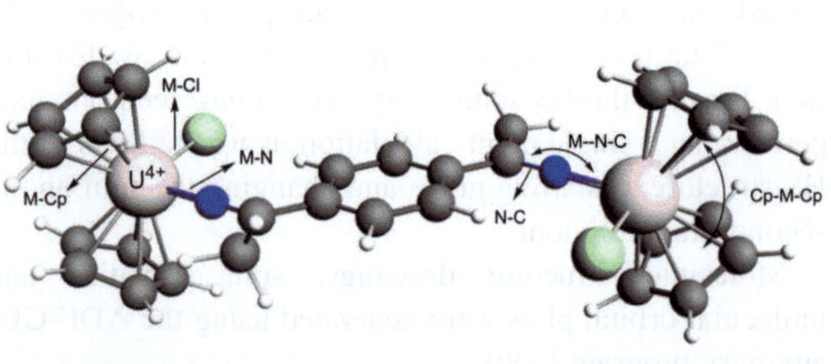

Fig. 2 Relevant structural parameters (values in Table 1)

From Table 1, it is noteworthy that U–N bond distances are in the range of multiple short bonds in agreement with experimental structural data of analogues ketimide or imido complexes found in the literature [74–77, 106, 115, 122]. The quasi-linear U–N–C coordination is also indicative of multiple bonding. Such coordination in the diuranium(IV) complex, throughout the ketimide (1,4-phenylenediketimide) bridging ligand, which appears also in bis(imido) diuranium(V) species [63, 90], is likely to maximize the possibility of electronic and magnetic communication between the two metallic centres. Different authors [106, 115, 120] attributed this peculiar linear U–N–C coordination, to a significant participation of the uranium 5*f* orbitals in metal–N bonding. This U–N(ketimide) coordination exhibits both σ and π interactions between the central metal and bis(ketimide) bridging ligand via the nitrogen electron pairs. We expect also that the 5*f* uranium orbitals are not only responsible for this peculiar geometry, but could play a key role in magnetic exchange interaction cooperatively with the π-conjugated aromatic spacer. This point will be supported by the MO analysis which follows.

3.2 DFT evaluation of the exchange coupling constant *J*

We study now the magnetic character of the complexes under consideration starting from the U_2ketimide model and $[Cp'_2UCl]_2(\mu$-ketimide) the X-ray characterized complex. We remind that the antiferromagnetic character of the latter was not clearly evidenced experimentally by variable temperature measurements [75]. The study of the hypothetical ThUketimide complex will come after. The U_2ketimide systems exhibiting the $5f^2$–$5f^2$ configuration will be considered in its HS(↑↑–↑↑) and BS(↑↑–↓↓) states.

On the basis of the HDvV Hamiltonian ($\hat{H} = -2J\hat{S}_1 \cdot \hat{S}_2$), exchange coupling constant J can be extracted from the energy difference $\Delta E = E_{BS} - E_{HS}$ of the HS and BS states [32–58]. According to the spin-projected (SP) approach [56], the energies of a two-centre A and B complex can be related to the J coupling as:

$$J_{AB} = (E_{BS} - E_{HS})/2S_AS_B \tag{1}$$

while in the non-spin-projected (NP) approach [32], the corresponding energies can be related to the J coupling as:

$$J_{AB} = (E_{BS} - E_{HS})/(2S_AS_B + S_B) \tag{2}$$

where S_A and S_B are the total local spin angular momenta. In our case, $S_A = S_B = 1$ so that Eqs. 1 and 2 lead,

Table 1 Relevant average optimized computed bond (Å) distances and angles (°) of U_2ketimide (U/U) and ThUketimide (Th/U) complexes at the ZORA/BP86/TZP level and X-ray values

M/M′	M–N	M–Cl	N–C	M–M	M–Cp	M–N–C	Cp–M–Cp
U/U	2.088	2.609	1.307	11.057	2.472	178.6	125.1
X-ray	2.250(2)	2.710(13)	1.260(3)	10.956	2.471–2.493	175.3	140.2
Th/U	2.219/2.104	2.666/2.580	1.275/1.296	11.034	2.564/2.458	179.8/179.2	126.8/125.5

Available X-ray structural parameters are for $[(C_5Me_4Et)_2UCl]_2(\mu$-ketimide)

respectively, to $J = \Delta E/2$ and $J = \Delta E/3$. Several authors argued that Eq. 1 represents the more physically meaningful mapping between the HDvV and the DFT/BS models [48, 92].

In addition, the Yamaguchi et al. spin-projected method [94] lead to the following expression of J (3):

$$J_{AB} = (E_{BS} - E_{HS})/\langle S^2\rangle_{HS} - \langle S^2\rangle_{BS} \quad (3)$$

where $\langle S^2\rangle_{HS}$ and $\langle S^2\rangle_{BS}$ are the mean values of the $\hat{S}^2$ squared spin operator, respectively, for the HS and BS states.

To investigate the effect of a small geometry modification on the computed properties, we computed, in addition to the U2ketimide model in its optimized structure, the actual $[Cp'_2UCl]_2(\mu$-ketimide) diuranium(IV) complex using its X-ray geometry [75]. In Table 2 are reported the computed $\Delta E = E_{BS} - E_{HS}$ energy differences, the mean values $\langle S^2\rangle$ of the squared spin operator and the computed exchange coupling constants J(cm^{-1}) using the three above-mentioned different expressions. In this table, computed values for the actual complex and the U2ketimide model have been obtained by means of the BP86 functional and the B3LYP one.

$\langle S^2\rangle$ exhibits the expected values, that is, of the order of 2 for the BS and 6 for the HS states; moreover, a weak spin contamination for the HS state can be noticed. An important observation from the results of Table 2 is that, contrarily to B3LYP, the BP86 functional leads to very high energy differences between the HS and BS states. As stressed by previous authors [32–36, 41–46, 48–55], the resulting values of the exchange coupling constant J, in the GGA case, are systematically larger in absolute values, than those expected for such systems [62, 63, 77, 78]. We noticed the same overestimation using another GGA functional.

We note that B3LYP predicts a rather weak antiferromagnetic character either for the actual complex (−1.7 to −3.4 cm^{-1}) or the U2ketimide model (from −2.3 to −4.6 cm^{-1}). To our knowledge, no coupling constant values have been published up to now for these complexes, so that a direct comparison is not possible. However, such small values of J are not unexpected for bimetallic uranium(IV) complexes [78]. Moreover, it is worth noting that the coupling constants in the case of a U(V)–U(V) $5f^1$–$5f^1$ binuclear complex like para-bis(imido) $[(C_5H_5)_3U]_2$ (μ-imido) uranium(V) are higher in absolute value than for the considered U(IV) $5f^2$–$5f^2$ complexes [60, 90].

The magnetic character of binuclear complexes can be well explained considering the spin density distributions in the HS and BS states, as already demonstrated by several authors [20, 23, 37–40, 72, 78]. Some mechanisms have been proposed to explain the exchange coupling between magnetic centres, that is, the spin polarization and spin delocalization effects, phenomena well known in the case of d-transition metals magneto-chemistry [37–40, 49, 93–97] described early by O. Kahn [47].

The obtained spin density distributions (differences between α and β electron densities) for the U_2ketimide model complex are displayed on Fig. 3.

We note that the HS state exhibits quasi-localized spin densities on the two magnetic uranium centres with non-negligible values on their nearest neighbours but no spin delocalization within the phenyl ring. On the contrary, the BS spin density distribution shows that the two magnetic

Table 2 ΔE (eV and cm^{-1}) energy differences, $\langle S^2\rangle$ values and computed exchange coupling J(cm^{-1}) constant for the U_2ketimide model (optimized geometry) and actual complex (X-ray structure)

Diuranium complex	$\Delta E_{(BS-HS)}$ cm^{-1}	$\langle S^2\rangle_{HS}$	$\langle S^2\rangle_{BS}$	$J^{(1)}$	$J^{(2)}$	$J^{(3)}$
U_2ketimide model						
BP86	−437.2	6.073	1.942	−218.1	−145.1	−105.8
B3LYP	−9.3	6.038	2.039	−4.6	−3.1	−2.3
Actual complex						
BP86	−680.4	6.095	2.044	−340.2	−226.8	−170.1
B3LYP	−6.9	6.046	2.047	−3.4	−2.3	−1.7

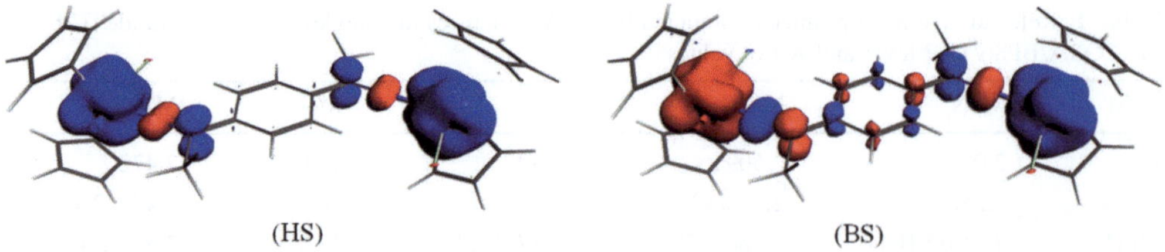

Fig. 3 ZORA/B3LYP spin density distributions for the HS (quintet) and BS states of U_2ketimide complex (blue colour: positive and red colour: negative spin density). The isodensity surface corresponds to a value of 0.0025 e bohr^{-3}

centres are antiferromagnetically coupled, with significant spin density contributions from all bis(ketimide) ligand atoms. The sign alternation of the spin densities along the path linking the two magnetic centres, due to spin polarization, proves the exchange coupling. It is worth noting that, due to the even number of atoms constituting the latter path, only an antiferromagnetic interaction due to spin polarization is possible. We observed the same behaviour in the case of $5f^1$–$5f^1$ bis (imido) uranium complexes [90]; the para isomer for which the bridging path between the magnetic centres contains an even number of atoms is antiferromagnetic, whereas the meta one, containing an odd number of linking atoms, is ferromagnetic. In the latter case, the sign alternation of the spin densities along the bridge is obtained for the HS state of the complex. So, the topology of the linking path determines the magnetic character driven by spin polarization, of such binuclear complexes.

We note that the Cp ligands are practically not involved in this coupling, clarifying partly why the U2ketimide model, that is, $[Cp_2UCl]_2(\mu\text{-ketimide})$ and the actual complex bearing substituted Cps, $[(C_5Me_4Et)_2UCl]_2$ (μ-ketimide) lead to very close J values. The antiferromagnetic coupling is also clarified by the alternating signs of the atomic Mulliken spin polarizations (difference of the α and β gross atomic charges) found in the BS state along the path linking the two magnetic metal centres as given in Table 3. Despite its known drawbacks, the Mulliken population analysis remains a useful tool to compare the variation of electronic populations in homologous series of compounds computed at the same level of theory.

In Table 3, atoms are numbered as indicated on Fig. 4; the computed values for the X-ray structure are given between parentheses.

As expected on the basis of the electron spin density distributions, the atomic spin polarizations of the atoms of the linking bridge increase significantly in absolute values when passing from the HS to BS state. This confirms the crucial rule of the bridging ligand to promote metal–metal electronic communications leading to exchange coupling interactions within an antiferromagnetic state. It is also interesting to note that that the model complex and the actual one exhibit almost the same spin populations in the HS state, whereas they differ slightly in the BS state.

Table 3 ZORA/B3LYP Mulliken spin populations for HS and BS states of U_2ketimide model complex (computed values for the actual complex given between parentheses)

U_2ketimide	HS		BS	
Atoms				
U1	2.148	(2.174)	2.146	(2.172)
Cl2	−0.043	(−0.042)	−0.041	(−0.022)
N3	−0.090	(−0.087)	−0.129	(−0.110)
C4	0.136	(0.131)	0.126	(0.114)
C5	−0.017	(−0.016)	−0.047	(−0.036)
C6	−0.013	(−0.010)	−0.057	(−0.039)
C7	0.018	(0.016)	0.037	(0.021)
C8	0.013	(0.011)	−0.043	(−0.022)
C9	−0.013	(−0.014)	0.057	(0.029)
C10	0.011	(0.012)	−0.037	(−0.021)
C11	0.013	(0.010)	0.043	(0.022)
C12	0.136	(0.110)	−0.126	(−0.114)
C13	−0.017	(−0.012)	0.017	(0.015)
N14	−0.090	(−0.087)	0.129	(0.110)
U15	2.149	(2.174)	−2.147	(−2.172)
Cl16	−0.044	(−0.042)	0.041	(0.022)

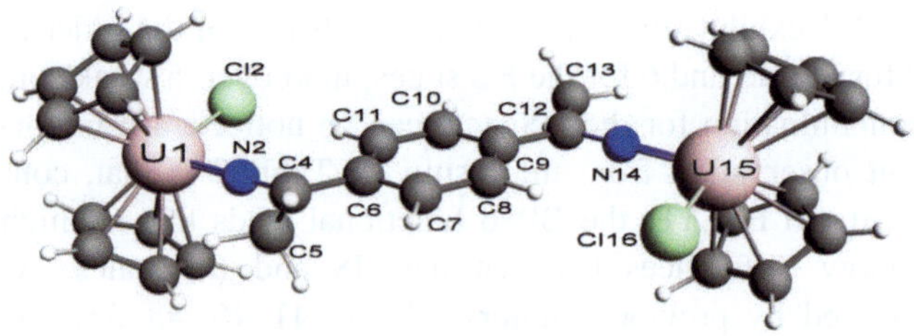

Fig. 4 Atom numbering of the U_2ketimide model

The effect of the replacement of one paramagnetic metal $U(5f^2)$ by a diamagnetic $Th(5f^0)$ one on the atomic spin populations appears in Table 4 (same atoms numbering as Fig. 3) where are reported these values for the Th^{IV}/U^{IV} ThUketimide hypothetical complex. We remind that such mixed metal complex, which has not yet been isolated experimentally, is designed to estimate the coupling constant in the U/U complex [75] by diamagnetic subtraction analysis [60].

Table 4 ZORA/B3LYP Mulliken spin populations of the ThUketimide model complex

Atoms	U1	Cl2	N3	C4	C5	C6	C7	C8	C9	C10	C11	C12	C13	N14	Th15	Cl16
	2.125	−0.023	−0.046	0.004	−0.002	−0.003	0.000	0.000	0.000	0.000	0.000	0.000	0.000	0.000	0.000	0.000

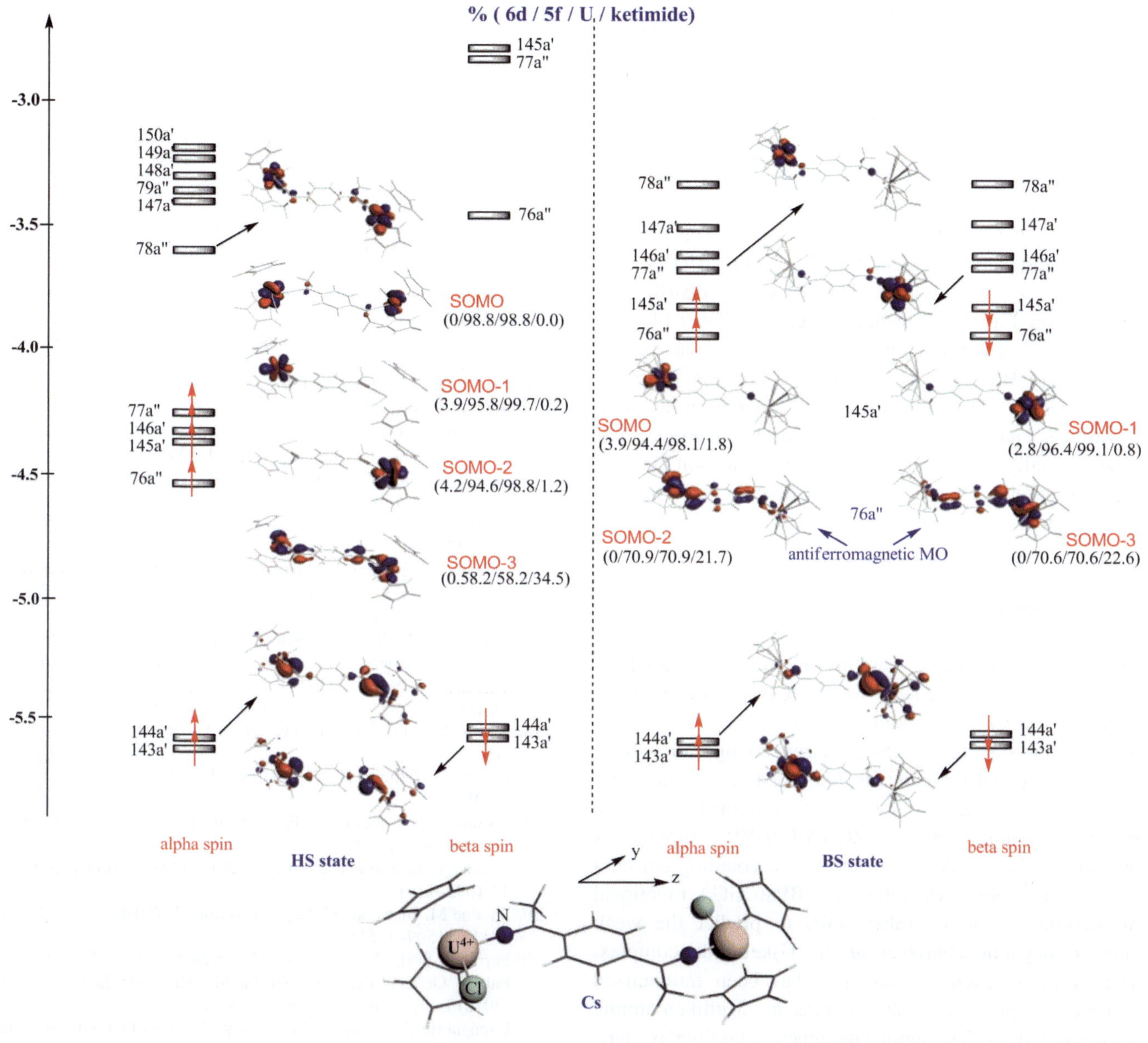

Fig. 5 B3LYP Frontier MO diagrams of HS and BS states for the U_2ketimide model complex in the C_s(yz) symmetry

From results of Table 4, it can be seen that the spin populations are equal to zero for thorium, the chlorine and nitrogen atoms directly bonded to it and for the atoms of the bridging ligand. Cl2 and N3 atoms, linked to the paramagnetic U(IV) ion, are the only ones for which the spin populations are relatively high with negative values owing to the spin polarization effect from the U(IV) spin centre. This mixed system illustrates well that replacing the paramagnetic centre U(IV) by its diamagnetic analogue Th(IV) affects drastically the spin polarization effect; the spin densities tend to zero beyond the first neighbours of the paramagnetic centre. Obviously, no magnetic exchange interaction could occur in such system.

The MO diagram of the U2ketimide model in C_s(yz) symmetry depicted on Fig. 5 displays the singly occupied molecular orbitals (SOMOs) of the HS state at the left side, with the corresponding BS SOMOs at the right side. In this diagram, %(6*d*/5*f*/U/ligand) represents the weight of, respectively, the $6d$, $5f$ orbitals, the uranium atoms and the bridging ligand within the displayed MOs.

In the HS quintet state, the four highest α SOMOs, SOMO-3 (76a″) excepted, that is, SOMOs #77a″, 145a′ and 146a′, are essentially $5f$ orbitals of the two uranium(IV) atoms with no contribution from the bridging ketimide ligand. On the contrary, in the BS state, SOMOs # α-76a″ and β-76a″ are significantly different and appear to be more delocalized with non-negligible contribution of the bis(ketimide) ligand as testified by the orbital percentages. In the BS state, two different pairs of molecular levels can be distinguished; the first one of a′ symmetry is essentially of localized $5f$ character, the second, with a″ symmetry and, combining α and β spin-parts, is rather well delocalized over the path linking the two spin centres U(IV). This picture agrees well with the spin density distribution, depicted on Fig. 4. Immediately below, MOs #144 and 143, with a′ symmetry, remain unchanged when passing from HS to BS state. These MOs of bonding character exhibit large π-bonding interactions, involving uranium $5f$ orbitals and π-orbitals of the N-bridging atoms. The experimentally observed electronic communication between the two uranium atoms is also well described by the MO description of the complex.

4 Conclusions

The magnetic exchange coupling interactions in bis(ketimide) $[(C_5H_5)_2(Cl)An]_2(\mu\text{-}\{N{=}CMe\text{-}(C_6H_4)\text{-}MeC{=}N\})$ and $[(C_5Me_4Et)_2UCl]_2(\mu\text{-ketimide})$ binuclear U^{IV}/U^{IV} complexes exhibiting the $5f^2$–$5f^2$ electron configuration have been investigated theoretically for a first time using relativistic DFT/ZORA computations combined with the broken symmetry approach. ZORA/B3LYP computations predict a rather weak antiferromagnetic coupling constant $|J| < 5\ cm^{-1}$. Not surprisingly, the BP86 GGA functional overestimating the $|J|$ values fails to predict the weak antiferromagnetic character of the U_2ketimide complex. The magnetic exchange coupling has been rationalized considering spin density distributions and Mulliken atomic spin populations. The antiferromagnetic coupling is clarified by the alternating signs of the atomic spin populations along the path linking the two magnetic metal centres occurring in the BS state. The fact that the number of atoms along the path linking the two metal ions is even leads to antiferromagnetic coupling, the latter being mainly due to spin polarization. Furthermore, from the MO point of view, the antiferromagnetic interaction between the two uranium(IV) ions mediated by the aromatic ketimide bridge is mainly due to the effective π-overlap between $5f$ orbitals and nitrogen atoms on bridging ligand groups along the path linking the two magnetic uranium(IV) centres.

Acknowledgments Financial support from the Algerian National Administration of Scientific Research NASR-ANDRU (PNR Grant No. 8/u250/4169) is gratefully acknowledged. Computing facilities were provided by IDRIS Computing Centre of CNRS.

References

1. Bencini A, Benelli C, Caneschi A, Carlin RL, Dei A, Gatteschi D (1985) J Am Chem Soc 107:8128–8136
2. Benelli C, Gatteschi D (2002) Chem Rev 102:2369–2387
3. Sessoli R, Gatteschi D (2003) Angew Chem Int Ed 42:268–297
4. Long JR (2003) Molecular cluster magnets. In: Yang P (ed) Chemistry of nanostructured materials. World Scientific, Hong Kong, pp 291–315
5. Sessoli R, Tsai HL, Schake AR, Wang SY, Vincent JB, Folting K, Gatteschi D, Christou G, Hendrickson DN (1993) J Am Chem Soc 115:1804–1816
6. Sessoli R, Gatteschi D, Caneschi A, Novak MA (1993) Nature 365:14–143
7. Osanai K, Okazawa A, Nogami T, Ishida T (2006) J Am Chem Soc 128:14008–14009
8. Atakol O, Boca R, Ercan I, Ercan F, Fuess H, Haase W, Herchel R (2006) Chem Phys Lett 423:192–196
9. Calzado CJ, Clemente-Juan JM, Coronado E, Gaita-Arino A, Suaud N (2008) Inorg Chem 47:5889–5901
10. Kim JI, Kwak HY, Yoon JH, Ryu DW, Yoo IY, Yang N (2009) Ki Cho B, Park JG, Lee H, Hong CS. Inorg Chem 48:2956–2966
11. Andruh M, Costes JP, Diaz C, Gao S (2009) Inorg Chem 48:3342–3359
12. Affronte M, Troiani F, Ghirri A, Carretta S, Santini P, Corradini V, Schuecker R, Muryn C, Timco G, Winpenny RE (2006) Dalton Trans:2810–2817
13. Milios CJ, Inglis R, Vinslava A, Bagai R, Wernsdorfer W, Parsons S, Perlepes SP, Christou G, Brechin EK (2007) J Am Chem Soc 129:12505–12511
14. Bogani L, Wernsdorfer W (2008) Nat Mater 7:179–186
15. Adamo C, Barone V, Subra R (2000) Theor Chem Acc 104:207–209
16. Barone V, Bencini A, Gatteschi D, Totti F (2002) Chem Eur J 8:5019–5027
17. Barone V, Cacelli I, Ferretti A (2009) J Chem Phys 130:094306–094309
18. Barone V, Bloino J, Biczysko M (2010) Phys Chem Chem Phys 12:1092–1101
19. Pavone M, Biczysko M, Rega N, Barone V (2010) J Phys Chem B 114:11509–11514
20. Ferrando-Soria J, Castellano M, Yuste C, Lloret F, Julve M, Fabelo O, Ruiz-Pérez C, Stiriba SE, Ruiz-García R, Cano J (2010) Inorg Chim Acta 363:1666–1678
21. Labéguerie P, Rohmer MM, Bénard M (2009) J Chin Chem Soc 56:22–25
22. Ismayilov RH, Wang WZ, Lee GH, Yeh CY, Hua SA, Song Y, Rohmer MM, Bénard M, Peng SM (2011) Angew Chem Int Ed 50:2045–2048
23. Gillon B, Mathonire C, Ruiz E, Alvarez S, Cousson A, Rajendiran TM, Kahn O (2002) J Am Chem Soc 124:14433–14441
24. Ciofini I, Daul CA (2003) Coord Chem Rev 238–239:187–209
25. Costuas K, Valenzuela ML, Vega A, Moreno Y, Pena O, Spodine E, Saillard JY, Diaz C (2002) Inorg Chim Acta 329:129–134
26. Kortus J (2007) C R Chimie 10:65–67
27. Castro I, Calatayud ML, Sletten J, Lloret F, Julve M (1997) Dalton Trans:811–818
28. Fink K, Fink R, Staemmler V (1994) Inorg Chem 33:6219–6229

29. Bencini A, Costes JP, Dahan F, Dupuis A, Garcia-Tojal J, Gatteschi D, Totti F (2004) Inorg Chim Acta 357:2150–2156
30. Albonico C, Bencini A (1988) Inorg Chem 27:1934–1940
31. Korzeniak T, Desplanches C, Podgajny R, Giménez-Saiz C, Stadnicka K, Rams M, Sieklucka B (2009) Inorg Chem 48:2865–2872
32. Ruiz E, Cano J, Alvarez S, Alemany P (1999) J Comput Chem 20:1391–1400
33. Ruiz E, Rodríguez-Fortea A, Cano J, Alvarez S, Alemany P (2003) J Comput Chem 24:982–989
34. Ruiz E, Rodriguez-Fortea A, Tercero J, Cauchy T, Massobrio CJ (2005) Chem Phys 123:074102–074110
35. Nunzi F, Ruiz E, Cano J, Alvarez S (2007) J Phys Chem C 111:618–621
36. Ruiz E, Cauchy T, Cano J, Costa R, Tercero J, Alvarez S (2008) J Am Chem Soc 130:7420–7426
37. Noh EAA, Zhang J (2006) Chem Phys 330:82–89
38. Noh EAA, Zhang J (2008) J Mol Struct THEOCHEM 867:33–38
39. Noh EAA, Zhang J (2009) J Mol Struct THEOCHEM 896:54–62
40. Yan F, Chen Z (2000) J Phys Chem A 104:6295–6300
41. Adamo C, Barone V, Bencini A, Totti F, Ciofini I (1999) Inorg Chem 38:1996–2004
42. Adamo C, Barone V, Bencini A, Broer R, Filatov M, Harrison NM, Illas F, Malrieu JP (2006) Moreira IdPR. J Chem Phys 124:107101–107103
43. Atanasov M, Comba P, Daul CA (2006) J Phys Chem A 110:13332–13340
44. Atanasov M, Comba P, Daul CA (2008) Inorg Chem 47:2449–2463
45. Atanasov M, Comba P, Hausberg S, Martin B (2009) Coord Chem Rev 253:2306–2314
46. Comba P, Hausberg S, Martin B (2009) J Phys Chem A 113:6751–6755
47. Kahn O (1993) Molecular Magnetism. VCH, New York
48. Moreira IdPR, Costa R, Filatov M, Illas F (2007) J Chem Theory Comput 3:764–774
49. Bencini A (2008) Inorg Chim Acta 361:3820–3831
50. Neese F (2009) Coord Chem Rev 253:526–563
51. Cramer CJ (2009) Truhlar D G. Phys Chem Chem Phys 11:10757–10816
52. Onofrio N, Mouesca JM (2011) Inorg Chem 50:5577–5586
53. Zhekova H, Seth M, Ziegler T (2011) J Chem Theory Comput 7:1858–1866
54. Peralta JE, Melo JI (2010) J Chem Theory Comput 6:1894–1899
55. Roy LE, Hughbanks T (2006) J Am Chem Soc 128:568–575
56. Noodleman LJ (1981) J Chem Phys 74:5737–5743
57. Noodleman LJ, Davidson ER (1986) Chem Phys 109:131–143
58. Noodleman LJ, Peng CY, Case DA, Mouesca JM (1995) Coord Chem Rev 144:199–244
59. Ephritikhine M (2006) Dalton Trans:2501–2516
60. Lukens WW, Walter MD (2010) Inorg Chem 49:4458–4465
61. Minasian SG, Krinsky JL, Rinehart JD, Copping R, Tyliszczak T, Janousch M, Shuh DK, Arnold J (2009) J Am Chem Soc 131:13767–13783
62. Rinehart JD, Harris TD, Kozimor SA, Bartlett BM, Long JR (2009) Inorg Chem 48:3382–3395
63. Rosen RK, Andersen RA, Edelstein NM (1990) J Am Chem Soc 112:4588–4590
64. Diaconescu PL, Arnold PL, Baker TA, Mindiola DJ (2000) Cummins C C. J Am Chem Soc 122:6108–6109
65. Fox AR, Bart SC, Meyer K, Cummins CC (2008) Nature 455:341–349
66. Jilek RE, Spencer LP, Kuiper DL, Scott BL, Williams UJ, Kikkawa JM, Schelter EJ, Boncella JM (2011) Inorg Chem 50:4235–4237
67. Roussel P, Errington W, Kaltsoyannis N, Scott P (2001) J Organomet Chem 635:69–74
68. Cloke FGN, Green JC, Kaltsoyannis N (2004) Organometallics 23:832–835
69. Gaunt AJ, Reilly SD, Enriquez AE, Scott BL, Ibers JA, Sekar P, Ingram KIM, Kaltsoyannis N, Neu MP (2008) Inorg Chem 47:29–41
70. Evans WJ, Montalvo E, Kozimor SA, Miller KA (2008) J Am Chem Soc 130:12258–12259
71. Kozimor SA, Bartlett BM, Rinehart JD, Long JR (2007) J Am Chem Soc 129:10672–10674
72. Monreal MJ, Carver CT, Diaconescu PL (2007) Inorg Chem 46:7226–7228
73. Rajaraman G, Totti F, Bencini A, Caneschi A, Sessoli R, Gatteschi D (2009) Dalton Trans 3153–3161
74. Schelter EJ, Veauthier JM, Thompson JD, Scott BL, John KD, Morris DE, Kiplinger JL (2006) J Am Chem Soc 128:2198–2199
75. Schelter EJ, Veauthier JM, Graves CR, John KD, Scott BL, Thompson JD, Pool-Davis-Tournear JA, Morris DE, Kiplinger JL (2008) Chem Eur J 14:7782–7790
76. Veauthier JM, Schelter EJ, Carlson CN, Scott BL, Da Re RE, Thompson JD, Kiplinger JL, Morris DE, John KD (2008) Inorg Chem 47:5841–5849
77. Spencer LP, Schelter EJ, Yang P, Gdula RL, Scott BL, Thompson JD, Kiplinger JL, Batista ER, Boncella JM (2009) Angew Chem Int Ed 48:3795–3798
78. Newell BS, Rapp AK, Shores MP (2010) Inorg Chem 49:1595–1606
79. Berthet JC, Ephritikhine M (1998) Coord Chem Rev 83:178–180
80. Borgne TL, Lance M, Nierlich M, Ephritikhine M (2000) J Organomet Chem 598:313–317
81. Le Borgne T, Rivière E, Marrot J, Thuéry P, Girerd JJ, Ephritikhine M (2002) Chem Eur J 8:773–783
82. Salmon L, Thuéry P, Rivière E, Girerd JJ, Ephritikhine M (2003) Dalton Trans:2872–2880
83. Salmon L, Thuéry P, Rivière E, Ephritikhine M (2006) Inorg Chem 45:83–93
84. Diaconescu PL, Cummins CC (2002) J Am Chem Soc 124:7660–7661
85. Evans WJ, Kozimor SA, Ziller JW, Kaltsoyannis N (2004) J Am Chem Soc 126:14533–14547
86. Monreal MJ, Diaconescu PL (2008) Organometallics 27:1702–1706
87. Nocton G, Horeglad P, Pécaut J, Mazzanti M (2008) J Am Chem Soc 130:16633–16645
88. Schelter EJ, Wu R, Scott BL, Thompson JD, Morris DE, Kiplinger JL (2008) Angew Chem Int Ed 47:2993–2996
89. Mills DP, Moro F, McMaster J, van Slageren J, Lewis W, Blake AJ, Liddle ST (2011) Nature Chemistry 3:454–460
90. Meskaldji S, Belkhiri A, Belkhiri L, Boucekkine A, Ephritikhine M (2011) C R Chimie (in press)
91. Caballol R, Castell O, Illas F, Moreira IP, Malrieu JP (1997) J Phys Chem A 101:7860–7866
92. Figgis BN, Hitchman MA (2000) Ligand field theory and its applications. Wiley, London
93. Fouqueau A, Casida ME, Daku LML, Hauser A, Neese F (2005) J Chem Phys 122:044110–044123
94. Mitani M, Mori H, Takano Y, Yamaki D, Yoshioka Y, Yamaguchi K (2000) J Chem Phys 113:4035–4051
95. Takano Y, Kitagawa Y, Onishi T, Yoshioka Y, Yamaguchi K, Koga N, Iwamura H (2002) J Am Chem Soc 124:450–461
96. Onishi T, Yamaki D, Yamaguchi K, Takano Y (2003) J Chem Phys 118:9747–9761

97. Shoji M, Koizumi K, Kitagawa Y, Kawakami T, Yamanaka S, Okumura M, Yamaguchi K (2006) Chem Phys Lett 432:343–347
98. Becke AD (1993) J Chem Phys 98:5648–5652
99. Lee C, Yang W, Parr RG (1988) Phys Rev B 37:785–789
100. ADF2010.02, SCM; Theoretical chemistry, Vrije University: Amsterdam, The Netherlands. http://www.scm.com
101. van Lenthe E, Baerends EJ, Snijders JG (1993) J Chem Phys 99:4597–4610
102. van Lenthe E, Baerends EJ, Snijders JG (1994) J Chem Phys 101:9783–9792
103. van Lenthe E, Ehlers A, Baerends EJ (1999) J Chem Phys 110:8943–8953
104. van Lenthe E, Snijders JG, Baerends EJ (1996) J Chem Phys 105:6505–6516
105. Elkechai A, Belkhiri L, Amarouche M, Clappe C, Boucekkine A, Hauchard D, Ephritikhine M (2009) Dalton Trans:2843–2849
106. Belkhiri L, Lissillour R, Boucekkine A (2005) J Mol Struct THEOCHEM 757:155–164
107. El Kechai A, Meskaldji S, Boucekkine A, Belkhiri L, Bouchet D, Amarouche M, Clappe C, Hauchard D, Ephritikhine M (2010) J Mol Struct THEOCHEM 954:115–123
108. Meskaldji S, Belkhiri L, Arliguie T, Fourmigué M, Ephritikhine M, Boucekkine A (2010) Inorg Chem 49:3192–3200
109. El Kechai A, Boucekkine A, Belkhiri L, Hauchard D, Clappe C, Ephritikhine M (2010) C R Chimie 13:860–869
110. Schreckenbach G, Shamov GA (2010) Acc Chem Res 43:19–29
111. Xiao H, Li J (2008) Chin J Struct Chem 27:967–974
112. García-Hernandez M, Lauterbach C, Krüger S, Matveev A, Rösch N (2002) J Comput Chem 23:834–846
113. Becke AD (1988) Phys Rev A 38:3098–3100
114. Perdew JP (1986) Phys Rev B 34:7406–7409
115. Graves CR, Yang P, Kozimor SA, Vaughn AE, Clark DL, Conradson SD, Schelter EJ, Scott BL, Thompson JD, Hay PJ, Morris DE, Kiplinger JL (2008) J Am Chem Soc 130:5272–5285
116. Fox AR, Creutz SE, Cummins CC (2010) Dalton Trans 39:6632–6634
117. Cavigliasso G, Kaltsoyannis N (2006) Inorg Chem 45:6828–6839
118. Shamov GA, Schreckenbach G (2005) J Phys Chem A 109:10961–10974
119. Schelter E J, Yang P, Scott BL, Thompson JD, Martin RL, Hay PJ, Morris DE, Kiplinger JL (2007) Inorg Chem 46:7477–7488
120. Clark AE, Martin RL, Hay PJ, Green JC, Jantunen KC, Kiplinger JL (2005) J Phys Chem A 109:5481–5491
121. Shannon RD (1976) Acta Crystallogr A 32:751–767
122. Brennan JG, Andersen R (1985) J Am Chem Soc 107:514–516

Theor Chem Acc (2012) 131:1154
DOI 10.1007/s00214-012-1154-6

REGULAR ARTICLE

Theoretical investigation of molecular excited states in polar organic monolayers via an efficient embedding approach

Aleksandrs Terentjevs · Eduardo Fabiano · Fabio Della Sala

Received: 28 June 2011 / Accepted: 13 September 2011 / Published online: 17 February 2012

Abstract In this work, we present a theoretical investigation on excitation energies of organic molecules embedded in a periodic monolayer. We use the self-consistent periodic-image-charges embedding approach, which takes into account all the electrostatic effects, to compute the perturbation on molecular orbitals and eigenvalues due to the presence of the surrounding periodic array of polar molecules. We considered vanadyl naphthalocyanine, mercaptobiphenyl, and tris-(8-hydroxyquinoline) aluminum (AlQ_3) at different coverages, and excitation energies computed using the time-dependent density-functional theory. We found a significant (0.1–0.2 eV) red- or blue-shift of the energies for different excited states, due to the different coupling of the molecule with the polarization field of the two-dimensional crystal.

Keywords Self-assembled monolayers · Time-dependent density-functional theory · QM/MM · Excited state · Polarizability · Workfunction shift

Dedicated to Professor Vincenzo Barone and published as part of the special collection of articles celebrating his 60th birthday.

A. Terentjevs · E. Fabiano · F. Della Sala (✉)
National Nanotechnology Laboratory (NNL), Istituto Nanoscienze-CNR, Via per Arnesano 16, 73100 Lecce, Italy
e-mail: fabio.dellasala@unisalento.it

F. Della Sala
Center for Biomolecular Nanotechnologies @UNILE, Istituto Italiano di Tecnologia (IIT), Via Barsanti, 73010 Arnesano, Italy

1 Introduction

Spontaneously assembled ordered single layers of polar organic molecules (SAMs) play an important role in many research areas [1–4]. In particular, they are largely employed in molecular (opto)electronics [5] to modify the effective work function of a surface and achieve a fine tuning of charge injection barriers and level alignment with additional organic adsorbates [6].

The work-function shift ($\Delta\Phi$) induced by a SAM of polar molecules is given (in atomic units) by the Helmholtz equation [7]

$$\Delta\Phi(n) = -4\pi n \mu_z(n) \tag{1}$$

where n is the number of molecules per unit area, μ_z is the z-component of the electronic molecular dipole, and we assumed the z-axis to be perpendicular to the surface. The molecular dipole can be further decomposed as $\mu_z(n) = \mu_z^{\mathrm{SAM}}(n) + \mu_z^{\mathrm{bond}}$ where μ_z^{SAM} is the z-component of the dipole moment of a molecule in the isolated SAM (i.e., without any substrate) and μ_z^{bond} is the so called bond-dipole [8–10]. The latter is determined by the interaction of the molecules with the surface and represents usually a rather small and local effect, quite independent on the structure of the SAM [8–10]. On the contrary, the dominating SAM's dipole depends strongly on the electronic properties of the molecules constituting the SAM and from the exact SAM's structure [7, 11].

Considering a SAM of polar molecules, in first approximation, as a two-dimensional (2D) infinite periodic array of point dipoles, μ_z^{SAM} is given by the Topping formula [12–14]

$$\mu_z^{\mathrm{SAM}}(n) = \frac{\mu_z^0}{1 + n^{3/2} f \alpha_{zz}} \tag{2}$$

with μ_z^0 the z-component of the dipole moment of the isolated molecule, α_{zz} the zz-component of the electric polarizability tensor, and f a numerical factor depending on the lattice periodic structure [12]. The Topping formula Eq. 2 provides an efficient qualitative tool to study depolarization effects in SAMs but, due to its approximations, suffers for several drawbacks, including the lack local-field and high-multipole effects, which are especially severe at high coverages [11].

An improved description of the problem is obtained performing full quantum simulations of the entire 2D system (see, e.g., Refs. [8–10]), using density-functional theory (DFT), with plane wave (PW) basis functions and periodic boundary conditions, within a supercell approach using a dipole-correction scheme [15]. Nevertheless, this approach has a high computational cost for low- and medium-coverage regimes [16] and, more importantly, is practically restricted to local or semilocal density functionals. This latter constrain implies several limitations in the description of the electronic structure of organic molecules [17–19], for example, the dipole moment and polarizability of conjugated systems [11, 20–22]. Alternative computational approaches, possibly including hybrid [23] and orbital-dependent treatments [24–26], are therefore desiderable.

Recently, we proposed an alternative approach to treat the depolarization in SAMs of polar organic molecules by introducing the self-consistent periodic-image-charges embedding (SPICE) method [11]. This is based on localized-orbitals quantum-mechanical calculations within a periodic-charge embedding scheme and allows to treat at any level of theory (including, e.g., coupled-cluster methods) the depolarization effect with good accuracy and high efficiency. In previous works [11, 22], the SPICE method was applied only to acquire information on the total ground-state density and the corresponding work-function shift. However, it is more general and can be applied also to excited states, similarly as other embedding methods [27]. In particular, it can be used to compute the perturbation within a SAM, due to the surrounding molecules, on the wavefunction and eigenvalues of different molecular orbitals and thus also modifications of the excitation energies (see Sect. 3).

In this paper, we consider this new aspect and we focus on the modifications of the excitation spectrum of molecular species when these are packed together to form a SAM extending infinitely in two dimensions. We will consider the case in which only a single molecule is excited, whereas all the other molecules are frozen in their ground state. Such a description is well suited to model optical experiments in molecular films and crystals in which the population of excited molecules is very low [28, 29]. In addition, as the lowest excited states of organic molecules are rather localized and we are interested in the low coverage regime, the electron-hopping probability among different molecules is small. In fact, we are not aiming to model the excitation energies of a molecular crystal or slab, whose excited states need to described in term of excitonic and/or charge-transfer bands [30–33]. Our approach is instead relevant to compute directly the polarization term of the excitonic band-structure [34, 35], to describe the excited-state depolarization in thin organic films [29], to study the dependence of single-molecule excited states on the coverage [36], and to describe the modifications of the photoluminescence of single molecule in a molecular crystal [34, 37].

To this end, we apply the SPICE method to study the depolarization effect and the single-molecule excitation spectrum in three typical SAMs with significant dipole moment and employed in molecular electronic and optoelectronic applications, namely those of vanadyl naphthalocyanine (VONc) [38], mercaptobiphenyl [39, 40], and tris-(8-hydroxyquinoline) aluminum (AlQ_3) [41] (for the latter both the facial and the meridional configurations were considered). The geometries and the unit cells of the all considered systems are reported in Fig. 1.

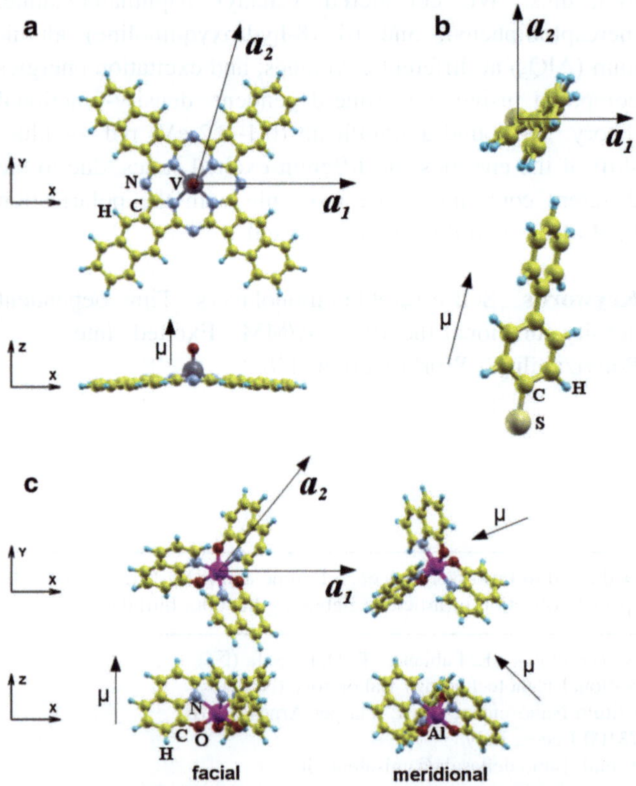

Fig. 1 Investigated systems: **a** VONc, **b** Biphenyl, **c** AlQ_3 facial and meridional. The dipole moment $\boldsymbol{\mu}$ and the lattice unit vectors $\mathbf{a}_1$ and $\mathbf{a}_2$ are also indicated

VONc has been largely investigated from an experimental point view [29, 36, 38, 42] due to its large dipole moment perpendicular to a large macrocyle; mercaptobiphenyl has been the subject of many theoretical studies as a model system of SAM/metal interfaces [10, 16, 39, 40]; AlQ_3 is a molecule often employed in the realization of optoelectronic devices [43–47].

2 Method

The ground-state electronic structure of a single, isolated molecule can be obtained, within the Kohn–Sham DFT or Hartree–Fock approach, solving

$$\hat{H}^0[\rho^0]\phi_i^0(\mathbf{r}) = \epsilon_i^0\phi_i^0(\mathbf{r}), \tag{3}$$

where H^0 is the Hamiltonian of a single isolated molecule, $\rho^0(\mathbf{r}) = \sum_{i=1}^{N/2} |\phi_i^0(\mathbf{r})|^2$ is the electron density (for simplicity we consider a closed-shell system with $N/2$ occupied orbitals), $\{\phi_i^0\}$ are the Kohn–Sham orbitals, and the superscript ^{0} denotes the ground-state unperturbed (isolated) quantities. The Hamiltonian contains the kinetic, the external potential, and the Coulomb-exchange-correlation operator. Therefore, it depends on ρ^0 (thus on $\{\phi_i^0\}$), requiring a self-consistent solution of Eq. 3.

To include the perturbation due to the 2D monolayer, we consider the perturbed Hamiltonian of a single unit cell (for simplicity we consider the case of one molecule per unit cell)

$$\hat{H}[\rho] = \hat{H}^0[\rho] + V_{\mathrm{emb}}[\rho](\mathbf{r}), \tag{4}$$

where ρ is now the perturbed molecular density, $H^0[\rho]$ is the Hamiltonian of a single isolated molecule but computed at the perturbed density, and V_{emb} is the embedding potential including the interaction of the molecule with the rest of the monolayer. The perturbed molecular orbitals (ϕ_i) and eigenvalues (ϵ_i) are then obtained from the equation

$$\hat{H}[\rho]\phi_i(\mathbf{r}) = \epsilon_i\phi_i(\mathbf{r}), \tag{5}$$

with $\rho(\mathbf{r}) = \sum_{i=1}^{N/2} |\phi_i(\mathbf{r})|^2$.

Within the SPICE method, the embedding potential V_{emb} is restricted to a Coulomb interaction and is given by

$$V_{\mathrm{emb}}[\rho](\mathbf{r}) = \sum_{(i,j)\neq(0,0)} \int d\mathbf{r}' \frac{q(\mathbf{r}' - i\mathbf{a}_1 - j\mathbf{a}_2)}{|\mathbf{r} - \mathbf{r}'|}, \tag{6}$$

with q the sum of the (fixed) nuclear and electronic charge densities. Equation 6 represents the electrostatic potential on the central molecule (i.e., with $i = j = 0$) due to all the surrounding molecules in the monolayer. It can be computed efficiently, introducing only a small overhead with respect to a single-molecule calculation, through the fast multipole method [48] after an expansion of q in localized charges (e.g., Mulliken charges). The embedding potential can be also generalized to include kinetic-exchange-correlation effects [49, 50], but these introduce additional computational complications and are only relevant for the high coverage regime. The coverage dependence is implicitly included in Eq. 6 through the lattice vectors $\mathbf{a}_1$ and $\mathbf{a}_2$.

Equations 4, 5, and 6 define the SPICE method and are solved in a double self-consistent loop requiring typically 3–5 steps for convergence.

It is worth to underline that (i) in the SPICE method, the full periodic structure of the monolayer is taken into account, and thus, it is different from other approaches that employ a finite cluster of molecules to model the environment [28, 51, 52] and are limited by the slow convergence of the electronic properties with system size; (ii) the SPICE method does not need any external parametrization, as it is required instead to model the QM/MM interaction with a solvent, nor the computation of the atom–atom polarizabilities [53] because the QM molecule is interacting with *images* of itself, which are computed at the same QM level of theory, and thus with the exact QM density and polarization.

The perturbed molecular orbitals (ϕ_i) and eigenvalues (ϵ_i) resulting from the solution of Eq. 5 differ from the ones of the isolated molecule (ϕ_i^0, ϵ_i^0) due to the electrostatic interaction with the surrounding molecules. They can be used to evaluate the ground-state electronic properties of the monolayer directly or through their use in post-Hartree–Fock wavefunction methods (e.g., CC2). In addition, excitation energies can then be evaluated via the linear response time-dependent DFT (TD-DFT) [54] formalism, using the perturbed orbitals and eigenvalues, or applying correlated wavefunction methods.

Note that in the present approach, no response of the environment to the excitation is included in the TD-DFT kernel. Therefore, we only model single-molecule excitations in a static embedding environment (analogous with static solvation models or biological environment). However, as discussed in the introduction, we are not aiming at describing excitonic splittings and long-range dielectric effects are expected to be quite small in a thin monolayer. Nevertheless, dynamic embedding effects can be possibly incorporated by considering the TD-DFT extension of the FDE theory [55–57].

3 Computational details

For all molecules, the gas-phase equilibrium geometry optimized at the APBE/def2-TZVP [58–60] level was considered. This is expected to be a fair choice for the relatively low coverages studied in this work. The APBE functional was chosen because of its ability to yield reasonably accurate geometries for both conjugated and metal-organic molecules [58].

The VONc monolayer was constructed considering high-resolution STM studies of Fe-phthalocyanine on an Au (111) surface [61], using a monoclinic cell with an angle of 81.8° between the primitive vectors (see Fig. 1a) and lattice constants varying from 18 Å (full coverage) to 53 Å. The use of experimental data referring to a Fe-phthalocyanine SAM was motivated by the lack of information in literature about the structure of SAMs of VONc molecules. This choice is expected anyway to have a negligible impact on the results of the present study.

For mercaptobiphenyl, a simple square cell was considered (see Fig. 1b). The molecule was assumed to be tilted with respect to the 2D plane by 10°, in agreement with PW-DFT calculations [40]. The cell-side dimensions were varied from 7.4 to 53 Å.

Finally, for both AlQ_3 configurations, we considered a monoclinic cell with an angle of 60° between the lattice vectors (see Fig. 1c). This was deduced from previous theoretical works considering the absorption of AlQ_3 on an aluminum (111) surface [41]. The dimensions of the cell were ranging from 11.6 to 53 Å.

All calculations were performed using the TURBOMOLE program package [62]. The APBE [58], BLYP [63, 64], B3LYP [23, 64], BHLYP [65], and PBE0 [66] functionals were considered, and the def2-TZVP basis set [59, 60] was used in all cases. In addition, approximate single and double coupled-cluster (CC2) [67, 68] calculations were performed for benchmark purposes using the cc-pVTZ [69] basis set.

4 Results

4.1 VONc

In this molecule, a large dipole moment is present due to the V–O group (see Fig. 1a). In Fig. 2, we report the work-function shift (computed with the SPICE method using Eq. 1) as a function of the VONc coverage and compare it with experimental data for VONc on highly oriented pyrolytic graphite [38].

The computed data agree well with experimental ones up to a coverage of 0.6 monolayers (ML) for all methods considered. The workfunction shift increases with the fraction of exact exchange in the series BLYP, B3LYP, BHLYP. APBE (not shown) almost coincides with BLYP. The increase of the work-function shift is directly related to the increase of the module of the V–O dipole moment that is 0.99 a.u., 1.05 a.u., and 1.15 a.u. at the BYLP, B3LYP, and BHLYP level, respectively. The comparison between the theoretical and experimental results seems to indicate, surprisingly, a better agreement for the BLYP functional. However, two main issues need to be considered: (i) The experimental data are expressed as a function of the monolayer (ML) fraction, while the theoretical results are obtained as a function of the number of molecules per unit surface. Since no information is available about the exact molecular density at 1 ML experimental conditions, the comparison is necessarily approximate. In plotting Fig. 2, we assumed for the full coverage limit a unit cell with a size of 18 Å. This is, however, a lower limit for the cell dimension, dictated by the steric repulsion between neighboring molecules, while the true cell dimension might be expected to be slightly larger. Increasing slightly the minimum cell dimension (i.e., expanding the x-axis of Fig. 2 for the theoretical data) causes the experimental curve to approach more the BHLYP one, thus making hybrid functionals in better agreement with the experiment. We recall in addition that in the present work, the monolayer was constructed using the experimental data for Fe-phthalocyanine rather than VONc. While this is not a problem at low coverages, where the exact structure of the SAM plays a minor role, it can lead to some small discrepancies at particularly high coverages where, due to different inter-molecular interactions, the arrangement of the molecules in the SAM might be different also for similar species.

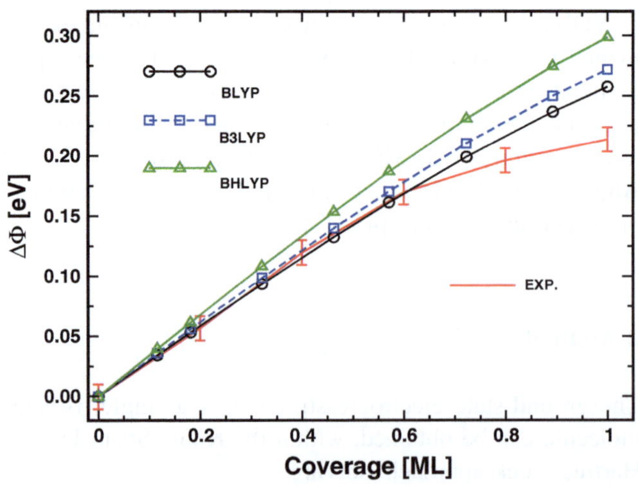

Fig. 2 Work-function shift for the VONc monolayer as a function of coverage, from SPICE calculations with different exchange-correlation functionals. Experimental data are taken from Ref. [38]

(ii) The experimental measurements were performed for VONc on a pyrolytic graphite surface, while simulations considered an isolated monolayer. Although the interaction of VONc with the graphite surface is small (certainly much smaller than, e.g., with metal substrates), an interfacial bond-dipole (μ^{bond}) cannot be excluded [70]. The latter, being in opposite direction to the V–O dipole, will reduce the computed work-function shift, making again BHLYP in better agreement with experiment.

At higher coverage, on the other hand, all functionals overestimate the experimental results, which show a clear reduction of the steepness of the curve. In addition to the above two points, we cannot rule out that other effects (e.g., kinetic-exchange-correlation) play a relevant role at the highest coverage, because these effects are not included in SPICE simulations. Moreover, at short intramolecular distances, the expansion of the molecular electron density in localized Mulliken charges is inappropriate to represent correctly the high-multipole moments and the strong local-field effects that contribute at these coverages. Despite all the above considerations, Fig. 2 clearly shows that up to 0.8 ML the SPICE method can describe very well the electronic properties of the VONc monolayer.

Concerning the variation of the excitation energy with the coverage (not reported), it is found to be negligible (smaller than 0.01 eV). This traces back to the fact that the VONc has its aromatic π system perpendicular to the molecular dipole and the latter is well localized on the V–O bond. Thus, the electrostatic field produced by the molecules in the monolayer is small and roughly uniform in the plane of the molecule. In fact, we found that all molecular orbitals display a similar shift in energy, and thus, the energy-gap and the excitation energies are not modified with the coverage: for example, the HOMO shifts by 0.17 eV (from isolated molecule to the coverage 0.9 ML) while LUMO shifts by 0.18 eV giving, an energy-gap difference of only 0.01 eV.

4.2 Mercaptobiphenyl

Larger shifts in excitation energies can be obtained when the dipole is parallel to the highly polarizable π-electron system: this is the case for mercaptobiphenyl. In this molecule, the depolarization is caused by the interaction of the dipole localized on the thiol group with the polarizable π-conjugated molecular backbone (see Fig. 1b). Because the dipole moment is directed in the same direction as the conjugated molecular backbone, and the π-electron density is highly polarizable in this direction, the depolarization is particularly relevant for this system and significantly larger than for the VONc molecule, where the dipole moment is pointing perpendicular to the π-plane.

Previous studies showed that an accurate description of the work-function shift can be achieved using hybrid or orbital-dependent functionals (see Fig. 2 of Ref. [11] and Fig. 4 of Ref. [22]). Local and semilocal functionals give instead a significant overestimation of both the dipole moment and the polarizability [22]. As a consequence, they yield a poor description of the depolarization at low- and medium-coverage, while they rely on an error cancelation at the highest coverages [11, 22].

In this paper, we use the mercaptobiphenyl as a model system to study the coverage dependence of singlet excitations in self-assembled monolayer. In Fig. 3, we report the lowest singlet excitations energies as a function of the number of molecules per unit area in the monolayer (the S_1 state lays much lower in energy and it shows the same behavior as the S_2 and S_3 states: it is not reported to keep the readability of Fig. 3). The calculations were performed at the PBE0 level, because this functional provides a good description of the depolarization effect and it includes nonlocality in the kernel.

Inspection of Fig. 3 shows that different excitations display different behaviors with the coverage. In particular, the lowest excited states show increasing energies with increasing coverage. On the contrary, the two highest singlets display a decrease of the excitation energy as the coverage increases. Interestingly, it is also possible to have different slopes for the lines so that two excited states can cross. This is the case of the S_4 (dark) and S_5 (with the highest oscillator strength) excited states that display an avoided crossing at a molecular density of about 0.004 Bohr^{-2}. This finding can open a new route in the design of optoelectronic devices. In fact, it can allow, for example, to control the radiative/nonradiative decay rates of the molecule inside the monolayer, as a function of the coverage.

To rationalize these results, it is useful to keep in mind the single-particle character of the excitations and look closer to the coverage dependence of the energies of the molecular orbitals involved in these excitations. They are

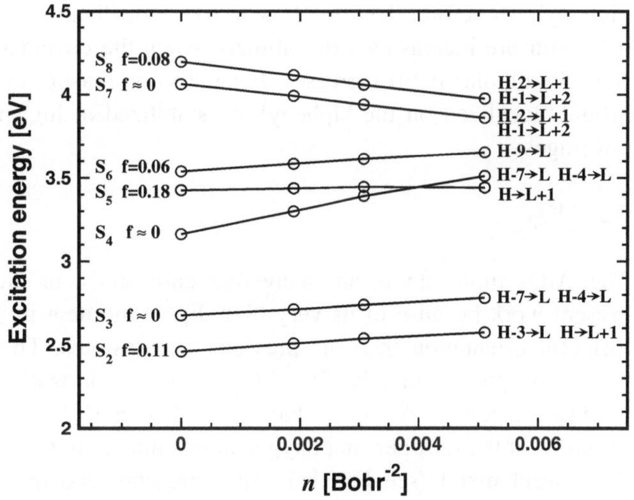

Fig. 3 TD-PBE0 singlet excitation energies of the mercaptobiphenyl as functions of the monolayer molecular density. The oscillator strength f of each state is also reported. S_1 (not shown) has a very low energy and $f \approx 0$. The single-particle character of the excitations is indicated on the right side (H HOMO, L LUMO)

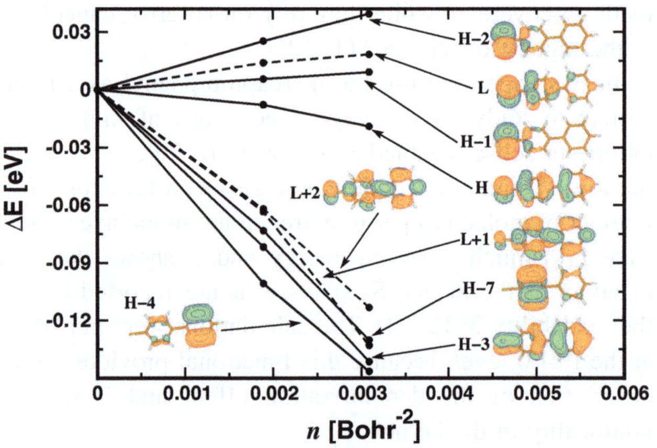

Fig. 4 Orbital energies (*H* HOMO, *L* LUMO) as a function of the molecular density for mercaptobiphenyl. Virtual orbitals are denoted with *dashed lines*

reported in Fig. 4. Orbitals with a significant localization on the thiol group (e.g., HOMO-2) increase their energy with increasing coverage, while those which are mainly localized on the biphenyl (e.g., H-4) lower their energy when the coverage increases. Thus, excitations that cause an increase of electron density on the thiol part and/or a reduction on the biphenyl (e.g., S_2) will increase their energy with increasing coverage, while the opposite will happen for excitations increasing the electron density on the molecular backbone and/or reducing it on the sulfur (e.g., S_7). A simple rationalization of the trend observed for the orbital energies is given by the observation that the field produced by the molecular dipole moments in the monolayer leads to a depolarization; hence, it tends to remove electron charge from the sulfur atom and to place it on the biphenyl. Therefore, the orbitals with some localization on the sulfur are increasingly destabilized when the coverage (i.e., the dipolar field) increases. For the same reason, the orbitals localized on the biphenyl are stabilized at higher coverages.

4.3 AlQ_3

The AlQ_3 molecule is an interesting case study in the present work because of its very high dipole moment in a different orientation than the previous two systems. This dipole originates mainly from the charge polarization around the central Al atom, due to the different electron affinities of the nitrogen and oxygen atoms interacting with the central metal (see Fig. 1c). AlQ_3 presents two main isomers, the meridional-AlQ_3 (mer-AlQ_3) and the facial-AlQ_3 (fac-AlQ_3), which differ for the position of oxygen and nitrogen atoms around the central metal Al atom. In fac-AlQ_3 in the central distorted octahedral coordinated site each oxygen atom faces a nitrogen so that three polarized O-Al-N linear bonds are formed, while in mer-AlQ_3 only one O-Al-N linear bond is observed and the dipole moment of fac-AlQ_3 is significantly larger than the one of mer-AlQ_3. Moreover, since for fac-AlQ_3, the molecular dipole moment is oriented perpendicular to the monolayer plane, while for mer-AlQ_3 it is tilted, very large differences in the magnitude of μ_z are observed. Both isomers present instead a very similar polarizability, thus a similar depolarization effect can be expected.

In the upper part of Fig. 5, we report the dipole moments of fac-AlQ_3 and mer-AlQ_3 as a function of the molecular density, calculated with the GGA PBE functional and the hybrid PBE0 functional. CC2 results are also shown for reference. For both isomers, a similar trend with the molecular density is found, as expected. The depolarization effect is described also rather similarly by both DFT methods. This relates well with the fact that the computed *zz*-polarizability are similar: 345.6 a.u. (366.9 a.u.) and 337.1 a.u. (358.0 a.u.) from PBE and PBE0, for fac-AlQ_3(mer-AlQ_3). Larger differences are found instead between PBE and PBE0 for what concerns the absolute value of the dipole moment. For the isolated fac-AlQ_3, PBE0 yields $\mu_z = 3.01$ a.u. while PBE gives a dipole moment of 2.74 a.u.. Similarly, for the mer-AlQ_3, dipole moments of 1.10 and 1.01 a.u. are obtained with PBE0 and PBE, respectively. A comparison with the CC2 calculations ($\mu_z = 2.90$ a.u. for fac-AlQ_3 and $\mu_z = 1.06$ a.u. for mer-AlQ_3) indicates that relatively accurate results are produced by DFT calculations employing the hybrid PBE0 functional, despite the polarizability (hence the depolarization) is overestimated at this level of theory. Poorer results are found with the PBE GGA functional, which yields a clear underestimation of the dipole moment.

Turning to excitation energies, we note preliminarily that a reasonably good description of the lowest singlet excited states can be obtained from TD-DFT calculations using the PBE0 functional. The calculations yield in fact a lower absorption band around 3 eV and a second band between 4 and 5 eV (not shown). These results compare well with those obtained in optical absorption measurements on amorphous AlQ_3 thin films (see Fig. 5 of Ref. [71]), except for a small underestimation of the excitation energies.

The evolution of excitation energies with the molecular density in the monolayer was simulated with the SPICE method at the TD-PBE0 level. Inspection of the bottom panels of Fig. 5 reveals two distinct behaviors. For fac-AlQ_3. all the examined excited states show decreasing energy with increasing coverage. For mer-AlQ_3, on the other hand, some excitation energies (S_1, S_4, S_5, S_6) increase with the coverage, while some of them (S_2, S_3, S_7, S_8, S_9) decrease. This makes the mer-AlQ_3 more interesting from the photophysical point of view because it offers the possibility of tuning the device performance through a

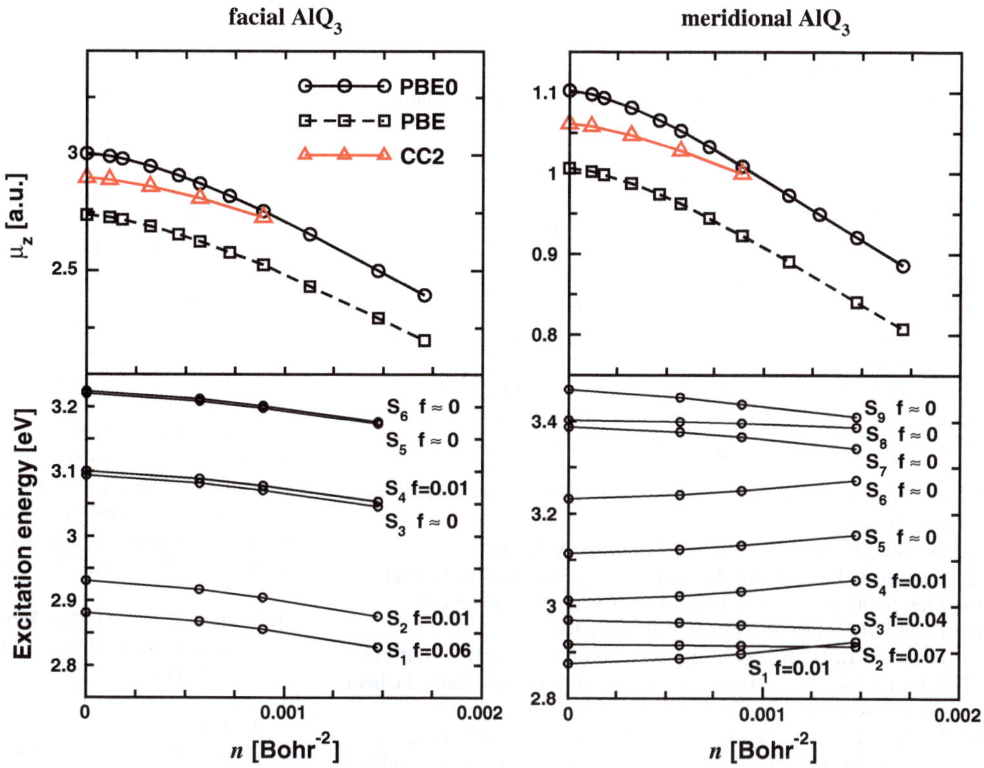

Fig. 5 Dipole moments (*top*) and excitation energies (*bottom*) of fac-AlQ_3 and mer-AlQ_3 as a function of the molecular density. Excitation energies were computed at the TD-PBE0 level of theory

modification of the excited-state energy gaps by varying the coverage. On the other hand, fac-AlQ_3 retains the isolated molecule properties.

The behavior of the different excitation energies can be rationalized in a similar way as for the biphenyl case, although here a more complex situation is faced due to the complicated structure of the AlQ_3 molecule. In general, molecular orbitals that present a stronger localization are more destabilized at high coverages than those that are rather delocalized. The latter can in fact relax more easily to respond to the external perturbation induced by the interaction in the monolayer.

5 Conclusions

The SPICE method is an efficient and powerful tool to perform theoretical investigations on periodic organic monolayers at low- and medium-coverages. It was in fact already successfully employed to study the depolarization effect and work-function shift in several systems as well as to assess the limitations of conventional DFT approaches [11].

In this work, the SPICE method has been extended to the calculation of the coverage dependence of excited states: single-molecule excitation spectra of VONc, mercaptobiphenyl, and AlQ_3 monolayers at different coverages were investigated.

We found that for molecules possessing a large dipole perpendicular to the monolayer plane and low-lying excitations characterized by single-particle transitions between orbitals with different characters (e.g., $n \rightarrow \pi^*$ transitions), large variations in the excitations energies are observed. This is the case for the example of mercaptobiphenyl and AlQ_3. The evolution of the excitation energies with the coverage was rationalized in terms of the increased stability of molecular orbitals with a large polarizability in the direction of the monolayer dipole. The opposite occurs for orbitals that cannot properly respond to the monolayer perturbing field, as in the case of rather localized orbitals.

The possibility of tuning the excitation-energy gaps by varying the coverage opens new ways to design and engineer molecular (opto)electronic devices with higher efficiency. The present work brings a fundamental contribution in this context by providing a rationale for the basic understanding of the phenomenon at the molecular level. Moreover, it proposes the SPICE method as an efficient tool for the simulation of single-molecule excitations in organic monolayers.

Acknowledgments We thank TURBOMOLE GmbH for providing us with the TURBOMOLE program package, and M. Margarito for technical support. This work was funded by the European Research Council (ERC) Starting Grant FP7 Project DEDOM, Grant Agreement No. 207441.

References

1. Ulman A (1996) Chem Rev 96:1533
2. Arima V, Fabiano E, Blyth RIR, Della Sala F, Matino F, Thompson J, Cingolani R, Rinaldi R (2004) J Am Chem Soc 126:16951
3. Love JC, Estroff LA, Kriebel JK, Nuzzo RG, Whitesides GM (2005) Chem Rev 105:1103
4. Heimel G, Romaner L, Zojer E, Brédas JL (2008) Acc. Chem. Res. 41:721
5. Tao N (2006) Nat Nanotech 1:173
6. Sushko ML, Shluger AL (2009) Adv Mater 21:1111
7. Romaner L, Heimel G, Ambrosch-Draxl C, Zojer E (2008) Adv Funct Mater 18:3999
8. Rusu PC, Brocks G (2006) J Phys Chem B 110:22628
9. Rusu PC, Brocks G (2006) Phys Rev B 74:073414
10. Heimel G, Romaner L, Zojer E, Brédas JL (2006) NanoLetters 7:932
11. Piacenza M, D'Agostino S, Fabiano E, Della Sala F (2009) Phys Rev B 80:153101
12. Topping J (1927) Proc R Soc Lond Ser A 114:67
13. Bagchi A, Barrera RG, Fuchs R (1982) Phys Rev B 25:7086
14. Maschhoff BL, Cowin JP (1994) J Chem Phys 101:8138
15. Bengtsson L (1999) Phys Rev B 59:12301
16. Romaner L, Heimel G, Zojer E (2008) Phys Rev B 77:045113
17. Johnson ER, Mori-Sánchez P, Cohen AJ, Yang W (2008) J Chem Phys 129:204112
18. Fabiano E, Della Sala F (2007) J Chem Phys 126:214102
19. Reimers JR, Cai ZL, Bili A, Hush NS (2003) Ann N Y Acad Sci 1006:235
20. Fuentealba P, Simn-Manso Y (1997) J Phys Chem A 101:4231
21. Kümmel S, Kronik L, Perdew JP (2004) Phys Rev Lett 93:213002
22. Della Sala F, Fabiano E, Laricchia S, D'Agostino S, Piacenza M (2010) Int J Quant Chem 110:2162
23. Becke AD (1993) J Chem Phys 98:5648
24. Della Sala F, Görling A (2001) J Chem Phys 115:5718
25. Kümmel S, Kronik L (2008) Rev Mod Phys 80:3
26. Della Sala F (2010) In: Springborg M (eds) Chemical modelling: applications and theory, vol 7, Royal Society of Chemistry, Cambridge, pp 115–161
27. Khait YG, Hoffmann MR (2010) J Chem Phys 133:044107
28. Kamiński R, Schmøkel MS, Coppens P (2010) J Phys Chem Lett 1:2349
29. Steele MP, Blumenfeld ML, Monti OLA (2010) J Phys Chem Lett 1:2011
30. Sariciftci, NS (eds) (1997) Primary photoexcitations in conjugated polymers: molecular exciton versus semiconductor band model. World Scientific, Singapore
31. Silbey R (1976) Ann Rev Phys Chem 27:203
32. Gigli G, Della Sala F, Lomascolo M, Anni M, Barbarella G, Di Carlo A, Lugli P, Cingolani R (2001) Phys Rev Lett 86:167
33. Yamagata H, Norton J, Hontz E, Olivier Y, Beljonne D, Brédas JL, Silbey RJ, Spano FC (2011) J Chem Phys 134:204703
34. Davidov A (1971) Theory of molecular excitons. Plenum Press, London
35. Müller M, Le Moal E, Scholz R, Sokolowski M (2011) Phys Rev B 83:241203
36. Steele MP, Blumenfeld ML, Monti OLA (2010) J Chem Phys 133:124701
37. Hochstrasser RM (1962) Rev Mod Phys 34:531
38. Blumenfeld ML, Steele MP, Monti OLA (2010) J Phys Chem Lett 1:145
39. Heimel G, Romaner L, Brédas JL, Zojer E (2006) Phys Rev Lett 96:196806
40. Fabiano E, Piacenza M, D'Agostino S, Della Sala F (2009) J Chem Phys 131:234101
41. Yanagisawa S, Lee K, Morikawa Y (2008) J Chem Phys 128:244704
42. Monti OLA, Steele MP (2010) Phys Chem Chem Phys 12:12390
43. Sirringhaus H, Tessler N, Friend RH (1998) Science 12:1741
44. Granström M, Petritsch K, Arias AC, Lux A, Andersson MR, Friend RH (1998) Nature 395:257
45. Gustafsson G, Cao Y, Treacy GM, Klavetter F, Colaneri N, Heeger AJ (1992) Nature 357:477
46. Friend RH, Gymer RW, Holmes AB, Burroughes JH, Marks RN, Taliani C, Bradley DDC, Santos DAD, Brdas JL, Lögdlund M, Salaneck WR (1999) Nature 397:121
47. Voss D (2000) Nature 407:442
48. Burow AM, Sierka M, Döbler J, Sauer J (2009) J Chem Phys 130:174710
49. Cortona P (1991) Phys Rev B 44:8454
50. Laricchia S, Fabiano E, Della Sala F (2010) J Chem Phys 133:164111
51. Cornil D, Olivier Y, Geskin V, Cornil J (2007) Adv Func Mater 17:1143
52. Norton JE, Brédas JL (2008) J Am Chem Soc 130:12377
53. Tsiper EV, Soos ZG (2001) Phys Rev B 64:195124
54. Casida ME (1995) Recent advances in density functional methods, vol 1. World Scientific, Singapore
55. Neugebauer J (2010) Phys Rep 489:1
56. Neugebauer J (2007) J Chem Phys 126:134116
57. Casida ME, Wesolowski TA (2004) Int J Quant Chem 96:577
58. Constantin LA, Fabiano E, Laricchia S, Della Sala F (2011) Phys Rev Lett 106:186406
59. Weigend F, Furche F, Ahlrichs R (2003) J Chem Phys 119:12753
60. Weigned F, Ahlrichs R (2005) Phys Chem Chem Phys 7:3297
61. Cheng ZH, Gao L, Deng ZT, Jiang N, Liu Q, Shi DX, Du SX, Guo HM, Gao HJ (2007) J Phys Chem C 111:9240
62. TURBOMOLE V6.1 2009: a development of University of Karlsruhe and Forschungszentrum Karlsruhe GmbH, 1989–2007, TURBOMOLE GmbH, since 2007; available from http://www.turbomole.com
63. Becke AD (1988) Phys Rev A 38:3098
64. Lee C, Yang W, Parr RG (1988) Phys Rev B 37:785
65. Becke AD (1993) J Chem Phys 98:1372
66. Adamo C, Barone V (1999) J Chem Phys 110:6158
67. Hättig C, Weigend F (2000) J Chem Phys 113:5154
68. Köhn A, Hättig C (2003) J Chem Phys 119:5021
69. Dunning TH Jr (1989) J Chem Phys 90:1008
70. Terentjevs A, Steele MP, Blumfeld ML, Ilyas N, Kelly LL, Fabiano E, Monti OLA, Della Sala F (2011) J Phys Chem C 115:21128
71. Brinkmann M, Gadret G, Muccini M, Taliani M, Masciocchi N, Sironi A (2000) J Am Chem Soc 122:5147

Theor Chem Acc (2012) 131:1155
DOI 10.1007/s00214-012-1155-5

REGULAR ARTICLE

Deactivation of Ru-benzylidene Grubbs catalysts active in olefin metathesis

Albert Poater · Luigi Cavallo

Received: 13 July 2011 / Accepted: 27 September 2011 / Published online: 22 February 2012

Abstract In this work, we explore the reactivity induced by coordination of a CO molecule trans to the Ru-benzylidene bond of a prototype Ru-olefin metathesis catalyst bearing a *N*-heterocyclic carbene (NHC) ligand. DFT calculations indicate that CO binding to the Ru center promotes a cascade of reactions with very low-energy barriers that lead to the final crystallographically characterized product, in which the original benzylidene group has attacked the proximal aromatic ring of the ligand leading to a cycloheptatriene ring through a Buchner ring expansion. In conclusion, the overall mechanism is best described as a carbene insertion into a C–C bond of the aromatic *N*-substituent of the NHC ligand, forming a cyclopropane ring. This cyclopropanation step is followed by a Buchner ring expansion reaction, leading to the experimentally observed product presenting a cycloheptatriene ring.

Keywords Homogeneous catalysis · Olefins metathesis · Deactivation · Cyclopropanation · *N*-heterocyclic carbene

Dedicated to Professor Vincenzo Barone and published as part of the special collection of articles celebrating his 60th birthday.

A. Poater
Catalan Institute for Water Research (ICRA),
Scientific and Technological Park of the University of Girona,
H2O Building, Emili Grahit 101, 17003 Girona, Spain

L. Cavallo (✉)
MoLNaC, Modeling Lab for Nanostructures and Catalysis,
Department of Chemistry and Biology, University of Salerno,
Via Ponte don Melillo, 84084 Fisciano, Italy
e-mail: lcavallo@unisa.it

1 Introduction

Ru-catalyzed olefin metathesis is on the verge to make the big leap from a flexible synthetic tool in the academy to a variety of large-scale applications of great economic return in the industry [1]. However, one of the drawbacks slowing this step still is the scarce stability of otherwise quite active *N*-heterocyclic carbene (NHC)-based second-generation catalysts [2–5]. The greater activity of NHC-based catalysts is now rather well understood [3, 6–10], while knowledge about their stability is still scarce [11–16].

One of the reasons for this limited understanding is that academic groups usually focus on the more rewarding improvement of activity and/or selectivity of a catalyst, since more or less rational strategies can be followed, rather than investing resources to follow catalyst deactivation along unexplored pathways. Nevertheless, seminal computational efforts have confirmed experimental insights, such as that one deactivation pathway starts with the activation of an aromatic ortho C–H bond of the N-substituents of the NHC ligand [17–19]. The emerging picture is that stable catalysts must be protected in the ortho position to avoid this deactivation pathway. This immediately explains the increased stability of catalysts with mesityl N-substituents. However, the increased stability is accompanied by inferior performances in the formation of tetrasubstituted C=C bonds. A solution to this impasse has been the synthesis of catalysts with *o*-tolyl N-substituents. The single o-methyl stabilizes the catalyst without degrading performances in challenging synthesis [20–22]. Another solution is the synthesis of NHC ligands tetramethyl substituted on the saturated NHC skeleton, with the precise idea that the bulky NHC skeleton should prevent free rotation around the N-aryl bond, thus shutting off the C–H deactivation pathway [23].

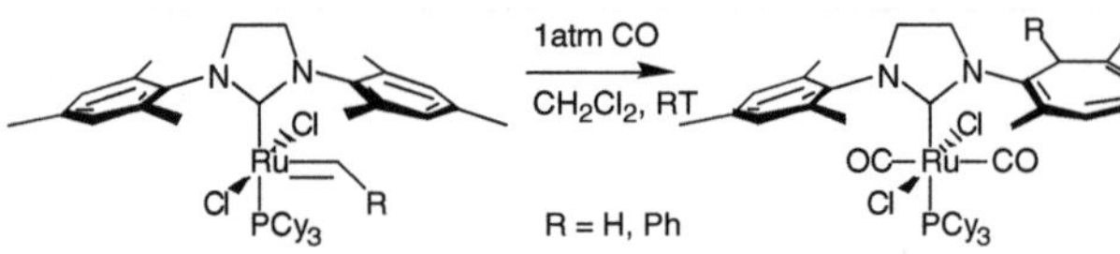

Scheme 1 Sketch of the deactivation reaction promoted by CO

Broadly speaking, deactivation pathways are those chemical transformations of the (pre)catalyst or the active species that remove them from productive metathesis. In this sense, Diver, Keister, and coworkers [24, 25] have clearly shown that addition of π acids such as CO or isocyanides to a solution of typical second-generation catalysts can switch a metathesis catalyst into a cyclopropanation catalyst [26–31], by activating the Ru-ylidene bond to a carbene-like reactivity. This promotes ylidene attack to the proximal N-bonded aromatic ring of the NHC ligand, which results in a Buchner type ring expansion reaction; see Scheme 1.

Focusing on the reactivity promoted by CO, for which accurate data are available, this decomposition reaction starts with coordination of a CO molecule trans to Ru-ylidene bond. In this respect, it is worth noting that the position trans to the Ru-ylidene bond has been scarcely considered during the years [24, 25, 32–34]. For example, Grubbs and coworkers have suggested that one of the ortho-F atoms of the NHC **A** shown in Scheme 2 could be engaged in an interaction with the Ru center that results in increased metathesis activity [35]. Furthermore, strategies for the synthesis of (pre)catalysts having labile ligands that could result in more stable and active catalysts, such as **B** and **C** of Scheme 2, have been also explored [36, 37]. In this framework, the deactivation reaction of Scheme 1 illustrates that the coordination position trans to the Ru-ylidene bond is not "innocent" and can trigger interesting reactivity. For this reason, we investigated the mechanism of deactivation shown in Scheme 1 in case of the Ru-methyidene bond [32].

Although it is difficult to call the reaction of Scheme 1 a deactivation reaction, since CO is not normally added during metathesis, understanding the chemistry behind this transformation is relevant for the following reasons. (1) π-acids are inevitably present during metathesis, and any organometallic textbook presents educative comparisons between the σ-basic/π-acid properties of CO, olefins, and phosphines. Clearly, olefins and phosphines are much weaker π-acids than CO, but a reduced acidity could still induce slow deactivation reactions as those shown in Scheme 1 for CO. (2) A clear understanding of the changes in the chemical behavior promoted by a σ-basic/π-acid ligand trans to the Ru-ylidene bond would expand our knowledge of the chemistry of Ru-based catalysts with possible consequences in the design of new ligands with tuned properties.

Along this line, we recently demonstrated that for a series of σ-basic/π-acid ligands coordinated trans to a Ru-methylidene bond of the systems, a quite acidic ligand was enough to produce such a decomposition of the olefin metathesis catalyst [24, 32]. To insert the methylidene moiety inside one of the aromatic rings of the SIMes ligand, the system must carry out a Buchner ring expansion reaction to allow the formation of a cycloheptatriene type ring. Calculations suggested that CO binding may weaken π-back-bonding between the Ru atom and the ylidene ligand [24, 32], making the latter more electrophilic and disengaging it from the metal center [38]. Electron density is donated from filled d orbitals on the Ru to the empty π orbital of the benzylidene. Differently, in the presence of the strong π-acid CO molecule, electron density from the Ru is more strongly donated to π-acid MOs of the CO. As suggested by Diver, this depletes electron density from the π orbital of the benzylidene, which is able to accept from the properly oriented π orbital of the C_{ipso} atom of the proximal ring, resulting in a very low barrier for formation of the $C_{benzylidene}$–C_{ipso} bond. However, this study was focused on the Ru-methylidene bond, which is known to be a rather reactive and unstable bond, while standard and commercial (pre)catalysts normally bear a Ru-benzylidene group. For this reason, we decided to extend our previous study to the reactivity induced by CO-coordinated trans to the Ru-benzylidene bond, see Scheme 1. Comparison between the behavior of the Ru-methylidene and Ru-benzylidene systems of Scheme 1 should increase our understanding of the basic properties of the Ru-ylidene bond and of the reactivity changes that could in principle be induced by changes in the nature of the ligands connected to the Ru atom. The only approximation we have made to the systems of Scheme 1 is replacing PCy_3 with PMe_3 to save computer time.

2 Computational details

All the DFT static calculations were performed at the GGA level with the Gaussian03 set of programs [39], using the BP86 functional of Becke and Perdew [40–42]. The electronic configuration of the molecular systems was described with the standard split-valence basis set with a polarization function of Ahlrichs and coworkers for H, C, N, O, P, F, and Cl (SVP keyword in Gaussian) [43]. For Ru, we used the small-core, quasi-relativistic Stuttgart/Dresden effective core potential, with an associated (8s7p6d)/[6s5p3d] valence basis set contracted according to a (311111/22111/411) scheme (standard SDD keywords in Gaussian03) [44–46]. The geometry optimizations were performed without symmetry constraints. The Synchronous Transit-Guided Quasi-Newton (STQN) Method, developed

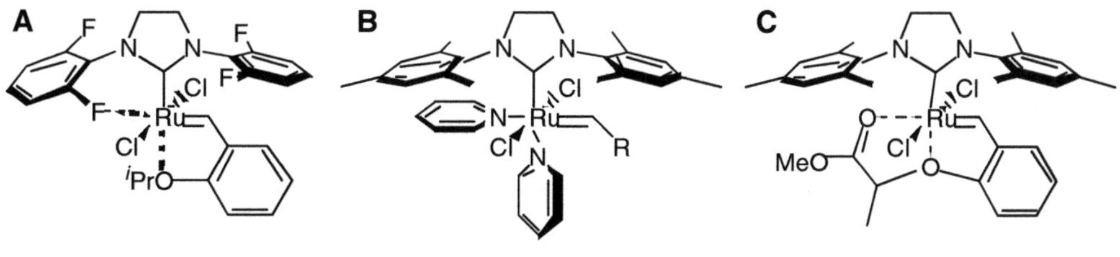

Scheme 2 Representation of Ru-complexes presenting a ligand coordinated trans to the Ru-alkylidene bond

by H. B. Schlegel and coworkers, was used to calculate the transition states. This method is requested with the QST3 keyword [47]. The characterization of the located stationary points was performed by analytical frequency calculations. Solvent effects including contributions of non-electrostatic terms have been estimated in single-point calculations on the gas-phase optimized structures, based on the polarizable continuous solvation model PCM using CH_2Cl_2 as a solvent [48, 49]. In conclusion, the energies reported correspond to the "Total free energy in solution: with all non-electrostatic terms" value in the Gaussian output. Previous studies have revealed that this scheme of calculation provides excellent results [50–52].

3 Results

To get insight into the decomposition for the so-called benzylidene olefin metathesis precatalyst, complex **1**, induced by CO, we first optimized complex **7**, for which we had available X-ray structural data. Although the optimization suffered the simplification of the tricyclohexyl phosphine by trimethyl phosphine, the agreement was excellent, with a rmsd of 0.039 Å for distances and of 0.9° for angles [53–55].[1] This indicates that the sterically demanding PCy_3 phosphine can be replaced with the less bulkier PMe_3 phosphine to shed light on this kind of reactions. Since no intermediate between the starting species **1** and the product **7** was experimentally characterized, we assumed a decomposition pathway similar to that we calculated for the similar Ru-methylidene complex. The complete energy profile for the benzylidene complex is shown in Fig. 1.

The first step in Fig. 1 shows that the coordination of a CO molecule trans to the Ru-benzylidene bond of **1** is an exoergonic process releasing 25.7 kcal/mol. Linear transit calculations using the Ru–CO distance as reaction coordinate revealed the absence of a transition state between complexes **1** and **2**, demonstrating CO coordination to be a barrierless step in terms of potential energy. The vacant position trans to the ylidene group reveals no sterical hindrance for the entering CO molecule; however, after bonding, the metal the Ru-benzylidene bond elongates from 1.85 Å in **1** to 1 99 Å in **2** and pushes the phosphine almost perfectly trans to the NHC ligand, with a small elongation (less than 0.05 Å) of the Ru-NHC and Ru–P bonds. CO coordination also results in the remarkable reduction of the NHC-Ru-benzylidene angle from 103.3° in **1** to 96.1° in **2**. These structural rearrangements increase the proximity of the benzylidene group with the proximal aromatic ring of the SIMes ligand, with the a sensible reduction of the distance between the $C_{benzylidene}$ and C_{ipso} atoms from 3.11 Å in **1** to 2.82 Å in **2**.

Starting from **2,** the cyclopropanation reaction between the benzylidene group and one of the C_{ipso}–C_{ortho} bonds of the nearby mesityl ring occurs in two steps. The former corresponds to attack of the benzylidene group to the C_{ipso} atom, which requires overcoming a barrier of 7.1 kcal/mol and leads to **3**, which still presents a Ru-benzylidene bond and is 6.8 kcal/mol less stable than **2** (see Fig. 1). It is worth noting that in the absence of a CO molecule coordinated trans to the Ru-methylidene bond, that is, in system **1**, the metallacycle structure **3** of Fig. 1 is not stable. All our attempts to locate a metallacycle species similar to **3** in the absence of the CO ligand failed, and the geometry optimization collapsed into the starting species **1**. Then, the second step of the cyclopropanation reaction requires the complete rupture of the previous Ru-benzylidene bond with the concerted formation of a cyclopropane ring by attack of the $C_{benzylidene}$ atom to one of the C_{ortho} atoms of the mesityl ring. This step requires the overcoming of a negligible barrier of 0.9 kcal/mol, after which the system collapses into the intermediate **4**. This species **4** is 4.7 kcal/mol below in energy with respect to species **2**.

Starting from intermediate **4**, with the newly formed cyclopropane ring, the next step involves the cleavage of the C_{ipso}–C_{ortho} bond of this tensioned small ring, evolving to intermediate **5** with a formal cycloheptatriene ring, with a small energy barrier, 4.6 kcal/mol, separating **4** from **5**. At this point, the decomposition reaction has completed, and the only steps required to reach the crystallographically characterized product are a rotation of nearly 180° around

[1] Standard deviations for distances and angles: $S_{n-1} = \sqrt{\frac{\sum_{i=1}^{N}(CV-EV)^2}{N-1}}$ where CV means calculated value, EV experimental value (X-ray data), and N is the number of distances or angles taken into account.

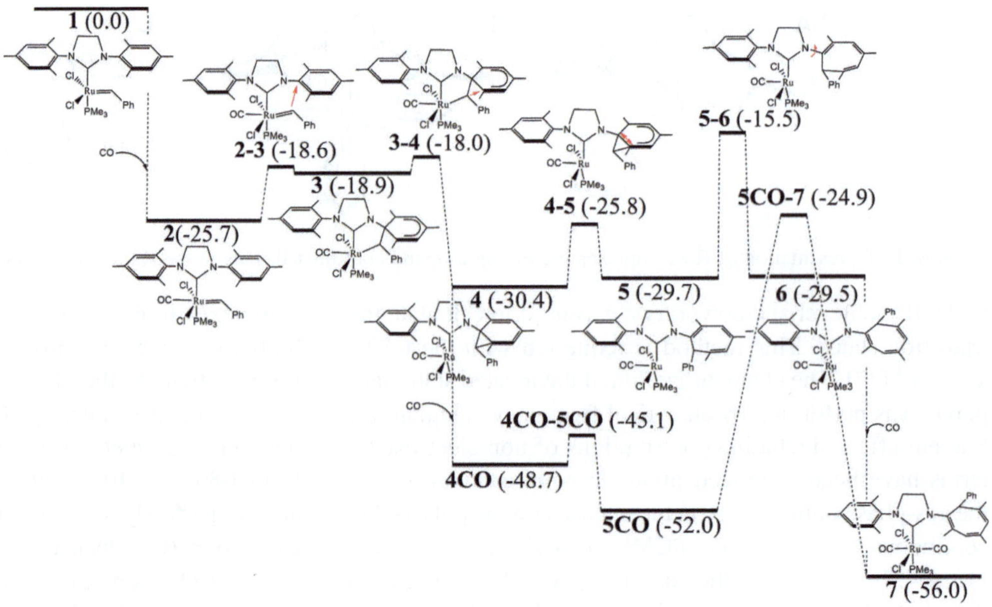

Fig. 1 Energy diagram of the complete **1–7** deactivation pathway. In *parentheses* is the energy of the various species, in kcal/mol, relative to the (pre)catalyst **1**

the N-cycloheptatriene bond, to reach a conformation in which the Ph substituent on the cycloheptatriene ring points away from the Ru atom, and coordination of a second CO molecule. Starting with the rotation around the *N*-cycloheptatriene bond, which corresponds to a conformational change from **5** to **6**, the transition state reported in Fig. 2 shows both the NHC and heptatriene rings almost coplanar. This transition state is 14.2 kcal/mol above **5**. It is worthy to note that this energy barrier could be somewhat higher when the model phosphine used in this work, PMe_3, is replaced by the experimentally used, and bulkier, PCy_3. Interestingly, intermediates **4**, **5**, and **6** are nearly isoenergetic with energy differences less than 1 kcal/mol. Coordination of a second CO molecule to the Ru atom of **6** leads to the experimental product **7** by release of 26.5 kcal/mol. This remarkable increase in the energy required for cycloheptatirene rotation is a consequence of the steric hindrance of the second CO molecule, which is coordinated just below the cycloheptatirene ring.

Although in our previous study on the CO-promoted decomposition in the presence of the Ru-methylidene group indicated that coordination of a second CO molecule is not needed to reach the experimental decomposition product, a result that is supported by the results presented in Fig. 1, we also investigated the energy profile in case the second CO molecule coordinates to the Ru atom of **4**, which means before the Buchner ring expansion step. Calculations indicate that CO coordination to 4, leading to **4CO**, releases 18.3 kcal/mol. This value is quite smaller than the energy gain associated with coordination of CO to **1** and **6**, 25.7 and 36.5 kcal/mol, respectively. From **4CO**, the next barrier corresponding to the opening of the cyclopropane ring, leading to structure **5CO**, is only

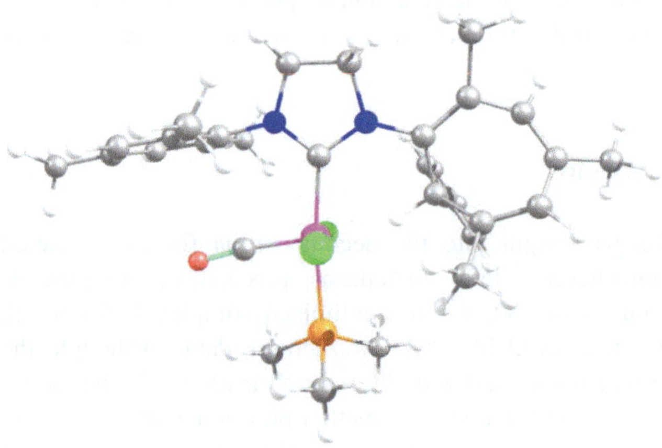

Fig. 2 DFT optimization of transition state **5–6**

1.9 kcal/mol. However, the barrier corresponding to a rotation of 180° around the N-cycloheptatriene bond, to reach the experimentally characterized conformer **7**, is now 27.1 kcal/mol. Thus, considering the small energy barrier for the **4** to **5** step, and the rather high barrier for the **5CO** to **7** step, it is likely that coordination of the second CO molecule might occur after the ring expansion step. This conclusion is also supported by considering that coordination of a second CO molecule would be further disfavored by an entropic contribution not included explicitly in the reported energy values. Comparison between these results and our previous results on the decomposition pathway in the presence of the Ru-methylidene bond indicates the nature of the Ru-ylidene bond has a very minor impact on the energetics of cyclopropane formation and of the following Buchner type ring expansion. Rather, that the main difference is in the rotation of 180° around the *N*-cycloheptatriene bond that, in case of the

Ru-benzylidene bond, is needed to place the Ph substituent on the cycloheptatriene ring away from the metal center.

4 Conclusions

In this contribution, we evaluated the complete energy profile corresponding to the deactivation, promoted by coordination of a CO molecule, of a NHC-Ru catalyst for olefins metathesis presenting a Ru-benzylidene bond. Consistent with our previous work on the deactivation of a similar NHC-Ru precatalyst presenting a Ru-methylidene bond, after CO coordination trans to the Ru-ylidene bond, the overall decomposition energy profile is characterized by very low-energy barriers. These results indicate that the nature of the Ru-ylidene bond has a minor impact on the energetics of this deactivation pathway. The main difference between the two decomposition mechanisms is that in case of the Ru-benzylidene system, an additional step, corresponding to rotation around the *N*-cycloheptatriene bond is needed in order to orientate the Ph substituent on the cycloheptatriene ring away from the metal atom.

Acknowledgments The research leading to these results has received funding from the European Community's Seventh Framework Programme (FP7/2007-2013) under grant agreement no CP-FP 211468-2 EUMET. LC thanks BSC (QCM-2010-2-0020) and the HPC team of Enea for using the ENEA-GRID and the HPC facilities CRESCO in Portici (Italy) for access to remarkable computational resources. AP thanks the Spanish MICINN for a Ramón y Cajal contract (RYC-2009-04170) and Generalitat de Catalunya for a BE-DGR fellowship.

References

1. Grubbs RH (2003) Handbook of olefin metathesis. Wiley, Weinheim
2. Scholl M, Ding S, Lee CW, Grubbs RH (1999) Org Lett 1:953
3. Huang J, Stevens ED, Nolan SP, Petersen JL (1999) J Am Chem Soc 121:2674
4. Weskamp T, Kohl FJ, Hieringer W, Gleich D, Herrmann WA (1999) Angew Chem Int Ed 38:2416
5. Bielawski CW, Grubbs RH (2000) Angew Chem Int Ed 39:2903
6. Ulman M, Grubbs RH (1998) Organometallics 17:2484
7. Adlhart C, Hinderling C, Baumann H, Chen P (2000) J Am Chem Soc 122:8204
8. Cavallo L (2002) J Am Chem Soc 124:8965
9. Correa A, Cavallo L (2006) J Am Chem Soc 128:13352
10. Poater A, Bahri-Lalehac N, Cavallo L (2011) Chem Commun 47:6674–6676
11. Hong SH, Chlenov A, Day MW, Grubbs RH (2007) Angew Chem Int Ed 46:5148
12. Abdur-Rashid K, Fedorkiw T, Lough AJ, Morris RH (2004) Organometallics 23:86
13. Scott NM, Dorta R, Stevens ED, Correa A, Cavallo L, Nolan SP (2005) J Am Chem Soc 127:3516
14. Hong SH, Wenzel AG, Salguero TT, Day MW, Grubbs RH (2007) J Am Chem Soc 129:7961
15. Vehlow K, Gessler S, Blechert S (2007) Angew Chem Int Ed 46:8082
16. Berlin JM, Campbell K, Ritter T, Funk TW, Chlenov A, Grubbs RH (2007) Org Lett 9:1339
17. Mathew J, Koga N, Suresh CH (2008) Organometallics 27:4666
18. Poater A, Cavallo L (2010) J Mol Catal A Chem 324:75
19. Poater A, Cavallo L (2010) In: Dragutan V, Demonceau A, Dragutan I, Finkelshtein ES (eds) NATO ASI series A. Springer, Dordrecht, p 275
20. Stewart IC, Ung T, Pletnev AA, Berlin JM, Grubbs RH, Schrodi Y (2007) Org Lett 9:1589
21. White DE, Stewart IC, Grubbs RH, Stoltz BM (2007) J Am Chem Soc 130:810
22. Stewart IC, Douglas CJ, Grubbs RH (2008) Org Lett 10:441
23. Chung CK, Grubbs RH (2008) Org Lett 10:2693
24. Galan BR, Gembicky M, Dominiak PM, Keister JB, Diver ST (2005) J Am Chem Soc 127:15702
25. Galan BR, Pitak M, Gembicky M, Keister JB, Diver ST (2009) J Am Chem Soc 131:6822
26. Peppers BP, Diver ST (2004) J Am Chem Soc 126:9524
27. Monnier F, Castillo D, Derien S, Toupet L, Dixneuf PH (2003) Angew Chem Int Ed 42:5474
28. Kitamura T, Sato Y, Mori M (2001) Chem Commun 1258–1259
29. Anciaux AJ, Demonceau A, Noels AF, Hubert AJ, Warin R, Teyssie P (1981) J Org Chem 46:873
30. Basato M, Tubaro C, Biffis A, Bonato M, Buscemi G, Lighezzolo F, Lunardi P, Vianini C, Benetollo F, Del Zotto A (2009) Chem Eur J 15:1516
31. Nishiyama H, Itoh Y, Matsumoto H, Park S-B, Itoh K (1994) J Am Chem Soc 116:2223
32. Poater A, Ragone F, Correa A, Cavallo L (2009) J Am Chem Soc 131:9000
33. Webster CE (2007) J Am Chem Soc 129:7490
34. Aharoni A, Vidavsky Y, Diesendruck CE, Ben-Asuly A, Goldberg I, Lemcoff NG (2011) Organometallics 30:1607
35. Ritter T, Day MW, Grubbs RH (2006) J Am Chem Soc 128:11768
36. Sanford MS, Love JA, Grubbs RH (2001) Organometallics 20:5314
37. Bieniek M, Bujok R, Cabaj M, Lugan N, Lavigne G, Arlt D, Grela K (2006) J Am Chem Soc 128:13652
38. Frenking G, Sola M, Vyboishchikov SF (2005) J Organomet Chem 690:6178
39. Frisch MJ, Trucks GW, Schlegel HB, Scuseria GE, Robb MA, Cheeseman JR, Montgomery JA, Vreven T, Kudin KN, Burant JC, Millam JM, Iyengar SS, Tomasi J, Barone V, Mennucci B, Cossi M, Scalmani G, Rega V, Petersson GA, Nakatsuji H, Hada M, Ehara M, Toyota K, Fukuda R, Hasegawa J, Ishida M, Nakajima T, Honda Y, Kitao O, Nakai H, Klene M, Li X, Knox JE, Hratchian HP, Cross JB, Adamo C, Jaramillo J, Gomperts R, Stratmann RE, Yazyev O, Austin AJ, Cammi R, Pomelli C, Ochterski JW, Ayala PY, Morokuma K, Voth GA, Salvador P, Dannenberg JJ, Zakrzewski VG, Dapprich S, Daniels AD, Strain MC, Farkas O, Malick DK, Rabuck AD, Raghavachari K, Foresman JB, Ortiz JV, Cui Q, Baboul AG, Clifford S, Cioslowski J, Stefanov BB, Liu G, Liashenko A, Piskorz P, Komaromi I, Martin RL, Fox DJ, Keith T, Al-Laham MA, Peng CY, Nanayakkara A, Challacombe M, Gill PMW, Johnson B, Chen W, Wong MW, Gonzalez C, Pople JA (2003) B. 1 ed, Gaussian, Inc., Pittsburgh
40. Becke AD (1988) Phys Rev A 38:3098
41. Perdew JP (1986) Phys Rev B 33:8822
42. Perdew JP (1986) Phys Rev B 34:7406
43. Schaefer A, Horn H, Ahlrichs R (1992) J Chem Phys 97:2571
44. Haeusermann U, Dolg M, Stoll H, Preuss H (1993) Mol Phys 78:1211

45. Kuechle W, Dolg M, Stoll H, Preuss H (1994) J Chem Phys 100:7535
46. Leininger T, Nicklass A, Stoll H, Dolg M, Schwerdtfeger P (1052) J Chem Phys 1996:105
47. Peng CY, Schlegel HB (1993) Isr J Chem 33:449
48. Barone V, Cossi M (1995) J Phys Chem A 1998:102
49. Tomasi J, Persico M (1994) Chem Rev 94:2027
50. Poater A, Ragone F, Mariz R, Dorta R, Cavallo L (2010) Chem Eur J 16:14348
51. Bosson J, Poater A, Cavallo L, Nolan SP (2010) J Am Chem Soc 132:13146
52. Luan X, Mariz R, Gatti M, Costabile C, Poater A, Cavallo L, Linden A, Dorta R (2008) J Am Chem Soc 130:6848
53. Poater A (2009) J Phys Chem A 113:9030
54. Poater A, Ribas X, Llobet A, Cavallo L, Sola M (2008) J Am Chem Soc 130:17710
55. Poater A, Cavallo L (2009) Inorg Chem 48:4062

Theor Chem Acc (2012) 131:1156
DOI 10.1007/s00214-012-1156-4

REGULAR ARTICLE

Is the dynamical polarization a significant part of the contribution of the triples to the correlation energy?

Jean-Paul Malrieu · Hongjiang Zhang · Jing Ma

Received: 14 July 2011 / Accepted: 22 September 2011 / Published online: 19 February 2012

Abstract One may call dynamical polarization of doubly excited configurations the energy lowering of these configurations under the response of the other electrons to the so-created fluctuation of the electric field. This contribution of triply excited configurations may be identified and calculated through a computation that only requires a computation time proportional to the sixth power of the number of molecular orbitals (MOs), instead of the seventh power for the total contribution of the triples. Its amplitude depends on the choice of the MOs and becomes important when localized MOs are used.

Keywords Dynamic polarization · Correlation energy · Triple excitations

1 Introduction

In the so-called closed shell systems, where a single determinant may be considered as a good approximation of the wave function, or a satisfactory zero-order reference to build a correlated description, the main part of the correlation energy is brought by the action of the double excitations on the top of the self-consistent determinant. This may be done through a configuration interaction of singles and doubles (CISD) [1]. This approach suffers a size-consistency problem, which can be easily corrected through coupled electron pair approximations (CEPA) [1–4] of various degrees of sophistication, depending on the way the exclusion principle violating (EPV) effects are taken into account, from full neglect (CEPA-0) [1–4] to full account [$(SC)^2$CISD] [5, 6]. The coupled cluster (CC) [7, 8] expansion involving single and double excitation operators (CCSD) [9] also introduce the linked effects of the quadruply excited determinants and a minor part of the effect of the triply excited ones. Actually, due to the Brillouin's theorem, the coefficients of the singly excited determinants are of 2nd order only, and the coefficients of the triply excited determinants in the CCSD wave function are only of 3rd order. In a perturbative expansion of the wave function, the triply excited determinants appear at the 2nd order, through their interaction with the doubly excited ones, and their effect on the energy is a fourth-order one. Their numerical impact is sometimes significant, and a widely used procedure, called CCSD (*T*) [10], consists in taking into account the perturbative correction of the triples in a CCSD scheme. The computational cost of CISD or CCSD scales as the 6th power of the number of electrons or molecular orbitals (MOs), while the perturbative treatment of the triples requires a N^7 (N being the basis set size) step. This is a severe increase in the computational cost, which limits the applicability of the method. The present paper examines the possible role of a fraction of the effect of the triples, the computation of which would only require a N^6 computational cost. Heully and one of us (JPM) have recently suggested [11] that this might be the leading contribution of the triples. It concerns the so-called

Dedicated to Professor Vincenzo Barone and published as part of the special collection of articles celebrating his 60th birthday.

J.-P. Malrieu (✉)
Laboratoire de Chimie et Physique Quantiques, IRSAMC, CNRS UMR 5626, Université Paul Sabatier, 118 Route de Narbonne, 31062 Toulouse, France
e-mail: jean-paul.malrieu@irsamc.ups-tlse.fr

H. Zhang · J. Ma
Key Laboratory of Mesoscopic Chemistry of MOE, School of Chemistry and Chemical Engineering, Institute of Theoretical and Computational Chemistry, Nanjing University, Nanjing 210093, People's Republic of China

dynamical polarization effect, the definition of which will be recalled in the next section, together with a discussion of its physical significance and the reasons why one might expect that its role is important. Numerical tests will be presented in Sect. 3, followed by a brief conclusion.

2 The dynamical polarization of the doubles

2.1 Recall

While in the mean field approximation each electron moves in a static field, the electronic correlation phenomenon introduces a fluctuation of the electric field. When a double excitation takes place on the top of the self-consistent zero-order determinant Φ_0, promoting two electrons from the occupied spin orbitals i and j to the vacant spin orbitals r and s, the other electrons feel an instantaneous field that is modified with respect to the mean field relative to Φ_0. One may call dynamical polarization the response to the fluctuation of the electric field.

In many problems, the dynamical polarization is an important phenomenon. For instance in magnetic systems, where two unpaired electrons are essentially localized in two distant magnetic orbitals a and b, the triplet configuration remains purely neutral, in the sense of valence bond (VB) theory [12, 13], $\Psi_T = |\text{core}\, ab|$ for the $M_s = 1$ component, or $\Psi_T = \left|\text{core}\,(a\bar{b} - b\bar{a})/\sqrt{2}\right|$ for the $M_s = 0$ component. On the contrary in the singlet state, two ionic components A^-B^+ and A^+B^- appear in the wave function,

$$\Psi_S = \lambda\left|\text{core}\left(a\bar{b} + b\bar{a}\right)/\sqrt{2}\right| + \mu|\text{core}\left(a\bar{a} + b\bar{b}\right)|/\sqrt{2}, \tag{1}$$

which introduces the non-dynamical (or valence left–right) correlation.

If one returns to the VB way of thinking, one understands easily that in one of these ionic components, say A^-B^+, two phenomena will take place:

- The doubly occupied orbital a is no longer adapted to a double occupancy, the orbital would prefer to be more diffuse and to hybrid in the direction of the positive center B^+. This is the so-called orbital breathing of the active orbitals [12, 13].
- The electrons of the core would tend to polarize in order to react to the field created by the active electrons in the instantaneous dipolar distribution A^-B^+.

Both effects are known to be responsible for a dramatic lowering of the effective energy of the ionic VB components of the wave function. The energy difference U between the neutral and ionic components of the wave function is typically reduced by a factor 3, going from 25 to 7 eV in copper dimer complexes [14, 15]. In a multireference configuration interaction (MRCI) picture, these effects are brought by single excitations on the complete active space (CAS of 2 electrons in 2 MOs) determinants.

2.2 Formal derivation of the dynamical polarization of the doubles

The electronic correlation problem for closed shell molecules, starting from the single-determinantal approximation Φ_0, is somewhat different, but in the doubly excited determinant $a_r^+ a_s^+ a_j a_i \Phi_0$, the field created by the electrons in the spin orbitals i and j is now replaced by that created by these electrons in the virtual MOs r and s. The response of the other electrons to the fluctuation of the electric field, created by the double excitation, may be considered as brought by the single excitations $a_t^+ a_k$ on the top of the double excitation.

Let us make explicit the corresponding correction to the energy, which is of fourth-order character and only requires a sextuple summation, three on the occupied MO indices, i, j and k, three on the virtual MOs r, s and t. The interaction between a doubly excited determinant Φ_{ij}^{rs} and a triply excited determinant Φ_{ijk}^{rst} (which only differ by a single excitation $a_t^+ a_k$) is

$$\left\langle \Phi_{ij}^{rs} \middle| H \middle| \Phi_{ijk}^{rst} \right\rangle = \langle k|J_r + J_s - J_i - J_j|t\rangle, \tag{2}$$

where the coulomb operators J_p incorporate the exchange counterpart K_p when the spins of the spin orbitals k and t are the same as the spin of the orbital p. The contribution to the correlation energy $\varepsilon_{\text{pol}\,T}^{(4)}$ is given by

$$\varepsilon_{\text{pol}\,T}^{(4)} = \sum_{i,j,k,r,s,t} \left(c_{ij}^{rs}\right)^2 \frac{\langle k|J_r + J_s - J_i - J_j|t\rangle^2}{\langle \Phi_0|H|\Phi_0\rangle - \left\langle \Phi_{ijk}^{rst}|H|\Phi_{ijk}^{rst}\right\rangle}, \tag{3}$$

where c_{ij}^{rs} is the coefficient of the doubly excited determinant Φ_{ij}^{rs}. This expression is based on the Epstein–Nesbet definition of the zero-order Hamiltonian [16]. It is worth noticing that this effect may be treated as an effective dressing of the diagonal energies of the doubly excited determinants

$$\left\langle \Phi_{ij}^{rs} \middle| H^{\text{eff}} \middle| \Phi_{ij}^{rs} \right\rangle = \left\langle \Phi_{ij}^{rs} | H | \Phi_{ij}^{rs} \right\rangle + \sum_{k\neq i,j,t\neq r,s} \frac{\langle k|J_r + J_s - J_i - J_j|t\rangle^2}{\langle \Phi_0|H|\Phi_0\rangle - \left\langle \Phi_{ijk}^{rst}|H|\Phi_{ijk}^{rst}\right\rangle}. \tag{4}$$

The sums may be calculated before the diagonalization of the CISD matrix, for each set of values of the indices i, j, r and s, and the resulting matrix may be diagonalized once for all. More efficient implementations are conceivable and will be evoked below.

2.3 Physical content of the dynamical polarization of doubles and its dependence to the MO choice

It is important now to consider the physical content of the dynamical polarization of doubles. It reflects a polarization of an occupied spin-orbital k, by mixing with a spin-orbital t, under the effect of a double excitation from i and j to r and s. This effect is not invariant under the unitary transformations of the orbitals of the same class (occupied between themselves, virtual between themselves). The amplitude of the coulomb operators decrease when the orbitals delocalize, and the effect is expected to be more pronounced when localized occupied MOs are used. Let us illustrate this point by considering a diatomic problem, with a covalent bond orbital k, an antibonding valence MO t and one lone pair on each atom, labeled a and b, respectively. From these localized orbitals, one may define a pair of delocalized MOs g and u

$$g = (a + b)/\sqrt{2}, \tag{5}$$

$$u = (a - b)/\sqrt{2}. \tag{6}$$

The ground-state determinant may be written $\Phi_0 = |k\bar{k}g\bar{g}u\bar{u}| = |k\bar{k}a\bar{a}b\bar{b}|$. Let us consider excitations from the MOs a and b or g and u to virtual rather diffuse MOs, r and s, whose coulomb operators J_r and J_s are weaker than those relative to the valence MOs. The single excitation $a_t^+ a_k$ is of u symmetry. Consider for instance the doubly excited determinant $\Phi_{gg}^{rs} = |k\bar{k}r\bar{s}u\bar{u}|$. The corresponding interaction with the triply excited determinant

$$\left\langle \Phi_{gg}^{rs} | H | \Phi_{ggk}^{rst} \right\rangle = \left\langle k | J_r + J_s - 2J_g | t \right\rangle \tag{7}$$

is zero for symmetry reasons, and this will remain true for the effect of the single excitations $a_t^+ a_k$ acting on the doubly excited determinants involving the other pairs of symmetry-adapted MOs, uu or gu. In the localized approach, the corresponding interactions, for instance,

$$\left\langle \Phi_{aa}^{rs} | H | \Phi_{aak}^{rst} \right\rangle = \left\langle k | J_r + J_s - 2J_a | t \right\rangle, \tag{8}$$

are important. Actually, the same effect is involved in the symmetry-adapted approach, but it is no longer through a single excitation operator but a simultaneous change of hole and single excitation. Actually,

$$\left\langle \Phi_{g\bar{g}}^{r\bar{s}} | H | \Phi_{g\bar{u}k}^{r\bar{s}t} \right\rangle = (kt, gu) \approx \left\langle k | (J_a - J_b)/2 | t \right\rangle \tag{9}$$

The effect that is clearly a polarization effect in the localized picture appears through a coupling of dipolar distributions and a 2 hole − 1 particle process in the delocalized picture. The physical effect is the same, but it is not located in the same place when one goes from localized to delocalized MOs. The preceding discussion suggests that it will be more easily harvested in a localized approach.

3 Numerical tests

In order to assess the performance of the dynamic polarization to the effect of triples, we first apply it to calculate the equilibrium properties of some small molecules, including the energies, bond lengths and harmonic vibrational frequencies. We also test its performance on the dissociation of single bonds and the symmetrical stretching of triangular Be_3 cluster. As described in Sect. 2.3, the results of the dynamic polarization of doubles depend on the choice of MOs. So in the energy calculations around equilibrium geometries, we examine all four possible combinations of MOs, that is, canonical occupied and virtual MOs ([CMO, CMO]), localized occupied and canonical virtual MOs ([LMO, CMO]), canonical occupied and localized virtual MOs ([CMO, LMO]) and localized occupied and virtual MOs ([LMO, LMO]). Then, we choose the MO combination of giving the largest part of the effect of triples to the remaining calculations. In all calculations, the CISD is carried out in the first stage through GAMESS package [17, 18], and then, the effect of the dynamic polarization is obtained by adding them to the diagonal elements of CISD matrix and diagonalizing the dressed matrix (named as CISD [t]) in the modified GAMESS program. The localized MOs (LMOs) are obtained through Pipek–Mezey localization method [19]. Notice that the CISD calculations are not size consistent, but we think that the present evaluation of the amplitude of the dynamical polarization effect would be approximately the same in a CCSD (T) approach.

3.1 Equilibrium properties

We examine the effect of dynamic polarization on the energetics for C_2, HF, F_2, Be_3, O_3 and C_4H_{10} molecules, in which the effect of triples plays an important role even around equilibrium geometries. The results are compared to the energy differences between CCSD [9] and CCSD (T) [10], which is a full fourth-perturbation treatment, as listed in Table 1. We also give the ratio of the correlation energy brought by the triples compared with the part due to the singles and doubles only for both the CI and CC methods (the last two columns in Table 1). At a first glance, the [LMO, CMO] combination gives the largest estimate of the full fourth-perturbation treatment, except HF molecule, in which the canonical occupied MOs are

Table 1 The numerical values (in a.u.) and percentages (under the numbers) retrieved by dynamical polarization to the effect of triples obtained through full fourth-perturbation treatment [the energy difference between CCSD (*T*) and CCSD]

Molecule	[CMO, CMO]	[LMO, CMO]	[CMO, LMO]	[LMO, LMO]	Ref.[a]	E_{corr} (T)/E_{corr} (SD)	
						CI[b] (%)	CC (%)
C_2^c	−0.001725	−0.005237	−0.003454	−0.003970	−0.028569	1.8	8.9
	(6.0%)	(18.3%)	(12.1%)	(13.9%)			
HF^d	−0.001808	−0.001689	−0.000786	−0.000637	−0.002060	0.8	1.0
	(87.8%)	(82.0%)	(38.2%)	(30.9%)			
F_2^e	−0.000704	−0.005526	−0.003310	−0.001333	−0.009322	1.4	2.3
	(7.5%)	(59.3%)	(35.5%)	(14.3%)			
Be_3^f	−0.001494	−0.003163	−0.002457	−0.001870	−0.013370	2.1	8.2
	(11.2%)	(23.7%)	(18.4%)	(14.0%)			
O_3^g	−0.005225	−0.013418	−0.009072	−0.005252	−0.033831	2.4	5.4
	(15.4%)	(39.7%)	(26.8%)	(15.5%)			
$C_4H_{10}^h$	−0.001472	−0.009625	−0.003039	−0.002480	−0.016401	1.7	2.5
	(9.0%)	(58.7%)	(18.5%)	(15.1%)			

All electrons are correlated, and Cartesian basis functions are used

[a] Energy difference between CCSD (*T*) and CCSD

[b] The ratio of correlation energy contributed by triples and that from singles and doubles for [LMO, CMO]

[c] Calculated with a cc-pVDZ basis set and at the equilibrium bond length of 1.27276 Å [21]

[d] Calculated with a cc-pVDZ basis set and at the equilibrium bond length of 0.92025 Å [21]

[e] Calculated with a cc-pVDZ basis set and at the equilibrium bond length of 2.66816 Bohr [22]

[f] Calculated with the ANO basis set [23] and at the distance of one Be atom to the center of mass of the equilateral triangle of 1.312998 Å [24]

[g] Calculated with a cc-pVDZ basis set and at the equilibrium bond length of 1.272 Å and bond angle of 116.8° [25]

[h] Calculated with a 6-31G* basis set [26, 27] and the equilibrium geometry from [28]

already localized. If we concentrate only on the use of [LMO, CMO], it appears that the dynamic polarization does not represent a systematic fraction of the total energy brought by the triples. For example, the dynamic polarization can retrieve as large as 82.0% of the total effect of triples for HF molecule with a cc-pVDZ basis set [20], but obtain only 18.3% of the total effect of triples for C_2 molecule. The largest fractions are obtained for HF, F_2 and C_4H_{10} molecules, in which more than half of the effect of triples is captured.

The calculated fraction also depends on the size of basis sets. For HF for instance when the cc-pVDZ basis set (totally 20 basis functions) is used, the dynamic polarization effect obtained with [LMO, CMO] can retrieve 82.0% of the effect of triples as listed in Table 1, but the percentage decreases to 54.8% when the cc-pVTZ basis set (with 50 basis functions) is used.

We also evaluate the equilibrium bond lengths and harmonic vibrational frequencies of HF and C_2 molecules (Table 2), as examined in [21]. Since [LMO, CMO] gives the best estimate of the effect of triples, the following tests used only such MO combination. The results are consistent with those of Table 1. The dynamic polarization retrieving the largest part of effect of triples for HF gives a great improvement for the bond length and harmonic vibrational frequencies, CISD [*t*] compares well with CCSD. But for C_2 molecule, the improvement is very limited, and the difference between bond lengths of CISD and CISD [*t*] is about 0.001 Å. CISD [*t*] gives an improvement of about 13 cm^{-1} to the harmonic vibrational frequency compared to CISD.

3.2 The dissociation of single bonds

3.2.1 HF

The full potential curves are investigated by considering the dynamic polarization effect with a 6-31G** basis set [26, 27]. The results are displayed in Fig. 1, along with the results of CCSD, CCSD (*T*) and full CI (FCI) [29]. CISD [*t*] does not present a divergent behavior in bond-breaking regions, although the dynamic polarization effect is incorporated by a dressing of perturbative character. CCSD (*T*) presents divergence when the bond length is longer than 1.8 Å. The dynamic polarization effect gives largest part of the effect of triples at 1.1 Å, amounting to 84.0%. This percentage decreases gradually when the bond length becomes larger than 1.1 Å, but generally the dynamic polarization effect can obtain no <50% of the total effect of triples when the bond length is no more than 1.8 Å.

Table 2 The equilibrium bond lengths (r_e) in Å and harmonic vibrational frequencies (ω_e) in cm^{-1} of HF and C_2 molecules obtained by fitting a polynomial using points around equilibrium bond length

	HF		C_2	
	r_e	ω_e	r_e	ω_e
CISD	0.91661	4212.1	1.26065	1899.3
CISD [*t*]	0.91884	4168.9	1.26143	1886.2
CCSD[a]	0.91895	4169.0	1.26616	1861.5
CCSD (*T*)[a]	0.91988	4150.2	1.27045	1828.1

The cc-pVDZ basis set and harmonic components of *d* orbitals are used. The frozen core approximation is adopted in post-Hartree–Fock calculations

[a] From [21]

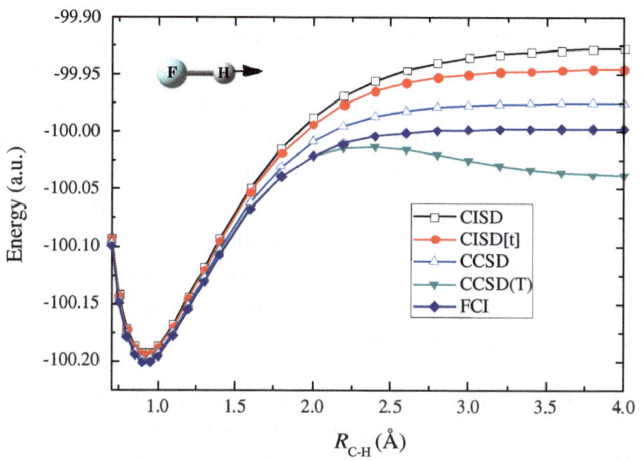

Fig. 1 The potential *energy curves* of HF molecule

The absolute value represented by dynamic polarization effect increases gradually from 1.225 m hartree (at 0.7 Å) to 17.899 m hartree (at 4.0 Å).

3.2.2 F_2

The dynamic correlation plays a significant role in F_2 molecule, especially the triples has an important effect on the correlation energy, which has been investigated by various types of CC methods [22]. The results calculated with a cc-pVDZ basis set are listed in Table 3, together with the results of CCSDT, CCSD and CCSD (*T*) taken from [22]. Different from HF molecule, CCSD (*T*) shows an overshooting behavior even at 1.25 R_e. The dynamic polarization captures 60.0% of the effect of triples at 1 R_e. The absolute numbers obtained by the dynamic polarization effect range from 4.324 m hartree (at 0.75 R_e) to 6.589 m hartree (at 1.5 R_e). It remains almost constant around 6.0 m hartree at large interatomic distances, while the full treatment of the triples in CCSDT recovers up to 50 m hartree at 5 R_e.

3.3 Triangular Be_3 cluster

The effect of connected triples plays an essential role in the triangular Be_3 cluster due to its obvious multireference character caused by the relatively small energy differences between the atomic 2 *s* and 2*p* orbitals [30]: CCSD cannot satisfactorily reproduce the binding energy, but CCSD (*T*) treatment gives greatly improved result [31, 32]. Although the dynamic polarization gives only 18.3% of the effect of total triples in Table 1, it is interesting to investigate its behavior in symmetrical stretching of three Be atoms. The ANO basis set [23] in a [3*s*2*p*1*d*] contraction was used for this system. The frozen core approximation was employed in all post-Hartree–Fock calculations. The results are displayed in Fig. 2, together with the results of CCSD, CCSD (*T*) and FCI [24]. Surprisingly, the CISD [*t*] indeed reproduces the symmetrical stretching process qualitatively (without the spurious hump of CCSD), although the dynamic polarization retrieves only about one-fifth of the effect of triples around equilibrium geometries. The dissociation energies from FCI, CCSD (*T*) and CCSD are 17.2, 14.5 and 6.4 kcal/mol, respectively. CISD [*t*] and CISD give values of 11.1 and 9.5 kcal/mol, respectively. Figure 3 displayed the percentages retrieved by the dynamic polarization effect relative to the energy difference between CCSD (*T*) and CCSD, which increase monotonically except the first two points.

Table 3 The total energy of F_2 molecule calculated by CCSDT and energy differences between other correlation methods and CCSDT at several geometries (R_e = 2.66816 Bohr)

Methods	0.75 R_e	1 R_e	1.25 R_e	1.5 R_e	1.75 R_e	2 R_e	3 R_e	5 R_e
CCSDT	−198.922138	−199.102796	−199.085272	−199.065882	−199.059433	−199.058201	−199.058511	−199.058586
CISD	22.937	35.881	59.429	88.571	111.737	125.492	139.798	142.466
CISD [*t*]	18.613	30.342	53.016	81.982	105.481	119.574	133.901	136.195
CCSD	4.504	9.485	19.917	32.424	41.184	45.638	49.425	49.816
CCSD (*T*)	0.102	0.248	−0.503	−5.711	−15.133	−23.596	−35.700	−39.348

The core orbitals are frozen in the correlated methods. The CCSDT, CCSD and CCSD (*T*) energies are taken from [22]

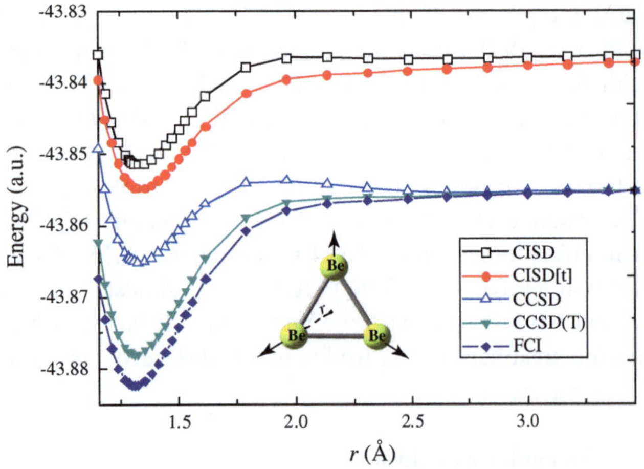

Fig. 2 The potential *curves* of the symmetrical *stretching* of *triangular* Be_3 cluster

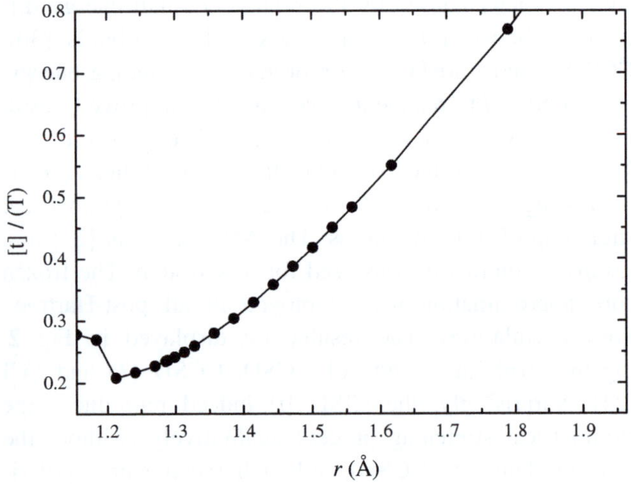

Fig. 3 The percentages retrieved by dynamic polarization effect relative to the energy difference between CCSD (*T*) and CCSD during the symmetrical *stretching* of *triangular* Be_3 cluster

4 Conclusion

Heully and one of us (JPM) had suggested [11] that the dynamical polarization of the doubly excited configurations might be a leading part of the contribution of the triples to the correlation energy. If this was true, one might avoid the bottleneck of CCSD (*T*) method, the computational cost of which scales as the 7th power of the number of electrons or MOs. This hypothesis is tested here for the first time. This work shows that the percentage of the effect of the triples brought by this effect strongly depends on the basis of occupied MOs. It is larger when LMOs are used and the reason why it is so has been clarified. The procedure we propose here, namely a dressing of the energies of the doubly excited determinants, may perfectly be employed in a CCSD formalism, which may always be expressed as a dressing of the CISD matrix [5, 6]. Moreover, the authors of [11] suggested treating this effect through the definition of energy-dependant modifications of the coulomb integrals between the pairs of spin orbitals m and n through a double summation

$$\delta J_{mn}(E) = \sum_{k,t} \frac{\langle k|J_m|t\rangle\langle t|J_n|k\rangle}{F_{kk} - F_{tt} - E}, \tag{10}$$

which might be performed before the CI process. In this case, the cost of the evaluation of the effect of the dynamical polarization of the doubles by the triples falls to N^4. The present numerical tests indicate that the so-calculated fraction of the contribution of the triples, although working in the right direction regarding the ground-state properties, is not sufficient to recover the overall effect of the triple excitations.

A surprising side result of the present work concerns the impact of the choice of the zero-order Hamiltonian. Our Epstein–Nesbet calculation goes through all the triples. The results do not exhibit a spurious overestimation of the calculated perturbative correction at large interatomic distances, in contrast with what is observed in CCSD (*T*) treatments, which use a Moller–Plesset zero-order Hamiltonian. Usually, the Epstein–Nesbet corrections are larger in absolute value than the Moller–Plesset ones. This point should deserve further analysis.

Acknowledgments Jean-Paul Malrieu thanks J.-L. Heully for stimulating discussions. The authors express their debt to the scientific achievements of V. Barone in the methodology of quantum chemistry. The support from the Chinese NSF (Grant No. 20825312) and National Basic Research Program (Grant No. 2011CB808600) is gratefully acknowledged.

References

1. Szabo A, Ostlund N (1982) Modern quantum chemistry: introduction to advanced electronic structure theory. MacMillan, New York
2. Wennmohs F, Neese F (2008) Chem Phys 343:217
3. Kutzelnigg W (1977) In: Schaefer HF (ed) Methods of electronic structure theory. Plenum, New York, p 172
4. Koch S, Kutzelnigg W (1981) Theor Chim Acta (Berl) 59:387
5. Daudey JP, Heully JL, Malrieu JP (1993) J Chem Phys 99:1240
6. Sanchez-Marin J, Nebot-Gil I, Malrieu JP, Heully JP, Maynau D (1997) Theor Chim Acta 95:215
7. Čížek J (1966) J Chem Phys 45:4256
8. Čížek J (1969) In: Hariharan PC (ed) Advances in chemical physics, vol 14. Wiley, New York, p 35
9. Purvis GD III, Bartlett RJ (1982) J Chem Phys 76:1910
10. Raghavachari K, Trucks GW, Pople JA, Head-Gordon M (1989) Chem Phys Lett 157:479
11. Heully JL, Malrieu JP (2006) J Mol Struct Theochem 758:53
12. Hiberty P, Shaik S (2002) Theor Chem Acc 108:255
13. Hiberty P, Humbel S, Byrman C, van Lenthe J (1994) J Chem Phys 101:5969
14. Calzado C, Caballol R, Malrieu JP (2002) J Chem Phys 116:3985

15. Calzado C, Angeli C, Taratiel D, Caballol R, Malrieu JP (2009) J Chem Phys 131:044327
16. Claverie P, Diner S, Malrieu JP (1967) Int J Quantum Chem 1:751
17. Schmidt MW, Baldridge KK, Boatz JA, Elbert ST, Gordon MS, Jensen JH, Koseki S, Matsunaga N, Nguyen KA, Su SJ, Windus TL, Dupuis M, Montgomery JA (1993) J Comput Chem 14:1347
18. Ivanic J (2003) J Chem Phys 119:9364
19. Pipek J, Mezey P (1989) J Chem Phys 90:4916
20. Dunning TH Jr (1989) J Chem Phys 90:1007
21. Evangelista FA (2011) J Chem Phys 134:224102
22. Kowalski K, Piecuch P (2001) Chem Phys Lett 344:165
23. Widmark PO, Malmqvist PA, Roos BO (1990) Theor Chim Acta 77:291
24. Junquera-Hernández JM, Sánchez-Marín J, Bendazzoli GL, Evangelisti S (2004) J Chem Phys 120:8405
25. Helgaker T, Jorgensen P, Olsen J (2004) Molecular electronic-structure theory. Wiley, Chichester
26. Hehre WJ, Ditchfield R, Pople JA (1972) J Chem Phys 56:2257
27. Hariharan PC, Pople JA (1973) Theor Chim Acta 28:213
28. Bonham RA, Bartell LS (1959) J Am Chem Soc 81:3491
29. Dutta A, Sherrill CD (2003) J Chem Phys 118:1610
30. Rendell AP, Lee TJ, Taylor PR (1990) J Chem Phys 92:7050
31. Watts JD, Cernusak I, Noga J, Bartlett RJ, Bauschlicher CW Jr, Lee TJ, Rendell AP, Taylor TR (1990) J Chem Phys 93:8875
32. Lee TJ, Rendell TJ, Taylor PR (1990) J Chem Phys 94:5463

Theor Chem Acc (2012) 131:1157
DOI 10.1007/s00214-012-1157-3

REGULAR ARTICLE

Modelling vibrational coupling in DNA oligomers: a computational strategy combining QM and continuum solvation models

Alessandro Biancardi · Roberto Cammi · Chiara Cappelli · Benedetta Mennucci · Jacopo Tomasi

Received: 20 July 2011 / Accepted: 30 September 2011 / Published online: 18 February 2012

Abstract In this paper, a computational strategy, based on DFT calculations at the M06-2X level, combined with the polarizable continuum model, the Hessian matrix reconstruction method and the Partial hessian vibrational approach is applied to evaluate inter- and intra-layer vibrational couplings between hydrogen bonded and stacked DNA base pairs. The present work demonstrates that this computational scheme can effectively predict and interpret the vibrational couplings of nucleic acids in solution. The effect of the environment described in a cluster or in a continuum manner is necessary in order to improve the agreement with the experimental values.

Keywords Vibrational coupling · Solvent effects · DNA pairs · DFT calculations · Hessian matrix reconstruction · Partial hessian vibrational approach · Guanine · Cytosine

1 Introduction

Two-dimensional infrared spectroscopy (2D-IR) [1–5] has been recently reported to provide valuable information on vibrational properties and dynamics of molecular systems, for which it is also a valuable tool to proof the local environment. 2D-IR somewhat overcomes limitations of conventional spectroscopic techniques, such as FTIR spectroscopy and circular dichroism, which need empirical qualitative relations in order to extract structural information from mono-dimensional spectra.

In addition to the usual diagonal peaks, the 2D-IR spectrum displays off-diagonal peaks that, similarly to the 2D-NMR spectroscopy, are produced by the coupling between oscillators and provide direct information about structure and dynamics. For these reasons, 2D-IR spectroscopy has been used to monitor the vibrational coupling in several molecular systems, such as hydrogen-bonded systems, nucleic acids [6–10], proteins and peptides.

Despite such potentialities, the extraction of structural information from 2D-IR spectra requires the coupling of experimental data with theoretical models, generally formulated in terms of quantum mechanical (QM) calculations, which permit to directly compute the vibrational properties of the molecular sub-units (frequencies and vibrational transition moments) and the vibrational coupling among them. The QM calculation of harmonic vibrational frequencies and transition dipole moments can be nowadays routinely performed, mainly due to the availability of analytical algorithms for the evaluation of energy second derivatives, for both isolated and solvated systems treated by means of continuum solvation models [11]. The theoretical modelling of the vibrational coupling is instead far from being a standard procedure and has been based on different approaches.

The simplest approach is the transition dipole coupling (TDC) model [12–14], based on the Coulomb interaction between the vibrational transition point dipoles associated with the interacting modes. However, due to its definition, the TDC is limited to describe vibrational couplings of electrostatic origin. A more accurate approach is the

Dedicated to Professor Vincenzo Barone and published as part of the special collection of articles celebrating his 60th birthday.

A. Biancardi · C. Cappelli · B. Mennucci · J. Tomasi (✉)
Dipartimento di Chimica e Chimica Industriale,
Università di Pisa, via Risorgimento, 35, 56126 Pisa, Italy
e-mail: tomasi@dcci.unipi.it

R. Cammi
Dipartimento di Chimica GIAF, Università di Parma,
viale delle Scienze, 43100 Parma, Italy

Hessian matrix reconstruction (HMR) method developed originally by Cho and co-workers [15–18], which is able to describe also the non-electrostatic components of the vibrational coupling. The basic formulation of the HMR requires an harmonic QM calculation of the vibrational modes of an appropriately chosen model system, followed by the computational vibrational analysis of the target system. The normal modes of the target system are then resolved in terms of vibrational modes localized on the molecular subunits, so that local modes frequencies and the inter-mode vibrational coupling can be computed. Both the TDC and the HMR methods have been originally used to describe the vibrational coupling for isolated systems, and only recently extended to solvated systems [19], by some of the present authors within the framework of the polarizable continuum model (PCM) [20], which describes the target solute at the QM level, while the solvent is described as a dielectric medium characterized by its macroscopic dielectric properties. The most recent implementations of the PCM manage to compute the molecular geometry and a large variety of properties of solvated systems, including vibrational normal modes and frequencies. In the formulation of the HMR within the PCM, the solvent effects are taken into account in terms of changes in the geometry of the system, of its normal modes and vibrational frequencies, due to the presence of the solvent. The extension of the TDC to PCM solutes relies instead on the evaluation of the vibrational transition dipoles of the local modes in the presence of the solvent and then explicitly includes medium screening effects on the Coulombic interaction between the transition dipoles. On the grounds of its formulation, the PCM–TDC model is suitable to model the vibrational coupling of weak interacting systems only.

In this paper, we will focus on the application of the PCM–HMR method to study the vibrational coupling between the nucleic acid bases in DNA double-stranded oligomers. It is in fact well known that infrared techniques are very sensitive to the structure of the polynucleotide helix [21, 22] and that DNA bases change in a unique way their vibrational properties after structural modifications, due to the different vibrational couplings and due to environment effects.

A good understanding of the structure and dynamics of a molecule is a key factor in order to correlate the change of couplings, to particular structural modifications [23]; in the case of DNA, this is fundamental, for instance, in order to develop more efficient DNA-binding drugs [24, 25].

QM calculations of the vibrational coupling in DNA oligomers have been reported by Zanni and co-workers [7] and by Cho and co-workers [26–29] adopting different computational strategies. In particular, Zanni and co-workers have studied the vibrational coupling between normal modes involving carbonyl stretching in Watson–Crick (WC) base pairs guanine (G) and cytosine (C) and between nearest-neighbour bases G and C, in two different DNA oligomers. QM calculations were performed on isolated G and C bases, by neglecting the deoxyribose and phosphate groups to reduce the computational costs, and the vibrational coupling was obtained by using the electrostatic TDC [12–14] and transition dipole density distribution [30] models, and a finite difference evaluation of the PES curvature. In particular, for the vibrational coupling between the WC pairs, a single pair was used as model, while for the evaluation of the vibrational coupling between nearest-neighbour bases G and C, two stacked WC base pairs were considered. The vibrational coupling for the same DNA oligomers has also been studied by Cho and co-workers [26–29], by performing QM calculations on G and C and computing the vibrational coupling through a modified HMR method. Large vibrational coupling models consisting of GC stacked base pairs up to quadruple GC stacked base pairs have been considered, but solvent effects have been only marginally studied in the particular case of a GC pair surrounded by the explicitly included adjacent base pairs.

The consideration of the local environment has been reported as crucial in order to realistically describe vibrational couplings in solution [10], principally because the interaction with the environment modifies both the single local vibrations and the coupling between them. Similarly to a previous work of some of the present authors on amide vibrational couplings in small peptides [19], in this paper, we will apply a computational strategy still based on the PCM combined with the HMR method and Partial hessian vibrational approach (PHVA) [31–37] to evaluate inter- and intra-layer couplings between hydrogen bonded and stacked DNA base pairs. In particular, the coupling between carbonyl vibrations located on G and C bases will be considered in the case of a A-form DNA of well-known structure, resolved with X-ray crystallography [38].

The paper is organized as follows. In the next section, the description of the systems investigated and of the methods used for the analyses is reported. Then, numerical results will be shown, by also putting into evidence how the results are influenced by the choice of the pair geometry and orientation, on the extension of the molecular model and on the accounting/discarding of solvent effects. Some conclusions end the presentation.

2 Methods and systems

2.1 Modelling DNA oligomers

In this study, we have investigated a sequence of guanine and cytosine in the double-stranded helix DNA $d(G_nC_n)$. In particular, we have used the structure of the DNA oligomer $d(G_4C_4)$ determined by McCall et al. [38] by single-crystal X-ray methods (PDB ID: 2ANA). In the structure, there are two strands in the asymmetric unit, and these coil about each other to form a right-handed *A*-type double helix with Watson–Crick hydrogen bonds between base pairs. The two d(G–G–G–G)·d(C–C–C–C) segments in the oligomer exhibit similar and uniform structures (see Fig. 1).

From this experimental structure, five different clusters were extracted, containing the coupled bases that from now on will be indicated as fragments 1–5. The borders of the helix were not considered.

To introduce the effect of the rest of DNA, two different models were used. In the first model (referred to as PCM), the environment was represented as a continuum dielectric, while in the second model (cluster), the adjacent base pairs were explicitly included in the QM calculation. More in details, for the H-bonded (HB) pairs, two additional bases were considered, while for the stacked pairs, four additional base pairs were included (see Fig. 2).

2.2 Methodologies

As pointed out by Lee and Cho [26], the application of the HMR approach to the study of coupled vibrations localized on base pairs presents critical points, due to the high delocalization of the vibrational modes. To overcome these critical points, we have combined the HMR approach with the PHVA [31–37].

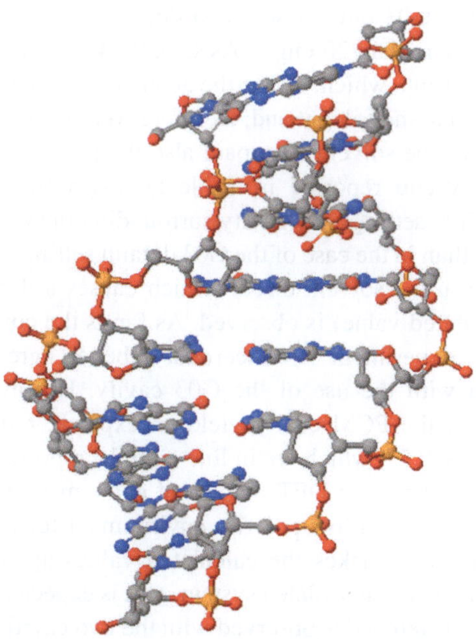

Fig. 1 Structure of the DNA oligomer investigated in this study. From this experimental structure, five different coupled bases were extracted, which from now on will be indicated as fragments 1–5

The PHVA approach is used both to perform the vibrational calculation on the coupled HB and stacked pairs and to localize the vibrational modes on each single basis. In this approach, the system is divided in two parts: a freely vibrating and a frozen part. The latter is taken frozen during the vibrational calculation and, therefore, does not directly contribute to the normal mode of the former. In agreement with the one-bond-distance rule [39], more than one bonds between the freely vibrating (the bases) and the fixed part (the furanoses and the adjacent base pairs in the cluster model) have been left in order not to bring the two parts too close (see Fig. 3).

The calculation of the vibrational coupling between two local modes located on different base pairs consists of the following steps:

- Normal modes calculation of the single uncoupled basis in the presence of the others kept frozen, using the PHVA approach. These are the local modes of the uncoupled basis, which are used as a basis for the determination of the vibrational coupling.
- Normal modes calculation of the coupled pair possibly in the presence of the other bases kept frozen, using the PHVA approach. These are the normal modes of the coupled pair.
- Application of the HMR using the results of the coupled and the uncoupled calculations to determine the vibrational coupling between the local vibrations.

2.3 Computational details

All calculations were performed by using the Gaussian 09 package [40], by employing DFT and using the M06-2X functional [41]. This functional was chosen for its ability to describe non-covalent interactions [42, 43]. To describe the continuum environment, the integral equation formalism [44] version of PCM was used. PCM cavities were built as a series of interlocking spheres centred on the atoms. Two different set of radii were tested: the UFF radii (all multiplied by a cavity size factor of 1.1), which are the Gaussian 09 default settings (from now on indicated as StCav), and a set of enlarged radii (employing a cavity size factor of 1.7). Since the distance between staked bases is about 3.4 Å [45], these enlarged radii are able to mimic the void space around the base pairs in the DNA double helix (from now on indicated as LCav). In the particular case of the PCM calculations for the fragments, in order to consider the possibility that in a DNA double helix the volume between the base pairs cannot be accessible to solvent, a larger radius (equal to that of the carbon atom) was used for the hydrogen atoms involved in hydrogen bonds between bases. Two different values of the dielectric permittivity were tested, corresponding to a weakly polar ($\varepsilon = 6$) and a non-polar ($\varepsilon = 2$) environment.

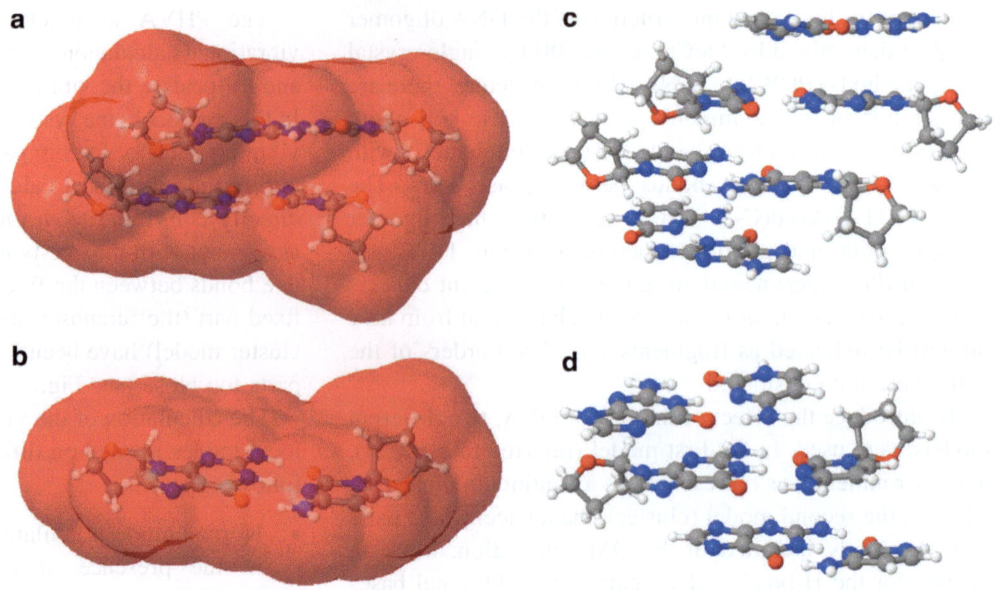

Fig. 2 Representations of the stacked and HB fragments within the PCM (**a** and **b**) and the explicitly included adjacent base pairs (**c** and **d**), respectively

The combination of three basis sets for the three different parts of the fragments was used, namely 6-311+G(d,p) for the freely vibrating bases, 6-31G(d) for the other vibrationally frozen bases and 3–21 g for the furanose rings. All the geometries were kept fixed to the experimental structure (taken from [38]), with the exception of the missing hydrogens in the experimental X-ray structure, that were optimized at the 6-311+G(d,p) level.

3 Results

3.1 Normal modes and frequencies of references bases in solution

Before discussing the couplings in the DNA oligomers, in this section, we will focus on the vibrational properties of the single bases composing the systems, that is, guanine (G) and cytosine (C), in their methylated form. The role of such results is twofold: in fact, besides being interesting per se, they can also give an idea of the accuracy of the results obtained in the following analysis on the DNA oligomers, since the single C and G are used as basis set for the PHVA and HMR analysis.

The vibrational properties of C and G have been previously studied in the literature, both from the experimental [7, 26, 46–50] and computational [7, 46, 51–54] points of view. In particular, values for the two selected modes, referred to in the following as Gs and Cs, have been reported in various environments. According to DFT calculations, these modes have predominately carbonyl stretch character, as shown in Fig. 4. Such modes are the ones on which the following discussion on vibrational couplings will be focused.

In Table 1, PCM calculated frequencies for G_s and C_s modes in different environments are reported, chosen according to the availability of experimental data. The outcomes of previous calculations reported in the literature are also shown for comparison's sake, as well as experimental values. In the case of PCM calculated data, two different choices of the definition of the PCM molecular cavity surrounding the solute in the dielectric are compared, that is, the cavity obtained by exploiting the default settings of Gaussian 03 [55] (G03,) and Gaussian 09 [40] (G09). In fact, the cavity size and shape are the only adjustable parameters in a PCM calculation, for a given solvent.

The calculated value of G_s in DMSO is sensitive to the choice of the PCM cavity. In particular, the use of a smaller cavity (the G03 default settings) decreases the calculated value of more than 20 cm^{-1}. As said, the Gs mode involves the C=O group, which, due to the geometry of the molecule, is exposed to the solvent and, therefore, very sensitive to the polarity of the solvent (compare also the previous calculations in vacuo reported in Table 1). However, with the default G09 settings, the cavity surrounding the C=O group is larger than in the case of the G03 default settings, so that a decrease in the solvent effect (which causes a decrease in the calculated value) is observed. As far as the comparison with the experiment is concerned, a better agreement is obtained with the use of the G03 cavity. However, both choices of the PCM cavity yield an exp/calc ratio in the range 0.96–0.97, which are in line with the reported scaling factors for the same DFT functional and similar basis sets [56]. The use of a low-polarity environment for the calculations ($\varepsilon = 2$) makes the calculated values approach the calculations for the isolated system, as it is expected. Also, a very low difference is observed with the two cavities, as the solvent effect being in both cases very low.

Also in the case of cytosine in water, the same trend as reported for DMSO as varying the cavity is observed. In

Fig. 3 Representations of the stacked (**a–f**) and HB fragments (**g**). The freely vibrating atoms are represented in "Ball and Stick", whereas the frozen atoms are shown in "wireframe". The systems **a–d** are obtained from the fragments 1, 2, 4, 5, whereas the systems represented in **d–f** are obtained from the fragment 3

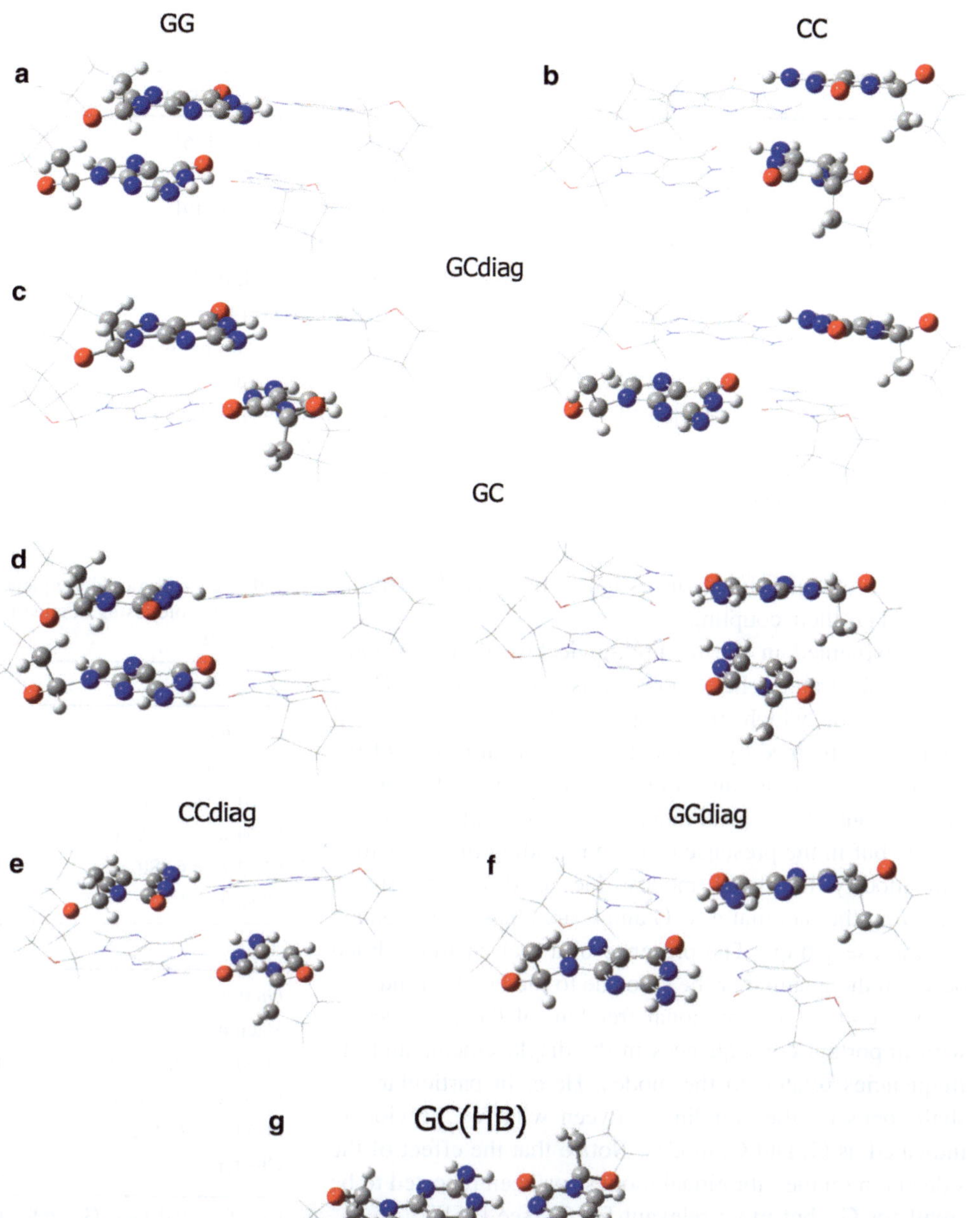

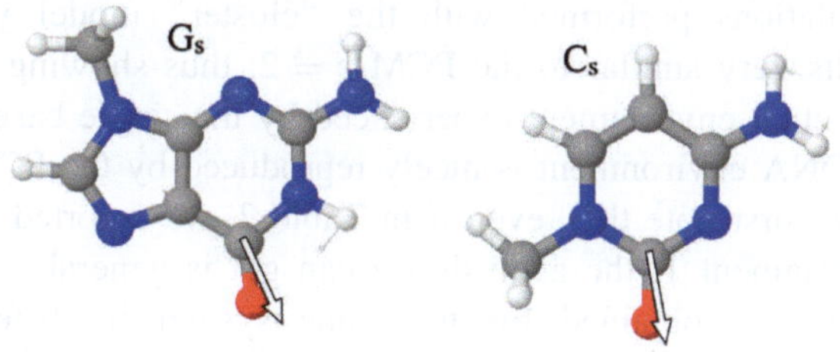

Fig. 4 Structure and pictorial view of the two selected normal modes of G (*left*) and C (*right*)

fact, a difference of about 20 cm^{-1} is reported for C_s, with the G03 once again approaching the experimental value, with an exp/calc ratio in the range 0.97–0.99, depending on the reference experimental value.

3.2 Normal modes and frequencies in DNA

The analysis reported in the previous section for the solvated G and C bases has shown the characteristics (including possible limits) of the QM model here adopted to describe vibrational motions in solvated systems. When moving to a more complex environment such as DNA, however, further difficulties appear due to the extremely inhomogeneous and anisotropic interactions between the bases and the embedding matrix. In particular, when focusing on vibrations, differences between solvated and embedded bases can appear not only in the frequencies of the investigated modes but also in the nature of the modes themselves. It is thus necessary to analyse the possible

Table 1 Calculated M06-2X/6-311+G(d,p) and experimental frequency values for a selected mode of C and G monomers in various environments

Mode	This work	Previous calculations in vacuo	Experimental
C_S	1747 G09 $\varepsilon = 2$	1750 (B3LYP/6-31G(d)) [26]	1712 (in N_2) [17, 46] 20 (in Ar) [52]
	1738 G03 $\varepsilon = 2$	1660 (BP86/6-311G) [46]	1730 (in Ne) [53]
	1702 G09 in H_2O	1722 (MP2/6-31G(d, p)) [49]	1645 in D_2O [7]
	1680 G03 in H_2O	1785 (B3LYP/6-31++G(d,p)) [7]	1660 in H_2O [54]
		1818 B3LYP/6-31G(d,p) [50]	
G_S	1799 G09 $\varepsilon = 2$	1686 (BP86/6-311G) [46]	1736, 1749 (in Ar) [51]
	1799 G03 $\varepsilon = 2$	1727 (B3LYP/6-31+G(d,p)) [47]	1742 (in N_2) [51]
	1760 G09 in DMSO	1762 (B3LYP/6-31G(d)) [26]	1693 in DMSO [7]
	1737 G03 in DMSO	1840 (B3LYP/6-31++G(d,p)) [7]	
		1833 (B3LYP/6-31G(d)) [48]	

All data are given in cm^{-1}

changes on the previously investigated modes before trying to simulate their couplings.

As explained in the methodological section, the computational strategy here adopted is based on the HMR approach, in which the couplings between vibrational modes are obtained by "projecting" the local modes of the uncoupled systems into those of the coupled pair. In present context, the local modes are those of the uncoupled G or C but in the presence of the embedding matrix, which will modify the local modes due to different effects, including the fact that now G and C are chemically bonded to furanose groups. The presence of a real chemical bond between the guanine or the cytosine to the DNA strand will in fact reduce the vibrational freedom of these molecules with important consequences in the displacements and the frequencies related to the modes. Here, in particular, we shall focus on the coupling between what we previously indicated as G_s and C_s modes. Notice that the effect of the side chain on the vibrational modes has been reported to be small for G_s, but more relevant for C_s (see [26]).

In Fig. 4, we previously reported the selected modes of G and C, whereas in Table 2, we report the corresponding frequencies obtained in vacuo and with the different models for mimicking the DNA environment.

With respect to the calculations in vacuo, the presence of the continuum dielectric, even if with a low dielectric constant, causes a decrease in the frequency absolute values, which is of the order of 6 cm^{-1} for C and 10 cm^{-1} for G. The use of a larger PCM cavity (i.e. LCav) causes the results to go towards the in vacuo limit, as expected. Overall, the differences among the results obtained with the two choices for the size of the PCM cavity are minor, in both cases of the order of 5–7 cm^{-1}, and independent on the choice of the fragment. The increase in the dielectric constant of the medium (moving from $\varepsilon = 2$ to $\varepsilon = 6$) also causes a shift of the same order. Noticeably, the

Table 2 Vibrational frequencies of uncoupled linked cytosine and guanine moieties as obtained with different solvation models, in the fragment *1*

	C_{S1}	C_{S2}	Average
Cytosine			
Vacuum	1709	1700	1704
PCM $\varepsilon = 2$, StCav	1703	1694	1698
PCM $\varepsilon = 2$, LCav	1707	1699	1703
PCM $\varepsilon = 6$, StCav	1696	1687	1691
Cluster	1703	1695	1699
	G_{S1}	G_{S2}	Average
Guanine			
Vacuum	1764	1755	1760
PCM $\varepsilon = 2$, StCav	1754	1745	1750
PCM $\varepsilon = 2$, LCav	1761	1752	1756
PCM $\varepsilon = 6$, StCav	1743	1734	1738
Cluster	1757	1749	1753

C_{S1}, C_{S2} and G_{S1}, G_{S2} refer to the local vibrations of cytosine and guanine in fragment 1. All values are given in cm^{-1}

calculations performed with the "cluster" model yield results very similar to the PCM $\varepsilon = 2$, thus showing that the actual environment experienced by the single bases in the DNA environment is nicely reproduced by the PCM.

We first note that, even if in Table 2, we reported data for fragment 1, the analysis we can get is general as the differences obtained for the same system in different fragments are always small, of the order of 10 cm^{-1}.

In order to get more insight into the trends just commented, it is relevant to recall here the actual molecular structures of the systems to which Table 2 refers. In particular, it is to note that all the calculations were performed on the same geometries, corresponding to what is extracted from the experimental X-ray structure taken from [38],

Table 3 Vibrational couplings calculated in vacuo and through different solvation models, in the fragment 1

	β_{GG}	β_{CC}	β_{GC}	β_{CG2}	β_{GC}(HB)
Vacuum	10.0	4.5	2.5	4.5	11.4
PCM $\varepsilon = 2$, StCav	9.1	3.7	0.9	4.0	11.8
PCM $\varepsilon = 2$, LCav	9.1	3.6	1.0	4.3	12.0
PCM $\varepsilon = 6$, StCav	8.1	3.5	0.5	2.9	13.7
Cluster	8.6	3.6	1.5	4.2	10.8

All values are in cm^{-1}

except for the positions of the hydrogen atoms, which were optimized. In addition, each basis is, in this case linked, to a partially frozen furanose group (see Fig. 3), and the corresponding frequencies have been obtained in the presence of the coupled bases, kept frozen. All these aspects make the comparison with what was previously shown in Table 1 for the "free" bases, quite difficult. In fact, in that case, the geometries were relaxed in each of the considered environments. As it can be seen, the differences are quite substantial: by referring to the PCM $\varepsilon = 2$ calculation, the calculated data for the "free" bases are about 45 cm^{-1} higher than the corresponding ones for the bonded bases in the frozen fragments.

3.3 Couplings

In Table 3, we report the vibrational coupling constants (β) obtained for the different pairs (stacked and HB) using different solvation models: the values reported in the different columns refer to G–G, C–C and G–C in the stacked pairs (see Fig. 3a–c) and G–C in the HB pair (see Fig. 3g), respectively. In the case of stacked G–C, two sets of values are reported corresponding to the two different diagonal arrangements (see Fig. 3c).

As far as the nomenclature in Table 3 is concerned, we note that the coupling terms β_{GG} and β_{CC} refer to stacked guanine and cytosine bases, respectively, and β_{GC}, β_{CG2} refer to the diagonal coupling between guanine and cytosine.

As it can be seen from the data reported in Table 3, the inclusion of the effects of the DNA environment always leads to a decrease in the coupling constants, with the only exception of the HB pair in a PCM environment.

Such a decrease in the β values can be ascribed to screening effects of the polarizable environment, and this is what found for all stacked pairs with all solvation models. As expected, the largest screening is found in the case of PCM with $\varepsilon = 6$ (i.e. the highest dielectric constant). For all stacked pairs, the cluster calculation is better reproduced by the PCM model with $\varepsilon = 2$, which can be defined a good continuum model to mimic the microscopic DNA-like environment here examined (which, however, does not fully consider the complexity of DNA, also comprising ribose, phosphate, surrounding waters and even counterions). In the following, the analysis of PCM environment will be limited to this description.

In the case of the HB pair, the analysis is less simple as the PCM description always leads to an increase of β also in the case of the smallest permittivity ($\varepsilon = 2$) and the largest cavity. This increase seems to be due to a large variation of the transition densities which is here opposite to the expected screening effect which is also diminished being the two coupled bases quite close [57].

Table 4 Calculated vibrational couplings for the investigated fragments in different environments

Fragment	β_{GG}			β_{CC}		
	Vac	$\varepsilon = 2$ LCav	$\varepsilon = 2$ StCav	Vac	$\varepsilon = 2$ LCav	$\varepsilon = 2$ StCav
1	10.0	9.1	9.1	4.5	3.7	3.6
2	11.7	10.7	9.9	3.5	3.4	3.4
3	6.7	6.7	6.5	2.9	2.6	2.5
4	13.9	12.9	12.0	3.4	3.3	3.3
5	10.6	9.8	9.0	2.9	2.7	2.7
$<\beta>$	10.6	9.8	9.3	3.4	3.1	3.1

Fragment	β_{GC}			β_{CG2}			β_{GC}(HB)		
	Vac	$\varepsilon = 2$ LCav	$\varepsilon = 2$ StCav	Vac	$\varepsilon = 2$ LCav	$\varepsilon = 2$ StCav	Vac	$\varepsilon = 2$ LCav	$\varepsilon = 2$ StCav
1	2.5	1.0	0.9	4.5	4.3	4.0	11.4	12.0	11.8
2	1.5	1.3	1.1	8.2	7.5	6.9	9.7	9.8	9.7
3	6.0	5.7	6.0	11.9	8.5	6.1			
4	1.5	1.3	1.3	3.4	3.6	5.0			
5	6.3	2.4	1.7	4.6	4.2	4.1			
$<\beta>$	3.6	2.3	2.2	6.5	5.6	5.2			

All values are given in cm^{-1}

Table 5 Internuclear C(O)–C(O) distances (in Å) and CO–CO dihedral angles (in degrees) in the coupled bases in the different fragments

Fragment	r_{GG}	r_{CC}	r_{GC}	r_{GC2}	r_{GC}(HB)	θ_{GG}	θ_{CC}	θ_{GC}	θ_{GC2}	θ_{GC}(HB)
1	4.2	4.4	7.3	4.2	4.5	24	24	174	162	174
2	3.7	4.7	7.2	4.5	4.4	32	40	168	154	170
3	4.6	5.5	4.7	4.2		47	47	91	96	
4	3.6	4.5	7.1	4.7		21	24	175	164	
5	3.8	4.4	7.1	4.5		27	31	178	156	
Avg	4.0	4.7	6.7	4.4	4.5	30	33	157	146	172

Moving now to the comparison among the different pairs, it is interesting to note that the GG stacked pair presents a coupling constant very similar to that of the GC HB pair, for which the largest interaction was expected.

In order to be properly compared with experiments, the results reported in Table 3 have to be further analysed so to estimate the effects of heterogeneity along the double strand. As explained in the methodological section, from the DNA oligomer, we have extracted five non-equivalent fragments (see Fig. 3). The corresponding β values obtained in vacuo and with PCM $\varepsilon = 2$ with the two cavities are reported in Table 4, whereas in Table 5, we collect the main geometrical parameters, namely the internuclear distances between the carbonyl carbons of the two coupled bases and the angles between the corresponding CO vectors.

We note that the significant differences found in the fragment n = 3 are due to the change in the sequence of bases. In fact, we note that, in fragments 1,2,4,5, the coupling terms β_{GG} and β_{CC} refer to stacked guanine and cytosine bases, respectively, and β_{GC}, β_{CG2} refer to the diagonal coupling between guanine and cytosine, whereas in fragment 3, the coupling terms β_{GC}, β_{CG2} refer to the coupling between stacked guanine and cytosine, respectively, and β_{GG} and β_{CC} refer to the coupling between diagonal guanine and cytosine (see Fig. 3d–f).

The comparison of the β values with the structural data reported in Table 5 reveals that the vibrational couplings are sensitive not only to the inter-layer distances, but also on the specific DNA base pair, that is, on the specific environment acting on the specific DNA base.

Finally, in Table 6, a comparison between our data and the outcomes of previous calculations are reported, as well as experimental data available in the literature.

We first note that previous calculations were performed by exploiting a different approach with respect to our. In fact, the data reported in [7] were obtained by using a finite difference (FD) approach, in which the coupling is calculated through the numerical evaluation of total energy second derivatives with respect to the local modes of the single nucleic acid bases. In this case, the local modes are normal modes of the isolated nucleic basis.

The calculations reported in [26] were performed with an extension of the HMR approach (Extended-HMR), which assumes the Hessian matrix in the atomic coordinate system to be divided into blocks, each associated with the local nucleic acid bases modes. Such an Hessian submatrix generates the corresponding local modes, which are used to transform the Hessian matrix from the global basis modes to the local modes. The same approach was also exploited to obtain the data reported in [29], where environmental effects are taken into account through molecular dynamics techniques.

The differences between the various approaches explain the variability of the results, being the set of values for the β_{GG} vibrational coupling the most significative under this comparative viewpoint. Also, due to the large differences between the computational approaches previously exploited and our method, it is not simple to clearly evidence the origins of the discrepancies. However, the trend reported for the extended-HMR/B3LYP approach by Lee and Cho [26] is in good agreement with our data, whereas the largest discrepancies are noticed for FD calculations [7], which is, however, the most different with respect to our model.

Table 6 also reports the comparison between the calculated and the fitted experimental vibrational couplings taken from [6, 7]. Overall, a general good agreement with experiments is found, especially if we consider that the experimental findings refer to a dG_5C_5 sequence that is not identical to the similar sequence dG_4C_4 used in the present study. The effect of the environment described in an explicit or in a continuum manner is not large, due to the low polarity of the environment to be modelled; however, a proper account of environmental effects improves the agreement with the experimental data.

4 Summary and conclusions

In this paper, the vibrational couplings of nucleic acids were investigated by carrying out QM calculations, combined with: a) the HMR approach in order to determine the coupling between vibrations, b) the PHVA in order to

Table 6 Comparison among experimental and calculated vibrational couplings

Method	Environment	β_{GG}	β_{CC}	β_{GC}	β_{CG2}	β_{GC}(HB)	References
HMR/M06-2X	Gas phase	10.0	4.5	2.5	4.5	11.4	This work
On fragment 1	$\varepsilon = 2$, LCav	9.1	3.7	1.0	4.3	12.0	This work
	$\varepsilon = 2$, StCav	9.1	3.6	0.9	4.0	11.8	This work
	Cluster	8.6	3.6	1.5	4.2	10.8	This work
HMR/M06-2X	$\varepsilon = 2$, LCav	9.8	3.1	2.3	5.6	10.9	This work
Averaged values on fragments 1–5	$\varepsilon = 2$, StCav	9.3	3.1	2.2	5.2	10.6	This work
FD/MP2	Gas phase	12.6	0.5	3.0	4.7	11.6	[7]
FD/B3LYP	Gas phase	21.8	1.0	1.6	5.0	10.4	[7]
Extended-HMR/M06-2X	Gas phase	8.4	1.3	0.4	2.5	13.9	[26]
	Cluster	8.3	1.0	0.2	2.3	13.7	[26]
	Classical MD simulations in D_2O solvent	17.4	3.2	1.8	4.3	12.8	[29]
Exp. (± 2 cm^{-1})		9.5	0.7	1.9	2.7	9.6	[7]
Exp.		9.7	2.3	>1.0	5.0		[6]

All values are given in cm^{-1}

accurately determine the vibrational local modes, c) an explicit description of the nearest parts of the DNA helix in order to account for a part of the DNA environment effect and d) a continuous description in order to account (in an average manner) for the effect of the rest of the DNA helix and of the solvent.

The present work demonstrates that this strategy can be effectively used in order to predict and interpret some features of the 2D-IR spectra of nucleic acids in solutions. The good correlation of our approach with both previous calculations and experimental findings shows that this method can indeed be used in combination with experimental measurements to gain an accurate simulation of the vibrational coupling between vibration localized on different DNA bases. However, many extensions of our work are possible. For example, the inclusion of water molecules potentially hydrogen bonded to the carbonyl groups could represent an important improvement towards a more realistic description: notice in fact that the authors of the X-ray structure [38] pointed out that "almost all exposed functional groups on the DNA are hydrated". Moreover, the notable change in the coupling term for fragment 3, which is attributed to the alteration in the base sequence, is an index on the important role played by the sequence which is known to affect the local structure and flexibility properties of DNA base pairs. Future studies could address this issue so to better understand the real impact of sequence on vibrational couplings.

References

1. Cho M (2008) Chem Rev 108:1331
2. Kim YS, Hochstrasser RM (2009) J Phys Chem B 113:8231
3. Jeon J, Yang S, Choi JH, Cho M (2009) Acc Chem Res 42:1280
4. Ganim Z, Chung HS, Smith AW, Deflores LP, Jones KC, Tokmakoff A (2008) Acc Chem Res 41:432
5. Jansen TIC, Knoester J (2009) Acc Chem Res 42:1405
6. Krummel AT, Mukherjee P, Zanni MT (2003) J Phys Chem B 107:9165
7. Krummel AT, Zanni MT (2006) J Phys Chem B 110:13991
8. Yang M, Szyc Y, Rottger K, Fidder H, Nibbering ETJ, Elsaesser T, Temps F (2011) J Phys Chem B 115:5484
9. Yang M, Szyc Y, Elsaesser T (2011) J Phys Chem B 115:1262
10. Szyc L, Yang M, Nibbering ETJ, Elsaesser T (2010) Angew Chem Int Ed 49:3598
11. Mennucci B, Cammi R, Tomasi J (1999) J Chem Phys 110:6858
12. Krimm S, Abe Y (1972) Proc Natl Acad Sci USA 69:2788
13. Torii H, Tasumi M (1992) Chem Phys 96:3379
14. Torii H, Tasumi M (1996) In: Mantsch HH, Chapman D (eds) Infrared spectroscopy of biomolecules. Wiley, New York
15. Gorbunov RD, Kosov DS, Stock G (2005) J Chem Phys 122:224904
16. Hamm P, Wouterson S (2002) Bull Chem Soc Jpn 75:985
17. Ham S, Cha S, Choi JH, Cho M (2003) J Chem Phys 119:1451
18. Hahn S, Ham S, Cho M (2005) J Phys Chem B 109:11789
19. Biancardi A, Cappelli C, Mennucci B, Cammi R (2010) J Phys Chem B 114:4924
20. Tomasi J, Mennucci B, Cammi R (2005) Chem Rev 105:2999
21. Taillandier E, Liquier J (1992) In: Lilley HJ, Dahlberg JJ (eds) Methods in enzymology. Academic Press, New York
22. Banyay M, Sarkar M, Gräslund A (2003) Biophys Chem 104:477
23. Hunt NeilT (2009) Chem Soc Rev 38:1837
24. Richards AD, Rodger A (2007) Chem Soc Rev 36:471
25. Biancardi A, Biver T, Marini A, Mennucci B, Secco F (2011) Phys Chem Chem Phys 13:12595
26. Lee C, Park K, Cho M (2006) J Chem Phys 125:114508
27. Lee C, Park K, Cho M (2006) J Chem Phys 125:114509
28. Lee C, Park K, Cho M (2006) J Chem Phys 125:114510
29. Lee C, Park K, Cho M (2007) J Chem Phys 126:145102
30. Moran A, Mukamel S (2004) Proc Natl Acad Sci USA 101:506
31. Besley NA, Metcalf KA (2007) J Chem Phys 126:035101
32. Besley NA, Bryan JA (2008) J Phys Chem C 112:4308
33. Jin S, Head JD (1994) Surf Sci 318:204
34. Calvin MD, Head JD, Jin S (1996) Surf Sci 345:161
35. Head JD (1997) Int J Quantum Chem 65:827

36. Head JD, Shi Y (1999) Int J Quantum Chem 75:815
37. Head JD (2000) Int J Quantum Chem 77:350
38. McCall M, Brown T, Kennard O (1985) J Mol Biol 183:385
39. Ghysels A, Van Speybroeck V, Verstraelen T, Van Neck D, Waroquier M (2008) J Chem Theory Comput 4:614
40. Frisch MJ, Trucks GW, Schlegel HB, Scuseria GE, Robb MA, Cheeseman JR, Scalmani G, Barone V, Mennucci B, Petersson GA, Nakatsuji H, Caricato M, Li X, Hratchian HP, Izmaylov AF, Bloino J, Zheng G, Sonnenberg JL, Hada M, Ehara M, Toyota K, Fukuda R, Hasegawa J, Ishida M, Nakajima T, Honda Y, Kitao O, Nakai H, Vreven T, Montgomery JJA, Peralta JE, Ogliaro F, Bearpark M, Heyd JJ, Brothers E, Kudin KN, Staroverov VN, Kobayashi R, Normand J, Raghavachari K, Rendell A, Burant JC, Iyengar SS, Tomasi J, Cossi M, Rega N, Millam NJ, Klene M, Knox JE, Cross JB, Bakken V, Adamo C, Jaramillo J, Gomperts R, Stratmann RE, Yazyev O, Austin AJ, Cammi R, Pomelli C, Ochterski JW, Martin RL, Morokuma K, Zakrzewski VG, Voth GA, Salvador P, Dannenberg JJ, Dapprich S, Daniels AD, Farkas Ö, Foresman JB, Ortiz JV, Cioslowski J, Fox DJ (2009) Gaussian 09, revision B.01, Gaussian, Inc., Wallingford
41. Zhao Y, Truhlar DG (2008) Theor Chem Acc 120:215
42. Hohenstein EG, Chill ST, Sherrill CD (2008) J Chem Theory Comput 4:1996
43. Rutledge LR, Wetmore SD (2010) Can J Chem 88:815
44. Cances E, Mennucci B, Tomasi J (1997) J Chem Phys 107:3032
45. Watson JD, Crick FH (1953) Nature 171:737
46. Santamaria R, Charro E, Zacarías A, Castro M (1999) J Comput Chem 20:511
47. Chandra AK, Nguyen MT, Uchimaru T, Zeegers-Huyskens T (1999) J Phys Chem A 103:8853
48. Pelmenschikov A, Hovorun DM, Shishkin OV, Leszczynski J (2000) J Chem Phys 113:5986
49. Ha TK, Gunthard HH (2001) Spectrochim Acta Part A 57:55
50. Kwiatkowski JS, Leszczynski J (1996) J Phys Chem 100:941
51. Szczepaniak K, Szczesniak M (1987) J Mol Struct 156:29
52. Szczesniak M, Szczepaniak K, Kwiatkowski JS, KuBulat K, Person WB (1988) J Am Chem Soc 110:8319
53. Nowak MJ, Lapinski L, Fulara J (1989) Spectrochim Acta Part A 45:229
54. Tena GN, Baranov VI (2005) J Appl Spectr 72:155
55. Frisch MJ, Trucks GW, Schlegel HB, Scuseria GE, Robb MA, Cheeseman JR, Zakrzewski VG, Montgomery JA Jr, Stratmann RE, Burant JC, Dapprich S, Millam JM, Daniels AD, Kudin KN, Strain MC, Farkas O, Tomasi J, Barone V, Cossi M, Cammi R, Mennucci B, Pomelli C, Adamo C, Clifford S, Ochterski J, Petersson GA, Ayala PY, Cui Q, Morokuma K, Malick DK, Rabuck AD, Raghavachari K, Foresman JB, Cioslowski J, Ortiz JV, Stefanov BB, Liu G, Liashenko A, Piskorz P, Komaromi I, Gomperts R, Martin RL, Fox DJ, Keith T, Al-Laham MA, Peng CY, Nanayakkara A, Gonzalez C, Challacombe M, Gill PMW, Johnson B, Chen W, Wong MW, Andres JL, Gonzalez C, Head-Gordon M, Replogle ES, Pople JA (2004) Gaussian 03. Gaussian, Inc., Wallingford
56. Alecu IM, Zheng J, Zhao Y, Truhlar DG (2010) J Chem Theory Comput 6:2872
57. Scholes GD, Curutchet C, Mennucci B, Cammi R, Tomasi J (2007) J Phys Chem B 111:6978

Theor Chem Acc (2012) 131:1165
DOI 10.1007/s00214-012-1165-3

REGULAR ARTICLE

Relativistic coupled cluster calculations of the electronic structure of KrH^+, XeH^+ and RnH^+

Francesco Ferrante · Giampaolo Barone · Dario Duca

Received: 28 June 2011 / Accepted: 13 September 2011 / Published online: 2 March 2012

Abstract Potential energy curves of NgH^+ cations (Ng = Kr, Xe, Rn) were obtained by using four-component relativistic CCSD(T) coupled cluster calculations. Dissociation energies, equilibrium bond lengths, electronic properties, such as dipole moments and electric field gradients at the nuclei, and the related spectroscopic parameters of the electronic ground state have been determined. The results obtained for KrH^+ and XeH^+ are in good agreement with available experimental data, while those for RnH^+ have been determined for the first time at this level of theory.

Keywords Relativistic coupled cluster · protonated noble gases · spectroscopic constants

1 Introduction

The formation, stability and physical–chemical properties of noble-gas (Ng) hydrides, as neutral, cationic or anionic species, have been subjects of several experimental and theoretical studies. It is well known, for example, that the NgH^- anions (Ng = He, Ne and Ar) are weakly bound systems, characterized by quite flat potential energy curves and long equilibrium distances, due to the interaction between the negative hydride ion H^- and the polarized Ng atom [1–3]. The van der Waals NgH neutral species (Ng = He, Ne, Ar, Kr, Xe) are also defined as Rydberg molecules, [4–6] because their excited states consist of one cationic NgH^+ core related to one Rydberg electron.

The protonated NgH^+ (Ng = He, Ne, Ar, Kr and Xe) species can be experimentally produced by electron impact on noble gas + hydrogen mixtures [7, 8]. Spectroscopic studies of neutral NgH molecules (Ng = Ne, Ar, Kr and Xe) [6, 9] revealed that the cationic species, generated at high energies, are characterized by relatively deep and harmonic-shaped curves, compared to the analogous curves of the ground state of the neutral species. These compounds are also of interest from an astrophysical point of view, since recent investigations suggest that the noble gas could be trapped by H_3^+ ions to form NgH_3^+ and that NgH^+ (Ng = Ar, Kr, Xe) cations could be formed in protoplanetary disks outside the solar system, following the dissociation reaction: $NgH_3^+ \rightarrow NgH^+ + H_2$ [10].

In the latter paper, it is claimed that the sequestration of He and Ne should not occur even at temperatures down to 30 K since they form very labile NgH_3^+ complexes. However, an additional explanation could be given by considering the easiness of the proton transfer reaction $NgH^+ + H_2 \rightarrow Ng + H_3^+$. As a matter of fact, mass spectrometry measurements have shown that the amount of NgH^+ species, present in Ng/H_2 or $Ng/H_2/C_2H_4$ gas mixtures, increases with increasing atomic mass of Ng [8, 7]. This evidence can be related to the strength of the proton transfer reaction of NgH^+ that apparently decreases by increasing the atomic mass along the group [8]. Furthermore, an exothermic proton transfer reaction of NgH^+ with ethylene was evidenced [7]: $NgH^+ + C_2H_4 \rightarrow Ng + C_2H_5^+$ The conversion rate constant of the reaction above and the amount of $C_2H_5^+$ monotonously decrease by increasing the atomic number of Ng. Such trend was explained by

Dedicated to Professor Vincenzo Barone and published as part of the special collection of articles celebrating his 60th birthday.

F. Ferrante (✉) · G. Barone · D. Duca
Dipartimento di Chimica "S. Cannizzaro",
Università degli studi di Palermo, Viale delle Scienze,
Parco d'Orleans II, Palermo 90128, Italy
e-mail: francesco.ferrante@unipa.it

considering that in the ion chamber, under similar conditions, the internal energy of NgH^+ decreases by increasing the atomic mass of Ng [7]. An interesting feature of the dissociation curve of protonated noble gas is related to the observation that while for Ng = He, Ne, Ar and Kr, the dissociation curve is described by the reaction: $NgH^+ \rightarrow Ng + H^+$ [8, 11, 12], for Ng = Xe and plausibly Rn, the dissociation curve is described by the reaction: $NgH^+ \rightarrow Ng^+ + H$ [12–14].

It is worth mentioning that several accurate quantum chemical calculations of the electronic structure of Ng hydrides (Ng = He, Ne, Ar, Kr), both ionic and neutral, have been reported in the literature [15–18]. On the other hand, those involving XeH^+ and RnH^+ have been performed by using relativistic pseudopotentials in nonrelativistic approaches [19] or by the consideration of scalar relativistic effects through the zeroth-order regular approximation [20] or by the inclusion of corrective terms to take into account spin-orbit effects [21].

In the present paper, we have revisited the electronic structure and the dissociation curve of KrH^+, XeH^+ and RnH^+ employing a fully relativistic coupled cluster approach, with the aim of verifying the relativistic effects influence on the rotovibrational spectroscopic parameters in their ground electronic state. The results obtained for KrH^+ and XeH^+ have been compared to the ones available by experimental studies concerning the dissociation curves and rotational and vibrational spectroscopic parameters [22, 23]. Such benchmarking studies are essential in order to validate the approach employed for characterizing RnH^+. It is to emphasize that calculations on the latter are especially interesting because it cannot be easily investigated by experimental procedures, due to the radioactive instability of Rn and because it can be expected that relativistic contributions in the RnH^+ species must be larger than those observed in the other Group-18 elements. In this respect, calculated rotovibrational spectroscopic properties, which take into account relativistic effects in RnH^+, could be, for example, helpful for its experimental characterization. Furthermore, the relativistic results obtained for RnH^+ can be employed as a reference for studying features of the group-18 superheavy $118H^+$ species [24, 25].

2 Computational details

The four-component relativistic Dirac-Coulomb CCSD(T) approach [26, 27] has been used to calculate the electronic structure and the potential energy curves of the protonated noble gases. The fully decontracted triple zeta quality basis set provided by Dyall [28] has been used for Kr, Xe and Rn, augmented with 1s1p1d1f diffuse functions and 2f1g core correlating functions. Final basis set schemes are as follows: Kr: 24s17p11d4f1g, Xe: 29s22p16d4f1g and Rn: 31s27p18d12f1g. The uncontracted functions of aug-cc-pVTZ basis set by Dunning, 6s3p2d, have been centered on the proton. Small basis sets for relativistic calculations have been generated by kinetic balance. For computational efficiency, SS integrals have not been included in the calculations. A total of 26 electrons have been considered in the coupled cluster calculations for all NgH^+ species, being correlated 162, 204 and 212 virtual spinors for KrH^+, XeH^+ and RnH^+, respectively. The dipole moments and the electric field gradients at nuclear positions have been calculated by using the MP2 gradient. The energy of the $NgH^+ \rightarrow Ng + H^+$ dissociation has been evaluated by calculating the energy of the isolated noble-gas atom, including BSSE correction (estimated *ca.* 0.15 eV, irrespective of the noble gas). Energies of the $NgH^+ \rightarrow Ng^+(^2P_{3/2}) + H$ dissociation have conversely been obtained by considering the first ionization energy, E_I, of the involved noble-gas atom. The E_I values have been calculated at the CCSD level by using the Fock space formalism [29]. Finally, in order to estimate how relativity affects the results, the same methods and basis sets have been used to perform nonrelativistic calculations with the Lévy–Leblond Hamiltonian [30] and a point charge model for the nuclei. All calculations have been performed by using the DIRAC 0.8 program [31].

3 Results and discussion

Before starting a detailed analysis of the potential energy curves, we in the following discuss some properties calculated at the equilibrium geometry and the dissociation behavior of the three NgH^+ species.

Structural and energetic parameters of the three title compounds are reported in Table 1. The dipole moment of the three protonated noble gases was calculated at relativistic level at the equilibrium geometry. The calculated dipole moment of KrH^+, 1.969 D, is in excellent agreement with that determined experimentally, 1.967 D [32], and markedly improves the one previously estimated, 1.944 D [33], obtained by *ab initio* calculations. Considering that the calculated equilibrium distance of KrH^+ (see Sect. 3.1) is slighter underestimated compared to the corresponding experimental value, the explanation for such good agreement between calculated and experimental dipole moments is not straightforward. This result has in fact to be considered cautiously, taking into account the intrinsic difficulties present in the experimental determination of dipole moments of charged species [34].

Table 1 Equilibrium distance (Å), dipole moment (D), electric field gradient on the Ng nucleus (a.u.), dissociation energy (eV), ionization energy of Ng (eV) and spin-orbit splitting (SOS) in Ng^+ (eV), obtained for the three protonated noble gases KrH^+, XeH^+ and RnH^+ at relativistic (*R*) and nonrelativistic (NR) coupled cluster level

	KrH^+			XeH^+			RnH^+	
	R	NR	Exp	*R*	NR	Exp	*R*	NR
r_e	1.414	1.415	1.421[a]	1.598	1.601	1.603[f]	1.710	1.700
μ	1.969	–	1.967[b]	1.512	–	–	1.025	–
q	8.8838	–	–	13.3819	–	–	28.0594	–
$D_e(Ng + H^+)$	4.58	4.58	4.51[c]	5.28	5.30	–	5.68	5.64
$D_e(Ng^+ + H)$[j]	4.79	5.03	–	3.64	4.09	3.81[g]	2.64	3.68
E_I	13.81	14.06	14.00[d]	11.96	12.40	12.13[h]	10.56	9.07
SOS	0.67	–	0.65[e]	1.31	–	1.312[i]	3.81	–

[a] Ref. [22], [b] Ref. [32], [c] Ref. [12], [d] Ref. [38, 39], [e] Ref. [40], [f] Ref. [13], [g] Ref. [14], [h] Ref. [41, 42], [i] Ref. [42], [j] $^2P_{3/2}$ state in the relativistic case

To our knowledge, the experimental value of the dipole moment of XeH^+ has not been reported in the literature. The value obtained in the present paper, 1.512 D, is smaller than that already calculated, 1.551 D, by Klein and Rosmus [35].

From Table 1, it is evident that for KrH^+ the dissociation energy to Kr + H^+ is lower than that leading to Kr^+ + H, both at relativistic and nonrelativistic levels. Hence, relativity seems to have no sensible effect, being the relativistic and nonrelativistic results equal, 4.58 eV, and within 1.5% from the experimental D_e value of 4.51 eV [8, 11, 12]. The relativistic effect is indeed very small also for the dissociation energy values to Ng + H^+ calculated for XeH^+ and RnH^+, because there is no spin-orbit coupling in these cases. The dissociation energy of XeH^+ experimentally determined, which produces $^2S_{1/2}$ H and $^2P_{3/2}Xe^+$, is equal to 3.81 eV [12–14]. The D_e value calculated at relativistic level for XeH^+, 3.64 eV, is lower of about 4.5% than the corresponding experimental value and is in worst agreement than the value previously calculated by Peterson and co-workers [19]. However, in the latter paper, the dissociation energy to $^2S_{1/2}$ H and $^2P_{3/2}Xe^+$ has been determined by using the experimental value of the ionization energy of Xe, in the following equation: $D_e(^2P_{3/2}Xe^+ + {}^2S_{1/2}) = D_e(Xe + H^+) + E_I(Xe) - E_I(H)$, where the value of 13.61 eV has been used for $E_I(H)$. Since the calculated E_I of Xe to $^2P_{3/2}Xe^+$, 11.96 eV, is lower of 0.17 eV than the experimental value, such error is propagated in the D_e value. If we use the E_I experimental value of Xe, we obtain 3.80 eV in perfect agreement with the experiment. As a matter of fact, while the ionization energy of Xe to $^2P_{3/2}Xe^+$ is strongly affected by spin-orbit coupling, the dissociation energy of XeH^+ to Xe and H^+ is not. Hence, the value of D_e that we have obtained, 3.64 eV, allows one to estimate the accuracy with which the effects of the spin-orbit coupling can be determined by relativistic *ab initio* calculations. Incidentally, the nonrelativistic calculation overestimates the experimental Xe E_I value of 0.27 eV. At the same time, the ionization energy of atomic radon is evaluated with an error of 0.19 eV with respect to the early experimental value of 10.748 eV reported by Rasmussen [36]. It is worth to note that, although the calculated effect of spin-orbit coupling is not extremely accurate on the ionization energy (which has been here evaluated without the inclusion of the perturbative correction for triple excitations), the $^2P_{3/2}$–$^2P_{1/2}$ spin-orbit splittings (SOS) in Kr^+ and Xe^+ have been determined with an excellent agreement with the experimental values (see Table 1). Interestingly, spin-orbit coupling is so large in RnH^+ that there is an inversion of the dissociation energy of $D_e(^2P_{1/2}\ Rn^+ + {}^2S_{1/2}H)$ with that of $D_e(Rn + H^+)$ and the order of the three possible dissociation energy values becomes: $D_e(^2P_{3/2}Rn^+ + {}^2S_{1/2}H) < D_e(Rn + H^+) < D_e(^2P_{1/2}Rn^+ + {}^2S_{1/2}H)$. On the other hand, for XeH^+, the order is the following: $D_e(^2P_{3/2}\ Xe^+ + {}^2S_{1/2}H) < D_e(^2P_{1/2}Xe^+ + {}^2S_{1/2}H) < D_e(Xe + H^+)$. The values of $D_e(^2P_{1/2}Ng^+)$ can be calculated by using the spin-orbit splitting data for the Ng^+ cation reported in Table 1.

The electric field gradient at the nucleus, *q*, is a useful parameter to determine the nuclear quadrupole moment (Q) once knowing the quadrupole coupling constant (eQq), for example, from the rotational line splitting. In particular, the calculated electric field gradient can be of particular importance for evaluating the quadrupole moment of a given radon isotope in RnH^+ from the registration of its spectrum. The fully relativistic treatment used in the present approach, together with the use of specifically tailored relativistic basis sets, allows us to think that the electric field gradients *q* reported in Table 1 are highly accurate. By using the nuclear quadrupole coupling constants for $^{83}KrD^+$ (549.1 MHz [11]) and $^{131}XeH^+$ (−369.5 MHz [14]), we obtain the following updated

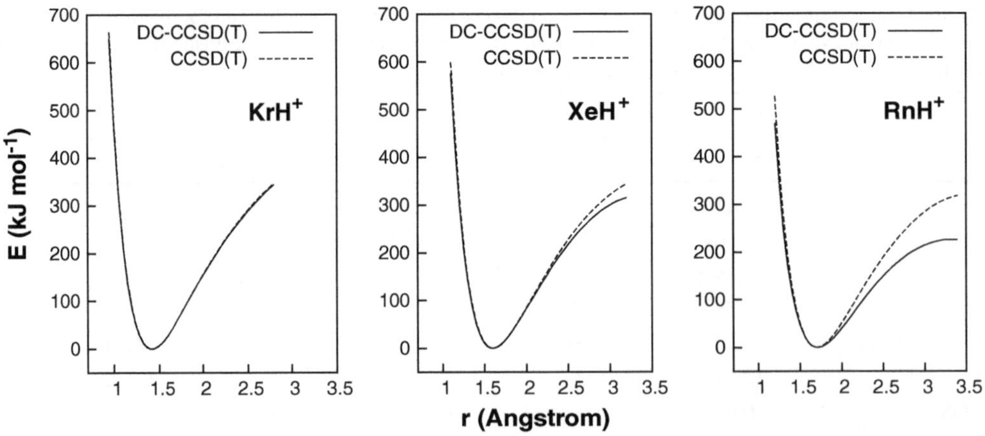

Fig. 1 The relativistic and nonrelativistic energy curves in the range $0.7r_e$–$2.0r_e$ for KrH^+, XeH^+ and RnH^+

values of nuclear quadrupole moments Q: +263.1 and −117.5 mbarn for ^{83}Kr and ^{131}Xe, respectively. Previous values of Q, +259 and −114 mbarn, were reported in the literature [37]. There is no experimental quadrupole coupling constant for RnH^+ or RnD^+. In this case, by using the nuclear quadrupole moments of ^{209}Rn reported in [37], Q = +311 mbarn, and the corresponding value of q in Table 1, we obtained the remarkably large value of eQq = 2.050 GHz for $^{209}Rn^+$.

3.1 Analysis of the potential energy curves

A Dunham analysis of the potential energy curves was performed in the bond length range $0.7r_e$–$2.0r_e$, being r_e the calculated NgH^+ equilibrium distance corresponding to the fundamental state. The potential energy curves have been obtained by a scan of the energy as a function of the internuclear distance, in steps of 0.05 Å. Near the energy minimum, the steps were reduced at 0.01 Å and in the harmonic region at 0.001 Å, for a total of about 110 points. The equilibrium distance was estimated from the analysis of the harmonic portion and was used as fixed parameter in the fitting procedure with the Dunham potential.

The analysis of the electronic structure along the dissociation curve of both XeH^+ and RnH^+ induces to think that the intersection between the (Ng + H^+) and the ($^2P_{3/2}Ng^+$ + H) potential energy curves occurs in the internuclear distance range 2.75–2.80 Å and 2.70–2.75 Å for Ng = Xe and Rn, respectively. According to the analysis of the spectroscopic constants determined in the present work (see below), the curve intersection should occur near the 11th and 8th vibration level of XeH^+ and RnH^+, respectively. The effect of the spin-orbit coupling is clear from the inspection of the dissociation curves obtained by relativistic and nonrelativistic calculations, as illustrated in Fig. 1. The latter shows that the difference between the two curves is negligible for KrH^+ and only a small effect on the values of the spectroscopic constants can be predicted, excepting the high-order ones (see Table 2). For XeH^+, a stabilization of the relativistic ($^2P_{3/2}Xe^+$ + H) compared to the nonrelativistic ($^2PXe^+$ + H) curve occurs at high r values. Such stabilization produces small curve shape variations also at internuclear distance values near the energy minimum. The effect is markedly more pronounced for RnH^+, because the spin-orbit coupling is by far larger. Such differences in the shapes of the dissociation curves of XeH^+ and RnH^+ lead to different values of the spectroscopic constants evaluated by including or not relativity in our calculations (see Tables 3 and 4).

Table 2 Dunham coefficients at relativistic (*R*) and nonrelativistic (NR) level for $^{84}KrH^+$ and their spectroscopic identification (SC)

	R	NR	Exp	SC
Y_{10}	2,551.1	2,567.1	2,494.7[a]	ω_e
Y_{20}	45.8	47.1	48.5[a]	$-\omega_e x_e$
Y_{30}	0.207	0.321		$\omega_e y_e$
Y_{40}	−0.017	−0.014		$\omega_e z_e$
Y_{01}	8.466	8.454	8.381[a]	B_e
Y_{11}	0.277	0.279	0.267[a]	$-\alpha_e$
$Y_{21} \times 10^3$	1.71	1.86		γ_e
$Y_{31} \times 10^4$	4.22	5.10		
$Y_{02} \times 10^4$	3.73	3.67		$-D_e$
$Y_{12} \times 10^6$	7.75	6.94		β_e
$Y_{22} \times 10^7$	−3.76	−1.86		
$Y_{03} \times 10^9$	6.74	6.11		H_e
$Y_{13} \times 10^{10}$	−2.01	−0.66		
$Y_{04} \times 10^{13}$	−3.02	−3.02		
B_0	8.328	8.315	8.248[b]	
$-D_0 \times 10^4$	3.69	3.63	3.76[b]	
$H_0 \times 10^9$	6.64	6.07	6.92[b]	

Dunham fit was performed in the range 0.95–2.80 Å. All values are in cm^{-1}

[a] Ref. [8], [b] Ref. [22]

Table 3 Dunham coefficients at relativistic (*R*) and nonrelativistic (NR) level for $^{132}XeH^+$ and their spectroscopic identification (SC)

	R	NR	Exp	SC
Y_{10}	2,286.6	2,320.3	2,270.0[a]	ω_e
Y_{20}	36.6	34.7	41.3[a]	$-\omega_e x_e$
Y_{30}	0.065	0.021		$\omega_e y_e$
Y_{40}	−0.021	−0.023		$\omega_e z_e$
Y_{01}	6.600	6.575	6.561[a]	B_e
Y_{11}	0.188	0.180	0.187[a]	$-\alpha_e$
$Y_{21} \times 10^3$	0.96	1.01		γ_e
$Y_{31} \times 10^4$	1.57	0.84		
$Y_{02} \times 10^4$	2.20	2.11		$-D_e$
$Y_{12} \times 10^6$	4.18	4.76		β_e
$Y_{22} \times 10^7$	−2.79	−3.77		
$Y_{03} \times 10^9$	2.94	2.90		H_e
$Y_{13} \times 10^{10}$	−1.17	−1.65		
$Y_{04} \times 10^{14}$	9.36	−6.39		
B_0	6.507	6.485	6.467[b]	
$-D_0 \times 10^4$	2.18	2.09	2.18[b]	
$H_0 \times 10^9$	2.89	2.82	2.57[b]	

Dunham fit was performed in the range 1.10–3.20 Å. All values are in cm^{-1}

[a] Ref. [12], [b] Ref. [23]

Table 4 Dunham coefficients at relativistic (*R*) and nonrelativistic (NR) level for $^{222}RnH^+$ and their spectroscopic identification (SC)

	R	NR	SC
Y_{10}	2,016.3	2,212.5	ω_e
Y_{20}	34.3	33.7	$-\omega_e x_e$
Y_{30}	−0.056	0.054	$\omega_e y_e$
Y_{40}	−0.031	−0.019	$\omega_e z_e$
Y_{01}	5.746	5.814	B_e
Y_{11}	0.172	0.155	$-\alpha_e$
$Y_{21} \times 10^4$	5.64	8.53	γ_e
$Y_{31} \times 10^4$	1.28	1.21	
$Y_{02} \times 10^4$	1.87	1.61	$-D_e$
$Y_{12} \times 10^6$	3.20	2.81	β_e
$Y_{22} \times 10^7$	−3.89	−1.82	
$Y_{03} \times 10^9$	1.81	1.51	H_e
$Y_{13} \times 10^{10}$	1.76	−0.57	
$Y_{04} \times 10^{14}$	−6.86	−4.02	
B_0	5.660	5.737	
$-D_0$	1.85	1.59	
H_0	1.72	1.50	

Dunham fit was performed in the range 1.20–3.20 Å. All values are in cm^{-1}

The a_i coefficients of the Dunham potential:

$$V(\xi) = a_0 \xi^2 \left(1 + \sum_{i=1}^{6} a_i \xi^i\right) \qquad \xi = \frac{r - r_e}{r_e} \tag{1}$$

have been obtained by the fit of the calculated potential energy curves. These have been used to determine the Dunham Y_{kj} coefficients [43], which are correlated to the position of the rotovibrational spectroscopic lines by the following equation:

$$E_{v,J} = \sum_{j=1}^{4} \sum_{k=0}^{5} \rho^{2j+k} Y_{kj} \left(v + \frac{1}{2}\right)^k [J(J+1)]^j \tag{2}$$

The factors $\rho = \mu/\mu_{iso}$ can be used to evaluate the isotopic effects on the spectroscopic constants, when considering species with different reduced mass μ_{iso} compared to that of the isotope used in the calculation. This consideration can be important for Xe, which consists of five principal isotopes, three of which with abundances larger than 20% and two with abundances about 10%. The values reported in Tables 2, 3 and 4 are referred to the molecular species having the following isotopic composition: $^{84}Kr^1H^+$, $^{132}Xe^1H^+$ and $^{222}Rn^1H^+$.

The rotational spectra of $^{84}KrH^+$ and $^{86}KrH^+$ are reported in [22], with a claimed accuracy of 40 kHz in the transition frequencies. The equilibrium distance of KrH^+ calculated at both relativistic and nonrelativistic levels is smaller than that obtained experimentally of 0.007 and 0.006 Å, respectively. An analogous but slighter difference was also observed for XeH^+. The underestimation of the calculated equilibrium bond length has a direct consequence on the rotational constants B_e and B_0, which are both about 1% larger than the corresponding experimental values. This occurs because the rotovibrational coupling constant α_e only marginally affects the ground vibrational state rotational constant B_0. The calculated values of the fourth- and sixth-order centrifugal distortion constants for the ground vibrational state, D_0 and H_0, are both smaller of 2 and 4% than the corresponding experimental values, respectively. Remarkably, such errors are about three times larger in the corresponding nonrelativistic case. The relativistic correction to the fundamental frequency, ω_e, is of only 16 cm^{-1}, and the experimental value is overestimated by about 2 and 3% at relativistic and nonrelativistic levels, respectively. The calculated first anharmonicity constant, $\omega_e x_e$, on the other hand, underestimates the experimental value of about 6 and 3% at relativistic and nonrelativistic levels, respectively. The agreement between calculated and experimental values of the spectroscopic parameters is worst than that obtained by Peterson et al. [19], likely because the equilibrium distance calculated in the latter paper is slightly larger than that obtained in the present work. However, it is interesting to note that in the paper by Peterson et al., the equilibrium bond lengths calculated at the CCSD(T)/aug-cc-pv5Z-PP level for all the diatomics there reported are systematically overestimated, of 0.005–0.011 Å, with respect to the corresponding experimental values. The only exception is KrH^+, for which experimental and calculated bond lengths agree within 0.001 Å. These findings, in our opinion, suggest the need of confirming the experimental result on the Kr–H distance in KrH^+.

A pure rotational spectrum of XeH+, with an estimated accuracy of 35 kHz on the transition frequencies, is reported by Ohtaki et al. [23], for all known isotopes and for both hydrogenated and deuterated species. Also in this case, the equilibrium bond distance is underestimated of 0.3%, with consequent overestimates of 0.6% on B_e and B_0. In comparison with KrH^+, however, the α_e and D_0 constants are more accurately determined, with an error of less than 0.5%. The calculated H_0 constant, on the other hand, is larger of about 12% than the corresponding experimental value. For XeH^+, the bond distance calculated at nonrelativistic level is in better agreement with the experimental value, and consequently the same occurs also for B_e and B_0. However, other constants show a larger deviation than that obtained at relativistic level, for example, $\alpha_e = -4\%$, $D_0 = -4\%$, $H_0 = -14\%$.

A slight deformation of the calculated potential energy curve, for example, due to spin-orbit coupling, could affect the relativistic fundamental vibration frequency that is 16 cm^{-1} (0.7%) and 50 cm^{-1} (2.2%) higher than the experimental value at relativistic and nonrelativistic levels, respectively. The relativistic fundamental vibration frequency of XeH^+ is comparable to that obtained by Runeberg et al. [21] by nonrelativistic CCSD(T) calculations, with corrections for the BSSE and spin-orbit coupling. On the other hand, the first anharmonicity constant is underestimated by 11 and 16% at relativistic and nonrelativistic level, respectively.

To the best of our knowledge, no experimental data have been reported in the literature concerning the structure and spectroscopic constants of RnH^+. The only available data are relative to CCSD(T) calculations with relativistic pseudopotential basis sets [19] and DFT calculations with scalar relativistic contributions evaluated by the ZORA approximation [20]. Being our calculations fully relativistic, the difference between the relativistic equilibrium distance evaluated in the present work and that previously obtained should be mainly attributable to the deformation of the potential energy curve due to the occurrence of a strong spin-orbit coupling at large internuclear distance, where the ($^2P_{3/2}Rn^+$ + H) character must be taken into account. Such curve deformation causes a lowering of the fundamental vibration frequency and a slight increase in the relativistic equilibrium distance compared to the nonrelativistic one. In fact, while the nonrelativistic equilibrium distance of RnH^+, 1.700 Å, is in good agreement with those previously calculated, 1.6957 Å [19] and 1.702 Å [20], the relativistic value is slightly larger, 1.710 Å. Analogously, the nonrelativistic vibration frequency is higher of about 200 cm^{-1} than the corresponding relativistic value (see Table 4) and of about 140 cm^{-1} compared to the result reported in Peterson et al. [19]. Of course, the consistent difference in the shape of the relativistic and nonrelativistic potential energy curves of RnH^+ produces large effects on the centrifugal and rotovibrational coupling constants, which show large percentage changes and also two sign variations. Considering that for XeH^+ the relativistic treatment leads to a by far better agreement with the experimental fundamental frequency, and considering also that relativistic effects increase with the atomic number, in our opinion the relativistic results obtained for the protonated radon are more reliable than those obtained at nonrelativistic level.

In Tables 2, 3 and 4, the high-order spectroscopic constants are also reported. They have not been discussed since neither experimental nor calculated corresponding data are reported in the literature. Such constants are so tiny that their values are strongly affected by the fit procedure. Indeed, they are only rarely used in the interpretation of biatomic molecule spectra.

4 Conclusions

In this work, the potential energy curves and the spectroscopic parameters of the protonated forms of the three heavier noble gases, KrH^+, XeH^+ and RnH^+, have been obtained by fully relativistic coupled cluster calculations. The relativistic contributions have been assessed by comparison with the values calculated at the same nonrelativistic level of theory. The obtained results are in good agreement with the available experimental data, though the calculations slightly underestimate the bond length in KrH^+ and, to a lesser extent, in XeH^+. A small change in the shape of the XeH^+ potential energy curve, due to the intersection of the (Xe + H^+) with the ($^2P_{3/2}Xe^+$ + H) curve at higher internuclear distances, improves the agreement with the experimental vibrational and rotovibrational constants with respect to the nonrelativistic results. The potential energy curve deformation is particularly evident for RnH^+, whose structural, electronic and spectroscopic properties are here reported for the first time at fully relativistic level. These data could be used for the experimental characterization of RnH^+, which is made difficult by the radon low abundance and high radioactive instability.

Acknowledgments We thank the "Fondazione Banco di Sicilia" for the financial support supplied to assemble the "Iride" computer cluster (convenzione PR13.b/07).

References

1. Li Y, Lin CD (1999) Phys Rev A 60:2009–2014
2. Robicheaux F (1999) Phys Rev A 60:1706–1709
3. Vallet V, Bendazzoli GL, Evangelisti S (2001) Chem Phys 263:33–40
4. Braga JP, Murrella JN (1984) Mol Phys 53:295–299
5. Douay M, Rogers SA, Bernath PF (1988) Mol Phys 64:425–436
6. Lo JMH, Klobukowski M, Bielinska-Waz D, Diercksen GHF, Schreiner EWS (2005) J Phys B: At Mol Opt Phys 38:1143–1159
7. Smith RD, Futrell JH (1976) Int J Mass Spect Ion Phys 20:71–76
8. Johns JWC (1984) J Mol Spectr 106:124–133
9. Moller T, Beland M, Zimmerer G (1987) Chem Phys Lett 136:551–556
10. Mousis O, Pauzat F, Ellinger Y, Ceccarelli C (2008) Astrophys J 673:637–646
11. Warner HE, Conner WT, Woods RC (1984) J Chem Phys 81:5413–5416
12. Hunter EP, Lias SG (1998) J Phys Chem Ref Data 27:413–656
13. Rogers SA, Brazier CR, Bernath PF (1987) J Chem Phys 87:159–162
14. Peterson KA, Petrmichl RH, Wood RC (1991) J Chem Phys 95:2352–2360
15. Liu DJ, Ho WC, Oka T (1987) J Chem Phys 87:2442–2446
16. Partridge H, Schwenke DW, Bauschlicher CW (1993) J Chem Phys 99:9776–9783
17. Schutte CJH (2002) Chem Phys Lett 353:389–395
18. Sebera CJ, Spirko V, Fiser J, Kraemer WP, Kawaguchi K (2004) J Mol Struct 695(696):5–11
19. Peterson KA, Figgen D, Goll E, Stoll H, Dolg M (2003) J Chem Phys 119:11113–11123
20. Pierrefixe SCAH, Poater J, Im C, Bickelhaupt FM (2008) Chem Eur J 14:6901–6911
21. Runeberg N, Seth M, Pyykkö P (1995) Chem Phys Lett 246:239–244
22. Linnartz H, Zink LR, Evenson KM (1997) J Mol Spectr 184:56–59
23. Ohtaki Y, Matsushima F, Odashima H, Takagi K (2001) J Mol Spectr 210:271–274
24. Nash CS (2005) J Phys Chem A 109:3493–3500
25. Pershina V (2009) Russ Chem Rev 78:1153–1171
26. Visscher L, Lee TJ, Dyall KG (1996) J Chem Phys 105:8769–8776
27. Pernpointner M, Visscher L (2003) J Comp Chem 24:754–759
28. Dyall KG (2002) Theor Chem Acc 108:335–340; erratum (2003) Theor Chem Acc 109:284; revision (2006) Theor Chem Acc 115:441. Basis sets available from the Dirac web site, http://dirac.chem.sdu.dk
29. Visscher L, Eliav E, Kaldor U (2001) J Chem Phys 115:9720–9726
30. Lévy-Leblond JM (1967) Commun Math Phys 6:286–311
31. DIRAC, a relativistic ab initio electronic structure program, Release DIRAC08 (2008), written by Visscher L, Jensen HJAa, Saue T, with new contributions from Bast R, Dubillard S, Dyall KG, Ekström U, Eliav E, Fleig T, Gomes ASP, Helgaker TU, Henriksson J, Ilias M, Jacob ChR, Knecht S, Norman P, Olsen J, Pernpointner M, Ruud K, Salek P, Sikkema J (see http://dirac.chem.sdu.dk)
32. Molski M (2002) Mol Phys 100:3545–3552
33. Rosmus P, Reinsch EA (1980) Z Naturforsch A 35:1066–1070
34. Linnartz H, Havenith M, Zwart E, Meerts WL, Ter Meulen JJ (1992) J Mol Spectr 153:710–717
35. Klein R, Rosmus P (1984) Z Naturforsch A 39:349–353
36. Rasmussen E (1933) Z Phys 80:726–734
37. Pyykkö P (2008) Mol Phys 106:1965–1974
38. Yoon S, Glab WL (1994) J Phys B: At Mol Opt Phys 27:1433–1441
39. Brandi F, Hogervorst W, Ubachs W (2002) J Phys B At Mol Opt Phys 35:1071–1084
40. Kaufman V (1993) J Res Natl Inst Stand Tech 98:717–724
41. Knight DL, Wang L (1985) J Opt Soc Am B 2:1084–1087
42. Brandi F, Velchev I, Horgervorst W, Ubachs W (2001) Phys Rev A 64:032505
43. Dunham JL (1932) Phys Rev 41:721–731

Theor Chem Acc (2012) 131:1166
DOI 10.1007/s00214-012-1166-2

REGULAR ARTICLE

$[Zn_{10}(\mu_4\text{-}S)(\mu_3\text{-}S)_6(Py)_9(SO_4)_3]$ as a molecular model of ZnS surfaces: an experimental and theoretical study

Francesco Avanzini · Maurizio Casarin · Daniel Forrer · Luciano Pandolfo · Mauro Sambi · Andrea Vittadini

Received: 5 July 2011 / Accepted: 11 October 2011 / Published online: 2 March 2012

Abstract Experimental and theoretical results pertaining to $[Zn_{10}(\mu_4\text{-}S)(\mu_3\text{-}S)_6(Py)_9(SO_4)_3]$, a possible molecular model of ZnS S-terminated polar surfaces, as well as a potential source of strictly monodispersed ZnS quantum dots, are presented and discussed. The results of density functional theory (DFT) calculations provided a rationale for the peculiar arrangement of $[Zn_{10}(\mu_4\text{-}S)(\mu_3\text{-}S)_6(Py)_9(SO_4)_3]$ clusters in the solid state, contemporarily indicating the unsuitability of the isolated species to mimic whatever (polar or non-polar) ZnS surface. Despite the fact that such a failure is further confirmed by time-dependent DFT and UV–Vis diffuse reflectance spectroscopy, the combined use of theoretical outcomes, DRIFT measurements, and literature data pertaining to the surface chemical properties of ZnS (Hertl in Langmuir 4:594, 1988) ultimately testifies that $[Zn_{10}(\mu_4\text{-}S)(\mu_3\text{-}S)_6(Py)_9(SO_4)_3]$ is perfectly suited to model the interaction of pyridine molecules with ZnS surface Lewis acid sites. The herein reported theoretical results are expected to be a useful reference for the interpretation of chemisorption experiments of Py-based Lewis bases on single crystal ZnS surfaces.

Keywords Zinc sulfide · Chemisorption · Quantum dots · Density functional calculations

Dedicated to Professor Vincenzo Barone and published as part of the special collection of articles celebrating his 60th birthday.

Electronic supplementary material The online version of this article (doi:10.1007/s00214-012-1166-2) contains supplementary material, which is available to authorized users.

F. Avanzini · M. Casarin (✉) · D. Forrer · L. Pandolfo · M. Sambi
Dipartimento di Scienze Chimiche DiSC, Università degli Studi di Padova, Padua, Italy
e-mail: maurizio.casarin@unipd.it

M. Casarin · D. Forrer · A. Vittadini
Istituto di Scienze e Tecnologie Molecolari, ISTM-CNR, V. Marzolo, 1, 35131 Padua, Italy

1 Introduction

When semiconductor nanocrystals have dimensions comparable to the Bohr exciton diameter, the spatial confinement of charge carriers gives rise to significant changes in the electronic properties of the material [1–4]. Further reductions in particle dimensions ultimately yield clusters of molecular size, which may provide, besides possible insights into the properties of the solid, an experimental validation of the cluster approach to the investigation of local phenomena in solids. Some of us have extensively used the molecular–cluster approach coupled to the density functional theory (DFT) to study solid-state phenomena [5–7], and, in this habit, we were able to single out molecular models of ZnO [8, 9],[1] ZnS [10], $Zn_4S(BO_2)_6$, and $Cd_4S(AlO_2)_6$ [11, 12], as well as of ZnS non-polar surfaces [13].

More than 10 years ago, Ali et al. [14] reported the synthesis and the structural characterization of a new type of zinc sulfide cluster ($[Zn_{10}(\mu_4\text{-}S)(\mu_3\text{-}S)_6(Py)_9(SO_4)_3]\cdot 3H_2O$, Py = pyridine) characterized by several fascinating peculiarities: (1) the occurrence of a neutral inorganic core ($[Zn_{10}(\mu_4\text{-}S)(\mu_3\text{-}S)_6(SO_4)_3]$, hereafter **IC**) constituted by a

[1] A molecular model of a solid is a molecule or an ion whose stereochemistry and electronic structure reasonably mimic the structural arrangement and the nature of the valence (conduction) band top (bottom) of the solid itself [8].

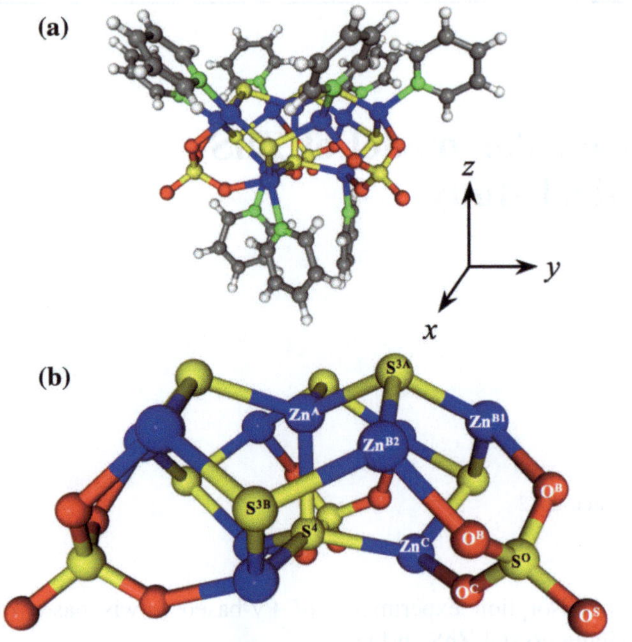

Fig. 1 Schematic representation of the $[Zn_{10}(\mu_4\text{-}S)(\mu_3\text{-}S)_6(Py)_9(SO_4)_3]$ cluster (**a**) and its kernel $Zn_{10}S_7$ capped by three tridentate SO_4^{2-} ligands (**b**). *Blue*, *yellow*, *red*, *green*, *gray*, and *white spheres* represent Zn, S, O, N, C, and H atoms, respectively; moreover, labels associated with the atoms of the $[Zn_{10}(\mu_4\text{-}S)(\mu_3\text{-}S)_6(SO_4)_3]$ inorganic core are the same proposed by Ali et al. [14] in their contribution

$Zn_{10}(\mu_4\text{-}S)(\mu_3\text{-}S)_6$ kernel capped by three SO_4^{2-} tridentate ligands (see Fig. 1) and distinguished by an unusual Zn–S connectivity neither corresponding to that of ZnS structural polymorphs (hexagonal wurtzite and cubic sphalerite) nor to the one of clusters formed by group 12 metals [15, 16]; (2) a singular supramolecular arrangement (see Fig. 2): the three Py molecules bonded to the Zn^C ions (Py^C) sink into the "basin" generated by the six Py ligands (Py^B) decorating the puckered ring of Zn^B ions of an adjacent cluster along the *a* axis[2]; (3) the presence of six $\mu_3\text{-}S^{2-}$ ions (S^{3A} and S^{3B} in Fig. 1b), each of them carrying a coordinative vacancy and trigonally coordinated to chemically complete Zn^{2+} ions which makes $[Zn_{10}(\mu_4\text{-}S)(\mu_3\text{-}S)_6(Py)_9(SO_4)_3]$ (hereafter **1**) a possible molecular model of ZnS, S-terminated, polar surfaces. Last but not least, **1** could represent a potential source of ZnS strictly monodispersed quantum dots having a diameter of ~1 nm,[3] that is, significantly smaller than the ZnS excitonic Bohr radius (2.5 nm) [17].

[2] The Zn^B set can be further split in two three-member subsets (Zn^{B1} and Zn^{B2}) according to their higher or lower *z* coordinates with respect to the selected origin placed in S^4 in the adopted framework (see Fig. 1).

[3] Depyridination reactions at different temperature are in progress in our laboratory to identify $[Zn_{10}(\mu_4\text{-}S)(\mu_3\text{-}S)_6(Py)_9(SO_4)_3]\cdot 3H_2O$ decomposition products.

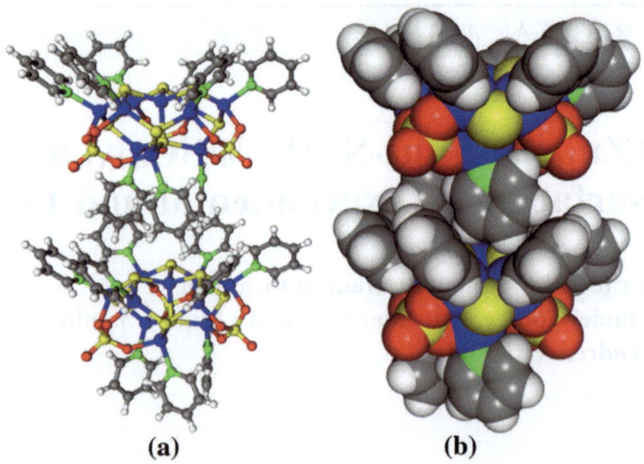

Fig. 2 Ball-and-stick (**a**) and space-filling (**b**) representations of two $[Zn_{10}(\mu_4\text{-}S)(\mu_3\text{-}S)_6(Py)_9(SO_4)_3]$ clusters along the *a*-axis. Water coordination molecules are not displayed for clarity

Despite all these features, neither experimental nor theoretical investigations of the electronic properties of **1** have been reported so far to our knowledge. In this contribution, the results of a series of DFT and time-dependent (TD) DFT numerical experiments on the isolated $[Zn_{10}(\mu_4\text{-}S)(\mu_3\text{-}S)_6(Py)_9(SO_4)_3]$ cluster are presented, discussed, and compared with the outcomes of diffuse reflectance spectroscopy measurements carried out in the UV–Vis and infrared wavelength ranges. The final goal of such a combined experimental and theoretical investigation is to obtain a rationale for the above described peculiarities.

2 Experimental and computational details

2.1 Chemicals, instrumentation, and syntheses

All solvents were distilled and degassed before use. PhSH, Et_3N, Me_4NCl, $ZnNO_3\cdot 6H_2O$, and S (Fluka) were used without further purification. GC/MS spectra of volatile products obtained in the thermal decomposition of $(Me_4N)_4[Zn_{10}(\mu_3\text{-}S)_4(\mu_2\text{-}SPh)_{12}(SPh)_4]$ (the precursor needed for the synthesis of $[Zn_{10}(\mu_4\text{-}S)(\mu_3\text{-}S)_6(Py)_9(SO_4)_3]\cdot 3H_2O$) were obtained on a FISONS MD 800 instrument equipped with a 30 ($\Phi = 0.25$ mm) capillary column, using He as carrier gas and EI ionization. Elemental analyses were provided by the Microanalysis Laboratory of the Department of Chemical Sciences of the University of Padova.

2.1.1 Synthesis of the $[Zn_{10}(\mu_3\text{-}S)_4(\mu_2\text{-}SPh)_{12}]$ precursor

The starting material to synthesize $[Zn_{10}(\mu_4\text{-}S)(\mu_3\text{-}S)_6(Py)_9(SO_4)_3]\cdot 3H_2O$ is the neutral cluster $[Zn_{10}(\mu_3\text{-}S)_4(\mu_2\text{-}SPh)_{12}]$, which has been prepared according to the synthetic route reported in [13, 18]: $[NMe_4][Zn_{10}(\mu_3\text{-}S)_4(\mu_2\text{-}SPh)_{12}(SPh)_4]$ (906 mg, 0.32 mmol) was heated at 250 °C under dynamic vacuum (0.2 mmHg) for 5 h. GC/MS of the volatile collected revealed the presence of Me_3N and PhSMe. The residual solid (661 mg) corresponded to $[Zn_{10}(\mu_3\text{-}S)_4(\mu_2\text{-}SPh)_{12}]$ with a yield of 99%.

2.1.2 Synthesis of $Zn_{10}(\mu_4\text{-}S)(\mu_3\text{-}S)_6(Py)_9(SO_4)_3]\cdot 3H_2O$

The synthesis of $[Zn_{10}(\mu_4\text{-}S)(\mu_3\text{-}S)_6(Py)_9(SO_4)_3]\cdot 3H_2O$ was carried out according to a procedure slightly modified with respect to the one reported by Ali et al. [14]: $[Zn_{10}(\mu_3\text{-}S)_4(\mu_2\text{-}SPh)_{12}]$ (156 mg, 0.075 mmol) was dissolved in 15 mL of a solution prepared by treating 100 mL of pyridine with 500 mg of anhydrous Na_2SO_4; 36 mg of sulphur were then added. The mixture was heated at 70 °C under stirring, and sulphur dissolved in ca. 2 min yielding a yellow solution, which was cooled at room temperature (RT) and layered with 3 mL of acetone. The solution was left undisturbed for 4 weeks, yielding a cream-white solid, which was decanted, washed with pyridine, and dried under vacuum at RT. The obtained light yellow powder (17 mg, 9×10^{-3} mmol) implied a 12% yield. Elemental analysis: calculated for $[Zn_{10}(\mu_4\text{-}S)(\mu_3\text{-}S)_6(Py)_9(SO_4)_3]\cdot 3H_2O$: C = 27.97, H = 2.66, N = 6.52, S = 16.59; found: C = 27.07, H = 2.44, N = 6.27, S = 16.43.

2.2 Diffuse reflectance UV–Vis spectra

UV–Vis measurements were carried out on a Cary 5E spectrophotometer (Varian Inc., Palo Alto, CA), equipped with an integrating sphere coated with polytetrafluoroethylene. The employed setup allowed the exclusion of the specular component of the reflected light. Spectra were collected within the 300–800 nm wavelength range with a spectral resolution of 1 nm by using a self-supporting pellet.

2.3 Diffuse reflectance infrared Fourier transform (DRIFT) spectra

Fourier transform IR spectra were recorded with a Bruker Tensor 27 spectrophotometer by loading few milligrams of $[Zn_{10}(\mu_4\text{-}S)(\mu_3\text{-}S)_6(Py)_9(SO_4)_3]\cdot 3H_2O$ in the accessory for diffuse reflectance (Harrick Scientific) using KBr as a reference, and accumulating 32 scans at a resolution of 4 cm^{-1}. The collected spectra are displayed in Kubelka–Munk units [19, 20].

2.4 Numerical experiments

All the numerical experiments have been carried out by using the Amsterdam density functional package [21]. The geometry of **1** has been fully optimized without any symmetry constraint (C_1 symmetry point group), by adopting the hybrid GGA corresponding to the Becke's three parameters hybrid exchange [22] and to the Lee–Yang–Parr's correlation functional [23], and by adding a dispersion correction [24] (hereafter, DFT-D) to take care of possible van der Waals interactions between Py rings. A further series of numerical experiments without dispersion corrections (hereafter, DFT) have also been carried out for comparison. A 20% Hartree–Fock exchange has been adopted [25]. A triple-ζ Slater-type basis set plus a polarization function (TZP) have been used for all the atoms; furthermore, the inner core of Zn and S (1s–2p), O (1s), N (1s), and C (1s) atoms have been kept frozen throughout the calculations in a fully occupied configuration, excluding them from the variational space. Finally, the electronic structures have been studied with the aid of density of states (hereafter DOS) and partial DOS (PDOS) curves. To this end, we have used the Mulliken's prescription for partitioning the overlap density [26], and we have applied a 0.25 eV Lorentzian broadening factor to the one-electron eigenvalues.

Harmonic frequencies have been analytically calculated by using the dispersion corrected BP86 exchange–correlation functional. In fact, on one hand, the B3LYP functional has not yet been included among the functional allowing the analytical evaluation of harmonic frequencies [21][4]; on the other hand, the BP86 functional is known to provide a satisfactory agreement between experimental and theoretical frequencies [27–34]. Finally, low-lying electronic transitions of $[Zn_{10}(\mu_4\text{-}S)(\mu_3\text{-}S)_6(Py)_9(SO_4)_3]\cdot 3H_2O$ have been evaluated by coupling the TD-DFT [35] to the B3LYP exchange–correlation potential. TD-DFT calculations have been run by adopting the atomic coordinates derived from the solid-state structure.[5]

3 Results and discussion

The geometry of the title cluster significantly changes on passing from the $[Zn_{10}(\mu_4\text{-}S)(\mu_3\text{-}S)_6(Py)_9(SO_4)_3]\cdot 3H_2O$ crystal structure [14] to the isolated species (DFT-D and

[4] The computational cost of a numerical evaluation of vibrational frequencies by using the B3LYP exchange–correlation potential (787 single point runs) resulted prohibitive.

[5] TD-DFT calculations have been run by using geometrical parameters corresponding to the $[Zn_{10}(\mu_4\text{-}S)(\mu_3\text{-}S)_6(Py)_9(SO_4)_3]\cdot 3H_2O$ solid-state structure rather than employing the optimized geometry of **1**, because corresponding outcomes have been compared with the results of diffuse reflectance measurements carried out on a powder sample. In this regard, it is worthwhile to mention that, on passing from $[Zn_{10}(\mu_4\text{-}S)(\mu_3\text{-}S)_6(Py)_9(SO_4)_3]\cdot 3H_2O$ to **1**, besides a widening of the HOMO–LUMO energy difference (3.18 eV in the former case, 3.69 eV in the latter), DOS and PDOS are very similar.

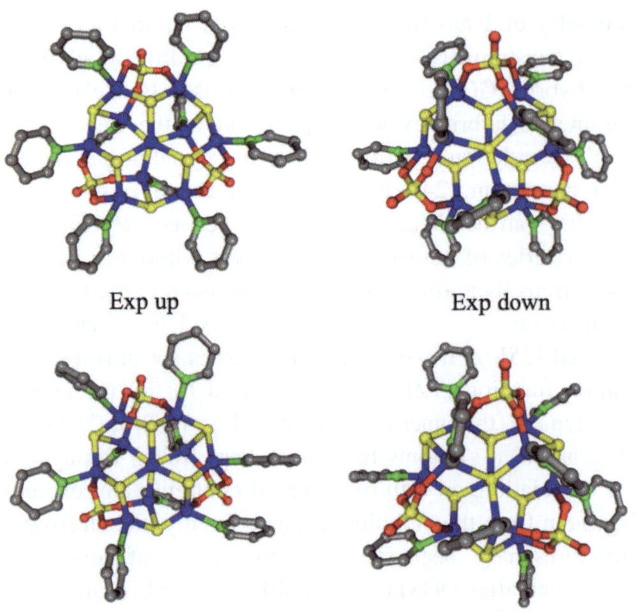

Fig. 3 Comparison of the $[Zn_{10}(\mu_4\text{-}S)(\mu_3\text{-}S)_6(Py)_9(SO_4)_3]$ experimental (Exp) and optimized (DFT-D) molecular arrangements along the *a*-axis

DFT optimized Cartesian coordinates of **1** are reported in the Online Resource 1 and 2, respectively), though the pseudo threefold axis is conserved. Variations mainly involve the relative orientation of Py^B molecules, which could be expected, given the supramolecular arrangement of $[Zn_{10}(\mu_4\text{-}S)(\mu_3\text{-}S)_6(Py)_9(SO_4)_3]\cdot 3H_2O$. On the other hand, both **IC** and the relative orientation of the three Py^C ligands undergo only minor perturbations upon optimization. Geometry modifications of **1** are better appreciated by referring to Fig. 3, which shows that at least three competing interactions contribute to determine the arrangement of Py^B ligands : (1) the H-bonding between C_α^{Py} H atoms and the unsaturated $\mu_3\text{-}S^{3A}$ ions (mainly involving Zn^{B1}-coordinated Py^B); (2) the H-bonding between C_α^{Py} H atoms and the SO_4^{2-} fragments (mainly involving Zn^{B2}-coordinated Py^B)[6]; (3) the π–π interaction between adjacent pyridine molecules. Obviously, the last interaction is absent in DFT calculations (the reader may refer to the Online Resource 2). Despite the fact that $\mu_3\text{-}S^{3B}$ ions represent a source of negative electrostatic potential comparable to $\mu_3\text{-}S^{3A}$ (see Fig. 4),[7] their role in determining the orientation of pyridine molecules seems to be negligible. In addition to that, the color-mapped molecular electrostatic potential (MEP) superimposed on the 3D map of the density matrix of **1** ultimately testifies that electrostatic interactions play an important role in determining the head–tail aggregation of **1** along *a* in $[Zn_{10}(\mu_4\text{-}S)(\mu_3\text{-}S)_6(Py)_9(SO_4)_3]\cdot 3H_2O$. As far as **IC** is concerned, bond distances and bond angles reported in Table 1 confirm that the inorganic core of **1** undergoes only minor perturbations on passing from the $[Zn_{10}(\mu_4\text{-}S)(\mu_3\text{-}S)_6(Py)_9(SO_4)_3]\cdot 3H_2O$ crystal structure [14] to the isolated cluster.

Before moving to the analysis of the electronic structure of **1**, a few words about the negligible effects of the geometry optimization on the relative arrangement of Py^C molecules are needed. As a matter of fact, a thorough inspection of the Py^C environment stresses that, both in the crystal structure and in the isolated cluster, each of them is constrained between two contiguous SO_4^{2-} ligands. More specifically, for a particular Zn^C ion (say Zn^{C1}), its four coordinative positions are occupied by S^4, S^{3C1}, O^{C1}, and the pyridine molecule Py^{C1} whose C_α^{Py} H atoms are H-bonded to the O atoms of two contiguous SO_4^{2-} ions: the former H-bonding involves O^{C1}, while the latter implies an O^B of the adjacent SO_4^{2-} group.[8]

DOS and PDOS of **1** are displayed in Figs. 5 and 6, where DOS main features have been alphabetically labeled. The inspection of both figures indicates that (1) the feature **C** is originated by **IC**-based occupied MOs, mainly localized on S and O atomic orbitals (AOs) and with negligible contributions from Py molecules; (2) the HOMOs, generating the evident shoulder on the higher energy side of **C**, account for linear combinations of S^{3A} lone pairs (n^{S3A}, see Fig. 6a); (3) the LUMOs, generating the feature **D**, correspond to ${}^{Py}\pi^*$ MOs. According to this picture, the low-energy region of the UV–Vis spectrum of **1** should be dominated by $n^S \rightarrow {}^{Py}\pi^*$ and $n^O \rightarrow {}^{Py}\pi^*$ electronic excitations.

In Fig. 7 the UV–Vis diffuse reflectance spectrum of $[Zn_{10}(\mu_4\text{-}S)(\mu_3\text{-}S)_6(Py)_9(SO_4)_3]\cdot 3H_2O$ is displayed. The experimental pattern in the wavelength range extending from 300 to 600 nm (the instrumental setup prevents any measurement at wavelengths shorter than 300 nm) is characterized by the presence of an evident shoulder **S**, extending from 550 to 420 nm, and the band **A** centered at ~310 nm. TD-DFT calculations indicate that the lowest 30 excitations[9] cover an energy range of ~0.6 eV, from 2.66 (466 nm) to 3.23 eV (384 nm), and mainly imply

[6] Geometrical parameters reported in the Online Resource 1 certainly correspond to a local minimum, and it seems reasonable to assume that the potential energy surface includes many of them.

[7] The mean values of the S^{3A} and S^{3B} Mulliken [26] (Hirshfeld) [36] gross atomic charges amount to -0.48 (-0.33) and -0.46 (-0.31), respectively.

[8] Experimental (optimized) $H(C_\alpha^{Py})\cdots O^C$and $H(C_\alpha^{Py})\cdots O^B$ internuclear distances are 2.668, 2.563, 2.551 Å (2.473, 2.450, 2.379 Å) and 2.752, 2.663, 2.619 Å (2.204, 2.174, 2.147 Å), respectively.

[9] The absence of any symmetry constraint and the use of a TZP basis set for all the atoms imply a very large number (475,881) of possible excitations to be computed. The available computational resources allowed us to evaluate the first 30.

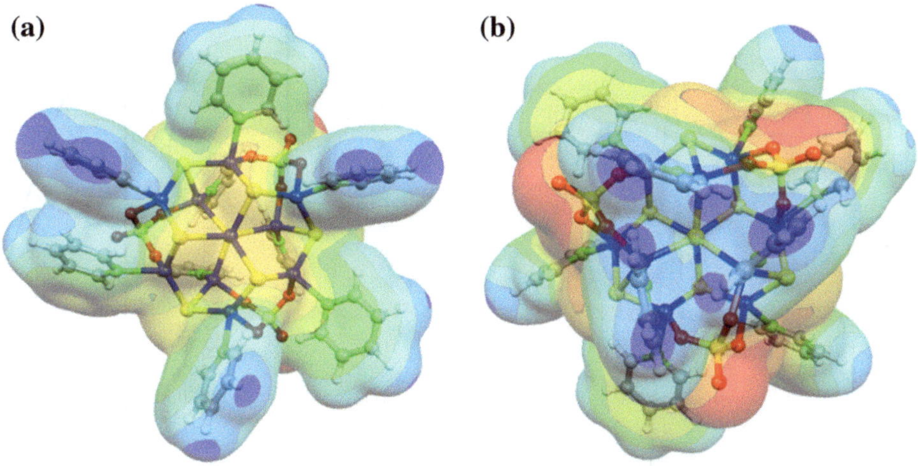

Fig. 4 Up (**a**) and down (**b**) 3D MEP map superimposed on the density matrix of **1**. The *red* (*blue*) MEP region corresponds to a negative (positive) potential of −0.09 (0.05) au. The reported isodensity surface is 0.001 Å^{-3}

Table 1 Selected mean internuclear distances (Å) and bond angles (°) of the $Zn_{10}S_7(SO_4)_3$ inorganic core of the $[Zn_{10}S_7(py)_9(SO_4)_3]$ cluster

	Theory	Experiment[a]
Zn^A–S^4	2.464	2.453
Zn^C–S^4	2.390	2.321
Zn^A–S^{3A}	2.381	2.339
Zn^B–S^{3A}	2.306	2.278
Zn^B–S^{3B}	2.317	2.285
Zn^B–N^B	2.165	2.060
Zn^C–N^C	2.111	2.059
Zn^C–O^C	2.038	2.006
Zn^B–O^B	2.070	2.033
S^4–Zn^A–S^{3A}	110.0	107.7
S^{3A}–Zn^A–S^{3A}	108.9	111.2
S^{3A}–Zn^B–S^{3B}	132.1	124.6
S^4–Zn^C–S^{3B}	119.8	120.2
Zn^A–S^4–Zn^C	105.8	103.3
Zn^C–S^{3B}–Zn^B	87.5	92.0
Zn^C–S^4–Zn^C	112.9	114.9
Zn^B–S^{3A}–Zn^B	93.1	101.9
Zn^A–S^{3A}–Zn^B	96.3	103.2
Zn^B–S^{3B}–Zn^B	101.1	106.8
S^4–Zn^C–O^C	113.0	106.4
S^4–Zn^C–N^C	106.2	108.9
S^{3B}–Zn^C–O^C	104.7	109.3
S^{3B}–Zn^C–N^C	118.9	113.8
O^C–Zn^C–N^C	90.3	95.2
O^B–Zn^B–N^B	91.2	95.6
S^{3B}–Zn^B–O^B	98.4	102.3
S^{3A}–Zn^B–O^B	116.2	111.7
S^{3B}–Zn^B–N^B	108.5	112.5
S^{3A}–Zn^B–N^B	103.1	106.0

[a] After Ref. [13]

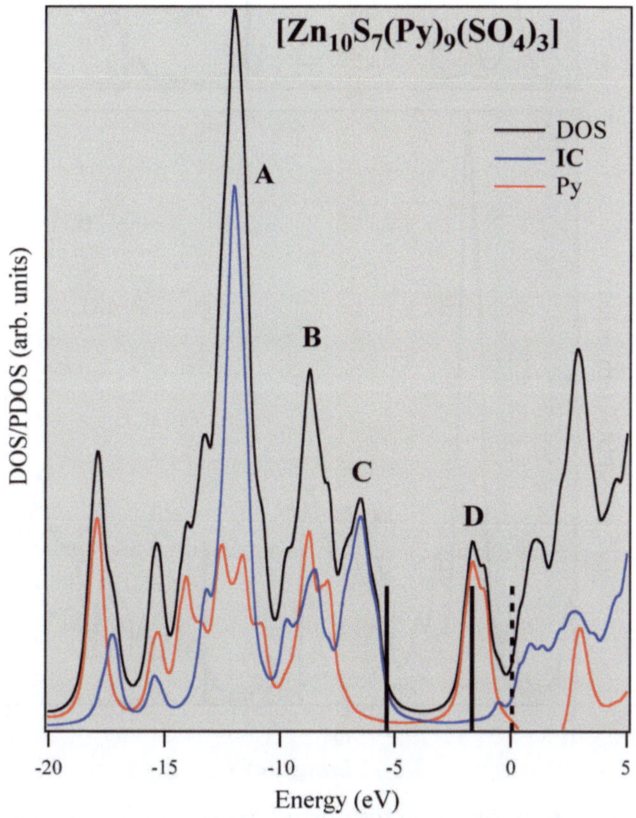

Fig. 5 DOS and PDOS of **1**. Vertical solid lines account for the HOMO and LUMO energies; the *dashed vertical line* corresponds to the energy of the lowest-unoccupied MO having a localization on Zn 4s empty AOs larger than 1.0%

$n^S \rightarrow {}^{Py}\pi^*$ transitions. This result, coupled with the Mulliken population analysis pertaining to the outermost occupied levels of **1** (see Figs. 5 and 6; see footnote 5), justifies the tentative assignment of spectral features **S** and **A** to $n^S \rightarrow {}^{Py}\pi^*$ and $n^O \rightarrow {}^{Py}\pi^*$ $[Zn_{10}(\mu_4\text{-}S)(\mu_3\text{-}S)_6(Py)_9(SO_4)_3]\cdot 3H_2O$ transitions, respectively.

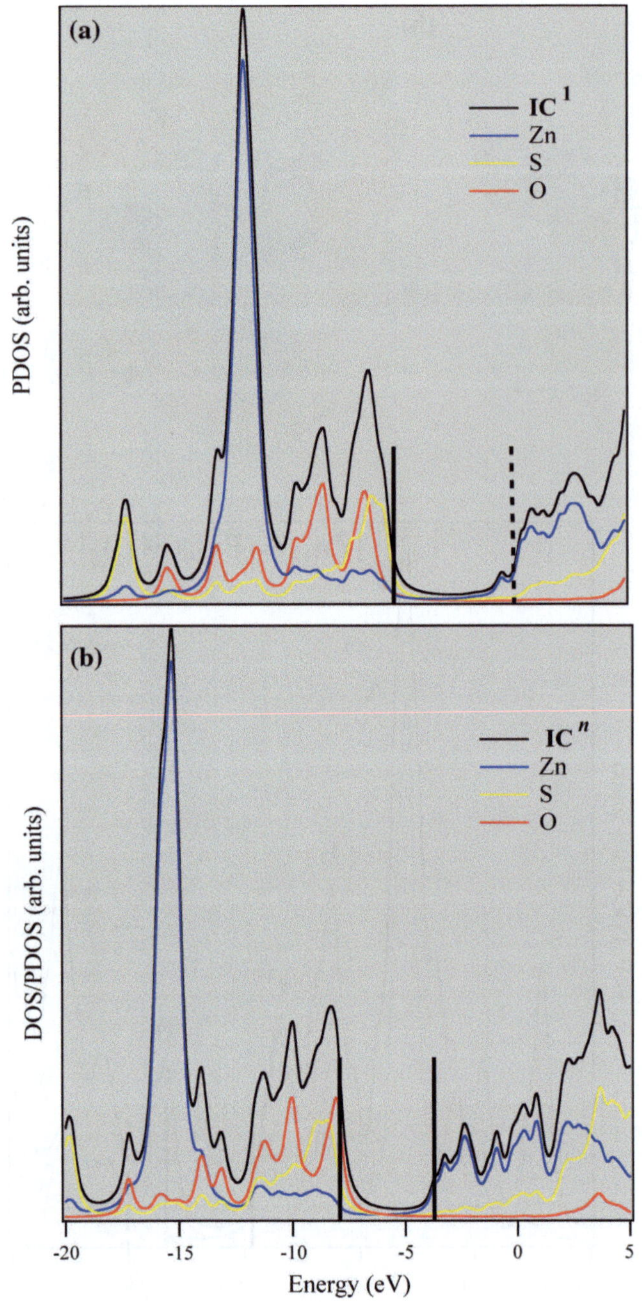

Fig. 6 **IC** PDOS (**a**) and **IC″** DOS (**b**). The physical meaning of the *vertical solid* and *dashed lines* is the same of Fig. 5

The strict application of the molecular model definition (see footnote 1) implies that **1** cannot be considered well tailored to mimic either polar or non-polar ZnS surfaces; however, if anything, a reasonable model of a ZnS surface characterized by the presence of Lewis acid sites (L_S^a, the surface cations Zn^B and Zn^C) and Lewis base sites (the surface anions S^{3A} and S^{3B}) with Py molecules coordinately bound to L_S^a. In this regard, it is noteworthy that, in the late 1988 [37], W. Hertl investigated by DRIFT spectroscopy the interaction of ZnS surfaces with different Lewis bases (Py and NH_3) with the aim of discriminating their interaction with proton-donating surface Brønsted acids and surface L_S^a. Peaks due to the pyridine coordinately bound to L_S^a were revealed at 1,448, 1,490, 1,570, and 1,609 cm^{-1} [37]. Interestingly, the DRIFT spectrum of $[Zn_{10}(\mu_4\text{-}S)(\mu_3\text{-}S)_6(Py)_9(SO_4)_3]\cdot 3H_2O$ in the same energy range reported by W. Hertl (1,800–1,300 cm^{-1}, see Fig. 8a) is characterized by four well evident and rather narrow peaks centered at 1,452, 1,489, 1,572, and 1,607 cm^{-1}[10] satisfactorily reproduced by dispersion corrected BP86 calculations[11] (see Fig. 8b). In this regard, the analysis of normal mode eigenvectors clearly indicates that each peak numerically labeled in Fig. 8b is generated by nine closely spaced normal modes corresponding to symmetric ($\nu_{1,588}$ and $\nu_{1,462}$) and antisymmetric ($\nu_{1,557}$ and $\nu_{1,452}$) stretchings of Py rings. Moreover, $\nu_{1,588}$ and $\nu_{1,462}$ ($\nu_{1,557}$ and $\nu_{1,452}$) symmetric (antisymmetric) vibrations involve the L_S^a–N stretchings (S–L_S^a–N and O–L_S^a–N bendings). As a whole, besides a confirmation of the assignment proposed by Hertl for DRIFT bands associated with Py molecules coordinately bound to ZnS L_S^a, the results herein presented ultimately proves that $[Zn_{10}(\mu_4\text{-}S)(\mu_3\text{-}S)_6(Py)_9(SO_4)_3]\cdot 3H_2O$ is a very well-tailored molecular model of Py molecules interacting with surface ZnS L_S^a.

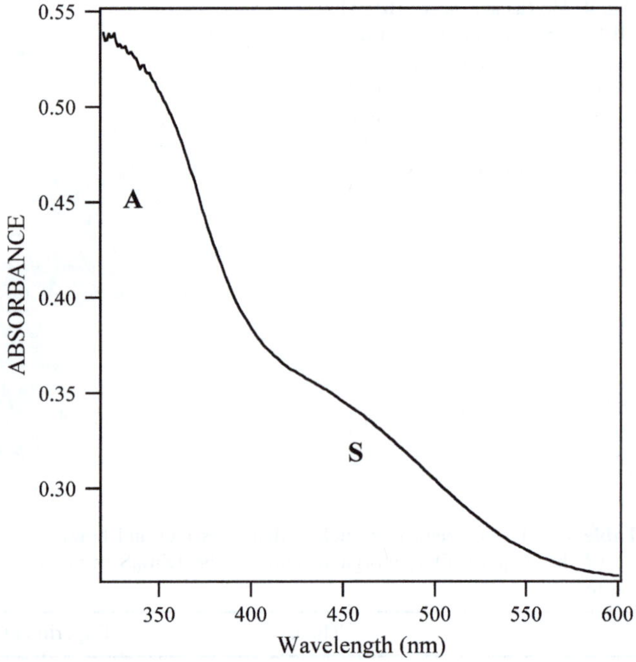

Fig. 7 Diffuse reflectance UV–Vis spectrum of $[Zn_{10}(\mu_4\text{-}S)(\mu_3\text{-}S)_6(Py)_9(SO_4)_3]\cdot 3H_2O$ in the wavelength range 300–600 nm

[10] The broad band centered at 1,655 cm^{-1} and exhibiting some structure is revealed in the H–O–H bending region.

[11] The optimized geometry of **1** at dispersion-corrected B3LYP and BP86 levels are almost indistinguishable.

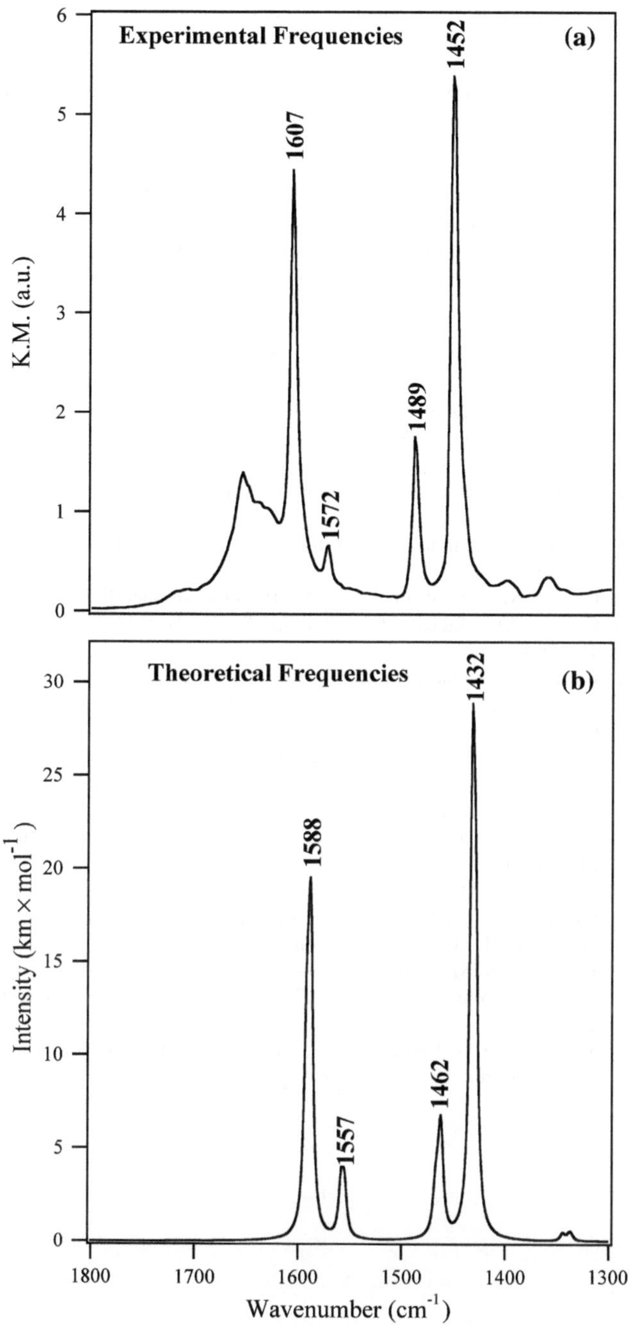

Fig. 8 **a** DRIFT spectrum of $[Zn_{10}(\mu_4\text{-}S)(\mu_3\text{-}S)_6(Py)_9(SO_4)_3]\cdot 3H_2O$ in the 1,300–1,800 cm^{-1} energy range; **b** theoretical frequencies, analytically evaluated, of $[Zn_{10}(\mu_4\text{-}S)(\mu_3\text{-}S)_6(Py)_9(SO_4)_3]$ in the 1,300–1,800 cm^{-1} energy range

It has been already mentioned that **1** could represent a potential source of ZnS strictly monodispersed quantum dots. In order to look into such a hypothesis from a theoretical point of view, the **IC** PDOS of **1** (**IC^1**) has been compared with the DOS of the naked-**IC** (**IC^n**) cluster.[12]

[12] A further series of B3LYP-D numerical experiments on **IC^n** (**1** with no Py ligand) have been carried out by using the optimized coordinates of **1**.

The inspection of Fig. 6 clearly testifies that the removal of Py molecules determines a general, but not uniform, shift toward more negative energies of the **IC**-based energy levels. In more detail, the largest shift (~3.5 eV) involves the Zn-based energy levels, while the smallest one (~1.5 eV) implicates the O-based states. Such a reorganization of the **IC** electronic structure implies: (1) a different localization of HOMOs on passing from **IC^1** to **IC^n**, mainly concentrated on n^{S3A} in the former case, strongly localized on n^{O} in the latter; (2) the narrowing of the ΔE between the **IC** HOMOs and the lowest states with a significant localization on Zn 4s AOs (the **IC^n** LUMOs). Incidentally, the **IC^n** HOMO–LUMO ΔE amounts to 4.2 eV. Now, the use of the relation proposed by Suyver et al. [38] to correlate nanoparticle radii (nm) and band gap ΔE (eV) for nanocrystalline ZnS,

$$r(\Delta E) = \frac{0.32 - 2.9 \times \sqrt{\Delta E - 3.49}}{2 \times (3.50 - \Delta E)}$$

would imply, with $\Delta E = 4.2$ eV, a $r(\Delta E) \sim 1.5$ nm, in qualitative agreement with the **IC** dimensions ($r = 0.7$–0.8 nm). Such a result corroborates the hypothesis that **1** could represent a potential starting material to get ZnS strictly monodispersed quantum dots with a diameter of the order of magnitude of 1 nm, contemporarily spurring the need of looking into its chemical reactivity (see footnote 3).

4 Conclusions

In this contribution, we have presented the results of an experimental and theoretical study of the electronic structure of $[Zn_{10}(\mu_4\text{-}S)(\mu_3\text{-}S)_6(Py)_9(SO_4)_3]$, a neutral zinc sulfide cluster characterized by the presence of two kinds of sulfide ions: a μ_4-S^{2-} tetrahedrally coordinated to four chemically complete Zn^{2+} cations and six μ_3-S^{2-} each of them trigonally coordinated to three chemically complete Zn^{2+} cations and contemporarily carrying a coordinative vacancy. Despite such a structural arrangement, the combined use of DFT-D, TD-DFT, and UV–Vis diffuse reflectance spectroscopy underlines the inability of $[Zn_{10}(\mu_4\text{-}S)(\mu_3\text{-}S)_6(Py)_9(SO_4)_3]$ to mimic the electronic properties of an extended ZnS surface, polar or non-polar. Nevertheless, theoretical outcomes and the results of DRIFT measurements ultimately testify that $[Zn_{10}(\mu_4\text{-}S)(\mu_3\text{-}S)_6(Py)_9(SO_4)_3]$ is perfectly suited to model the interaction of Py molecules with ZnS surface L_S^a, thus confirming the assignment proposed by Hertl for DRIFT bands associated to Py molecules coordinately bound to ZnS L_S^a. Theoretical outcomes herein reported might then be useful in the future to drive the rationalization of experimental evidences concerning the chemisorption of Lewis bases, such as pyridine, on single crystal ZnS surfaces. As a whole, the comparison with literature data [10, 11, 13] confirms and reinforces our belief that identical or very similar cluster skeletons do not necessarily imply equivalent or very similar cluster electronic properties; the suitability of specific systems to be molecular models of extended solids has to be verified for each single case.

Acknowledgments Computational resources and assistance were provided by the "Laboratorio Interdipartimentale di Chimica Computazionale" (LICC) at the Department of Chemistry of the University of Padova. Molecular graphics has been generated by Molekel [39].

References

1. Wang Y, Herron N (1991) J Phys Chem 95:525
2. Weller H (1993) Angew Chem Int Ed Engl 32:41
3. Henglein A (1997) Ber Bunsenges Phys Chem 101:1562
4. Hamad S, Woodley SM, Catlow CRA (2009) Mol Simulat 35:1015
5. Casarin M, Maccato C, Vittadini A (1997) Surf Sci 377:587
6. Casarin M, Maccato C, Vittadini A (1998) Inorg Chem 37:5842
7. Casarin M, Falcomer D, Glisenti A, Vittadini A (2003) Inorg Chem 42:436
8. Bertoncello R, Bettinelli M, Casarin M, Gulino A, Tondello E, Vittadini A (1992) Inorg Chem 31:1558
9. Casarin M, Tondello E, Calderazzo F, Vittadini A, Bettinelli M, Gulino A (1993) J Chem Soc, Faraday Trans 89:4363
10. Albinati A, Casarin M, Eisentraeger F, Maccato C, Pandolfo L, Vittadini A (2000) J Organomet Chem 593–594:307
11. Albinati A, Casarin M, Maccato C, Pandolfo L, Vittadini A (1999) Inorg Chem 38:1145
12. Casarin M, Maccato C, Vittadini A (2002) J Phys Chem B 106:2569
13. Bertoncello R, Bettinelli M, Casarin M, Maccato C, Pandolfo L, Vittadini A (1997) Inorg Chem 36:4707
14. Ali B, Dance IG, Craig DC, Scudder ML (1998) J Chem Soc Dalton Trans 1661
15. Dance IG, Fisher KJ (1994) Prog Inorg Chem 41:637
16. Archibald SJ (2003) In: McCleverty JA, Meyer TJ (eds) Comprehensive coordination chemistry II, vol 6. p 1148
17. Biggs MM, Ntwaeaborwa OM, Terblans JJ, Swart HC (2009) Phys B 404:4470
18. Farneth WE, Herron N, Wang Y (1992) Chem Mater 4:916
19. Kubelka P, Munk F (1931) Z Tech Phys 12:593
20. Kortum G (1969) Reflectance spectroscopy. Springer, New York
21. Amsterdam Density Functional (ADF) version 2010. http://www.scm.com
22. Becke AD (1993) J Chem Phys 98:5648
23. Lee C, Yang W, Parr RG (1988) Phys Rev B 37:785
24. Grimme S (2006) J Comput Chem 27:1787
25. Stephens PJ, Devlin FJ, Chabalowski CF, Frisch MJ (1994) J Phys Chem 98:11623
26. Mulliken RS (1955) J Chem Phys 23:1833
27. Casarin M, Maccato C, Vittadini A (1998) J Phys Chem B 102:10745
28. Casarin M, Pandolfo L, Vittadini A (2001) Organometallics 20:754
29. Casarin M, Pandolfo L, Sassi A (2002) Organometallics 21:2235
30. Casarin M, Maccato C, Vittadini A (2002) J Phys Chem B 106:795
31. Casarin M, Vittadini A (2003) Phys Chem Chem Phys 5:2461
32. Casarin M, Falcomer D, Glisenti A, Vittadini A (2003) Inorg Chem 42:436
33. Casarin M, Corvaja C, Di Nicola C, Falcomer D, Franco L, Monari M, Pandolfo L, Pettinari C, Piccinelli F, Tagliatesta P (2004) Inorg Chem 43:5865
34. Casarin M, Pandolfo L, Vittadini A (2009) Phys Chem Chem Phys 11:94
35. Gross EKU, Dobson JF, Petersilka M (1996) In: Nalewajski RF (ed) Density functional theory. Springer, Heidelberg
36. Hirshfeld FL (1977) Theor Chim Acta 44:129
37. Hertl W (1988) Langmuir 4:594
38. Suyver JF, Wuister SF, Kelly JJ, Meijerink A (2001) Nano Lett 1:429
39. Varetto U MOLEKEL 5.4.0.8, Swiss National Supercomputing Centre: Manno (Switzerland)

Theor Chem Acc (2012) 131:1167
DOI 10.1007/s00214-012-1167-1

REGULAR ARTICLE

Validation of a hybrid MD-SCF coarse-grained model for DPPC in non-lamellar phases

Antonio De Nicola · Ying Zhao · Toshihiro Kawakatsu · Danilo Roccatano · Giuseppe Milano

Received: 18 July 2011 / Accepted: 4 December 2011 / Published online: 2 March 2012

Abstract In the framework of a recently developed scheme for a hybrid particle-field simulation technique where self-consistent field theory and molecular dynamics simulation method are combined, specific coarse-grained models for aqueous solutions of phospholipids have been validated. In particular, the transferability of the model in the correct reproduction of non-lamellar phases has been validated against reference particle–particle simulations. By varying the water content, the proposed model is able to correctly describe the different morphologies that are experimentally observed such as micelles and reverse micelles. The lower computational costs of the hybrid techniques allow us to perform simulations of large-scale systems that are needed to investigate the applications of self-assembled structures of lipids in nanotechnologies.

Keywords Coarse-graining · Molecular dynamics · Self-consistent field theory · Lipids

Dedicated to Professor Vincenzo Barone and published as part of the special collection of articles celebrating his 60th birthday.

A. De Nicola · Y. Zhao · G. Milano (✉)
Dipartimento di Chimica e Biologia, Università di Salerno, via Ponte don Melillo, 84084 Fisciano, SA, Italy
e-mail: gmilano@unisa.it

A. De Nicola · G. Milano
IMAST Scarl-Technological District in Polymer and Composite Engineering, P.le Fermi 1, 80055 Portici, NA, Italy

T. Kawakatsu
Department of Physics, Tohoku University, Aoba, Aramaki, Aoba-ku, Sendai 980-8578, Japan

D. Roccatano
Jacobs University Bremen, Campus Ring 1, 28759 Bremen, Germany

1 Introduction

Phospholipids are an important class of compounds having both important biological functions (e.g., cell membranes) and technological applications (e.g., liposomes and micelles).

They can undergo different phase transition in aqueous environments characterized by different morphologies (mesomorphism and polymorphism).

They are an important class of biomolecules because they form lipid bilayer structures that are fundamental building blocks of cellular membranes. For this reason, lipid bilayers have been attracting the interest of computational biophysics community, where people perform atomistic and coarse-grained molecular dynamics simulations of these systems for long time [1–13].

Mesomorphism and polymorphism can occur when lipids are not included in the cell biomembranes. Despite their importance in the biological systems, the molecular structures of biological cell membranes, especially the structures of ion channels and transmembrane proteins, are not well understood yet.

When phospholipids are no longer under the constraints that are imposed by intermolecular interactions that are present within the biomembrane, they can form non-lamellar (non-bilayer) phases, where the non-lamellar phases include the hexagonal and cubic phases as well as the diluted micellar phases.

Hexagonal phases are characterized by tubular aggregates, and they can be composed of either normal or reverse aggregates. Cubic phases are composed of curved bilayers or micelles. Depending on the water concentration, micelles change their aggregation form from normal ('oil in water') to reverse ('water in oil') micelles.

Due to their relevance to the biological systems, most of the simulation works on lipids have been focused on the bilayer phase [10, 14–17]. With the advent of nanotechnology, self-assembled structures of lipids started to be exploited in different ways. Lipids can be used as templates for the assembly of nanoparticles. For example, gold nanoparticles can be prepared using templates formed by the tubular nanodomains of lipids [18, 19]. Lipid micelles can be used as dispersant for carbon nanotubes in aqueous media [20]. Liposomes can be used for drug delivery [21] and as nanoreactors [22]. Reverse micelles can be used as templates for the nanofabrication of tubes of conductive polymers [23].

Although atomistic simulations can provide very detailed and chemically consistent models of biological phospholipids, they are very expensive way of studying the assembling processes involving nanomaterials occurring on the mesoscopic time (>μs) and length scales (>100 nm).

In order to bridge this gap between the length and time scales covered by the atomistic simulations and the actual scales of the relevant phenomena, different types of coarse-grained (CG) models have been developed for phospholipids. For example, implicit solvent models and explicit solvent models for lipid bilayers have been reported [24–26]. Among different approaches for developing such CG models, specific models that retain some chemical details are useful in determining the connection between the chemical structure on the atomistic scales and the nanostructures that are obtained by self-assembly in different conditions.

In particular, these CG models usually employ several different types of beads (not just hydrophobic and hydrophilic), with several parameters for the force fields derived from simulations based on atomistic models. A successful and very widely explored example of this approach is the MARTINI CG model developed by Marrink and co-workers [27]. In the MARTINI force field, the phospholipids are described by several types of beads interacting via Lennard–Jones-type potentials with different interaction parameters, with which one can describe a smooth variation from the hydrophobic to hydrophilic characters. Despite its simplicity, the MARTINI force field is able to reproduce with surprisingly good accuracy the properties of the self-assembly of lipid molecules in the bilayers and the micellar phases [15, 28, 29].

A different coarse-graining approach is based on a field representation and has been proposed to model soft matter systems. In particular, in the framework of the self-consistent field (SCF) theory, the model systems are not represented by particles but by density fields and the mutual interactions between segments (beads) are decoupled and replaced by interactions with static external fields [30]. Numerous applications of this approach to block copolymers [31–35], proteins [36], polymer composites [37] and colloidal particles [38, 39] for studying the large-scale phenomena in soft matter have been proposed. There have been several approaches to phospholipids and water mixtures using approaches based on the field descriptions [34, 40–43].

More recently, Müller and Smith [44] introduced a hybrid approach in the framework of SCF theory by combining it with a Monte Carlo simulation of a coarse-grained model of polymer chains to study the phase separation in binary polymer mixtures. This approach has been widely and successfully applied by Daoulas et al. [45] to coarse-grained models of diblock copolymer thin films and by Detcheverry et al. [46, 47] to polymer nanocomposites. One of the advantages of this hybrid approach is the absence of any limitation in treating complex molecular architectures and/or intramolecular interactions. With these preceding studies, very recently, a hybrid particle-field approach, where the molecular dynamics (MD) simulation method is combined with SCF description (MD-SCF), was proposed and an implementation suitable for the treatment of atomistic force fields and/or specific coarse-grained models has been reported [48, 49]. In particular, specific coarse-grained models for phospholipids and water, suitable for MD-SCF simulations, have been developed. These models and the set of parameters needed to evaluate the interactions between the particles and the density fields have been optimized to reproduce structural properties of reference full-atomistic MD simulations of lipid bilayers. It was also proved that this MD-SCF model is able to reproduce correctly the physical behavior of the lipid bilayers [50]. The aim of the present study was to validate these models for non-bilayer phases. In particular, the transferability of the model to systems different water contents has been validated against reference simulations.

The paper is organized as follows: in Sect. 2, we describe the basic concept of the SCF theory, which is useful for the reader to understand the present investigation. We also give a brief description of the computational scheme for the hybrid particle-field MD-SCF simulations. In Sect. 3, the validation of the model by comparison with classical MD simulations using the MARTINI force field is reported. In the following, these full-atomistic simulations will be referred as particle–particle (PP) simulations. Finally, simulation results on the non-lamellar-type self-assembly of dipalmitoylphosphatidylcholine (DPPC) molecules in its aqueous solutions with various concentrations are shown.

2 Computational methods

2.1 MD-SCF theory and implementation

In this section, a brief exposition of the recently developed hybrid MD-SCF simulation scheme is reported. This section is intended to quickly guide the reader to get the basis of the methodology and to understand the framework of the present investigation. In order to obtain this approach in more detail, the readers should refer to Ref. [48] where the complete derivation and the implementation are described and to references [30, 33, 51–53] for general reviews of SCF methods.

The main feature of the hybrid MD-SCF approach is that the evaluation of the non-bonded force and its potential between atoms of different molecules, that is, the most computationally expensive part of MD simulations, can be replaced by the evaluation of an external potential that is dependent on the local density at position $\mathbf{r}$ in MD-SCF simulations. According to the spirit of SCF theory, a many body problem like molecular motion in many molecule systems is reduced into the problem of deriving the partition function of a single molecule in an external potential $V(\mathbf{r})$. Then, non-bonded force between atoms of different molecules can be obtained from a suitable expression of the $V(\mathbf{r})$ and its derivatives.

In the frame of SCF theory, a molecule is regarded to be interacting with the surrounding molecules not directly but through a mean field. To derive such a mean field picture, we start from a Hamiltonian of a system that is composed of M molecules. The Hamiltonian is split into two parts $\hat{H}(\Gamma) = \hat{H}_0(\Gamma) + \hat{W}(\Gamma)$, where Γ specifies a point in the phase space, which is used as shorthand for a set of positions of all atoms in the system. Here and also in the following, the symbol ^ (hat) indicates that the associated physical quantity is a function of the microscopic states described by the phase space Γ.

$\hat{H}_0(\Gamma)$ is the Hamiltonian of a reference ideal system composed of non-interacting chains but with all the intra-molecular interaction terms (bond, angle and non-bonded interactions) that are taken into account in the standard MD simulations. The term $\hat{W}(\Gamma)$ is the deviation from the reference system and taking into account of the intermolecular non-bonded interactions.

Assuming the canonical (*NVT*) ensemble, the partition function of this system is given by the following equation:

$$Z = \frac{1}{M!}\int d\Gamma \exp\{-\beta[\hat{H}_0(\Gamma) + \hat{W}(\Gamma)]\}, \tag{1}$$

where $\beta = \frac{1}{k_B T}$.

From a microscopic point of view, the number density of segments is defined as a sum of delta functions placed at the center of mass of each segment as [30, 54]

$$\hat{\phi}(\mathbf{r};\Gamma) = \sum_{p=1}^{M}\sum_{i=0}^{S(p)} \delta\left(\mathbf{r} - \mathbf{r}_i^{(p)}\right), \tag{2}$$

where M is the total number of molecules in the system, $S(p)$ is the number of particles contained in pth molecule, and $\mathbf{r}_i^{(p)}$ is the position of the ith particle in pth molecule. The deviation $\hat{W}(\Gamma)$, according to Eq. 1, from the reference state $\hat{H}_0$ originates from the interactions between molecules. Several assumptions are introduced to calculate the interaction term $\hat{W}(\Gamma)$. First of all, we assume that $\hat{W}(\Gamma)$ depends on Γ only through the segment number density $\hat{\phi}(\mathbf{r};\Gamma)$ as

$$\hat{W}(\Gamma) = W[\hat{\phi}(\mathbf{r};\Gamma)], \tag{3}$$

where $W[\hat{\phi}(\mathbf{r};\Gamma)]$ means that W is a functional of $\hat{\phi}(\mathbf{r};\Gamma)$. Using the identity for the δ-functional

$$\delta[\hat{\phi}(\mathbf{r};\Gamma) - \phi(\mathbf{r})] = \int D\{w(\mathbf{r})\} \exp\left[i\int w(\mathbf{r})\left\{\hat{\phi}(\mathbf{r};\Gamma) - \phi(\mathbf{r})\right\}d\mathbf{r}\right],$$

we can rewrite the partition function in Eq. 1 as

$$Z = \frac{1}{M!}\int D\{\phi(\mathbf{r})\}\int D\{w(\mathbf{r})\}\exp\left\{-\beta\left[-\frac{M}{\beta}\ln z + W[\phi(\mathbf{r})] - \frac{1}{i\beta}\int w(\mathbf{r})\phi(\mathbf{r})d\mathbf{r}\right]\right\}, \tag{4}$$

where z is the single-molecule partition function, and $w(\mathbf{r})$ is a conjugate field of $\phi(\mathbf{r})$ which appeared in the Fourier representation of the δ-functional, and is a complex field [53].

For evaluating this partition function approximately, the integrals over $\phi(\mathbf{r})$ and $w(\mathbf{r})$ in Eq. 4 are replaced with a Gaussian integral around the most probable state that minimizes the argument of the exponential function on the right side of Eq. 4 (so-called saddle point approximation).

The minimization conditions in the form of functional derivatives result in

$$\begin{cases} V(\mathbf{r}) \equiv \frac{1}{i\beta}w(\mathbf{r}) = \frac{\delta W[\phi(\mathbf{r})]}{\delta\phi(\mathbf{r})} \\ \phi(\mathbf{r}) = -\frac{M}{\beta z}\frac{\delta z}{\delta V(\mathbf{r})} = \left\langle \hat{\phi}(\mathbf{r};\Gamma)\right\rangle, \end{cases} \tag{5}$$

where $\phi(\mathbf{r})$ is the coarse-grained density at position $\mathbf{r}$ and $V(\mathbf{r})$ is the external potential conjugate to $\phi(\mathbf{r})$ which is a real field now.

In term of Eq. 5, it is possible to acquire an expression for a density-dependent external potential acting on each segment.

Next, we assume that the density-dependent interaction potential W, where each component species is specified by the index K, takes the following form:

$$W[\{\phi_K(\mathbf{r})\}] = \int \mathrm{d}\mathbf{r} \left(\frac{k_B T}{2} \sum_{KK'} \chi_{KK'} \phi_K(\mathbf{r}) \phi_{K'}(\mathbf{r}) + \frac{1}{2\kappa} \left(\sum_K \phi_K(\mathbf{r}) - 1 \right)^2 \right), \tag{6}$$

where $\phi_K(\mathbf{r})$ is the coarse-grained density of the species K at position $\mathbf{r}$ and $\chi_{KK'}$ are the mean field parameters for the interaction of a particle of type K with the density fields due to the particles of type K'. Here, we redefined $\phi_K(\mathbf{r})$ as the number density of the species K normalized by the total average number density of the segments. This change in the definition of $\phi_K(\mathbf{r})$ only affects the scale of the conjugate field $V(\mathbf{r})$.

The second term of the integrand on the right-hand side of Eq. 6 is the relaxed incompressibility condition, and κ is the compressibility that is assumed to be sufficiently small [54]. Then, the corresponding mean field potential is given by

$$V_K(\mathbf{r}) = \frac{\delta W[\{\phi_K(\mathbf{r})\}]}{\delta \phi_K(\mathbf{r})} = k_B T \sum_{K'} \chi_{KK'} \phi_{K'}(\mathbf{r}) + \frac{1}{\kappa} \left(\sum_K \phi_K(\mathbf{r}) - 1 \right). \tag{7}$$

Taking the case of a mixture of two components A and B as an example, the mean field potential acting on a particle of type A at position $\boldsymbol{r}$ is given by

$$V_A(\mathbf{r}) = k_B T[\chi_{AA} \phi_A(\mathbf{r}) + \chi_{AB} \phi_B(\mathbf{r})] + \frac{1}{\kappa} (\phi_A(\mathbf{r}) + \phi_B(\mathbf{r}) - \phi_0). \tag{8}$$

Thus, the force acting on the particle A at position $\mathbf{r}$ imposed by the interaction with the mean field is

$$F_A(\mathbf{r}) = -\frac{\partial V_A(\mathbf{r})}{\partial r} = -k_B T \left(\chi_{AA} \frac{\partial \phi_A(\mathbf{r})}{\partial \mathbf{r}} + \chi_{AB} \frac{\partial \phi_B(\mathbf{r})}{\partial \mathbf{r}} \right) - \frac{1}{\kappa} \left(\frac{\partial \phi_A(\mathbf{r})}{\partial \mathbf{r}} + \frac{\partial \phi_B(\mathbf{r})}{\partial \mathbf{r}} \right). \tag{9}$$

The main advantage of hybrid MD-SCF scheme is that the most computationally expensive part of the MD simulations, that is, the evaluation of the non-bonded force between atoms of different molecules, is replaced by the evaluation of forces between single molecules with an external potential. In order to connect particle and field models, for the proposed hybrid MD-SCF scheme, it is necessary to obtain a smooth coarse-grained density function directly from the particle positions Γ. Let us denote this procedure as

$$\bar{S}\{\hat{\phi}(\mathbf{r};\Gamma)\} = \phi(\mathbf{r}), \tag{10}$$

where $\bar{S}$ is a symbolic name of the mapping from the particle positions to the coarse-grained density. In order to obtain a smooth spatial density from particle positions, the simulation box is divided into several cells. In particular, particles are sorted and, according to their positions, assigned to $ncell = n_x n_y n_z$ (where n_x, n_y and n_z are the number of cells in the x, y and z directions, respectively). Furthermore, according to the position of each particle inside a cell, a fraction of it will be assigned to each vertex of the cell [46, 48–50, 55–57]. In order to explain the procedure, a simpler two-dimensional case is described. In Fig. 1a, the structure of a phospholipid and the corresponding density field is schematized.

As described in Fig. 1b, the fraction of a particle assigned to a given lattice point is proportional to the area of a rectangle shown in the figure. For example, for a particle with coordinates $\underline{x}$ and $\underline{y}$, a fraction $(l - \underline{x})(l - \underline{y})/l^2$ will be assigned to the mesh point 1 and a fraction of $\underline{xy}/l^2$ at mesh point 4 in Fig. 1b (for simplicity, l is the length of the cell in both x and y directions). Thus, the density at every mesh point is the sum of all fractions assigned from all the cells that share a given lattice point. According to the procedure described above, the size of the cell l is a parameter defining the density coarse-graining. Larger is the value of l, more particles will be included in every cell and coarser will be the calculated density. Once the coarse-grained density has been calculated from particle positions, the spatial derivatives of the density field can be evaluated. Spatial derivatives can be obtained by the differentiation of the density lattice. In this way, the lattice where the derivatives are defined is staggered with respect to the lattice where the density is defined. As schematized in Fig. 1b, the squares indicate the lattice points where the density is defined. Correspondingly, the density gradients are defined on the center of each edge (staggered lattice points indicated by crosses in Fig. 1b) of the square surrounding the density lattice points.

Once that both density and derivatives have been computed on their corresponding lattices, the potential energy and forces acting on the particles can be calculated using values obtained by the interpolation of the density and of its spatial derivatives in Eqs. 8 and 9.

The iteration algorithm used in MD-SCF approach is explained in the following. According to the initial configurations of the system (at time t_0), a starting value of the coarse-grained density is obtained, where the coarse-grained density is defined on a lattice and the values of the density and density gradients at the particles positions are

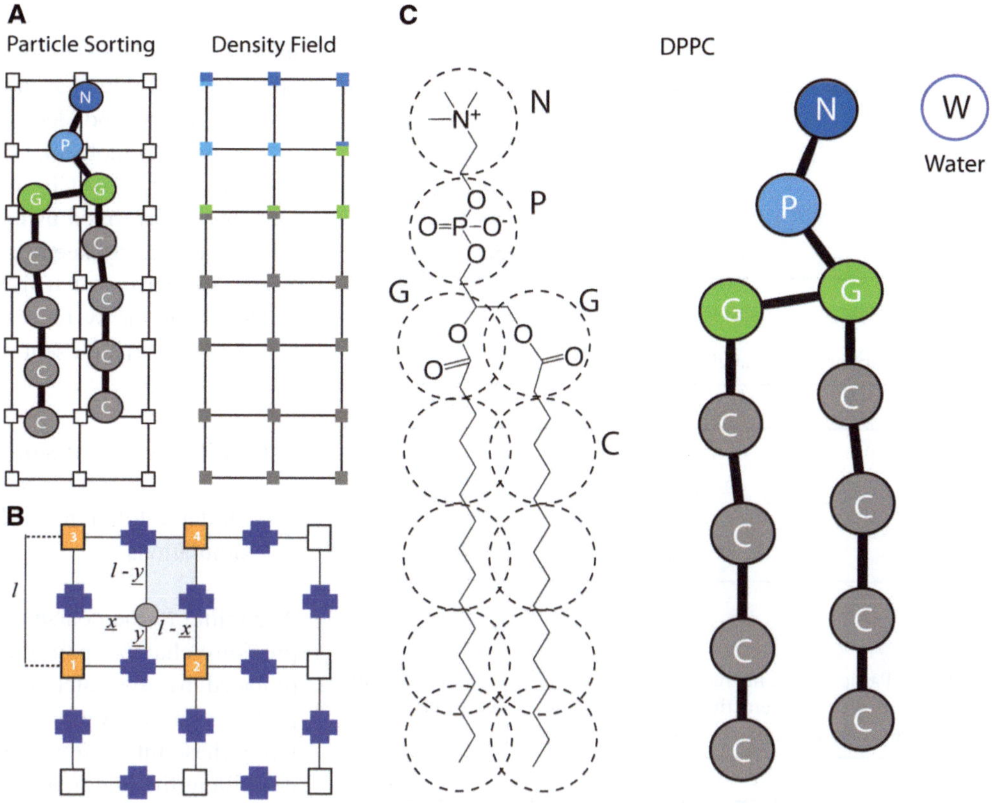

Fig. 1 **a** Construction of coarse-grained density for a phospholipid. **b** Criterion for the assignment of particle fractions to lattice points. **c** Adopted coarse-graining scheme for DPPC

calculated by linear interpolation. Then, from the density gradients, forces acting on the particles at position **r** due to the interaction with the density fields are computed according to Eq. 9. The total force acting on the particles will be the sum of the intramolecular forces (bonds, angles and intramolecular non-bonded forces calculated as in classical MD simulations) and the forces due to the interactions of particles with density fields. After the force calculation, a new configuration will be then obtained by the integration of the equation of motion. In principle, for every new configuration, an update of the CG density calculated from the new coordinates should be performed. Test simulations have shown that, due to the collective nature of the density fields, it is possible to define an update frequency of the coarse-grained densities without the loss of accuracy [48, 49]. These results are in agreement with the concepts behind the quasi-instantaneous field approximation discussed by Daoulas et al. [58] in the framework of Single Chain in Mean Field (SCMF) Monte Carlo simulations.

In other words, the values of the coarse-grained density at lattice points are not updated at every timestep but only at every prefixed density update time (Δt_{update}). Then between two updates, the values of the densities on the lattice used to interpolate both density and its derivatives will be constant. When an update of density is performed, a new coarse-grained density will be obtained and the iteration algorithm converges when the coarse-grained density and the particle-field potential become self-consistent.

2.2 Models and parameters

For the intramolecular interactions, parameters of the coarse-grained (CG) MARTINI model have been considered [59]. In this model, bonds are described by a harmonic-type potential of the form:

$$V_{\text{bond}}(R) = \frac{1}{2}K_{\text{bond}}(R - R_{\text{bond}})^2 \tag{11}$$

where R_{bond} is the equilibrium bond length and K_{bond} is the force constant of the bond. These parameters of the bond potential used in the present study are reported in Table 1.

The stiffness of the chains is also taken into account by a harmonic bending potential $V_{\text{angle}}\ (\theta)$ that depends on the cosine of the angle θ between two successive bonds.

$$V_{\text{angle}}(\theta) = \frac{1}{2}K_{\text{angle}}\{\cos(\theta) - \cos(\theta_0)\}^2 \tag{12}$$

where K_{angle} is the force constant and θ_0 is the equilibrium bond angle. Angle parameters adopted for the models of this paper are reported in Table 2.

Intramolecular non-bonded interactions are modeled using the parameters of the Lennard–Jones potentials and have been set for all non-bonded pairs $\varepsilon = 2.0$ kJ/mol and

Table 1 Parameters for bonding energetic terms

Bond type	b_o (nm)	K_b (kJ mol^{-1} nm^{-2})
N–P	0.470	1,250
P–G	0.470	1,250
G–G	0.370	1,250
G–C	0.470	1,250
C–C	0.470	1,250

Table 2 Parameters for angle energetic terms

Angle type	θ_o (°)	K_θ (kJ mol^{-1})
P–G–G	120	25.0
P–G–C	180	25.0
G–C–C	180	25.0
C–C–C	180	25.0

Table 3 Particle–field interaction parameters $\chi_{KK'} \times RT$ (kJ/mol) for particles of type K interacting with density fields due to particle of type K' are reported

	N	P	G	C	W
N	0.00	−1.50	6.30	9.00	−8.10
P	−1.50	0.00	4.50	13.50	−3.60
G	6.30	4.50	0.00	6.30	4.50
C	9.00	13.50	6.30	0.00	33.75
W	−8.10	−3.60	4.50	33.75	0.00

$\sigma = 0.47$ nm (which are repulsive according to the MARTINI nomenclature) [59].

Particle-field $\chi_{KK'}$ parameters (Eqs. 6 and 7) needed to calculate the interactions between a particle of type K and the density fields due to the particles of type K' are listed in Table 3. These parameters have been tuned to reproduce structural properties of lipid bilayers. In particular, electron density profiles of reference particle–particle simulations and bilayer thicknesses can be well reproduced for DPPC and several other lipids [50]. Furthermore, in order to determine the value of the parameter κ that regulates the strength of the incompressibility condition imposed in Eq. 7, we analyzed the behavior of density fluctuations in reference PP simulation. More detailed information about the model parameters can be found in Ref. [50].

2.3 Simulations details

For the use of a reference system for the CG simulations, classical PP-MD simulations have been performed using the program GROMACS (ver. 3.3) [60]. The timestep used for the integration of equation of motion was 0.03 ps. The temperature was kept constant using the weak coupling method with $\tau_T = 0.1$ ps, where the target temperatures are listed in Table 3. A cut-off of 1.5 nm has been used to truncate the non-bonded interactions.

The parallel molecular dynamics program OCCAM [61] was used for MD-SCF simulations. MD-SCF simulations have been performed using a timestep of 0.03 ps, with the *NVT* ensemble by keeping the temperature constant using Andersen thermostat with a collision frequency of 5 ps^{-1}.

Details on the systems size and composition used in the simulations in the present study are summarized in Table 4.

3 Results and discussion

3.1 Model validation: PP-MD versus MD-SCF simulations

According to the coarse-graining strategy explained in the previous chapter, pair interactions between particles are replaced by the calculation of the interactions of single particles in an external field. The main parameters regulating these interactions are the mean field parameters $\chi_{KK'}$ describing the interaction between particles of type K with the density fields due to the particles of type K'. As described in the previous chapters, a set of parameters that are able to reproduce structural properties of lipid bilayers for DPPC and several other lipids has been recently reported [50].

In principle, the parameters of a coarse-grained model are not always transferable. For example, χ parameters can be temperature and composition dependent. This dependency cannot be known a priori and needs to be investigated for every coarse-grained model. This feature of coarse-grained models is general and is relevant also for particle–particle coarse-grained models [62].

The main aim of the present paper was to validate the hybrid MD-SCF models proposed for lipid bilayers in non-lamellar phases corresponding to high or low water content.

In this subsection, in order to validate the MD-SCF simulations, the results of MD-SCF simulations are compared with those of the classical PP-MD simulations.

In Fig. 2, self-assemblies of DPPC/water systems simulated using PP and MD-SCF models at two different water contents (systems **1** and **2** of Table 4, respectively) are compared. In the reference PP model, each CG water bead corresponds to 4 real water molecules. In the present article, this factor has been taken into account. For example, ratio between CG water beads and DPPC molecules reported in Table 4 is for systems **1** is 182 and for system **2** is 0.8. In terms of real water molecules, these ratios would be 728 and 3.2.

Table 4 Details about simulated systems

System	Box size (nm)	Composition				T (K)
	$x = y = z$	No. of particles	No. of DPPC	No. of water	Water/DPPC[a] ratio	
1	7.923	3,876	20	3,636	182	325
2	8.176	4,096	320	256	0.8	325
3	12.964	21,216	1,664	1,248	0.75	318
4	13.486	24,128	1,664	4,160	2.5	318
5	12.376	18,600	300	15,000	50	318
6	17.309	42,588	208	40,092	192.7	318

[a] In the reference PP model, each CG water bead corresponds to 4 molecules; in the text of the manuscript and in the figures, this factor has been taken into account

For both of these simulations, the initial configuration and the simulation conditions are the same. The starting configuration used in both simulations is a random mixture of DPPC and water molecules.

For high water concentration (system **1**), in both PP and MD-SCF simulations, the formation of a micelle is observed. In the case of MD-SCF simulation, the self-assembling process of the micelle takes about 15 ns, which is faster than the PP simulation (see Fig. 2a), where the micelle is obtained after 25 ns (see Fig. 2b). In Fig. 3, the radial density profiles of the micelle obtained in the MD-SCF (Fig. 3a) and PP (Fig. 3b) simulations are compared. Radial density profiles obtained in the MD-SCF and PP simulations are similar. The main difference is observed in the slopes of the curves. A further validation of the proposed MD-SCF models can be done by calculating the average number of lipid per micelle. These calculations can be reliable only using larger systems in which two or more micelles are in equilibrium with free lipid molecules. These results will be discussed in Sect. 3.3.

The system at low water concentration (system **2**) shows a different behavior; in both PP and MD-SCF simulations, a formation of reverse micelles is observed. In particular, from Fig. 2c, it is clear that at about 12 ns the system reaches a stable reverse micellar phase. In this reverse micellar phase, polar head groups of the lipids and the water molecules form cylinders included in the hydrophobic majority phase made of lipid tails. This result is in agreement with the formation of a hexagonal reverse-cylindrical phase. The formation of this phase structure is more clearly observed in Fig. 4a where structures obtained from MD-SCF and PP simulations are compared.

Differently from MD-SCF simulations, the cylinders obtained in the PP simulation are less regular and the water beads are not included in the cylinders but they form clusters with different sizes inside the hydrophobic phase formed by the lipid tails. We expect that, in this case, the simulated system is "trapped" in a metastable phase and the slower dynamics of PP simulations does not allow the system to escape from such a metastable phase to become more stable hexagonal structure. In order to confirm this point, the equilibrium structure obtained in the MD-SCF simulation is used as the initial state for the PP simulation after a short energy minimization is performed. In this case, as shown in Fig. 4b, the non-bonded interaction energy is lower and the structure is stable during all the simulations. In Fig. 4c, this structure is depicted in a view showing the hexagonal arrangement.

3.2 Effects of density coarse-graining on structure and dynamics

As described in Sect. 2, coarse-grained density fields $\phi_K(\mathbf{r})$, obtained from the particle positions for every particle type K, are used to calculate MD-SCF potentials and forces using Eqs. 8 and 9.

According to the scheme described in Sect. 2, two parameters, that is, the cell size l and the update frequency Δt_{update}, regulate the degree of coarse-graining of the density fields. Larger cell sizes lead to more collective density fields. As for the value of the update frequency, it has to be chosen in a way that the approximation of slow variation in the field with respect to the displacements of the particles is valid between two density updates. As already pointed out by Daoulas et al. in the context of SCMF Monte Carlo simulations, the update of the density field introduces the correlations between different molecules. On the other side, in the timesteps between two updates, the molecules are decoupled and move independently in the external density field. The validity of the so-called *quasi-instantaneous field approximation* [58] will be considered in the following by comparing simulation results using different update frequencies. It is worth to note that the concept of using different time-integration steps for "stiff" and "soft" degrees of freedom has been exploited also for PP simulations. Popular examples of this can be Molecular Dynamics algorithms with multiple time scales [63].

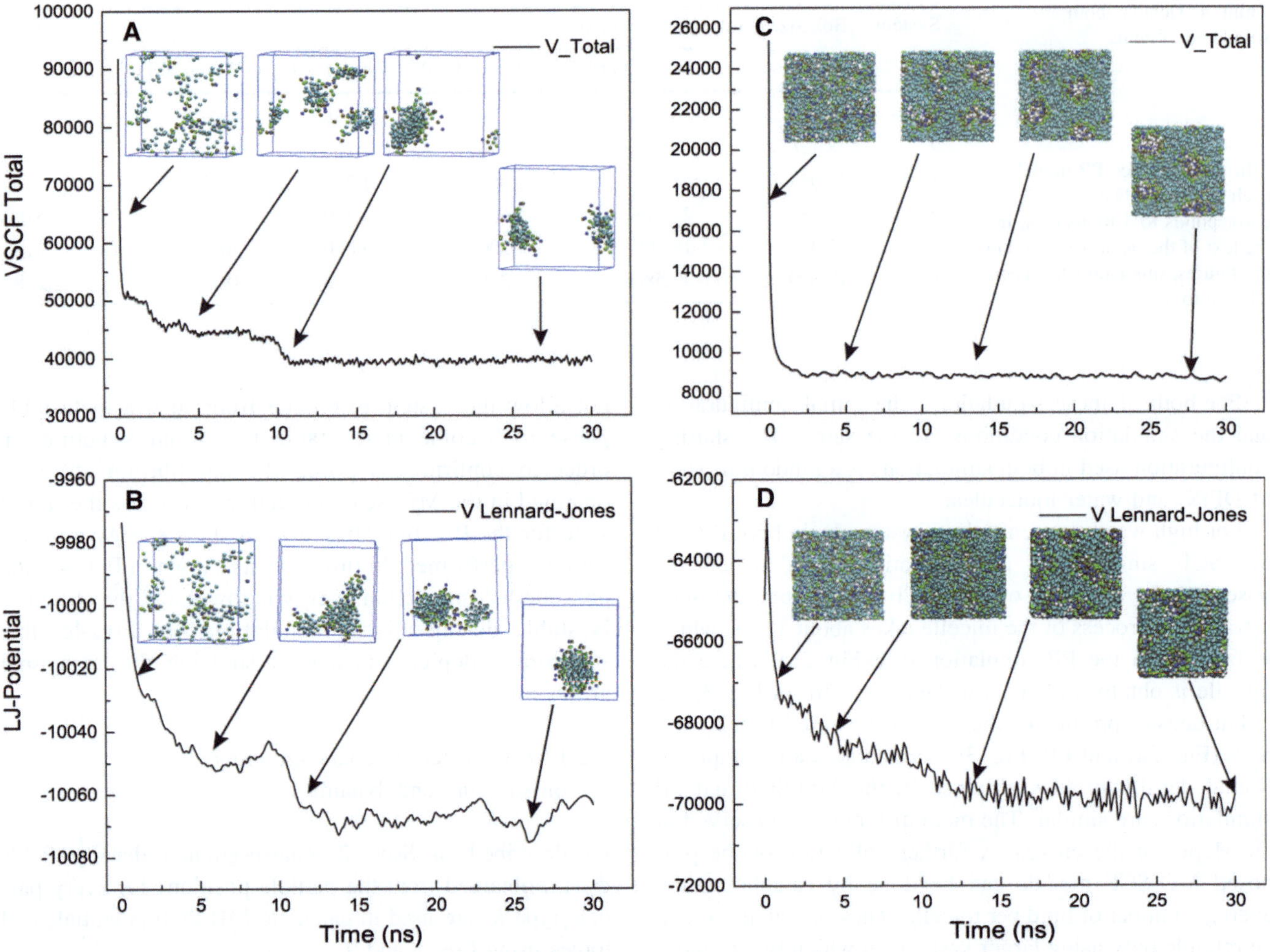

Fig. 2 Comparison of the self-assembling process of a DPPC micelle in water for MD-SCF (**a**) and PP simulations (**b**). Comparison of self-assembling process of a reverse micellar phase for MD-SCF (**c**) and PP simulations (**d**) is given. In the figures, the time behavior of particle-field intermolecular potential in the MD-SCF simulation is compared with the behavior of the non-bonded Lennard–Jones potential in the PP-MD simulation. Potential units are kJ/mol

In Fig. 5, the structures obtained using different density update frequencies and cell sizes are summarized. In particular, for high and intermediate water concentrations, the formation of a micelle and lipid bilayer is always observed using large values of field update interval (up to 900 timesteps) and the grid size (up to $2.5\sigma = 1.175$ nm). Differently, the hexagonal phase expected at low water concentration can be obtained for grid sizes smaller than 2σ ($l \approx 1$ nm) and update interval shorter than 500 timesteps. For larger values of the update interval or of the grid size, the cylinders are not formed and instead irregular reverse micelles are obtained as shown in Fig. 5 (labeled as im).

This behavior can be explained by comparing the size of the grid with the lengthscale of the self-assembling structure. The diameter of the micelle and the bilayer thickness are both about 4 nm, while the diameter of the tubes present in the hexagonal phase is smaller (≈ 1 nm). Then in this case when the size of the grid used for the density coarse-graining starts to approach the size of the cylindrical tubes, these structures cannot be described correctly. Similar considerations can be made for the density update interval. In Fig. 6 is reported the behavior of the mean square displacement as a function of time for different values of the density update intervals for the two different systems reported in Fig. 2. In particular, in order to compare the displacements with the size of the structures, the square root of the mean square displacement (MSD) normalized by the grid size is reported.

This is a quantitative way to understand the validity of the approximation of slow variation in the field with respect to the particle displacement between two density updates. In fact, the plots shown in Fig. 6 quantify how many cells a particle can cross in a given amount of simulation time. From Fig. 6, it is clear that for update intervals between 500 and 900 steps (corresponding to 15 and 36 ps), both water and DPPC beads undergo a

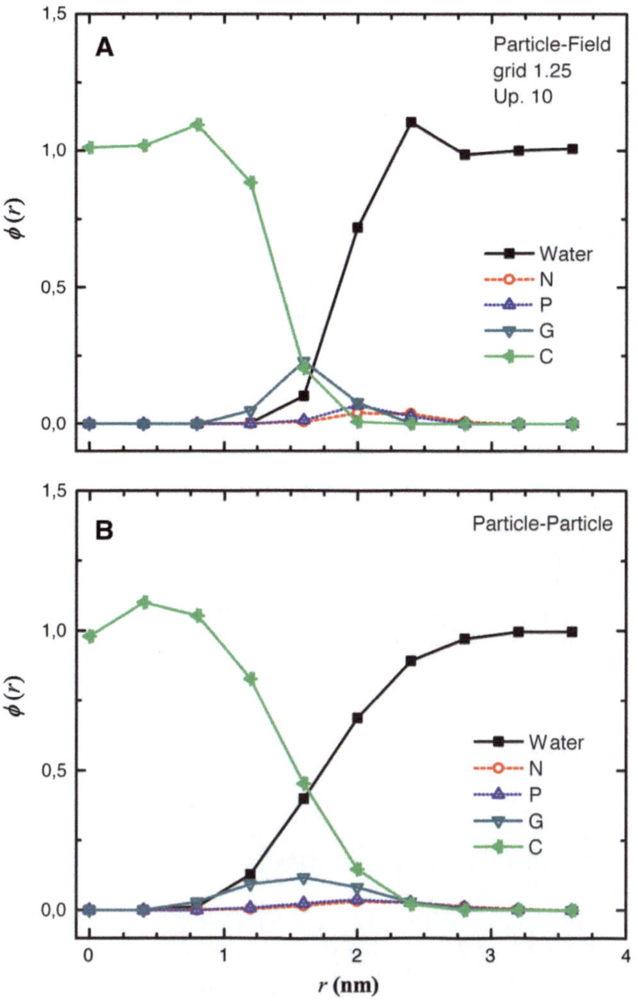

Fig. 3 Radial density profiles of the micelle obtained in the MD-SCF (**a**) and PP (**b**) simulations on system **1**

displacement equal to or larger than a cell size (i.e., larger than 0.6 nm).

From the comparison of the self-assembling processes obtained in the simulations shown in Fig. 2, it is clear that the dynamics of the system simulated in the MD-SCF method is faster than in the PP method. This is due to the smoother potentials and forces characterizing the MD-SCF Hamiltonian. In particular, the models used in the MD-SCF simulation include the effect of excluded volume interactions between particles by using incompressibility condition as described in Eq. 7. Then, forces depend on the derivatives of the density fields with a change much more smoother than distances between particles pairs.

In order to compare more quantitatively the different dynamics in PP and PF models, diffusion coefficients have been calculated from the behavior of the MSD for water and DPPC particles versus time. In particular in Fig. 7, the ratio D^* between diffusion coefficients calculated in the MD-SCF simulations (update interval of 300 timesteps and grid size $l = 2.0\sigma$) and the one calculated in the reference PP simulation is reported. From the figure, it is clear that for all considered systems, the diffusion coefficients calculated in the MD-SCF simulations are always larger than ones calculated in the PP simulations. In particular, for DPPC, they are from 4 to 6 times larger than those in the PP simulations, and for water, they are from 1.25 to 2.5 larger. Absolute values of diffusion coefficients calculated using different grid sizes and update interval for both water and DPPC are reported in Tables 5, 6, 7, 8.

The ratios between the diffusion coefficients obtained in the MD-SCF and PP simulations can be regarded as scaling factors to connect the dynamics of the MD-SCF simulations with that of the reference PP ones. This kind of comparison has been made also for the reference PP simulations to connect their dynamics with atomistic ones [59], and according to this comparison, we can estimate that the reference PP models have a dynamics that is about 4 times faster than atomistic simulations. Considering this point, the dynamics in MD-SCF simulations should be about 20–25 times faster than that in the atomistic simulations. It is interesting to note that the scaling factors between the PP and MD-SCF simulations are functions of water concentrations. In particular, they become larger as the water content becomes smaller. This behavior, from practical purposes, is very convenient because we can simulate with a largely improved efficiency the very slow dynamics in atomistic systems whose equilibration is difficult with the PP coarse-grained or the atomistic simulations.

3.3 Simulations on larger systems

On the basis of the results presented in the previous subsection, MD-SCF simulations on larger systems have been performed using a grid size $l = 1.5\sigma$ and the density update interval of 300 timesteps. In particular, we simulated the spontaneous self-assembling processes in several DPPC/water systems for 1.2 µs using a cubic box with the side lengths that are about double of those used in the simulations reported in the previous subsections. For all considered systems, the starting configuration is made up of randomly mixed lipid and water molecules.

In Fig. 8, the time behaviors of MD-SCF potential together with some snapshots for four DPPC/water systems at different water concentration have been shown (systems **3**–**6**), while the details of these systems are shown in Table 4.

For all of the systems shown in Fig. 8 after 500 ns, the equilibrium is achieved and a stable phase-separated structure is formed. In particular for the system at lower water concentration shown in Fig. 8a (system **3**, 3 water/lipid) after 400 ns, a stable reverse hexagonal phase is formed. At intermediate water content (system **4**, 10 water/

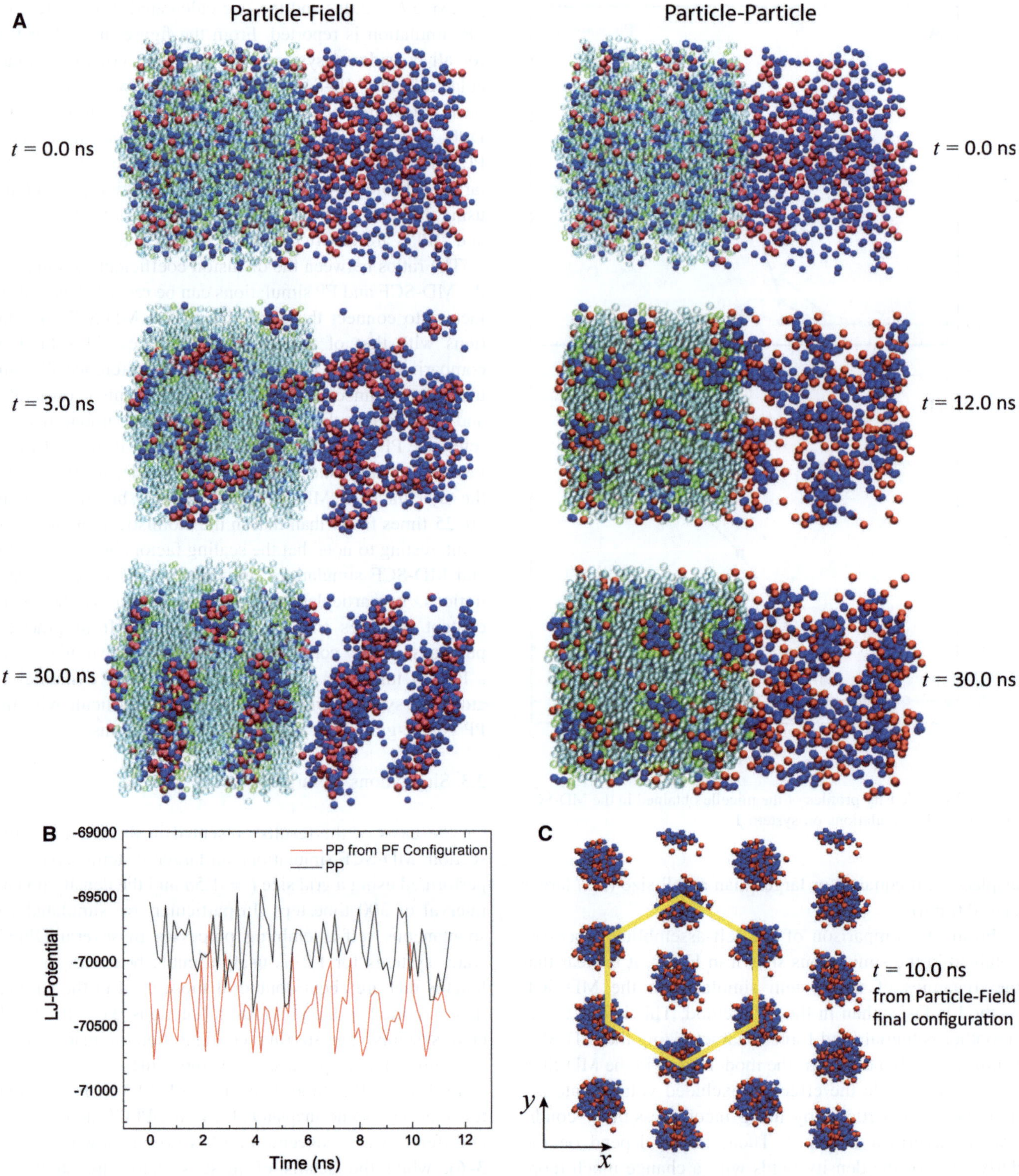

Fig. 4 Comparison of self-assembling process of cylinders in the reverse micellar phase obtained in the MD-SCF and PP simulations (**a**). Comparison between time behavior of Lennard–Jones non-bonded potential (kJ/mol) of PP simulations obtained in a spontaneous assembling process from a uniformly mixed state (*black curve*) and that of PP simulation starting from the self-assembled structure obtained in the MD-SCF simulation (*red curve*) (**b**). Snapshot of the system showing the hexagonal arrangement of cylinders in the reverse micellar phase (**c**)

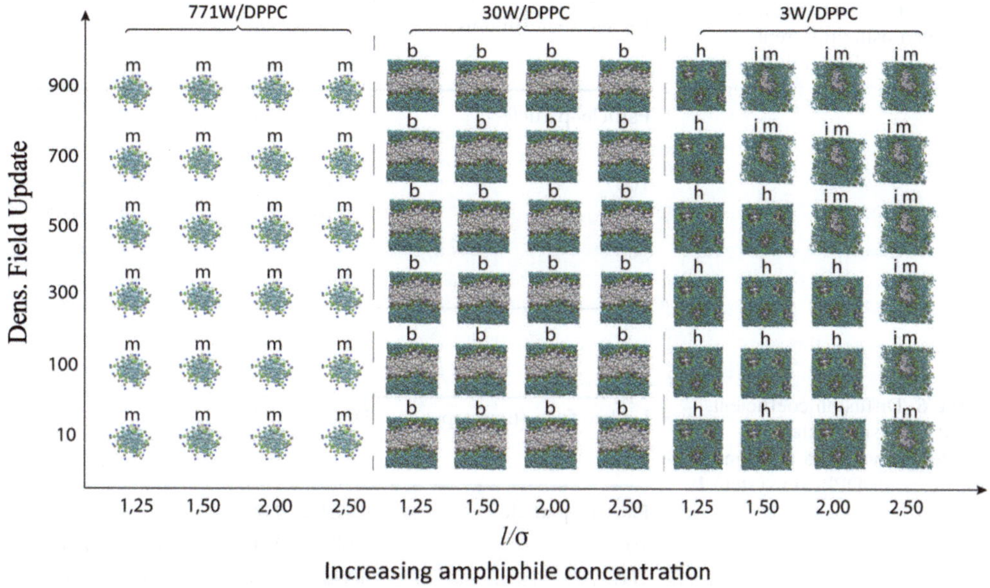

Fig. 5 Graphical matrix summarizing the structures obtained using different density updates intervals (*y* axis, timesteps unit) and cell sizes *l* (*x* axis, unit of σ)

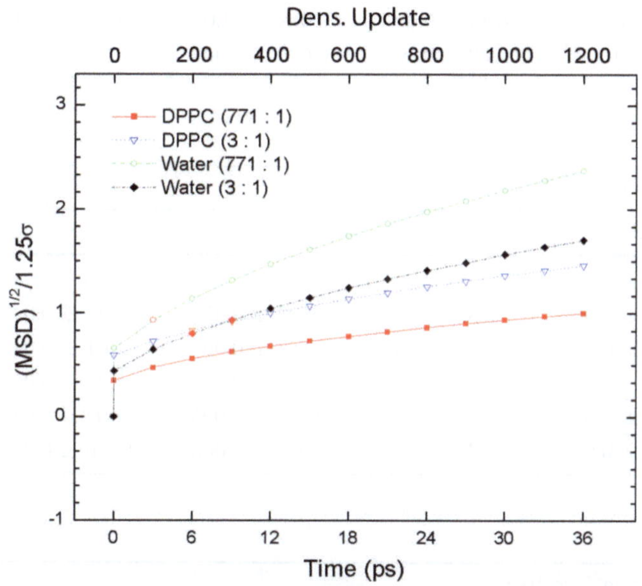

Fig. 6 Square root of normalized mean square displacement of water, DPPC and water beads as a function of time

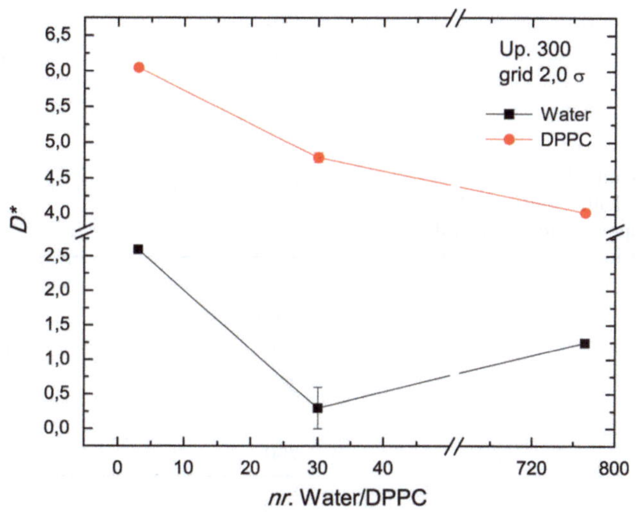

Fig. 7 Ratio between MD-SCF and PP diffusion coefficients as a function of water/DPPC ration calculated for water (*black curve*) and DPPC (*red curve*)

lipid), a stable bilayer phase is formed before 500 ns (Fig. 8b).

Differently, in system **5** (Fig. 8c), where the water content is higher, after a rapid initial local clustering after 100 ns, there is a coalescence to a bilayer structure with curved edges (bicelle) leaving a small single spherical micelle beside it. Between 100 and 200 ns, a process of fusion starts to give a stable bilayer structure with curved edges involving all 300 DPPC molecules present in the simulation box. Finally, the simulation of the system **6** with the water content of 771 water/lipid, after an initial clustering, the formation of two nearly spherical micelles is obtained.

A further validation of the proposed models can be done by considering the average number of lipids/micelle. In the Fig. 9a is reported the time evolution of the number of clusters together with the average number of lipid per cluster for a system having 771 water/lipids. Two lipids are considered to be in the same cluster if at least one distance between their beads is smaller than 1.2 nm. According to this choice, the average number of lipids/cluster (micelle) is about 80. This value has been averaged starting from 400 ns, when the system shows to be in equilibrium. As for available experimental data, sodium dodecyl sulfate (SDS) in physiological conditions forms micelles with an aggregation number ranging from 50 to 80 [64]. In Fig. 9b, the micelle size distribution, averaged from 400 to 600 ns, has

Table 5 Diffusion coefficients calculated using different update intervals and different grid sizes for water in system **1**

Update freq. (timesteps)	Water (cm^2/s × 10^5)			
	$l = 1.25\sigma$	$l = 1.50\sigma$	$l = 2.0\sigma$	$l = 2.5\sigma$
Particle–particle	2.45 ± 0.06	–	–	–
10	3.04 ± 0.06	3.02 ± 0.02	3.04 ± 0.02	2.93 ± 0.07
100	3.06 ± 0.07	3.07 ± 0.01	2.9 ± 0.1	2.94 ± 0.06
300	3.05 ± 0.01	3.12 ± 0.02	3.05 ± 0.05	3.04 ± 0.03
500	3.15 ± 0.03	3.18 ± 0.05	3.08 ± 0.04	3.06 ± 0.01
700	3.23 ± 0.02	3.02 ± 0.05	3.12 ± 0.05	3.08 ± 0.05

Table 6 Diffusion coefficients calculated using different update intervals and different grid sizes for DPPC in system **1**

Update freq. (timesteps)	DPPC (cm^2/s × 10^5)			
	$l = 1.25\ \sigma$	$l = 1.50\ \sigma$	$l = 2.0\ \sigma$	$l = 2.5\ \sigma$
Particle–particle	0.024 ± 0.003	–	–	–
10	0.085 ± 0.004	0.090 ± 0.003	0.132 ± 0.002	0.093 ± 0.001
100	0.12 ± 0.01	0.118 ± 0.008	0.091 ± 0.002	0.992 ± 0.008
300	0.09 ± 0.02	0.12 ± 0.04	0.14 ± 0.05	1.01 ± 0.03
500	0.076 ± 0.004	0.14 ± 0.02	0.089 ± 0.003	1.03 ± 0.03
700	0.089 ± 0.006	0.088 ± 0.003	0.096 ± 0.001	1.14 ± 0.09

Table 7 Diffusion coefficients calculated using different update intervals and different grid sizes for water in system **2**

Update freq. (timesteps)	Water (cm^2/s × 10^5)			
	$l = 1.25\ \sigma$	$l = 1.50\ \sigma$	$l = 2.0\ \sigma$	$l = 2.5\ \sigma$
Particle–particle	0.3813 ± 0.0007	–	–	–
10	1.02 ± 0.06	0.85 ± 0.02	0.98 ± 0.05	1.42 ± 0.02
100	0.957 ± 0.005	0.61 ± 0.08	0.84 ± 0.06	1.61 ± 0.04
300	0.88 ± 0.09	0.90 ± 0.07	0.6 ± 0.3	1.36 ± 0.06
500	0.87 ± 0.08	0.8 ± 0.2	0.859 ± 0.003	1.02 ± 0.04
700	1.07 ± 0.04	0.46 ± 0.03	0.936 ± 0.008	1.340 ± 0.007

Table 8 Diffusion coefficients calculated using different update intervals and different grid sizes for DPPC in system **2**

Update freq. (timesteps)	DPPC (cm^2/s × 10^5)			
	$l = 1.25\ \sigma$	$l = 1.50\ \sigma$	$l = 2.0\ \sigma$	$l = 2.5\ \sigma$
Particle–particle	0.032 ± 0.002	–	–	–
10	0.12 ± 0.01	0.14 ± 0.02	0.20 ± 0.01	0.24 ± 0.05
100	0.132 ± 0.002	0.148 ± 0.004	0.196 ± 0.004	0.24 ± 0.01
300	0.135 ± 0.002	0.155 ± 0.008	0.183 ± 0.001	0.25 ± 0.01
500	0.153 ± 0.007	0.158 ± 0.002	0.16 ± 0.02	0.24 ± 0.01
700	0.16 ± 0.01	0.156 ± 0.003	0.186 ± 0.001	0.25 ± 0.04

been also reported. From the figure, it is clear that the distribution is trimodal, showing three peaks. The first one corresponds to free lipid molecules, the second one corresponds to a smaller micelle composed of about 50 lipids, and the biggest one composed of about 150 lipids. This distribution can be also visualized by looking at some representative simulation snapshot like the one depicted in Fig. 9c. Analogous behavior has been found from PP simulations, where very similar structures can be obtained as shown in Fig. 9c.

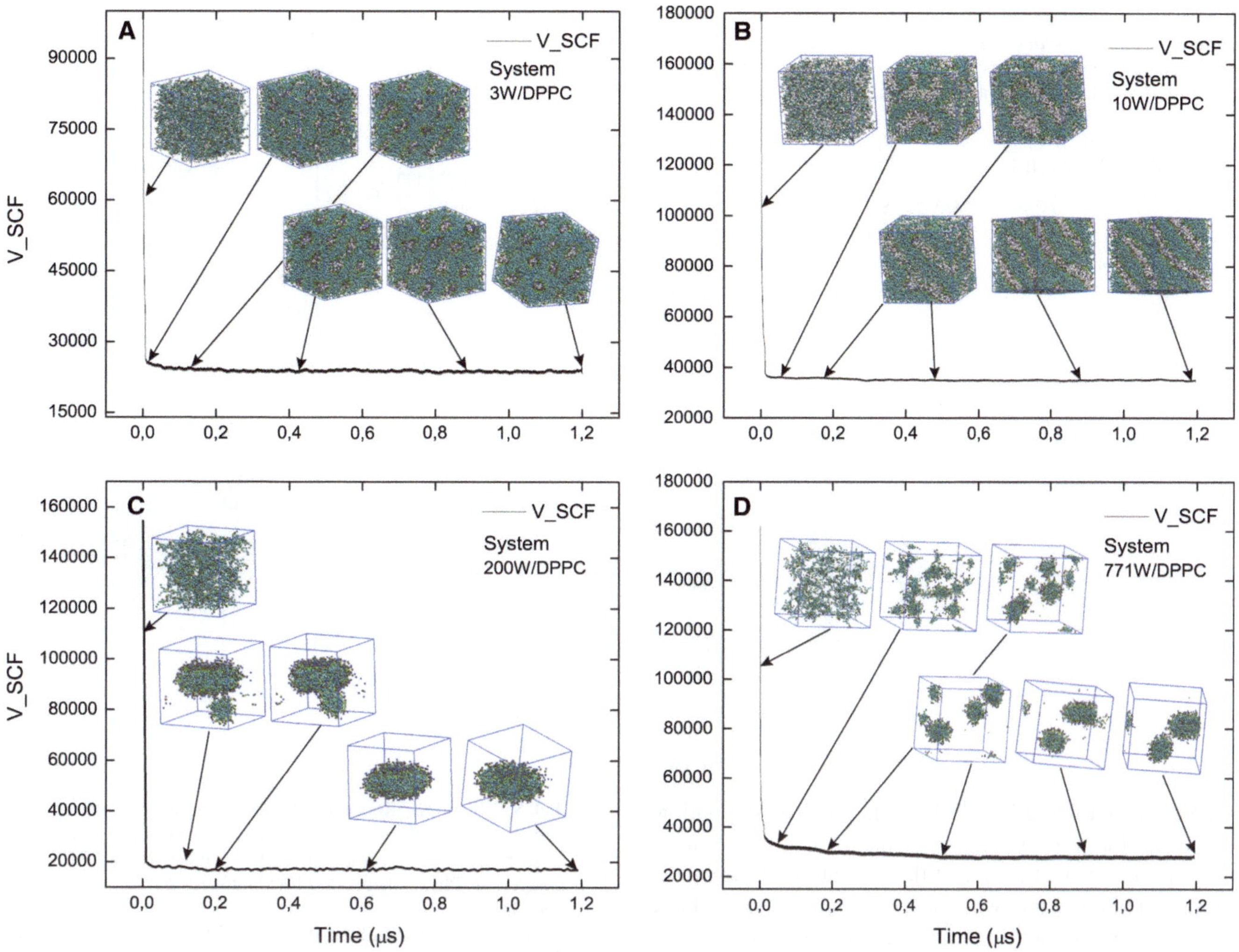

Fig. 8 Time behaviors of the MD-SCF potential together with some snapshots for **a** system **3** forming a reverse micellar hexagonal phase, **b** system **4** forming a lipid bilayer phase, **c** system **5** forming a single bicelle and **d** system **6** forming a micellar phase

4 Conclusions

The validation of a coarse-grained model for phospholipids suitable for a recently developed hybrid methodology combining particles and density fields (MD-SCF) has been reported for DPPC in non-bilayer phases.

According to the mean field coarse-graining strategy, pair interactions between particles are replaced by the calculation of the interactions of single particles in an external field. In principle, the parameters of a coarse-grained model are not always transferable. The comparison between PP and MD-SCF results for non-lamellar phases (stable at low or high water concentration) gave a first confirmation that the proposed models reproduce the correct phase behavior at different compositions. These results encouraged to compare at a more detailed level hybrid particle-field and the reference particle–particle models. In particular, the density profiles of micelles simulated with MD-SCF and conventional MD methods are very similar. Furthermore, the average number of lipids/micelle is well reproduced with respect to reference particle–particle simulations and available experimental data.

As for dynamical properties, the self-assembly process is faster in MD-SCF simulations. Using a systematic comparison between diffusion coefficients, a quantitative scale factors between the dynamics of the coarse-grained hybrid models and the dynamics of corresponding coarse-grained or atomistic particle–particle models have been obtained.

The lower computational costs of the hybrid techniques together with a faster dynamics due to the smoothness of the potentials and forces enable us to perform simulations with a considerably improved efficiency. To have an idea about simulations presented in this paper, for systems **3–6** of Fig. 8, it is possible to calculate about 1 μs/day using 8 processors (Intel E7330, 2.40 GHz). The hybrid MD-SCF

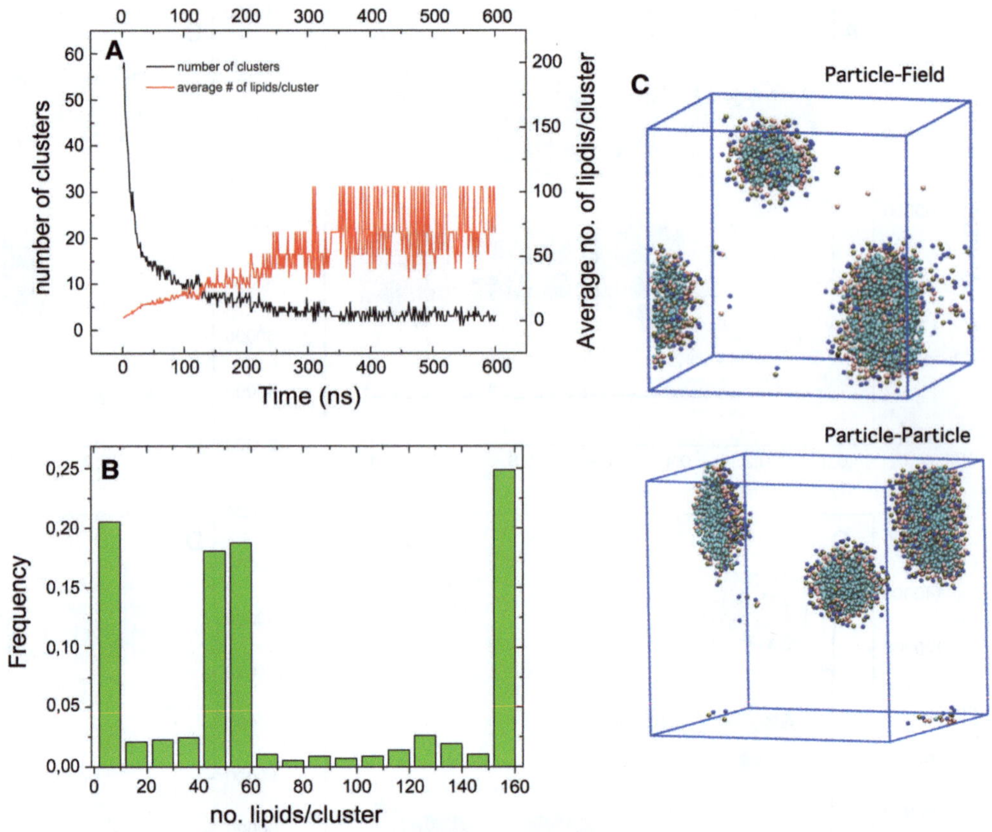

Fig. 9 Time behaviors of number of clusters (*black line*) and average number of lipids per cluster/micelle (*red line*) (**a**). Number of lipids/cluster distribution system (**b**) Snapshots of MD-SCF an PP micellar systems (**c**)

scheme is particularly efficient in parallel simulations, especially for large systems when the use of a large number of CPUs is efficient [61], and the validation of the models presented in this paper will allow their application to large-scale systems needed to study the application of the self-assembled structures of lipids to nanotechnologies.

Acknowledgments This paper is dedicated to Vincenzo Barone for his 60th birthday. G. M. is particularly beholden to Vincenzo who drove his first steps into the fascinating world of theoretical chemistry. G. M. thanks MIUR (PRIN2008 and FIRB "RETE ITALNANONET") for financial support and the HPC team of Enea (http://www.enea.it) for using the ENEA-GRID and the HPC facilities CRESCO (http://www.cresco.enea.it) in Portici, Italy. D. R. and G. M. thank Deutschen Forschungsgemeinschaft (DFG) for funding in the framework of the project "*The study of detailed mechanism of polymers/biological membrane interactions using computer simulation*" (RO 3571/3-1). T. K. thanks the Grant-in-Aid for Science from the Ministry of Education, Culture, Sports, Science and Technology, Japan on the Priority Area "Soft Matter Physics".

References

1. Bandyopadhyay S, Tarek M, Klein ML (1999) Molecular dynamics study of a lipid-DNA complex. J Phys Chem B 103(46):10075–10080. doi:10.1021/jp9927496
2. Saiz L, Klein ML (2002) Computer simulation studies of model biological membranes. Acc Chem Res 35(6):482–489. doi:10.1021/ar010167c
3. Faller R, Marrink S-J (2004) Simulation of domain formation in DLPC-DSPC mixed bilayers. Langmuir 20(18):7686–7693. doi:10.1021/la0492759
4. Pal S, Milano G, Roccatano D (2006) Synthetic polymers and biomembranes. How do they interact?: Atomistic molecular dynamics simulation study of PEO in contact with a DMPC lipid bilayer. J Phys Chem B 110(51):26170–26179. doi:10.1021/jp063418d
5. Bennett WFD, MacCallum JL, Hinner MJ, Marrink SJ, Tieleman DP (2009) Molecular view of cholesterol flip-flop and chemical potential in different membrane environments. J Am Chem Soc 131(35):12714–12720. doi:10.1021/ja903529f
6. Katsaras J, Tristram-Nagle S, Liu Y, Headrick RL, Fontes E, Mason PC, Nagle JF (2000) Revisiting the ripple phase using fully hydrated, aligned DPPC multibilayers. Biophys J 78(1):20A
7. Ayton G, Smondyrev AM, Bardenhagen SG, McMurtry P, Voth GA (2002) Calculating the bulk modulus for a lipid bilayer with non-equilibrium molecular dynamics simulation. Biophys J 82:1226–1238
8. Ayton G, Voth GA (2004) Mesoscopic lateral diffusion in lipid bilayers. Biophys J 87(5):3299–3311
9. Ayton G, Voth GA (2007) Multiscale simulation of transmembrane proteins. J Struct Biol 157(3):570–578
10. Gurtovenko AA, Anwar J, Vattulainen I (2010) Defect-mediated trafficking across cell membranes: insights from in silico modeling. Chem Rev 110(10):6077–6103. doi:10.1021/cr1000783
11. Izvekov S, Voth GA (2006) Multiscale coarse-graining of mixed phospholipid/cholesterol bilayers. J Chem Theory Comput 2(3):637–648. doi:10.1021/ct050300c
12. Shi Q, Voth GA (2005) Multi-scale modeling of phase separation in mixed lipid bilayers. Biophys J 89(4):2385–2394
13. Tepper HL, Voth GA (2006) Mechanisms of passive ion permeation through lipid bilayers: insights from simulations. J Phys Chem B 110(42):21327–21337. doi:10.1021/jp064192h
14. Venturoli M, Sperotto MM, Kranenburg M, Smit B (2006) Mesoscopic models of biological membranes. Phys Rep Rev Sect Phys Lett 437(1–2):1–54. doi:10.1016/j.physrep.2006.07.006

15. Marrink SJ, de Vries AH, Tieleman DP (2009) Lipids on the move: simulations of membrane pores, domains, stalks and curves. Biochim Biophys Acta 1788(1):149–168. doi:10.1016/j.bbamem.2008.10.006
16. Psachoulia E, Marshall DP, Sansom MSP (2010) Molecular dynamics simulations of the dimerization of transmembrane alpha-helices. Acc Chem Res 43(3):388–396. doi:10.1021/ar900211k
17. Lyubartsev AP, Rabinovich AL (2011) Recent development in computer simulations of lipid bilayers. Soft Matter 7(1):25–39. doi:10.1039/c0sm00457j
18. Burkett SL, Mann S (1996) Spatial organization and patterning of gold nanoparticles on self-assembled biolipid tubular templates. Chem Commun 3:321–322
19. Bhattacharya S, Srivastava A (2003) Synthesis and characterization of novel cationic lipid and cholesterol-coated gold nanoparticles and their interactions with dipalmitoylphosphatidylcholine membranes. Langmuir 19(10):4439–4447. doi:10.1021/la0269513
20. Wallace EJ, Sansom MSP (2009) Carbon nanotube self-assembly with lipids and detergent: a molecular dynamics study. Nanotechnology 20(4):045101. doi:10.1088/0957-4484/20/4/045101
21. Lian T, Ho RJY (2001) Trends and developments in liposome drug delivery systems. J Pharm Sci 90(6):667–680
22. Bolinger PY, Stamou D, Vogel H (2004) Integrated nanoreactor systems: triggering the release and mixing of compounds inside single vesicles. J Am Chem Soc 126(28):8594–8595. doi:10.1021/ja049023u
23. Jang J, Yoon H (2005) Formation mechanism of conducting polypyrrole nanotubes in reverse micelle systems. Langmuir 21(24):11484–11489. doi:10.1021/la051447u
24. Müller M, Katsov K, Schick M (2006) Biological and synthetic membranes: what can be learned from a coarse-grained description? Phys Rep 434(5–6):113–176
25. Sintes T, Baumgaertner A (1998) Interaction of wedge-shaped proteins in flat bilayer membranes. J Phys Chem B 102(36):7050–7057
26. Lenz O, Schmid F (2005) A simple computer model for liquid lipid bilayers. J Mol Liq 117(1–3):147–152. doi:10.1016/j.molliq.2004.08.008
27. Marrink SJ, de Vries AH, Mark AE (2003) Coarse grained model for semiquantitative lipid simulations. J Phys Chem B 108(2):750–760. doi:10.1021/jp036508g
28. Marrink S-J, Mark AE (2004) Molecular view of hexagonal phase formation in phospholipid membranes. Biophys J 87(6):3894–3900
29. Marrink SJ, Mark AE (2003) The mechanism of vesicle fusion as revealed by molecular dynamics simulations. J Am Chem Soc 125(37):11144–11145. doi:10.1021/ja036138+
30. Kawakatsu T (2004) Statistical physics of polymers. Springer, Berlin
31. Matsen MW, Schick M (1994) Stable and unstable phases of a diblock copolymer melt. Phys Rev Lett 72(16):2660–2663
32. Drolet F, Fredrickson GH (1999) Combinatorial screening of complex block copolymer assembly with self-consistent field theory. Phys Rev Lett 83(21):4317–4320
33. Fredrickson GH, Ganesan V, Drolet F (2002) Field-theoretic computer simulation methods for polymers and complex fluids. Macromolecules 35(1):16–39. doi:10.1021/ma011515t
34. Lauw Y, Leermakers FAM, Stuart MAC (2006) Self-consistent-field analysis of the micellization of carboxy-modified poly(ethylene oxide)-poly(propylene oxide)-poly(ethylene oxide) triblock copolymers. J Phys Chem B 110(1):465–477
35. Ly DQ, Honda T, Kawakatsu T, Zvelindovsky AV (2008) Hexagonally perforated lamella-to-cylinder transition in a diblock copolymer thin film under an electric field. Macromolecules 41(12):4501–4505
36. Dickinson E, Pinfield VJ, Home DS, Leermakers FAM (1997) Self-consistent-field modelling of adsorbed casein interaction between two protein-coated surfaces. J Chem Soc, Faraday Trans 93(9):1785–1790
37. Balazs AC, Singh C, Zhulina E (1998) Modeling the interactions between polymers and clay surfaces through self-consistent field theory. Macromolecules 31(23):8370–8381
38. Roan JR, Kawakatsu T (2002) Self-consistent-field theory for interacting polymeric assemblies. I. Formulation, implementation, and benchmark tests. J Chem Phys 116(16):7283–7294
39. Roan JR, Kawakatsu T (2002) Self-consistent-field theory for interacting polymeric assemblies. II. Steric stabilization of colloidal particles. J Chem Phys 116(16):7295–7310
40. Marcelja S (1973) Molecular model for phase transition in biological membranes. Nature 241(5390):451–453
41. Leermakers FAM, Rabinovich AL, Balabaev NK (2003) Self-consistent-field modeling of hydrated unsaturated lipid bilayers in the liquid-crystal phase and comparison to molecular dynamics simulations. Phys Rev E 67(1):011910
42. Leermakers FAM, Scheutjens J (1988) Statistical thermodynamics of association colloids. 1. Lipid bilayer-membranes. J Chem Phys 89(5):3264–3274
43. Ayton G, McWhirter JL, Voth GA (2006) A second generation mesoscopic lipid bilayer model: connections to field-theory descriptions of membranes and nonlocal hydrodynamics. J Chem Phys 124(6):12. doi:10.1063/1.2165194
44. Muller M, Smith GD (2005) Phase separation in binary mixtures containing polymers: a quantitative comparison of single-chain-in-mean-field simulations and computer simulations of the corresponding multichain systems. J Polym Sci, Part B: Polym Phys 43(8):934–958
45. Daoulas KC, Muller M, Stoykovich MP, Park SM, Papakonstantopoulos YJ, de Pablo JJ, Nealey PF, Solak HH (2006) Fabrication of complex three-dimensional nanostructures from self-assembling block copolymer materials on two-dimensional chemically patterned templates with mismatched symmetry. Phys Rev Lett 96(3):036104
46. Detcheverry FA, Kang HM, Daoulas KC, Muller M, Nealey PF, de Pablo JJ (2008) Monte Carlo simulations of a coarse grain model for block copolymers and nanocomposites. Macromolecules 41(13):4989–5001
47. Detcheverry FA, Pike DQ, Nealey PF, Müller M, de Pablo J (2010) Simulations of theoretically informed coarse grain models of polymeric systems. Faraday Discuss 144:111–125
48. Milano G, Kawakatsu T (2009) Hybrid particle-field molecular dynamics simulations for dense polymer systems. J Chem Phys 130(21):214106
49. Milano G, Kawakatsu T (2010) Pressure calculation in hybrid particle-field simulations. J Chem Phys 133(21):214102. doi:10.1063/1.3506776
50. De Nicola A, Zhao Y, Kawakatsu T, Roccatano D, Milano G (2011) Hybrid particle-field coarse-grained models for biological phospholipids. J Chem Theory Comput 7(9):2947
51. Schmid F (1998) Self-consistent-field theories for complex fluids. J Phys: Condens Matter 10(37):8105–8138. doi:10.1088/0953-8984/10/37/002
52. Matsen MW (2002) The standard Gaussian model for block copolymer melts. J Phys Condens Matter 14(2):R21. doi:10.1088/0953-8984/14/2/201
53. Fredrickson GH (2005) The equilibrium theory of inhomogeneous polymers, vol 1. Oxford University Press, Oxford
54. Helfand E, Tagami J (1972) Theory of the interface between immiscible polymers. II. J Chem Phys 56(7):10. doi:10.1063/1.1677735
55. Eastwood JW, Hockney RW, Lawrence DN (1980) P3M3DP—the three-dimensional periodic particle–particle/particle-mesh program. Comput Phys Commun 19(2):215

56. Deserno M, Holm C (1998) How to mesh up Ewald sums. I. A theoretical and numerical comparison of various particle mesh routines. J Chem Phys 109(12):7678
57. Norizoe Y, Daoulas KC, Muller M (2010) Measuring excess free energies of self-assembled membrane structures. Faraday Discuss 144:369–391
58. Daoulas KC, Muller M (2006) Single chain in mean field simulations: Quasi-instantaneous field approximation and quantitative comparison with Monte Carlo simulations. J Chem Phys 125(18):184904
59. Marrink S-J, Risselada HJ, Yefimov S, Tieleman DP, de Vries AH (2007) The MARTINI force field: coarse grained model for biomolecular simulations. J Phys Chem B 111(27):7812–7824. doi:10.1021/jp071097f
60. Van der Spoel D, Lindahl E, Hess B, Groenhof G, Mark AE, Berendsen HJC (2005) GROMACS: fast, flexible, and free. J Comput Chem 26(16):1701–1718
61. Zhao Y, De Nicola A, Kawakatsu T, Milano G (2012) Hybrid particle-field molecular dynamics simulations: parallelization and benchmarks. J Comput Chem 33:868–880. doi:10.1002/jcc.22883
62. Carbone P, Varzaneh HAK, Chen X, Muller-Plathe F (2008) Transferability of coarse-grained force fields: the polymer case. J Chem Phys 128(6):064904–064911
63. Tuckerman ME, Glenn JM, Bruce JB (1990) Molecular dynamics algorithm for condensed systems with multiple time scales. J Chem Phys 93(2):5. doi:10.1063/1.459140
64. Moroi Y (1992) Micelles. Theoretical and applied aspects. Springer, Berlin

Theor Chem Acc (2012) 131:1177
DOI 10.1007/s00214-012-1177-z

REGULAR ARTICLE

Thermal and environmental effects on Oligothiophene low-energy singlet electronic excitations in dilute solution: a theoretical and experimental study

Massimiliano Aschi · Andrea Amadei · Andrea Pellegrino · Nicola Perin · Riccardo Po'

Received: 20 August 2011 / Accepted: 19 October 2011 / Published online: 2 March 2012

Abstract The absorption spectra in dilute dichloromethane solution at 300 K of four Oligotiophenes (OT), namely 2,2′:5′,2″-Terthiophene, 2,2′:5′,2″:5″,2‴-Quaterthiophene, 4,4‴-Didodecyl-2,2′:5′,2″:5″,2‴-quaterthiophene and 5,5′′′′′-Dihexyl-2,2′:5′,2″:5″,2‴:5‴,2′′′′:5′′′′,2′′′′′-sexithiophene, have been studied both experimentally and theoretically by using a combination of molecular dynamics simulations, time-dependent density functional theory (TD-DFT) and perturbed matrix method calculations. A deep analysis of the theoretical results, affected by a systematic although not dramatic underestimation of the absorption maxima due to the well-documented TD-DFT limitations, clearly indicates that both the environmental (solvent) and thermal effects significantly alter the Oligotiophenes photophysical properties mainly because of the wide repertoire of the S_0–S_1 energy gaps and electronic densities experienced in solution by the different conformers actually populated. In particular, all the investigated OT display a very high flexibility resulting in a very high repertoire of sampled conformations. The comparison of the calculated and experimental lineshape of the S_0–S_1 electronic transition has clearly indicated that for a correct modeling of OT spectral features, the lack of an exhaustive sampling of semiclassical configurational space of the overall system, i.e., solute in interaction with the solvent, might result in an incomplete picture even in the presence of well-documented important aspects such as reliable definition of excited electronic states and the inclusion of quantum vibronic effects.

Keywords Oligothiophenes excited states modeling · Solar cell · Molecular dynamics · Quantum chemistry

Dedicated to Professor Vincenzo Barone and published as part of the special collection of articles celebrating his 60th birthday.

Electronic supplementary material The online version of this article (doi:10.1007/s00214-012-1177-z) contains supplementary material, which is available to authorized users.

M. Aschi (✉)
Dipartimento di Chimica, Ingegneria Chimica e Materiali
Universita' di l'Aquila, Via Vetoio, 67100 Coppito I,
l'Aquila, Italy
e-mail: aschi@caspur.it

A. Amadei
Dipartimento di Scienze e Tecnologie Chimiche, Universita' di Roma 'Tor Vergata', Via della Ricerca Scientifica,
00100 Rome, Italy

A. Pellegrino · N. Perin · R. Po'
Solar Energy Research Center for Non-Conventional Energies
Istituto ENI Donegani Eni S.p.A, Via Giacomo Fauser 4,
28100 Novara, Italy

1 Introduction

Oligothiophenes (OT) are conjugated polycyclic thiophene oligomers extensively used in organic electronics [1] as light-emitting diodes [2] and solar cells [3] which have attracted, in the last years, a huge amount of experimental and theoretical work [4]. One of the most investigated aspects, considered to be fundamental for characterizing the photophysics of OT, is the full characterization of the low-lying singlet and triplet excited states involved in many photochemical processes, including electron transfer, taking place upon low-energy OT electronic excitations. Many studies have addressed the actual localization of the vertical and adiabatic singlet excited electronic states, the relative position of the first singlet excited state S_1 and the first triplet state T_1 [5–7], their dependency with respect the length of the OT chain [8–10], and the energetics and dynamics related to the intra and intermolecular electron transfer upon excitation [11, 12].

Theoretical studies have also revealed that electronic-correlation, multi-reference wave-function, and multi-excitation character seem to be important ingredients to correctly reproduce OT electronic excitations [7, 13–18].

Accurate, and more recent, investigations carried out with different combinations of experimental and theoretical approaches have also revealed that a full understanding of OT spectral bandshapes requires a proper exploration of vibronic coupling [10, 19, 20].

As a matter of fact, OT species typically display, at low temperature, absorption spectra characterized by mirror symmetry.

However, temperature increase has been shown to produce spectral broadening [10] clearly indicating that thermal excitations of torsional modes producing different OT conformations [21–25] in interaction with the solvent [20, 26, 27] play a very important role in OT photophysics whose modeling should also take into account these effects.

As a consequence, several studies have recently proposed advanced theoretical modeling of OT spectral features including both vibronic coupling, torsional modes excitations and solvent effects [10, 19, 20].

However, the same studies [19, 20] have also clearly stated that OT have to be considered as soft materials hence plausibly characterized by a rather complicated conformational repertoire at relatively high temperature and in the presence of solvent inhomogeneity. In this respect, an unbiased and exhaustive sampling of the OT-solvent semiclassical conformational space—carried out with molecular dynamics (MD) simulations—associated with a proper treatment of the quantum degrees of freedom might greatly help in improving our understanding of the photophysics of these systems.

In line with our ongoing interest in the theoretical–computational modeling of electronic properties of molecular systems in solution, in the present investigation, we attempt to shed some light on the low-lying OT singlet excited states in solution in thermal conditions, i.e., explicitly taking into account thermal effects both on the OT conformational repertoire and on the solvent fluctuations.

In particular, we have considered the absorption spectra of four species in dilute CH_2Cl_2 solution at 300 K, namely 2,2′:5′,2″-Terthiophene (TT), 2,2′:5′,2″:5″,2‴-Quaterthiophene (QT), 4,4‴-Didodecyl-2,2′:5′,2″:5″,2‴-quaterthiophene (DQT), and 5,5′′′′′-Dihexyl-2,2′:5′,2″:5″,2‴:5‴,2′′′′:5′′′′,2′′′′′-sexithiophene (DET), schematically reported in Fig. 1. The experimental spectra of the above species have been compared with the corresponding spectra calculated in the same conditions utilizing a theoretical–computational scheme, based on MD simulations, Quantum-Chemical (QM) calculations, and the Perturbed Matrix Method (PMM) [28], extensively utilized in our laboratory in the last years for calculating the photophysical equilibrium properties, e.g., UV absorption spectrum, of molecules in solution [29–31].

Such a methodology, which might be considered somewhat complementary to more popular QM/MM approaches [32, 33] has been usually applied, so far, for characterizing the perturbed quantum states of rigid or semi-rigid molecules in solution. However, its extension to very flexible species such as OT [21] is relatively straightforward and, in the present investigation, its systematic implementation is presented for the first time. We wish to further remark that our goal in this work is not to provide an exact reproduction of OT low-lying singlet excited states in solution. In fact, the already cited effects such as multireference character of the wavefunction as well as inclusion of vibronic coupling [10, 19, 20], although already incorporated in our theoretical–computational setup [34], have been disregarded in the present investigation whose main aim is to specifically enucleate the effects of the solute and solvent conformational fluctuations onto the vertical low-lying singlet electronic excitations of solvated OT. Hence, well aware of its limitations for describing low-energy electronic excitations in OT [6, 15] and, more in general in π-conjugated species [35], Time-Dependent Density Functional Theory (TD-DFT) [36] has been adopted for characterizing OT unperturbed singlet electronic states. As a matter of fact, TD-DFT provides, at the moment, a very good compromise between costs and benefits when a huge number of geometries must be considered, as in the present case. In the first part of the study, the theoretical–computational method—hereafter concisely termed as PMM-MD—and the experimental setup are described in detail. Then, after a brief outline of the theoretical aspects of the work part, the experimental and computational results are presented and commented.

2 Theoretical and experimental methods

PMM-MD approach, similarly to many of the QM-MM methods [37], is based on the determination of the quantum states of a portion of a complex system (hereafter termed as Quantum Centre, QC), in interaction with the environment, e.g., the solvent molecules. The main differences with the other commonly employed approaches might be essentially summarized in two points:

- PMM-MD main goal is to maintain as much as possible, at a proper level of description, *both* the explicit definition of QC quantum degrees of freedom in interaction with the environment *and* an exhaustive

Fig. 1 Schematic representation of the species investigates in the present study

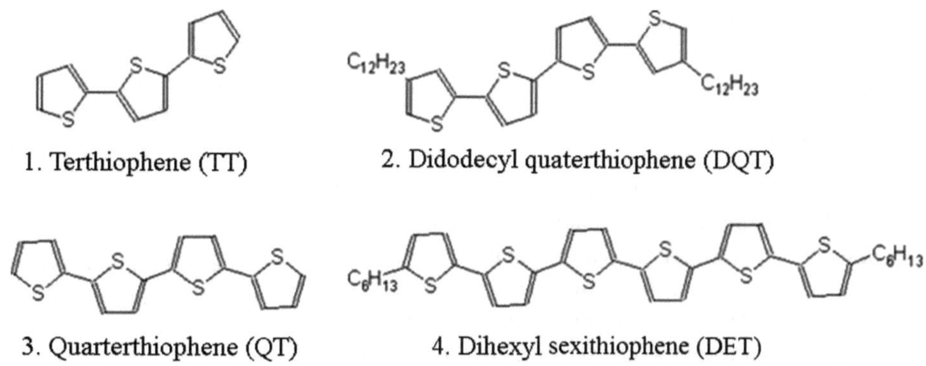

1. Terthiophene (TT)
2. Didodecyl quaterthiophene (DQT)
3. Quarterthiophene (QT)
4. Dihexyl sexithiophene (DET)

and significant sampling of the semiclassical phase-space of the overall system;

- differently from many of the commonly employed QM/MM methods, PMM-MD is based on a homogenous definition of the configurational space of the *overall* system, i.e., both the solute and the solvent are simulated according to classical MD making use of the *same* Hamiltonian. Subsequently, the configurations sampled by the overall system along MD simulation are utilized a posteriori to applying the PMM procedure for the evaluation of the QC ground to excited quantum (e.g., electronic) states interacting with the rest of the simulation box acting as a classical (e.g., electrostatic) perturbation. In the next three paragraphs, details of PMM, MD simulations, and the Quantum-Chemical (QM) calculations are reported.

2.1 Perturbed matrix method

The determination of the perturbed quantum (electronic) properties of the QC interacting with a classical perturbation along the MD simulation is achieved by defining, in matrix terms, a perturbed Hamiltonian (electronic) operator constructed on the basis set provided by the unperturbed (i.e., evaluated in vacuum) Born–Oppenheimer QC geometry, assumed for the moment to be characterized by a rigid covalent framework.

Hence, indicating with $\mathbf{r}_n$ the QC coordinates and with $\mathbf{x}(t)$ the atomic coordinates of the perturbing environment at a generic MD frame, we can write the perturbed electronic Hamiltonian matrix as

$$\tilde{H}[\mathbf{r}_n, \mathbf{x}(t)] = \tilde{H}^o(\mathbf{r}_n) + q_T V[\mathbf{r}_{\rm com}, \mathbf{x}(t)] + \tilde{Z}_1[\mathbf{E}(\mathbf{r}_{\rm com}, \mathbf{x}(t))] + \tilde{Z}_2\left[\tilde{\Theta}(\mathbf{r}_{\rm com}, \mathbf{x}(t))\right] + \Delta\varsigma\tilde{I} \qquad (1)$$

where $\tilde{H}^o(\mathbf{r}_n)$ is the unperturbed (gas-phase) electronic Hamiltonian calculated for the selected QC geometry (see Sect. 2.3), q_T is the QC total charge, $V[\mathbf{r}_{\rm com}, \mathbf{x}(t)]$ and $\mathbf{E}$ are the electrostatic potential and electrostatic field exerted by perturbing molecules present in the simulation box at their instantaneous $\mathbf{x}(t)$ configuration and acting at the QC centre of mass ($\mathbf{r}_{\rm com}$), and $\tilde{I}$is the identity matrix. The terms $\tilde{Z}_1[\mathbf{E}(\mathbf{r}_{\rm com}, \mathbf{x}(t))]$ and $\tilde{Z}_2\left[\tilde{\Theta}(\mathbf{r}_{\rm com}, \mathbf{x}(t))\right]$ are the perturbed Hamiltonian matrices whose generic l, l' elements are defined as:

$$\left[\tilde{Z}_1\right]_{l,l'} = -\mathbf{E} \cdot \langle \Phi_l^o | \hat{\mu} | \Phi_{l'}^o \rangle \qquad (2)$$

and

$$\left[\tilde{Z}_2\right]_{l,l'} = \frac{1}{2} Tr[\tilde{\Theta}\tilde{Q}_{l,l'}] \qquad (3)$$

where $\Theta_{k,k'} = \left(\frac{\partial E_k}{\partial r_{k'}}\right)_{\mathbf{r}=\mathbf{r}_{\rm com}}$ is the electric field gradient evaluated at QC centre of mass and $\tilde{Q}_{l,l'}$ is the unperturbed l, l' transition quadrupole matrix, and $\Delta\varsigma$ is scalar term which approximates the perturbation due to all higher-order terms to simple short-range potential [28].

Note that in the case of dipolar approximation, i.e., first order, Eq. 1 can be rewritten as

$$\tilde{H}[\mathbf{r}_n, \mathbf{x}(t)] = \tilde{H}^o(\mathbf{r}_n) + q_T V[\mathbf{r}_{\rm com}, \mathbf{x}(t)] + \tilde{Z}_1[\mathbf{E}(\mathbf{r}_{\rm com}, \mathbf{x}(t))] + \Delta\varsigma'\tilde{I} \qquad (4)$$

with the term (3), approximated in this case as a scalar term, merged together with $\Delta\varsigma$ into the same scalar term, $\Delta\varsigma'$. Both the transition dipole and transition quadrupole in (2) and (3) are derived by using the Φ_1^0 unperturbed electronic Hamiltonian eigenfunctions evaluated only once in correspondence of the QC (rigid) geometry. Diagonalization of the perturbed electronic Hamiltonian (Eq. 1 or 4) at each MD frame produces a trajectory of perturbed eigenvalues and eigenvectors ($\mathbf{c}_i$) of QC which can be used, in the present work, to evaluate at each configuration of the perturbing environment the perturbed excitation energies (i.e., the difference between the ith eigenvalue and the lowest eigenvalue of operator 1) and transition dipoles, $\boldsymbol{\mu}_{i,j}$, to be used for calculating the absorption spectra intensities [30], according to Eq. 5

$$\mu_{i,j} = \mathbf{c}_i^{*T}\tilde{\Lambda}_x^0\mathbf{c}_j\mathbf{i} + \mathbf{c}_i^{*T}\tilde{\Lambda}_y^0\mathbf{c}_j\mathbf{j} + \mathbf{c}_i^{*T}\tilde{\Lambda}_z^0\mathbf{c}_j\mathbf{k} \tag{5}$$

where

$$[\tilde{\Lambda}_k^0]_{l,l'} = \langle\Phi_l^0|\hat{\mu}_k|\Phi_{l'}^0\rangle \tag{6}$$

The above PMM-MD procedure might be concisely described using the chart in Scheme 1 for a rigid QC.

These considerations are valid when the QC does not undergo conformational transitions, i.e., for a rigid covalent framework. Hence, it should be modified when, for the QC, several conformational states come out from MD simulation. In this case, we need to derive a set of unperturbed electronic Hamiltonian states ($\Phi_{l'}^{0-j}$) for each of the jth sampled conformation. Consequently, the PMM scheme modifies according to the Scheme 2.

The only additional problem, in this second case with respect to the rigid QC case, derives from the difficulty associated with the definition of the QC conformational transitions which becomes a relatively complicated task in the presence of more than one or two QC internal coordinates. This has been overcome using Essential Dynamics as described in the next paragraph dedicated to MD simulations details.

2.2 Molecular dynamics simulations and essential dynamics analysis

As already remarked, the overall system configurations are sampled by using a MD simulation with a properly selected (at a properly sophisticated level of theory) force field. The solute, i.e., in the case described in this paper each of the species 1–4 in Fig. 1, is put at the center of a rectangular box filled with dichloromethane (CH_2Cl_2) at the typical density of .786 g/ml at 300 K and 1.0 bar and simulated for

Scheme 1 Approximate flow chart for PMM application in the case of QC rigid geometry

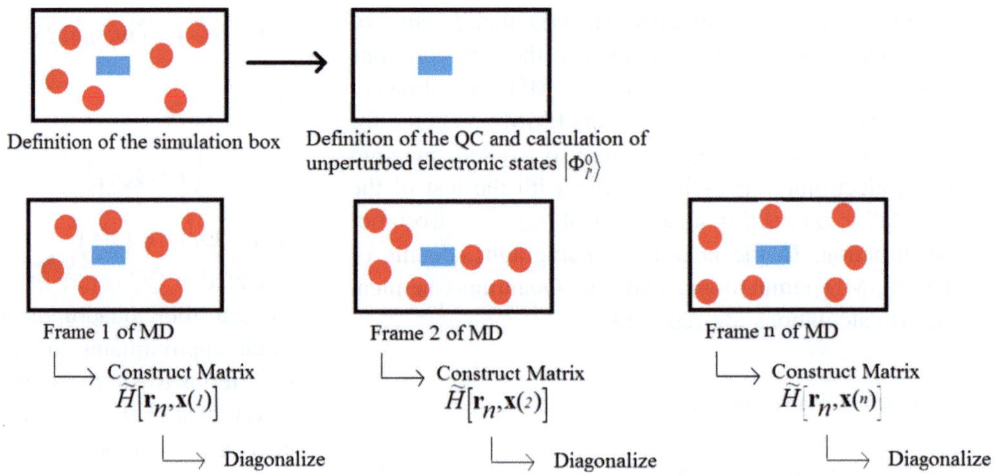

Scheme 2 Approximate flow chart for PMM application in the case of QC geometry transition. Note that the *blue square* and *triangle* correspond to different QC geometries

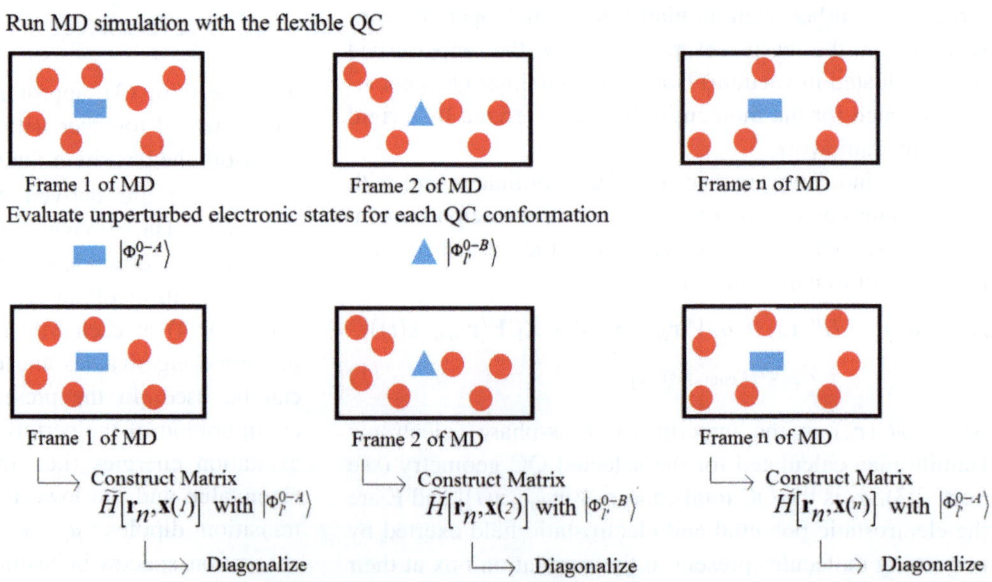

40 ns. For the species 1–4, we have adopted the force field recently developed by one of us [38]. For CH_2Cl_2, we used a force field adapted from opls-AA force field [39]. The following protocol was used for all the simulations: an energy minimization was first carried out and, after a solvent relaxation, the system was gradually heated from 50 to 300 K using short (typically 200.0 ps) MD simulations; the trajectories were then propagated up to 40 ns in a NVT ensemble using an integration step of 2 fs removing the solute roto-translation motions [40] (note that in the case of DQT and DET, only the OT moiety was used for the roto-translational fitting, see below); temperature was kept constant at 300 K by the Berendsen thermostat [41] with the τ equal to the time-step used for the integration algorithm. All bond lengths were constrained using LINCS algorithm [42]. Long-range electrostatics were computed by the Particle Mesh Ewald method [43] with 34 wave vectors in each dimension and a fourth-order cubic interpolation. Gromacs package [44] was used for all the simulations.

Essential Dynamics (ED) [45] was used for analyzing OT conformational transition. Briefly: using the equilibrated MD trajectory, the all-atoms OT covariance matrix was constructed and diagonalized. Such a transformation produces a set of eigenvectors which represent the directions along which OT fluctuates. Eigenvectors showing large eigenvalues define the space (essential space) in which the actual conformational transitions take place and are used for identifying the most stable (most sampled) conformations. In correspondence of these structures, unperturbed QM calculations are carried out as described in detail in the next Sect. 2.3.

2.3 Quantum-chemical calculations

On each of the *j*th conformation previously extracted from the essential space, we first carried out a geometrical relaxation of the quantum internal degrees of freedom (all bond lengths and angles) by keeping fixed all the dihedral angles using the B3LYP [46, 47] hybrid functional in conjunction to the 6–31 + G(d, p). These locally minimized structures were then used to obtain the ground and the first eight vertical excited states taken as basis set ($\Phi_{l'}^{0-j}$, $l' = 0, 1,\ldots,8$ in Scheme 2) utilizing TD-DFT calculations with the same functional and basis set. For these calculations, we used the Gaussian03 program [48].

2.4 Experimental setup

2,2′:5′,2″-Terthiophene (TT), 2,2′:5′,2″:5″,2‴-Quaterthiophene (QT), and 5,5‴‴′-Dihexyl-2,2′:5′,2″:5″,2‴:5‴,2‴′:5‴′,2‴‴′-sexithiophene (DET) were purchased from Sigma-Aldrich Co., and used without purification. 4,4‴-Didodecyl-2,2′:5′,2″:5″,2‴-quaterthiophene (DQT) was prepared by Stille coupling of 2-tributylstannyl-4-dodecylthiophene and 2,2′-bithiophene-5,5′-dibromide and purified by column chromatography.

UV–Vis absorption spectra were recorded at room temperature with a Lambda 950 spectrophotometer (Perkin-Elmer Inc. Waltham, MA, USA). Samples in solution were prepared by dissolving the products in $CHCl_3$ (TT, QT, DQT, or DET) or CH_2Cl_2 (TT, QT, or DQT) solutions in cell with 10 mm optical path length and optical density of typically .5 at max. absorption in the visible region of spectrum.

3 Results

3.1 Experimental UV–Vis absorption spectra

The experimental UV–Visible spectra are reported in Fig. 2. All compounds present a band with maximum at 253 nm (only HST band is slightly red shifted). The main absorption band instead is red shifted with increase in π-conjugation of molecules. In particular, the maximum of absorption is centered at 355 nm for TT, 393 nm for QT, 398 nm for DQT, and 415 nm for DET. Note that DET was analyzed in $CHCl_3$ due to solubility problem anyway an experimental verification confirmed that for TT, QT, DQT the spectra in $CHCl_3$ and the spectra in CH_2Cl_2 are substantially the same.

Increasing the number of thiophene rings, the π-conjugation is enhanced and the maximum of absorption shifts to red. With the same number of thiophene rings, the alkyl side chains have influence on the maximum absorption band position; in fact, DQT is slightly red shifted compared

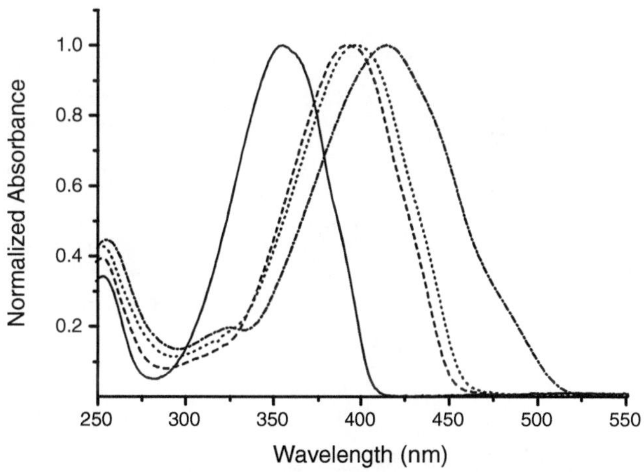

Fig. 2 UV–Visible absorption spectra of TT (*solid curve*), QT (*dash curve*), DQT (*dot curve*), and DET (*dash-dot curve*). All spectra are normalized at maximum of absorption and were recorded in CH_2Cl_2 solutions except DET that were recorded in $CHCl_3$ solutions

with QT due to inductive effect of alkyl side chains. In DET UV–Vis absorption spectrum, the shoulder on the red side might be due to some impurities, as confirmed by ^{1}H-NMR and mass spectra.

3.2 Extraction of the relevant conformations of OT moieties from MD simulation

For theoretically modeling the above spectra in this work, we assume that the electronic transitions we are looking at, i.e., essentially the S_0–S_1 and S_0–S_2 ones, essentially involve only the OT moieties in all the investigated species in Fig. 1. This approximation has been repeatedly shown, by the analysis of the molecular orbitals predominantly involved in low-lying valence excited excitations, to be acceptable in molecules similar of the ones addressed in this study [49]. It follows that in DQT and DET, the alkyl side chains have been assumed not to directly contribute to the electronic spectrum but their role is only relevant for the MD conformational sampling.

In the first part of this section, we analyze the OT moieties conformations sampled along the MD simulations to be used in the following PMM analysis.

In Fig. 3, we report the root mean square fluctuation (RMSF) of the four investigated systems in order to compare the internal mobility of the four species.

From Fig. 3, it is evident that, when present, alkyl side chains undergo a relatively large fluctuation and that all the systems are characterized by a rather high internal flexibility of the OT only slightly increasing with the dimension of the system. This means that all the species are expected to show a not negligible repertoire of OT conformations which should be in principle taken into account, with their own statistical weight, to construct the overall observed spectrum. A further aspect emerging from Fig. 3, which will turn out to be important for subsequent analyses, is the strong similarity between the RMSFs of OT moieties in QT and DQT systems.

The significant OT conformations sampled along the simulation have been extracted using ED as follows. First of all, it is important to underline that the construction and the diagonalization of the OT covariance matrix has produced for all the species a very steep spectrum of the eigenvalues indicating that the first and second eigenvector, accounting for almost the 70% of the overall fluctuation, would properly be used as the essential space for a correct conformational analysis.

Hence, the first step toward the extraction of OT geometries is represented by the projection of the trajectory onto such an essential space (i.e., essential plane, EP) producing differently populated portions (spots), hereafter termed as basins. An example of these basins is reported in Fig. 4a as a result of the ED analysis for TT species. Moreover, assuming the trajectory at the equilibrium, it is also possible to derive the relative stability, i.e., the relative

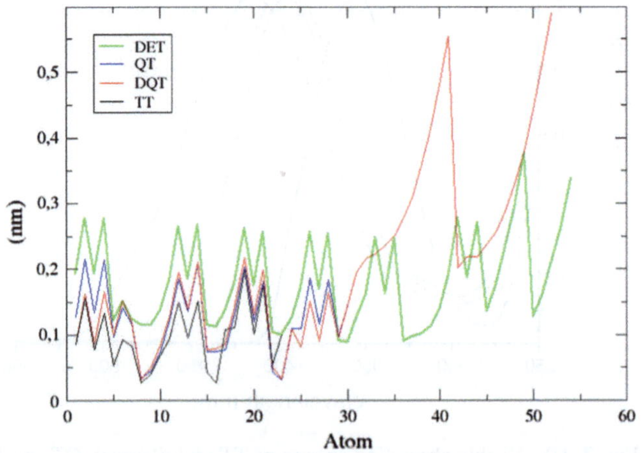

Fig. 3 Root mean square fluctuation for TT (*black*), QT (*blue*), DQT (*red*), and DET (*green*). Atoms starting from 31 for DQT and starting from 45 for DET correspond to alkyl side chains

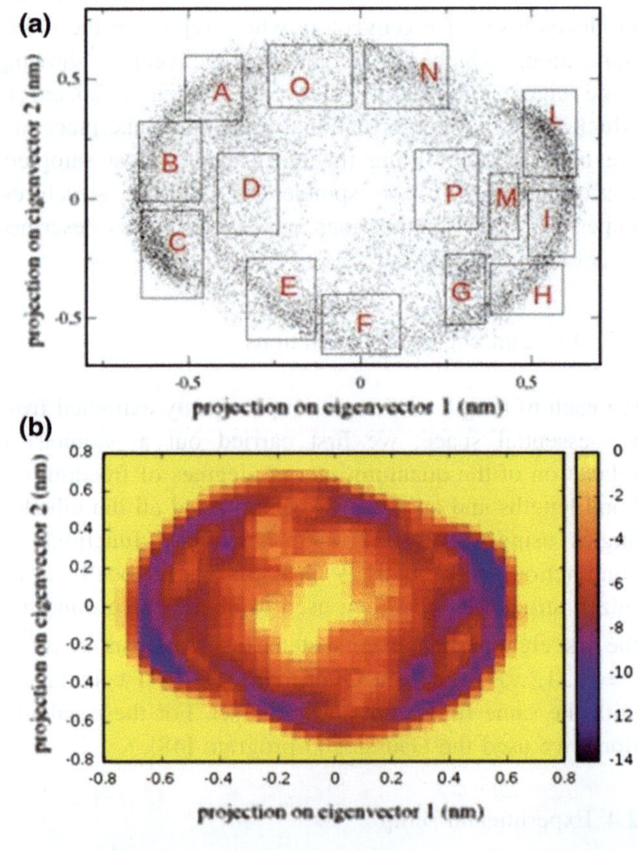

Fig. 4 **a** Identification of the conformational basins (*A–P*) on the TT essential plane and **b** corresponding 300 K Helmholtz free energy (kJ/mol) landscape from Eq. 7. Conformational basins are indicated with *red letters* and the representative structures are reported in the Supplementary Information *yellow* regions correspond to not sampled regions of the essential plane

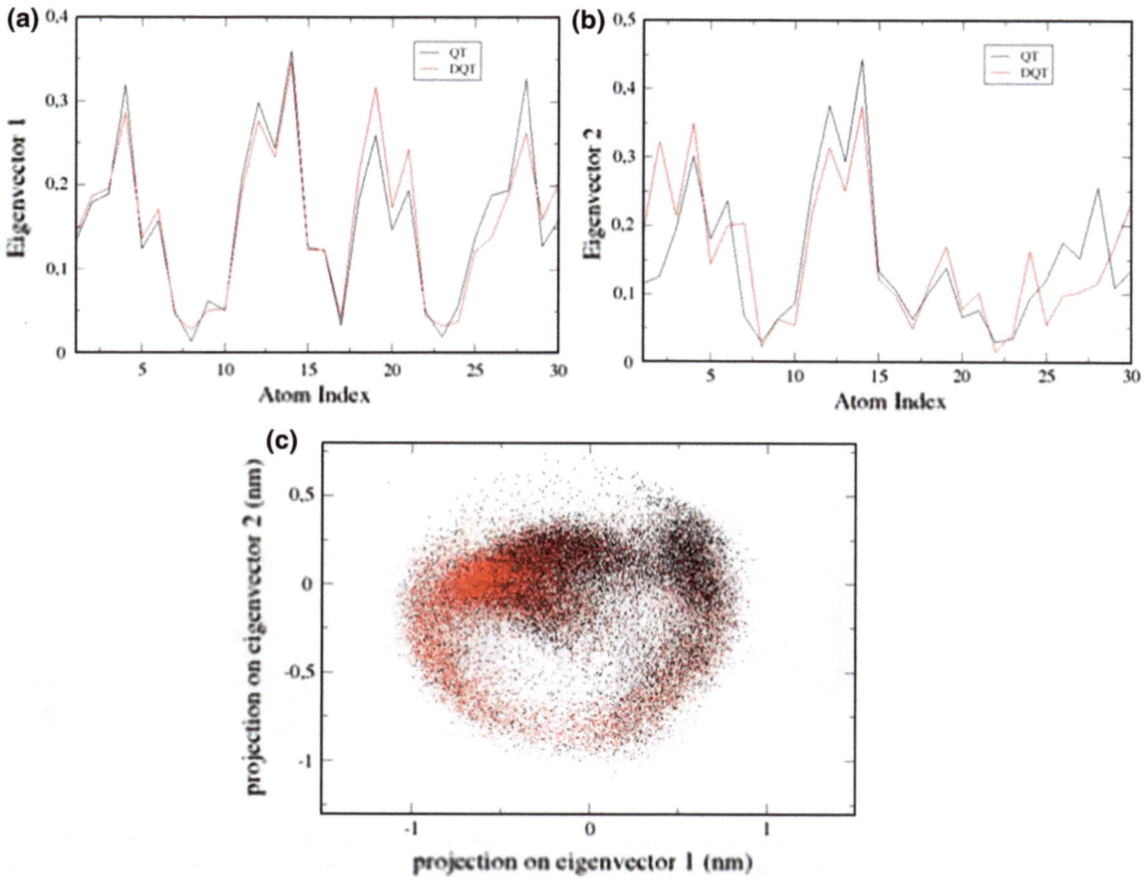

Fig. 5 Atom component of the first (**a**) and second (**b**) eigenvector from diagonalization of OT all-atom covariance matrix from MD simulation of QT (*black line*) and DQT (*red line*) in solution. **c** Projection of the QT (*black dots*) and DQT (*red dots*) trajectories on the corresponding essential plane

300 K Helmholtz free energy, for each OT basin using the standard formula

$$\Delta A_{i-ref} = A_i - A_{\text{ref}} = -\text{RTln}\frac{P_i}{P_{ref}} \quad (7)$$

where P_i and P_{ref} are the probability of finding the projected trajectory onto the basin *i* and *ref*, respectively. Note that the *ref* basin is the one showing the largest probability, in the case of TT the basin L. The resulting free energy landscape reported in Fig. 4b for TT indicates that all the basins lie within a free energy interval not exceeding 10 kJ/mol.

In correspondence of each basin, we then extract the OT conformations. This is obviously an approximation as a single geometrical conformation would strictly correspond to a single point on EP. However, as far as all the points (corresponding to different OT conformations) falling within a single basin display, as in the present case, geometrical deviations well below 1.0 Angstrom, this approximation might be considered as acceptable. As a consequence, a single OT structure might be used as *representative* of the basin. For the sake of brevity, pictures and additional details about the representative OT structures for TT simulations are reported in the Supplementary Information. Here, it suffices to remark that only the region corresponding to basins I and L do actually correspond to structures somewhat similar to all-anti all-planar moieties. It follows that most stable conformations, i.e., most sampled TT geometrical states, show clear torsional excitations between vicinal thiophene rings as well as out-of-plane collective deviations of the whole polithiophene moiety.

The same analysis was then carried out for QT and DQT systems for which, as previously pointed out from the comment of Fig. 2, we may expect an almost identical conformational repertoire. As a matter of fact, the atomic components of the first two eigenvectors obtained from diagonalization of OT moiety covariance matrix from QT and DQT simulation, reported in Fig. 5a, b, show an almost perfect overlap. Moreover, see Fig. 5c, the projection of the QT and DQT trajectories in the plane constituted by the first two eigenvectors from QT covariance matrix also produces a very good overlap.

As a consequence of these observation, we have decided to carry out the extraction of the representative OT

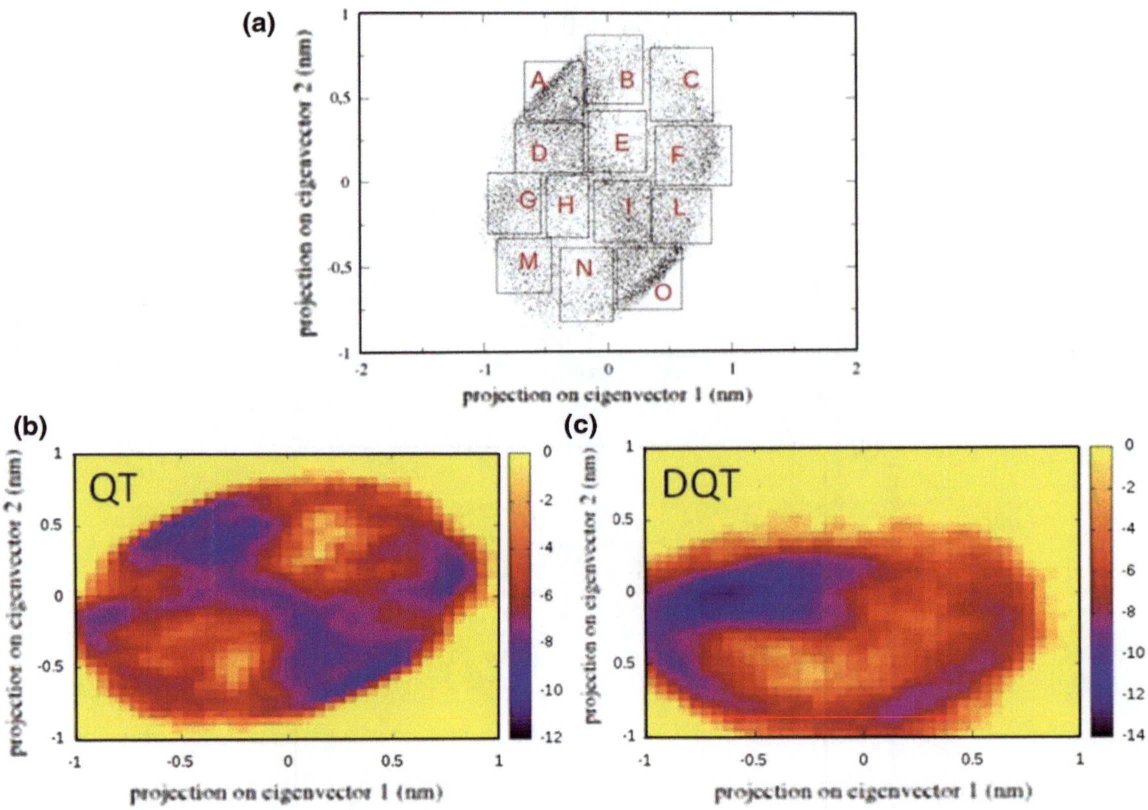

Fig. 6 **a** Identification of the conformational basins (*A–O*) in the OT essential plane from QT (DQT) MD simulation and **b** corresponding 300 K free energy landscapes (kJ/mol) for QT and DQT, from Eq. 7. Conformational basins are indicated with *red letters* and the representative structures are reported in the Supplementary Information. *Yellow* regions identify not sampled regions of the essential plane

structures for QT and DQT by using the same essential plane. The results presented below refer to the QT essential plane. Note, however, that the same procedure performed utilizing the DQT plane provided an almost identical result. The result is reported in the Fig. 6.

Similarly to TT, also for QT and DQT, many OT conformations—whose details are reported in the Supplementary Information—are populated a 300 K in solution with the *all-anti* all-planar structures (representative of basins A and O), representing the most stable ones. It is also interesting to note that the almost perfect symmetry emerged in the essential plane of TT and QT (unsubstituted species) is slightly but sharply broken in the case of DQT suggesting that the alkyl side chains do provide some differences also on the relative stabilities of OT conformations. The above analyses were finally repeated for the largest DET species.

The results, collected in Fig. 7, with the structural details also in this case reported in the Supplementary Information, indicate that also for DET, the *all-anti* structure (in this case falling in the basin A) is the most stable one with other conformations, however, significantly sampled during the trajectory. At this point, we are ready to utilize all the representative structures extracted for TT, QT, DQT, and DET for evaluating, by means of PMM-MD, the related spectra as described in the next two subparagraphs.

3.3 Calculation of UV spectra for TT, QT, DQT, and DET

Using the Schemes 1 and 2, within first-order approximation, i.e., Eq. 4, we evaluated the PMM-derived UV spectrum for each species. In particular, we analyzed the region between 300 and 600 nm. In Table 1, we schematically summarize the pre-selected QCs and the perturbing environment for each simulation.

Within the present model, in all the simulations, the solvent molecules produce an electrostatic (fluctuating) perturbation on the QCs. In the case of DQT and DET, beyond the solvent molecules, also the fluctuating alkyl side chains produce the perturbation.

Note that the actual spectra should arise from the convolution of the spectra of all the conformers with their own weight provided by the simulation ensemble as already reported in Scheme 2.

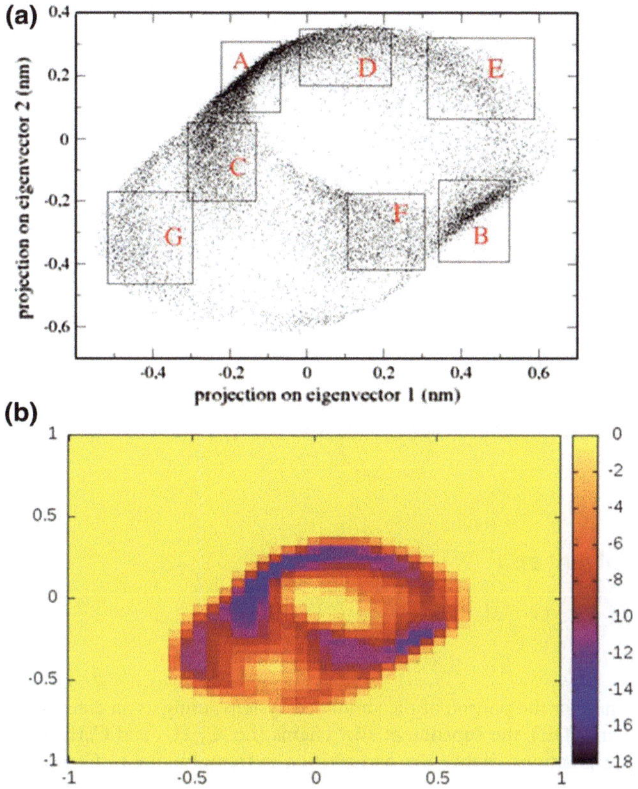

Fig. 7 **a** Identification of the conformational basins (*A–G*) in the OT essential plane from DET MD simulation and **b** corresponding 300 K free energy landscapes (kJ/mol) for DET from Eq. 7. Conformational basins are indicated with *red letters* and the representative structures are reported in the Supplementary Information. *Yellow* regions identify not sampled regions of the essential plane

By a practical point of view, Scheme 2 has been applied as follows.

Step 1 The QC conformations are extracted from their own basins (Sect. 3.2, Figs. 4, 6, 7 and Supplementary Information). We wish to point out that the size of each of these portions is arbitrary and its effect on the spectrum will be addressed later (see Step 5).

Step 2 According to Scheme 2 in correspondence of each structure representative of the different basins (Sect. 3.2, Figs. 4, 6, 7 and Supplementary Information), the unperturbed set of ground and eight excited states wavefunctions $\Phi_{l'}^{o-m}$ were obtained in vacuo (superscript 'o') using TD-DFT with B3LYP functional and 6–31 + G(d, p), with $l' = 0,1, 2,\ldots,8$ (note: 0 stands for ground electronic state) and $m = A, B, C, \ldots$. It is important to further remark that in the case of DQT and DET, the methyl groups were included in the unperturbed calculations in the 4–4‴ and 4–4‴″ positions, respectively.

Step 3 PMM is applied along the trajectory selecting the $\Phi_{l'}^{o-m}$ unperturbed set on the basis of the actual location on the essential plane evaluated by a root mean square deviation (RMSD)-based procedure (see Scheme 2) as reported in the Fig. 8.

More precisely: when the trajectory (blue line in Fig. 8) enters a selected area, e.g., basin H, the calculation of the RMSD carried with respect to all the structures will produce the minimal value in correspondence of the structure representative of basin H. Consequently, as far as this condition is hold along the trajectory, PMM is applied with the unperturbed basis for the QC structure H (i.e., $\Phi_{l'}^{o-H}$with $l' = 0, 1, 2,\ldots,8$). As soon as the trajectory escapes from the basin H, the RMSD minimal value will turn out to fall in correspondence, let's say, of basin I whose unperturbed basis set (i.e., $\Phi_{l'}^{o-I}$with $l' = 0, 1, 2,\ldots,8$) is then utilized for PMM.

Step 4 The spectrum of each species is evaluated following the Step 3 procedure, and the relative statistical weight of each species is automatically included.

Step 5 The choice of using a limited number of basins of an arbitrary area will obviously produce a non-continuous spectrum for which the calculation of *all* the points on the essential plane (all the conformations and not only the representative structures) would be required. In order to circumvent this practical drawback and whatever systematic deficiency arising from the arbitrariness of the size of the basin, we extracted four different structures in each of three different basins of TT simulation (namely: four structures in basin L, four in basin N and four in basin C). In each basin, we extracted the structures showing the largest root mean square deviation (which, however, as previously stated, does not exceed 1.0 Angstrom). On these structures, we calculated, in vacuum, the absorption frequency at the same level of theory then obtaining the corresponding standard deviation for each conformational basin. In this way, assuming no relevant variation for the standard deviation due to solvent effects and considering that for each basin, the corresponding dispersion was almost the same, we estimated the proper frequency standard deviation to be used for all the representative structures of the related basins as σ^2 equal to 3.1×10^{-6} a.u.^{2}. This value was then utilized to reconstruct the frequency fluctuation in each conformational basin via a Gaussian distribution centered at the basin mean frequency. This has been accomplished, in practice, by multiplying in each frequency bin, the spectrum value as obtained by steps 3 and 4 by such a Gaussian distribution and the bin size itself.

The S_0–S_1 spectra coming out from steps 1–4 are shown in Fig. 9 for TT, QT, and DQT. Note that in this case, the S_0–S_2 transition has been not included as producing a signal negligibly small.

Table 1 PMM system

SPECIES	QC	PERTURBATION
TT	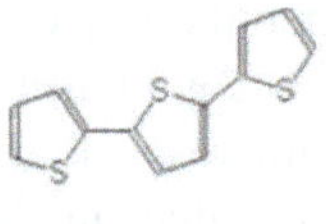	Solvent
QT	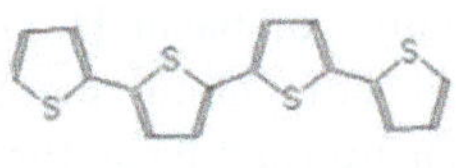	Solvent
DQT	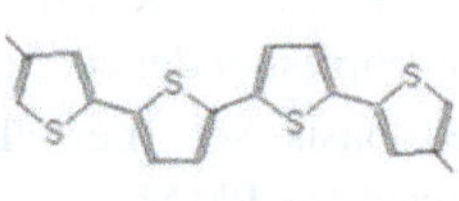	Solvent and 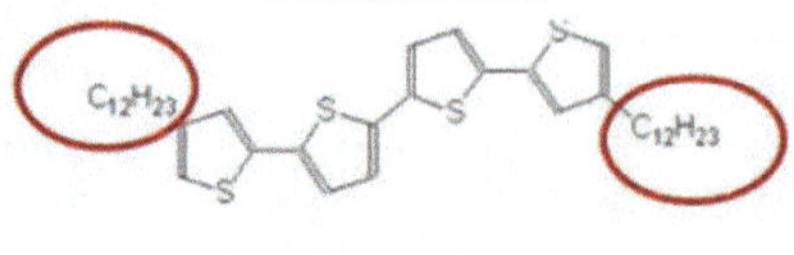
DET	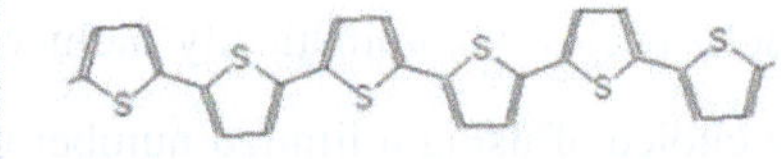	Solvent and 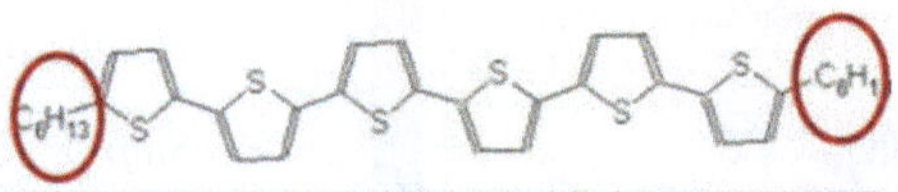

QC stands for the region explicitly treated at electronic level. Perturbation stands for the portion of the simulated system acting as an electrostatic perturbation on the QC. We wish to further remark that in the case of DQT and DET the substituent alky chains (i.e. $C_{12}H_{23}$ and C_6H_{13}) have been substituted in the unperturbed calculation with CH_3 groups in the same positions. Hence the perturbation is formally due to $C_{11}H_{20}$ and C_5H_{10}, respectively

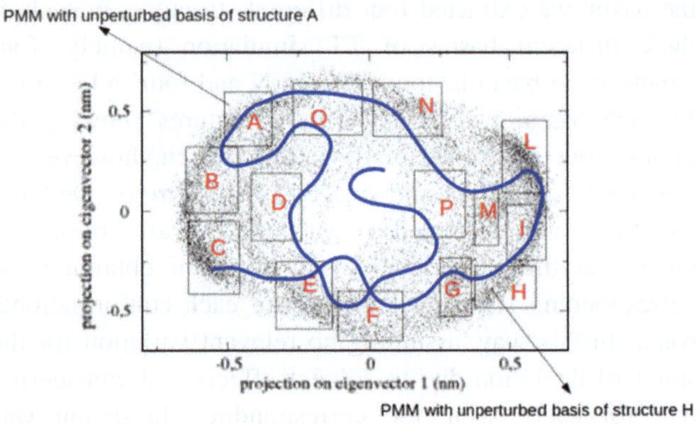

Fig. 8 Schematic representation of the time evolution of the trajectory onto the essential plane

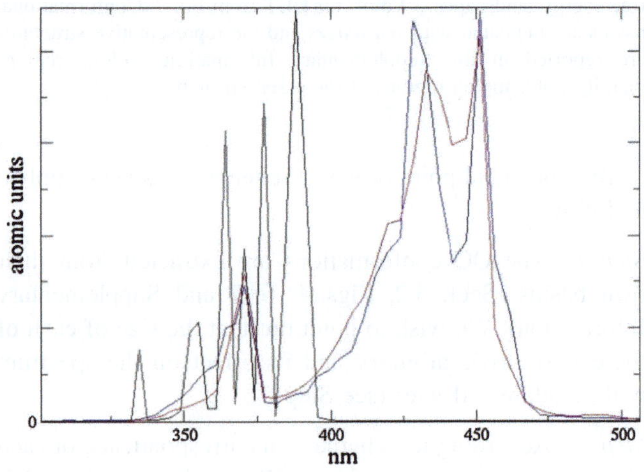

Fig. 9 Superposition of S_0–S_1 transitions (spectrum) resulting from steps 1–4 for TT (*black*), QT (*blue*), and DQT (*red*)

The spectra resulting from the steps 1–5 are then reported in the Fig. 10a–c for TT, QT, and DQT for the first and second electronic transition.

Note also that S_0–S_2 have been reported only for TT and QT as we did not appreciate any difference for this signal in the case of QT and DQT.

In Fig. 11, we have finally reported the spectrum for DET, compared with the one of QT.

As already remarked in the Theoretical section, the use of the first-order approximation in the construction of the electronic perturbed Hamiltonian (Eq. 4), utilized for the spectra previously shown, may produce artifacts and unrealistic electrostatic QC-environment interactions in particular when long-shaped QCs are concerned. For this reason, we have evaluated the spectral features also using the second-order approximation (Eqs. 1–3). This has been carried out for all the systems and shown for the largest one, i.e., DET.

The result, reported in Fig. 12, shows that in the case of DET, the presence and the absence of the second-order

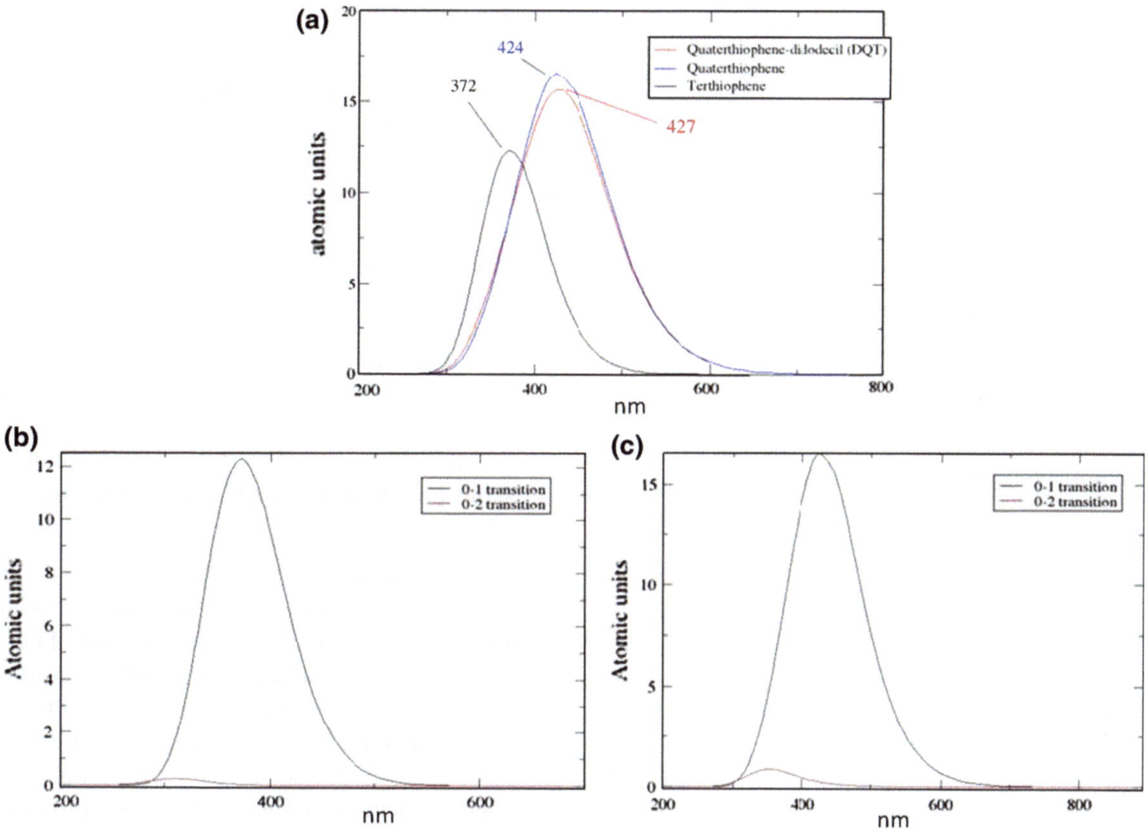

Fig. 10 **a** Comparison of 0–1 transition for TT, DQT, and QT. **b** Comparison of S_0–S_1 and S_0–S_2 transition for TT. **c** Comparison of S_0–S_1 and S_0–S_2 transition for QT

term in the operator do not appreciably alter the result. The strong similarity between all the reported points, probably due to the relatively large value of the ground-state dipole confining within the noise the effect of the quadrupole-electric field gradient interaction, clearly suggests that the description up to the dipolar interaction is sufficient in this case in the species addressed in this work.

4 Discussion

The previously described computational and experimental results have been summarized in the following Table 2.

As already remarked in the Introduction, the above data do not include two important aspects: the basis set for Perturbed Hamiltonian is not based on multi-reference calculations and the vibronic effects are not included. It is interesting to observe that despite the above limitations, the experimental sequence for S_0–S_1 transition, i.e., from TT to DET, as well as the experimental bandshapes are both rather well reproduced. The only severe discrepancy arising from the comparison between calculated and experimental spectra is the systematic under-estimated od the simulated spectral maxima. This effect, having already excluded systematic deficiencies of the first-order definition of the Hamiltonian operator (Fig. 12), might be principally ascribed to the already cited limitations of TD-DFT unperturbed calculations. It follows that other important plausible source of error, i.e., the lack of vibronic coupling, seems not to dramatically alter in this case the final picture. This might be partially ascribed to the fact that the large spectral broadening induced by conformational transitions of the chromophore in solution, i.e., by excitation of classical internal degrees of freedom and inhomogeneous interaction with solvent electric field, almost completely overwhelms the vibronic fine structure. This is obviously a supposition which deserves more accurate deep investigation. Hence, well aware of the previously remarked intrinsic limitations of the computational setup, let us illustrate some key points emerging from our results.

First of all, we wish to remark the presence of the S_0–S_2 signal red shifted with respect the S_0–S_1 one. This result, very difficult to be proved in solution, nicely fits early experimental observations carried out for OT with 2 up to 5 monomers [5].

The most interesting aspect emerging from our study resides in the importance of the inclusion of different conformations in the evaluation of the spectral UV

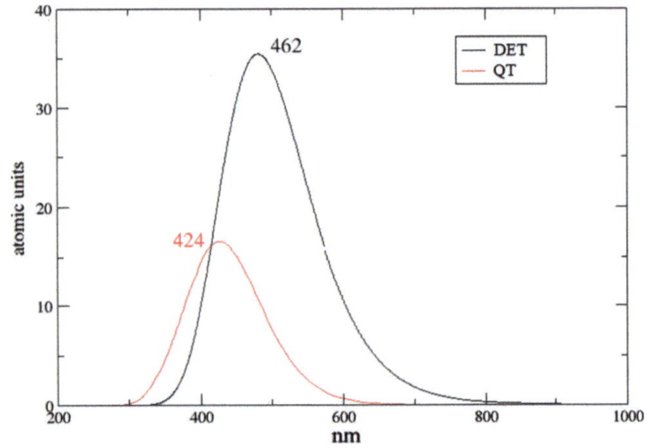

Fig. 11 Comparison of S_0–S_1 transition for DET and QT

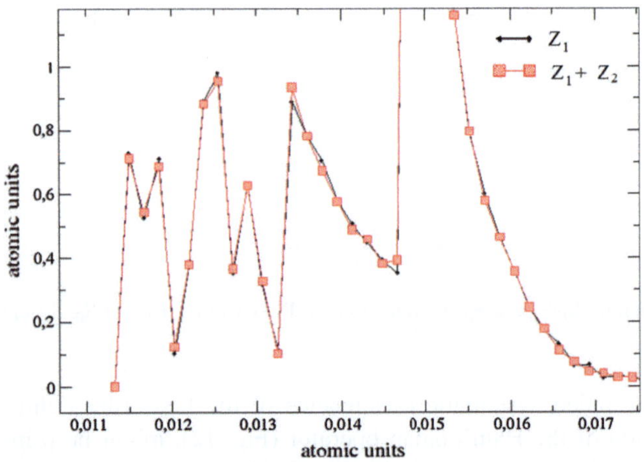

Fig. 12 Comparison between DET spectrum (without application of step 5) within first order, Z_1 (*black diamonds*, Eq. 4) and second order, $Z_1 + Z_2$ (*red squares*, Eq. 1) definition of the perturbed Hamiltonian operator. Note that the linear (energy) scale has been used in this case and not the wavelength

transition in solution which can be shortly summarized in Fig. 13 where we have reported, for each species (free energy basin) the excitation energies experienced during the simulation and the excitation energy in vacuum.

The above findings clearly indicate that:

- we do not observe any red spot (i.e., the value of 0–1 transition in vacuum) outside the black regions (the distribution of the 0–1 transitions perturbed along MD simulation).
- the black regions span a rather large energy range, in some cases (i.e., for some OT conformations) *very* large (more than 100 nm).
- The red spots may vary even in a relevant fashion when passing from one conformer to the other one.

This implies that:

- the maximum of the absorbance (roughly speaking the average absorption wavelength) is not dramatically influenced by the solvent (or more in general by the electrostatic perturbation).
- the solvent (or more in general the electrostatic perturbation) does have a strong effect on the *shape* of the spectrum resulting in sharp enlargements of the absorbing range even on a single conformation immersed in the fluctuating solvent.
- Different conformations of the same molecule may not only result in a different (sometimes *very* different) ground- to excited-state energy (shift of the red spots) but also in a very different *response* to the electrostatic perturbation (black spots distributed over long 'chains').

As a consequence of these finding, it turns out that when one has to model the OT electronic properties in condensed state and in thermal conditions, i.e., the typical OT working conditions, it is important to carefully consider all the conformations sampled in solution and also their interaction with the solvent. As a matter of fact, different OT conformations in interaction with the solvent do actually represent molecular systems with very different photophysical properties and, at least in principle, an external action leading to the alteration of the OT conformational equilibrium may produce dramatic alterations of the overall electronic properties.

In order to provide further insights into the OT photophyics and how it is influenced by thermal and environmental effects, we have more carefully investigated the changes in the charge distribution upon OT excitation. This has been accomplished by evaluating the *perturbed* electric dipole (Eq. 6) of the OT first excited electronic state, $\boldsymbol{\mu}_{\text{first excited}}$ and its projection the *perturbed* electric dipole of OT ground state, $\boldsymbol{\mu}_{\text{ground}}$. We indicate such a projection as $\boldsymbol{\mu}_{\text{ground}} \cdot \boldsymbol{\mu}_{\text{firstexcited}}$ It is first important to remark that $\boldsymbol{\mu}_{\text{ground}}$ undergoes a scarce fluctuation in solution. As a matter of fact (see Supplementary Information), its projection along the simulation onto the unperturbed ground-state dipole produce a value essentially always equal to one. Hence, a value of $\boldsymbol{\mu}_{\text{ground}} \cdot \boldsymbol{\mu}_{\text{firstexcited}}$ very different to 1 would imply an intra-molecular (intra OT moiety) charge transfer, i.e., a substantial reorganization of the perturbed electronic distribution upon excitation.

The resulting picture is reported in Fig. 14 where we have correlated $\boldsymbol{\mu}_{\text{ground}} \cdot \boldsymbol{\mu}_{\text{firstexcited}}$ values with the corresponding excitation energy along the overall trajectory.

In this Figure, a projection equal to one for a given excitation wavelength implies that upon such a ground- to first-excited-state vertical excitation, the OT first-excited-state

Table 2 Comparison of the calculated and experimental absorption maxima

Species	Calculated S_0–S_1 absorbance maximum (nm) ($\Delta\lambda$ with respect to TT)	Experimental S_0–S_1 absorbance maximum (nm) ($\Delta\lambda$ with respect to TT)	Calculated S_0–S_1 half maximum width (nm)	Experimental S_0–S_1 half maximum width (nm)
TT	372 ± 2 (0)	355 (0)	90	80
DQT	427 ± 2 (+55)	398 (+43)	110	100
QT	424 ± 2 (+52)	393 (+38)	120	100
DET	462 ± 2 (+90)	415 (+60)	100	100

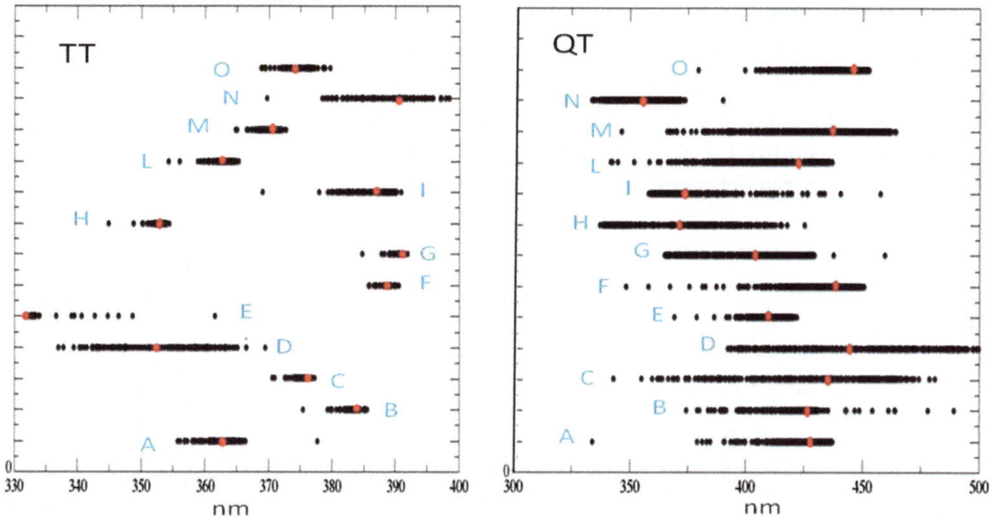

Fig. 13 Gas-phase (*red spots*) and solvent-perturbed (*black spots*) S_0–S_1 transition wavelengths along TT and QT simulations for each free energy basin (representative conformations). Note that basin P (Fig. 5) for TT has not been reported as scarcely sampled

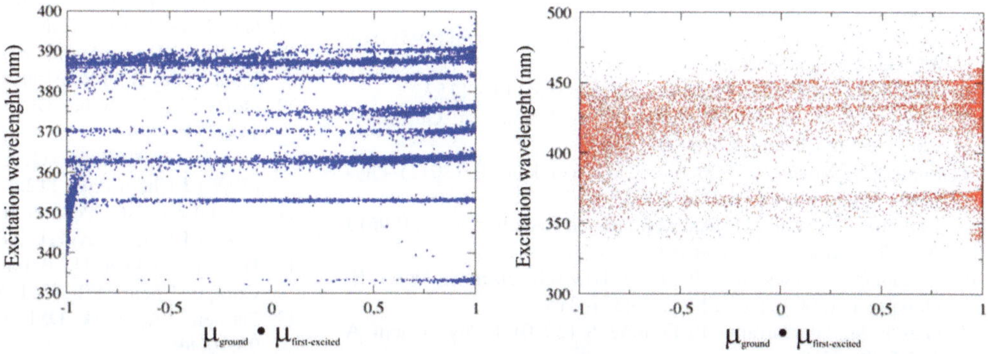

Fig. 14 Correlation between electronic energy excitation and charge fluctuation in the first excited state of TT (*left panel*) and QT (*right panel*). Value of the abscissa equal to 1

dipole moment—i.e., OT first-excited-state electronic density—does not undergo appreciable variation. Consequently, all the other values of the projection indicate intramolecular charge fluctuation upon the specific excitation. Our result shows that when OT vertically accesses the S_1 electronic state the occurrence of intra-molecular charge transfer strongly depends both on the instantaneous field exerted by the solvent on the QC and on the QC geometry itself. As a matter of fact, at a given value of excitation wavelength, the electronic response of OT moiety ($\mu_{\text{ground}} \cdot \mu_{\text{firstexcited}}$) does span the whole domain from -1 to 1. However, e.g., in the case of TT depicted in the left-side panel of Fig. 14, it is possible to extract narrow photon energies between 340 and 350 nm and between 355 and 360 nm corresponding to $\mu_{\text{ground}} \cdot \mu_{\text{firstexcited}}$ values equal to -1, i.e., indicating an intramolecular charge transfer.

Such a result, somewhat speculative, might provide a good starting point for modeling the possibility of designing perturbing field for controlling the possibility of intramolecular electron transfer in complex environments [50].

5 Conclusions

The S_0–S_1 and S_0–S_2 electronic excitations for a series of polycyclic thiophene oligomers (OT), extensively used in organic electronics and solar cells, have been studied experimentally and theoretically in dilute dichloromethane solution. It emerges that the electronic properties of the OT low-lying excited states might result very different when different conformations, even energetically very close, are sampled in solution at 300 K. This finding confirms that the thermal and environmental effects, beyond an advanced description of the electronic excited states and the inclusion of the vibronic effects, are very important for better understanding the photophysics of these systems in liquid solutions. Further aspects, essentially related to the characterization of OT low-lying excited states in solution, also indicate that a subtle combination of solvent electric field and OT conformational transition might induce intramolecular electron transfer and virtually open the possibility of designing a proper OT-environment condition for controlling electron transfer processes.

References

1. Horowitz G, Garnier F, Yassar A, Hajlaoui R, Kouki F (1996) Adv Mater 8:52
2. Pal AJ, Ostergard T, Paloeimo J, Stubb H (1997) Phys Rev B Condens Matter 55:1306–1309
3. Alem S, Pandey AK, Unni KNN, Nuzi JM, Blanchard P (2006) J Vac Sci Technol A 24:645–648
4. Taliani C, Gebauer W (1999) In: Fichou D (ed) Handbook of oligo and polythiophenes. Wiley, Weinheim
5. Becker RS, Seixas de Melo J, Macanita AL, Elisei F (1995) Pure & Appl Chem 67:9–16
6. Salzner U (2007) J Chem Theory Comput 3:1143–1157
7. Fabiano E, Della Sala F, Cingolani R, Weimer M, Gurling A (2005) J Phys Chem A 109:3078–3085
8. Siegert S, Vogelert F, Marian CM, Weinkauf R (2011) Phys Chem Chem Phys 13:10350–10363
9. Beljonne D, Cornil J, Friend RH, Janssen RAJ, Bredas L (1996) J Am Chem Soc 118:6453–6461
10. Giershner J, Mack HG, Egelhaaf HJ, Schweizer S, Doser B, Oelkrug D (2003) Synth Met 138:311–315
11. Garcia M, Guadarrama P, Fomine S (2010) J Phys Chem A 114:5406–5413
12. Fujitsuka M, Nakatani T, Sakamoto M, Sugimoto A, Majima T (2010) J Phys Chem A 114:10789–10794
13. Colditz R, Grebner D, Helbig M, Rentsch S (1995) J Phys Chem 201:309–320
14. Telesca R, Bolink H, Yunoki S, Hadziioannou G, Th. Van Duijen P, Snijders JG, Jonkman HT, Sawatzky GA (2011) Phys Rev B Condens Matter 63:155112
15. Della Sala F, Heinze HH, Gorling A (2001) Chem Phys Lett 339:343–350
16. Rubio M, Merchan M, Pou-Amerigo R, Orti E (2003) ChemPhysChem 4:1308–1315
17. Rubio M, Merchan M, Orti E (2005) ChemPhysChem 6:1357–1368
18. Risko C, McGehee MD, Bredas JL (2011) Chem Sci 2:1200–1218
19. Macchi G, Miliàn Medina M, Zambianchi M, Tubino R, Cornil J, Barbarella G, Giershner J, Meinardi F (2009) Phys Chem Chem Phys 11:984–990
20. Milian Medina B, Wassenberg D, Meskers SCJ, Janssen RAJ, Mena-Osterit E, Bauerle P, Grierchner J (2008) J Phys Chem A 112:13282–13286
21. Gierschner J, Cornil J, Egelhaaf HJ (2007) Adv Mater 19:173–191
22. Aleman C, Julià L (1996) J Phys Chem 100:1524
23. Aleman C, Julià L (1996) J Phys Chem 100:14661
24. Roncali J (1992) Chem Rev 92:711
25. Barbarella G, Bongini G, Zambianchi M (1994) Macromolecules 27:3039
26. Egelhaaf HJ, Gierschner J, Oelkrug D (2002) Synth Met 127: 221–227
27. Barbarella G, Bongini G, Zambianchi M, Antolini L (1992) Adv Mater 4:282
28. Amadei A, D'Alessandro M, D'Abramo M, Aschi M (2009) J Chem Phys 130(8):08410
29. Amadei A, D'Abramo M, Zazza C, Aschi M (2003) Chem Phys Lett 381:187
30. Zazza C, Amadei A, Sanna N, Grandi A, Chillemi G, Di Nola A, D'Abramo M, Aschi M (2006) Phys Chem Chem Phys 8:1385
31. Aschi M, Fontana A, Di Meo E, Zazza C, Amadei A (2010) J Phys Chem B 114(5):1915–1924
32. Improta R, Barone V (2004) J Am Chem Soc 126:14320–14321
33. Marini A, Munoz-Losa A, Biancardi A, Mennucci B (2010) J Phys Chem B 114:17128–17135
34. Amadei A, Marinelli F, D'Abramo M, D'Alessandro M, Anselmi M, Di Nola A, Aschi M (2005) J Chem Phys 122:124506
35. Parac M, Grimme S (2003) Chem Phys 292:11
36. Gross EKU, Dobson JF, Petersilka M (1996) In: Nalewajski RF (ed) Density functional theory II, Springer series in topics in current chemistry, band 181. Springer, Heidelberg
37. Gao J (1996) Rev Comp Chem 7:119–185
38. Moreno M, Casalegno M, Raos G, Meille SV, Po' R (2010) J Phys Chem B 114(4):1591–1602
39. Duffy EM, Severance DL, Jorgensen WL (1992) J Am Chem Soc 114:7535–7542
40. Amadei A, Chillemi G, Ceruso MA, Grottesi A, Di Nola A (2000) J Chem Phys 112:9–23
41. Berendsen HJC, Postma JPM, van Gunsteren WF, Di Nola A (1984) J Chem Phys 81:3684–3690
42. Hess B, Bekker H, Berendsen HJC, Fraaije JGEM (1997) J Comput Chem 18:1463–1472
43. Darden TA, York DM, Pedersen LG (1993) J Chem Phys 98:10089
44. Van der Spoel D, van Buren AR, Apol E, Meulenhoff PJ, Tieleman DP, Sijbers ALTM, van Drunen R, Berendsen HJC (1996) Gromacs user manual version 1.3
45. Amadei A, Linssen ABM, Berendsen HJC (1993) Proteins Struct Funct Genet 17:412–425
46. Becke AD (1993) J Chem Phys 98(2):1372–1377
47. Lee C, Yang W, Parr RG (1998) Phys Rev B 37(2):785–789
48. GAUSSIAN Frish MJ et al (2004) Gaussian 03, Revision C.02 Gaussian Inc. Wallingford, CT
49. Yang L, Feng JK, Ren AM, Sun JZ (2006) Polymer 47:1397–1404
50. Zhao J (2002) Chem Phys Lett 351:481–485

Theor Chem Acc (2012) 131:1179
DOI 10.1007/s00214-012-1179-x

REGULAR ARTICLE

Improving the study of proton transfers between amino acid side chains in solution: choosing appropriate DFT functionals and avoiding hidden pitfalls

Pedro J. Silva · Marta A. S. Perez · Natércia F. Brás · Pedro A. Fernandes · M. J. Ramos

Please note an erratum was published subsequent to publication of the original article. For reasons of consistency with the electronic record, the article and the corresponding erratum are reproduced here as published in the journal.

Received: 24 September 2011 / Accepted: 2 November 2011 / Published online: 2 March 2012

Abstract We have studied the influence of implicit solvent models, inclusion of explicit water molecules, inclusion of vibrational effects, and density functionals on the quality of the predicted pK_a of small amino acid side chain models. We found that the inclusion of vibrational effects and explicit water molecules is crucial to improve the correlation between the computed and the experimental values. In these micro-solvated systems, the best agreement between DFT-computed electronic energies and benchmark values is afforded by BHHLYP and B97-2. However, approaching experimental results requires the addition of more than three explicit water molecules, which generates new problems related to the presence of multiple minima in the potential energy surface. It thus appears that a satisfactory ab initio prediction of amino acid side chain pK_a will require methods that sample the configurational space in the presence of large solvation shells, while at the same time computing vibrational contributions to the enthalpy and entropy of the system under study in all points of that surface. Pending development of efficient algorithms for those computations, we strongly suggest that whenever counterintuitive protonation states are found in a computational study (e.g., the presence of a neutral aspartate/neutral histidine dyad instead of a deprotonated aspartate/protonated histidine pair), the reaction profile should be computed under each of the different protonation microstates by constraining the relevant N–H or O–H bonds, in order to avoid artifacts inherent to the complex nature of the factors contributing to the pK_a.

Keywords DFT · Benchmarking · Acid/base reactions · Amino acid side chains

1 Introduction

The ubiquitous presence of protonation/deprotonation steps in the catalytic mechanisms of organic and biochemical reactions is responsible for the experimentally observed large effects of pH on many reaction rates. A good theoretical description of these reaction mechanisms, therefore, depends on the ability to predict accurately the solution pK_a of the reacting functional groups. Several protocols for computational determination of pK_a have been developed by several authors [1–6]. They all rely on the determination of gas phase reaction energies, followed by computation of solvation energies of the basic and acidic forms of the molecular species under study using implicit solvent models [1–3, 5, 6] or the Poisson equation [4], in the presence [2, 3, 5, 6] or absence [1, 4, 6] of a few explicit water molecules. The computed energy values are then converted to pK_a using appropriate thermodynamic cycles.

Dedicated to Professor Vincenzo Barone and published as part of the special collection of articles celebrating his 60th birthday.

Electronic supplementary material The online version of this article (doi:10.1007/s00214-012-1179-x) contains supplementary material, which is available to authorized users.

P. J. Silva
REQUIMTE/Faculdade de Ciências da Saúde, Universidade Fernando Pessoa, Rua Carlos da Maia, 296, 4200-150 Porto, Portugal

M. A. S. Perez · N. F. Brás · P. A. Fernandes · M. J. Ramos (✉)
REQUIMTE/Departamento de Química e Bioquímica, Faculdade de Ciências, Universidade do Porto, Rua do Campo Alegre, s/n, 4169-007 Porto, Portugal
e-mail: mjramos@fc.up.pt

Clearly, the quality of the results obtained critically depends on the accuracy of the individual computations performed in each step of the protocol, and this is reflected, for example, on the different correlations between computed and experimental values obtained for different classes of acids [7]. We have recently performed a thorough benchmarking study of the ability of 62 density functionals to accurately describe the proton affinity of large amino acid models [8]. However, to the best of our knowledge, no systematic study of the influence of each factor on the quality of the predicted pK_a values has been performed to date.

In this report, we describe the influence of: the implicit solvent model, inclusion of explicit water molecules, neglect of vibrational effects, and density functional used on the quality of predicted pK_a. We found that the inclusion of vibrational effects and explicit water molecules is crucial to improve the correlation between the computed and the experimental values. However, achieving convergence of the results requires the addition of too many explicit water molecules, which introduce new problems related to the presence of multiple minima in the potential energy surface. A satisfactory resolution of this problem will likely require ab initio methods capable of simultaneously addressing the presence of large solvation shells and the accurate computation of enthalpic/entropic effects.

2 Computational methods

The geometries of every molecule described were optimized at the MP2 level and with each of the tested density functionals. Autogenerated delocalized coordinates [9] were used in geometry optimizations performed with a medium-sized basis set, 6-31 + G (d), as increasing the basis sets to triple-ζ quality gives very small additional corrections to the geometries while dramatically increasing the computational cost [10–12]. Accurate DFT energies of the optimized geometries obtained with each density functional were then calculated using the same functional with several triple-ζ quality basis sets: 6-311 + G (d, p), 6-311 + G (2d, p), 6-311 + G (2d, 2p), and 6-311 + G (3d, 2p). We used 18 functionals in total—three GGA functionals (PBE96 [13, 14], PBEPW91 [13, 14] and PW91 [13]), eight hybrid-GGA functionals (B3LYP [15–17], B3PW91 [13, 15], B97-1 [18], B97-2 [19], BHHLYP (50% HF exchange + 50% B88 [20] exchange, with LYP correlation), PBE0 [21], PBE1PW91 [13, 14], and X3LYP [22]), three meta-GGA functionals (TPSS [23, 24], TPSSm [25], and M06-L [28]), and four meta-hybrid-GGA functionals (TPSSh [26], M06 [27], M06-2X [27], and M06-HF [29]). Functionals with dispersion corrections have been tried elsewhere to describe the proton affinity of large amino acid models [8] and found to generate very small dispersion effects. Computations involving CCSD (T), the B97-1, -2 and the M06 and TPSS families of functionals were performed with Gamess (US) [30]. All other computations were performed with the Firefly [31] quantum chemistry package. CCSD (T) and MP2 single-point energies were computed on the MP2 geometries using 6-31 + G (d), aug-cc-pVDZ and -pVTZ basis sets and extrapolated to the complete basis set limit as described by Schwenke [32] [for CCSD (T) level] or Truhlar [33, 34] (at the MP2 level). Comparison of these values to experimental free energies of protonation required the inclusion of the contribution of H^+ to the Gibbs free energy of the protonation reactions ($G_{gas,H+}= -6.28$ kcal mol^{-1} at 298 K and 1 atm) and the evaluation of zero-point and thermal effects on these geometries (computed from the MP2/6-31 + G (d) frequencies using a scaling factor of 0.967 [35]).

Since the computation of gas phase Gibbs free energies through any computational method entails the determination of the electronic energies and zero-point/vibrational energies (ZPVE) of each reactant and each product, the performance of a functional depends not only on the accuracy of the computed electronic energies but also on the quality of the vibrational frequencies provided by the method. In this work, we decided to focus on the electronic reaction energies *only*, to prevent spurious results arising from mutual cancellation of errors in energy/ZPVE, and because no ZPVE scaling factors are available for many of the method/basis set combinations tested. Basicities were computed as E_{acid}–E_{base}. DFT solvation energies were computed using the polarizable continuum model [36] implemented in each package [D-PCM [37, 38] in Firefly and C-PCM [39, 40] in Gamess (US)]. Average DFT solvation energies were used as estimates of MP2 solvation energies. For the solvation free energy of H^+, $G_{solv,H+}$, we used the value of -265.9 kcal mol^{-1}, obtained by converting the experimental value of -263.98 kcal mol^{-1} (Tissandier et al. [41]) to the appropriate thermodynamic standard state conventions as recommended by Kelly et al. [42].

Guanidine was selected as model for arginine acidities, and aspartate/glutamate and lysine were modeled as acetate and methylamine (respectively). Methanethiol was chosen to represent the cysteine side chain and phenol as tyrosine model. Histidine was modeled as methylimidazole. Both δ- and ε-deprotonated forms of methylimidazole were computed; all methods predicted the δ-deprotonated form to lie slightly lower in energy than the ε-deprotonated form. The average value for these protonations was used in all method comparisons, to avoid lending more weight to the histidine protonation versus other amino acids.

3 Results and discussion

3.1 Benchmarking selected density functionals in the gas phase

The energies of the selected set of model reactions range from −223.6 to −364.1 kcal mol^{-1} at the CCSD (T)/CBS//MP2/6-31 + G (d) level (Table 1). Energies predicted by MP2/CBS lie a few kcal mol^{-1} above CCSD (T)/CBS in all instances. Both theory levels agree very well with experimental results (Table 1) and can, therefore, be used as appropriate benchmarks for the density-functional theory computations. Interestingly, MP2/CBS agrees the most with experiment in the reactions involving a protonated and a neutral species, whereas CCSD (T) outperforms MP2/CBS in reactions involving a neutral and an anionic species.

Density-functional theory generally afforded geometries in very good agreement with MP2 geometries (root-mean-squared differences were most often below 0.02 Å). Large differences between functionals were only observed for the protonated arginine model: TPSS, TPSSm, PBE96, PBEPW91, and PW91 agree almost perfectly with the MP2 geometry (RMSD <0.015 Å), whereas BHHLYP places the hydrogens closer to the plane defined by the heavier atoms than the MP2 reference (RMSD = 0.077 Å); other functionals show RMSD versus MP2 between 0.02 and 0.05 Å. All tested functionals correctly predict the relative ordering of the gas phase reaction energies in this test set as dictated by experiment and CCSD (T)/CBS (with the exception of PBE96, PBEPW91, and PW91, which erroneously predict the protonation of the tyrosine model to be marginally less favorable than that of the aspartate model). The reaction energies obtained with the DFT functionals are generally overestimated relative to the CCSD (T)/CBS energies; B97-2 and BHHLYP are the major exceptions to this trend, as they underestimate all protonation energies (see Supporting Information).

3.2 Computing basicities in solution

Gas phase benchmark results are not directly applicable to the computation of reaction energies in solution because the amount of stabilization provided by the solvent depends on the amount of delocalized charge in the tested molecule. We, therefore, computed the reaction energies in solution for each of the reaction with every density functional, using the polarizable continuum model. As expected from simple charge considerations, the inclusion of solvent strongly favored the protonation of neutral species and disfavored the protonation of charged (anionic) species and, therefore, dramatically reduced the large spread of protonation energies, from −220/−360 kcal mol^{-1} in gas phase to −280/−306 kcal mol^{-1} in solution. However, the computed values correlate very poorly (or not at all) with the experimental trends: for example, aspartate is still predicted to be more basic than histidine (though much less so than in gas phase), irrespective of the polarizable continuum model used (either C-PCM [39, 40] or D-PCM [36] using an escaped charge compensation scheme [37, 38]) (Fig. 1, upper panel). Inclusion of zero-point and thermal effects does not improve these results.

Such poor performance of the PCM models conflicts with the usually good estimates of solvation energies they are able to provide [45]. Further analysis showed that this problem arises from the neglect of solvent-dependent stabilizing effects, since the inclusion of two explicit water molecules to the MP2 optimization of every system (or three water molecules in the case of the lysine and arginine models) greatly improves the correlation with experiment, though *only* upon addition of zero-point and thermal effects. Under these conditions, the correlation between computed and experimental values increases considerably (especially with the D-PCM solvation model) but still remains far below the desirable accuracy (Fig. 1, lower panel). Our attempts at systematic improvement of these results by expanding the micro-solvation spheres were

Table 1 Computed gas phase protonation energies of the amino acid side chain models

	Modeled amino acid	MP2/CBS	CCSD (T)/CBS	Experimental value of the model molecules
1.	Arginine	−239.9 (**−226.6**)	−242.6 (**−229.3**)	−226.9 [43]
2.	Aspartate	−352.6 (**−338.3**)	−356.1 (**−341.8**)	−341.4 ± 1.2 [44]
3.	Cysteine	−360.5 (**−348.3**)	−364.1 (**−351.8**)	−350.6 ± 2.0 [44]
4.	Histidine (δ-protonation)	−234.1 (**−218.4**)	−237.4 (**−221.7**)	−220.1 ± 2.0 [43]
5.	Histidine (ε-protonation)	−234.5 (**−219.6**)	−237.8 (**−222.9**)	
6.	Lysine	−221.6 (**−206.2**)	−223.6 (**−208.2**)	−206.6 ± 0.5 [43]
7.	Tyrosine	−354.4 (**−340.1**)	−358.1 (**−343.8**)	−342.9 ± 1.4 [44]

All values in kcal mol^{-1}. Bolded data in parentheses include zero-point energy, enthalpic and entropic contributions at 298.15 K computed at the MP2/6-31 + G (d) level and can be compared to the experimental ΔG values; conformational sampling was not needed to achieve experimental-quality results

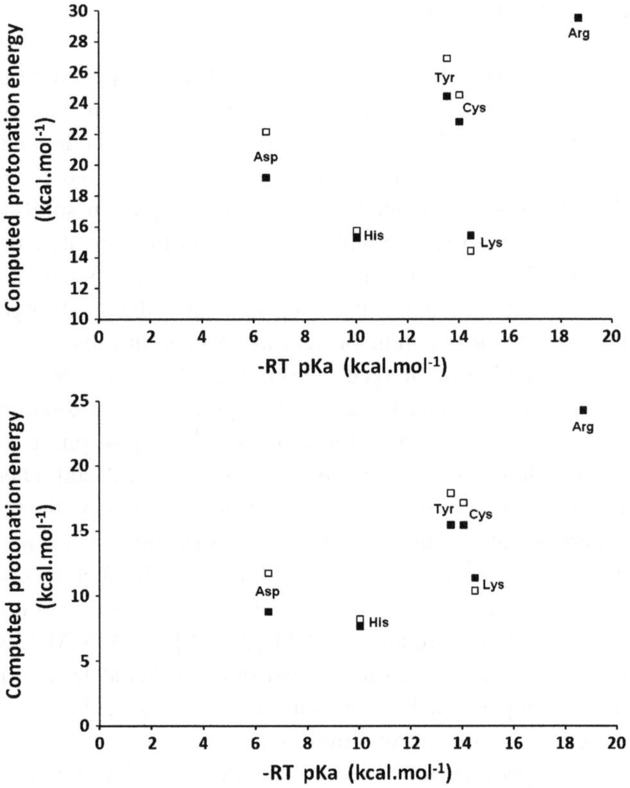

Fig. 1 Computed MP2/CBS protonation energy in solution versus experiment. *Filled square* solvation energies computed with D-PCM. *Open square* solvation energies computed with C-PCM. *Upper panel* without micro-solvation, ZPVE nor thermal effects. *Lower panel* including micro-solvation, ZPVE, and thermal effects

thwarted by the onset of difficulties regarding conformational sampling. This occurs because as the number of water molecules around the solute increases, so does the number of water–water interactions. In these small solute-water clusters, the change in the number of hydrogen bonds between the solute and the water as the solute becomes protonated often causes a dramatic reorganization of the water–water hydrogen bonds in the cluster, leading to large geometric rearrangements between the acidic and basic forms of the solute and misleading energy differences, that depend not only on the protonation energy itself, but also on the different regions of the potential energy surface of the solvent micro-cluster analyzed in each protonation state. The results above clearly show that attaining acceptable accuracy in the ab initio prediction of pK_a is prevented by the need to include a large solvation shell around the solutes, zero-point and vibrational effects throughout the potential energy surface, and sufficient conformational sampling to take into account each of the many local minima present in the potential energy surface. However, at least for biological systems in general, such large ab initio studies still lie beyond the reach of current computational technologies. It is, therefore, useful to identify the density functional(s) that more closely approach the MP2 results in micro-solvated systems for eventual application in DFT-based applications. We compared the MP2 micro-solvation behavior of the amino acid systems with density-functional theory by performing additional geometry optimizations with the explicitly solvated systems with each density functional, followed by computation of high-quality single-point energies in a polarizable continuum model. In all cases, the hydrogen bond lengths remain quite constant among methods, but due to the rotational flexibility of the added water molecules, the agreement between the obtained geometries and the MP2 geometries is substantially lower than observed in the water-free models. Root-mean squared deviations range from 0.17 Å (average value obtained with B3LYP or X3LYP) to 0.39 Å (average value obtained with M06). We found that whereas the precise implementation of the polarizable continuum model (either D- or C-PCM) affected the magnitude of solvation effect on the reaction energies, within each implementation, its effect was mostly independent of the precise geometry of the models (standard deviation of the continuum contribution to the reaction energies <0.7 kcal mol^{-1}). Since the increase in system size caused by the addition of explicit water molecules made CCSD (T) computations impractical, DFT-derived reaction energies were compared (Table 2) with the MP2/CBS benchmark, which has been shown above (Table 1) to provide overall deviations from experiment almost identical to those afforded by CCSD (T). Although most functionals afforded errors above 1.4 kcal mol^{-1} ($\approx$1 pK_a unit), and two (B97-2 and BHHLYP) yielded large errors >5 kcal mol^{-1} (due to the systematic introduction of very similar errors in all reaction energies), three functionals (M06, M06-2X and PW91) yielded results very similar to the MP2 benchmark. Further analysis shows that the surprisingly large errors arise mostly from systematic underestimation (M06-HF, PBE0, PBE96, PW91, PBEPW91, and PBE1PW91) or overestimation (all other functionals, except M06-2X) of reaction energies. Many of these systematic errors cancel when the deprotonation energies of the amino acid models are compared with each other. Table 3 depicts the mean unsigned errors for the *differences* in reaction energies (which we call *relative* reaction energies) for all 6 × 5/2 reactions obtained by coupling any two acid/base pairs tested for each method, with the 6-311 + G (3d, 2p) basis sets. BHHLYP (MUE = 1.20 kcal mol^{-1}) and B97-2 (MUE = 1.35 kcal mol^{-1}) now emerge clearly as the best overall choices (closely followed by TPSSh). Much higher accuracy (MUE <0.4 kcal mol^{-1}) may, however, be obtained in selected cases involving proton transfer between only two amino acid side chains by using other functionals (Supporting information: Table S4).

Table 2 Mean unsigned errors in the computation of reaction energies with micro-solvated models using the 6-311 + G (3d, 2p) basis set

		MUE versus MP2/CBS
PBE96	GGA	2.11 ± 1.90
PBEPW91	GGA	2.01 ± 1.83
PW91	GGA	**1.34 ± 1.13**
B3LYP	Hybrid-GGA	1.91 ± 1.27
B3PW91	Hybrid-GGA	3.62 ± 1.35
B97-1	Hybrid-GGA	2.50 ± 1.43
B97-2	Hybrid-GGA	5.16 ± 1.11
BHHLYP	Hybrid-GGA	5.24 ± 1.12
PBE0	Hybrid-GGA	2.31 ± 1.28
PBE1PW91	Hybrid-GGA	1.96 ± 1.74
X3LYP	Hybrid-GGA	1.56 ± 1.17
TPSS	Meta-GGA	2.16 ± 1.13
TPSSm	Meta-GGA	2.32 ± 1.17
M06	Meta-hybrid-GGA	**1.34 ± 0.89**
M06-2X	Meta-hybrid-GGA	**1.35 ± 1.43**
M06-HF	Meta-hybrid-GGA	2.78 ± 2.35
M06-L	Meta-hybrid-GGA	3.72 ± 1.69
TPSSh	Meta-hybrid-GGA	3.24 ± 1.13

All values in kcal mol^{-1}. ZPVE not included. Error values below 1.5 kcal mol^{-1} are highlighted in bold

Table 3 Mean unsigned errors in the computation of reaction energy *differences* with micro-solvated models using the 6-311 + G (3d, 2p) basis set

		MUE versus MP2/CBS
PBE96	GGA	2.92 ± 1.76
PBEPW91	GGA	2.86 ± 1.73
PW91	GGA	1.97 ± 1.23
B3LYP	Hybrid-GGA	1.52 ± 0.99
B3PW91	Hybrid-GGA	1.65 ± 1.01
B97-1	Hybrid-GGA	1.85 ± 1.21
B97-2	Hybrid-GGA	**1.35 ± 0.83**
BHHLYP	Hybrid-GGA	**1.20 ± 1.06**
PBE0	Hybrid-GGA	2.00 ± 1.25
PBE1PW91	Hybrid-GGA	2.78 ± 1.69
X3LYP	Hybrid-GGA	**1.48 ± 1.00**
TPSS	Meta-GGA	1.52 ± 1.22
TPSSm	Meta-GGA	1.51 ± 1.21
M06	Meta-hybrid-GGA	1.77 ± 1.15
M06-2X	Meta-hybrid-GGA	2.18 ± 1.97
M06-HF	Meta-hybrid-GGA	2.73 ± 1.97
M06-L	Meta-hybrid-GGA	1.98 ± 1.40
TPSSh	Meta-hybrid-GGA	**1.26 ± 1.02**

All values in kcal mol^{-1}. ZPVE not included. Error values below 1.5 kcal mol^{-1} are highlighted in bold

4 Conclusions

It is clear from the results that, although in most cases a very satisfactory agreement between density functional theory and MP2 is found, the accuracy of the computed MP2 protonation energies in solution is still far from perfect, even with explicit micro-solvation of the chemical systems under study (Fig. 1). Preliminary trials revealed that improvement is unlikely to be achieved simply by a small increase in the number of explicit water molecules. Very large increases on the size of the solvent cluster, on the other hand, quickly take the problem out of reach of pure QM methodologies and further compound the problems due to the very large number of possible minima in the potential energy surface. The large effect of the zero-point and thermal effects on the quality of the correlations should also be kept in mind whenever proton transfer between amino acid side chains is studied, as these effects may reverse the computed relative basicities (e.g., using only electronic energies, lysine is predicted to be more acidic than histidine, whereas the inclusion of ZPVE and thermal effects results in the prediction of higher acidity of histidine relative to lysine). Since the computation of these vibrational contributions from the molecular hessian is only possible at stationary points, the full ab initio description of the free energy surface landscape must necessarily involve other methods of estimating entropic contributions at an ab initio level.

These confounding effects are probably at play in a number of theoretical studies that predict abnormal protonation behaviors, such as the predictions of a neutral aspartate/neutral histidine dyad instead of an anionic aspartate/cationic histidine [46–48], the proposal of an anionic histidine/neutral aspartate dyad in the reaction mechanism of Ca^{2+}-dependent phospholipase [49], or the prediction of neutral arginine/aspartic acid pairs instead of arginine/aspartate salt bridges [50]. We strongly suggest that whenever counterintuitive protonation states are found in a computational study, the reaction profile should be computed under the each of the different protonation micro-states by constraining the relevant N–H or O–H bonds. This strategy will provide a fuller understanding of the influence of each specific state of the relevant amino acid dyad on the reaction mechanism and avoid artifacts inherent to the complex nature of the factors contributing to the basicities of amino acid side chains.

References

1. da Silva CO, da Silva EC, Nascimento MAC (1999) J Phys Chem A 103:11194–11199

2. Adam KR (2002) J Phys Chem A 106:11963–11972
3. Pliego JR Jr, Riveros JM (2002) J Phys Chem A 106:7434–7439
4. Schmidt am Busch M, Knapp EW (2004) Chem Phys Chem 5:1513–1524
5. Kelly CP, Cramer CJ, Truhlar DG (2006) J Phys Chem A 110:2493–2499
6. Ho J, Coote ML (2010) Theor Chem Acc 125:3–21
7. Peräkylä M (1999) Phys Chem Chem Phys 1:5643–5647
8. Brás NF, Perez MAS, Fernandes PA, Silva PJ, Ramos MJ J Chem Theor Comp (submitted)
9. Baker J, Kessi A, Delley B (1996) J Chem Phys 105:192
10. Siegbahn PEM, Eriksson L, Himo F, Pavlov M (1998) J Phys Chem B 102:10622
11. Fernandes PA, Ramos MJ (2003) J Am Chem Soc 125:6311
12. Riley KE, Op't Holt BT, Merz KM Jr (2007) J Chem Theory Comput 3:407
13. Perdew JP (1991) Unified theory of exchange and correlation beyond the local density approximation. In: Ziesche P, Eschig H (eds) Electronic structure of solids '91. Akademie Verlag, Berlin, pp 11–20
14. Perdew JP, Burke K, Ernzerhof M (1996) Phys Rev Lett 77:3865
15. Becke AD (1993) J Chem Phys I(98):5648
16. Lee C, Yang W, Parr R (1998) J Phys Rev B 37:785
17. Hertwig RW, Koch W (1995) J Comp Chem 16:576
18. Hamprecht FA, Cohen AJ, Tozer DJ, Handy NC (1998) J Chem Phys 109:6264
19. Wilson PJ, Bradley TJ, Tozer DJ (2001) J Chem Phys 115:9233
20. Becke AD (1988) Phys Rev A 38:3098
21. Adamo C, Barone V (1999) J Chem Phys 110:6158
22. Xu X, Zhang Q, Muller RP, Goddard WA (2005) J Chem Phys 122:014105
23. Perdew JP, Tao J, Staroverov VN, Scuseria GE (2003) Phys Rev Lett 91:146401
24. Perdew JP, Tao J, Staroverov VN, Scuseria GE (2004) J Chem Phys 120:6898
25. Perdew JP, Ruzsinszky A, Tao J, Csonka GI, Scuseria GE (2007) Phys Rev A 76:042506
26. Staroverov VN, Scuseria GE, Tao J, Perdew JP (2003) J Chem Phys 119:12129 (erratum in J Chem Phys 121:11507)
27. Zhao Y, Truhlar DG (2008) Theor Chem Acc 120:215
28. Zhao Y, Truhlar DG (2006) J Chem Phys 125:194101
29. Zhao Y, Truhlar DG (2006) J Phys Chem A 110:13126
30. Schmidt MW, Baldridge KK, Boatz JA, Elbert ST, Gordon MS, Jensen JJ, Koseki S, Matsunaga N, Nguyen KA, Su S, Windus TL, Dupuis M, Montgomery JA (1993) J Comput Chem 14:1347
31. Granovsky AA Firefly version 7.0. http://classic.chem.msu.su/gran/gamess/index.html
32. Schwenke DW (2005) J Chem Phys 122:014107
33. Truhlar DG (1998) Chem Phys Lett 294:45
34. Zhao Y, Truhlar DG (2005) J Phys Chem A 109:6624
35. Merrick JP, Moran D, Radom L (2007) J Phys Chem A 111:11683–11700
36. Tomasi J, Persico M (1994) Chem Rev 94:2027–2094
37. Mennucci B, Tomasi J (1997) J Chem Phys 106:5151–5158
38. Cossi M, Mennucci B, Pitarch J, Tomasi J (1998) J Comput Chem 19:833–846
39. Barone V, Cossi M (1998) J Phys Chem A 102:1995–2001
40. Cossi M, Rega N, Scalmani G, Barone V (2003) J Comput Chem 24:669–681
41. Tissandier MD, Cowen KA, Feng WY, Gundluach E, Cohen MH, Earhart AD, Coe JV, Tuttle TR (1998) J Phys Chem A 102:7787–7794
42. Kelly CP, Cramer CJ, Truhlar DG (2006) J Phys Chem B 110:16066–16081
43. Hunter EP, Lias SG (1998) J Phys Chem Ref Data 27:413–656
44. Linstrom P, Mallard W (eds) (2003) NIST chemistry web book, NIST standard reference database number 69. National Institute of Standards and Technology, Gaithersburg. http://webbook.nist.gov
45. Barone V, Cossi M, Tomasi J (1997) J Chem Phys 107:3210–3221
46. Gómez PC, Pacios LF (2005) Chem Phys Phys Chem 7:1374–1381
47. Wang J, Dong H, Li H, He H (2005) J Phys Chem B 109:18664–18672
48. Schiøtt B (2004) Chem Commun, 498–499
49. Leopoldini M, Russo N, Toscano M (2010) J Phys Chem B 114:11584–11593
50. Nagy PI, Erhardt PW (2010) J Phys Chem B 114:16436–16442

Theor Chem Acc (2013) 132:1307
DOI 10.1007/s00214-012-1307-7

ERRATUM

Erratum to: Improving the study of proton transfers between amino acid side chains in solution: choosing appropriate DFT functionals and avoiding hidden pitfalls

Pedro J. Silva · Marta A. S. Perez · Natércia F. Brás · Pedro A. Fernandes · M. J. Ramos

Published online: 30 December 2012

**Erratum to: Theor Chem Acc (2012) 131:1179
DOI 10.1007/s00214-012-1179-x**

The authors would like to correct an error in the final paragraph of the original publication. The error is an incorrect reference citation to the paper by Nagy and Erhardt. The correct sentence should read as "[…] the prediction of neutral lysine/aspartic acid pairs instead of lysine/aspartate salt bridges [50]". The authors thank Peter Nagy for making them aware of this mistake.

Reference

50. Nagy PI, Erhardt PW (2010) J Phys Chem B 114:16436–16442

The online version of the original article can be found under doi:10.1007/s00214-012-1179-x.

P. J. Silva
REQUIMTE/Faculdade de Ciências da Saúde, Universidade Fernando Pessoa, Rua Carlos da Maia, 296, 4200-150 Porto, Portugal

M. A. S. Perez · N. F. Brás · P. A. Fernandes · M. J. Ramos (✉)
REQUIMTE/Departamento de Química e Bioquímica, Faculdade de Ciências, Universidade de Porto, Rua do Campo Alegre, s/n, 4169-007 Porto, Portugal
e-mail: mjramos@fc.up.pt

Theor Chem Acc (2012) 131:1181
DOI 10.1007/s00214-012-1181-3

REGULAR ARTICLE

Franck–Condon factors in curvilinear coordinates: the photoelectron spectrum of ammonia

Amedeo Capobianco · Raffaele Borrelli · Canio Noce · Andrea Peluso

Received: 1 November 2011 / Accepted: 3 February 2012 / Published online: 2 March 2012

Abstract An approach to the calculation of Franck–Condon factors in curvilinear coordinates is outlined. The approach is based on curvilinear normal coordinates, which allows for an easy extension of Duschinsky's transformation to the case of curvilinear coordinates, and on the power series expansion of the kinetic energy operator. Its usefulness in the case of molecules undergoing large displacements of their equilibrium nuclear configurations upon excitation is then demonstrated by an application to the vibrational structure of the photoelectron spectrum of ammonia, using an anharmonic potential only for the symmetric stretching and bending coordinates of the radical cation.

Keywords Franck–Condon factors · Curvilinear coordinates · Photoelectron spectra

1 Introduction

Ammonia is a very interesting molecule for spectroscopists and theoreticians. In the ground electronic state, it exhibits two distinct equilibrium nuclear configurations that, by interconverting each other along the large amplitude motion, the so-called umbrella motion, cause the splitting of the rovibrational energy levels. Furthermore, ammonia undergoes significantly large displacements of its equilibrium nuclear geometry upon excitation and photoionization, which reflect into broad absorption bands, with a well-resolved vibrational and rovibrational structure, as shown in Fig. 1 for the $\tilde{X}^2A_2'' \leftarrow \tilde{X}^1A_1'$ transition. The analyses of both the vibrational pattern [1] and of the rotational fine structure of some vibronic peaks [2] of the UV absorption band occurring at 217 nm have led to the conclusion that the excited electronic state has a planar equilibrium nuclear configuration, belonging to D_{3h} point group. A very similar vibrational pattern has also been observed in the lower-energy region of its photoelectron spectrum, leading to the conclusion that even the $\tilde{X}^2A_2''$ ionic state is planar [3–5].

Both the splitting of the energy levels of the rovibrational spectrum and the well-resolved vibrational pattern of the electronic and photoelectronic spectra provide important experimental data for theoreticians involved in the application of sophisticated theoretical models for treating strong anharmonic effects [6, 7], and in the development of efficient methodologies for the calculations of spectroscopic band shapes, that is, Franck–Condon (FC) factors. It is indeed well-known that in the case an electronic transition takes place between two electronic states exhibiting a large displacement of their equilibrium positions, the calculation of the FC factors may pose problems, especially when the Cartesian representation of normal modes and the harmonic approximation are adopted [8, 9]. In fact, in rectilinear coordinates, a large displacement along a bending coordinate always implies a motion along the two bond distances of the angular coordinate, Fig. 2, so that on carrying out the projection of the normal modes of one

Dedicated to Professor Vincenzo Barone and published as part of the special collection of articles celebrating his 60th birthday.

A. Capobianco · C. Noce
Dipartimento di Fisica E.R. Caianiello, Università di Salerno, Via Ponte don Melillo, 84084 Fisciano, SA, Italy

R. Borrelli · A. Peluso (✉)
Dipartimento di Chimica e Biologia, Università di Salerno, Via Ponte don Melillo, 84084 Fisciano, SA, Italy
e-mail: apeluso@unisa.it

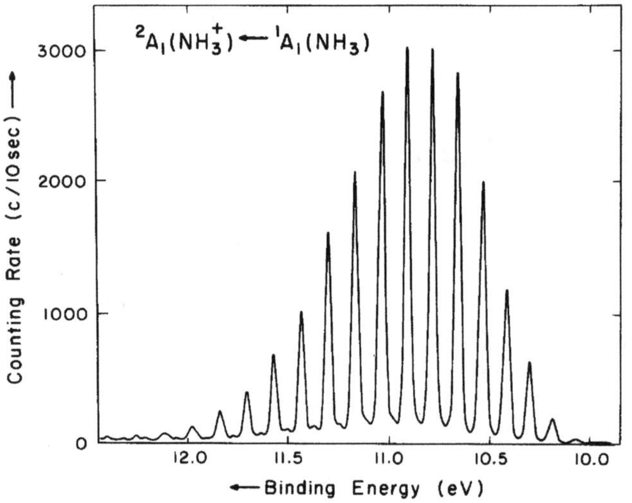

Fig. 1 The lowest energy band of the photoelectron spectrum of ammonia, reproduced with permission from Rabalais et al. [5]

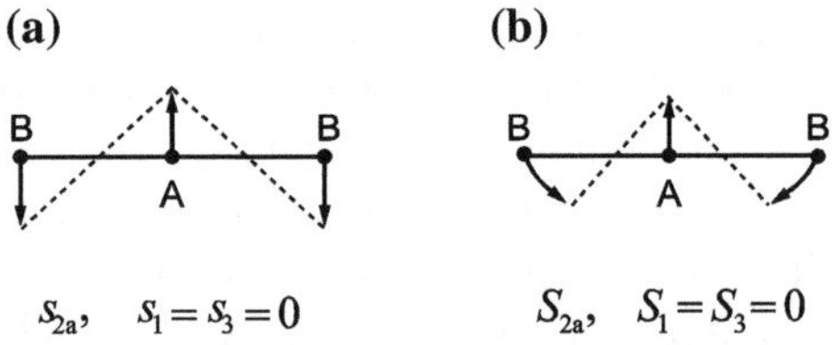

Fig. 2 Representations of the bending motion (at arbitrary low speed) in rectilinear (**a**) and curvilinear (**b**) coordinates for the linear molecule AB_2. *S* denotes curvilinear coordinates, *s* rectilinear ones

electronic state onto those of the other—the Duschinsky's transformation—those stretching modes come out to be significantly displaced, causing the appearance in the computed spectrum of several vibrational progressions, due to excitations of both the large amplitude and the stretching modes, which have no counterparts in the observed spectrum [9, 10].

The most successful attempts to rationalize the photoelectron spectrum of ammonia through the computation of Franck–Condon factors was based on procedures that overcome Duschinsky's transformation, avoiding the use of normal modes of both electronic states [11–14]. Recently, a very satisfying reproduction of the photoelectron spectrum of ammonia, based on Duschinsky's transformation and the Cartesian representation of the normal coordinates, has been obtained by using a two-dimensional approach, including the symmetric stretching and bending coordinates. A high-order polynomial for describing the strong anharmonic couplings between the two modes was necessary for correcting the spurious displacement of the symmetric stretching mode predicted by Duschinsky's transformation in the Cartesian representation of the normal modes [10], which causes the appearance in the computed spectrum of vibrational progressions that are not present in the observed one. The adopted methodology is therefore quite cumbersome, requiring the knowledge of the potential energy of NH_3^+ on a large grid of nuclear coordinates. In the present work, with the aim of setting up an easier and general procedure to compute the FC factors for floppy molecules, those undergoing large displacements of their equilibrium positions upon light excitation, we afford the computation of the ammonia photoelectron spectrum by using curvilinear coordinates, with the hope of relieving the high computational effort of computing anharmonic potential energy terms. Because our main interest is in finding out practical ways for computing FC factors for difficult cases, anharmonic terms in the potential energy expression have been considered only for the two vibrational coordinates, the symmetric stretching and bending, which in the photoelectron spectrum of ammonia play a predominant role.

2 Theoretical approach

Curvilinear internal coordinates such as bond elongations (stretching) or variations of valence angles (bending) are potentially the best suited to treat molecular vibrations [15]. Figure 2 gives the classical description of the bending motion for an AB_2 linear molecule [16], both in linearized internal coordinates (a) and in true curvilinear internal coordinates (b). In curvilinear coordinates, nuclei are allowed to follow curved trajectories, whereas linearized internal coordinates force nuclear motion along straight lines, defined by the projections of the exact internal coordinates onto fixed Cartesian axes. A linearized internal coordinate s_j coincides with the curvilinear one S_j in the limit of infinitesimal displacements of atoms from their equilibrium positions. Therefore, in the range of validity of the harmonic approximation, there is no distinction between linear and curvilinear coordinates, but when large displacements are involved, such as those observed in going from the equilibrium geometry of NH_3 to that of NH_3^+, the two sets of coordinates are no longer equivalent. In the example of Fig. 2, the motion along a linearized bending coordinate involves an elongation of the A-B bond. Even if the potential energy was strictly quadratic in curvilinear coordinates, anharmonic terms would be needed in the rectilinear representation to obtain an equivalent description of the nuclear motion [17]. Of course, both the rectilinear and the curvilinear representation must lead to identical results when an exact treatment of the vibrational problem is carried out, but they yield different results at lower orders of approximation. In particular, the expression of the potential energy in curvilinear coordinates is generally simpler than that obtained by adopting rectilinear coordinates [16–18], but the linear momentum and the kinetic energy operators in curvilinear coordinates are much more involved [19–22].

The quantum vibrational Hamiltonian in curvilinear coordinate is [19]:

$$\hat{H}_S = -\frac{\hbar^2}{2} G^{1/4} \left(\frac{\partial}{\partial \mathbf{S}}\right)^T G^{-1/2} \mathbf{G} \left(\frac{\partial}{\partial \mathbf{S}}\right) G^{1/4} + V, \tag{1}$$

where $\mathbf{S}$ is the column vector of curvilinear internal coordinates, V is the potential energy, $\mathbf{G}$ and G are the metric matrix and its determinant, respectively, both depending on $\mathbf{S}$. The subscript S in the operator indicates normalization with respect to the $\mathbf{S}$ coordinates and the superscript T denotes transposition. By performing the differentiation, Eq. 1 can be rewritten as:

$$\hat{H}_S = -\frac{\hbar^2}{2} \sum_{j,k} \frac{\partial}{\partial S_j} G_{jk}(S) \frac{\partial}{\partial S_k} + V(S) + V_{\text{kin}}(S), \tag{2}$$

where

$$V_{\text{kin}} = \frac{5\hbar^2}{32G^2} \sum_{j,k} G_{jk} \frac{\partial G}{\partial S_j} \frac{\partial G}{\partial S_k} - \frac{\hbar^2}{8G} \sum_{j,k} \frac{\partial G_{jk}}{\partial S_j} \frac{\partial G}{\partial S_k} - \frac{\hbar^2}{8G} \sum_{j,k} G_{jk} \frac{\partial^2 G}{\partial S_j \partial S_k} \tag{3}$$

collects the terms of the kinetic energy operator which depend only on coordinates, thus acting like a potential energy term. The easiest way of taking properly into account the dependence of G_{jk} on $\mathbf{S}$ coordinates is that of expanding them into a power series. That yields:

$$\hat{H}_S = -\frac{\hbar^2}{2} \left(\sum_{j,k} G^0_{jk} \frac{\partial^2}{\partial S_j \partial S_k} + \sum_{j,k,\ell} G'_{jk\ell} \frac{\partial}{\partial S_j} S_\ell \frac{\partial}{\partial S_k} + \frac{1}{2} \sum_{j,k,\ell,m} G''_{jk\ell m} \frac{\partial}{\partial S_j} S_\ell S_m \frac{\partial}{\partial S_k} + \cdots \right) + V_{\text{kin}}(S) + V(S) \tag{4}$$

where

$$G^0_{jk} = G_{jk}(0) \tag{5}$$

$$G'_{jk\ell} = \left(\frac{\partial G_{jk}}{\partial S_\ell} \right)_0 ; \tag{6}$$

$$G''_{jk\ell m} = \left(\frac{\partial^2 G_{jk}}{\partial S_\ell \partial S_m} \right)_0 . \tag{7}$$

The Hamiltonian in Eq. 4 can be cast into a more manageable form by introducing curvilinear normal coordinates [18, 19, 23–26]. First, the so-called linearized internal coordinates, which are linear combinations of Cartesian displacements [15, 27, 28], are introduced:

$$\mathbf{s} = \mathbf{B}_0 \boldsymbol{\sigma}, \tag{8}$$

where $\mathbf{s}$ and $\boldsymbol{\sigma}$ are the column vectors of the linearized internal and the Cartesian displacement coordinates, respectively, and $\mathbf{B}_0$ is the Jacobian matrix whose elements depend only on the equilibrium geometry of the molecule. Normal coordinates $\mathbf{Q}$ are linear combinations of linearized internal coordinates:

$$\mathbf{s} = \mathbf{L}_0 \mathbf{Q}, \tag{9}$$

in which:

$$\mathbf{L}_0 = \frac{\partial \mathbf{s}}{\partial \mathbf{Q}} = \mathbf{B}_0 \mathbf{M}^{-1/2} \mathbf{L}, \tag{10}$$

where $\mathbf{M}$ is the diagonal matrix of atomic masses and $\mathcal{L}$ is the normalized matrix of normal modes in Cartesian coordinates, whose elements are not a function of the coordinates. The linear relationship between $\mathbf{Q}$ and $\mathbf{s}$ ensures that the elements of matrix of the effective masses:

$$\mathbf{G}^0 = \mathbf{L}_0 \mathbf{L}_0^T, \tag{11}$$

do not depend on the coordinates $\mathbf{s}$.

Then, "curvilinear normal coordinates" $\bar{\mathbf{Q}}$, defined as:

$$\mathbf{S} = \mathbf{L}_0 \bar{\mathbf{Q}}, \tag{12}$$

are introduced so that the vibrational Hamiltonian Eq. 2 assumes the form:

$$\hat{H} = -\frac{\hbar^2}{2} \sum_{r,s} \frac{\partial}{\partial \bar{Q}_r} g_{rs}(\bar{Q}) \frac{\partial}{\partial \bar{Q}_s} + V(\bar{Q}) + V_{\text{kin}}(\bar{Q}), \tag{13}$$

where

$$\mathbf{g} = \mathbf{R}^T \mathbf{G} \mathbf{R}; \quad \mathbf{R} = \left(\mathbf{L}_0^{-1}\right)^T. \tag{14}$$

In the limit of infinitesimal vibrational amplitudes, curvilinear internal coordinates coincide with linearized ones; therefore, (Eqs. 9, 12) curvilinear normal coordinates coincide with the linear ones. This in turn implies that (1) $\mathbf{g}^0$, that is, the metric matrix over curvilinear normal coordinates at the zero order of expansion, coincides with the unit matrix and (2) anharmonic terms vanish in the potential energy expressed as a function of $\bar{Q}$. Therefore, for finite amplitudes, a power series expansion provides a kinetic energy operator of the form:

$$\hat{T} = -\frac{\hbar^2}{2} \sum_r \frac{\partial^2}{\partial \bar{Q}_r^2} + \Delta\hat{T} + V_{\text{kin}}. \tag{15}$$

$\Delta\hat{T}$ and V_{kin} being the kinetic energy terms originated by the curvilinear nature of $\bar{Q}$:

$$\Delta\hat{T}\left(\bar{Q}, \frac{\partial}{\partial \bar{Q}}\right) = -\frac{\hbar^2}{2} \left(\sum_{r,s,t} g'_{rst} \frac{\partial}{\partial \bar{Q}_r} \bar{Q}_t \frac{\partial}{\partial \bar{Q}_s} + \frac{1}{2} \sum_{r,s,t,u} g''_{rstu} \frac{\partial}{\partial \bar{Q}_r} \bar{Q}_t \bar{Q}_u \frac{\partial}{\partial \bar{Q}_s} + \cdots \right), \tag{16}$$

where, as before (Eq. 6) [26]:

$$g'_{rst} = \left(\frac{\partial g_{rs}}{\partial \bar{Q}_t}\right)_0 = \sum_{j,k,l} G'_{jk\ell} R_{jr} R_{ks} L^0_{\ell t}; \quad \text{etc.} \tag{17}$$

and a potential energy in the form:

$$V = 2\pi^2 c^2 \sum_r \tilde{\nu}_r^2 \bar{Q}_r^2 + \Delta V(\bar{Q}) \tag{18}$$

where ΔV collects the anharmonic terms of the potential and $\tilde{\nu}_r$ are the harmonic frequencies expressed as wavenumbers. Combining Eqs. 15 and 18, the vibrational Hamiltonian may be written as:

$$\hat{H} = \hat{H}_0 + \Delta\hat{T} + \Delta V + V_{\text{kin}}, \tag{19}$$

where $\hat{H}_0$ is the harmonic Hamiltonian:

$$\hat{H}_0 = -\frac{\hbar^2}{2}\sum_r \frac{\partial}{\partial \bar{Q}_r^2} + 2\pi^2 c^2 \sum_r \tilde{\nu}_r^2 \bar{Q}_r^2, \tag{20}$$

Approximate eigenfunctions of $\hat{H}$ can be obtained by means of perturbation theory [29] or variational method, using the basis set of the eigenfunctions of $\hat{H}_0$.

Although both $\bar{\mathbf{Q}}$ and $\mathbf{S}$ are curvilinear coordinates, they are linearly related (Eq. 12), this is the key-point for the extension of Duschinsky's transformation [30] in curvilinear normal coordinates. Reminding that internal coordinates $\mathbf{S}$ represent displacements from equilibrium positions:

$$\mathbf{S} = \zeta - \zeta_0 = \mathbf{L}_0\bar{\mathbf{Q}} \tag{21}$$

and denoting with a prime one of the electronic states involved in the transition:

$$\zeta = \zeta_0 + \mathbf{L}_0\bar{\mathbf{Q}} \tag{22}$$

$$\zeta = \zeta'_0 + \mathbf{L}'_0\bar{\mathbf{Q}}'. \tag{23}$$

By Eqs. 22 and 23 and eliminating $\bar{\mathbf{Q}}$:

$$\begin{aligned}\bar{\mathbf{Q}} &= \mathbf{L}_0^{-1}\mathbf{L}'_0\bar{\mathbf{Q}}' + \mathbf{L}_0^{-1}(\zeta'_0 - \zeta_0) \\ &= \mathbf{J}\bar{\mathbf{Q}}' + \mathbf{K}.\end{aligned} \tag{24}$$

3 Results

Several sets of internal coordinates for describing the vibrational motion of ammonia in both the neutral and the cationic $\tilde{X}^2A''_2$ state have been proposed [7, 16, 31–33]. We have adopted the set suggested by Hoy et al. [16] and developed by Handy and coworkers [7], which, although not orthogonal in the whole range of nuclear coordinates as that of Ref. [33], offers the advantage that the elements of the $\mathbf{G}$ matrix are already tabulated. It consists of the three N–H stretching coordinates (r_1, r_2, r_3), the angle (β) that each N-H bond forms with the trisector, that is, the axis forming the same angle with each bond vector, see Fig. 3, and any two of the three angles (α_i) obtained by projecting ammonia onto a plane perpendicular to the trisector. The adopted set of coordinate is shown in Fig. 3. Instead of r_1, r_2, r_3, we have used the symmetry adapted stretching coordinates R_1, R_2, R_3 defined as:

$$\begin{cases} R_1 = \frac{1}{\sqrt{3}}(r_1 + r_2 + r_3) \\ R_2 = \frac{1}{\sqrt{6}}(2r_1 - r_2 - r_3) \\ R_3 = \frac{1}{\sqrt{2}}(r_2 - r_3) \end{cases} \tag{25}$$

Harmonic vibrational frequencies and equilibrium geometries of the ground electronic states of NH_3 and NH_3^+ are reported in Table 1. They have been computed at two levels of computation in order to allow for comparison with both our previous FC computations [10] and with some of the high-level computations published up to now [34, 35]. The fourth order of the Møller Plesset perturbation theory, including single, double and quadruple excitations, in conjunction with the 6-311++G(3df,3pd) basis set (hereafter MP4/TZ) and the CCSD(T) method in conjunction with the aug-cc-pCVQZ core-valence basis set (hereafter CC/QZ) were used.

The $\mathbf{B}_0$ and $\mathbf{G}$ matrices have been evaluated following the procedure given in Ref. [7] and transformed according to Eq. 25. Analytical expressions of the elements of $\mathbf{G}$ are given in the Appendix. The coefficients g'_{rst} and g''_{rstu} appearing in the $\Delta\hat{T}$ term are reported in Table 2 both for MP4/TZ and for CC/QZ equilibrium geometries.

At the MP4/TZ level of approximation, the potential energy surface of NH_3^+ as a function of $\bar{Q}_1$ and $\bar{Q}_2$, the symmetric stretching and bending curvilinear normal coordinates, respectively, has been calculated over a grid of 431 points, covering the range $r_{\text{eq}} - 0.05\,\text{Å} \leq r \leq r_{\text{eq}} +$

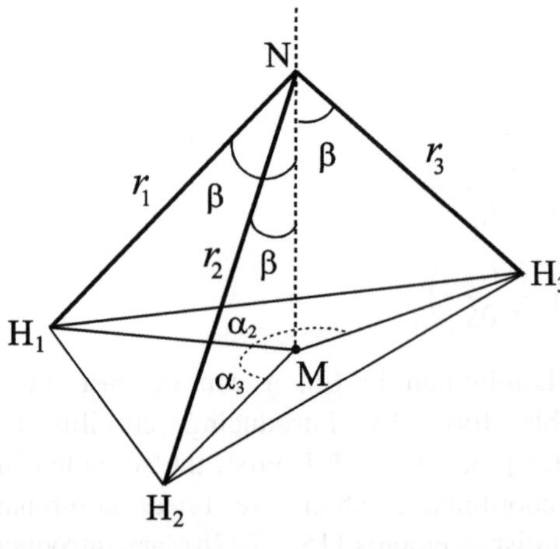

Fig. 3 Trisector axis (NM) at any nuclear configuration of ammonia: M belongs to the H_1–H_2–H_3 plane and is chosen in such a way that $\angle H_1NM = \angle H_2NM = \angle H_3NM = \beta$

Table 1 Predicted –MP4/TZ and CC/QZ– bond lengths (r_{eq}, Å), valence angles ∠HNH (θ_{eq}, degrees) and harmonic frequencies of the two symmetric modes ($\tilde{\nu}_i$ as wavenumbers, cm^{-1}) of NH_3 ($\tilde{X}^1A_1'$) and NH_3^+ ($\tilde{X}^2A_2''$)

	r_{eq} [a]	θ_{eq}	$\tilde{\nu}_1$ [b]	$\tilde{\nu}_2$
MP4[c]				
$\tilde{X}^1A_1'$	1.0114	106.76	3523.94	1056.72
$\tilde{X}^2A_2''$	1.0203	120.00	3423.10	876.67
CC				
$\tilde{X}^1A_1'$	1.0115	106.71	3484.26	1053.01
$\tilde{X}^2A_2''$	1.0206	120.00	3376.51	865.87

[a] Experimental values are $r_{eq} = 1.0124$ Å, $\theta_{eq} = 106.67°$ for NH_3, Ref. [36], and $r_{eq} = 1.0145$ Å for NH_3^+, Ref. [37]

[b] Herzberg notation, $\tilde{\nu}_{1(2)}$ refers to symmetric stretching (bending)

[c] From Ref. [10]

Table 2 First-($uma^{-1/2}$ $Å^{-1}$) and second-order (uma^{-1} $Å^{-2}$) power expansion non-null coefficients of the kinetic energy of the $\tilde{X}^2A_2''$ electronic state of cationic ammonia

	MP4	CCSD(T)
g'_{212}	0.12170	0.12167
g'_{221}	−1.12732	−1.12703
g''_{2222}	−0.13720	−0.13713
g''_{2211}	1.90627	1.90530
g''_{2121}	−0.06860	−0.06856
g''_{1122}	0.16682	0.16674

1(2) is the symmetric stretching (bending)

0.05 Å and $\beta_{eq} - 10° \leq \beta \leq \beta_{eq} + 10°$. The potential energy has been fitted by different polynomials, one without coupling terms between the $\bar{Q}_1$ and the $\bar{Q}_2$ coordinates, the others with coupling terms having maximum order 4 in each variable. Since in the D_{3h} point group the totally symmetric stretching and the out-of-plane bending belong to different irreducible representations, all the terms in odd powers of $\bar{Q}_2$ are zero. The coefficients of the fitted polynomials are given in Table 3. It must be remarked that, in the case of MP4/TZ computations, we have deliberately chosen to use a small grid of points, much smaller than those used previously in the literature [10, 34, 35], in order to test the effectiveness of a good choice of internal curvilinear coordinates in relieving the computational efforts of computing anharmonic potential energy hypersurfaces for the calculation of FC factors. On the contrary, CC/QZ potential energy hypersurface has been calculated over a grid of 1,653 points, covering the range 0.9 Å ≤ r ≤ 1.2 Å and 88.6° ≤ θ ≤ 120.0°. The potential energy was fitted, in that case, by a polynomial of fourth degree in the dimensionless Morse variable:

Table 3 The coefficients of different polynomial fits of the MP4/TZ potential energy surface of the $\tilde{X}^2A_2''$ electronic state of cationic ammonia in terms of curvilinear normal coordinates (coefficients in $cm^{-1}/(amu^{1/2}$ Å$)^n$, rms error in cm^{-1})

	P1	P2	P3
$\bar{Q}_2^2$	11319.66	11319.66	11318.95
$\bar{Q}_2^4$	11885.66	11886.66	11799.86
$\bar{Q}_1^2$	173774.00	173774.00	173766.66
$\bar{Q}_1^3$	−279077.51	−206229.84	−206229.85
$\bar{Q}_1^4$	161562.75	161562.75	152225.84
$\bar{Q}_1\bar{Q}_2^2$		−14873.10	−14873.09
$\bar{Q}_1^2\bar{Q}_2^2$			3196.46
rms error	22.2	0.3	0.2

$$\xi_1 = 1 - \exp(-a\bar{Q}_1),$$

and of 12th degree in $\bar{Q}_2$, with coupling terms up to 11th power:

$$V = \sum_{i=2}^{4} a_i \xi_1^i + \sum_{j=1}^{6} b_j \bar{Q}_2^{2j} + \sum_{k=1}^{4}\sum_{\ell=1}^{5} c_{k\ell}\xi_1^k \bar{Q}_2^{2\ell}; \quad k + 2\ell \leq 11 \tag{26}$$

The coefficients of the fitted polynomial are given in Table 4.

The parameter a in ξ_1 was obtained by minimizing the RMS error of the monodimensional fit of the potential energy over the stretching coordinate (for $\theta = \theta_{eq}$) by using a potential of the form:

$$V_{str}(\bar{Q}_1) = \sum_{m=2}^{4} d_m [1 - \exp(-a\bar{Q}_1)]^m, \tag{27}$$

and was not re-optimized in the bidimensional fit.

The photoelectron spectrum of ammonia consists of two transitions, one corresponding to the removal of an electron from the HOMO, the nitrogen lone pair $3a_1$ orbital, the $\tilde{X}^2A_2'' \leftarrow \tilde{X}^1A_1'$ transition, the other corresponding to the removal of an electron from the 1e orbital, the $\tilde{X}^2E \leftarrow \tilde{X}^1A_1'$ transition. The ground-state electronic configuration is $1a_1^2 2a_1^2 1e^4 3a_1^2$ [4]. The $\tilde{X}^2A_2'' \leftarrow \tilde{X}^1A_1'$ lowest energy absorption exhibits a long vibrational progression, extending over about 2 eV and consisting of sixteen well-resolved vibrational peaks see Fig. 1 [5]. The highest intensities occur for the 7th and 8th peaks, whose intensity are nearly the same [5, 38], but the origin of the band at lowest energy is still uncertain, it could be either the $0 \leftarrow 0'$ transition (hereafter a prime will be used for the states of neutral ammonia) or a hot band. The observed vibrational spacing is *ca.* 0.12 eV (970 cm^{-1}), so that the long vibrational progression has been assigned to the ν_2 mode

Table 4 Parameters of the CC/QZ potential energy hypersurfaces of NH_3^+, Eq. 26

a_2	166968.647510294	c_{15}	−817.398711959657
a_3	−35289.0487979425	c_{21}	−1594.69984392527
a_4	11953.0683296960	c_{22}	−1951.90198081176
b_1	11121.4657867805	c_{23}	1191.88081069391
b_2	13245.5214498090	c_{24}	−258.787556100943
b_3	−7562.83554727964	c_{31}	3333.02647681526
b_4	5229.27984543509	c_{32}	−3721.84786330654
b_5	−2421.23199626368	c_{33}	21.1138010388612
b_6	547.255930418840	c_{34}	597.278450112574
c_{11}	−14062.5970311914	c_{41}	1044.63030334237
c_{12}	2027.23704747984	c_{42}	−1565.01400236844
c_{13}	−4463.71091704471	c_{43}	−401.619141894396
c_{14}	2826.24603526959	a	1.0063077647

Energy in cm^{-1}; ξ_1 dimensionless; $\bar{Q}_2$ in uma$^{1/2}$ Å; a in uma$^{-1/2}$ Å^{-1}, rms error = 0.1 cm^{-1}

(Herzberg's notation [39]), the so-called umbrella mode. In the high-resolution spectrum recorded by Edvardsson et al. [38], the strong progression due to the symmetric bending mode is accompanied by a much weaker but well-resolved one, falling at higher energy, which was tentatively assigned to the asymmetric bending mode with one more quantum on the symmetric bending mode.

As concerns neutral ammonia, a good representation of the two quasi-degenerate vibrational states is all we need for computing the photoelectron spectrum. A few test computations, carried out with an analytical anharmonic potential that reproduces the height of the inversion barrier (1,882 cm^{-1}), showed that the lowest energy states are well described by the symmetric and antisymmetric linear combinations of the two lowest energy vibrational states of each nuclear configuration, no significant contributions of higher-energy harmonic wavefunctions have been found. From the symmetric vibrational ground state, transitions to vibrational states with even quantum numbers are allowed, whereas states with odd quantum numbers are excited from the antisymmetric combination. Since the symmetric and antisymmetric vibrational modes of neutral ammonia are separated by only 0.79 cm^{-1} [32], Boltzmann populations of the two levels are equal and that allows to compute FC factors using only one equilibrium configuration of neutral ammonia. The initial state in all the FC calculations is therefore the harmonic ground vibrational state of the neutral molecule, the only one significantly populated at room temperature.

Among the six normal modes of the $\tilde{X}^2A_2''$ cationic state, only the symmetric stretching and the symmetric bending inversion modes play a role in the photoelectron spectrum if vibronic couplings are neglected [10, 38]. The stronger progression observed in the photoelectronic spectrum extends over an energy range of about 2 eV, so that a good representation of at least the lowest sixteen states associated with the inversion mode is needed. Thus, both potential anharmonic terms and the $\Delta\hat{T}$ kinetic terms are in principle important for this electronic state. Anharmonic wavefunctions of NH_3^+ have been computed by using the variational method, truncating the expansion of Eq. 16 to second order and neglecting the V_{kin} term, which is known to give very small contributions to the energies of the vibrational states [18, 19]. Forty harmonic oscillator basis functions for each mode centered in the minimum energy nuclear configuration of the cationic state have been used; that choice ensures convergence on the excited vibrational states in an energy region much larger than that of interest. Eigenstates have been computed by an implicitly restarted Arnoldi procedure. Analytical integrals have been used [15, 40, 41]. The computed energies of the vibrational states are reported in Table 5 together with the available experimental values [37, 42, 43]. Energies are predicted with reasonable accuracy by both the simple P1 potential of Table 3 and by the more accurate potential of Table 4; the latter one gives a significant improvement only for states with high quantum numbers on the ν_2 mode. The agreement is not as good as that obtained by previous theoretical works [34], but it must be remarked that in the present work, the potential energy couplings of $\bar{Q}_1$ and $\bar{Q}_2$ with the other four normal coordinates have been neglected.

The $\tilde{X}^2A_2'' \leftarrow \tilde{X}^1A_1'$ photoelectronic spectrum of ammonia computed by considering the harmonic potential obtained at MP4/TZ level of computation, and the kinetic energy operator $\Delta\hat{T}$ of Eq. 16 is shown in Fig. 4.

Table 5 Computed and experimental energies (cm^{-1}) of the vibrational states of NH_3^+

Assignment	P1	CCSD(T)	Obs.[a]	Ref. [10]	Ref. [34]
$\tilde{\nu}_2$	918	908	903[b]	921	899
$2\tilde{\nu}_2$	1,874	1,854	1,844	1,873	1,837
$3\tilde{\nu}_2$	2,865	2,831	2,813	2,852	2,804
$\tilde{\nu}_1$	3,304	3,329	3,232[c]	3,373	3,231
$4\tilde{\nu}_2$	3,885	3,832	3,807	3,855	3,796
$\tilde{\nu}_1 + \tilde{\nu}_2$	4,221	4,231	4,127[c]	4,286	4,123
$5\tilde{\nu}_2$	4,931	4,855	4,821	4,878	4,809
$6\tilde{\nu}_2$	6,002	5,895	5,852	5,920	5,839
$\tilde{\nu}_1 + 3\tilde{\nu}_2$	6,165	6,140	6,022[c]	6,204	6,015
$7\tilde{\nu}_2$	7,093	6,952	6,899	6,979	6,885
$8\tilde{\nu}_2$	8,203	8,022	7,958	8,052	7,944
$9\tilde{\nu}_2$	9,331	9,104	9,030	9,139	9,015

[a] Ref. [42], except where otherwise noted

[b] Ref. [37]

[c] Ref. [43]

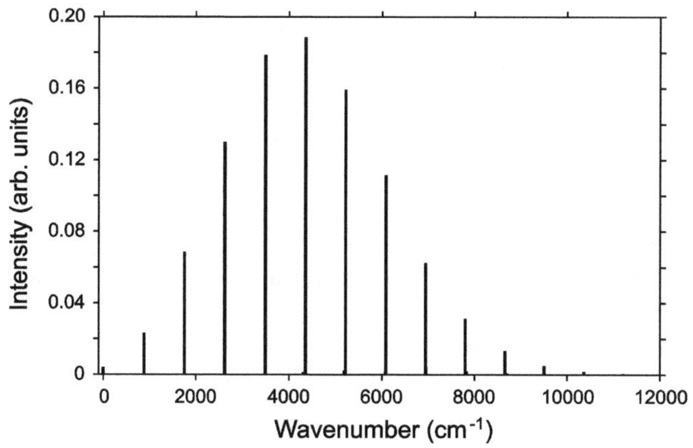

Fig. 4 Photoelectron spectrum of NH_3 computed using the MP4 harmonic potential and the kinetic energy operator $\Delta\hat{T}$

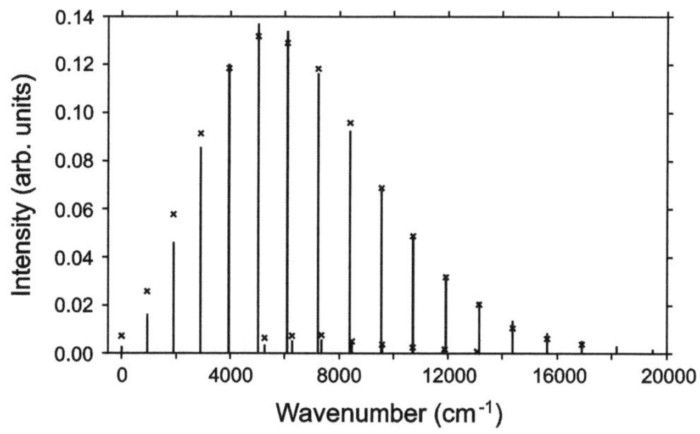

Fig. 5 Computed photoelectron spectrum of NH_3. *Verticle line* anharmonic potential P1 of Table 3, *Multiplication symbol* relative experimental intensities from Ref. [38]

Although the main features of the experimental spectrum are reproduced by FC computations, the predicted intensity distribution is significantly different from the observed one. The maximum peak height occurs for the $5 \leftarrow 0'$ transition, but the $6 \leftarrow 0'$ one exhibits a significantly lower intensity than the $5 \leftarrow 0'$ transition. More important discrepancies are observed in the high-wavenumber region, where the relative intensities of the computed spectrum decay much faster than the experimental ones. The intensities of the higher-energy transitions are significantly underestimated, and therefore, the spectrum exhibits only 13 peaks, whereas in the experimental one at least 16 peaks are clearly observed. The results obtained by using a harmonic potential are thus in many aspects similar to those previously obtained by employing the internal coordinate representation but using a simplified expression of $\hat{T}$, without including the $\Delta\hat{T}$ terms [10, 44]. Indeed, the $\Delta\hat{T}$ kinetic operator only provides corrections of a few tens of cm^{-1} on the computed energies, without significantly affecting the computed eigenvectors.

The computed spectrum obtained by using the P1 anharmonic potential of Table 3 together with kinetic $\Delta\hat{T}$ is shown in Fig. 5 and compared with the experimental intensities of Ref. [38]. The latter ones have been obtained from the peak areas of the deconvoluted photoelectron spectrum, which have been then normalized to unity for comparison with the theoretical one. The whole spectrum is well reproduced, especially as concerns its decay in the longer-wavenumber region, which had posed problems in previous theoretical investigations [12]. The experimental bandwidth is also well reproduced by FC calculations; Rabelais' experimental spectrum shows sixteen well-resolved peaks, with a little not resolved tail, whereas the theoretical one exhibits one more. The computed spectrum also shows the presence of a second much weaker progression, which the calculation assigns to transitions to vibrational states with excitations on the bending mode ν_2 and a single quantum on the totally symmetric stretching mode ν_1, in agreement with previous theoretical results [10, 13, 14]. The weaker progression is very well resolved in the high-resolution spectrum of Edvardsson et al. [38] who tentatively assigned it to transitions to a vibrational progression of the asymmetric bending mode with one more quantum on the symmetric bending mode ν_2. The asymmetric bending mode is of E symmetry, and therefore, vibronic couplings should be invoked to justify the presence of such combination bands in the experimental spectrum. In the experimental spectrum, the weak progression starts just after the sixth peak of the stronger progression, about 5,400 cm^{-1} after the $0 \leftarrow 0'$ transition and exhibits a maximum for the third peak. In the computed spectrum the weak progression starts after the fifth peak and the maximum peak height falls at the forth peak. The adopted scaling was the same as for the main progression; thus, the good agreement between the relative intensities of the weak and the strong progression further supports our assignment.

The computed spectra obtained by using the other potential energy functions of Table 3 are very similar to that obtained by using the simplest P1 potential as concerns the stronger progression; a slight detriment on the intensities and width of the weaker progression is observed when couplings between $\bar{Q}_1$ and $\bar{Q}_2$ are introduced in the potential energy expressions, possibly because of the small grid used in the electronic calculations, not sufficiently large, especially for the totally symmetric stretching mode, to provide the coupling terms with the required accuracy for calculations of the FC factors.

The deviation from the experimental relative intensities is more pronounced for low-energy levels, whereas the high region of the spectrum is better reproduced. That unexpected behavior—the quality of the computed states at low energy should be higher than those at higher energy—can be due to the presence of overlapping bands in the

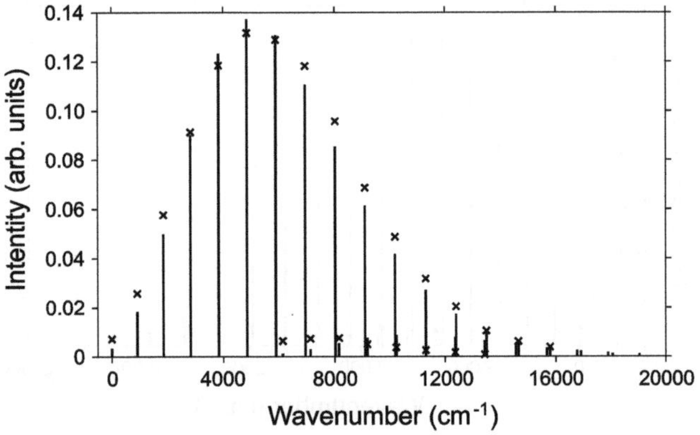

Fig. 6 Computed photoelectronic spectrum of NH_3. *Verticle line* anharmonic potential of Table 4, *Multiplication symbol* relative experimental intensities from Ref. [38]

experimental spectrum, which makes the estimation of FC factors by the deconvolution technique difficult. Indeed, on the assumption that the first observed band is a hot band, as experimental and theoretical evidence leads to suppose [13, 38], a possible explanation of the low intensity is that other peaks of the hot progression could be hidden under the lower-energy peaks. Indeed previous calculations of FC factors from the $|0'1'\rangle$ state of neutral ammonia indicates that the transition $00 \leftarrow 0'1'$ is not the most intense one, but it should be accompanied by at least four transitions of the type $0n \leftarrow 0'1'$ $(n = 1, 2, \ldots)$, which have still higher intensities [10]. Actually FC computations yield a quite long progression for the hot band with relative intensities 0.02, 0.08, 0.14, 0.13 and 0.05 for n = 0, 1, 2, 3 and 4, respectively. Those hot transitions could affect the intensities of the peaks of the main progression making the comparison of the computed and the observed spectrum even more difficult.

The spectrum obtained by using the CC/QZ potential energy of Table 4 is shown in Fig. 6. It is quite similar to that obtained by using the simplest P1 potential, exhibiting a slight improvement in the low-energy region of the spectrum, but to detriment of the middle energy region, and of the second less intense progression, which, within the limits of the deconvolution procedure used by us for assigning experimental intensities, is slightly longer than the observed one.

4 Conclusion

Prediction of the band shapes for radiative transitions between electronic states with significantly different minimum energy nuclear configurations along an angular coordinate (floppy molecules) requires caution. Gribov's definition of curvilinear normal coordinates allows the generalization of Duschinsky's transformation to the case of curvilinear coordinates. Those coordinates can be extremely useful for floppy molecules, inasmuch they could allow for decoupling the displaced modes, making it easier the computation of the potential energy hypersurface. Here we have shown that the photoelectron spectrum of ammonia, the prototype of floppy molecules, can be reproduced with sufficient accuracy, by using the curvilinear internal coordinate representation of the normal modes of the neutral and ionic state, providing that anharmonic potential energy terms for the two coordinates that play the dominant role in the photoelectron spectrum are properly included. The treatment of anharmonic effects at the full six-dimensional level does not appear to be necessary, although high accuracy on vibrational frequencies, comparable to that of previous work [34], cannot be obtained within the two-dimensional anharmonic model adopted here. The computational effort for the characterization of anharmonic potential energy terms is significantly relieved by using curvilinear coordinates, both because in the case of ammonia curvilinear coordinates allows for decoupling the two active modes with a good accuracy, and because of the limited number of electronic computations to be carried out for the characterization of the potential energy hypersurface. Indeed in the Cartesian coordinate representation, a grid of about 4300 points, covering the range $0.57\,\text{Å} \leq r \leq 1.65\,\text{Å}$ and $40° \leq \theta \leq 120°$, was necessary for obtaining reasonable results, whereas in the present work, a much smaller grid of 431 points, covering the range $r_{eq} - 0.05\,\text{Å} \leq r \leq r_{eq} + 0.05\,\text{Å}$ and $\beta_{eq} - 10° \leq \beta \leq \beta_{eq} + 10°$, has been sufficient. In the case of ammonia, probably because of the good choice of internal coordinates, the dependence of the **G** matrix terms upon the curvilinear coordinates plays a minor role, it can be neglected if a high accuracy on frequencies is not required.

5 Computational details

Equilibrium geometries, harmonic frequencies and Cartesian normal coordinates for NH_3 and NH_3^+ were computed by using the CFOUR program at both MP4 and CCSD(T) levels [45]. The unrestricted formalism was used for NH_3^+. All the electrons –included core 1s of nitrogen– were explicitly correlated both in CCSD(T) and MP4 computations. The MP4 potential energy surface of NH_3^+ was computed at the UMP4(SDQ)/6-311++G(3df,3pd) level by using the Gaussian package [46]. The CCSD(T) potential energy surface of NH_3^+ has been calculated by using the Molpro suite of programs [47]. The UCCSD(T) level of theory in conjunction with the aug-cc-pCVQZ core-valence basis set was used. Franck–Condon factor computations were carried out by using a development version of the MolFC package [48–50].

Acknowledgments The financial support of the University of Salerno is gratefully acknowledged.

Appendix: The G matrix

In the following:

$$\alpha_1 = 2\pi - \alpha_2 - \alpha_3; \tag{28}$$

$$\psi_i = \frac{\alpha_i}{2}; \quad i = 1, 2, 3; \tag{29}$$

$$\mu = \frac{m_N m_H}{m_N + m_H}. \tag{30}$$

$$\begin{aligned} G_{\beta\beta} = & \frac{\cos^2\psi_1}{4\mu(\sin^2\psi_2)(\sin^2\psi_3)\left(\frac{R_1}{\sqrt{3}} + \frac{2R_2}{\sqrt{6}}\right)^2} \\ & + \frac{\cos^2\psi_2}{4\mu(\sin^2\psi_1)(\sin^2\psi_3)\left(\frac{R_1}{\sqrt{3}} - \frac{R_2}{\sqrt{6}} + \frac{R_3}{\sqrt{2}}\right)^2} \\ & + \frac{\cos^2\psi_3}{4\mu(\sin^2\psi_1)(\sin^2\psi_2)\left(\frac{R_2}{\sqrt{6}} - \frac{R_1}{\sqrt{3}} + \frac{R_3}{\sqrt{2}}\right)^2} \\ & + \frac{(\cot\psi_2)(\cot\psi_3)\left[2(\sin^2\psi_1)(\cos^2\beta) - 1\right]}{2m_N(\sin^2\psi_1)\left(\frac{R_2}{\sqrt{6}} - \frac{R_1}{\sqrt{3}} + \frac{R_3}{\sqrt{2}}\right)\left(\frac{R_1}{\sqrt{3}} - \frac{R_2}{\sqrt{6}} + \frac{R_3}{\sqrt{2}}\right)} \\ & + \frac{(\cot\psi_1)(\cot\psi_2)\left[2(\sin^2\psi_3)(\cos^2\beta) - 1\right]}{2m_N(\sin^2\psi_3)\left(\frac{R_1}{\sqrt{3}} + \frac{2R_2}{\sqrt{6}}\right)\left(\frac{R_2}{\sqrt{6}} - \frac{R_1}{\sqrt{3}} - \frac{R_3}{\sqrt{2}}\right)} \\ & + \frac{(\cot\psi_1)(\cot\psi_3)\left[2(\sin^2\psi_2)(\cos^2\beta) - 1\right]}{2m_N(\sin^2\psi_2)\left(\frac{R_1}{\sqrt{3}} + \frac{2R_2}{\sqrt{6}}\right)\left(\frac{R_2}{\sqrt{6}} - \frac{R_1}{\sqrt{3}} + \frac{R_3}{\sqrt{2}}\right)} \end{aligned} \tag{31}$$

$$G_{R_1R_1} = \frac{1}{\mu} - \frac{4\sin^2\beta(\sin^2\psi_1 + \sin^2\psi_2 + \sin^2\psi_3) - 6}{3m_N} \tag{32}$$

$$\begin{aligned} G_{\beta R_1} = & \frac{\sqrt{3}\sin(2\beta)}{6m_N\sin\psi_2}\left(\frac{\cos\psi_3\sin\psi_1}{\frac{R_2}{\sqrt{6}} - \frac{R_1}{\sqrt{3}} + \frac{R_3}{\sqrt{2}}} - \frac{\cos\psi_1\sin\psi_3}{\frac{R_1}{\sqrt{3}} + \frac{2R_2}{\sqrt{6}}}\right) \\ & - \frac{\sqrt{3}\sin(2\beta)}{6m_N\sin\psi_3}\left(\frac{\cos\psi_1\sin\psi_2}{\frac{R_1}{\sqrt{3}} + \frac{2R_2}{\sqrt{6}}} + \frac{\cos\psi_2\sin\psi_1}{\frac{R_1}{\sqrt{3}} - \frac{R_2}{\sqrt{6}} + \frac{R_3}{\sqrt{2}}}\right) \\ & - \frac{\sqrt{3}\sin(2\beta)}{6m_N\sin\psi_1}\left(\frac{\cos\psi_2\sin\psi_3}{\frac{R_1}{\sqrt{3}} - \frac{R_2}{\sqrt{6}} + \frac{R_3}{\sqrt{2}}} - \frac{\cos\psi_3\sin\psi_2}{\frac{R_2}{\sqrt{6}} - \frac{R_1}{\sqrt{3}} + \frac{R_3}{\sqrt{2}}}\right) \end{aligned} \tag{33}$$

References

1. Walsh AD, Warsop PA (1961) Trans Faraday Soc 57:345
2. Douglas AE (1963) Discuss Faraday Soc 35:158
3. Branton GR, Frost DC, Herring FG, McDowell CA, Stenhouse LA (1969) Chem Phys Lett 3:581
4. Weiss MJ, Lawrence GM (1970) J Chem Phys 53:214
5. Rabalais JW, Karlsson L, Werme LO, Bergmark T, Siegbahn K (1973) J Chem Phys 58:3370
6. Leonard C, Handy NC, Carter S, Bowman JM (2002) Spectrochim Acta A 58:825
7. Handy NC, Carter S, Colwell SM (1999) Mol Phys 96:477
8. Reimers JR (2001) J Chem Phys 115:9103
9. Borrelli R, Peluso A (2006) J Chem Phys 125:194308
10. Peluso A, Borrelli R, Capobianco A (2009) J Phys Chem A 113:14831
11. Domcke W, Cederbaum LS, Köppel H, Von Niessen W (1977) Mol Phys 34:1759
12. Ågren H, Reineck I, Veenhuizen H, Maripuu R, Arneberg R, Karlsson L (1982) Mol Phys 45:477
13. Viel A, Eisfeld W, Neumann S, Domcke W, Manthe U (2006) J Chem Phys 124:214306
14. Viel A, Eisfeld W, Evenhuis CR, Manthe U (2008) Chem Phys 347:331
15. Wilson EBJ, Decius JC, Cross PC (1955) Molcular vibrations. McGraw Hill, New York
16. Hoy AR, Mills IM, Strey G (1972) Mol Phys 24:1265
17. Kuchitsu K, Morino Y (1965) Bull Chem Soc Japn 38:805
18. Sibert EL, Hynes JT, Reinhardt WP (1983) J Phys Chem 87:2032
19. Gribov LA, Orville-Thomas W (1988) Theory and methods of calculation of molecular spectra. Wiley, Chichester
20. Gribov LA (1992) J Struct Chem (Engl. Transl.) 33:478
21. Villaseñor-González P, Cisneros-Parra J (1981) Am J Phys 49:754
22. Nauts A, Chapuisat X (1985) Mol Phys 55:1287
23. Gribov LA (1970) Opt Spectrosc 30:1247
24. Pavlyuchko AI, Gribov LA (1983) Opt Spectrosc 54:644
25. Seidner L, Stock G, Sobolewski AL, Domcke W (1992) J Chem Phys 96:5298
26. Isaacson AD (2006) J Phys Chem A 110:379
27. Cederbaum LS, Domcke W (1976) J Chem Phys 64:603
28. Cederbaum LS, Domcke W (1976) J Chem Phys 64:612
29. Quade CR (1976) J Chem Phys 64:2783 (See also erratum *ibid.* 79, 4089 (1983))
30. Dushinsky F (1937) Acta Physicochim. URSS 7:551
31. Špirko V (1983) J Mol Spectrosc 101:30
32. Špirko V, Kraemer WP (1989) J Mol Spectrosc 133:331
33. Ragni M, Lombardi A, Pereira Barreto PR, Peixoto Bitencourt AC (2009) J Phys Chem A 113:15355
34. Yurchenko SN, Thiel W, Carvajal M, Jensen P (2008) Chem Phys 346:146
35. Léonard C, Carter S, Handy NC (2002) Phys Chem Chem Phys 4:4087
36. Morino Y, Kuchitsu K, Yamamoto S (1968) Spectrochim Acta A 24:335
37. Lee SS, Oka TT (1991) J Chem Phys 94:1698
38. Edvardsson D, Baltzer P, Karlsson L, Wannberg B, Holland DMP, Shaw DA, Rennie EE (1999) J Phys B 32:2583
39. Herzberg G (1945) Infrared and raman spectra of polyatomic molecules. Van Nostrand Co. Inc., Princeton, NJ
40. Califano S (1976) Vibrational states. Wiley, New York
41. Morales J, Palma A, Sandoval L (1986) Int J Quantum Chem 29:211
42. Reiser G, Habenicht W, Müller-Dethlefs K (1993) J Chem Phys 98:8462
43. Bahng MK, Xing X, Baek SJ, Ng CY (2005) J Chem Phys 123:084311
44. Harshbarger WR (1970) J Chem Phys 53:903
45. CFOUR, a quantum chemical program package written by Stanton JF, Gauss J, Harding ME, Szalay PG with contributions from Auer AA, Bartlett RJ, Benedikt U, Berger C, Bernholdt DE, Bomble YJ, Christiansen O, Heckert M, Heun O, Huber C, Jagau

T-C, Jonsson D, Jusélius J, Klein K, Lauderdale WJ, Matthews DA, Metzroth T, O'Neill DP, Price DR, Prochnow E, Ruud K, Schiffmann F, Stopkowicz S, Tajti A, Vázquez J, Wang F, Watts JD and the integral packages MOLECULE (Almlöf J and Taylor PR), PROPS (Taylor PR), ABACUS (Helgaker T, Jensen HJ Aa, Jørgensen P, and Olsen J), and ECP routines by Mitin AV, van Wüllen C. For the current version, see http://www.cfour.de
46. Frisch MJ, Trucks GW, Schlegel HB, Scuseria GE, Robb MA, Cheeseman JR, Scalmani G, Barone V, Mennucci B, Petersson GA, Nakatsuji H, Caricato M, Li X, Hratchian HP, Izmaylov AF, Bloino J, Zheng G, Sonnenberg JL, Hada M, Ehara M, Toyota K, Fukuda R, Hasegawa J, Ishida M, Nakajima T, Honda Y, Kitao O, Nakai H, Vreven T, Montgomery JA Jr., Peralta JE, Ogliaro F, Bearpark M, Heyd JJ, Brothers E, Kudin KN, Staroverov VN, Kobayashi R, Normand J, Raghavachari K, Rendell A, Burant JC, Iyengar SS, Tomasi J, Cossi M, Rega N, Millam JM, Klene M, Knox JE, Cross JB, Bakken V, Adamo C, Jaramillo J, Gomperts R, Stratmann RE, Yazyev O, Austin AJ, Cammi R, Pomelli C, Ochterski JW, Martin RL, Morokuma K, Zakrzewski VG, Voth GA, Salvador P, Dannenberg JJ, Dapprich S, Daniels AD, Farkas, Foresman JB, Ortiz JV, Cioslowski J, Fox DJ (2009) Gaussian 09 revision A.02. Gaussian Inc. Wallingford, CT
47. Werner HJ, Knowles PJ, Knizia G, Manby FR, Schütz M, Celani P, Korona T, Lindh R, Mitrushenkov A, Rauhut G, Shamasundar KR, Adler TB, Amos RD, Bernhardsson A, Berning A, Cooper DL, Deegan MJO, Dobbyn AJ, Eckert F, Goll E, Hampel C, Hesselmann A, Hetzer G, Hrenar T, Jansen G, Köppl C, Liu Y, Lloyd AW, Mata RA, May AJ, McNicholas SJ, Meyer W, Mura ME, Nicklass A, O'Neill DP, Palmieri P, Pflüger K, Pitzer R, Reiher M, Shiozaki T, Stoll H, Stone AJ, Tarroni R, Thorsteinsson T, Wang M, Wolf A (2010) Molpro, version 2010.1, a package of ab initio programs. See http://www.molpro.net
48. Peluso A, Santoro F, Del Re G (1997) Int J Quantum Chem 63:233
49. Borrelli R, Peluso A (2003) J Chem Phys 119:8437
50. Borrelli R, Peluso A. MolFC: a program for Franck-Condon integrals calculation. Package available online at http://www.theochem.unisa.it

Theor Chem Acc (2012) 131:1182
DOI 10.1007/s00214-012-1182-2

REGULAR ARTICLE

Computational modelling of de novo synthesis of Dibenzofuran: oxidative pathways of Pyrene and Benzodibenzofuran

Ugo Cosentino · Demetrio Pitea · Giorgio Moro

Received: 20 July 2011 / Accepted: 9 November 2011 / Published online: 2 March 2012

Abstract The dominating route to polychlorinated Dibenzo-*p*-dioxin and Dibenzofuran formation in the "cold zones" of flue gas cleaning systems of municipal solid waste incinerators is the so-called de novo synthesis, that is, carbonaceous matrix burnoff with simultaneous oxidation and chlorination reactions. Pyrene (**1**) and Benzodibenzofuran (**2**) were chosen as the model compounds of carbonaceous material present in fly ash. Possible routes of Dibenzofuran formation by oxidative pathways of compounds (**1**) and (**2**) were investigated by theoretical calculations at the density functional theory level. The key intermediate peroxy radical, formed by reaction with molecular oxygen, can follow three main paths leading to Dibenzofuran. In the kinetically favourite path, the highest energetic barriers (25–30 kcal mol^{-1}) are encountered in the steps where CO molecules are released from ketene-like structures. These findings agree with previously reported temperature-programmed desorption results on CO desorption. Moreover, along this path, phenanthrene and biphenyl intermediates are formed, in agreement with the detection of these products in previously reported experimental Pyrene oxidation. Along the preferred path, different steric constraints in compounds (**1**) and (**2**) play a role in determining the relative stability of the intermediates, while they have less influence on the energetic barriers. As a consequence, compounds (**1**) and (**2**) should present similar kinetic behaviour as they present similar energetic barriers.

Keywords Oxidation mechanism · Benzodibenzofuran · Pyrene · Dibenzofuran · DFT calculations

Dedicated to Professor Vincenzo Barone and published as part of the special collection of articles celebrating his 60th birthday.

Electronic supplementary material The online version of this article (doi:10.1007/s00214-012-1182-2) contains supplementary material, which is available to authorized users.

U. Cosentino (✉) · D. Pitea
Dipartimento di Scienze dell'Ambiente e del Territorio, University of Milano-Bicocca, Piazza della Scienza 1, 20126 Milan, Italy
e-mail: ugo.cosentino@unimib.it

G. Moro
Dipartimento di Biotecnologie e Bioscienze, University of Milano-Bicocca, Piazza della Scienza 2, 20126 Milan, Italy

1 Introduction

Incineration and thermal decomposition of waste material frequently lead to the production of trace quantities of the pollutants Polychlorinated Dibenzo-*p*-dioxins (PCDD) and Dibenzofurans (PCDF). In order to minimize the emission of these pollutants, it is important to develop appropriate combustion technologies based on a deep knowledge of the detailed mechanisms of their formation and destruction.

PCDD/F yield in thermal systems is a balance between destruction and formation pathways. Since the activation energies of the destruction reactions are higher than those of the formation reactions, the net rate of production has maxima at certain temperatures [1].

On the basis of numerous experimental studies, it is now widely accepted that PCDD/Fs are formed through two main pathways, one homogeneous and the other heterogeneous [2–4].

The homogeneous pathway involves the reaction, at temperatures between 400 and 800 °C, of structurally related precursors in the gas phase, the most important and direct precursors being monocyclic aromatic compounds such as chlorinated phenols and chlorobenzenes. Heterogeneous pathways act in the 200–400 °C range on solid fly-ash surfaces and involve two routes. The first, the so-called de novo synthesis, proceeds through carbonaceous matrix burnoff with simultaneous oxidation and chlorination; the second takes place by the coupling of precursors. Both routes are catalyst-assisted, mainly due to copper and iron compounds.

The dominating route in PCDD/F formation in the "cold zones" of flue gas cleaning systems of municipal solid waste (MSW) incinerators and secondary not-ferrous metal smelters and sinter plants is de novo synthesis [3, 5].

The carbon source for de novo synthesis is the solid carbon matrix of fly ash, which has imperfect or degenerate graphitic structures, called turbostratic carbon crystallite, linked by amorphous regions, that is, aliphatic and alicyclic bridges [3]. Although the carbon layers in turbostratic carbon crystallite are similar to those in graphite, the layers are stacked randomly and disorientedly with larger inter-layer spacing and are thus more easily attacked by chemical reactions. The main chemical fate of a carbonaceous matrix is carbon gasification, PCDD/F de novo synthesis being only a trace reaction in the low-temperature regime.

Dibenzofuran (DF) and PCDF formation has been investigated in several experimental studies using a wide variety of polycyclic aromatic hydrocarbons (PAH), and these have lead to the hypotheses on the species and mechanisms involved in DF formation. Perylene has been proposed as the basic molecular structure able to explain the PCDF pattern in fluidized bed incinerators [6, 7], and analyses of PAH emissions, particularly of DF, from the high-temperature oxidation of anthracene have been carried out [8]. From the CuO catalysed oxidation of different PAHs [9], it was found that DF formation prevailed in oxidative pathways where phenanthrene is the reagent or an intermediate.

Also, the oxidative destruction of DF was experimentally investigated [10–12], leading to the proposal of two different DF oxidation mechanisms, one in high regime temperatures (>1,200 K) and the other in low.

Theoretical methods were used to investigate the mechanistic aspects of PCDD/F formation and destruction. Formation mechanisms from structurally related precursors were investigated at the ab initio level in both the gas-phase and heterogeneous catalytic–assisted reactions [2]. Furthermore, oxidative [13, 14] and pyrolytic [15] decomposition routes of DF were investigated.

For a theoretical investigation into the mechanisms of PCDD/F formation and destruction, advantage can be taken of available detailed knowledge, both experimental and theoretical, for oxidation mechanisms in the combustion process of aromatic and unsaturated hydrocarbons. In the initial steps, the combustion reaction of these systems involves the formation of a radical species with the loss of a hydrogen atom, through reaction with other radical species present in the environment (O, OH, O_2H, etc.) or with molecular oxygen. Following this, the newly formed radical species reacts with the molecular oxygen to form a peroxy radical R-OO˙. This is the key intermediate that can evolve following different pathways. The potential energy surface of vinyl [16], phenyl [17] and Dibenzofuranyl radical [13, 14] oxidation pathways was carefully investigated at the ab initio level, addressing the overall similarities of these systems with respect to reactions with O_2 [14]. Thermochemical parameters of the species involved in the mechanisms were also calculated [18–20]. The oxidation of molecular PAHs and one-layer graphite by H, HO, NO, NO_2, and NO_3 species was investigated to model soot particle formation [21].

The general aim of our research was a theoretical investigation into the mechanism of PCDD/PCDF formation by de novo synthesis, in order to support experimental studies focused on preventing and/or minimizing PCDD/F formation.

In the present study, we chose Pyrene (**1**) and Benzodibenzofuran (**2,** BDF) as the model compounds of carbonaceous material present in fly-ash and involved in de novo synthesis (Scheme 1). Solid-state ^{13}C NMR experiments have shown [22] that char has structures of randomly connected graphene clusters, consisting of 12–25 aromatic carbon atoms, corresponding to 3–7 benzene rings. Mass spectra analysis of PAHs in flames has shown a significant presence of species consisting of a highly condensed system of 6-membered rings [23]. On the basis of these findings, compound (**1**), consisting of four condensed aromatic rings, can be considered a "small size" molecular model representative of carbonaceous material involved in de novo synthesis. Moreover, also compound (**2**) was considered due to the presence of partially oxidized compounds in the fly-ash carbon matrix.

Here, we present the results of our computational study on the reaction pathways of Pyrene (**1**) and Benzodibenzofuran (**2,** BDF) involved in the de novo synthesis of DF (Scheme 2). Indeed, any investigation into possible pathways involved in oxidative decomposition reactions of

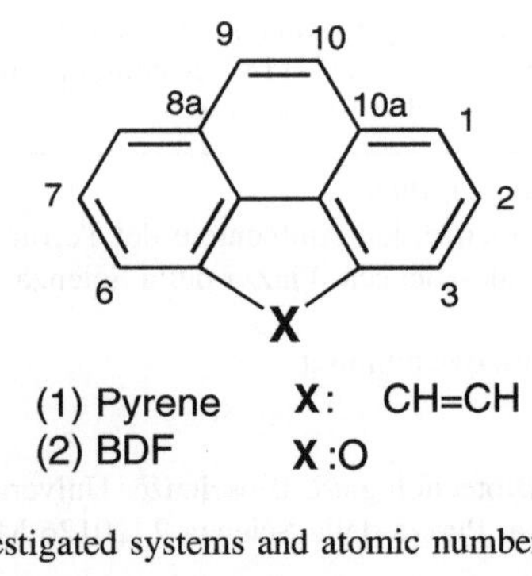

Scheme 1 Investigated systems and atomic numbering

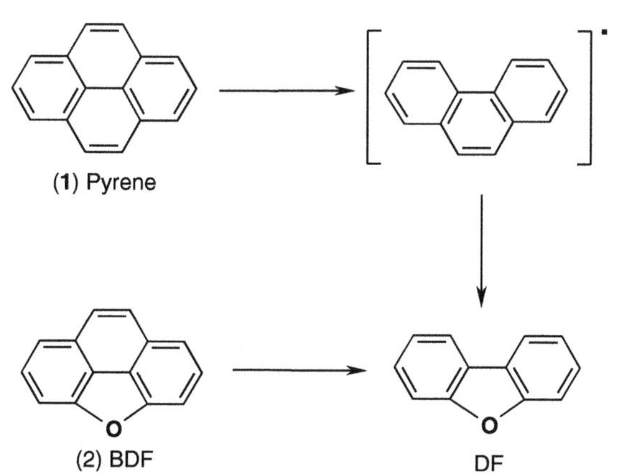

Scheme 2 Overall oxidative pathways for compounds (**1**) and (**2**)

aromatic hydrocarbons is a very demanding task, also for the simplest case, that is, the phenyl radical system [17]. In compounds (**1**) and (**2**) the presence of more than one ring significantly complicates the system, further increasing the number of possible reactive channels. Thus, given all the possible pathways, focusing only on those pathways involved in DF formation allows other competitive demolition pathways to be ignored and/or disregarded.

2 Computational details

All the calculations were performed with the Gaussian09 program [24]. Optimized geometries and harmonic vibrational frequencies of reactants, intermediates, products and transition states were calculated using the density functional theory (DFT) with the hybrid functional PBE1PBE [25] and the 6-311G** basis set. Spin-unrestricted calculations were performed for the radical systems. DFT densities and energies are known to be less affected by spin contamination than the corresponding unrestricted Hartree–Fock quantities [26]. In all the investigated radical systems, spin contamination was found to be negligible, the only exception being molecular systems made up by (Ar–O$^{\cdot}$) and the ground-state oxygen atom, O(^{3}P), deriving from the cleavage of the O–O bond in (Ar–OO$^{\cdot}$). Some caution is warranted in interpreting the results in these cases.

Intrinsic Reaction Coordinate (IRC) calculations were used to link reactants and products with their transition states. All the energy values reported include electron energies and Zero Point Energy (ZPE) corrections.

3 Test calculations on the phenyl peroxy system

The reliability of the PBE1PBE functional was tested by performing a preliminary investigation on the starting reaction channels of the phenyl peroxy radical, using the 6-311++G** basis set. According to a previous investigation [17], the starting reaction channels of the phenyl peroxy radical (**P$_1$**) can undergo three main reactions (Scheme 3). The (**P$_1$**) radical can isomerize to the (**P$_8$**) radical through the closure of a 3-membered ring (path A) or to the (**P$_{19}$**) through the closure of a 4-membered ring (path B). Path C involves the O–O cleavage and formation of the phenoxy radical and the ground-state oxygen atom, O(^{3}P).

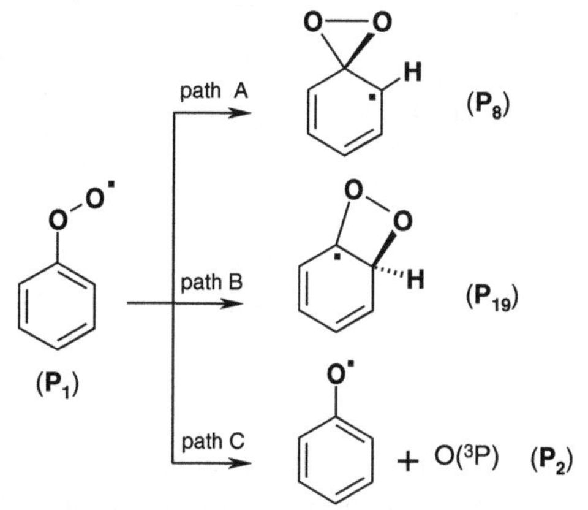

Scheme 3 Entry channel reactions of the phenyl peroxy radical. Molecular numbering as in Ref. [17]

Table 1 PBE1PBE/6-311++G** barriers (E_{att}, kcal mol^{-1}) and reaction energies ($E_{react} = E_{products} - E_{reagent}$, kcal mol^{-1}) for the starting reactions of (**P$_1$**) (Scheme 3). The B3LYP and G2 M(MP2) results from Ref. [17]

	PBE1PBE[a]		B3LYP[b]		G2 M(MP2)[b]	
	E_{att}	E_{react}	E_{att}	E_{react}	E_{att}	E_{react}
$P_1 \rightarrow P_8$	27.3	20.3	27.1	22.6	24.3	14.7
$P_1 \rightarrow P_{19}$	45.6	42.3	45.7	44.5	41.3	38.0
$P_1 \rightarrow P_2$	35.6	35.4	33.7	33.6	36.3	35.9

[a] PBE1PBE/6-311++G** geometries, present work

[b] B3LYP/6-311++G** geometries

Comparison of our results (Table 1) with those previously obtained [17] at the B3LYP and G2 M(MP2) levels [27] on B3LYP/6-311++G** geometries shows that the two functionals provide similar barriers and reaction energies as well as transition state geometries, with the PBE1PBE results in better agreement with the G2 M(MP2) results. Thus, the PBE1PBE functional is certainly suitable for modelling the reactivity of these systems.

4 Results

4.1 (Ar–OO$^{\cdot}$) formation

The initial step in the oxidative pathway of compounds (**1**) and (**2**) involves the formation of a radical species (Ar$^{\cdot}$)

Table 2 PBE1PBE/6-311G** relative energies (ΔE, kcal mol^{-1}) of the radical species of compounds (**1**) and (**2**), with respect to the parent compound

	ΔE (kcal mol^{-1})
$(\mathbf{1}^{\cdot})_{10}$	107.7
$(\mathbf{1}^{\cdot})_{1}$	108.1
$(\mathbf{1}^{\cdot})_{2}$	107.4
$(\mathbf{2}^{\cdot})_{10}$	107.4
$(\mathbf{2}^{\cdot})_{1}$	108.6
$(\mathbf{2}^{\cdot})_{2}$	107.0
$(\mathbf{2}^{\cdot})_{3}$	110.1

through the loss of a hydrogen atom by reaction with radical species ($X^{\cdot}$) present in the environment such as H, OH, O, alkyl groups or molecular oxygen. Depending on the environmental conditions, the barrier to overcome in this initial step will be quite different. To avoid the problem of having to choose a specific route to (**Ar˙**) formation, we assumed the (**Ar˙**) species to be already formed, regardless of the reactions involved.

The radicals obtained by the extraction of a hydrogen atom from compounds (**1**) and (**2**) present close relative energies (Table 2), and it can be expected that all of them will form, thus providing different reaction channels. However, as just the attack of molecular oxygen to C_{10} (or equivalent) position can lead to DF formation, only the $(\mathbf{1}^{\cdot})_{\mathbf{10}}$ and $(\mathbf{2}^{\cdot})_{\mathbf{10}}$ radicals are considered in the following (see Scheme 1 for numbering).

A barrier-less reaction between the radical species and the molecular oxygen leads to the (**Ar–OO˙**) peroxy radicals. These peroxy radicals can follow different reaction pathways, involving intramolecular rearrangements of the peroxy radical or O–O bond scission, as in the case of the phenoxy radical [17]. However, of all these, only paths A–C, reported in Schemes 4, 5, 6, can lead to DF formation, and were considered for (**1-OO˙**) and (**2-OO˙**) evolution. Paths A and B involve intramolecular rearrangements of the peroxy radical, while path C involves O–O bond scission. All three paths (A–C) converge to a common intermediate (**A5**), and further evolution of (**A5**) finally leads to DF (Scheme 7).

Moreover, each of the considered paths can present several collateral paths, two converging to intermediates of the A–C paths, and thus investigated (Schemes 8, 9), the others evolving towards products other than DF and thus not considered.

4.2 (Ar–OO˙) decomposition

Whenever possible, intermediates and transition states are named with the same label, without reference to the parent compound, (**1**) or (**2**). Relative energies with respect to (**Ar–OO˙**) of the species involved along the different paths are reported in Tables 3 and 4 and Figs. 1, 2, 3.

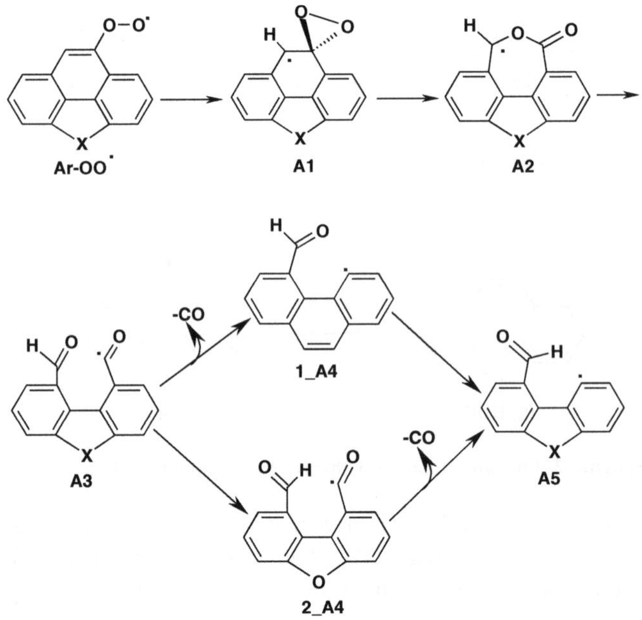

Scheme 4 Path A

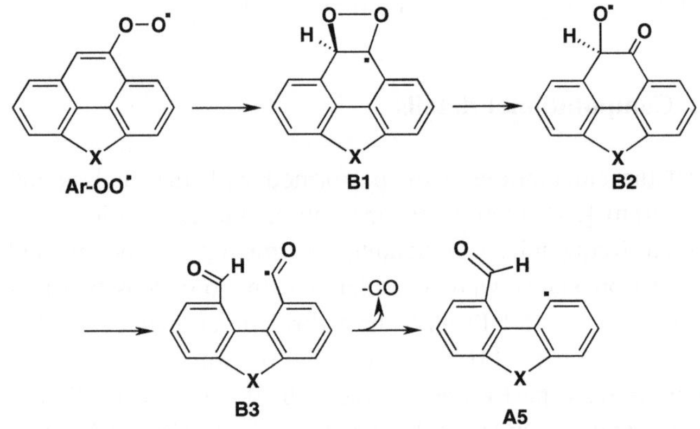

Scheme 5 Path B

Path A. The first step in this pathway (Scheme 4, Table 3, Fig. 1) consists in the isomerization of (**Ar–OO˙**) to the dioxiranyl radicals (**A1**) through the closure of the 3-membered COO ring. The involved energetic barriers are reported in Table 3. In the next step, one of the oxygen atoms inserts into the C_9 and C_{10} bond to produce the 7-membered lactonic radical (**A2**), significantly more stable than (**A1**). Even though two isomers could be expected from the (**A1**) rearrangement (the oxygen atom could insert into the C_9–C_{10} or the C_{10}–C_{10a} bond), only the (**A2**) product is obtained, as confirmed by IRC calculations. Moreover, the stability of (**1_A2**) and (**2_A2**) differs notably. The difference can be explained on the basis of the different molecular geometries. Indeed, in (**2_A2**) the lactonic ring is planar, while in (**1_A2**) the phenanthrenic moiety constrains the carbonyl group outside the molecular

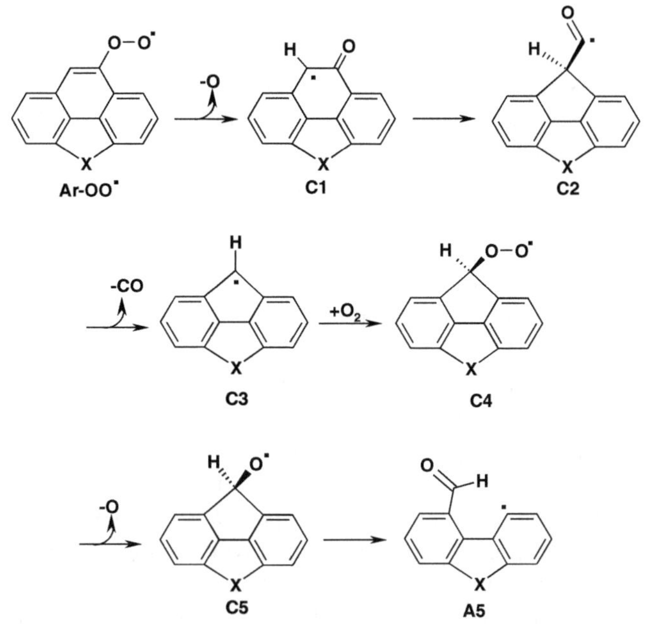

Scheme 6 Path C

plane. Along the (**A2**)–(**A5**) path, all intermediates and TSs from (**1**) present non-planar geometry due to the steric hindrance between the ortho substituents. On the contrary, derivatives from compound (**2**) are always planar; as a consequence, the path of compound (**2**) is at lower energies than that of compound (**1**).

Successively, the lactonic ring can be opened by stretching the O–C_{10} bond, providing the (**A3**) dicarbonyl intermediate. In the next two steps the paths of compounds (**1**) and (**2**) differ slightly. The (**1_A3**) intermediate evolves, losing one CO molecule and forming (**1_A4**) that undergoes a conformational rearrangement of the aldehydic group, providing (**1_A5**). In the case of (**2_A3**), the conformational rearrangement of the aldehydic group to (**2_A4**) is required before losing CO and yielding (**2_A5**). A van der Waals adduct between (**A5**) and the leaving CO molecule is observed in both cases.

A collateral path (Scheme 8, Table 4) starts from (**A2**) and converges to the (**C3**) intermediate along path C (see below). The lactonic radical (**A2**) fuses into the bicyclic intermediate (**A2$_1$**), a planar structure for compound (**1**) and a highly bent structure in (**2**), overcoming significant barriers due to the formation of a 4-membered ring. The (**C3**) intermediate is then obtained through the loss of one CO_2 molecule from (**A2$_2$**). Contrary to the (**A2**)–(**A5**) steps, here the path of compound (**2**) is at higher energy than that of compound (**1**), due to the constraints imposed by two condensed 5-membered rings.

Path B. The first step in this path (Scheme 5, Table 3, Fig. 2) involves an intramolecular cyclization in (**Ar–OO˙**) by the terminal oxygen atom of the peroxy group,

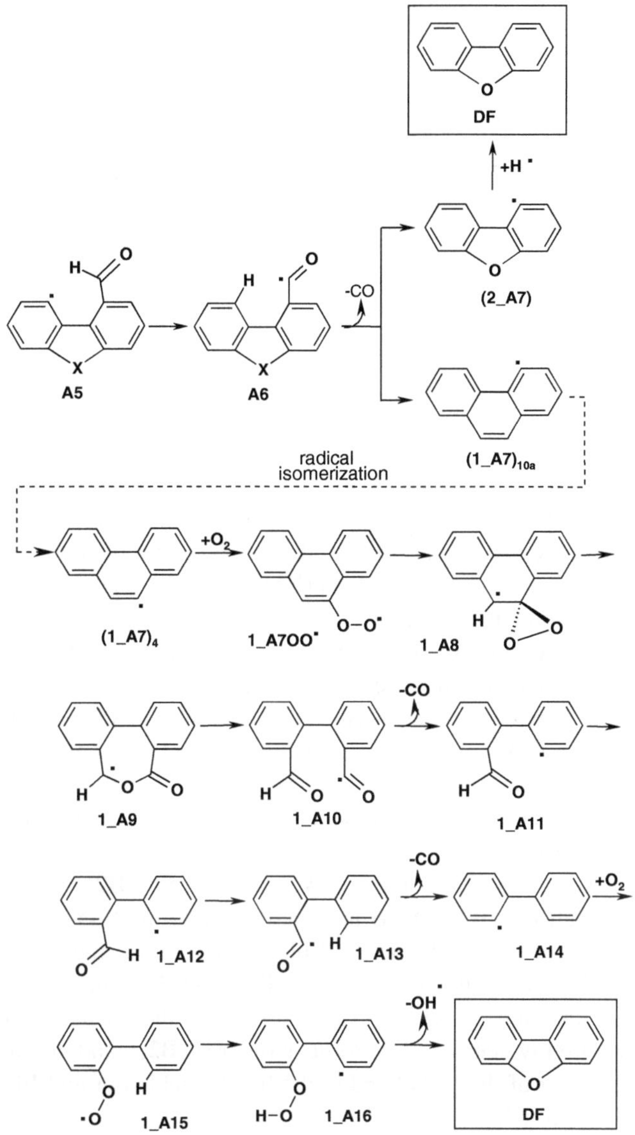

Scheme 7 Path from (A5) to DF

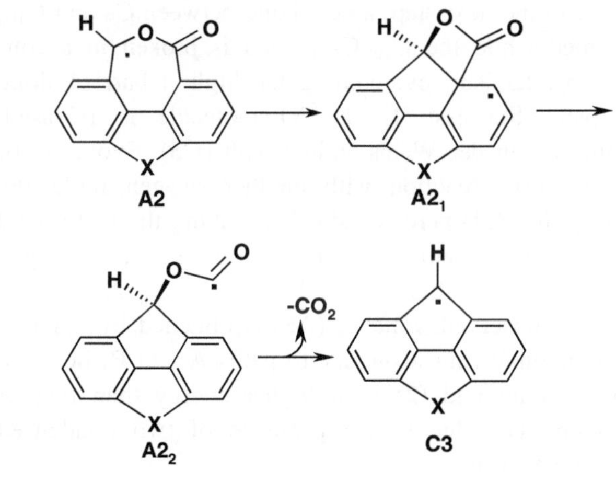

Scheme 8 Collateral path to A

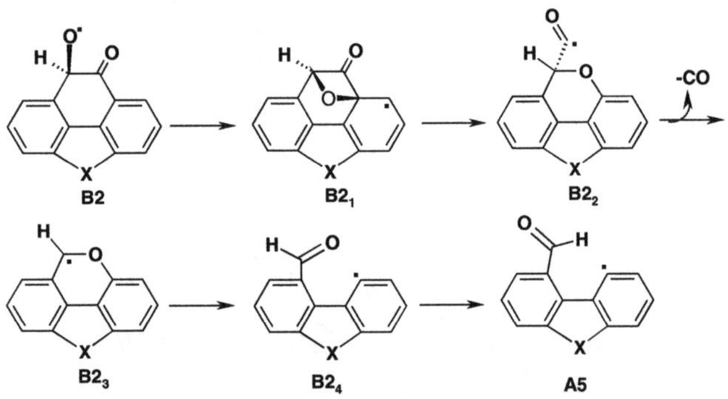

Scheme 9 Collateral path to B

producing the bicyclic structure (**B1**). The reaction involves the overcoming of a significant energetic barrier. Then, the O–O bond is cleaved: The (**B2**) intermediates are significantly more stable than (**B1**), due to the release of the steric hindrance imposed by the 4-membered ring. Successively, the C_9–C_{10} bond is cleaved, providing the dicarbonyl intermediates (**B3**) that, losing one CO molecule, provide van der Waals adducts between (**A5**) and the leaving CO molecule. As in path A, the intermediates and TSs between (**B2**) and (**A5**) of compound (**1**) are at higher energies than that of compound (**2**) due to their non-planar geometries.

An alternative route to (**A5**) starts from (**B2**), passes through the (**B2$_1$**)-(**B2$_4$**) intermediates and the loss of one CO molecule (Scheme 9, Table 4). The most significant barriers along this route involve (**B2$_1$**) and (**B2$_4$**) formation. The extent of these two barriers is due to the formation of the highly tensioned bicyclic structure (**B2$_1$**) and to the breaking of the C–O bond in the highly stable pyranyl-like structure (**B2$_3$**).

Path C. Path C involves O–O cleavage in (**Ar–OO$^·$**) and the formation of the aryloxy radical (**C1**) and the ground-state oxygen atom, O(^{3}P) (Scheme 6, Table 3, Fig. 3). In the next step, a new bond between C_9 and C_{10a} is formed while the C_{10}–C_{10a} bond is broken in a concerted mechanism, overcoming the highest barrier along this path. Successively, a CO molecule is released, forming a van der Waals adduct with (**C3**). Evolution of (**C3**) involves reaction with another oxygen molecule, yielding the (**C4**) peroxy radical. Breaking the O–O bond yields the (**C5**) aryloxy radical and a ground-state oxygen atom, O(^{3}P). For this high endothermic step, no discrete TSs were observed. Finally, (**A5**) is obtained by opening the pentatomic ring. Contrary to paths A and B, here the path of compound (**2**) is at higher energy than that of compound (**1**), due to the presence of two condensed 5-membered rings.

As mentioned above, (**C3**) can also be obtained from (**A2**) through the path collateral to path A.

4.3 From (A5) to DF

At this point, the three paths, A, B, and C, converge to the common (**A5**) intermediate that continues in evolution towards the DF final product (Scheme 7, Table 5, Fig. 4). After a hydrogen transfer from the aldehydic moiety and CO expulsion, a van der Waals adduct between (**A7**) and CO is obtained. In the case of (**1_A5**), the step involves a variational transition state, with a lower energy, by 0.9 kcal mol^{-1}, than the reagent.

The (**2_A7**) intermediate can react with a hydrogen donor species (HX), providing the final DF product. Instead, the (**1_A7**) phenanthrenic radical requires further oxidative decomposition steps. Indeed, only the attack of a new O_2 molecule at the C_3 or C_4 positions can provide the DF final product; thus, only this path was considered. The starting step of this process must involve radical isomerization between (**1_A7**)$_{10a}$ and (**1_A7**)$_4$; such isomerization involves hydrogen migration that can occur via consecutive intra-molecular paths or via inter-molecular paths assisted by a HX species. Whatever the isomerization process, only the steps subsequent to (**1_A7**)$_4$ formation are analysed in the following.

The (**1_A7**)$_4$ undergoes the attack of O_2, forming the (**1_A7OO$^·$**) peroxy radical that can follow three pathways analogous to those previously discussed for the phenyl and the (**Ar–OO$^·$**) peroxy radicals. Those results show that path A involves the overcoming of lower energetic barriers with respect to paths B and C (see Discussion). For this reason, only the A-type path was investigated to characterize the route from (**1_A7OO$^·$**) to DF.

The steps from (**1-A7OO$^·$**) to the (**1_A14**) biphenyl radical have already been described for the case of Pyrene/BDF, and are not further discussed here. From the (**1_A14**) intermediate, the oxidation process to DF continues with the formation of the (**1_A15**) peroxy radical, by reaction with another oxygen molecule, and of the (**1_A16**) hydroperoxy radical, by transfer of a hydrogen atom. After conformational rearrangement of the C–C–O–O torsion, the closure of the furanic ring and the simultaneous expulsion of an OH radical leads to the final Dibenzofuran product.

5 Discussion

The oxidative pathway of compounds (**1**) and (**2**) to DF initially requires the formation of a radical species (Ar$^·$), through the loss of a hydrogen atom. This occurs through reaction with molecular oxygen or radical species (X$^·$) present in the environment and involves very different activation energies. In the case of the benzene reaction with molecular oxygen, a barrier of about 61 kcal mol^{-1} was

Table 3 PBE1PBE/6-311G** relative energies (ΔE, kcal mol^{-1}) calculated with respect to the peroxy radicals (**1-OO**·) or (**2-OO**·) for transition states and intermediates involved along paths A, B, and C

	Pyrene				BDF		
	ΔE^{a}	E_{att}	E_{react}		ΔE^{a}	E_{att}	E_{react}
1-OO·	0			**2-OO**·	0		
Path A							
(1OO-A1)$_{TS}$	22.2	22.2		**(2OO-A1)**$_{TS}$	22.8	22.8	
1_A1	11.4		11.4	**2_A1**	10.9		10.9
1_(A1-A2)$_{TS}$	27.9	16.5		**2_(A1-A2)**$_{TS}$	27.1	16.2	
1_A2	−44.6		−56.1	**2_A2**	−68.6		−79.6
1_(A2-A3)$_{TS}$	−19.5	25.1		**2_(A2-A3)**$_{TS}$	−49.1	19.5	
1_A3	−20.6		24.1	**2_A3**	−50.0		18.7
1_(A3A4)$_{TS}$	4.55	25.1		**2_(A3-A4)**$_{TS}$	−38.5	11.5	
1_W(A4-CO)	−0.22		20.4	**2_A4**	−47.8		2.2
1_A4	1.1		1.3	**2_(A4-A5)**$_{TS}$	−19.1	28.7	
1_(A4-A5)$_{TS}$	5.00	3.9		**2_W(A5-CO)**	−21.8		25.9
1_A5	−2.0		−3.0	**2_A5**	−21.2		0.6
Path B							
(1OO-B1)$_{TS}$	40.5	40.5		**(2OO-B1)**$_{TS}$	38.3	38.3	
1_B1	33.8		33.8	**2_B1**	29.8		29.8
1_(B1-B2)$_{TS}$	34.0	0.2		**2_(B1-B2)**$_{TS}$	30.3	0.4	
1_B2	−33.3		−67.1	**2_B2**	−36.7		−66.6
1_(B2-B3)$_{TS}$	−17.6	15.8		**2_(B2-B3)**$_{TS}$	−36.1	0.7	
1_B3	−18.6		14.7	**2_B3**	−47.8		−11.0
1_(B3-A5)$_{TS}$	3.0	21.7		**2_(B3-A5)**$_{TS}$	−19.1	28.7	
1_W(A5-CO)	−3.2		15.4	**2_W(A5-CO)**	−21.8		26.0
1_A5	−2.0		1.3	**2_A5**	−21.2		0.5
Path C							
(1OO-C1)$_{TS}$	31.4	31.4		**(2OO-C1)**$_{TS}$	31.8	31.8	
1-W(C1-O)	28.5		28.5	**2_W(C1-O)**	28.2		28.2
1_C1	30.9		2.4	**2_C1**	30.9		2.7
1_(C1-C2)$_{TS}$	91.3	60.4		**2_(C1-C2)**$_{TS}$	118.3	87.4	
1_C2	65.2		34.4	**2_C2**	97.3		66.4
1_(C2-C3)$_{TS}$	72.4	7.2		**2_(C2-C3)**$_{TS}$	104.1	6.8	
1_W(C3-CO)	60.1		−5.2	**2_W(C3-CO)**	93.0		−4.3
1_C3	60.5		0.4	**2_C3**	93.4		0.4
1_W(C3-O$_2$)	59.8		−0.7	**2_W(C3-O$_2$)**	92.7		−0.7
1_(C3-C4)$_{TS}$	64.9	5.1		**2_(C3-C4)**$_{TS}$	98.2	5.5	
1_C4	51.2		−8.6	**2_C4**	83.0		−9.7
1_C5	108.7		57.5	**2_C5**	139.4		56.4
1_(C5-A5)$_{TS}$	125.1	16.4		**2_(C5-A5)**$_{TS}$	145.0	5.6	
1_A5	117.4		8.7	**2_A5**	98.1		−41.3

Barriers (E_{att}, kcal mol^{-1}) and reaction energies ($E_{react} = E_{products} - E_{reagents}$, kcal mol^{-1}) for each step, are also reported. Transition states are labelled with TS subscript; van der Waals adducts includes a W in the name

[a] As an example: $\Delta E_{(1\_A5)} = [E_{(1\_A5)} + E_{(CO)}] - E_{(1\text{-}OO)}$ in path A and B; $\Delta E_{(1\_A5)} = [E_{(1\_A5)} + E_{(CO)} + 2\,E_{(O)}] - [E_{(1\text{-}OO)} + E_{(O2)}]$ in path C

found [13]. Hydrogen abstraction reactions in acenaphthylene, phenanthrene, naphthalene and biphenyl by H or CH_3 radicals involve activation energies in the range 7–10 kcal mol^{-1} [28]. Moreover, it is well known that copper and iron, as the metal itself or their oxides and salts, catalyse gas-carbon reactions, acting as a dissociation centre for O_2 chemisorption [3]. To overcome the problem of choosing a specific route to (**Ar**·) formation, we assumed

Table 4 PBE1PBE/6-311G** relative energies (ΔE, kcal mol^{-1}) calculated with respect to the peroxy radicals (**1-OO**˙) or (**2-OO**˙) for transition states and intermediates involved along collateral paths to A and B (see text for description)

	Pyrene				BDF		
	ΔE	E_{att}	E_{react}		ΔE	E_{att}	E_{react}
Collateral path to A							
1_A2	−44.6			**2_A2**	−68.6		
1_(A2-A2$_1$)$_{TS}$	−17.8	26.8		**2_(A2-A2$_1$)$_{TS}$**	−9.9	58.8	
1_A2$_1$	−32.1		12.5	**2_A2$_1$**	−12.4		56.2
1_(A2$_1$-A2$_2$)$_{TS}$	−20.0	12.2		**2_(A2$_1$-A2$_2$)$_{TS}$**	6.0	18.5	
1_A2$_2$	−29.9		2.3	**2_A2$_2$**	2.1		14.6
1_(A2$_2$-C3)$_{TS}$	−12.5	17.4		**2_(A2$_2$-C3)$_{TS}$**	21.6	19.5	
1_W(C3-CO$_2$)	−71.5		−41.6	**2_W(C3-CO$_2$)**	−38.6		−40.8
1_C3	−70.3		1.2	**2_C3**	−37.4		1.3
Collateral path to B							
1_B2	−33.3			**2_B2**	−36.7		
1_(B2-B2$_1$)$_{TS}$	−9.9	23.4		**2_(B2-B2$_1$)$_{TS}$**	−4.0	32.7	
1_B2$_1$	−18.4		14.9	**2_B2$_1$**	−7.2		29.5
1_(B2$_1$-B2$_2$)$_{TS}$	−12.4	6.0		**2_(B2$_1$-B2$_2$)$_{TS}$**	−2.1	5.1	
1_B2$_2$	−33.0		−14.6	**2_B2$_2$**	−28.8		−21.5
1_(B2$_2$-B2$_3$)$_{TS}$	−29.8	3.2		**2_(B2$_2$-B2$_3$)$_{TS}$**	−24.6	4.2	
1_W(B2$_3$-CO)	−41.5		−8.5	**2_W(B2$_3$-CO)**	−35.3		−6.5
1_B2$_3$	−41.4		0.1	**2_B2$_3$**	−34.6		0.6
1_(B2$_3$- B2$_4$)$_{TS}$	6.3	47.7		**2_(B2$_3$-B2$_4$)$_{TS}$**	0.2	34.9	
1_B2$_4$	1.1		42.5	**2_B2$_4$**	−21.0		13.6
1_(B2$_4$-A5)$_{TS}$	5.00	3.9		**2_(B2$_4$-A5)$_{TS}$**	−12.4	8.6	
1_A5	−2.0		−3.0	**2_A5**	−21.2		−0.2

Barriers (E_{att}, kcal mol^{-1}) and reaction energies ($E_{react} = E_{products} - E_{reagents}$, kcal mol^{-1}) for each step, are also reported. Transition states are labelled with TS subscript; van der Waals adducts includes a W in the name

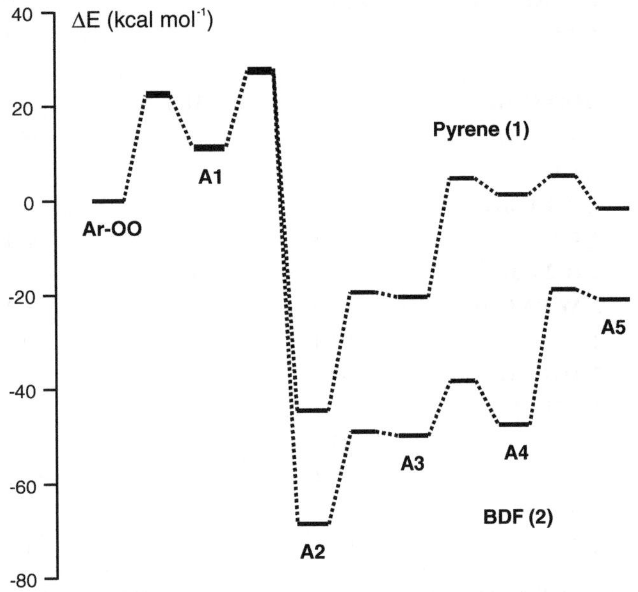

Fig. 1 Relative energies (ΔE, kcal mol^{-1}), calculated with respect to (**Ar–OO**˙), along path A for Pyrene (**1**) and Benzodibenzofuran (**2**)

the (**Ar**˙) species to be already formed, regardless of the reactions involved.

The attack of molecular oxygen to non-equivalent positions in (**Ar**˙) provides different (**Ar–OO**˙) peroxy radicals that undergo different oxidative paths, as in the case of the phenyl radical [17]. However, only some of these paths lead to the formation of DF, and it is these that were investigated.

Once formed, the (**Ar–OO**˙) peroxy radical can follow three possible paths (A, B and C) to the (**A5**) common intermediate. It is noteworthy that the first step along the three paths presents similar activation energies to the corresponding steps in vinyl, phenyl and Dibenzofuranyl peroxy radical reactions. This finding further supports analogies already observed for olefinic and aromatic systems [14].

Compounds (**1**) and (**2**) present similar energetic profiles in the three paths, but the paths show different features.

Along path A, the highest energetic barrier corresponds to the CO release step, with a barrier of 25.1 and 28.7 kcal mol^{-1} for compounds (**1**) and (**2**), respectively. From (**A2**), the path of compound (**2**) is always at lower energy than that of compound (**1**), which presents non-planar intermediates due to the steric hindrance between the orto substituents.

In path B, the highest barrier is found in the first step involving intramolecular cyclization to (**B1**), with barriers of 40.5 and 38.3 kcal mol^{-1} for compounds (**1**) and (**2**), respectively. Also in path B, compound (**2**) intermediates are lower in energy than those of compound (**1**), because of the non-planarity of Pyrene intermediates.

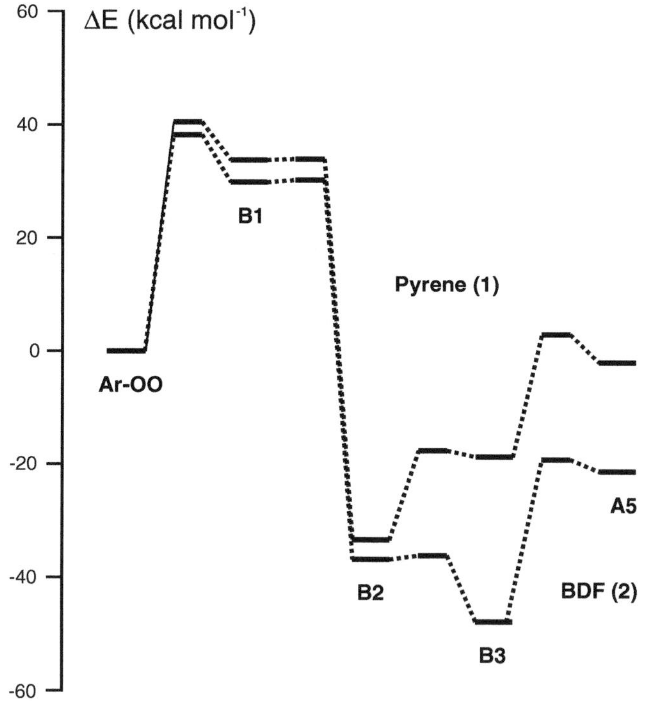

Fig. 2 Relative energies (ΔE, kcal mol^{-1}), calculated with respect to (**Ar–OO·**), along path B for Pyrene (**1**) and Benzodibenzofuran (**2**)

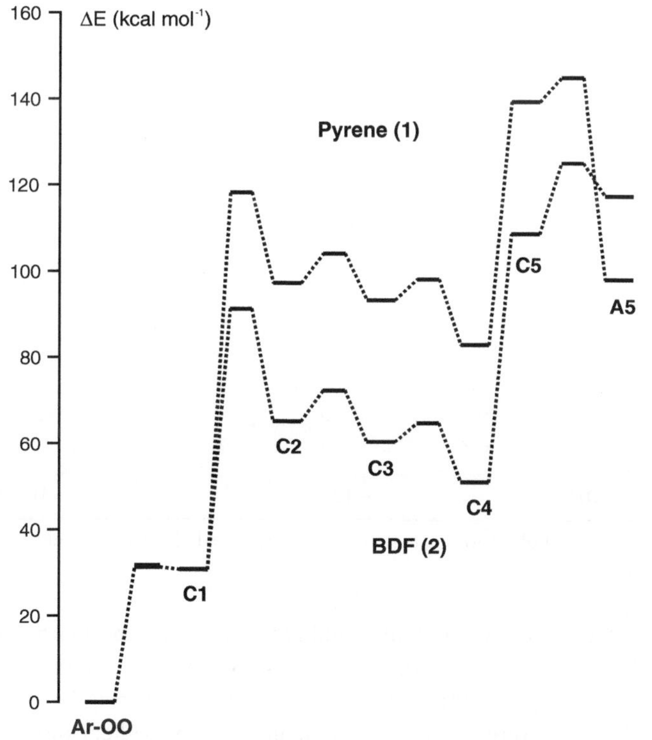

Fig. 3 Relative energies (ΔE, kcal mol^{-1}), calculated with respect to (**Ar–OO·**), along path C for Pyrene (**1**) and Benzoidbenzofuran (**2**)

Along path C, both compounds encounter the highest barrier in the formation of the 5-membered ring, (**C1**) to (**C2**) step, with barriers of 60.4 and 87.4 kcal mol^{-1}, respectively. Here, the Pyrene intermediates are planar while those of (**2**) are bent due to the presence of two condensed penta-atomic rings. Thus in this case compound (**1**), intermediates lie at a lower energy than those of compound (**2**).

Two possible collateral paths were also identified. The first starts from (**A2**) and leads to the (**C3**) intermediate, losing a CO_2 molecule. This route to (**C3**) is competitive with path C as it involves the overcoming of lower barriers, 26.8 and 58.8 kcal mol^{-1} for compounds (**1**) and (**2**), respectively. The second collateral path starts from (**B2**) and leads to the formation of the (**A5**) intermediate. This route is not competitive with path B as it involves the overcoming of a significant barrier (47.7 and 34.9 kcal mol^{-1}, for compounds (**1**) and (**2**), respectively) when the C–O bond in the (**B2$_3$**) pyranyl-like structure is broken.

As a general remark, the different steric constraints in compounds (**1**) and (**2**) imposed by the presence of a benzenic or a furanic ring condensed to the phenanthrenic system affects the planarity of the intermediates along the three paths. This molecular characteristic plays a role in determining the relative stability of the intermediates but affects the energetic barriers to a lesser extent, at least in the case of path A.

Comparison between the different paths of compounds (**1**) and (**2**) to (**A5**) shows that path A is always the kinetically favoured one, and the highest barrier is encountered when the CO molecule is released. In the case of compound (**1**), the highest barrier (26.8 kcal mol^{-1}) in the collateral channel involving CO_2 release and (**C3**) formation is comparable with that involving CO release in path A (25.1 kcal mol^{-1}). However, during the reaction steps from (**C3**) to (**A5**), a significant barrier of 57.5 kcal mol^{-1} has to be overcome.

Evolution of the (**A5**) intermediate involves the release of another CO molecule, overcoming barriers of 26.5 and 30.1 kcal mol^{-1} for the respective compounds (**1**) and (**2**), with the formation of the Phenanthrene (**1_A7**) and Dibenzofuran (**2_A7**) radicals. While compound (**2**) completes its oxidative path to DF, the (**1_A7**) radical must undergo further steps towards the final DF product. The first step involves a radical isomerization between C_{10a} and C_4. As previously noted for (**Ar·**) formation, the presence of a hydrogen donor HX species in the reaction environment can assist the isomerization process, lowering the high barriers (about 60 kcal mol^{-1}) expected for an intramolecular mechanism. Once the (**1_A7**)$_4$ radical has been formed, four high energetic barriers (23–29 kcal mol^{-1}) are encountered towards DF formation; two of these barriers involve the loss of a CO molecule.

Our results can be compared with temperature-programmed desorption (TPD) spectra from disperse

Table 5 PBE1PBE/6-311G** relative energies (ΔE, kcal mol^{-1}) calculated with respect to the peroxy radicals (**1-OO$^{\cdot}$**) or (**2-OO$^{\cdot}$**) for transition states and intermediates from (**A5**) to DF

	Pyrene				BDF		
	ΔE	E_{att}	E_{react}		ΔE	E_{att}	E_{react}
1-A5	−2.0			**2_A5**	−21.2		
1_(A5-A6)$_{TS}$	−2.9	−0.9		**2_(A5-A6)$_{TS}$**	−8.3	13.0	
1-A6	−21.6		−19.7	**2_A6**	−43.9		−22.7
1_(A6-A7)$_{TS}$	4.8	26.5		**2_(A6-A7)$_{TS}$**	−13.9	30.1	
1_W(A7-CO)	2.0		23.7	**2_W(A7-CO)**	−14.4		29.5
(1_A7)$_{10a}$	2.8		0.8	**2_A7**	−13.7		0.8
(1_A7)$_4$	4.6		−0.1				
1_A7OO$^{\cdot}$	−39.5						
1_(A7OO-A8)$_{TS}$	−16.9	22.6					
1_A8	−27.1		12.4				
1_(A8-A9)$_{TS}$	−10.8	16.3					
1_A9	−86.4		−59.2				
1_(A9-A10)$_{TS}$	−63.2	23.1					
1_A10	−65.9		20.4				
1_(A10-A11)$_{TS}$	−39.7	26.3					
1_W(A11-CO)	−41.1		24.8				
1_A11	−39.8		1.3				
1-(A11-A12)$_{TS}$	−35.5	4.4					
1_A12	−42.8		−3.0				
1_(A12-A13)$_{TS}$	−42.1	0.7					
1-A13	−64.9		−22.1				
1_(A13-A14)$_{TS}$	−36.2	28.7					
1_W(A14-CO)	−37.8		27.0				
1_A14	−37.1		0.8				
1_W(A14-O$_2$)	−38.0		−1.0				
1_(A14-A15)$_{TS}$	−37.9	0.1					
1_A15	−75.6		−37.6				
1_(A15-A16)$_{TS}$	−50.3	25.3					
1_A16	−49.1		26.5				
1_(A16-A16′)$_{TS}$	−48.5	0.6					
1_A16′	−49.6		−0.4				
1_(A16′-DF)$_{TS}$	−42.6	6.5					
1_W(DF-OH)	−118.1		−69.0				
DF	−113.5		4.6	**DF**	−121.9		−108.2

Barriers (E_{att}, kcal mol^{-1}) and reaction energies ($E_{react} = E_{product} - E_{reagents}$, kcal mol^{-1}) for each step, are also reported. Transition states are labelled with TS subscript; van der Waals adducts includes a W in the name

polycrystalline graphite oxidized with CO [29]. These spectra show two main broad CO evolution features in the 973–1,253 K and 400–700 K range. Activation desorption energies (64–83 kcal mol^{-1}) derived from high-temperature measurements have been associated with the desorption of a CO group belonging to 6-membered ring systems (usually referred to in the literature as semiquinone). In the low-temperature range, activation desorption energies (25–44 kcal mol^{-1}) have been associated with CO desorption from ketene structures. Indeed, in our calculations, along path A and along the route from (**A5**) to DF, the highest energetic barriers are always encountered in the steps where a CO molecule is released from ketene-like structures, the values spanning 25–30 kcal mol^{-1}. Moreover, along path C, the rate determining step in CO desorption is the rearrangement of the (**C1**) semiquinone structure, involving a barrier of 60.4 and 87.4 kcal mol^{-1} in compounds (**1**) and (**2**), respectively. Thus, our results compare well with the reported TPD energies and structural features proposed by Marchon et al. [29].

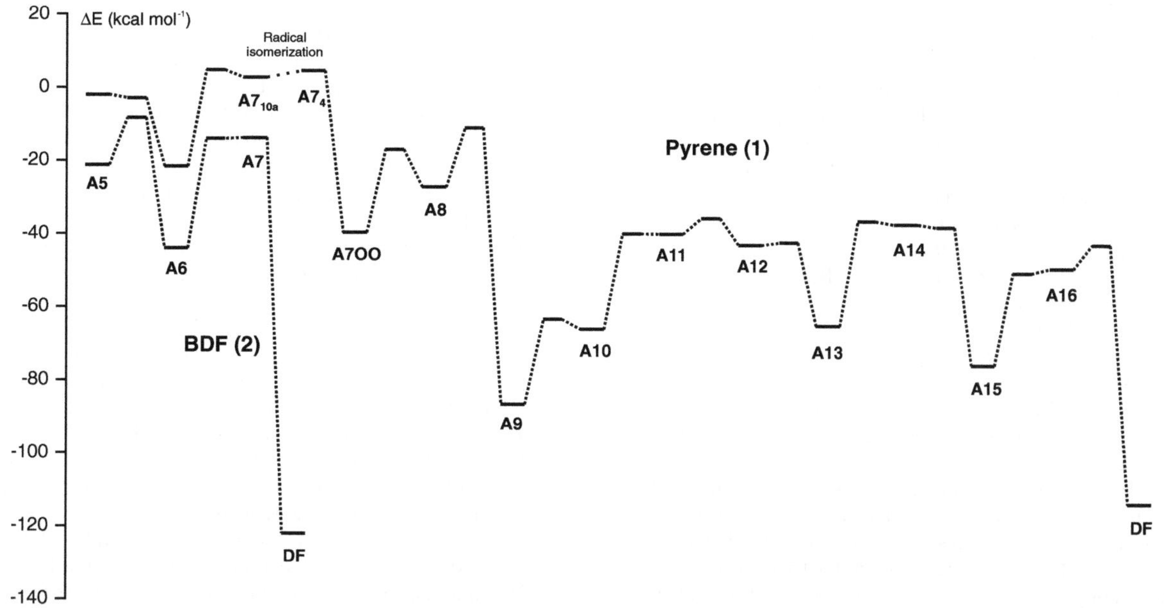

Fig. 4 Evolution from (**A5**) to DF: relative energies (ΔE, kcal mol^{-1}) calculated with respect to (**Ar–OO˙**), for Pyrene (**1**) and Benzodibenzofuran (**2**)

Our results also agree with the experimental evidence obtained for the oxidation of Pyrene [9] that shows the formation, among other trace compounds, of DF, phenanthrene and biphenyl. Indeed, along our calculated path, we observed the formation of both (**1_A7**) phenanthrene and (**1_A14**) biphenyl radicals.

6 Conclusion

The present work is a first attempt to shed light on the mechanism of PCDD/F formation by de novo synthesis, a trace reaction that occurs in the "cold zones" of flue gas control devices of MSW incinerators. The aim of studying such formation mechanisms is to find a way to prevent and/or minimize PCDD/F formation.

Here we propose a model for the de novo synthesis of Dibenzofuran from Pyrene and Benzodibenzofuran, chosen as model compounds of carbonaceous material present in fly-ash. Our calculations have led to the identification of a kinetically preferred pathway, though the temperature dependence of energetic barriers could increase the relevance of other higher energetic paths.

Along the preferred pathway, different steric constraints in compounds (**1**) and (**2**) play a role in determining the relative stability of the intermediates; however, such constraints have less influence on the energetic barriers. Thus, we believe that compounds (**1**) and (**2**) should present similar kinetic behaviour as the energetic barriers to be overcome are similar.

Future work will investigate the reciprocal influence of chlorination and oxidation reactions in polychlorinated Dibenzofuran formation.

Acknowledgments We gratefully acknowledge financial support from the University of Milano-Bicocca (FAR2010). We wish to thank Prof. V. Barone, Scuola Normale di Pisa, Italy, for his constant encouragement and stimulating discussions.

References

1. Grandesso E, Ryan S, Gullett B, Touati A, Collina E, Lasagni M, Pitea D (2008) Kinetic modeling of polychlorinated Dibenzo-*p*-Dioxin and dibenzofuran formation based on carbon degradation reactions. Environ Sci Technol 42:7218–7224
2. Altarawneh M, Dlugogorski BZ, Kennedy EM, Mackie JC (2009) Mechanisms for formation, chlorination, dechlorination and destruction of polychlorinated Dibenzo-*p*-Dioxins and dibenzofurans (PCDD/Fs). Prog Energy Combust Sci 35:245–274
3. Huang H, Buekens A (1996) De novo synthesis of polychlorinated Dibenzo-*p*-Dioxins and dibenzofurans. Proposal of a mechanistic scheme. Sci Total Environ 193:121–141
4. Stanmore BR (2004) The formation of dioxins in combustion systems. Combust Flame 136:398–427
5. Pitea D, Bortolami M, Collina E, Cortili G, Franzoni F, Lasagni M, Piccinelli E (2008) Prevention of PCDD/F formation and minimization of their emission at the stack of a secondary aluminum casting plant. Environ Sci Technol 42:7476–7481
6. Iino F, Imagawa T, Takeuchi M, Sadakata M (1999) De novo synthesis mechanism of polychlorinated dibenzofurans from polycyclic aromatic hydrocarbons and the characteristic isomers of polychlorinated naphthalenes. Environ Sci Technol 33:1038–1043

7. Weber R, Iino F, Imagawa T, Takeuchi M, Sakurai T, Sadakata M (2001) Formation of PCDF, PCDD, PCB, and PCN in de novo synthesis from PAH: mechanistic aspects and correlation to fluidized bed incinerators. Chemosphere 44:1429–1438
8. DeCoster J, Ergut A, Levendis YA, Richter H, Howard JB, Carlson JB (2007) PAH emissions from high-temperature oxidation of vaporized anthracene. Proc Combust Inst 31:491–499
9. Fullana A, Sidhu SS (2005) Fate of PAHs in the post-combustion zone: partial oxidation of PAHs to dibenzofuran over CuO. J Anal Appl Pyrolysis 74:479–485
10. Marquaire P, Worner R, Rambaud P, Baronnet F (1999) Organohalogen Compd 40:419–422
11. Marquaire PM, Bounaceur R, Baronnet F (2003) Detailed mechanism of dibenzofuran oxidation. Organohalogen Compd 63:377–380
12. Marquaire PM, Wörner R, Baronnet F (2002) Dioxins abatement by total oxidation. IT3'02 conference, May 13–17, New Orleans, Louisiana
13. Altarawneh M, Dlugogorski BZ, Kennedy EM, Mackie JC (2006) Quantum chemical study of low temperature oxidation mechanism of dibenzofuran. J Phys Chem A 110:13560–13567
14. Sebbar N, Bockhorn H, Bozzelli JW (2008) Thermochemical similarities among three reaction systems: Vinyl + O_2 – Phenyl + O_2 – Dibenzofuranyl + O_2. Combust Sci Tech 180:959–974
15. Altarawneh M, Dlugogorski BZ, Kennedy EM, Mackie JC (2007) Theoretical study of reaction pathways of dibenzofuran and Dibenzo-*p*-Dioxin under reducing conditions. J Phys Chem A 111:7133–7140
16. Mebel AM, Diau EWG, Lin MC, Morokuma K (1996) Ab initio and RRKM calculations for multichannel rate constants of the C_2H_3 + O_2 reaction. J Am Chem Soc 118:9759–9771
17. Tokmakov IV, Kim-Sue G, Kislov VV, Mebel AM, Lin MC (2005) The reaction of phenyl radical with molecular oxygen: a G2 M study of the potential energy surface. J Phys Chem A 109:6114–6127
18. Sebbar N, Bockhorn H, Bozzelli J (2008) Thermodynamic properties of the species resulting from the phenyl radical with O_2 reaction system. Int J Chem Kinet 40:583–604
19. Sebbar N, Bockhorn H, Bozzelli JW (2005) Thermochemical properties, rotation barriers, and group additivity for unsaturated oxygenated hydrocarbons and radicals resulting from reaction of vinyl and phenyl radical systems with O_2. J Phys Chem A 109:2233–2253
20. Sebbar N, Bockhorn H, Bozzelli JW (2004) Thermochemical properties, rotation barriers, bond energies, and group additivity for vinyl, phenyl, ethynyl, and allyl peroxides. J Phys Chem A 108:8353–8366
21. Ghigo G, Maranzana A, Tonachini G, Zicovich-Wilson CM, Causà M (2004) Modeling soot and its functionalization under atmospheric or combustion conditions by density functional theory within molecular (Polycyclic-Aromatic-Hydrocarbon-like) and periodic methodologies. J Phys Chem B 108:3215–3223
22. Perry ST, Hambly EM, Fletcher TH, Solum MS, Pugmire RJ (2000) Solid-state 13C NMR characterization of matched tars and chars from rapid coal devolatilization. Proc Combust Inst 28:2313–2319
23. Homann KH (1998) Fullerenes and soot formation: new pathways to large particles in flames. Angew Chem Int Ed 37:2434–2451
24. Frisch MJ, Trucks GW, Schlegel HB, Scuseria GE, Robb MA, Cheeseman JR, Scalmani G, Barone V, Mennucci B, Petersson GA, Nakatsuji H, Caricato M, Li X, Hratchian HP, Izmaylov AF, Bloino J, Zheng G, Sonnenberg JL, Hada M, Ehara M, Toyota K, Fukuda R, Hasegawa J, Ishida M, Nakajima T, Honda Y, Kitao O, Nakai H, Vreven T, Montgomery JrJA, Peralta JE, Ogliaro F, Bearpark M, Heyd JJ, Brothers E, Kudin KN, Staroverov VN, Kobayashi R, Normand J, Raghavachari K, Rendell A, Burant JC, Iyengar SS, Tomasi J, Cossi M, Rega N, Millam NJ, Klene M, Knox JE, Cross JB, Bakken V, Adamo C, Jaramillo J, Gomperts R, Stratmann RE, Yazyev O, Austin AJ, Cammi R, Pomelli C, Ochterski JW, Martin RL, Morokuma K, Zakrzewski VG, Voth GA, Salvador P, Dannenberg JJ, Dapprich S, Daniels AD, Farkas Ö, Foresman JB, Ortiz JV, Cioslowski J, Fox DJ (2009) Gaussian 09, Revision A.02. Gaussian Inc. Wallingford CT
25. Adamo C, Barone V (2002) Physically motivated density functionals with improved accuracy: the modified Perdew-Burke-Ernzerhof model. J Chem Phys 116:5933–5940
26. Menon AS, Radom L (2008) Consequences of spin contamination in unrestricted calculations on open-shell species: effect of Hartree–Fock and Møller–Plesset contributions in hybrid and double-hybrid density functional theory approaches. J Phys Chem A 112:13225–13230
27. Mebel AM, Morokuma K, Lin MC (1995) Modification of the Gaussian-2 theoretical model: the use of coupled-cluster energies, density-functional geometries, and frequencies. J Chem Phys 103:7414–7421
28. Carissan Y, Klopper W (2010) Hydrogen abstraction from biphenyl, acenaphthylene, naphthalene and phenanthrene by atomic hydrogen and methyl radical: DFT and G3(MP2)-RAD data. J Mol Struct THEOCHEM 940:115–118
29. Marchon B, Tysoe WT, Carrazza J, Heinemann H, Somorjai GA (1988) Reactive and kinetic properties of carbon monoxide and carbon dioxide on a graphite surface. J Phys Chem 92:5744–5749

Theor Chem Acc (2012) 131:1183
DOI 10.1007/s00214-012-1183-1

REGULAR ARTICLE

Modelization of vibrational spectra beyond the harmonic approximation from an iterative variation–perturbation scheme: the four conformers of the glycolaldehyde

Philippe Carbonniere · Claude Pouchan

Received: 19 July 2011 / Accepted: 19 October 2011 / Published online: 6 March 2012

Abstract This paper presents the computed anharmonic frequencies and IR intensities in the mid-infrared region for the four conformers of glycolaldehyde (Cis cis, Trans trans, Trans gauche and Cis trans forms). The fundamental transitions and their connected overtones and combination bands through strong anharmonic couplings (Fermi resonances) are provided. The results are stemmed from an iterative variational–perturbational resolution of the vibrational problem implemented in the VCI-P code. The four potential electronic surfaces are built as a Taylor series truncated to the fourth order around each minimum geometry. The second derivatives with respect to the normal coordinates were computed at the CCSD(T)/cc-pVTZ level, while the third and fourth derivatives were estimated with the B3LYP/6-31 + G(d,p) model chemistry. For the most stable Cc form, an average deviation of about 10 cm^{-1} is obtained with respect to the unambiguous experimental values. Furthermore, some of the transitions observed in the CH stretchings region were reassigned. The theoretical values calculated for the Tt and Tg forms are compared to the experimental data obtained from the irradiation of the Cc conformer isolated in Ar matrix with an IR source.

Keywords Ab initio · DFT · Anharmonic vibrational spectra · Glycolaldehyde · IR intensities

Dedicated to Professor Vincenzo Barone and published as part of the special collection of articles celebrating his 60th birthday.

P. Carbonniere (✉) · C. Pouchan (✉)
Groupe Chimie Théorique et Réactivité, ECP, IPREM UMR CNRS 5254, Université de Pau et des Pays de l'Adour 2, Rue Jules Ferry, 64053 Pau, France
e-mail: philippe.carbonniere@univ-pau.fr

C. Pouchan
e-mail: claude.pouchan@univ-pau.fr

1 Introduction

The vibrational fingerprint of a molecular system that infrared (IR) or Raman spectroscopy allows to observe is a widely used information for substance recognition and for the understanding of its behavior within a chemical environment. Thus, it is also used for applications such as structural and conformational analysis [1–3] reactivity monitoring [4, 5] and molecular design [6].

The subtle interplay between the different effects that induce the experimental picture of such fingerprint makes the identification task non-trivial, particularly when the molecular system presents strong anharmonic couplings giving rise to IR (or Raman)-active overtones and combination bands. From a theoretical point of view, the need to go beyond the harmonic approximation for an accurate modelization of a vibrational spectrum requires to face the computational cost that makes the treatment impracticable in its most complete formalism.

In the framework of the time independent formalism, several models have been developed and implemented to explicitly take into account anharmonicity. For small molecules (3–5 atoms), converged rovibrational levels can be obtained by fully variational methods, *that is,* the discrete variable representation (DVR) [7] or the vibrational configuration interaction (VCI) [8] methods. For larger molecules, four approximate schemes have been proposed. The vibrational self-consistent field (VSCF) approach represents the total vibrational wave function by a separable product of single-mode wave functions optimized separately using an effective mean field potential [9]. For a better accuracy, the correlation between modes is commonly treated by (i) the vibrational Moller–Plesset perturbation theory (VMP) [10] that is computationally cheap but overestimates the strong anharmonic complings, (ii) a vibrational configuration interaction (VCI) [8] that treats properly the strongest interactions but proves much more time consuming, (iii) the vibrational mean field configuration interaction (VMFCI) [11] that plays on the partitioning of the vibrational modes and encompasses both VSCF and VCI as particular case and (iv) the vibrational coupled cluster (VCC) [12] level of theory. Several strategies to strongly reduce the computational cost of such treatment were proposed in the literature [13–16] (see Ref. [16] for a more detailed description). A possible way which is used in the present study is to take advantage of both perturbative and variational approaches [17]. This variational–perturbational scheme in its most advanced version has been implemented in the VCI-P code [16] that uses small VCI matrices to treat the strongest interactions while the myriads of the weakest interactions are treated perturbationally (see the section computational details and method). Such a recipe was used to investigate the vibrational properties of the four conformers of glycolaldehyde (CH_2OHCHO).

The glycolaldehyde is the first monosaccharide detected in the interstellar medium which has increased the interest in the study of isomerism in interstellar chemistry (see Ref. [18] and therein). Among its four conformers [18, 19] so-called Cc, Tt, Tg and Ct (see Fig. 1), only the Cc conformer was detected from IR/Raman spectroscopy in gas phase [20–23] or in matrix isolation [24, 25], while the Tt form is observed from IR-induced conformer interconversion process in low-temperature matrices.

In the present study, we report the computed anharmonic fundamental transitions and their connected overtones or combination transitions through Fermi resonances of the four conformers of glycolaldehyde at least for two reasons (i) because of the presence of strong anharmonic couplings in the CH stretching region in which the assignment of the vibrational transitions appears non-univocal for the Cc and Tt conformers and (ii) because no experimental data are reported for the Tg and Ct forms.

2 Computational details and method

Computations at the CCSD(T) [26] level of theory were performed with the CFOUR [27] program, while Gaussian 09 [28] was used for the DFT computations. The anharmonic vibrational treatment relies on a hybrid [29] quartic force field in which the second derivatives are computed at the CCSD(T)/cc-pVTZ [30] level of theory, while the third and fourth derivatives are computed at the DFT level using the B3LYP functional [31, 32] with a valence double zeta Pople basis set including diffuse and polarization functions, 6–31+G(d,p) [33]. This model chemistry was chosen since it has been previously shown [34, 35] that for the prediction of harmonic and anharmonic force constants and in the case of small organic systems, it is able to approach the results obtained using the more expensive CCSD(T)/cc-pVTZ level of theory by an average value of 10 cm^{-1} on the fundamental transitions.

The analytic model of the potential functions is determined as follows [36]: from a minimum energy structure of a given system, a quartic force field is built in which the third and fourth derivatives are computed by 6 N–11 (N = number of atoms) numerical differentiation of analytical second derivatives. In the quartic approximation, the potential provided by the Gaussian code is a Taylor series in normal coordinates limited to the 3-mode interactions. From a technical point of view, it can be shown [36] that the best setup is obtained using a step size of 0.01 Å for the numerical differentiation of harmonic frequencies, tight convergence criteria for structural optimizations and fine grids for integral evaluation (that is, at least 99 radial and 590 angular points).

The Hamiltonian used is the pure vibrational Hamiltonian as detailed in Ref. [37]:

$$\hat{H} = \frac{1}{2}\sum_i \omega_i p_i^2 + V_{(q_1,\ldots,q_M)} + \sum_\alpha B_\alpha \left(\sum_{i,j\neq i,k,l\neq k} \zeta_{ij}^\alpha \zeta_{kl}^\alpha q_i p_j q_k p_l \sqrt{\frac{\omega_j \omega_l}{\omega_i \omega_k}} \right) \quad (1)$$

where q_i and p_i are, respectively, the dimensionless normal coordinates and their conjugate momenta. $V(q1,\ldots, qM)$ is a polynomial expansion of the PES in terms of normal coordinates q_i truncated to the fourth order The last term represents the major component of the rotational contribution to the anharmonicity in which B_α is the rotational constant of the system with respect to the Cartesian axis α, and ζ_{ij}^α is the Coriolis constant that couples q_i and q_j through the rotation about the α axis.

Cc	Tt	Tg	Ct

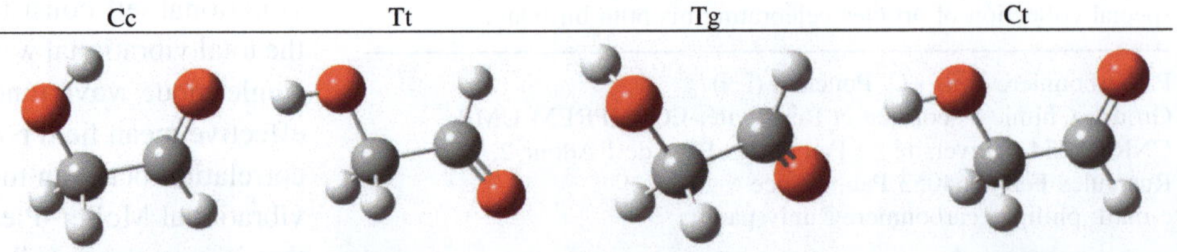

Cc	Tt	Tg	Ct
0	13,3[a]	14,2[a]	21,2[a]
0	15,3[b]	15,5[b]	22,3[b]

Fig. 1 The four conformers of glycolaldehyde and their corresponding relative energy (in $kJ\cdot mol^{-1}$) computed at the [a]CCSD(T)/cc-pVTZ level of theory (our work) and [b]MP4(SDTQ)/cc-pVQZ (see Ref. [18])

Table 1 Calculated wavenumbers (cm^{-1}) for the glycolaldehyde in its Cc form with the VCI-P method from a CCSD(T)/cc-pVTZ//B3LYP/6-31G(d,p) hybrid quartic force field

Description	Mode	Experimental			Calculations				
		Gas phase[a]	Ar[b]	Ar[c]	ν^{d}_{VMP2}	ν^{d}_{VCI-P}	I_ν[d]	Description[d]	ν^{e}
OH stretch	ν_{18}	3,549	3,541	3,542	3,542	3,570	54	81% ω_{18} + 7% $2\omega_{18}$	3,624
as. CH_2 stretch	ν_{17}	2,881	2,895	2,855	2,876	2,898	13	74% ω_{17} + 16% $\omega_{16} + \omega_{17}$	2,921
	$\nu_9 + \nu_{12}$				2,649	2,650	1	83% $\omega_9 + \omega_{12}$ + 4% ω_{17}	
	$\nu_9 + \nu_{10}$				2,503	2,496	1	83% $\omega_9 + \omega_{10}$ + 4% ω_{17}	
	$2\nu_{13}$				2,798	2,913	14	58% $2\omega_{13}$ + 15% ω_{16}	
	$\nu_{12} + \nu_{13}$				2,882	2,883	9	54% $\omega_{12} + \omega_{13}$ + 16% $2\omega_4 + \omega_{12}$ + 10% ω_{16}	2,836
	$2\nu_4 + \nu_{13}$					2,874	11	32% $2\omega_4 + \omega_{13}$ + 27% $\omega_{12} + \omega_{13}$ + 13% ω_{16}	
s. CH_2 stretch	ν_{16}	2,840	2,845	2,895	2,999				2,967
	$2\nu_4 + \nu_{11}$					2,794	6	55% $2\omega_4 + \omega_{11}$ + 16% $4\omega_4$ + 4% ω_{16}	
	$\nu_{10} + \nu_{12}$				2,689	2,681	2	76% $\omega_{10} + \omega_{12}$ + 10% $\omega_{10} + \omega_{11}$ + 4% ω_{16}	
CH stretch	ν_{15}	2,820	2,907	2,845	2,822	2,854	19	49% ω_{15} + 26% $\omega_{11} + \omega_{12}$	2,870
	$2\nu_{12}$				2,831	2,816	14	43% $2\omega_{12}$ + 33% $\omega_{11} + \omega_{13}$ + 4% ω_{15} + 7% ω_{16}	
	$\nu_{11} + \nu_{12}$				2,775	2,767	5	62% $\omega_{11} + \omega_{12}$ + 16% ω_{15}	
	$2\nu_{11}$	2,710	2,712	2,713	2,720	2,707	6	78% $2\omega_{11}$ + 10% ω_{15}	2,718
C=O stretch	ν_{14}	1,754	1,747	1,747	1,770	1,761	108	64% ω_{14} + 25% $2\omega_6$	1,753
	$2\nu_6$	1,706	1,697	1,707	1,702	1,716	33	58% $2\omega_6$ + 27% ω_{14}	1,711
	$2\nu_4$				1,441	1,412	10	73% $2\omega_4$ + 12% ω_{13} + 9% ω_{11}	
CH_2 scissor	ν_{13}	1,456	1,443	1,430	1,460	1,454	11	81% ω_{13} + 13% $2\omega_4$	1,462
		1,425	1,429	1,424					
			1,421						
CH_2 wag, ip OH bend	ν_{12}	1,378	1,399	1,400	1,426	1,421	27	92% ω_{12}	1,420
ip CH bend	ν_{11}	1,356	1,367	1,366	1,367	1,363	31	94% ω_{11}	1,378
ip OH bend, CH_2 wag,	ν_{10}	1,275	1,267	1,268	1,279	1,273	41	94% ω_{10}	1,265
CH_2 twist	ν_9		1,229	–	1,229	1,225	3	95% ω_9	1,235
	$2\nu_8$			2,209	2,229	2,233	1	86% $2\omega_8$ + 1% ω_8	2,222
C–O stretch	ν_8	1,112	1,110	1,110	1,120	1,120	80	93% ω_8	1,116
CH_2 twist, op CH bend	ν_7		1,130	–	1,087	1,081	0	95% ω_7	1,090
CC stretch	ν_6	859	858	856	865	865	49	91% ω_6	867
O=CC bend, CCO bend	ν_5	752	749	751	757	758	9	95% ω_5	740
CH_2 rock, op CH bend	ν_4		–	–	722	713	0	96% ω_4	715

Comparison with the experimental values obtained in gas phase and in Ar matrix and VPT2 results from a MP2/6-311++G(d,p) quartic force field [24]. Calculated intensities are in km·mol^{-1}

[a] Taken from Ref. [23], [b] taken from Ref. [24], [c] taken from Ref. [25], [d] our work, [e] taken from Ref. [24]

The anharmonic vibrational treatment is based on an iterative variational–perturbational scheme implemented in the VCI-P code [16]. This code computes the anharmonic wavenumbers and the anharmonic IR intensities of the fundamental transitions, overtones and combination bands which are IR (or Raman) actives through strong anharmonic couplings (Fermi or Darling Dennisson resonances).

Here, we recall [16] that the VCI-P method consists in 3 N–5 independent VCI computations (one for the fundamental state and one for each mono-excitation). For each VCI computation, an automated procedure builds iteratively an active space by selecting, through the second-order perturbational formula, the vibrational configurations that lead to the strongest couplings with the states of interest. Each active space gathers from some dozens to some hundreds of configurations while the weakest interactions, up to several hundred of thousands, are discarded from the variational process and contribute perturbationally to the anharmonicity:

$$E_{i,L}^{VCI-P} = E_{i,L}^{VCI} + \sum_{j_{weak}}^{N_{weak}} \frac{\left\langle \Psi_{i,L-1}^{n_i} \middle| \hat{H} \middle| \Phi_{j_{weak},L}^{n_j} \right\rangle^2}{E_i^0 - E_{j_{weak},L}^0} \quad (2)$$

The process is ended at the Lth iteration when the convergence on the VCI-P energy of the state i is reached. It has been set to 0.2 cm^{-1} in the present study. Note that in the above expression, $E_{i,n}^{VCI}$ corresponds to the VCI energy obtained by the diagonalization of an active space containing the N_{strong} configurations $\Phi_k^{n_k}$, $\Phi_k^{n_k}$ being the n_kth excitations of the configuration i. The second term represents the perturbational contribution of the weakest configurations $\Phi_j^{n_j}$ interacting with the state $\Psi_i^{n_i}$ determined variationally at the iteration *L-1*.

Thus, any state of interest $\Psi_i^{n_i}$ appears as a linear combination of the most pertinent configurations $\Phi_k^{n_k}$ that belong to the final active space:

$$\Psi_i^{n_i} = \sum_{k}^{Nstrong} c_{ik} \Phi_k^{n_k} \quad (3)$$

Note that any configuration $\Phi_k^{n_k}$ for which c_{ik}^2 is greater than a threshold (4% in the present study) is considered as a strong resonant configuration. As

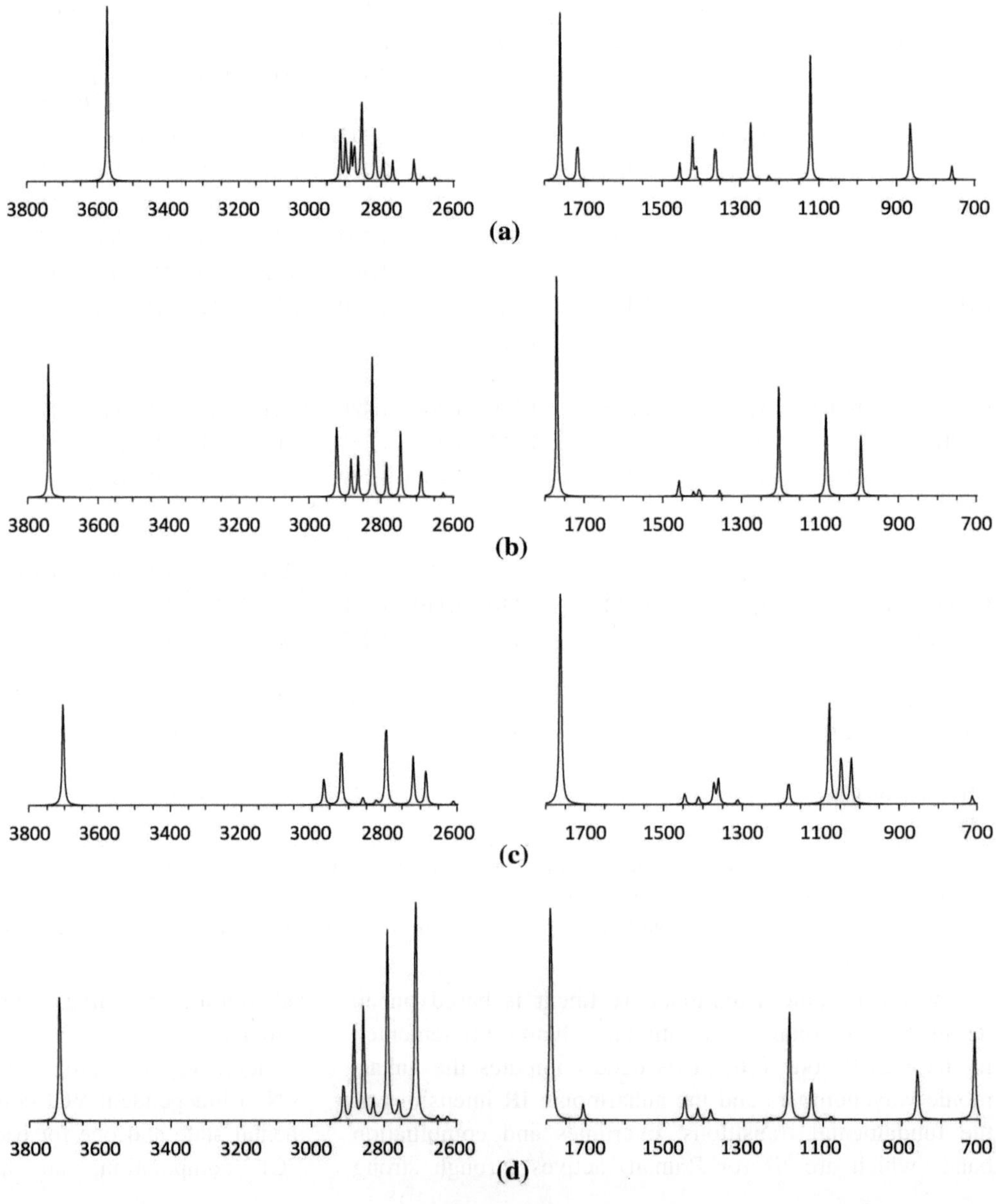

Fig. 2 anharmonic spectra of the four conformers of glycolaldehyde in the 3,800–2,600 cm^{-1} region and the 1,800–700 cm^{-1} region with the VCI-P method from a CCSD(T)/cc-pVTZ//B3LYP/6-31G(d,p) quartic force field. **a** Cc form, **b** Tt form, **c** Tg form, **d** Ct form

consequence, their multiexcitations are generated from the procedure detailed above up to the convergence of their VCI-P energy. This ensures for each CI matrix the orthogonality between the state $\Psi_i^{n_i}$ and its connected states through Fermi resonances for which the corresponding anharmonic intensities can be determined more safely.

Furthermore, this program allows the computation of anharmonic intensities [16]: starting from the quadratic expansion of the dipole surface, the second derivatives of dipole moment, d_{ij} (where α is a Cartesian component of the dipole moment vector d_j referred to the Eckart axes) are calculated in a manner analogous to the cubic and quartic force constants detailed above:

$$D_\alpha = D_\alpha(0) + \sum_i d_{\alpha_i} q_i + \frac{1}{2} d_{\alpha_{ij}} q_i q_j \quad (4)$$

The anharmonic intensity between an initial state i and a final state j is then evaluated as follows:

$$I_{i,j} = \frac{8\pi^3 N_A}{3hc(4\pi\varepsilon_0)} \nu_{i,j} \sum_\alpha \left\langle \Psi_{i,L} \middle| D_\alpha \middle| \Psi_{j,L'} \right\rangle^2 (N_i - N_j). \quad (5)$$

3 Results and discussion

Table 1 reports the vibrational transitions in the mid-infrared region of the glycolaldehyde in its Cc form. The first column corresponds to the description of the modes in terms of functional groups involved in the vibrational transition as reported in Ref. [25]. In the column 3, are reported data stemmed from a gas phase IR spectrum recorded at 300 K [23], while the data in columns 4 and 5 are issuing from low-temperature experiments in Ar matrix [24, 25]. These are compared to our theoretical results comprising the anharmonic wavenumbers, their corresponding anharmonic intensities and their theoretical descriptions. In the last columns are reported the second-order perturbative results available in the literature from a MP2/6-311++G(d,p) quartic force field [25].

Globally, the VCI-P results achieve an average convergence of 11 cm^{-1} with the overall of the non-ambiguous experimental values (values reported in Table 1 excepted ν_7, ν_{13}, ν_{15}, ν_{16}, as discussed). This mean absolute

Table 2 Calculated wavenumbers (cm^{-1}) for the glycolaldehyde in its Tt form with the VCI-P method from a CCSD(T)/cc-pVTZ//B3LYP/6-31G(d,p) hybrid quartic force field

Description	Mode	exp[a]	ν_{VMP2}[b]	ν_{VCIP}[b]	Iν[b]	Description[b]
OH stretch	ν_{18}	3,668	3,707	3,740	36	82% ω_{18} + 6% $2\omega_{18}$
as. CH_2 stretch	ν_{17}	2,940	2,903	2,921	15	74% ω_{17} + 4% $\omega_{16} + \omega_{17}$
	$\nu_{10} + \nu_{12}$		2,621	2,625	1	83% $\omega_{10} + \omega_{12}$ + 4% ω_{17}
	$\nu_8 + \nu_{13}$		2,553	2,534	1	65% $\omega_8 + \omega_{13}$ + 4% ω_{17}
	$2\nu_{13}$		2,849	2,921	8	64% $2\omega_{13}$ + 19% $2\omega_5 + \omega_{13}$ + 13% ω_{16}
	$2\nu_5 + \nu_{13}$			2,882	10	34% $2\omega_5 + \omega_{13}$ + 14% ω_{16}
s. CH_2 stretch	ν_{16}	2,935	2,971	2,863	10	33% ω_{16} + 27% $2\omega_{13}$
CH stretch	ν_{15}	2,810	2,807	2,823	35	34% ω_{15} + 30% $2\omega_{12}$
	$2\nu_{12}$		2,795	2,783	9	52% $2\omega_{12}$ + 21% ω_{15} + 8% ω_{16}
	$\nu_{11} + \nu_{12}$		2,748	2,743	20	67% $\omega_{11} + \omega_{12}$ + 16% ω_{15} + 6% ω_{16}
	$2\nu_{11}$		2,701	2,686	9	76% $2\omega_{11}$ + 15% ω_{15}
C=O stretch	ν_{14}	1,747	1,767	1,770	145	91% ω_{14}
CH_2 scissor	ν_{13}	1,439	1,472	1,458	12	76% ω_{13} + 16% $2\omega_5$
	$2\nu_5$		1,441	1,420	3	64% $2\omega_5$ + 17% ω_{13}
CH_2 wag, ip OH bend	ν_{12}		1,410	1,406	6	91% ω_{12}
ip CH bend	ν_{11}	1,353	1,357	1,354	4	94% ω_{11}
CH_2 twist	ν_{10}	–	1,220	1,217	0	94% ω_{10}
ip OH bend, CH_2 wag,	ν_9	1,203	1,214	1,203	68	93% ω_9
op CH bend	ν_8	1,129	1,088	1,083	2	95% ω_8
C–O stretch	ν_7	1,065	1,082	1,083	63	90% ω_7
CC stretch	ν_6	998	996	994	46	92% ω_6
CH_2 rock, op CH bend	ν_5	–	728	719	0	95% ω_5
O=CC bend, CCO bend	ν_4	538	539	540	6	91% ω_4

Comparison with the experimental values obtained in Ar matrix. Calculated intensities are in km·mol^{-1}

[a] Taken from Ref. [24], [b] our work

deviation (M.A.D) is confronted to the VPT2 results [25] stemmed from a MP2/6-311G(d,p) quartic force field for which a mean deviation of about 17 cm^{-1} is observed on the same set.

The experimental values in the 1,800–700 cm^{-1} region are particularly well reproduced by the VCI-P approach, namely for the remarkable case of Fermi resonance that occurs between the C=O stretching and the first overtone of the C–C stretching mode. Indeed, a gap of 45–50 cm^{-1} is observed between the two transitions both experimentally and theoretically with an intensity ratio $I_{\nu_{C=O}} : I_{2\nu_{C-C}}$ of about 5:1 in gas phase [23], 2:1 in Ar matrix [24] and 3:1 from our calculations.

Here, the most ambiguous assignment concerns the fundamental transition of the CH_2 scissoring mode for which two bands of about the same intensity [23, 24] were observed. From the experiment in gas phase [23], the two values (1,425 and 1,456 cm^{-1}) were assigned to the fundamental transition even though it was pointed out that one of the two positions could correspond to the first overtone of ν_4 (CH_2 rocking mode). From low-temperature experiments in Ar matrix, a triplet is reported in Ref. [24] (1,421–1,429–1,443 cm^{-1}) invoking possible site effects in this material, while the Ref. [25] mentions only a doublet (1,424–1,430 cm^{-1}). These observations are consistent with our calculations concerning both the positions and the intensity ratio. Thus, the values around 1,430 and 1,450 cm^{-1} are assigned to the $2\nu_4$ and ν_{13} (CH_2 scissoring), respectively.

More problems are encountered in the region above 2,600 cm^{-1} since the symmetric CH_2 and CH stretching modes are affected by strong anharmonic couplings, giving rise to several IR-active overtones and combination bands. Starting from the CH stretching mode region, the presence of a Fermi resonance between this mode (ν_{15}) and the first overtone of the op CH bending mode ($2\nu_{11}$) was observed in the three experiments [23–25], but different values of ν_{15} were reported (2,820, 2,907 and 2,845 cm^{-1}), while the

Table 3 Calculated wavenumbers (cm^{-1}) for the glycolaldehyde in its Tg and Ct forms with the VCI-P method from a CCSD(T)/cc-pVTZ// B3LYP/6-31G(d,p) hybrid quartic force field

Tg				Ct		
Description	Mode	CC//B3	Iν	Mode	CC//B3	Iν
OH stretch	ν_{18}	3,703	27	ν_{18}	3,715	36
as. CH_2 stretch	ν_{17}	2,965	7	ν_{17}	2,883	20
				$\nu_{10}+\nu_{12}$	2,650	1
				$\nu_{10}+\nu_{11}$	2,624	1
				$\nu_7+\nu_{13}$	2,531	1
s. CH_2 stretch	ν_{16}	2,918	16	ν_{16}	2,857	29
	$2\nu_5+\nu_{13}$	2,859	2	$2\nu_5+\nu_{13}$	2,884	5
	$\nu_{10}+\nu_{12}$	2,683	9			
	$2\nu_{13}$	2,822	1	$2\nu_{13}$	2,912	10
				$4\nu_5$	2,757	5
				$\nu_{11}+\nu_{12}$	2,830	5
CH stretch	ν_{15}	2,794	23	ν_{15}	2,710	62
	$2\nu_{12}$	2,718	12	$2\nu_{12}$	2,790	49
				$2\nu_{11}$	2,763	1
	$\nu_{10}+\nu_{12}$	2,682	13			
C=O stretch	ν_{14}	1,764	137	ν_{14}	1,790	151
				$2\nu_6$	1,707	10
CH_2 scissor	ν_{13}	1,446	7	ν_{13}	1,449	16
	$2\nu_5$	1,411	5			
CH_2 wag, ip OH bend	ν_{12}	1,372	13	ν_{12}	1,416	7
ip CH bend	ν_{11}	1,361	16	ν_{11}	1,384	7
ip OH bend, CH_2 wag,	ν_{10}	1,312	3	ν_{10}	1,233	1
CH_2 twist	ν_9	1,182	15	ν_9	1,184	68
C–O stretch	ν_8	1,079	66	ν_8	1,129	26
CH_2 twist, op CH bend	ν_7	1,049	32	ν_7	1,080	2
CC stretch	ν_6	1,023	29	ν_6	860	35
O=CC bend, CCO bend	ν_5	712	5	ν_5	718	0
CH_2 rock, op CH bend	ν_4	519	2	ν_4	713	56

Calculated intensities are in km·mol^{-1}

value of $2\nu_{11}$ is determined around 2,710 cm^{-1}. The theoretical description reveals a more subtle interplay involving four transitions, in particular between the $2\nu_{12}$ and the ν_{15}, calculated at 2,816 and 2,854 cm^{-1}, respectively, for which the intensities are of the same magnitude. Thus, the experimental value of 2,820 cm^{-1} corresponds probably to the $2\nu_{12}$ transition. Note, therefore, that the deviation between theory and experiment on the $2\nu_{11}$, ν_{15} and $2\nu_{12}$ are of 9, 4 and 3–5 cm^{-1}, respectively.

The assignment of the CH_2 stretching mode (ν_{16}) is not straightforward. According to our theoretical description, the harmonic character ω_{16} is widely spread within the states $2\nu_{13}$, $\nu_{12} + \nu_{13}$ and $2\nu_4 + \nu_{13}$ and no major component of ω_{16} appears, according to the form of the PES used. However, Fig. 2a that displays the anharmonic spectrum of the Cc form shows a multiplet shape in the CH stretching region very close to the pictures reported in Ref. [20, 23].

If we turn our attention on the asymmetric CH_2 and OH stretching modes, the observed deviation is about 3–17 cm^{-1} and 21–29 cm^{-1} with the VCI-P results, respectively, while it is about 5–19 cm^{-1} and 0–7 cm^{-1} with the VMP2 results. Here again [38], in the absence of Fermi resonance, the VMP2 treatment is closer to the experimental counterpart for a mode having a well-defined Morse potential even though the analytical form of the PES used has no Morse behavior far from the local minimum (quartic force field). Nevertheless, the VCI-P wavenumber, based on such a force field, is closer to the experimental value than its VCI counterpart (see Eq. 2) that yields some discrepancies of 20–34 cm^{-1} and 39–50 cm^{-1} for the asymmetric CH_2 stretching mode and OH stretching mode, respectively.

Table 2 reports the anharmonic frequencies and activities calculated in the mid-infrared region for the glycolaldehyde in its Tt form. These results are compared with the experimental data obtained when irradiating the Cc conformer isolated in matrix with an IR source [24]. In the region below 1,800 cm^{-1}, the very strong absorptions ($\nu_{C\text{–}C}$, $\nu_{C\text{–}O}$, ip OH bending) are in good agreement with the theoretical computations, while the very weak absorption (op CH bending) is too far from its theoretical counterpart. In the region above 2,600 cm^{-1}, the experimental transition observed at 2,935 cm^{-1} probably does not correspond to $\nu_{a,CH2}$ since it appears 72 cm^{-1} higher than the corresponding theoretical value and the observed ν_{OH} mode at 3,668 cm^{-1} is overestimated by 72 cm^{-1} by our computations. According to Table 3, it should be noted that the experimental values pointed out here (2,935 and 3,668 cm^{-1}) are closer to their theoretical counterpart belonging to the Tg form which is expected very close in energy to the Tt form.

References

1. Sevegney M, Kannan R, Siedle A, Naik R, Naik V (2006) Vib Spectrosc 40:246
2. Gagarinov A, Degtyareva O, Khodonov A, Terpugov E (2006) Vib Spectrosc 42:231
3. Schweitzer-Stenner R (2006) Vib Spectrosc 42:98
4. Zhao W, Gao X, Hao L, Huang M, Huang T, Wu T, Zhang W, Chen W (2007) Vib Spectrosc 44:388
5. Portnov A, Ganot Y, Bespachiansky E, Rosenwaks S, Bar I (2006) Vib Spectrosc 42:147
6. Xie W, Ye Y, Shen A, Zhou L, Lou Z, Wang X, Hu J (2008) Vib Spectrosc 47:119
7. Bramley MJ, Carrington T Jr (1993) J Chem Phys 99:8519
8. Dunn KM, Boggs JE, Pulay P (1986) J Chem Phys 85:5838
9. Bowman J, Christoffel K, Tobin F (1979) J Phys Chem 83:905
10. Christiansen O (2003) J Chem Phys 119:5773
11. Cassam-Chenai P, Lievin J (2006) J Comput Chem 27:627
12. Christiansen O (2004) J Chem Phys 120:2149
13. Iung C, Gatti F, Meyer HD (2004) J Chem Phys 120:6992
14. Culot F, Lievin J (1994) Theor Chim Acta 89:227
15. Culot F, Laruelle F, Lievin J (1995) Theor Chim Acta 92:211
16. Carbonniere P, Dargelos A, Pouchan C (2009) Theor Chem Acc 125:543
17. Pouchan C, Zaki K (1997) J Chem Phys 107:342
18. Senent ML (2004) J Phys Chem A 108:6286
19. Ratajczyk T, Pecul M, Sadlej J, Helgaker T (2004) J Phys Chem A 108:2758
20. Niki H, Maker PD, Savage CM, Breitenbach LP (1981) Chem Phys Lett 80:499
21. Michelsen H, Klaboe P (1969) J Mol Struct 4:293
22. Marstokk KM, Mollendal H (1969) J Mol Struct 16:259
23. Jetzki M, Luckhaus D, Signorell R (2004) Can J Chem 82:915
24. Aspiala A, Murto J, Stén P (1986) Chem Phys 106:399
25. Ceponkus J, Chin W, Chevalier M, Broquier M, Limongi A, Crepin C (2010) J Chem Phys 133:094502
26. Raghavachri K, Trucks GW, Pople JA, Head-Gordon M (1989) Chem Phys Lett 157:479
27. CFOUR, a quantum chemical program package written by Stanton JF, Gauss J, Harding ME, Szalay PG, with contributions from Auer AA, Bartlett RJ, Benedikt U, Berger C, Bernholdt DE, Bomble YJ, Cheng L, Christiansen O, Heckert M, Heun O, Huber C, Jagau TC, Jonsson D, Jusélius J, Klein K, Lauderdale WJ, Matthews DA, Metzroth T, O'Neill DP, Price DR, Prochnow E, Ruud K, Schiffmann F, Schwalbach W, Stopkowicz S, Tajti A, Vázquez J, Wang F, Watts JD and the integral packages MOLECULE (Almlöf J and Taylor PR), PROPS (Taylor PR), ABACUS (Helgaker T, Jensen HJ Aa, Jørgensen P, and Olsen J), and ECP routines by Mitin AV and van Wüllen C. For the current version, see http://www.cfour.de
28. Gaussian 09, Revision A.1, Frisch MJ, Trucks GW, Schlegel, HB, Scuseria GE, Robb MA, Cheeseman JR, Scalmani G, Barone V, Mennucci B, Petersson GA, Nakatsuji H, Caricato M, Li X, Hratchian HP, Izmaylov AF, Bloino J, Zheng G, Sonnenberg JL, Hada M, Ehara M, Toyota K, Fukuda R, Hasegawa J, Ishida M, Nakajima T, Honda Y, Kitao O, Nakai H, Vreven T, Montgomery Jr J. A, Peralta JE, Ogliaro F, Bearpark M, Heyd JJ, Brothers E, Kudin KN, Staroverov VN, Kobayashi R, Normand J, Raghavachari K, Rendell A, Burant JC, Iyengar SS, Tomasi J, Cossi M, Rega N, Millam NJ, Klene M, Knox JE, Cross JB, Bakken V, Adamo C, Jaramillo J, Gomperts R, Stratmann RE, Yazyev O, Austin AJ, Cammi R, Pomelli C, Ochterski JW, Martin RL, Morokuma K, Zakrzewski VG, Voth GA, Salvador P, Dannenberg JJ, Dapprich S, Daniels AD, Farkas Ö, Foresman JB, Ortiz JV, Cioslowski J, Fox DJ, Gaussian, Inc., Wallingford CT, 2009

29. Begue D, Carbonniere P, Pouchan C (2005) J Phys Chem A 109:4611
30. Dunning TH Jr (1989) J Chem Phys 90:1007
31. Lee C, Yang W, Parr R (1988) Phys Rev B 37:785
32. Becke AD (1993) J Chem Phys 98:5648
33. Ditchfield R, Hehre WJ, Pople JA (1971) J Chem Phys 54:724
34. Carbonniere P, Barone V (2004) Chem Phys Lett 399:226
35. Carbonniere P, Lucca T, Pouchan C, Rega N, Barone V (2005) J Comput Chem 26:384
36. Barone V (2005) J Chem Phys 122:014108
37. Papousek D, Aliev MR (1982) Molecular vibrational rotational spectra. Elsevier, Amsterdam 139
38. Burcl R, Carter S, Handy NC (2003) Chem Phys Lett 373:357

Theor Chem Acc (2012) 131:1184
DOI 10.1007/s00214-012-1184-0

REGULAR ARTICLE

Generalized gradient exchange functionals based on the gradient-regulated connection: a new member of the TCA family

Éric Brémond · Diane Pilard · Ilaria Ciofini · Henry Chermette · Carlo Adamo · Pietro Cortona

Received: 19 July 2011 / Accepted: 14 October 2011 / Published online: 11 March 2012

Abstract A new type of generalized gradient approximation exchange functional coupled with the TCA correlation model has been proposed. Based on gradient-regulated connection, this exchange functional is able to mix performances of a modified PBE for the bulk region, to those of the PW91 for the asymptotic one, leading to a significant improvement on the modelization of weak interacting systems, while keeping a good accuracy for the atomization energies.

Keywords Exchange-correlation functional · Gradient-regulated connection · Tognetti–Cortona–Adamo functional

Electronic supplementary material The online version of this article (doi:10.1007/s00214-012-1184-0) contains supplementary material, which is available to authorized users.

Dedicated to Professor Vincenzo Barone and published as part of the special collection of articles celebrating his 60th birthday.

É. Brémond · D. Pilard · I. Ciofini · C. Adamo (✉)
Laboratoire d'Électrochimie, Chimie des Interfaces et Modélisation pour l'Énergie, UMR 7575, Chimie ParisTech, 11 rue P. et M. Curie, 75231 Paris Cedex 05, France
e-mail: carlo-adamo@chimie-paristech.fr

H. Chermette
Institut de Sciences Analytiques-Laboratoire de Chimie Physique Théorique, bat. Dirac, Université de Lyon, Université Lyon 1 and CNRS UMR 5280, 43 bd du 11 novembre 1918, 69622 Villeurbanne Cedex, France

P. Cortona
Laboratoire Structure, Propriété et Modélisation des Solides, UMR 8580, École Centrale Paris, Grande Voie des Vignes, 92295 Châtenay-Malabry, France
e-mail: pietro.cortona@ecp.fr

1 Introduction

Density functional theory (DFT) is, nowadays, probably the most widely used first principle approach, both in solid-state physics and in chemistry. The main reason is its good balance between performances (*e.g.*, accuracy in computed geometrical and electronic structure features) and computational cost. Thus, it is not surprising that both the development of new exchange-correlation functionals and their benchmarking are currently extremely flourishing, the aim being a more and more accurate description of system's properties.

Functionals are normally ranked using the Perdew ladder, which focuses on the degree of non-locality included in the functional expression [1]. In such a way, functionals are classed from strictly local, such as the local density approximation, to non-local, for example, global hybrid functionals containing a fraction of exchange computed in a Hartree-Fock fashion using Kohn-Sham orbitals. In between, there are the semi-local functionals belonging to the generalized gradient approximation (GGA) or the meta-GGA classes, which offer a compromise between accuracy and efficiency. During the last decades, a great variety of GGAs have been proposed and tested. Some of them are nowadays extremely popular: for instance, the Perdew-Wang PW86 [2] and PW91 [3] or the Perdew, Burke, and Ernzerhof (PBE) [4] functional.

Nonetheless, from their first proposal [5], global hybrid functionals marked a real breakthrough for applications to chemical systems, since they normally outperform the GGAs for thermochemical properties. Among them, B3LYP [6, 7] and PBE0 [8, 9] probably best represent the two different philosophies used to develop new hybrid functionals, which is the introduction in the functional form of fitted parameters (B3LYP) and the "parameter-free" approach, based on the fulfillment of a number of theoretical conditions (PBE0). Indeed, if such functionals definitely changed the DFT world at molecular level, their use and implementation in codes using plane waves, that is the very large majority of codes for solid state or first principle dynamics, are still far from efficient, thus making their use often unaffordable [10].

In this context, recently, some of us proposed an exchange-correlation functional, obtained by combining the PBE exchange, which is known to lead pretty accurate estimations of atomization energies and activation barriers, with a new GGA correlation functional [11]. The latter, usually referred to as TCA, was derived by including gradient-dependent terms in the local correlation functional proposed by Ragot and Cortona [12]. Successively, a revised version of the TCA correlation functional (called revTCA) was derived [13] by taking into account a supplementary physical constraint, that is, the vanishing correlation energy of one-electron systems. The revTCA correlation, combined with a modified PBE exchange proposed by the same authors [13] (revPBE, which should not be confused with the similar, but not identical, revised PBE exchange proposed by Zhang and Yang [14]), led to a GGA exchange-correlation functional whose performances are not far from those of hybrid functionals for several properties [15, 16], thus constituting a net improvement with respect to standard GGAs. Nevertheless, this functional still fails to model weak interactions, a failure of which can be attributed to the exchange part of the functional, particularly for hydrogen-bonded systems. Since the PW91 exchange has proven to yield satisfactory results for the description of weakly interacting systems, while the PBE (or revPBE) functional provides a good description of iono-covalent systems, it is reasonable to conceive that a new GGA exchange functional, combining PBE (or revPBE) for the description of the bulk region, that is, for small reduced gradients, and the PW91 for the asymptotic region, responsible for the modeling of a weak interaction, could overcome the limitations of the original revTCA functional.

In this paper, we focus on a possible approach, the gradient-regulated connection (GRAC) [17], allowing a smooth merge of two different exchange functionals. The resulting exchange will be combined with the revTCA correlation, which has previously proved to outperform other GGA correlation functionals for classical thermochemical benchmarks. The GRAC approach follows the same philosophy of those proposed by Pople [18], by Goddard and coworkers [19], or of that recently proposed by Truhlar [20]. However, in contrast with all these approaches where two analytical expressions for the exchange functional are mixed in a uniform way, here the ratio between the two contributions is regulated by a continuous function.

The structure of the paper is the following. After a brief description of the GRAC approach and of the computational setup, the optimal range of switching between the bulk and asymptotic regions will be investigated. Once this range defined, the properties of the newly defined GRAC-revTCA functional will be tested over a series of standard benchmark sets, also including weakly interacting systems.

2 Gradient-regulated connection based exchange functionals

Originally developed by Grüning et al. [17], the GRAC was proposed to correct the asymptotic behavior of an exchange potential, so as to allow the description of the bulk density region by a potential and the density tail region by another one. The link between these two potentials is made by a switching function f varying from zero (in the region where the reduced density gradient is null) to one (in the region where the reduced density gradient tends to infinity).

Using a similar approach in order to join functionals instead of potentials, the new exchange functional $\mathfrak{F}^{b\to a}$ developed in this paper combines a first exchange functional $\mathfrak{F}^{b}$, recognized to be accurate in the bulk region, with a second one $\mathfrak{F}^{a}$, which reproduces well the asymptotic behavior:

$$\mathfrak{F}^{b\to a}[\rho(\mathbf{r})] = [1 - f[x(\mathbf{r})]] \cdot \mathfrak{F}^{b}[\rho(\mathbf{r})] + f[x(\mathbf{r})] \cdot \mathfrak{F}^{a}[\rho(\mathbf{r})] \tag{1}$$

In this equation, ρ is the electron density and x is the reduced density gradient defined by:

$$x : \mathbb{R}^3 \to [0, +\infty[\quad \mathbf{r} \mapsto \frac{|\nabla \rho(\mathbf{r})|}{\rho(\mathbf{r})^{\frac{4}{3}}} \tag{2}$$

The switching function f is chosen to be the Fermi-like function (Fig. 1):

$$f_{\alpha,\beta}[x(\mathbf{r})] = \frac{1}{1 + \exp(-\alpha(x(\mathbf{r}) - \beta))} \tag{3}$$

where α and β are parameters governing how the bulk and asymptotic regions are linked. In particular, α regulates the switching slope, whereas β indicates the position of the switch region, that is, the value of the reduced density gradient for which $\mathfrak{F}^{b\to a}$ is equal to the average of $\mathfrak{F}^{a}$ and $\mathfrak{F}^{b}$.

Considering the conversion completed when $f_{\alpha,\beta}[x(\mathbf{r})] = M$, the functional domain can be split in three parts (Fig. 1): $[0, \beta - \beta_s]$, $[\beta - \beta_s, \beta + \beta_s]$ and $[\beta + \beta_s, +\infty[$, named $b, b \to a$, and a, respectively. The β_s value is given

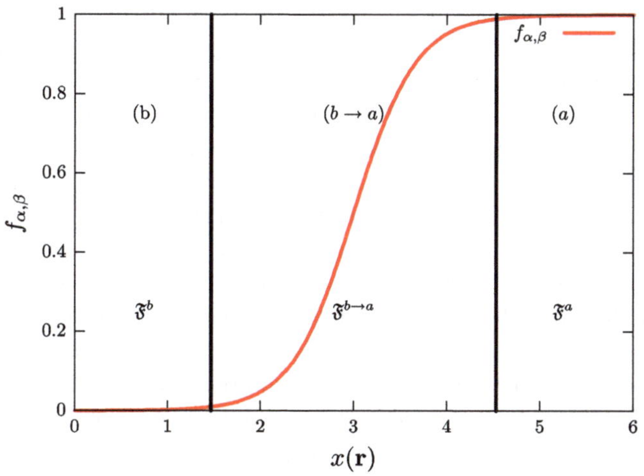

Fig. 1 The representation of the Fermi function

by: $\beta_s = \alpha^{-1} \cdot \ln\left(\frac{M}{1-M}\right)$. Clearly, the steeper is the switching slope, the more the value of β_s tends to zero.

In order to define the β value, a first approach could be to use the definition of the "physical interval" given by Perdew et al. [21, 22]. Following these authors, the bulk region is restrained to the interval $x(\mathbf{r}) \in [0, 3 \times 2 \cdot (3\pi^2)^{\frac{1}{3}}]$. It is expected that values of β not too different from this one do not produce significant modifications of the results. Similarly, as α determine how quickly the switch from one exchange to the other is done, its precise value should not influence too much the results, if the switch region is not too large.

3 Computational details

The GRAC exchange functional was implemented in the Gaussian Development Version [23], including its analytical first to third derivatives. The correlation functional was always fixed to revTCA [13]. Several reference datasets were used in order to analyze the behavior in function of the α and β parameters of the Fermi function and to assess the performances of the resulting exchange-correlation functional. The result changes in function of the α and β values were studied by computing the mean absolute error (MAE) for the atomization energies of the G2-148 dataset [24] and for the NCB31 test set [25]. The latter is a dataset that was especially conceived in order to test the functional performances for systems characterized by weak interactions. For both datasets, the Pople's triple-ζ basis set 6-311+G(3*df*, 2*p*) was used.

The performances of the so obtained GRAC-revTCA method were compared with those of standard GGA exchange functionals (PBE [4], revPBE [13], PW91 [3]) using the same revTCA correlation counterpart, and to those of the global hybrid PBE0 [8, 9], considered here as a reference for the performances of hybrid functionals.

The performances of all the above mentioned functionals were tested on barrier heights (from the DBH24/08 test set [26, 27]), weak interactions (from the NCB31 dataset [25]), structural parameters (from the G2-32 set [28]), or hydrogen-bond length optimizations and dissociation energies (from the HB10 bench test [28]). As described in literature, systems belonging to the DBH24/08 test set were described by a 6–31+G(*d*, *p*) basis set.

For the HB10 database, structures were fully optimized using 6–311+ G(3*df*, 2*p*) basis set, while dissociation energies were computed using the larger Dunning quadruple-ζ basis set aug-cc-pVQZ, in order to minimize the error due to the basis set superposition (BSSE).

Finally, all the calculations were performed using a tight criterion for the self-consistent field (SCF) convergence and a standard integration grid.

4 Results and discussions

4.1 Analysis of the influence of the α and β parameter values

Let us start by considering the atomization energies of the G2-148 dataset. As previously discussed, on the basis of the "physical interval" concept introduced by Perdew and coworkers [21, 22], reasonable values of β should be close to $3 \times 2 \cdot (3\pi^2)^{\frac{1}{3}} \approx 18.6$. Let us set $\beta = 19$ and study the influence of α.

In Fig. 2 are reported the enhancement functions $F(s)$ of the original revPBE and PW91 exchange functionals as well as the GRAC-revTCA $F(s)$ for different α values. The matching of the functionals is very smooth for $\alpha = 0.5$ and

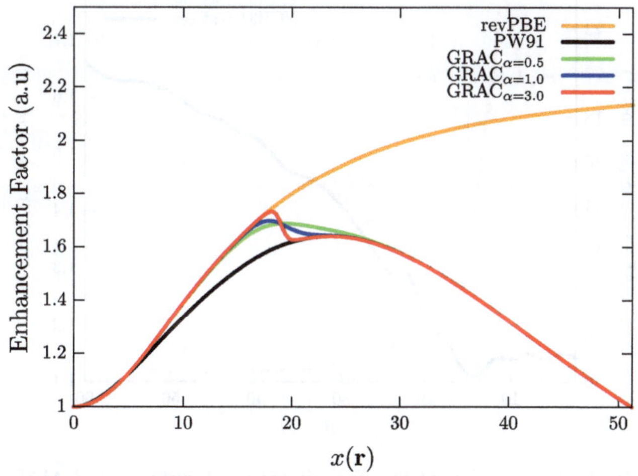

Fig. 2 The representation of the revPBE, PW91, and GRAC functional for β equal to 19 and different values of α

becomes progressively sharp (but always keeping a "regular" character in mathematical sense) increasing the α value. Variations of the α value within the interval $\alpha \in [1.0, 4.5]$ produce MAE changes smaller than 0.5 kcal mol^{-1}.

A second piece of information is obtained by studying the MAE in function of α and β. The latter is assumed to vary in an interval of values around $\beta \approx 18.6$. The results are reported in Fig. 3, where it can be seen that the MAE surface, in a quite large range of the parameter values ($\alpha \in [1.0, 4.5], \beta \in [17.5, 19.5]$), is indeed flat. There is no MAE minimum in this region (for a reason that will become clear in the following), and the flat character of the surface actually means that any pair of values could be arbitrarily chosen with only minor consequences on the results.

A further, important, piece of information is obtained by studying the MAE changes in a largest range of β values for a fixed value of α ($\alpha = 1$). The results of this study, for both the G2-148 and the NCB31 datasets, are reported in Fig. 4. This figure confirms some known features of the parents functionals and furnishes some interesting new insights. Increasing the weight of the revPBE exchange (that is increasing the β value) improves the G2-148 atomization energies, while the description of the weak interactions worsens. The contrary happens reducing the β value. A more surprising result is that relatively high values of the reduced gradient are still important for the calculation of the atomization energies. The revPBE exchange seems to be (when combined with the revTCA correlation) the optimum one on the overall range of the reduced gradient values. Thus, any choice of the β value will give rise to a G2-148 MAE greater than the one of the parent revPBE functional. However, our aim is to improve the description of weakly interacting systems, keeping the errors for the atomization energies reasonably low. Figure 4 shows that this can be done by choosing β on the basis of the "physical interval" concept. Changes of its value, within a range that can be deduced by inspection of Fig. 4, have the only consequence of slightly improving the results for one dataset at the expense of the other. For this reason, the choice of the β value is relatively arbitrary.

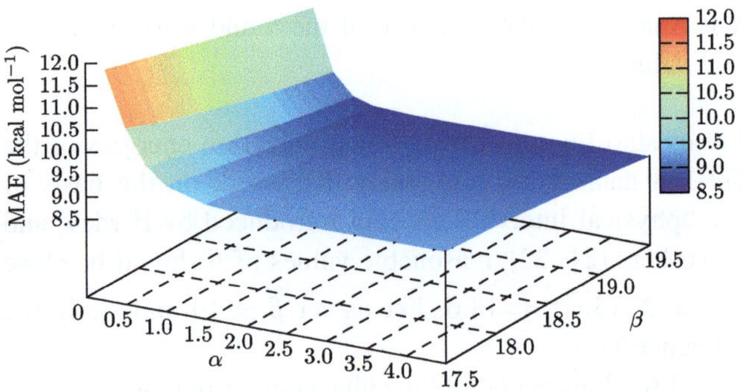

Fig. 3 Mean average error (MAE) surface for the atomization energy of the G2-148 test set function of the α and β parameters

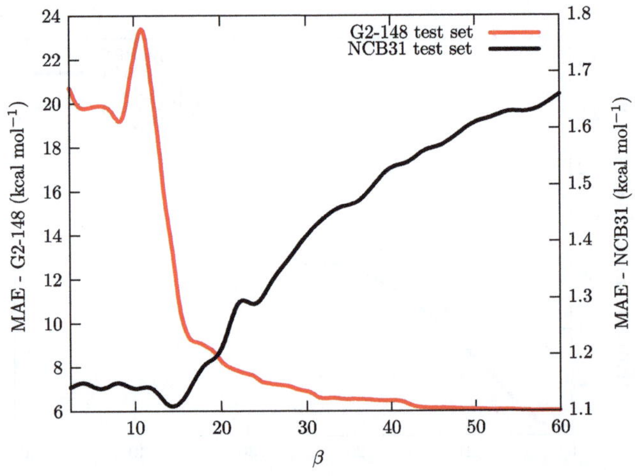

Fig. 4 Cross section of the three dimensional surface (α, β, MAE) at $\alpha = 1$ of the G2-148 and NCB31 test sets as a function of the β parameter

Our calculations have been performed by setting $\beta = 19$ and $\alpha = 1$. This latter value corresponds to $\beta_s \approx 2.9$ and was chosen in order to have a $F(s)$ function as smooth as possible, keeping the transition from one exchange to the other sufficiently fast. Thus, the GRAC-revTCA functional used in the following has mainly a revPBE character for $x(\mathbf{r}) \in [0, 16]$, a PW91 character for $x(\mathbf{r}) \in [22, +\infty[$, while the interval $x(\mathbf{r}) \in [16, 22]$ can be considered the switching region.

4.2 Benchmarks

The performances of the above defined GRAC-revTCA functional were tested on different databases and compared to those of PBE-revTCA, PW91-revTCA, revPBE-revTCA, PBE0, and the GRAC0-revTCA functional. The last one is obtained by using the same α and β parameter values as in the GRAC-revTCA case, but the PBE exchange functional in the bulk region instead of revPBE, that is $\mathfrak{F}^{\mathfrak{PBE}\rightarrow\mathfrak{PW}}$ instead of $\mathfrak{F}^{\mathrm{rev}\mathfrak{PBE}\rightarrow\mathfrak{PW}}$ as used for GRAC-revTCA.

The main results are summarized in Fig. 5, while detailed information for each database is reported in the Supporting Information.

4.2.1 Atomization energies and structural optimizations

The atomization energy datasets, G2-X (with X equal to the number of molecules), are the reference test sets routinely used both to optimize the functional parameters (if any) and to assess the functional performances [29].

Analyzing the results obtained for the G2-148 set, it appears that the PBE0 global hybrid is just slightly better

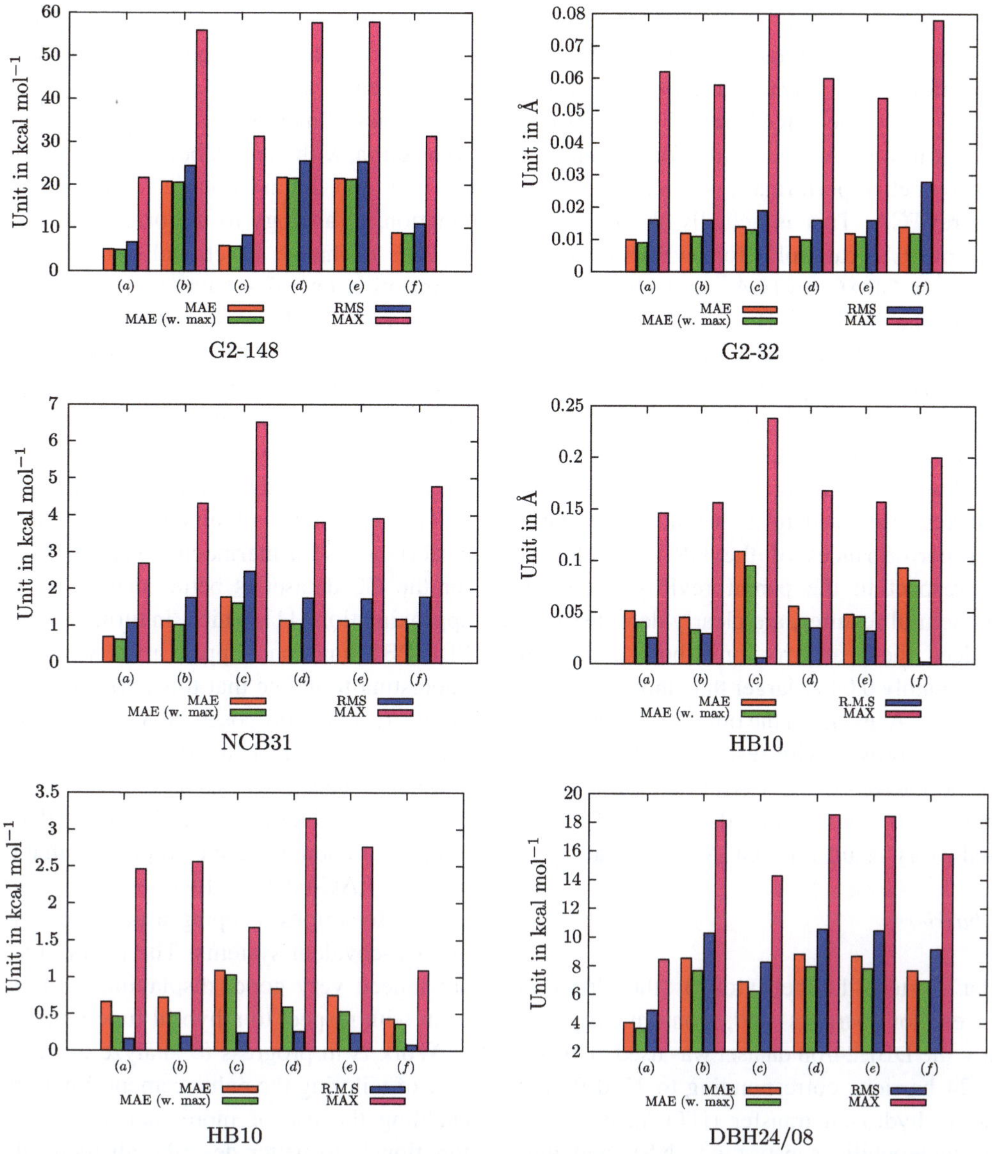

Fig. 5 Histogram representation of the error computed on different test sets. (*a*) PBE0, (*b*) PBE-revTCA, (*c*) revPBE-revTCA, (*d*) PW91-revTCA, (*e*) $\mathfrak{F}^{\mathfrak{PBE}\rightarrow\mathfrak{PW}}$-revTCA (GRAC0-revTCA), (*f*) $\mathfrak{F}^{\mathrm{rev}\mathfrak{PBE}\rightarrow\mathfrak{PW}}$-revTCA (GRAC-revTCA)

than the revPBE-revTCA GGA functional (the MAE is 0.8 kcal mol^{-1} smaller). The performances of both are clearly better than those of all the other GGA exchange functionals (PBE, PW91) coupled with the revTCA correlation. The GRAC-revTCA functional provides a satisfactory MAE, intermediate between that of revPBE-revTCA and PW91-revTCA, of only 8.9 kcal mol^{-1}, while a much higher MAE (21.5 kcal mol^{-1}) is computed for GRAC0-revTCA. These results follow the evolution of the error computed going from revPBE-revTCA to PBE-revTCA and thus clearly underly the importance of the use of the revPBE exchange functional instead of the PBE one in conjunction with the revTCA correlation.

The second test set, G2-32, was used to evaluate the performances of the different functionals for structural optimizations. Actually, all functionals show a very similar behavior. The computed MAE is always lower than 0.02 Å, and the differences among the various functionals being smaller than 0.004 Å. In the case of the GRAC-revTCA, the worst result is obtained for the Li_2 compound.

4.2.2 *Weak interactions*

The first dataset we have used, in order to test the new functional on weakly bonded systems, was the NCB31 one. This dataset is composed of 31 "dimers", that is, six hydrogen-bonding dimers (HB6), seven charge transfer complexes (CT7), six dipole interaction complexes (DI), seven weak-interaction pairs (WI7), and five π-π stacking complexes (PPS5). Since the GRAC-revTCA functional was devised in order to improve the results for this kind of systems, we expect better performances with respect to those of revPBE-revTCA. This is actually the case: the MAE is reduced from 1.8 kcal mol^{-1} for revPBE-revTCA, to 1.2 kcal mol^{-1} for GRAC-revTCA, very close to the values computed for the PW91-revTCA (1.14 kcal mol^{-1}), and the PBE-revTCA (1.13 kcal mol^{-1}) functionals. Although the errors are still larger than those obtained at the PBE0 level (0.70 kcal mol^{-1}), the accuracy is satisfactory for a pure GGA functional.

The HB10 dataset, containing ten hydrogen-bonded molecular complexes, was used for a second benchmark. Again the better performances of the GRAC exchange functional, with respect to the parent revPBE, both on energetics and structural features, are clearly demonstrated by our results. Indeed, the mean errors on H-bond optimized distances are still roughly 0.04 Å larger than those obtained at PBE0 or PBE-revTCA level. On the other hand, energetic features are greatly improved with respect to all the other functionals, and the GRAC-revTCA errors on the dissociation energies being two times smaller than those of PBE0, a more than satisfactory result for a GGA functional.

4.2.3 Reaction barriers

Finally, in order to test the behavior of the different functionals on reaction barriers and, consequently, on chemical kinetics, the DBH24/08 dataset was used. This set is composed by 24 barriers, corresponding to 12 different reactions, including hydrogen transfer (HT), heavy atom transfer (HAT), nucleophilic substitution (NS), and unimolecular and association (UA) reactions. The performances of the GRAC-revTCA functional are better than those of PBE-revTCA, PW91-revTVA, or GRAC0-revTCA, but slightly worst than those of the parent revPBE-revTCA functional. For barriers, the best results are obtained by using the hybrid PBE0 functional, whose MAE is only 4.0 kcal mol$^{\wedge-1}$, to be compared with 7.7 kcal mol^{-1} for GRAC-revTCA.

5 Conclusion

A new exchange functional was designed to allow, when coupled with the GGA revTCA correlation, a balanced description of both iono-covalent and weak interacting systems. The main idea is to combine two different exchange functionals in such a way that their strengths are kept and their weaknesses are reduced. A similar idea has been followed by other authors [30, 31] in order to derive functionals, which give good performances simultaneously for molecules and solids.

The new functional is based on the gradient-regulated connection, with the link between the functionals describing the bulk and the asymptotic regions regulated by a Fermi function, in analogy to the original GRAC approach. The revPBE exchange functional, which provides results for atomization energies having an accuracy close to that of hybrid functionals, was used in the bulk region, while the PW91 one, reputed for its performances on weakly interacting systems, was chosen for the asymptotic region. The values of the two parameters, required to define the Fermi switch function, can be chosen quite arbitrarily in a relatively large region with only minor consequences on the results.

The functional obtained (GRAC-revTCA) has good performances for thermochemistry. For example, the MAE for the G2 dataset is better than those of the PBE-TCA approximation [11], which, in turn, reduces the errors of the PBE-PBE functional approximately by a factor of two. It is interesting to notice that this performance is similar to that of the recently proposed SOGGA11 approximation [20]. This is an empirical functional, and its ability to calculate atomization energies was tested by using the MGAE109/05 dataset [32], obtaining a MAE smaller than the PBE one for a factor of about two. We emphasize that the main strength of the GRAC-revTCA functional is its ability to describe weak interactions, keeping a good accuracy for properties of iono-covalent systems. The results for hydrogen bonds are indeed very good, displaying improved performances even with respect to a hybrid functional such as PBE0.

Work is in progress to analyze more in details the best way of tailoring the enhancement function $F(s)$, eventually enabling the use of more than two exchange-correlation functionals to better describe all reduced density gradient domains.

6 Supporting informations

Detailed data on the G2-148, G2-32, NCB31, HB10, and DBH24 dataset as well as MAE maps computed for the NCB31 set.

Acknowledgments This work was funded by the ANR agency under the project DinfDFT ANR 2010 BLANC n. 0425, and by Sanofi-Aventis.

References

1. Perdew JP, Ruzsinszky A, Constantin LA, Sun J, Csonka GI (2009) J Chem Theory Comput 5:902–908

2. Perdew JP (1986) Phys Rev B 33:8822–8824
3. Perdew JP, Wang Y (1992) Phys Rev B 45:13244–13249
4. Perdew JP, Burke K, Ernzerhof M (1996) Phys Rev Lett 77:3865–3868
5. Becke AD (1992) J Chem Phys 96:2155–2160
6. Becke AD (1993) J Chem Phys 98:5648–5652
7. Barone V, Adamo C (1994) Chem Phys Lett 224:432–438
8. Adamo C, Barone V (1999) J Chem Phys 110:6158–6170
9. Ernzerhof M, Scuseria GE (1999) J Chem Phys 110:5029–5037
10. Guidon M, Schiffmann F, Hutter J, VandeVondele J (2008) J Chem Phys 128:214104
11. Tognetti V, Cortona P, Adamo C (2008) J Chem Phys 128:034101
12. Ragot S, Cortona P (2004) J Chem Phys 121:7671–7680
13. Tognetti V, Cortona P, Adamo C (2008) Chem Phys Lett 460:536–539
14. Zhang Y, Yang W (1998) Phys Rev Lett 80:890
15. Tognetti V, Cortona P, Adamo C (2009) Theor Chim Acta 122:257–264
16. Tognetti V, Adamo C, Cortona C (2010) Inter Sci 2:163–168
17. Grüning M, Gritsenko OV, van Gisbergen SJA, Baerends EJ (2001) J Chem Phys 114:652–660
18. Adamson RD, Gill PMW, Pople JA (1998) Chem Phys Lett 284:6
19. Xin X, Zhang Q, Muller RP, Goddard WA (2005) J Chem Phys 122:014105
20. Peverati R, Zhao Y, Thrular DG (2011) Phys Chem Lett 2:1991–1997
21. Zupan A, Burke K, Ernzerhof M, Perdew JP (1997) J Chem Phys 106:10184–10193
22. Zupan A, Perdew JP, Burke K, Causà M (1997) Int J Quant Chem 61:835–845
23. Frisch MJ et al (2010) Gaussian DV, revision H.10. Gaussian, Inc., Wallingford, CT
24. Johnson BG, Gill PMW, Pople JA (1993) J Chem Phys 98:5612–5626
25. Zhao Y, Schultz NE, Truhlar DG (2006) J Chem Theory Comp 2:364–382
26. Zheng J, Zhao Y, Truhlar DG (2009) J Chem Theory Comp 5:808–821
27. Lynch BJ, Truhlar DG (2003) J Phys Chem A 107:8996–8999
28. Boese AD, Martin JML (2004) J Chem Phys 121:3405–3416
29. Adamo C, Barone V (1998) J Chem Phys 108:664–675
30. Haas P, Tran F, Schwarz K (2011) Phys Rev B 83:205117
31. Fabiano E, Constantin LA, Della Sala F (2010) Phys Rev B 82:113104
32. Zhao Y, Schultz NE, Thrular DG (2005) J Chem Theory Comput 2:364–382

Theor Chem Acc (2012) 131:1185
DOI 10.1007/s00214-012-1185-z

REGULAR ARTICLE

Giant correlation effects in the photoelectron spectrum of $Ni(C_3H_5)_2$: clues from accurate calculation of ionization cross-sections

P. Decleva · G. Fronzoni · M. Stener

Received: 21 July 2011 / Accepted: 14 December 2011 / Published online: 10 March 2012

Abstract Several general issues relating to the strong correlation effects present in the ionic states and the interpretation of molecular photoelectron spectra are discussed in the context of the $Ni(C_3H_5)_2$ molecule, and the problematic assignment of its spectrum is considered. Accurate calculations of photoionization cross-sections relative to all the valence ionizations are computed and compared with the available experimental evidence. It is shown that most ambiguities are resolved by a careful analysis of cross-section profiles and branching ratios, which reveal a wealth of details on the complex electronic structure of this molecule.

Keywords Photoelectron spectroscopy · Time-dependent density functional theory · Transition metal compounds · Correlation effects

1 Introduction

The accurate description of correlation effects in transition metal compounds has remained an important challenge for theoretical approaches [1, 2]. In the domain of ab initio methods such effects are often unusually large, especially for excited and ionized states relevant to spectroscopy. The situation is much alleviated with DFT approaches, which are able to better describe the differential correlation associated with metal d electrons [3]. However, in close situations, the accuracy of present DFT approaches may not be sufficient to give an unambiguous answer, and one has to resort to pushing ab initio approaches to a high level of correlation.

One of the most direct probes of the electronic structure is the measurement of ionization potentials by photoelectron spectroscopy. In fact since the earliest studies, it has become apparent that huge deviation from the predictions of Koopmans Theorem (KT), which often works reasonably well for main group elements, occur in organometallic compounds, although by no means always, and the origin of this widely varying behavior is not well understood. Actually one of the first and most simple systems investigated, the nickel bis (π-allyl) molecule [$Ni(\eta^3$–$C_3H_5)_2$] [4], has proved one of the most difficult examples, revealing successive layers of complexity, so that it is still not completely understood. Besides comparison with similar compounds, a decisive clue to the assignment of the spectrum, and in particular, to the nature of the outermost ionization, has been provided by cross-sections and branching ratios (BR) profiles obtained with synchrotron radiation [5, 6], which have supplemented earlier suggestions originating from HeI/HeII intensity variations [4, 7], eliminating all the ambiguity inherent in previous studies, like accidental ratios due to the many non-monotonic features present in molecular cross-sections, like delayed onsets, shape and autoionization resonances, Cooper minima, further multi-electron excitations [8–10]. Unfortunately, at the time of the experimental studies, only rather primitive theoretical approaches were available (basically

Dedicated to Professor Vincenzo Barone and published as part of the special collection of articles celebrating his 60th birthday.

P. Decleva · G. Fronzoni · M. Stener (✉)
Dipartimento di Scienze Chimiche, Università di Trieste,
Via L. Giorgieri 1, 34127 Trieste, Italy
e-mail: stener@univ.trieste.it

P. Decleva · G. Fronzoni · M. Stener
Consorzio Interuniversitario Nazionale per la Scienza e Tecnologia dei Materiali, Unita' di Trieste and CNR-IOM DEMOCRITOS, Trieste, Italy

the Gelius model [11] and the MS-Xα approach [12–14]) to supplement and interpret the experimental data, for molecules of such complexity. Since then significant advances have been made in the calculation of molecular cross-sections, and it is presently feasible to compute full TDDFT continuum properties of organometallic compounds obtaining convergent results without further approximations [15–17].

As a contribution to the clarification of the electronic structure of this deceptively simple prototypical system, we have undertaken a TDDFT calculation of the photoionization parameters, to check the consistency and interpretation of the SR data, gaining further evidence and confirmation of the assignment of the spectrum and the sequence of ionization potentials. Additional interest comes to the clarification of the possible occurrence of resonances, and in particular, the characterization of the expected "giant resonance" associated with the 3p → 3d excitation in the metal center [18], which has hitherto escaped experimental detection. Moreover, it is expected that finer details revealed by improved theoretical description may stimulate further experimental investigations of photoionization parameters, both angular distributions, hitherto neglected and cross-sections with the much improved resolution and wider energy range now currently available, which could shed light on the most difficult details of the ordering of the outermost ionizations, still not unambiguously settled. At the same time, the challenge of an accurate description of the electronic structure of this basic organometallic compound and its lower ionic states, notably the characterization of the HOMO, should give a stringent test of the accuracy of the most highly correlated approaches presently available.

2 Theoretical method

In this work, we have employed the non-iterative TDDFT B-spline LCAO formalism for the calculation of the continuum. The formalism, its implementations and the numerical details are fully described in [16], and will be summarized here.

A conventional ground state calculation is first performed, employing the ADF program [19]. Then the KS (Kohn–Sham) hamiltonian matrix is calculated, employing the LCAO B-spline basis set. Occupied orbitals are then obtained as bound eigenvectors:

$$H_{KS}\varphi_i = \varepsilon_i\varphi_i \quad i = 1,\ldots,n.$$

While continuum photoelectron orbitals are extracted as eigenvectors with minimum modulus eigenvalue of the energy-dependent matrix $\mathbf{A}^+\mathbf{A}$:

$$\mathbf{A}^+\mathbf{A}(E)c = ac, \quad \mathbf{A}(E) = \mathbf{H} - E\mathbf{S}$$

where $\mathbf{H}$ and $\mathbf{S}$ are the Hamiltonian and overlap matrices, respectively, E is the photoelectron kinetic energy, c the eigenvectors and a the minimum modulus eigenvalues. In the TDDFT formalism, the linear response of an electronic density to an external time-dependent electromagnetic field with frequency ω is evaluated following the original scheme of Zangwill and Soven [20]; an effective Self-Consistent Field (SCF) potential Φ^{SCF} is introduced, which consists of the external dipole potential Φ^{EXT} plus the Coulomb and exchange correlation screening due to the first-order change in the electron density $\delta\rho$ induced by the field

$$\delta\rho = \chi(\omega)\,\Phi^{SCF}$$

$$\delta\Phi = K\,\delta\rho \quad \Phi^{SCF} = \Phi^{EXT} + \delta\Phi$$

Here $\chi(\omega)$ is the frequency-dependent electric susceptibility and K the linear response of the field to a density perturbation. Together they afford a linear equation for the effective potential Φ^{SCF}

$$(K\chi - 1)\,\Phi^{SCF} = -\,\Phi^{EXT}$$

Representing the functional equation with the same B-spline LCAO basis set leads to the solution of a linear system of algebraic equations. Once Φ^{SCF} is obtained, all the parameters that describe the photoionization process, namely the cross-section and the asymmetry parameters, are calculated via transition moments using Φ^{SCF} instead of dipole operator. Introducing an additional subscript λ_r to distinguish among the three dipole components that transform like the standard spherical harmonics with $l = 1$ and m $= \lambda_r$, that is, $\Phi^{EXT}_{\lambda_r}(\bar{r},\omega) = \sqrt{\frac{4\pi}{3}}rY_{1,\lambda_r}$. The dipole transition moments are finally obtained, employing the $\Phi^{SCF}_{\lambda_r}$ potential instead of the dipole operator:

$$D^{\lambda\mu-}_{lh}(\lambda_r) = \left\langle \varphi^{\lambda\mu-}_{lh} \middle| \Phi^{SCF}_{\lambda_r} | \varphi_i \right\rangle,$$

where φ_i is the one-electron KS orbital, which is currently ionized, and $\varphi^{\lambda\mu-}_{lh}$ is the continuum solution for channel lh in final symmetry $\lambda\mu$, normalized to incoming wave asymptotic boundary condition.

A DZP basis set of Slater type orbitals (TZP for Ni), taken from the ADF database [19], has been employed in the initial SCF calculation. The B-spline basis [15] employed in subsequent calculation is defined by the radial interval [0, 20] a.u., with step 0.2 a.u. and maximum angular expansion up to $L = 16$, supplemented by short expansions around the nuclei, with maximum $L = 2$ for C and $L = 1$ for H. Convergence tests at the DFT level have shown that the basis employed is completely convergent up to 2.5 a.u. electron kinetic energy, covering the entire range reported. LB94 exchange correlation potential [21] has been employed in H_{KS}, while the exchange correlation kernel in the TDDFT response equation has been approximated at the ALDA level [22].

3 Results and discussion

A qualitative orbital scheme, as obtained by the present DFT calculation, relative to $Ni(C_3H_5)_2$ and the two separate fragments, $(C_3H_5)_2$ and Ni ($3d^9 4$ s), is reported in Fig. 1. Metal d orbitals (percentage of d contribution in parenthesis) are $13a_g$ (75%, mostly d_{xy}), $12a_g$ (90%, $d_{x2\text{-}y2}$), $11a_g$ (87%, d_{z2}), $6b_g$ (85%, 54% d_{yz} + 34% d_{xz}), $5b_g$ (57%, 34%d_{yz} + 23% d_{xz}). Moreover, 15% of d_{xy} and d_{xz} are in $9a_g$ and $4b_g$, respectively. A HeI photoelectron spectrum of the outer valence ionization is shown in Fig. 2 (from Ref. [7]). The 9 bands resolved (actually the next band at 17.9 eV is also weakly apparent), together with the three following bands, revealed in the SR study [6], comprise 22 main

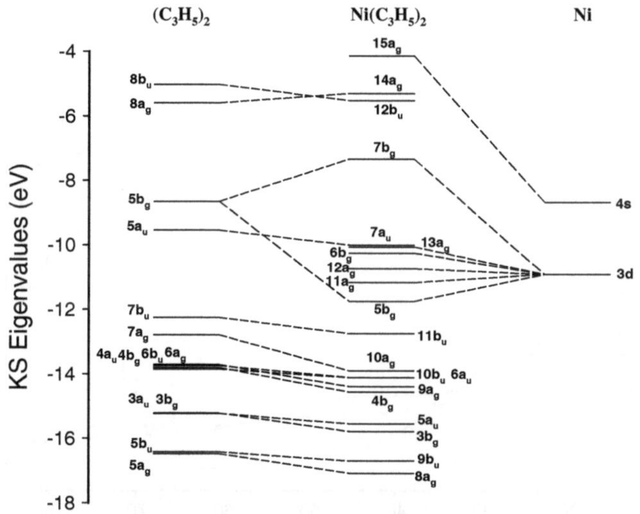

Fig. 1 Outer valence molecular orbital *diagram* for $Ni(C_3H_5)_2$ from *eigenvalues* of present ground state DFT calculation

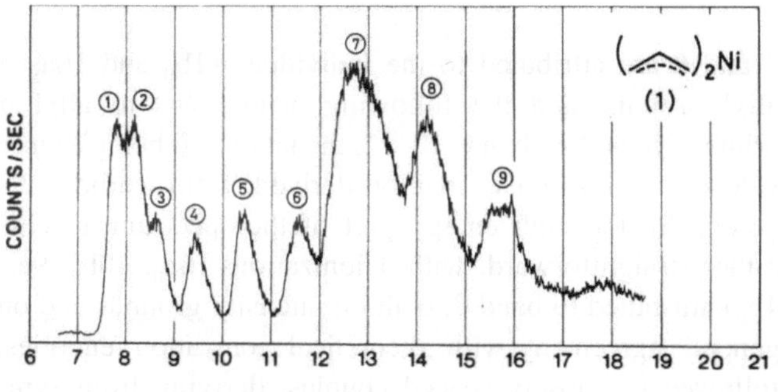

Fig. 2 He I photoelectron spectrum of $Ni(C_3H_5)_2$. Adapted with permission from Ref. [7]. Copyright (1976) American Chemical Society

ionizations, so several bands are composite. Actually the spectrum separates neatly into 3 regions. The first, comprising bands 1–4, are attributed to the mainly Ni 3d molecular orbitals (11–13 a_g, 5–6 b_g) and the highest π orbital of the allyl moiety, $7a_u$. Bands 5 ($11b_u$) and 6 ($10a_g$) are the ionizations due to the remaining two allyl π orbitals, while the remaining bands comprise the allyl σ ionizations. The assignment is by no means unambiguous and will be discussed below. A summary of different proposals and theoretical results is reported in Table 1.

As already mentioned, the interest in the electronic structure and ionization energies of $Ni(C_3H_5)_2$ was spurred by the strong disagreement between experimental evidence and the results of Hartree–Fock (HF) calculations [1, 23], that is, the huge breakdown of the Koopmans Theorem [24]. This is primarily due to very large relaxation effects, about 10 eV, for the metal d ionizations, which are buried deep in the valence shell at the HF level, but are assigned to the outermost ionizations (bands 1–4) on experimental grounds. Apart from relaxation, additional correlation effects, notably strong correlation within the 3d shell, and differential correlation between metal 3d and ligand orbitals come into play, so that even highly correlated calculations [25–28] have proved inadequate for a clear-cut assignment of the spectrum. In particular, already the nature of the HOMO ionization has been debated, with most calculations (as well as comparison with Pd and Pt analogues) favoured assignment to the allyl π orbital [25, 26, 28–32], while one of the first careful experimental analysis [7], later amply confirmed by a synchrotron radiation study [5, 6], assign it to metal d ionization.

An important issue regards also the nature of the orbital being ionized. Due to the large relaxation and additional correlation effects, the nature of the orbitals change dramatically upon ionization, as all ab initio calculations have clearly indicated [25–28, 32]. This renders essentially useless the use of canonical HF orbitals to describe ionization and has broad implications about describing molecular properties and reactivity in terms of HF HOMO, in such compounds. In fact, it is the Dyson orbital:

$$\varphi_I(\mathbf{x}) = \int \Psi_I^*(\mathbf{x}_2, \ldots, \mathbf{x}_N)\Psi_0(\mathbf{x}, \mathbf{x}_2, \ldots, \mathbf{x}_N)d\mathbf{x}_2 \ldots d\mathbf{x}_N$$

which is well defined, as a transition property between the initial neutral and the final ionic state, Ψ_0 and Ψ_I, respectively, and does not rely on any specific form of the initial and final wave functions. Note also that it is the Dyson orbital, which is in principle observable, for instance, by electron momentum spectroscopy [33], or the imaging techniques recently developing from high harmonic generation or ionization yields with strong IR laser pulses [34, 35]. In any case, also ionization cross-sections are basically described as dipole transitions from the Dyson

Table 1 IP values (eV) and assignments of the valence photoelectron spectrum of $Ni(C_3H_5)_2$

Band	Experiment			Theory					
	IP[a]	IP[b]	Assignment[a]	Ref.[c,d]	2ph-TDA	CI	2 h-1p	4 h-3v-2p	LB94
1	7.76	7.64/7.79	d + d	$7a_u$	$7a_u$	$13a_g$, $6b_g$	$13a_g$, $5b_g$	$7a_u$	$7a_u$, $13a_g$
2	8.19	7.95/8.15	$7a_u$ (π) + d	$12a_g$, $5b_g$, $13a_g$	$13a_g$, $6b_g$	$7a_u$	$11a_g$, $10a_g$	$13a_g$, $5b_g$	$6b_g$, $12a_g$
3	8.58	8.52	d	$11a_g$	$11a_g$, $5b_g$		$7a_u$	$11a_g$, $10a_g$	$11a_g$
4	9.40	9.38	d_{xz} (b_g)	$6b_g$	$10a_g$		$6b_g$	$6b_g$	$5b_g$
5	10.38	10.36	b_u (π)	$11b_u$	$11b_u$		$11b_u$	$11b_u$	$11b_u$
6	11.55	11.48	a_g (π)	a_g (π + d)	$9a_g$			$9a_g$	$10a_g$
7	12.7	12.6	a_g + b_u (σ)	$6a_u$, $10b_u$, $12a_g$, $4b_g$	$6a_u$, $10b_u$, $12a_g$, $4b_g$				$6a_u$, $10b_u$, $9a_g$, $4b_g$
8	14.2	14.2	b_g + a_u (σ)		$5a_u$, $3b_g$				$5a_u$, $3b_g$
9	15.6	15.6			$9b_u$, $8a_g$				$8a_g$, $9b_u$
10		17.9							$7a_g$, $8b_u$
11		20.8							$4a_u$, $2b_g$
12		24.0							$6a_g$, $7b_u$

[a] Ref. [7]
[b] Ref. [6]
[c] Ref. [28]
[d] Ref. [30]

orbital to the photoelectron continuum and can therefore afford a clear signature of the difference between the different descriptions. It is remarkable that in cases where HF and Dyson orbitals are very different, DFT Kohn–Sham orbitals appear much closer to the latter, underlying a deep difference between the two, which has not been adequately investigated.

With all these caveats, the computed cross-sections at the TDDFT level will be presently discussed.

A strong important clue to the assignment of the metal d ionizations as distinct from the ligand ones is the behavior of the cross-section. MOs derived from 2p orbitals of the first row atoms show cross-sections that decay much faster than those derived from metal 3d AO's, although they are larger at low energy [8–10]. It is well exemplified in the computed cross-section data of the $13a_g$–$9a_g$ orbitals reported in Fig. 3. $13a_g$–$11a_g$ are predominantly metal 3d, while $10a_g$ and $9a_g$ are essentially ligand orbitals. This has been successfully employed already in the earlier HeI/HeII studies [7] and is confirmed much more clearly with the SR studies [5, 6]. This information alone strongly supports the attribution of the metal d ionizations to the first four experimental bands in the photoelectron spectrum. All calculations agree in placing the allyl π ionization $7a_u$ among the outermost ones, so that it is also firmly comprised in the same experimental bands. Moreover, all calculations as well as qualitative arguments place immediately below the two orbitals deriving from the first allyl π orbital. Therefore, also on intensity grounds, bands

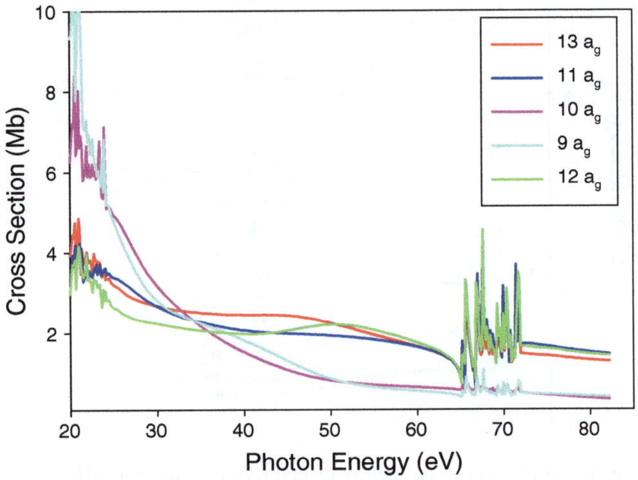

Fig. 3 Calculated cross-sections for the valence orbitals of a_g symmetry of $Ni(C_3H_5)_2$, $9a_g$–$13a_g$

5 and 6 are attributed to the individual $11b_u$ and $10a_g$ π allyl orbitals, and the following bands, 7–9 to allyl σ orbitals (including bands 10–12, the latter two being "inner valence" ionizations, mainly derived from carbon 2 s AO's). So the high energy part of the spectrum appears rather straightforward, with 4 ionizations ($6a_u$, $10b_u$, $9a_g$, $4b_g$) attributed to band 7, both on intensity grounds and on general agreement with theoretical ionization energies, followed by closely spaced couples, deriving from symmetric and antisymmetric combinations of individual allyl orbitals, either (a_u, b_g) or (a_g, b_u): bands 8 to $5a_u$, $3b_g$

ionizations, bands 9 to $8a_g$ and $9b_u$, bands 10 to $7a_g$, $8b_u$, bands 11 to $4a_u$, $2b_g$ and bands 12 to $6a_g$ and $7b_u$.

4 Ligand bands

The branching ratios of the ligand bands 5–9 are reported in Fig. 4. They are clearly proportional to the number of single ionizations under each band, 1 for bands 5–6, 2 for bands 8–9, 4 for band 7, giving a neat evidence for this assignment. One can observe a much larger contribution of the 3p → 3d resonance in band 7, a clear indication of some metal 3d contribution to these states. Moreover, from individual cross-sections reported in Fig. 5, it appears that this contribution is dominated by that of $9a_g$ and $4b_g$ ionizations, as it is expected on symmetry grounds. The d participation leads also to a distinct increase in cross-section for these two ionizations in comparison with the other two, $6a_u$ and $10b_u$, at high energy, because of the larger d atomic cross-section. The resonances are, however, weakly present in all channels, even though symmetry forbids d participation, so that rather strong interchannel coupling is active in this case.

A second feature superimposed to the general decreasing behavior of the cross-sections is the presence of oscillations that give rise to large-scale structures, sufficiently strong that they can be observable, giving therefore a precise characterization of the individual ionizations. As an example, bands 5 and 6 which both derive from the ionization of two very similar orbitals, the even and odd combinations of the 1π orbital of the allyl molecule, $11b_u = \pi_+$ and $10a_g = \pi_-$, show a strong oscillation in their respective cross-sections, which is even more clearly apparent taking their ratio, reported in Fig. 6. The similarity with the analogous oscillations well known in C_{60} [36, 37] and $MgCp_2$ [38] definitely point to their attribution to interference phenomena due to coherent emission from equivalent centers, in their case from the two 1π orbitals on the two allyl fragments, with opposite phase for the π_u and π_g combinations. Oscillations are present also in the remaining bands, notably with doubled frequency for the bands 8 and 9, although with smaller amplitude. This is because of the interplay of two overlapped ionizations within each experimental band.

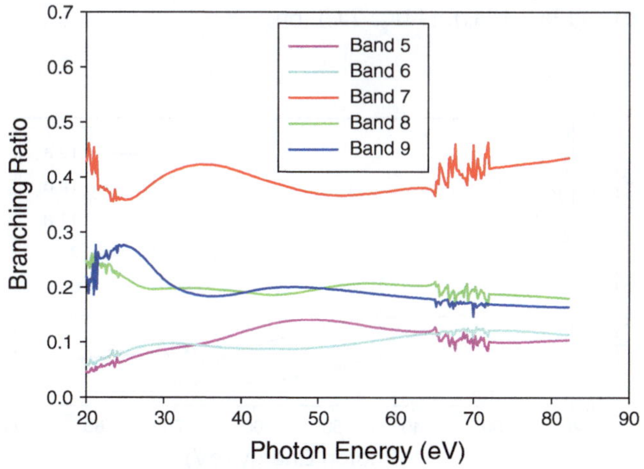

Fig. 4 Calculated branching ratios for the ligand photoelectron bands (bands 5–9)

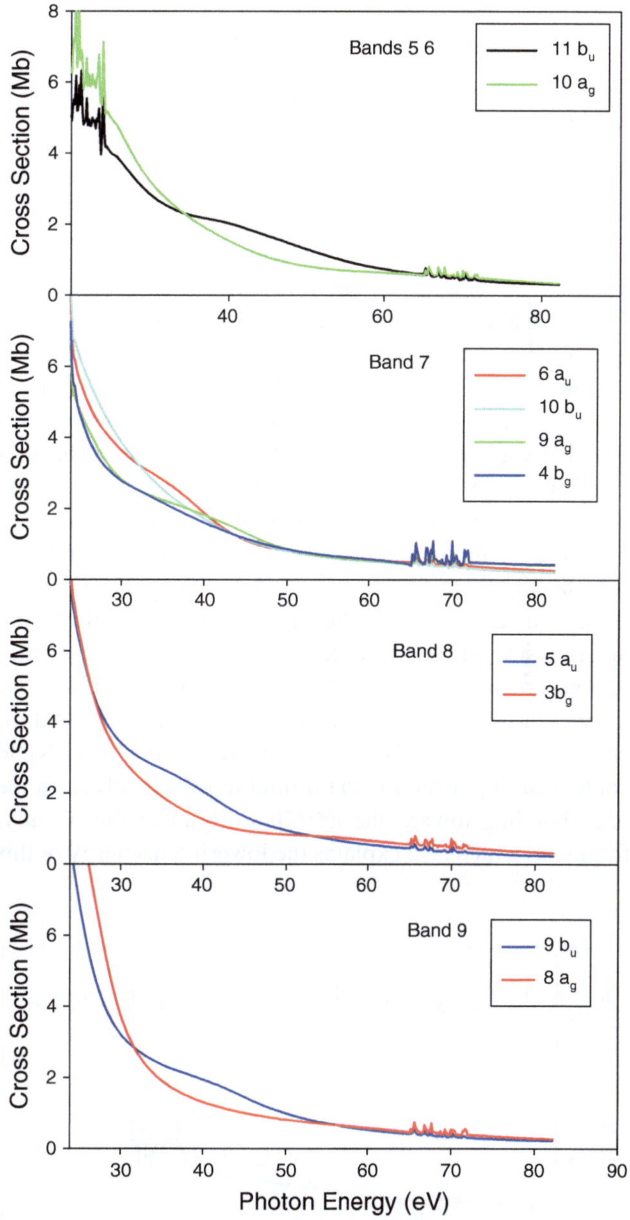

Fig. 5 Cross-sections for individual ionizations comprising the ligand bands, from *upper to lower panel*: *bands 5* ($11b_u$) and *6* ($10a_g$); *band 7* ($6a_u$, $10b_u$, $9a_g$, $4b_g$); *band 8* ($5a_u$, $3b_g$); *band 9* ($8a_g$, $9b_u$)

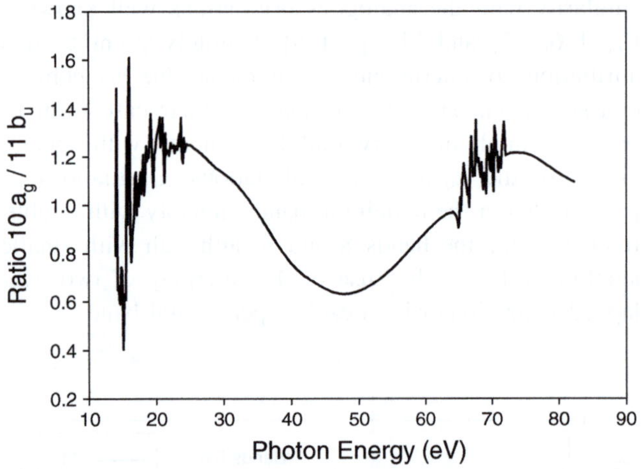

Fig. 6 Cross-section ratio of the $10a_g$ to $11b_u$ ionization (*band 6 to band 5*)

5 Bands 1–4

A first problem is the assignment of band 4, which on intensity grounds is attributed to a single ionization, with metal d character. In fact, assignment to the $7a_u$ (π) ionization is easily discarded, both on the basis of computed ionization potentials, which give a much larger separation with the other π ionizations (bands 5 and 6) and on the intensity ratios at high energy. The experimental ratio [5, 6] between bands 4 and 5 at $h\nu = 80$ eV is around 2, which is by far too large to attribute the $7a_u$ ionization to band 4. The two metal d orbitals available at higher energy are $11a_g$ and $5b_g$, and their calculated ratio to $11b_u$ in band 5 is shown in Fig. 7. As can be seen their ratio is similar up to about h$\nu = 50$ eV, but diverges strongly at high energy past the 3p → 3d resonance region, around $h\nu \sim 70$–80 eV. Here the ratio of the $5b_g$ ionization is a little larger than 2 and that relative to $11a_g$ becomes closer to 4. This is partially due to the lower Ni 3d content of $5b_g$, which is the orbital mostly involved in the back bonding toward the π^* ($7b_g$) orbital of the bis-allyl fragment, which also explains the lowering in energy of this orbital with respect to the other Ni 3d-derived ionizations. Again this ratio gives a strong support for the attribution of the $5b_g$ ionization to band 4, as several previous studies have suggested [7, 27–31], although the high level ab initio 2ph TDA approach gives the reverse assignment to 11 a_g [25, 26, 32]. The most difficult part is the assignment of bands 1–3. As already mentioned, they should comprise 5 ionizations, four mainly metal d orbitals, $11a_g$–$13a_g$ and $6b_g$, and the $7a_u$ π allyl orbital. The relative cross-sections are reported in Fig. 8. Again the much faster decrease in $7a_u$ cross-section compared to that of metal 3d ionizations is apparent. It is about 50% larger close to threshold, and it becomes lower by a factor of 5 at 80 eV. At this energy, the spread of metal 3d ionizations varies from 1.325 Mb for $13a_g$ to 1.648 Mb for $6b_g$. The attribution of the first peak to the $7a_u$ ionization, favored by most of the latter studies [25–32], comprising some of the highest level ab initio calculations is therefore completely untenable, as pointed out by the SR study [5, 6].

Again the behavior of the experimental intensities at high energies [5, 6], $\sigma1 > \sigma2 > \sigma3$, but of similar magnitudes, suggests the attribution of two metal d ionizations to the first band, 1 metal d + $7a_u$ to the second and 1 metal d to the third.

In fact the intensity ratios exclude the assignment of the $7a_u$ alone to any band (Fig. 9). Moreover, branching ratios associating $7a_u$ with one metal d ionization invariably show σ (2 metal d) > σ (1 metal d + 7au) > σ (1 metal d), so that the only sensible assignment is two metal d ionizations to the first band, 1 metal d + $7a_u$ to the second and 1 metal d to the third.

Individual assignment is more difficult.

The third band should be attributed either to $11a_g$ or to $6b_g$. That leaves the three possibilities for the first three bands:

(a) ($13a_g$, $12a_g$), ($6b_g$, $7a_u$), $11a_g$
(b) ($13a_g$, $6b_g$), ($12a_g$, $7a_u$), $11a_g$
(c) ($13a_g$, $12a_g$), ($11a_g$, $7a_u$), $6b_g$

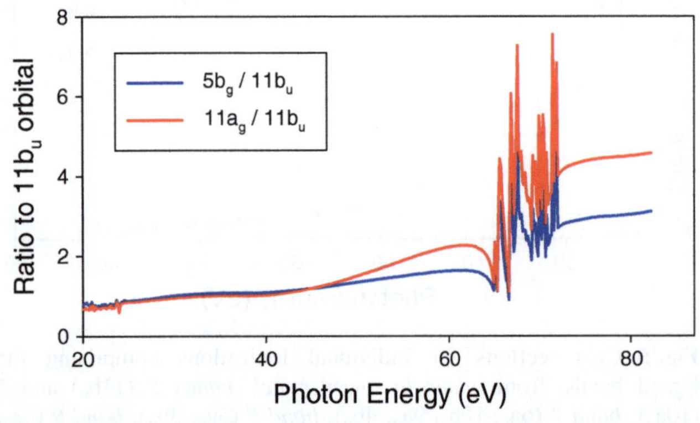

Fig. 7 Cross-section ratio of the $11a_g$ and $5b_g$ to $11b_u$ ionization

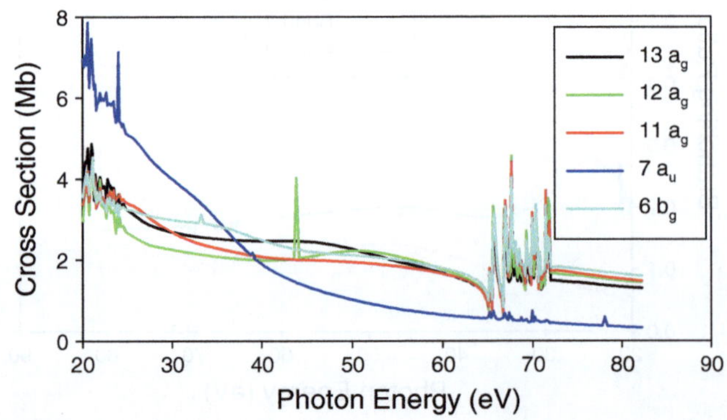

Fig. 8 Cross-sections for the ionizations comprised in the three outermost *bands*: $11a_g$–$13a_g$, $6b_g$, $7a_u$

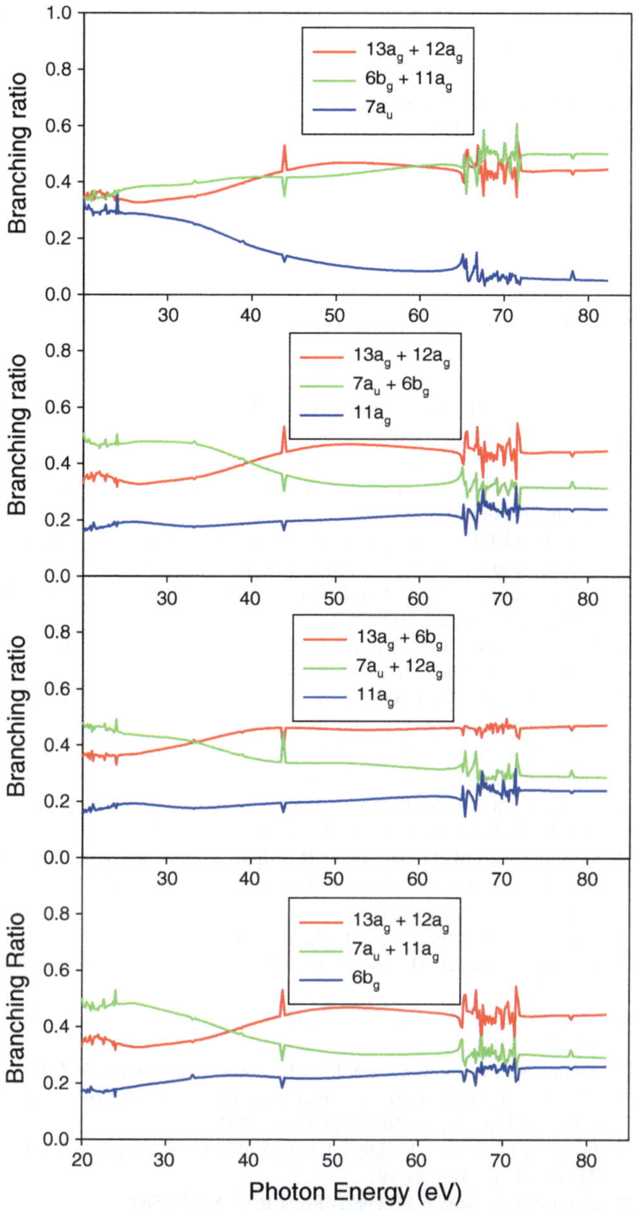

Fig. 9 Branching ratios relative to different possible assignments for the three outermost bands, see text for discussion

The relevant branching ratios, together with one BR relative to $7a_u$ considered alone, are reported in Fig. 9.

It will be probably hard to distinguish experimentally between the three possibilities, not to mention inadequacies of the theoretical results in comparison with experiment. One may only point out that assignment (a) leads to a significantly more even distribution of relative intensities at high energy, which appear quantitatively more in line with the experimental results available [5, 6], while the other two, especially (c), give a first peak intensity much larger than the other two, with a small difference between the second and the third.

A possible significant feature of assignment (b) is the much reduced amplitude of the 3p → 3d resonant structure in the first BR, compared to that of the other two. Possible additional information could come from the behavior of the cross-sections at still higher energies, or the angular distributions, in particular for band 3 which could distinguish between $11a_g$ and $6b_g$ ionizations, or more exotic techniques, like photoionization from the oriented molecule [39, 40]. Also a very high-resolution study may be able to further slice bands 1–3, as suggested from high-resolution spectra reported in Ref. [6], and possibly give the full ordering of ionizations within the first three bands.

Moreover, investigation of geometry dependence [41] and the possible presence of the cis conformer [42] have to be assessed.

At the present stage, assignment (a) seems slightly more favored, which is in complete agreement with that of the last SR study [6]. It is remarkable, however, how by a careful consideration of cross-sections alone a rather complete and firm assignment of this difficult spectrum can be achieved, the only ambiguity being confined to the three possibilities just considered for the three outermost bands.

Finally, it is interesting to point out the remarkably low intensity of the 3p → 3d resonance peaks in this region. A comparison of total cross-section for $Ni(C_3H_5)_2$ with that for $Ni(CO)_4$ is reported in Fig. 10. A striking difference is apparent. While in $Ni(CO)_4$, the spectrum shows a single large-intensity Fano profile with weak Rydberg-like structures superimposed in $Ni(C_3H_5)_2$, the large structure is absent while low intensity peaks are distributed over the whole resonant region. This indicates that a single virtual orbital carries most of the available 3d content, which is sizable, in the former molecule, while little 3d participation is available in the virtual space of $Ni(C_3H_5)_2$, and it is distributed over several virtual states. That is a clear

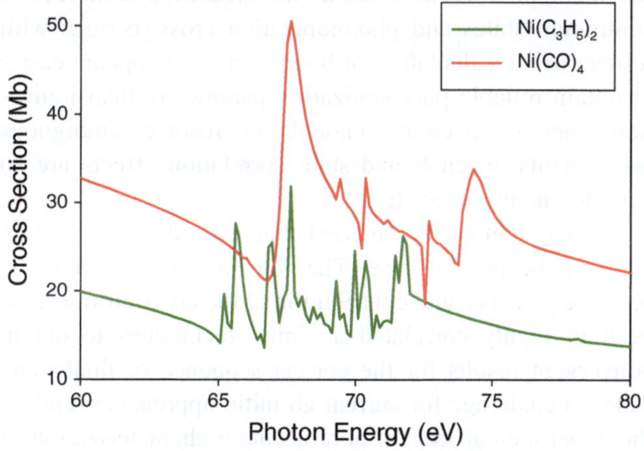

Fig. 10 Total ionization cross-section for $Ni(C_3H_5)_2$ and $Ni(CO)_4$ in the region of the 3p → 3d resonance

indication of the rather weak back bonding in $Ni(C_3H_5)_2$, which has a metal configuration closer to $3d^{10}$, while a much stronger backbonding is present in $Ni(CO)_4$, which is in fact a very stable system, pushing a significant 3d component in the corresponding antibonding orbital.

6 Conclusions

The $Ni(C_3H_5)_2$ molecule is a remarkable compound, a model for organometallic chemistry and olefine transition metal bonding, which despite its apparent simplicity shows a very complex electronic structure, dominated by strong correlation effects. It represents therefore a prototypical molecule for the study of electron correlation in transition metal compounds, and of the theoretical approaches to the computation of ionization potentials. Not only is the sequence of ionization energies widely different from that afforded by HF eigenvalues, but notably the Dyson orbitals, physically associated with the various final ionic states, are much different from the canonical HF orbitals, which challenge the use of such orbitals for the description of the electronic structure, and in particular of the HF HOMO in discussing chemical properties and reactivity in this class of molecules.

On the contrary, it appears that DFT has the ability to give orbitals much closer to Dyson than to HF in cases where their difference is large, notably in transition metal compounds [43], a remarkable property that remains to be further investigated.

In this context, the study of dynamical photoionization properties, cross-sections and possibly angular distributions give substantial and unique information on the electronic structure of such compounds, yielding a precise characterization of the orbitals associated with the various ionizations, and in particular, strong clues for the assignment of difficult spectra. It is remarkable that despite the additional complexities involved in the calculation of molecular continuum states and photoionization cross-sections with respect to the calculation of bound states, it appears easier to obtain reliable photoionization parameters than ionization energies accurate enough to resolve ambiguous assignments, when bound state correlation effects are so complex as in this system.

Finally, it must be stressed that several issues remain open in the present case: The role of geometry and the possible presence and contribution of the cis conformer, the push of highly correlated ab initio techniques to obtain convergent results for the correct sequence of final ionic states, a challenge for current ab initio approaches, and on the experimental side, a more accurate characterization of the photoionization parameters, including angular distributions and over a larger energy range, exploiting the much higher fluxes and energy resolution presently available at SR facilities.

Acknowledgments A generous amount of computer time at CINECA through ISCRA B project TRAMCREP "Transition metal compounds: resonances in photoionization" is gratefully acknowledged.

References

1. Veillard A, Demuynck J (1977) In: Schaefer HF (ed) Modern theoretical chemistry, vol 4. Plenum Press, New York, p 187
2. Neese F, Petrenko T, Ganyushin D, Olbrich G (2007) Coord Chem Rev 251:288
3. Decleva P, Fronzoni G, Lisini A (1991) In: Labanowski JK, Andzelm JW (eds) Density functional methods in chemistry. Springer, New York, p 323
4. Lloyd and DR, Lynaugh N (1972) In: Shirley DE (ed) Electron spectroscopy. North-Holland, Amsterdam, p 445
5. Li X, Bancroft GM, Puddephatt RJ, Hu YF, Liu Z, Tan KH (1992) Inorg Chem 31:5162
6. Li X, Bancroft GM, Puddephatt RJ, Liu ZF, Hu YF, Tan KH (1994) J Am Chem Soc 116:9543
7. Batich CD (1976) J Am Chem Soc 98:7585
8. Green JC (1994) Acc Chem Res 27:131
9. Li X, Bancroft GM, Puddephatt RJ (1997) Acc Chem Res 30:213
10. Green JC, Decleva P (2005) Coord Chem Rev 249:209
11. Gelius U, Siegbahn K (1972) Faraday discuss. Chem Soc 54:257
12. Dill D, Dehmer JL (1974) J Chem Phys 64:692
13. Dehmer JL, Dill D (1979) In: Rescigno T, McKoy V, Schneider B (eds) Electron-molecule and photon-molecule collisions. Plenum, New York, p 225
14. Chermette H (1992) New J Chem 16:1081
15. Toffoli D, Stener M, Fronzoni G, Decleva P (2002) Chem Phys 276:25
16. Stener M, Fronzoni G, Decleva P (2005) J Chem Phys 122:234301
17. Stener M, Fronzoni G, Decleva P (2009) Chem Phys 361:49
18. Martins M, Godehusen K, Richter T, Wernet P, Zimmermann P (2006) J Phys B: At Mol Opt Phys 39:R79
19. Fonseca Guerra C, Snijders JG, te Velde G, Baerends EJ (1998) Theor Chem Acc 99:391
20. Zangwill A, Soven P (1980) Phys Rev A 21:1561
21. Van Leeuwen R, Baerends EJ (1994) Phys Rev A 49:2421
22. Gross EKU, Kohn W (1990) Adv Quantum Chem 21:255
23. Rohmer MM, Demuynck J, Veillard A (1974) Theor Chim Acta 36:93
24. Cederbaum LS, Domcke W, Schirmer J, von Niessen W (1986) Adv Chem Phys 65:115
25. Moncrieff D, Hillier IH, Saunders VR, von Niessen W (1985) J Chem Soc Chem Commun 779
26. Moncrieff D, Hillier IH, Saunders VR, von Niessen W (1985) Inorg Chem 24:4247
27. Decleva P, Fronzoni G, De Alti G, Lisini A (1988) J Mol Struct Theochem 184:49
28. Decleva P, Fronzoni G, Lisini A (1989) Chem Phys 134:307
29. Böhm MC, Gleiter R, Batich CD (1980) Helv Chim Acta 63:990
30. Böhm MC, Gleiter R (1980) Theor Chim Acta 57:315
31. Böhm MC, Gleiter R (1986) Chem Phys Lett 123:87
32. Moncrieff D, Hillier IH, Saunders VR, von Niessen W (1986) Chem Phys Lett 131:545
33. Ning CG, Hajgato B, Huang YR, Zhang SF, Liu K, Luo ZH, Knippenberg S, Deng JK, Deleuze MS (2008) Chem Phys 343:19

34. Itatani J, Levesque J, Zeidler D, Niikura H, Pépin H, Kieffer JC, Corkum PB, Villeneuve DM (2004) Nature 432:867
35. Pavičić D, Lee KF, Rayner DM, Corkum PB, Villeneuve DM (2007) Phys Rev Lett 98:243001
36. Liebsch T, Plotzke O, Heiser F, Hergenhahn U, Hemmers O, Wehlitz R, Viefhaus J, Langer B, Whitfield SB, Becker U (1995) Phys Rev A 52:457
37. Decleva P, Fronzoni G, Furlan S, Stener M (2001) Chem Phys Lett 348:363
38. Decleva P, Fronzoni G, Stener M, de Simone M, Coreno M, Green JC, Hazari N, Plekan O (2005) Phys Rev Lett 95:263401
39. Yamazaki M, Adachi J, Teramoto T, Yagishita A, Stener M, Decleva P (2009) J Phys B: At Mol Opt Phys 42:051001
40. Yamazaki M, Adachi J, Kimura Y, Stener M, Decleva P, Yagishita A (2010) J Chem Phys 133:164301
41. Goddard R, Krüger C, Mark F, Stansfield R, Zhang X (1985) Organometallics 4:285
42. Casarin M, Pandolfo L, Vittadini A (2001) Organometallics 20:754
43. Gritsenko OV, Braïda B, Baerends EJ (2003) J Chem Phys 119:1937

Theor Chem Acc (2012) 131:1186
DOI 10.1007/s00214-012-1186-y

REGULAR ARTICLE

Copper coordination to the putative cell binding site of angiogenin: a DFT investigation

Luca Bertini · Maurizio Bruschi · Marco Romaniello · Giuseppe Zampella · Matteo Tiberti · Valentina Barbieri · Claudio Greco · Diego La Mendola · Raffaele P. Bonomo · Piercarlo Fantucci · Luca De Gioia

Received: 29 July 2011 / Accepted: 7 December 2011 / Published online: 3 March 2012

Abstract We present a DFT study of the structural and spectroscopic properties of the complex formed by Cu^{2+} with the peptide fragment Ac-PHREN-NH_2, which encompasses the putative cell binding domain of angiogenin, as well as with its Ac-PHRQN-NH_2 variant. Analysis of structures, energies and spectroscopic parameters has allowed to conclude that the metal coordination environment at pH 8 is formed by a nitrogen atom of His, two deprotonated amide groups, and an oxygen atom from the COO^- side chain of Glu, in nice agreement with recent experimental results (La Mendola et al. in Dalton Trans, 39:10678, 2010). Moreover, DFT results allowed to reveal that the Glu side chain of the Ac-PHREN-NH_2 peptide is coordinated in equatorial position, in a tetrahedrically distorted square planar arrangement, fully disclosing the effects of Cu^{2+} binding on the structural properties of this key angiogenin portion. In the Ac-PHRQN-NH_2 variant, the carboxylate group is replaced by a H_2O molecule in a coordination arrangement similar to that of the wild-type system.

Keywords DFT · Bioinorganic chemistry · Copper · Angiogenin

Dedicated to Professor Vincenzo Barone and published as part of the special collection of articles celebrating his 60th birthday.

L. Bertini · M. Romaniello · G. Zampella · M. Tiberti · V. Barbieri · C. Greco · P. Fantucci · L. De Gioia (✉)
Dipartimento di Biotecnologie e Bioscienze,
Università degli Studi di Milano Bicocca,
Piazza della Scienza, 2, 20126 Milan, Italy
e-mail: luca.degioia@unimib.it

M. Bruschi
Dipartimento di Scienze Ambientali, Università degli Studi di Milano Bicocca, Piazza della Scienza, 1, 20126 Milan, Italy

D. La Mendola
Istituto di Biostrutture e Bioimmagini,
CNR, Viale A. Doria 6, 95125 Catania, Italy

R. P. Bonomo
Dipartimento di Scienze Chimiche,
Università degli Studi di Catania, Viale A. Doria 6,
95125 Catania, Italy

1 Introduction

Angiogenesis is a set of functional processes responsible for the formation of new blood vessels from existing ones and is also a key component of the homeostatic process that regulates the distribution of oxygen to tissues [1]. Angiogenesis occurs in several tightly regulated stages that orchestrate a network of cooperative interactions and can be divided in several phases: (1) an initiation phase, characterized by increased vasopermeability; (2) a progression phase, in which proteolytic enzymes that degrade the extracellular matrix and promote endothelial cell migration are produced; (3) a final phase, in which differentiation into new vessels takes place. The latter phase is mediated by molecules that recruit mesenchymal cells to vessel walls [2].

Angiogenin (Ang) is a protein that undergoes nuclear translocation in endothelial cells, where it accumulates in the nucleolus and stimulates rRNA transcription, a rate limiting step in ribosome biogenesis, protein translation, and cell growth [3]. Ang was originally isolated from the conditioned medium of HT-29 human colon adenocarcinoma cells [4]. The mature Ang is a basic, single-chain protein containing 123 amino acids with a molecular weight of about 14.400 Da, and it is a homologue of bovine pancreatic ribonuclease A. Indeed, its ribonucleolytic activity is rather low, but Ang is essential for angiogenesis and other functions [5]. In fact, Ang belongs to the ribonuclease superfamily, showing 33% sequence identity to the pancreatic ribonuclease A. However, although the crystal structures of human Ang and pancreatic ribonuclease A are highly similar, a notable difference is evident in the ribonucleolytic active site: the pyrimidine binding site of Ang is "obstructed" by the Gln117 residue, explaining its very weak ribonucleolytic activity (about 10^5–10^6 lower than that of RNase A).

Besides its ribonucleolytic activity, binding of Ang to the endothelial cells surface is needed for its biological functions, and amino acid residues from 60 to 68 are critical in this process. Notably, affinity of Ang for endothelial cells is largely increased in the presence of copper ions [6]. In fact, copper has been recognized to be an angiogenic factor, but the mechanism whereby it exerts this function is still not well understood [7, 8]. In particular, Ang binds 2.4 mol of copper per mol of protein, and the metal ion is important for Ang binding to calf pulmonary artery endothelial cells, which increases by 4.3-fold in the presence of Cu^{2+}.

A putative endothelial cell binding domain of Ang is located in a loop exposed to the solvent and largely unstructured in the native protein [9]. The related amino acid sequence (hAng60–68 = KNGNPHREN) contains the prolyl-histidyl (PH) dyad, which is reminiscent of that present in other copper binding proteins, such as prion proteins [10] and the Wilson's and Menkes' ATPases [11]. In this respect, the comparison between Ang and RNase A structures already suggested possible copper binding sites [12]. In fact, RNase A binds copper at several sites including His-12, His-105 and His-119 at pH 5 [13], but Ang features three additional histidine residues that are not present in RNase A: His-8, His-65 and His-84 [14, 15]. The similarity between the copper binding plasma tripeptide Gly–His–Lys and the copper binding site of albumin and α-fetoprotein, where copper binds to a histidyl residue adjacent to a basic residue (Arg or Lys), suggests that His-65, which is adjacent to Arg-66 in Ang, might be involved in copper binding.

Very recently La Mendola et al. [16] reported the synthesis and characterization of the complex formed by the peptide fragments encompassing the sequence hAng64–68 (Ac-PHREN-NH_2; Scheme 1) and the whole sequence hAng60–68 (Ac-KNGNPHREN-NH_2) with copper.

In particular, combined potentiometric and spectroscopic investigations allowed to reveal the species distribution and the coordination environments of the Cu(II) complexes. It turned out that both peptides coordinate Cu^{2+} in a similar fashion. Moreover, thermodynamic and spectroscopic data indicated that the side chains of Glu and

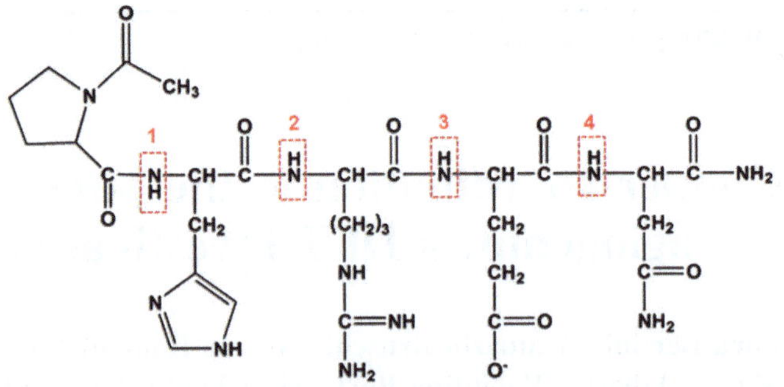

Scheme 1 Structure of the Ac-PHREN-NH_2 peptide. The four NH group of the peptide bonds have been highlighted in *red boxes*

His residues are involved in copper binding at physiological pH. The Cu(II) interaction with the peptide fragment Ang64–68(E67Q) (Ac-PHRQN-NH_2), in which glutamate was substituted by a glutamine residue, was also studied in order to unveil the role of glutamate carboxylate group on Cu(II) coordination. The comparison between results obtained studying the Cu(II) complexes formed by Ac-PHREN-NH_2 and its E67Q variant provided further evidence of the presence of a carboxylate oxygen atom in the copper coordination sphere. On the ground of such results, it was concluded that at pH 8 the metal coordination environment in the complex of Cu(II) with Ac-PHREN-NH_2 is formed by a nitrogen atom of His, two deprotonated amide groups, a water molecule and an oxygen atom from the COO^- side chain of Glu. However, it was not possible to reveal whether the carboxylate oxygen atom was coordinated in equatorial or apical position, hindering the full disclosure of the effects of Cu(II) binding on the structural properties of the Ang60–68 protein portion.

With the aim of complementing available experimental data in the elucidation of the nature of the copper first coordination environment, as well as the apical/equatorial disposition of oxygen ligands, we have carried out a DFT investigation of the Cu(II) complexes formed by Ac-PHREN-NH_2, as well as by the single point mutated peptide Ac-PHRQN-NH_2.

2 Computational details

The generation of the starting structures for DFT optimization was carried out using a combined MM/MD approach. In particular, starting structures, differing for coordination geometry and/or orientation of aminoacid side chains, were initially optimized using the MMFF94x forcefield [17, 18] as implemented in the MOE suite (MOE Molecular Operating Environment, version 2008.10; Chemical Computing Group Inc.: Montreal, Canada, 2008). Then, short 1,000 ps molecular dynamic simulations were performed at 100 K to relax the systems and allow it to escape from metastable high-energy minima. Simulation temperature was kept very low because the aim of such MM/MD simulations was local (and not global) sampling of the potential energy hypersurface.

Quantum chemical geometry optimizations were carried out using the pure Generalized Gradient Approximation (GGA) BP86 DFT functional [19, 20] and the Resolution of Identity (RI) technique [21], as implemented in the TURBOMOLE suite of programs [22]. Basis sets of triple-zeta plus polarization split valence quality (TZVP hereafter) [23] were adopted for all atoms in the complexes. The DFT grid-size was set to standard m3 value.

The computation of the EPR g tensors and hyperfine coupling constants (hcc) for the ^{63}Cu nucleus has been performed using the B3LYP hybrid functional on the geometries optimized at the RI-BP86/TZVP level. In fact, as shown in a previous DFT investigation about copper coordination to prion protein peptides [24], the structural features (bond distances and angles) of Cu–peptide complexes obtained with B3LYP and BP86 are almost identical. However, BP86 calculations are faster, due to the possibility of using the RI technique, and therefore more suited to test a large number of species, such as in the present work. On the other hand, EPR g tensors and hyperfine coupling constants obtained with B3LYP are in better agreement with experimental data, when compared to BP86 results. An extended basis set (14s, 10p, 5d) [25], augmented by a set of diffuse s, p and d functions (with exponents equal to 0.01, 0.03087, and 0.1, respectively) and contracted to (9s, 7p, 4d), was adopted for the Cu atom. The IGLO-II basis was adopted for all other atoms [26]. EPR hyperfine coupling constants (hcc) were calculated, explicitly taking Spin–Orbit (SO) contributions into account, with the one-centre and mean field approximation (AMFI) [27] for the two electron terms (see Ref. [28] for a complete discussion of the SO operators). For the calculations of g tensors, the gauge origin has been set to the centre of electronic charge. Calculations of EPR properties have been carried out using the ORCA suite of programs [29].

3 Results and discussion

3.1 Geometries and relative energies of isomers of [Cu(Ac-PHREN-NH_2)(H_2O)] and [Cu(Ac-PHRQN-NH_2)(H_2O)]$^{+1}$ complexes

According to the experimental evidences, Cu(II) coordination to the Ac-PHREN-NH_2 peptide at pH 8 involves two N^- anions from the deprotonated amide groups of the peptide bonds, the carboxylate group of Glu and a nitrogen atom from the His side chain. In addition, a water molecule should also be coordinated to the metal ion [16]. Before discussing which deprotonated N^- groups are actually involved in Cu(II) coordination, some considerations are in order. Previous results obtained studying coordination compounds formed by Cu(II) with fragments of the prion protein [24, 30–33], as well as simple considerations about the preferential formation of 5- and 6-member rings when dealing with chelate ligands, lead to the conclusion that the two N^- anions involved in Cu(II) coordination must be contiguous. Therefore, only three configurations are possible, namely those involving the **1–2**, **2–3** or **3–4** deprotonated amide groups (see Scheme 1 for amide groups labeling). We preliminary analyzed the conformation of these three configurations at Molecular Mechanics (MM) level, in order to quickly understand which is the most likely Cu^{2+} coordination geometry. The preliminary MM analysis revealed that, when the **1–2** or **3–4** configurations are taken into account, the histidine (in the **1–2** configuration) or the glutamic acid (in the **3–4** configuration) side chains occupy the apical position in the metal coordination sphere. In fact, since the side chains of His (in the **1–2** configuration) or Glu (in the **3–4** configuration) are in between the deprotonated N^- groups that form the planar 5-member cycle (Scheme 2), the aminoacid side chain must necessarily coordinate Cu^{2+} in apical position due to steric restrictions. As a consequence, in complexes characterized by **1–2** or **3–4** binding mode, the water molecule must occupy an equatorial position in the Cu(II) coordination environment.

MM results also indicate that the **1–2** and **3–4** binding modes induce some strain in the peptide chain. In particular, the side chain of the coordinating aminoacid which does not belong to the five-member ring (Glu in **1–2** complexes, His in **3–4** structures) is far from the metal center and its coordination to the metal atom is always accompanied by some strain of the peptide backbone. MM analysis of the **2–3** binding mode reveals that the imidazole ring of His and the carboxylate group of Glu can occupy equatorial positions in the Cu(II) coordination sphere without inducing steric strain in the molecule. These results suggest that the structures featuring the **2–3** binding mode

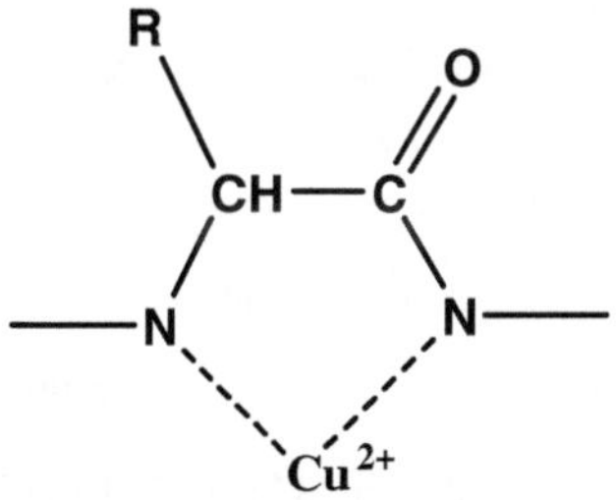

Scheme 2 Five-member ring formed in Ac-PHREN-NH_2 upon coordination to Cu(II) of the two adjacent deprotonated amide groups

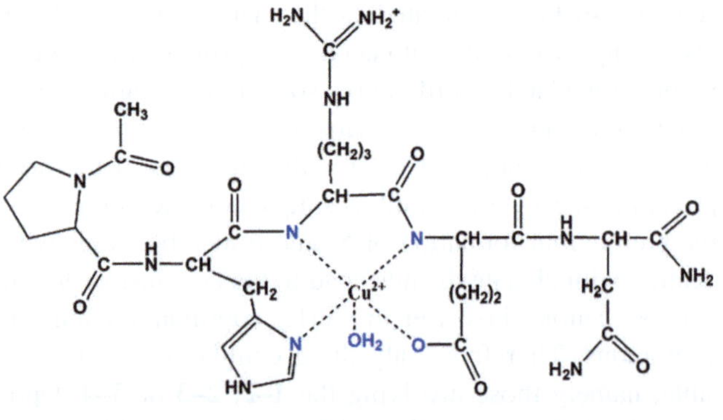

Scheme 3 Schematic representation of the **2–3** binding mode in complexes between Cu(II), the peptide Ac-PHREN-NH_2 and a water molecule

most likely correspond to the lowest-energy isomers (see Scheme 3).

Therefore, we have carried out a thorough DFT investigation into the conformational properties of coordination compounds featuring the **2–3** binding mode. However, for the sake of completeness, we have evaluated by DFT also some structures characterized by the **1–2** or **3–4** binding mode, to quantify the strain of the peptide chain.

To select reasonable starting structures for DFT optimization, we have carried out classical MD simulations (see "Computational details") of the [Cu(Ac-PHREN-NH_2)(H_2O)] complex featuring the **2–3** binding mode, selecting snapshots along the MD trajectory which correspond to low-energy conformations. The protocol adopted to explore the configurational space of the system is the following. The structures characterized by the **2–3** binding mode can be ideally divided into two families: the first containing structures featuring an apical water molecule (and consequently an equatorial Glu side chain), and the second containing structures featuring an equatorial water molecule (Scheme 4).

We initially generated "ideal" (dihedral angle of the base of the pyramid equal to zero, molecular axes of the apical ligand exactly perpendicular to the base) structures characterized by the N_3O_2 coordination environment and

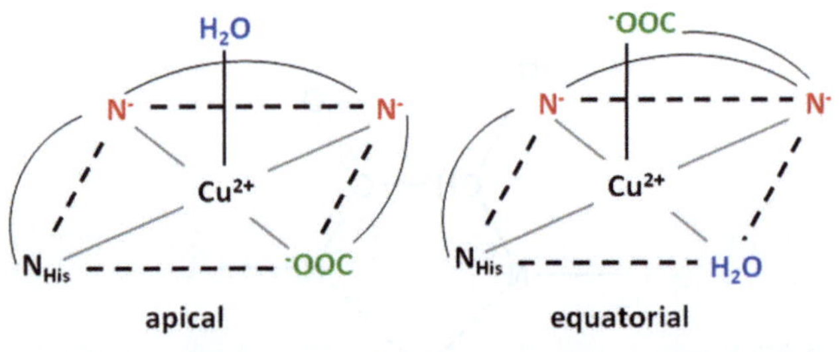

Scheme 4 The two families of [Cu(Ac-PHREN-NH_2)(H_2O)] complexes characterized by the **2–3** binding mode. Apical and equatorial labels are referred to the position of the water molecule in the Cu^{2+} coordination sphere

featuring an apical or equatorial water molecule. The value of Cu–N^- distances has been set to 2.000 Å, according to the BP86 optimized Cu–N distances in the [Cu(HGGG)] system [33]. For the Cu–N_{His} and Cu-O^-_{Glu} distances, we adopted the value of 1.894 and 2.007 Å obtained, respectively, from BP86/TZVP gas-phase optimization of the Cu-His and Cu-Glu adducts. Regarding the Cu-O distance to describe Cu–H_2O binding, we considered as experimental reference the XRD value of 2.4 Å in *cis*-Cu(Gly)$_2$·H_2O, which fits well also with the analysis of Sobolović et al. [34], which measured a mean value of 2.4 ± 0.1 Å for Cu-O distances in 10 aminoacid/water Cu(II) complexes.

Using the two "ideal" structures as templates, we generated apical and equatorial conformations from molecular dynamics simulations in which the Cu(II) coordination distances were constrained to the above-mentioned reference values. Then, for each structure obtained, we first optimized its geometry at the MM level of theory, and the corresponding minimum energy structure was used as starting point of a two-step DFT optimization. In the first step, the Cu(II) coordination was constrained, allowing only the peptide moiety to relax. In the second step, the whole geometry was optimized removing the constraints.

The relevant optimized geometry parameters, as well as computed relative energies of all structures characterized by the **2–3** binding mode are collected in Table 1. It must be noted that in structures characterized by an apical water molecule, the N_3O_2 coordination can be realized only in one way, whereas in the case of equatorial water binding, the isomer shown in Scheme 4 is not the only plausible possibility, but other three ligands arrangements, referred to as the A, B and C, can in principle exist (Scheme 5).

Before discussing the fine details of the various structures optimized, it is interesting to highlight some trends obtained by our DFT investigation. Except for a few structures, geometry optimizations usually led to tetrahedral distortion of the initial square pyramidal coordination. In addition, the H_2O ligand often resulted loosely bound or non-bound to Cu(II), yielding 4-coordinated N_3O structures. It is also worth noting that when starting from structures featuring an equatorial water ligand, the H_2O molecule always left the Cu(II) coordination sphere, and it was replaced in the equatorial position by the oxygen atom of the glutamic acid residue. Therefore, structures characterized by equatorial coordination of the H_2O molecule should not correspond to stable forms of the [Cu(Ac-PHREN-NH_2)(H_2O)] complex.

In the exploration of the conformation space of [Cu(Ac-PHREN-NH_2)(H_2O)], we often found that structures differ only for the number of intramolecular H-bonds. As shown in the following, when a complex is explicitly solvated by a reasonable number of H_2O molecules, intramolecular interactions are often lost because the side chains of the

Table 1 Selected geometrical parameters (in Å and degree) and energy differences relative to the lowest-energy form (in kcal mol^{-1}) of the [Cu(Ac-PHREN-NH_2)(H_2O)] isomers

Isomer	Distances								
	CN	ΔE	$Cu{-}O_{H_2O}$	Cu-O_{CO}	Cu-N_1^-	Cu-N_2^-	Cu-N_{His}	δ Dihedral	H-bonds
2–3 Binding mode									
1	5	0.0	2.645	2.039	2.016	2.019	2.039	0.4	6
2	4	2.0	3.777	2.001	2.017	2.001	2.000	16.6	5
3	4	2.1	3.863	2.012	2.020	2.000	2.006	12.4	5
4	4	2.1	3.796	2.010	2.012	1.999	1.994	13.5	5
5	4	5.5	3.565	2.001	1.996	1.948	2.016	38.5	2
6	4	6.0	3.374	1.973	1.965	2.034	2.003	9.2	4
7	4	6.2	3.354	1.999	2.029	2.033	2.032	17.6	4
8	5	6.3	2.488	2.014	2.017	2.056	2.029	0.9	3
9	4	8.4	3.684	1.983	1.986	2.022	2.012	8.7	4
10	4	12.2	3.655	1.982	1.932	2.074	1.986	42.1	8
11	5	12.7	2.492	2.020	1.994	2.070	2.028	4.3	5
12	5	13.7	2.348	2.044	2.037	2.023	1.971	3.4	4
13	4	14.0	3.390	2.002	1.963	1.929	2.100	37.3	3
14	5	14.8	2.753	2.009	2.032	1.993	2.047	1.3	6
15	4	15.3	3.912	1.954	1.954	2.016	2.021	35.3	5
16	4	15.5	3.755	2.010	1.990	2.019	2.011	18.2	5
17	4	16.0	3.924	1.962	1.949	2.023	2.020	34.9	5
18	4	17.7	4.073	1.932	1.972	2.002	2.038	35.2	6
19	5	24.6	2.889	1.997	2.000	2.013	2.054	19.6	4
20	5	32.3	2.407	1.997	2.000	2.004	1.994	0.4	6
2–3 Binding mode									
21	5	13.7	2.031	2.042	1.984	2.150	2.281	11.0	8
22	5	13.8	2.023	2.042	1.982	2.130	2.326	14.4	8
23	5	20.0	2.020	2.050	2.050	2.004	2.302	9.6	4
24	4	22.3	1.994	2.041	1.977	1.976		44.0	5
25	4	25.4	1.979	2.070	1.987	1.968		47.2	6
26	5	26.7	2.014	2.112	2.029	2.038	2.237	38.9	6
27	4	26.8	1.973	1.998	1.998	1.985		30.6	3
28	5	31.2	2.017	1.989	2.107	2.063	2.084	47.5	4
3–4 Binding mode									
29	5	8.1	2.450	2.242*	1.931	1.999	2.050	31.0	5
30	3	9.7	3.075		1.915	2.023	1.978		6
31	5	16.4	2.544	2.178*	2.012	1.927	2.024	19.7	5
32	5	26.1	2.170	2.010	2.000	2.078	2.168	47.5	5
33	5	26.2	3.414	2.574	1.935	1.986	2.003	53.7	6
34	5	35.2	3.641	2.711* 2.085 1.996*	1.911	1.993		28.8	5
35	5	36.1	2.035	2.054	1.999	2.073	2.294	14.3	5
36	5	37.3	2.289	2.627*	2.071	1.923	2.018	16.3	5
37	5	38.1	2.142	2.088	2.003	2.054	2.227**	21.0	5
38	5	39.4	2.041	2.142	1.984	2.058	2.287	55.6	4
39	5	41.6	5.610	1.995	1.969	2.038	2.033	36.9	6
40	5	42.4	2.289	2.267*	2.020	1.940	2.035	16.3	3
41	5	50.6	2.165	2.306	2.000	2.023	2.025	20.6	3
42	5	51.2	3.889	1.993	1.983	2.053	2.222	58.3	3

In the table are reported the five Cu-L distances, the dihedral angle that defines the base of the square pyramid of the N_3O_2 coordination sphere and the number of hydrogen bonds (H···O distance less than 2.0 Å). *CN* stands for coordination number

* The oxygen atom coordinated to Cu^{2+} does not belong to the COO^- group of the Glu side chain residue

** Histidine in apical position

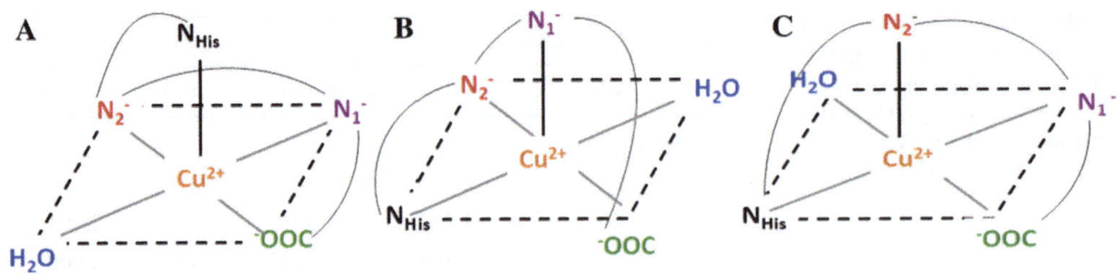

Scheme 5 The three subfamilies of [Cu(Ac-PHREN-NH_2)(H_2O)] complexes characterized by the **2–3** binding mode and the water molecule in equatorial position

residues not involved in Cu(II) coordination (in particular arginine) prefer to form H-bonds with the solvent. Nevertheless, even if we are aware of the limits of such approach, we carried out most DFT calculations neglecting solvent effects. In fact, the explicit description of a shell of water molecules surrounding the [Cu(Ac-PHREN-NH_2)(H_2O)] complex would be computationally too demanding. Indeed, since one of the main purposes of this study was the elucidation of the position of the H_2O ligand in the Cu(II) coordination sphere, in principle, solvent effects due to bulk water could be described using an implicit solvent model. However, it was already observed that DFT tends to underestimate H_2O–Cu(II) binding in similar complexes [33], and we noted that the adoption of an implicit solvent model makes the H_2O–Cu(II) bond even weaker (vide infra), not allowing to properly study H_2O coordination to Cu(II) in this class of compounds.

The lowest-energy structure characterized by the **2–3** binding mode features an apical water molecule (**1**; Fig. 1). In **1**, the N_3O_2 square pyramidal Cu(II) coordination environment is characterized by a dihedral angle δ (defined by the atoms N_{HIS}–N_1^-–N_2^-–O^- forming the base of the square pyramid) equal to 0.4°, and a Cu–O_{H_2O} distance = 2.645 Å. Other five structures are characterized by a Cu–O_{H_2O} distance lower than 3.0 Å (**8**, **11**, **12**, **14**, **19**; Table 1). When compared to **1**, these structures are characterized by a smaller number of H-bonds, except for structure **14**, which, however, is characterized by a loosely bound H_2O molecule (Cu–O_{H_2O} = 2.753 Å). The structure with the shortest Cu–O_{H_2O} distance (**12**) features an apical water molecule and is 13.7 kcal mol^{-1} higher in energy than **1**. In this context, it is also worth noting that **1** and **12** differ for the number of intramolecular H-bonds (4 in **12**, 6 in **1**; d(H···O) $\leq$ 2.0 Å), underlying the role played by this type of interactions for the relative energies of the structures considered.

The electronic spectra of Ac-PHREN-NH_2 and the lowest-energy [Cu(Ac-PHREN-NH_2)(H_2O)] form (structure **1**) have been computed at BP86/TZVP TDDFT level and are shown in Fig. 2. The first 100 transitions have been

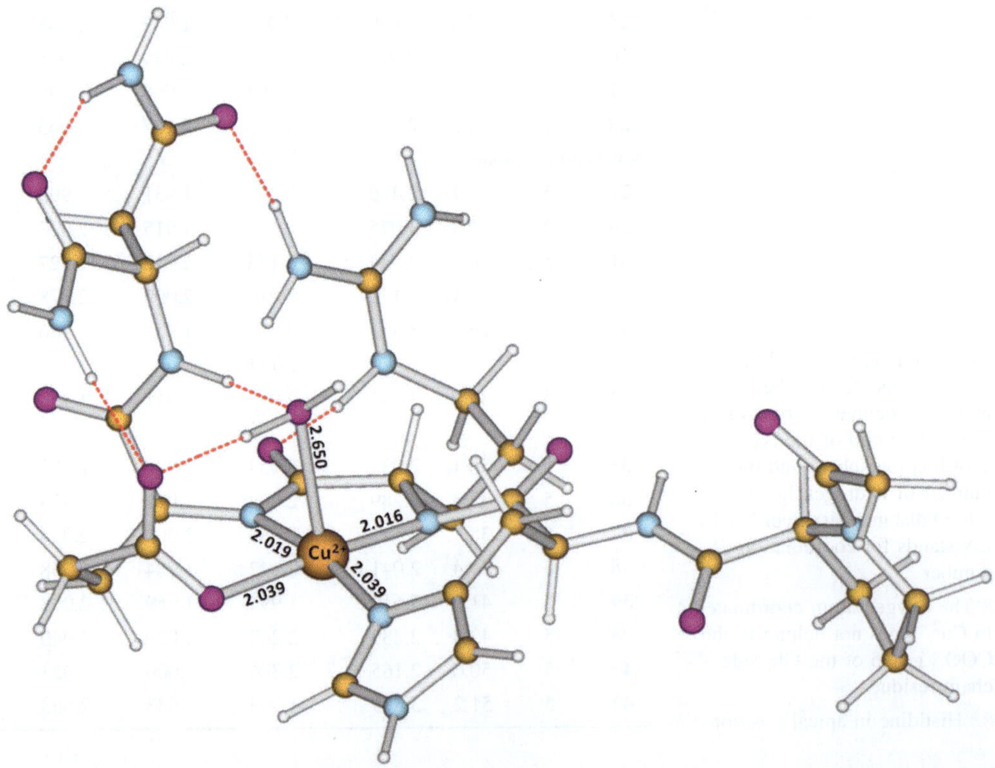

Fig. 1 Lowest-energy structure of the [Cu(Ac-PHREN-NH_2)(H_2O)] complex characterized by the **2–3** binding mode (**1**). Distances in Å

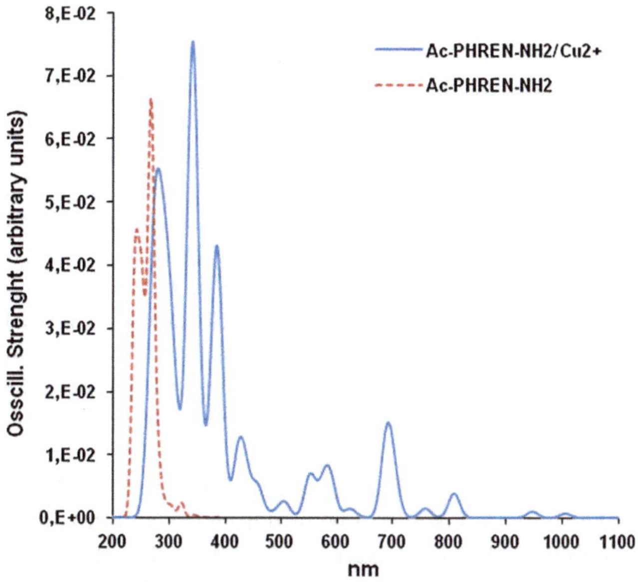

Fig. 2 TDDFT computed UV–Vis spectra for Ac-PHREN-NH_2 and the lowest-energy structure of the [Cu(Ac-PHREN-NH_2)(H_2O)] complex (**1**)

computed, and excitation energies, corresponding oscillation strengths and MO compositions for the first 30 transitions are collected in Table 2. It must be noted that the BP86 functional is not the optimal choice to reproduce excitation energies, in particular when metal-to-ligand charge transfer (MLCT) bands are considered, but the large size of the systems investigated did not allow us to adopt more suited functional, as PBE0 [35]. In any case, it can be concluded that computed UV–Vis bands of Ac-PHREN-NH_2 are mainly due to the histidine residue and correspond to two intense absorptions at 266 and 240 nm, and one weak band at 320 nm. In the corresponding Cu(II) complex, the two intense bands are red-shifted by about 100 nm. The bands in the range 1,000–500 nm (Table 2) are of SOMO-n(β) → SUMO(β) type (n = 0–14, Singly Occupied and Singly Unoccupied MO, respectively) and essentially correspond to the Cu(II) d–d transitions. In particular, the first five transitions that involve the d-Cu(II) MOs are of d–d type. The most intense band is at 689 nm (excitation 6), in fairly good agreement with the 640 nm absorption reported by La Mendola et al. [16].

As already pointed out, **2–3** isomers characterized by equatorial disposition of the water molecule do not correspond to minima on the potential energy surface, suggesting an intrinsic instability of this coordination mode. In order to quantify the destabilization of structures featuring an equatorially coordinated water molecule, we have also optimized a structure (**20**), which is obtained from **1** after exchange of the positions of the water molecule and the carboxylate group. In order to maintain the Cu(II) coordination environment, the position of the two coordinated oxygen atoms from the water molecule and the carboxylate group have been constrained to the Cu-O distances observed in **1**. Although this evaluation is qualitative, we found that structure **20** is more than 30 kcal mol^{-1} higher in energy than **1**.

Table 2 TDDFT BP86/TZVP electronic spectrum of **1**

Transition	nm	f	$a \rightarrow i$
1	1,005	7×10^{-4}	202β → 203β
2	947	1×10^{-3}	201β → 203β
3	808	4×10^{-3}	200β → 203β
4	757	2×10^{-3}	199β → 203β
5	702	5×10^{-3}	198β → 203β (55%)
			198β → 203β (45%)
6	689	1×10^{-2}	197β → 203β (45%)
			198β → 203β (33%)
7	685	2×10^{-3}	196β → 203β (83%)
8	637	2×10^{-4}	195β → 203β
9	623	1×10^{-3}	194β → 203β
10	597	2×10^{-4}	193β → 203β
11	588	5×10^{-3}	192β → 203β
12	575	4×10^{-3}	191β → 203β
13	552	6×10^{-3}	190β → 203β
14	507	2×10^{-3}	189β → 203β
15	497	5×10^{-4}	188β → 203β
16	485	5×10^{-4}	187β → 203β
17	462	9×10^{-4}	185β → 203β (62%)
18	457	4×10^{-3}	186β → 203β (87%)

For each transition, the excitation energy (nm), the oscillation strength (f) and the main ($a \rightarrow i$) monoelectronic excitations with the corresponding percentage composition are reported

In order to evaluate whether an extra water molecule might be bound to Cu(II), leading to a pseudo-octahedral coordination environment, we further investigated complex **1** after addition of a second H_2O molecule in apical position. However, the optimization of such a species led to a square pyramidal structure in which the extra water molecule moved away from the metal atom (final Cu–O_{H_2O} distance = 3.567 Å).

Going back to the issue of the computational approach adopted in this study to describe solvation effects, we also optimized the lowest-energy structure (**1**) including solvent effects through the implicit solvent model COSMO (COnductor-like Screening MOdel) [36–38]. The main effect observed in COSMO calculations was the removal of the H_2O molecule from the Cu(II) coordination sphere, which consequently becomes tetrahedrally distorted (δ moves from 0.4 to 30.7°) .

Finally, we have also evaluated the lowest-energy complex **1** when explicitly solvated (**1 W**). To address this point, structure **1** was solvated with a cluster of 55 H_2O

molecules, whose position was optimized by mean of a short MD simulation (1,500 ps) and successive MM geometry optimization. The solvated structure obtained in this way was successively optimized at DFT level using the same level of theory adopted for the non-solvated models (BP86/TZVP). Upon geometry optimization, the Cu(II) coordination does not change significantly. The H_2O molecule initially at 2.645 Å gets closer to the metal center (2.445 Å) and the δ dihedral angle increases of only 2°. Notably, no intramolecular H-bonds are found because the aminoacid side chains are all involved in H-bond with the solvent.

DFT results obtained studying structures characterized by the **1–2** and **3–4** binding mode indicate that these isomers are higher in energy with respect to the lowest-energy **2–3** isomer (**1**). In particular, most of the **1–2** structures (**21**, **22**, **23**, **26** and **28**; Table 1) feature a Cu(II) coordination with the histidine residue in apical position (see Scheme 6) and are higher in energy with respect to isomer **1** by at least 13.7 kcal mol^{-1}.

The three lowest-energy **1–2** structures feature square pyramid coordination with a small tetrahedral distortion. The lowest-energy **1–2** form (structure **21**) is shown in Fig. 3. It is interesting to point out that the H_2O ligand is coordinated to Cu(II) with a much shorter Cu-O distance (2.031 Å) compared with the **2–3** structures. In fact, in **21**, the Cu(II)–H_2O binding is "assisted" by the formation of two H-bonds formed with the two CO groups of the Glu side chain and the peptide bond between Pro and His (Scheme 7).

During the conformational sampling of structures characterized by the **1–2** binding mode, we have found three 4-coordinated structures (**24**, **25** and **27**), in which the histidine side chain is not coordinated to Cu(II). However, these isomers are at least 8.6 kcal mol^{-1} higher in energy than the lowest-energy **1–2** form (**21**).

The analysis of the MD simulations of the **3–4** configuration revealed a larger number of possible copper coordinations. In particular, several structures obtained from the MD simulations can be classified in two families (Scheme 8).

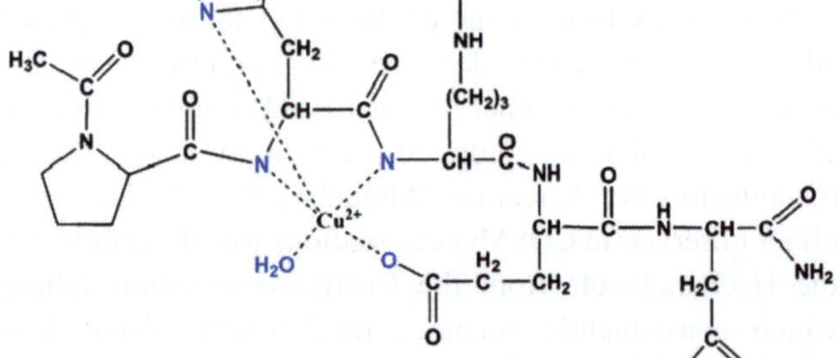

Scheme 6 Schematic representation of the **1–2** binding mode in complexes between Cu(II), the peptide Ac-PHREN-NH_2 and a water molecule

In structures **29**, **31**, **36** and **40** (Scheme 8; family *a*), the carboxylate of the glutamic acid forms an electrostatic interaction with the side chain of arginine, and consequently the carbonyl group of the C-terminal amide group enters the Cu(II) coordination sphere in equatorial position. Among the family *a* species, structure **29** is the lowest in energy (+8.1 kcal mol^{-1} with respect to **1**).

In structures **32**, **35**, **38**, **41** and **43** (Scheme 8; family *b*), the carboxylate group of Glu is coordinated to Cu(II) in apical position, and therefore the water molecule occupies the equatorial position. All these structures are genuine equatorial forms, but, however, are even higher in energy than the constrained **2–3** equatorial form (structure **20**).

The other **3–4** isomers obtained from the conformational sampling are higher in energy and feature a different Cu(II) coordination environment. It is worth noting that the 3-coordinated structure **30** ($2N^-$ and N_{His}) lies only 1.6 kcal mol^{-1} higher in energy with respect to the lowest-energy **3–4** structure (**29**) because it is stabilized by the formation of 6 intramolecular H-bonds. Structures **33** and **34**, which are characterized by the substitution of the coordinated H_2O with a CO group of the peptide chain, lie at least 29.6 kcal mol^{-1} higher in energy with respect to **1**. In **34**, also the histidine side chain leaves the Cu(II) coordination sphere, giving a 4-coordinated form. Structure **37** is the only **3–4** form that features the histidine side chain in apical position, similar to the **1–2** forms.

In order to better understand the role of the glutamate side chain in Cu(II) binding, and prompted by the previous observation that cleavage of the E67–N68 peptide bond compromises the angiogenic activity of Ang [39], La Mendola et al. [16] investigated also Cu(II) binding to the $E \rightarrow Q$ variant of the Ac-PHREN-NH_2 peptide ([Cu(Ac-PHRQN-NH_2)(H_2O)]$^{+1}$) observing at pH 8 a value of the $A_{\parallel}$ component of the ^{63}Cu hyperfine coupling constant slightly larger than that measured for the wild-type species.

In principle, four different types of coordination can be observed in complexes between Cu(II) and Ac-PHRQN-NH_2 (Scheme 9), in which the glutamine residue can be coordinate to Cu(II) via the NH_2 or CO group of its side chain, and the H_2O ligand can occupy apical or equatorial position. As well know from experimental evidences at neutral pH, the amidic NH_2 group should not coordinate the Cu(II) ion. However, to test the computational approach employed in this investigation, it is interesting also to consider this type of coordination. Sampling of the configuration space of the [Cu(Ac-PHRQN-NH_2)(H_2O)]$^+$ complex has been carried out using the same approach adopted for the wild-type peptide. In particular, starting from the lowest-energy forms of the Ac-PHREN-NH_2 complex, we introduced the E → Q substitution and successively carried out

Fig. 3 Lowest-energy structure of the [Cu(Ac-PHREN-NH_2)(H_2O)] complex characterized by the **1–2** binding mode (**21**). Distances in Å

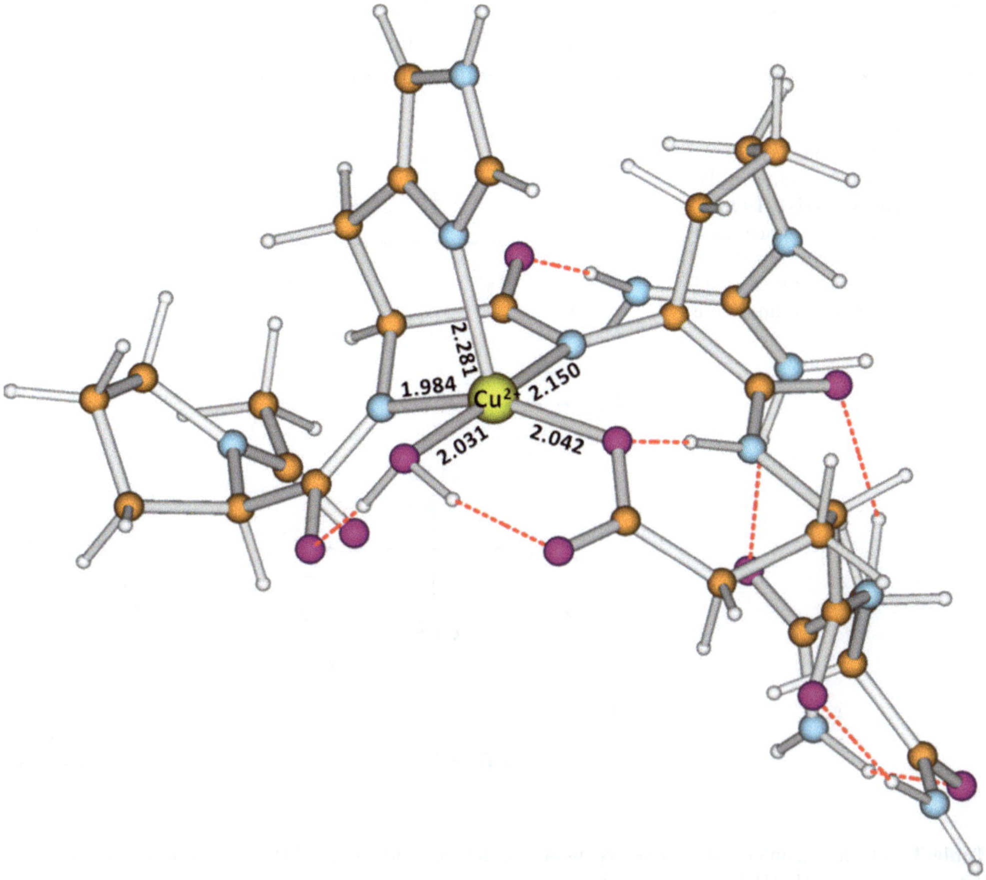

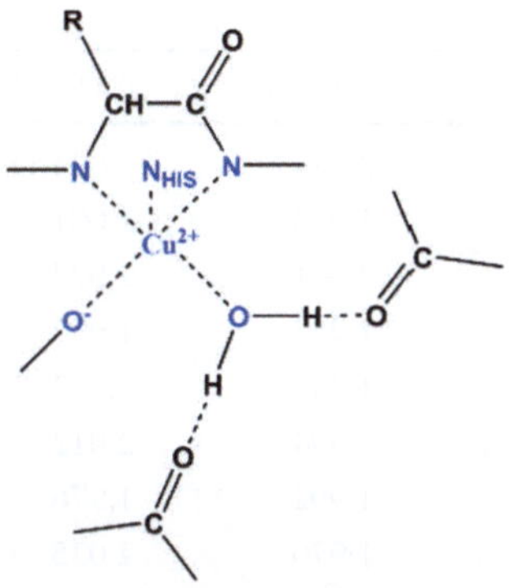

Scheme 7 Schematic representation of the Cu(II) coordination environment in the lowest-energy complex characterized by the **1–2** binding mode (structure **21**)

MD simulations to obtain starting point structures for DFT optimizations. We extracted from the MD trajectories 8 isomers characterized by apical (4 structures) or equatorial (4 structures) H_2O coordination, and for each structure, we evaluated both CO and NH_2 coordination of the glutamine side chain, obtaining a total number of 16 starting point structures for DFT optimization.

In general, in structures characterized by coordination of the NH_2 group of Gln, the Cu–N distance is about 2.3 Å and the NH_2 group is in equatorial position. The corresponding forms in which the CO amidic group is coordinated in equatorial position to Cu^{2+} feature a Cu-O distance within 2.1–2.2 Å.

Scheme 8 Schematic representation of the two families of complexes between Cu(II), the peptide Ac-PHREN-NH_2 and a water molecule, characterized by the **3–4** binding mode

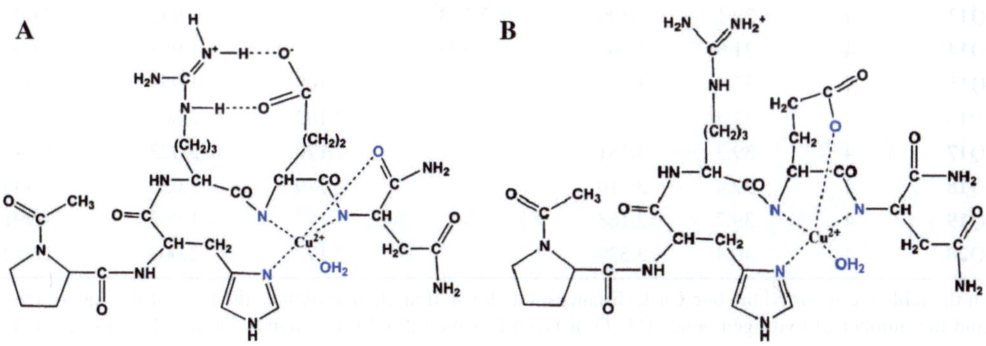

Relative energies of the various DFT optimized structures, as well as relevant bond distances for $[Cu(Ac\text{-}PHRQN\text{-}NH_2)(H_2O)]^{+1}$ complexes are collected in Table 3. Notably, all $[Cu(Ac\text{-}PHRQN\text{-}NH_2)(H_2O)]^{+1}$ low-energy structures are characterized by a significant tetrahedral distortion. The lowest-energy [Cu(Ac-PHRQN-

Scheme 9 The four conformation subfamilies of the $[Cu(Ac\text{-}PHRQN\text{-}NH_2)(H_2O)]^{+1}$ complex. The labels apical and equatorial are referred to the position of the water molecule in the Cu^{2+} coordination sphere

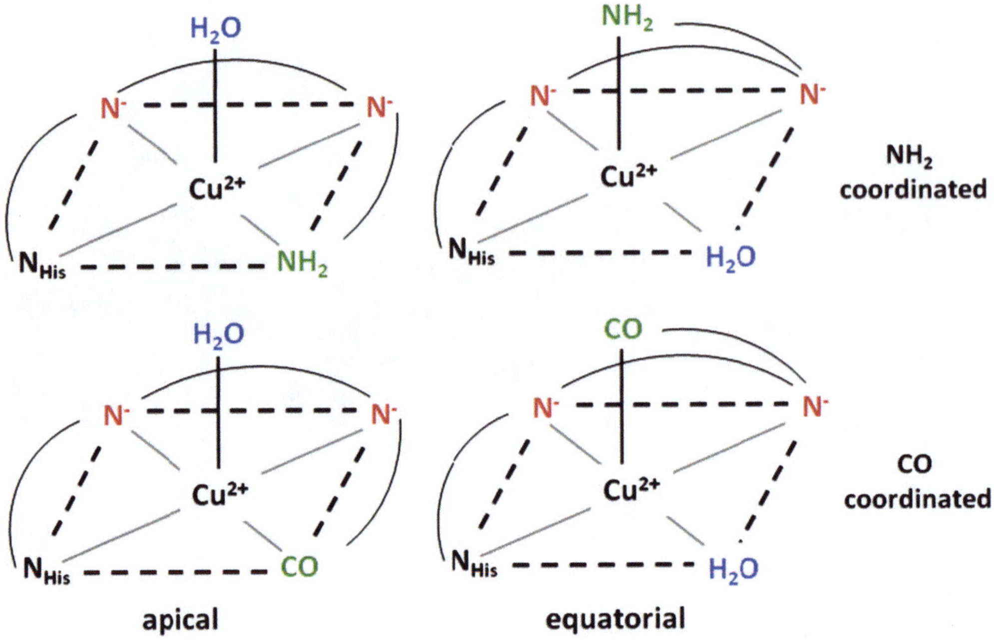

Table 3 Selected geometrical parameters (in Å and degree) and energy differences relative to the lowest-energy form (in kcal mol^{-1}) of the $[Cu(Ac\text{-}PHRQN\text{-}NH_2)(H_2O)]^{+1}$ complexes

Isomer	Distances								
	CN	ΔE	$Cu–O_{H_2O}$	$Cu\text{-}O_{CO}$	$Cu–N_1^-$	$Cu–N_2^-$	$Cu–N_{His}$	δ Dihedral	H-bonds
Q1	5	0.0	2.198		2.278	1.969	2.044	2.024	25.2
Q2	5	1.3	2.301		2.341	2.013	1.973	2.001	20.4
Q3	4	2.2	2.124			1.992	2.001	2.027	16.6
Q4	5	7.1	2.257		2.348	2.049	1.954	1.997	29.5
Q5	4	10.3	2.174			1.997	1.947	1.992	40.8
Q6	5	11.0	2.242		2.252	2.022	1.994	2.012	39.4
Q7	4	13.2	3.495		2.201	1.928	1.992	1.976	37.5
Q8	5	15.2	2.484	2.317		2.026	1.970	2.035	21.4
Q9	5	15.2	2.537	2.302		1.966	2.023	2.031	20.7
Q10	4	16.8	4.652		2.134	1.998	1.931	1.968	29.6
Q11	5	17.9	2.738		2.203	2.060	1.932	1.978	16.3
Q12	3	19.2	4.123			1.972	1.915	1.939	
Q13	4	20.1	4.199	2.273		1.955	2.000	1.998	9.0
Q14	4	21.9	3.668	2.305		1.953	2.009	1.981	48.0
Q15	4	37.4	3.544		2.108	1.957	2.007	2.108	6.1
Q16	4	37.4	3.521		2.107	2.006	1.958	1.983	5.7
Q17	4	39.3	3.181		2.176	2.023	1.948	1.973	19.3
Q18	4	39.4	3.310		2.179	2.023	1.946	1.977	19.1
Q19	4	39.7	2.168			1.962	1.991	1.983	32.7
Q20	4	40.4	3.539		2.172	2.008	1.952	1.989	14.7

In the table are reported the five Cu-L distances, the dihedral angle that defines the base of the square pyramid of the N_3O_2 coordination sphere and the number of hydrogen bonds (H···O distance less than 2.0 Å). *CN* stands for coordination number

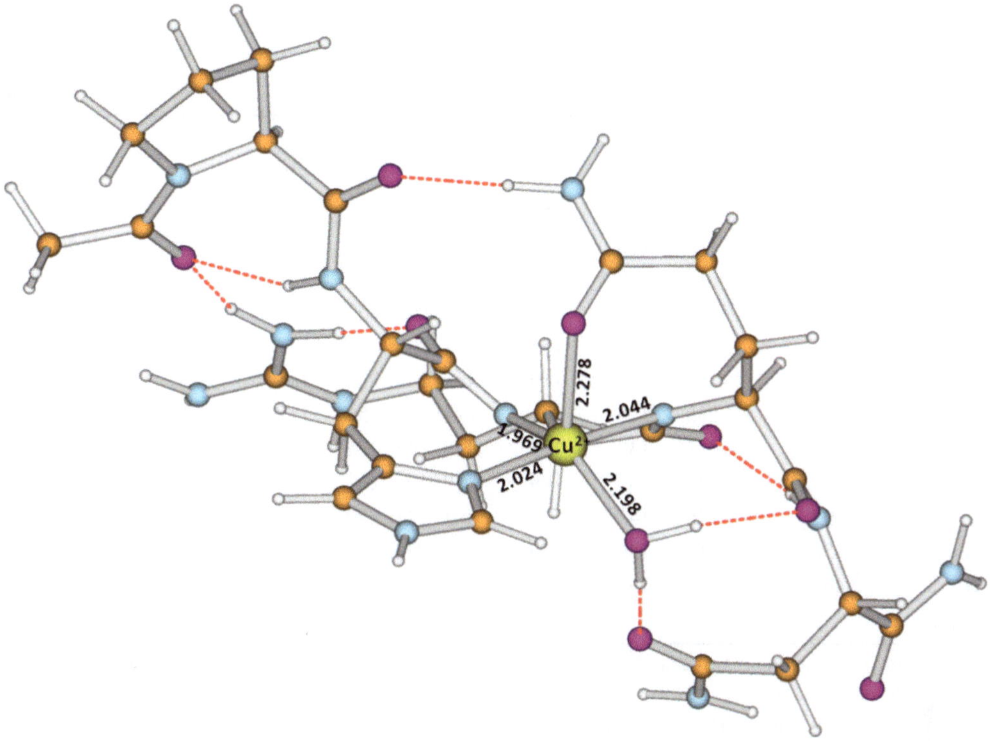

Fig. 4 Lowest-energy structure of the [Cu(Ac-PHRQN-NH_2)(H_2O)]$^{+1}$ complex (**1Q**). Distances in Å

NH_2)(H_2O)]$^{+1}$ isomer (**Q1**, see Fig. 4) features the H_2O molecule in equatorial position, whereas the carbonyl group of the glutamine side chain is coordinated to the Cu(II) atom in apical position. The high stability of **Q1** may be due to the large number of intramolecular H-bond (8). In fact, the isomer **Q2**, which is characterized by a coordination environment similar to that of **Q1**, but with a lower number of H-bond (5), is only 1.3 kcal mol^{-1} higher in energy than **Q1**. **Q3** is also close in energy to **Q1** ($\Delta E = 2.2$ kcal mol^{-1}). The latter isomer is a four-coordinated species in which the glutamine side chain has left the Cu^{2+} coordination sphere and the equatorial H_2O molecule is coordinated to the Cu atom with the shortest O-Cu distance (2.12 Å). Similarly, the isomer **Q5** is also 4-coordinated, but it is 10.3 kcal mol^{-1} higher in energy, possibly due to a smaller number of intramolecular H-bond interactions.

The lowest-energy isomer in which the carbonyl group of the glutamine side chain is coordinated to the Cu(II) atom in the equatorial position is **Q7**, which is 13.2 kcal mol^{-1} higher in energy than **Q1**. Interestingly, **Q7**, as well as all other isomers with the glutamine carbonyl group coordinated at the equatorial position (**Q15**–**Q18**), are 4-coordinated species, because the H_2O molecule leaves the Cu(II) coordination sphere.

The structures **Q8** and **Q9** are similar and are the lowest in energy among those in which the NH_2 group of glutamine is coordinated to Cu(II) atom. However, their energy is 15.2 kcal mol^{-1} higher than in **Q1**. These two isomers are also 3.0 kcal mol^{-1} less stable than **Q7** even if two more H-bond interactions are present. **Q8** and **Q9** are penta-coordinated species with the H_2O molecule weakly coordinated in apical position at about 2.5 Å from the Cu(II) atom. On the other hand, the other two isomers in which the NH_2 group of glutamine is coordinated to Cu(II) (**Q13** and **Q14**) are fully tetra-coordinated species.

3.2 EPR hyperfine coupling constants and g values of isomers of [Cu(Ac-PHREN-NH_2)(H_2O)] and [Cu(Ac-PHRQN-NH_2)(H_2O)]$^{+1}$ complexes

In the previous section, it was pointed out that the energy criterion is not sufficient to confidently predict the coordination mode of Cu(II) in the [Cu(Ac-PHREN-NH_2)(H_2O)] and [Cu(Ac-PHRQN-NH_2)(H_2O)]$^{+1}$ complexes, because of the different number and strength of intramolecular H-bonds in the isolated isomers, which affect their energy ranking. Therefore, in order to get a better insight into the coordination mode of these complexes, EPR ^{63}Cu hyperfine coupling constants (^{63}Cu-hcc) and g tensors of the most relevant isomers have been calculated, and compared with the experimental $A_{\parallel}$ and $g_{\parallel}$ values reported in Ref. [16]. Results are collected in Tables 4 and 5. As well known, for the Cu^{2+} ion the spin–orbit coupling (relativistic), contributions are important, due to its relatively large nuclear charge. Therefore, the Fermi contact (A_{FC}) and the principal values of the dipolar

Table 4 EPR hyperfine coupling constants (MHz) for the Cu atom in relevant isomers of the $[Cu(Ac\text{-}PHREN\text{-}NH_2)(H_2O)]$ complex, computed using the B3LYP functional on geometries optimized at the RI-BP86/TZVP level of theory

Complex	$Cu{-}O_{H_2O}$	A_{FC}^{a}	A_{PC}^{a}	A_{iso}	$A_{dip}^{a,b}$	$A_{dip2}^{a,b}$	Total dipolar[b]	$A_{\parallel}^{c}$	A_{xx}^{c}	A_{yy}^{c}	δA_{dip}^{d}
1	2.645	−308	139	−170	−488; 243; 245	114; −58; −57	−373; 186; 189	−543	16	19	3
1w	2.445	−303	144	−159	−489; 244; 245	118; −61; −57	−371; 183; 188	−530	25	29	4
2	3.777	−315	132	−183	−478; 236; 241	113; −54; −60	−364, 183, 181	−547	0	−1	−1
3	3.863	−319	128	−191	−478; 238; 240	114; −55; −58	−364; 183; 182	−555	−8	−9	1
4	3.796	−310	127	−183	−478; 240; 238	112; −58; −55	−366; 182; 183	−549	−1	0	1
5	3.565	−251	140	−111	−456; 203; 252	115; −50; −65	−340; 153; 187	−451	42	76	34
6	3.374	−291	135	−155	−461; 230; 232	112, −61; −52	−349; 169; 180	−504	13	25	12
7	3.359	−320	131	−189	−468; 249; 219	113; −61; −51	−355; 188; 168	−544	−1	−21	20
8	2.488	−286	145	−141	−495; 239; 256	119; −58; −62	−376; 182; 194	−517	41	53	12
9	3.684	−321	130	−190	−479; 243; 235	114; −58; −55	−366; 185; 180	−556	−5	−10	−5
10	3.655	−244	135	−109	−432; 171; 262	109; −45; −47	−323; 126; 197	−432	16	88	72
11	2.492	−276	143	−133	−491; 241; 250	117; −57; −59	−374; 184; 191	−507	51	58	7
16	3.755	−313	134	−179	−476; 238; 239	115; −56; −60	−361; 182; 179	−540	3	0	−3
20	2.407	−230	152	−78	−468; 107; 361	111; −14; −100	−357; 93; 261	−435	17	184	167
Expt.[e]								462			

[a] A_{FC} and A_{dip} are non-relativistic Fermi contact and traceless dipolar coupling tensor, respectively. A_{PC} and A_{dip2} are the corresponding relativistic corrections

[b] The components are reported in the order A_{33}, A_{11}, A_{22}

[c] $A_{\parallel} = A_{iso} + A_{33}$; $A_{xx} = A_{iso} + A_{11}$; $A_{yy} = A_{iso} + A_{22}$

[d] δA_{dip} is computed as $\delta A_{dip} = A_{22} - A_{11}$ and may be considered as the deviation from the axial symmetry of the tensor

[e] Experimental values taken from Ref. [16]

Table 5 g tensors of relevant isomers of the $[Cu(Ac\text{-}PHREN\text{-}NH_2)(H_2O)]$ complex, computed using the B3LYP functional on the geometries optimized at the RI-BP86/TZVP level of theory

	g_{iso}	g_{xx}	g_{yy}	$g_{\parallel}$
1	2.098	2.055	2.058	2.182
1w	2.103	2.058	2.061	2.190
2	2.092	2.048	2.053	2.175
3	2.090	2.047	2.051	2.171
4	2.088	2.046	2.050	2.170
5	2.097	2.050	2.058	2.184
6	2.095	2.049	2.055	2.180
7	2.091	2.047	2.055	2.172
8	2.106	2.059	2.060	2.192
9	2.091	2.047	2.053	2.172
10	2.093	2.044	2.058	2.177
11	2.102	2.058	2.060	2.189
16	2.093	2.049	2.054	2.177
20	2.109	2.036	2.102	2.190
Expt.[a]				2.238

[a] Experimental values taken from Ref. [16]

tensor (A_{dip}) are discussed in connection with relativistic corrections (Pseudo-contact (A_{PC}), and A_{dip2}).

All of the complexes investigated, with the exception of **5** and **10**, are characterized by a roughly planar N_3O disposition of ligands ($\delta < 20$; see Table 1). ^{63}Cu-hcc tensor components of these complexes feature an axial symmetry, as deduced from the observation that the A_{xx} ($A_{xx} \equiv A_{iso} + A_{11}$) and A_{yy} ($A_{yy} \equiv A_{iso} + A_{22}$) elements of tensor do not differ by more than 20 MHz. However, for these complexes the $A_{\parallel}$ ($A_{\parallel} \equiv A_{iso} + A_{33}$), component is significantly overestimated with respect to the experimental value (see Table 4). In this respect, it is worth noting that the $A_{\parallel}$ component appears to be suppressed by the axial coordination of the H_2O molecule, as deduced by the observation that in the 4-coordinated complexes ($Cu{-}O_{H_2O} > 3$ Å) $A_{\parallel} = 540$–556 MHz, whereas in **8** and **11** ($Cu{-}O_{H_2O} \approx 2.49$ Å) $A_{\parallel} = 507$–517 MHz. For the latter complexes, the $A_{\parallel}$ value is in better agreement with the experiment. Therefore, the comparison between experimental and computational EPR parameters might suggest a N_3O_2 (slightly distorted) square pyramidal Cu(II) coordination environment, with the H_2O molecule weakly coordinated in axial position.

Isomers **5** and **10**, which are 5.5 and 12.2 kcal mol^{-1}, respectively, higher in energy than **1**, are characterized by a significant tetrahedral distortion in the N_3O arrangement of ligands. For these two isomers, the $A_{\parallel}$ component is in very good agreement with the experimental value. In fact, the calculated tensor of **5** slightly deviates from the axial symmetry ($\delta A_{dip} = 34$) by a value which may not be resolved in the experimental spectrum. A larger deviation ($\delta A_{dip} = 72$) is observed for the complex **10**. It is interesting to note that **5** and **10** are both characterized by one of the two N$^-$–Cu distances significantly shorter than in the other complexes, a feature that could also account for the depression of the $A_{\parallel}$ component when moving from the square planar to the distorted configuration.

The $A_{\parallel}$ value computed for isomer **20**, which features an equatorial H_2O molecule constrained to 2.407 Å from the Cu(II) atom, is significantly lower than the $A_{\parallel}$ value obtained in isomers characterized by axial coordination of the H_2O molecule, and apparently in fairly good agreement with the experimental value. However, equatorial coordination of the H_2O molecule can be ruled out since the calculated tensor components significantly deviate from ideal axial symmetry (see δA_{dip} in Table 4), a feature that is not compatible with the experimental EPR spectrum.

The computed values of g components are only slightly affected by a particular coordination mode, as reported for other copper–peptide complexes [33], and are, therefore, of little help in discriminating different isomeric forms. In general, we observe for the calculated g tensors that the trend is similar to that discussed for the ^{63}Cu-hcc parameters (see Table 5). The g tensor computed for isomer **20** exhibits a significant rhombic distortion, whereas all other isomers are characterized by an almost ideal axial symmetry of the g tensor. In addition, the $g_{\parallel}$ component of the penta-coordinated complexes **8** and **11** is slightly larger than that of the tetra-coordinated complexes, and in closer agreement with the experimental value.

The experimental $A_{\parallel}$ and $g_{\parallel}$ values obtained studying [Cu(Ac-PHRQN-NH$_2$)(H_2O)]$^{+1}$ differ from those measured for the [Cu(Ac-PHREN-NH$_2$)(H_2O)] complex, leading to the suggestion that the carboxylate group is involved in Cu(II) coordination in the latter [16].

EPR parameters of selected [Cu(Ac-PHRQN-NH$_2$)(H_2O)]$^{+1}$ complexes are reported in Tables 6 and 7. As a general remark, we observe for most of the species investigated that, contrary to the experimental data, the $A_{\parallel}$ component of the ^{63}Cu-hcc tensor is lower than that calculated for the isomers of the [Cu(Ac-PHREN-NH$_2$)(H_2O)] complex. Inspection of the ^{63}Cu-hcc tensor contributions reported in Table 6 shows that this behavior is mainly due to the isotropic term. The lowest-energy isomers **Q1** and **Q2** feature a similar distorted square pyramidal arrangement of ligands, with the O_{H2O} and the O_{Gln} atoms coordinated in basal and apical positions, respectively. The main difference in the coordination sphere between the two isomers lies in the O_{H2O}-Cu and O_{Gln}-Cu distances, which in **Q1** are about 0.1 and 0.04 Å shorter than in **Q2**. The $A_{\parallel}$ component of the ^{63}Cu-hcc tensor of **Q1** and **Q2** is smaller than the experimental value by about 90 and 40 MHz, respectively. In addition, for both isomers, the tensor is characterized by a large rhombic distortion, a feature that is not compatible with the experimental spectrum. **Q3** and **Q5** are 4-coordinated isomers in which the N_3O coordination sphere is completed by the H_2O molecule. The **Q3** complex, which is roughly planar and only 2 kcal mol^{-1} higher in energy than **Q1**, is characterized by a ^{63}Cu-hcc tensor with axial symmetry. The $A_{\parallel}$ component of the calculated tensor is about 60 MHz larger than the experimental value. In addition, the calculated $A_{\parallel}$ value is very similar to that of the square planar species in the [Cu(Ac-PHREN-NH$_2$)(H_2O)] complex, even if the carboxylate ligand in the equatorial position is replaced by the H_2O molecule. The arrangement of ligands in **Q5**, differently from **Q3**, is characterized by a significant tetrahedral distortion. Notably, the $A_{\parallel}$ component of the ^{63}Cu-hcc tensor is smaller than in **Q3** and matches within 10 MHz the experimental value. In addition, the tensor deviates only slightly from the ideal axial symmetry ($\delta A_{dip} = 34$). Furthermore, this is the only isomer for which the replacement of the carboxylate ligand with the H_2O molecule increases the $A_{\parallel}$ value by 30 MHz, a value in very good agreement with that observed in the experiment when replacing the E residue with Q. **Q4** and **Q6** are 5-coordinated species in which the arrangement of the ligands approaches the trigonal bipyramidal symmetry. The calculated hcc tensor for these two tensors is both quantitatively and qualitatively very different from the experimental one, further indicating that the structure of the complex in solution should not deviate significantly from the square planar or square pyramidal symmetry. **Q8** is the lowest-energy isomer characterized by the equatorial coordination of four nitrogen atoms and the axial coordination of the H_2O molecule. The ^{63}Cu-hcc tensor of this isomer has axial symmetry, and the $A_{\parallel}$ value is about 40 MHz lower than the experimental one, indicating that this species cannot be excluded only on the basis of the computed EPR parameters. Finally, we have also calculated the ^{63}Cu-hcc tensor of **Q7**, **Q11** and **Q16** in which the carbonyl group of the glutamine side chain is coordinated to the Cu(II) atom in the equatorial position, and featuring different tetrahedral distortions ($\delta = 37.5°$, 16,3° and 5,7° for **Q7**, **Q11** and **Q16**, respectively). In **Q11**, the H_2O molecule is also weakly coordinated to the apical position. As shown in Table 6, the $A_{\parallel}$ component in this latter species is underestimated by more than 100 MHz with respect to the experimental value, and the tensor is characterized by a large rhombic distortion. In **Q7**, the $A_{\parallel}$ component is about 50 MHz smaller than in the experiment and the tensor is less affected by the rhombic distortion. In **Q16**, the $A_{\parallel}$ component of the ^{63}Cu-hcc tensor is in very good agreement with the experiment, and the tensor further approaches the axial

Table 6 EPR hyperfine coupling constants (MHz) for Cu atoms in relevant isomers of the $[Cu(Ac\text{-}PHRQN\text{-}NH_2)(H_2O)]^{+1}$ complex, computed using the B3LYP functional on geometries optimized at the BP86/TZVP level of theory

Complex	$Cu{-}O_{H_2O}$	A_{FC}^{a}	A_{PC}^{a}	A_{iso}	$A_{dip}^{a,b}$	$A_{dip2}^{a,b}$	Total dipolar[b]	$A_{\parallel}^{c}$	A_{xx}^{c}	A_{yy}^{c}	δA_{dip}^{d}
Q1		−206	144	−62	−452; 81; 371	111; 11; −100	−341; 70; 271	−403	8	109	101
Q2		−234	145	−89	−465; 123; 342	107; −21; −86	−358; 102; 256	−447	13	167	154
Q3		−319	133	−186	−471; 235; 236	109; −53; −56	−362; 182; 180	−548	−4	−7	−3
Q4		−120	149	29	406; 30; −436	−113; 9; 104	293; 39; −332	322	68	−304	−372
Q5	2.174	−255	126	−129	−450; 196; 254	98; −42; −57	−352; 154; 198	−481	25	69	44
Q6	2.242	−253	128	−125	−441; 80; 362	101; −11; −90	−317; 69; 272	−465	−56	147	203
Q7	3.495	−214	124	−90	−443; 178; 265	97; −37; −60	−346; 142; 205	−436	52	115	63
Q8		−234	144	−90	−461; 215; 246	111; −50; −60	−350; 165; 186	−441	76	96	−20
Q11		−165	131	−34	−456; 167; 289	110; −36; −72	−346; 131; 217	−380	97	183	96
Q16		−251	124	−127	−462; 214; 248	109; −50; −59	−353; 165; 188	−480	28	68	40
Expt.[c]								489			

[a] A_{FC} and A_{dip} are non-relativistic Fermi contact and traceless dipolar coupling tensor, respectively. A_{PC} and A_{dip2} are the corresponding relativistic corrections

[b] The components are reported in the order A_{33}, A_{11}, A_{22}

[c] $A_{\parallel} = A_{iso} + A_{33}$; $A_{xx} = A_{iso} + A_{11}$; $A_{yy} = A_{iso} + A_{22}$

[d] δA_{dip} is computed as $\delta A_{dip} = A_{22} - A_{11}$ and may be considered as the deviation from the axial symmetry of the tensor

[e] Experimental values taken from Ref. [16]

Table 7 g tensors of relevant isomers of the $[Cu(Ac\text{-}PHRQN\text{-}NH_2)(H_2O)]^{+1}$ complex, computed using the B3LYP functional on geometries optimized at the RI-BP86/TZVP level of theory

	g_{iso}	g_{xx}	g_{yy}	$g_{\parallel}$
Q1	2.106	2.023	2.100	2.184
Q2	2.102	2.036	2.088	2.181
Q3	2.091	2.048	2.055	2.171
Q4	2.102	2.030	2.094	2.182
Q5	2.085	2.041	2.056	2.159
Q6	2.090	2.026	2.083	2.161
Q7	2.084	2.037	2.058	2.157
Q8	2.098	2.051	2.063	2.179
Q11	2.091	2.039	2.065	2.169
Q16	2.084	2.043	2.050	2.160
Expt.[a]				2.240

[a] Experimental values taken from Ref. [16]

symmetry. However, **Q16** is 37.4 kcal mol^{-1} higher in energy than **Q1**, an energy difference which allow to exclude that such isomer is present in aqueous solution.

As observed for the corresponding ^{63}Cu-hcc parameters, the g tensors computed for **Q1**, **Q2**, **Q4** and **Q11** are characterized by a significant rhombic distortion, whereas in **Q3**, **Q8** and **Q11**, the tensors approach to the axial symmetry. It is worth noting that the $g_{\parallel}$ component of **Q16**, which features the glutamine carbonyl ligand in the equatorial position, is smaller than that of the species with the H_2O molecule in the equatorial position, a trend opposite with respect to the experimental results.

4 Conclusions

In this paper, we have reported results obtained studying Cu^{2+} coordination to the Ac-PHREN-NH_2 peptide that represents the Cu^{2+} binding portion of angiogenin, a protein playing a key role in the process of angiogenesis. Using as a reference the recent experimental results reported by La Mendola et al. [16], we have used classical MM/MD calculations followed by DFT optimizations to explore the configurational space of the $[Cu(Ac\text{-}PHREN\text{-}NH_2)(H_2O)]$ complex.

EPR experiments suggested that Cu(II) is coordinated in a N_3O square planar environment characterized by a significant tetrahedral distortion, where the Cu(II) coordination sphere is composed by $2N^-$ of deprotonated peptide bonds, the nitrogen atom of the histidine group and the carboxylate group of the glutamic residue. An alternative N_3O_2 (less distorted) square pyramidal coordination in which the oxygen atom of a weakly coordinated water molecule occupies the apical position cannot be excluded. In addition, a detailed comparison of computed minimum energy structures and EPR parameters suggests that the binding of Cu(II) to the deprotonated amide groups of the peptide bonds takes place between the residues H-R and R-E.

We have also investigated the configuration space of the E → Q mutated system $[Cu(Ac\text{-}PHRQN\text{-}NH_2)(H_2O)]^{+1}$. In this case, computational results led to the conclusion that, similar to the $[Cu(Ac\text{-}PHREN\text{-}NH_2)(H_2O)]$ complex, the coordination mode is described by a N_3O square planar environment with a large tetrahedral distortion, in which the carboxylate group is replaced by a H_2O molecule. Finally, the structures that better fit the experimental EPR parameters do not correspond to the most stable isomers, further indicating that the energy ranking of isomers, derived in the absence of solvent water molecules, is not a reliable criterion in the study of speciation of metal–peptide complexes in aqueous solution.

Acknowledgments This work was supported by the PRIN Project N 200875WHMR.

References

1. Folkman J (1989) J Natl Cancer Inst 82:4
2. Bussolino F, Mantovani A, Persico G (1997) Trends Biochem Sci 22:251
3. Kishimoto K, Liu S, Tsuji T, Olson KA, Hu G (2005) Oncogene 24:445
4. Fett JW, Strydom DJ, Lobb RR, Alderman EM, Bethune JL, Riordan JF, Vallee BL (1985) Biochemistry 24:5486
5. Gao X, Xu Z (2008) Acta Biochim Biophys Sin 40:619
6. Badet J, Soncin J, Guitton JD, Lamare O, Cartwright T, Barritault D (1989) Proc Natl Acad Sci USA 86:8427
7. McAuslan BR, Reilly W (1980) Exp Cell Res 130:147
8. Hu G-F (1998) J Cell Biochem 69:326
9. Lequin O, Thuring H, Robin M, Lallemand J-Y (1997) Eur J Biochem 250:712
10. Millhauser GL (2007) Annu Rev Phys Chem 58:299
11. Solioz M, Vulpe C (1996) Trends Biochem Sci 21:237
12. Soncin F, Guitton JD, Cartwright T, Badet J (1997) Biochem Biophys Res Commun 236:604
13. Joyce BK, Cohn M (1969) J Biol Chem 244:811
14. Acharya KR, Shapiro R, Allen SC, Riordan JF, Vallee BL (1994) Proc Natl Acad Sci USA 91:2915
15. Smyth DG, Stein WH, Moore S (1963) J Biol Chem 238:227
16. La Mendola D, Magri A, Vagliasindi LI, Hansson O, Bonomo RP, Rizzarelli E (2010) Dalton Trans 39:10678
17. Halgren TA (1996) J Comput Chem 17:490
18. Halgren TA (1999) J Comput Chem 20:720
19. Becke AD (1988) Phys Rev A 38:3098
20. Perdew JP (1986) Phys Rev B 33:8822
21. Eichkorn K, Weigend F, Treutler O, Ahlrichs R (1997) Theor Chem Acc 97:119
22. Ahlrichs R, Bar M, Haser M, Horn H, Kolmel C (1989) Chem Phys Lett 62:165
23. Schafer A, Huber C, Ahlrichs R (1994) J Chem Phys 100:5829
24. Bruschi M, De Gioia L, Mitric R, Bonacic-Koutecky V, Fantucci P (2008) Phys Chem Chem Phys 10:4573
25. Schafer A, Horn H, Ahlrichs R (1992) J Chem Phys 97:2571
26. Kutzelnigg W, Fleischer U, Schindler M (1990) The IGLO-method: Ab initio calculation and interpretation of NMR chemical shifts and magnetic susceptibilities. Springer, Berlin
27. Hess BA, Marian CM, Walhgren U, Gropen O (1996) Chem Phys Lett 251:365
28. Kaupp M, Buhl M, Malkin VG (2004) Calculation of NMR and EPR parameters: theory and applications. Wiley-VCH
29. Neese F (2003) J Chem Phys 117:3939
30. Marino T, Russo N, Toscano M (2007) J Phys Chem B 111:635
31. Marino T, Russo N, Toscano M (2011) Int J Quantum Chem 111:1152
32. Pushie MJ, Rauk A (2003) J Biol Inorg Chem 8:53
33. Franzini E, De Gioia L, Fantucci P, Zampella G, Bonacic-Koutecky V (2003) Inorg Chem Commun 6:650
34. Sabolovic J, Tautermann CS, Loerting T, Liedl KR (2003) Inorg Chem 42:2268
35. Adamo C, Scuseria G, Barone V (1999) J Chem Phys 111:2889
36. Klamt A (1995) J Phys Chem A 99:2224
37. Klamt A (1996) J Phys Chem A 100:3349
38. Klamt A, Schüürmann G (1993) J Chem Soc Perkin Trans 2:799
39. Hallahan TW (1991) Proc Natl Acad Sci USA 85:5061

Theor Chem Acc (2012) 131:1188
DOI 10.1007/s00214-012-1188-9

REGULAR ARTICLE

Impact of DFT functionals on the predicted magnesium–DNA interaction: an ONIOM study

José P. Cerón-Carrasco · Alberto Requena · Denis Jacquemin

Received: 29 June 2011 / Accepted: 9 September 2011 / Published online: 10 March 2012

Abstract We investigate the descriptions of several density functional theory functionals in modeling the interactions between DNA and magnesium, the divalent cation with the largest intracellular concentration. Precisely, the metal-induced proton transfer (PT) in the guanine-cytosine base pairs is analyzed within an hybrid two-layers QM/QM approach (ONIOM) by combining several functionals: BP86, B3LYP, B97-D, ωB97x-D and M06-2X, as well as the MP2 approach. The aqueous environment is simulated through, on the one hand, explicit solvent molecules present in the first hydration shell and, on the other hand, the well-known polarizable continuum model for the effects of solvation beyond this first shell. Calculations with all methods indicate that the Mg^{2+} cation coordination to DNA promotes a single PT reaction from the guanine to the cytosine base, characterized by a low back-reaction barrier. M06-2X and ωB97x-D functionals provide consistent results for the simulation of PT reactions. These methods can be combined with B97-D functional in ONIOM partition for the description of the stacking effects.

Dedicated to Professor Vincenzo Barone and published as part of the special collection of articles celebrating his 60th birthday.

J. P. Cerón-Carrasco · D. Jacquemin (✉)
CEISAM UMR CNRS 6230, Université de Nantes,
2, rue de la Houssinière, BP 92208,
44322 Nantes Cedex 3, France
e-mail: denis.jacquemin@univ-nantes.fr

J. P. Cerón-Carrasco
e-mail: jose.ceron@univ-nantes.fr

A. Requena
Departamento de Química Física,
Facultad de Química, Universidad de Murcia,
30100 Campus de Espinardo, Murcia, Spain

Keywords DNA · Magnesium · Tautomerism · Density functional calculations · ONIOM

1 Introduction

Although DNA is protected by the surrounding cellular environment, it is constantly exposed to external agents that may induce variations of its nature and hence alter its biological functions. In that framework, Prof. Barone and collaborators have demonstrated that theoretical chemistry provides a useful and deep understanding of the interactions between UV–visible radiation and DNA. Their studies unravel the nature of the excited state of nucleic acids [1–5] and the subsequent implications for life that they rationalize as radiation-induced changes in DNA canonical structure [6–8]. Interestingly, not only external factors might tune the biological functions of DNA but natural components such as metallic cations (mainly Na^+, K^+, Ca^{2+}, and Mg^{2+}) also play a major role in the stabilization (or unstabilization) of the double-helix structure [9]. Among these compounds, magnesium can be considered as one of the most important because it is present in most of nucleic acids activation processes [10–12]. Furthermore, recent studies suggested that the Mg^{2+}–DNA interaction is a prerequisite for the activity and therapeutic efficiency of some antitumorals drugs, such as mithramycin [13–19]. However, at high concentrations, metals, including Mg^{2+}, may play an opposite role by distorting the DNA structure or by inducing tautomeric changes in the base pairs, which might eventually yield an error in the genetic information [20–23]. It is therefore unsurprising that the influence of cations in life tissues has attracted a considerable attention and has initiated numerous experimental [24–28] and theoretical studies [29–37] aiming to unravel the biological consequences of such metal–nucleic acids interactions.

For DNA, it is well-recognized that metal cations are predominantly coordinated to the guanine–cytosine (GC) base pairs and more precisely on the N7(G) site [38, 39] (see Fig. 1). As a possible consequence of this binding, the amount of so-called rare tautomers, that is, structures arising from exchanging protons with respect to the original canonical form, might be increased [23]. Indeed, though proton transfer (PT) reactions in DNA are energetically unfavorable in the absence of external perturbations [40–50], metalation at N7(G) can possibly activate the transfer of the H1 proton from the N1(G) position to N3(C) leading to a rare tautomer form [51–57]. For the Mg^{2+} cation, Šponer and co-workers have evaluated the importance of PT mechanism on the global mutagenicity of DNA [58]. In agreement with experiments [52, 53], their calculations on deoxyribonucleotide monophosphate–GC model indicate that Mg^{2+} cation promotes the transfer of H1 proton between GC base pair. However, as these authors pointed out, a more refined DNA model is necessary to quantify the Mg^{2+}-induced PT energetic profile [58]. Indeed, for free DNA, we have recently demonstrated that this mechanism is qualitatively changed when the base stacking and confinement effects are considered [59].

In this work, we assess density functional theory (DFT) approaches in the framework of simulations of the interactions between the DNA molecule and the Mg^{2+} cation. Specifically, we test the impact of the functional in the predicted metal-induced H1 PT in the tautomeric equilibrium connecting the canonical system to the rare tautomer, hereafter denoted simply as Mg–GC and Mg–GC1, respectively. This paper is divided as follows: in Sect. 2, we describe the theoretical methods employed for the treatment of metal–DNA interactions; in Sect. 3, we explore the results of different functionals for describing the induced PT reaction and the biological implications of the metal–DNA interaction.

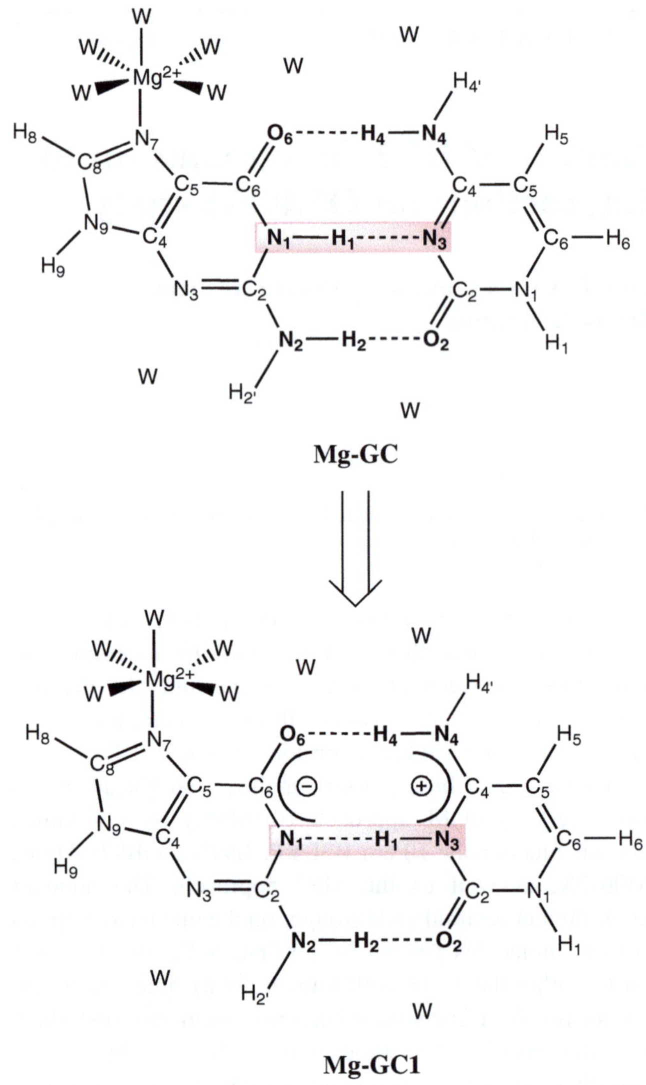

Fig. 1 Structure of the hydrated Mg^{2+}–GC complex and atomic numbering. Mg–GC and Mg–GC1 refers to the canonical and rare tautomer forms, respectively. The transferred proton in the central hydrogen bond is highlighted in *pink*. The explicit water molecules of the first hydration shell around the magnesium cation and the GC pair are shown as W

2 Computational methods

The DNA structure and properties are mainly governed by two non-covalent interactions: interbase bonds and base stacking [60–62]. As recently demonstrated by Acosta-Silva et al. [63], these two interactions are not independent. Consequently, to obtain reliable predictions of the DNA changes induced by the Mg^{2+}, an adequate treatment of both interactions is needed. The minimum DNA fragment able to cover the stacking influence on interbase bonds is a three base pair sequence, where a central base pair is under the influence of two neighboring bases pairs, located below and above the investigated moiety [64]. Here, a DNA trimer model is built up by inserting the Mg^{2+}–GC system into a double-stranded B-form trimer d(5′-GGG-3′)d(3′-CCC-5′) inspired by the study of DNA's radical anion of Chen et al. [65]. As solvent affects both the metal bonding and the GC tautomeric equilibrium [66, 67], we have combined the first hydration shell of Mg^{2+} cation with the first hydration shell for the GC base pair. For the former, the Mg^{2+} cation exists in its hexahydrated form in bulk water, but it looses one water molecule when bonded to N7(G) of the nucleotide, and therefore five explicit water molecules are included to saturate the first hydration shell of the cation [68]. For the latter, four water molecules have been placed in the vicinity of the N3(G), O6(G), O2(C), and N4(C), which is a suitable representation of the experimentally observed hydration shell of the GC base pair embedded in DNA [69] and is also consistent with previous molecular dynamics simulations [70, 71]. Accordingly, nine explicit water molecules are included in our model as shown in Fig. 1. Since lateral sugar-phosphate backbones are known to have a trifling influence on the PT reactions [59, 64], only the base pairs are included in the model, subsequently allowing to test several theoretical schemes. The resulting model is shown in Fig. 2 and allows a straightforward partition of the molecular system in two well-defined regions: (1) the central GC base pair, the Mg^{2+} cation and all explicit water molecules (in ball-and-sticks), and (2) the two GC neighboring bases (in tube). These two regions have been described within the partitioning approach developed by Morokuma and co-workers (ONIOM) [72] as implemented in Gaussian09 [73]. When the ONIOM scheme is adopted the choice of the combined methods become a crucial factor. From a theoretical point of view, CCSD(T) is the reference method [74]. Based on the rather similar dependences with respect to the basis set of CCSD(T) and MP2 energies, Hobza and Šponer proposed to use the difference between CCSD(T) and MP2 energies computed with medium basis to extrapolate the CCSD(T)/CBS, where CBS refers to the complete basis set limit [75]. Unfortunately, even within this approach, the CCSD(T) calculations are beyond reach for large systems as the one required in our investigation (115 atoms). As alternative, the DFT methods provide a computational tractable approach to biological systems. However, though early functionals provide a quite acceptable description for the covalent bonds, they fail to model correctly dispersion energy that is essential to describe the stacking effect and are not always very efficient for H-bonds [76, 77]. To partially overtake these limitations, Grimme proposed a dispersion corrected DFT scheme (DFT-D), where an empirical C_6R^{-6} term is added to usual DFT equations [78]. With this correction, DFT methods can be successfully applied to study intermolecular interactions in biological systems [79–82]. Additionally, novel *meta*-hybrid functionals developed by Zhao and Truhlar, and more precisely the M06-2X variation, have also been shown to adequately model non-covalent interactions [83–85].

To evaluate the impact of the selected functional in the study of magnesium–DNA interaction, pure (BP86 [86, 87]), pure dispersion corrected (B97-D [88]), hybrid (B3LYP [89, 90]), *meta*-hybrid (M06-2X [91]), and range-separated dispersion corrected (ωB97X-D [92]) functionals are combined as summarized Table 1. We should note that the inclusion of high percentage of Hartree–Fock exchange in the functional might be problematic in certain cases as transition metals [85]. However, they can be successfully used for the study of light element like Mg. For instance, the good performance of M06-2X functional (54% of exact exchange) has been demonstrated in a study of the absorption of CO on MgO surfaces [93]. The listed functionals are combined with the 6-311++G(d,p) and 6-31+G(d) basis sets for the description of the atoms in the high and low layers, respectively. For each scheme, we follow the H1 proton exchange by scanning the H1–N3(C) bond distance. More specifically, the ONIOM potential energy curves are computed through the partial optimization of the high layer (the H1–N3(C) distance is fixed along the scanning), whereas atoms in the low layer are frozen in space because their relaxation is expected to have a minor effect on the whole energy profile [64, 65]. In a second step, the resulting three critical points in the energy curves, namely the two minima and the maximum, are selected as starting point for a full optimization of the high layer, keeping the restriction in the low layer. Although the

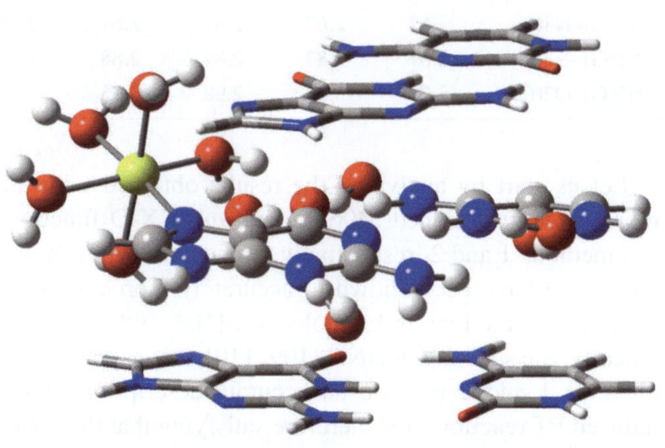

Fig. 2 Designed DNA fragment. The central Mg^{2+}–GC complex corresponding to the high layer level in the ONIOM scheme is represented with balls-and-sticks. The border base pairs confining the central GC included in the low layer, are displayed in tube

Table 1 Theoretical methods for the description of the high and low layer within the two-layers ONIOM approach

Method	High layer[a]	Low layer[b]	PCM
1	M06-2X	B97-D	–
2	ωB97X-D	B97-D	–
3	B97-D	B97-D	–
4	B3LYP	B97-D	–
5	BP86	B97-D	–
6[c]	MP2/6-31G(d)(0.25)		–
7	M06-2X	M06-2X	–
8	M06-2X	PM6	–
9[c]	M06-2X	B97-D	Single-point
10	M06-2X	B97-D	Optimization

[a] Combined with the 6-311++G(d,p) basis set

[b] Combined with the 6-31+G(d) basis set

[c] Single-point calculation using the geometry of method 1

vibrational calculations in non-stationary points (frozen low layer) should be analyzed cautiously, the absence of imaginary frequency for minima (Mg–GC and Mg–GC1) and presence of a single imaginary mode for the transition state (Mg–GC1‡) related to the H1–N3(C) stretching confirm the nature of the structures. The vibrational corrections obtained in the same approximate way have been used to estimate the Gibbs energies. Additional MP2 calculations have been carried out, that is the energy of hydrogen bonding and stacking interactions was calculated through a MP2/6-31G(d)(0.25) single-point calculation on the geometries obtained through method 1. In the 6-31G(d)(0.25) basis set, originally proposed by Hobza and Šponer [94, 95], a more diffuse polarization *d*-functions are used for the element belonging to the first period by replacing the exponent of 0.80 by 0.25 in the standard Pople's basis set to account for dispersion attraction. This choice provides an improved description for base stacking in nucleic acids at a reasonable computational cost [96, 97] but remains less potent than 6-311++G(d,p). Finally, since solvent affects both the metal bonding and the GC tautomeric equilibrium [98–100], the influence of aqueous environment beyond the first hydration shell has been also assessed by including the well-known polarizable continuum model (PCM) [101–103] in our model following two strategies. On the one hand, PCM corrections are included through single-point energy calculations performed on the optimized geometries, and on the other hand, PCM optimizations have been carried out.

3 Results and discussion

3.1 The high-layer method

As shown Table 1, methods 1–5 have been designed to assess the impact of the high-layer method (M06-2X, ωB97X-D, B97-D, B3LYP, and BP86) on the global PT reaction, the low layer being treated through the dispersion corrected B97-D functional. The computed potential energy curves along H1 PT are represented in Fig. 3 whereas the hydrogen bond distances and relatives energies for the optimized high layer are collated in Table 2. All selected theoretical methods predict the canonical Mg–GC structure to be more stable than the rare tautomer form, Mg–GC1, and do provide a similar picture for the changes in the interbase bonds pattern related to the PT. The major effect of the H1 PT is a shortening of the O6(G)–N4(C) and N1(G)–N3(C), whereas the N2(G)–O2(C) bond is slightly elongated. However, the inspection of Fig. 3 reveals significant differences in the energetic profiles showing the impact of the selected DFT functionals on the description of DNA interactions.

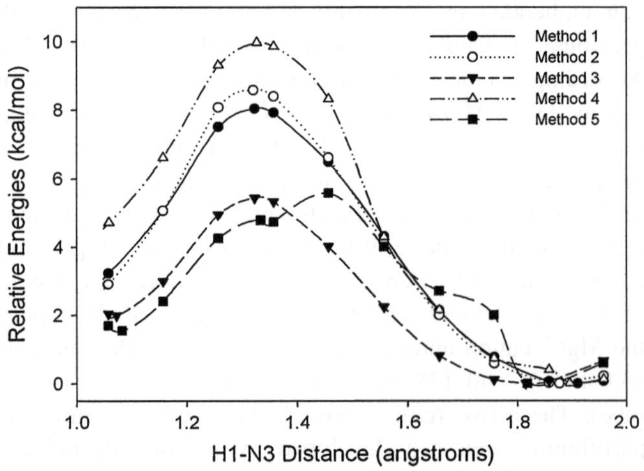

Fig. 3 ONIOM potential energy curves along the H1 proton transference

Table 2 Computed relative energies (Δ*E* in kcal/mol) along H1 proton transfer with selected high-layer methods. The interbase bonds (in Å) are also listed

	Method				
	1	2	3	4	5
Mg–GC					
Δ*E*	0.00	0.00	0.00	0.00	0.00
O6(G)–N4(C)	2.84	2.80	2.79	2.83	2.79
N1(G)–N3(C)	2.95	2.92	2.87	2.94	2.88
N2(G)–O2(C)	2.98	2.96	2.91	2.96	2.91
Mg–GC1‡					
Δ*E*	8.04	8.58	5.44	9.97	5.58
O6(G)–N4(C)	2.65	2.63	2.65	2.65	2.65
N1(G)–N3(C)	2.69	2.68	2.68	2.69	2.68
N2(G)–O2(C)	2.91	2.92	2.90	2.93	2.87
Mg–GC1					
Δ*E*	3.24	2.91	1.99	4.71	1.55
O6(G)–N4(C)	2.67	2.67	2.67	2.67	2.66
N1(G)–N3(C)	2.87	2.87	2.84	2.88	2.84
N2(G)–O2(C)	3.03	3.03	2.98	3.05	3.00

Let us start by analyzing the results obtained with the more recently designed M06-2X and ωB97X-D functionals, methods 1 and 2, respectively. Since both M06-2X and ωB97X-D have been shown to accurately reproduce non-covalent interactions in biomolecules [104–108] and, specifically, the stacking in DNA [109, 110], it is expected that methods 1 and 2 provide an accurate description of the induced PT reaction. It is therefore satisfying that these two functionals yield extremely similar results, as evidenced by the potential energy curves in Fig. 3 and the relative energies in Table 2. When both layers are described with the B97-D functional (method 3), the generated potential energy curve significantly differs from the ones produced by the M06-2X and ωB97X-D functionals. Indeed, while M06-2X and ωB97X-D locate the transition state ca. 8 kcal/mol above the canonical structure, the corresponding B97-D value is ca. 5 kcal/mol. A similar trend is found for the relative energy of the rare tautomer, which lies ca. 3 kcal/mol above the canonical structure with M06-2X and ωB97X-D, in contrast with the ca. 2 kcal/mol prediction of B97-D. This dissimilarity can be analyzed in terms of the definition of these three models. B97-D is a local functional, whereas in M06-2X and ωB97X-D, exact exchange is added which generally yields larger (and more realistic) transition energies. Indeed, as Barone et al. demonstrated that the inclusion of exact exchange improves the description of hydrogen bonds and provides more accurate PT energies [111, 112].

Turning to the most widely used functional, namely B3LYP (method 4), sizable energetic barrier (9.97 kcal/mol) and rare tautomer energies (4.72 kcal/mol) are reached, ca. 2 kcal/mol larger than the predicted by M06-2X and ωB97X-D. As Bickelhaupt and coworkers discussed, B3LYP often fails to reproduce the reaction barrier [77] so that this outcome is unsurprising. Additionally, these authors demonstrated that B3LYP is not adequate for mimicking biological systems involving π-stacking interactions [82]. Interestingly, B3LYP seems to overestimate the reactions barrier for the Mg-induced PT, an unexpected error sign. B3LYP is probably not adequate for the study of stacked DNA. We have also evaluated the performance of BP86 (method 5). Unlike methods 1–4, where two well-defined minima and one transition state are found, method 5 produces a rather erratic dependence of the energy along the H1 PT. This markedly different behavior arises from the changes in the hydration pattern of the central GC base pair. As shown in Fig. 4, the water molecule close to O2(C) site (see Fig. 1) "escapes" from the central base pairs and goes toward one of the borders in the Mg–GC1 form. As a consequence, the observed energy oscillation results from the combination of two effects (the tautomeric equilibrium and the change in the hydration shell), which prevents an adequate comparison of its energies. Of course, we could have frozen the position of this water molecule, but we wished to illustrate the difficulty to reach valuable results. It should also be noted that, though BP86 has been shown to adequately describe interbase bonds in gas-phase and microsolvated single base pairs [113–116], it fails in the description of stacking interactions in DNA [76, 82]. This finding clearly reflects that the performance of DFT for DNA is connected to the nature of the selected chemical model. Our ONIOM results suggest that both M06-2X and ωB97X-D functionals provide a consistent description of the induced Mg–PT process.

Eventually, we compare ONIOM results with single-point calculations at MP2/6-31G(d)(0.25). The MP2

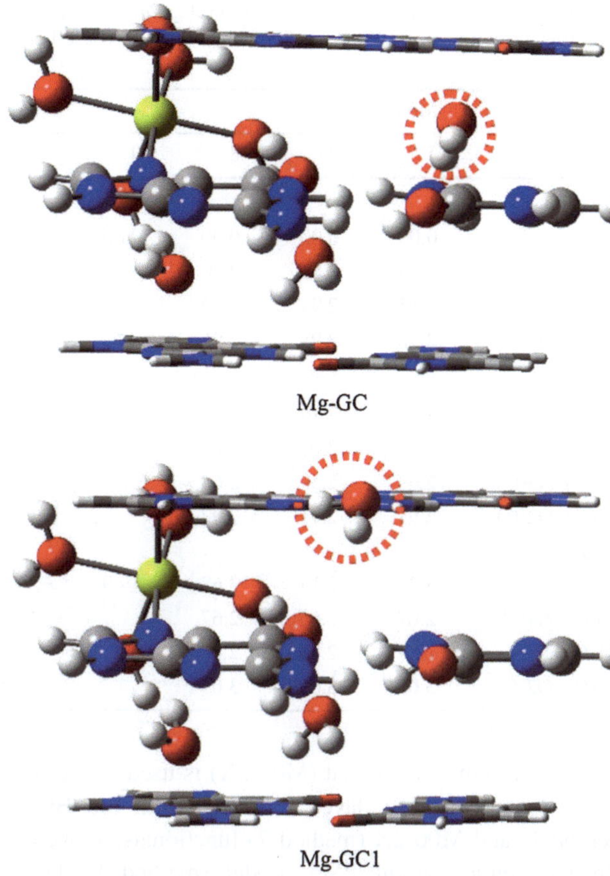

Fig. 4 ONIOM(BP86:B97-D) optimized structures for the canonical and rare tautomer forms, Mg-GC and Mg-GC1, respectively. The *circles* highlight the position of the "moving" water molecule

relative energies for the transition state and rare tautomer are 2.47 and 0.34 kcal/mol, respectively. These values are significantly below than their DFT counterparts. Precisely, the predicted energy for the transition state is ca. 6 kcal/mol smaller than the M06-2X and ωB97X-D data. Furthermore, according to this MP2 approach, the rare tautomer is practically isoenergetic with the canonical form, a rather unexpected result. Although the MP2/6-31G(d)(0.25) approach accurately describes the stacking effect [94, 95], we attribute the origin of such discrepancy to the size of the basis set. Indeed, MP2 calculations show a larger dependence with respect to basis set size than the DFT approaches when applied to stacked systems [117]. To verify this point, MP2 calculations with larger basis set would be necessary but are, unfortunately, computationally impossible due the size of the system.

3.2 The low layer method

Comparison between methods 1, 7, and 8 allow to assess the influence of the selected method in the description of the border base pairs effects. As shown in Table 1, in these

Table 3 Computed relative energies (ΔE in kcal/mol) along H1 proton transfer with selected low layer methods. The interbase bonds (in Å) are also listed

	Method				
	1	7	8	9	10
Mg–GC					
ΔE	0.00	0.00	0.00	0.00	0.00
O6(G)–N4(C)	2.84	2.82	2.80		2.87
N1(G)–N3(C)	2.95	2.94	2.92		2.93
N2(G)–O2(C)	2.98	2.98	2.96		2.90
Mg–GC1‡					
ΔE	8.04	7.74	3.69	6.90	8.02
O6(G)–N4(C)	2.65	2.65	2.63		2.67
N1(G)–N3(C)	2.69	2.69	2.68		2.68
N2(G)–O2(C)	2.91	2.91	2.92		2.86
Mg–GC1					
ΔE	3.24	3.14	3.68	1.49	2.99
O6(G)–N4(C)	2.67	2.65	2.67		2.70
N1(G)–N3(C)	2.87	2.86	2.87		2.87
N2(G)–O2(C)	3.03	3.03	3.03		2.99

methods, the same functional (M06-2X) is used in the high layer, whereas the low layer is treated with the B97-D (method 1) and M06-2X (method 7) functionals, as well as with the semiempirical PM6 model (method 8) [118]. According to the values listed in Table 3 and plotted in Fig. 5, methods 1 and 7 provide very similar results for both optimized geometries and relative energies. Indeed, the predicted interbase bonds distances are practically the same, whereas the difference in the relative energies is smaller than 0.30 kcal/mol. Therefore, M06-2X and B97-D similarly describe the stacking effects and, consequently, both methods could be used for the description of the low layer, though B97-D is apparently not adequate to model the region where PT reaction takes place as discussed above.

Moving to the PM6 description of border base pairs, we observe that the relative energy of the rare tautomer agrees with the more refined DFT calculations, while the transition state is dramatically stabilized. Consequently, the transition state and the rare tautomers structures are nearly isoenergetic. Like for method 5, this is a consequence of the changes in the hydration pattern: one of the water molecules moves from the $N4(C)H_{4'}$ site to the metal hydration shell. Since the evolution of the relative energies does not reflect exclusively the H1 proton transference, the energy of the calculated Mg–GC1‡ structure cannot be used for the study of the metal-induced PT. For this reason, the complete PM6 scan was not performed.

Our calculations reveal an inherent limitation of the local microhydrated model: the optimized positions for the

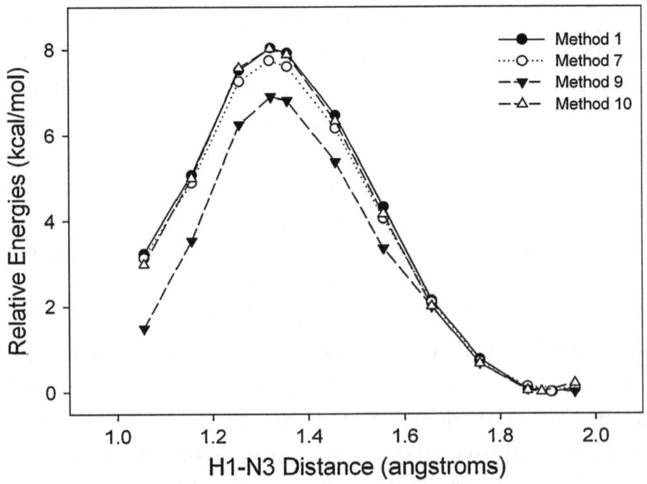

Fig. 5 ONIOM potential energy curves along the H1 proton transfer with methods 1, 7 and 9

water molecules are deeply connected to the selected methods, which eventually affect the stability of the entire system. Such limitation could be overtaken by increasing the number of water molecules, for example, including the first hydration water molecules for the border basis pairs or by adding constraints, but at the price of more demanding calculations or less satisfying simulations. Alternatively, one can rely on a partial hydration model as the one of Fig. 2, but it is necessary to systematically check that there are no changes in the hydration pattern resulting from the PT.

3.3 Continuum solvent models

We have evaluated the impact of the aqueous environment beyond the first hydration by combining the explicit nine water molecules to the PCM model. First, PCM corrections are included through single-point energy calculations (method 9) performed on the geometries of method 1. As shown Fig. 5, consideration of the continuum solvent gives a small stabilization (ca. 2 kcal/mol) for both Mg–GC1‡ and Mg–GC1 with respect to the PCM-free model, which is consistent with the larger charge separation in these two cases compared to the canonical structure. However, when PCM corrections are included during the optimization (method 10), the bulk solvent effects are in fact less important. As shown in Table 3, the differences between method 1 (without PCM correction) and method 10 (PCM optimization) are less than 0.30 kcal/mol. Furthermore, practically the same hydrogen bond distances are obtained with both methods 1 and 10, and only a slight increase of the O6(G)–N4(C) and decrease of N1(G)–N3(C) and N2(G)–O2(C) bond lengths are observed. Overall, the PCM optimization provides the same picture for the structural changes induced by PT reaction as in the original method, as evidenced by Fig. 5. This finding suggests that explicit solvent model could be enough to reproduce the main effects of solution in the induced PT, at least for the comparison of the relative energies between different tautomers of the same base pairs.

3.4 Biological implications

To clarify the actual impact of the theoretical approaches in the predicted magnesium–DNA interactions, we use methods 1–5 to calculate the PT equilibrium constant (K_{eq}) as well as the forward (k_f) and reverse (k_r) rate constants at 298.15 K. For the latter, we considered tunneling effects by introducing the Wigner term correction [119–121] in the conventional transition state theory:

$$k = \left(1 + \frac{1}{24}\left[\frac{h|\nu|}{k_B T}\right]^2\right)\frac{k_B T}{h}e^{-\Delta G^{\ddagger}/RT} \tag{1}$$

where $|\nu|$ is the (imaginary) frequency characteristic of the transition state, h is the Planck's constant, k_B is de Boltzmann's constant, and $\Delta G^{\ddagger}$ is the activation Gibbs free energy.

As showed in Table 4, the order of the predicted equilibrium constant range between 10^{-1} and 10^{-4}, depending on the method, which significantly exceeds the natural mutation rate, between 10^{-8} and 10^{-10} [122] when no bonding between DNA and metals cations is considered. Moreover, the forward rate constants, which lies in the range of 10^7–10^{11} s^{-1}, are large enough to allow the formation of the Mg–GC1 structure during the cell life [40]. Particularly, the extreme values agrees with our previous discussion: on the one hand, smallest k_f ($\sim 10^7$ s^{-1}) correspond to the method 4 (B3LYP), which overestimates the potential energy barrier, and on the other hand, method 5 (BP86) provides the largest k_f ($\sim 10^{11}$ s^{-1}) due to the change in the hydration pattern. Accordingly, all methods (including the most accurate methods 1 and 2) predict a high mutagenicity effect of the Mg^{2+}. This conclusion seems to contrast with the reality: magnesium is essential for the life and its adverse effects appears only at large doses. To explain this a priori conflicting result between theory and experience it should be noted the short life time of the Mg–GC1 form that can be quantified by means of the reverse rate constant (k_r). As Florián and Leszczyński demonstrated, only rare tautomers associated to a enough stable minimum of energy may yield permanent genetic errors [40]. According to the energetics requirements related to the time for DNA replication process, if the reverse rate constant exceeds ca. 10^{10} s^{-1}, the lifetime of the rare tautomer will be too short to induce a mutation [40]. The reverse constants listed in Table 4 rise above that limit regardless the selected high-layer method, and therefore, Mg–GC1 will reverts so quickly to the canonical structure that its impact on life is rather negligible. To confirm our results, we have also considered the tunneling correction method by Skodje and Truhlar, which depends not only on the imaginary frequency of the transition state but also on the energy barrier [123]. It is observed that both corrections predict reverse rate constants with the same order of magnitudes (e.g. 2.35×10^{11} and 2.64×10^{11} s^{-1} for method 2 and 3, respectively), with the only exception of method 1, for which the Skodje and Truhlar correction gives rise to $k_r = 1.36 \times 10^{11}$ s^{-1}. However, this finding is consistent with the good agreement between both tunneling corrections since the Wigner-corrected value is close to this order of magnitude (9.72×10^{10} s^{-1}). Finally, we notice that none of the methods are in total agreement in the magnitudes of rate and equilibrium constant, which clearly shows that the selected methods have a decisive impact on the study of metal–DNA interactions, and therefore, DFT functionals need to be cautiously combined. Among selected ONIOM schemes, methods 1 and 2 provide a similar picture of the magnesium effects, showing the reliability of the M06-2X and ωB97X-D functionals to perform these calculations.

Table 4 Calculated activation and reaction Gibbs free Energies at 298.15 K ($\Delta G^{\ddagger}$ and ΔG, respectively, in kcal/mol), forward and reverse rate constants (k_f and k_r, respectively, in s^{-1}) and equilibrium constants (K_{eq}) along the PT reaction

Method	$\Delta G^{\ddagger}$	ΔG	k_f	k_r	K_{eq}
1	5.02	1.98	3.44×10^9	9.72×10^{10}	3.54×10^{-2}
2	6.48	3.51	3.42×10^8	1.29×10^{11}	2.65×10^{-3}
3	3.25	1.16	6.51×10^{10}	4.63×10^{11}	1.41×10^{-1}
4	7.19	4.68	9.75×10^7	2.66×10^{11}	3.67×10^{-4}
5	2.54	2.00	2.01×10^{11}	5.93×10^{12}	3.39×10^{-2}

4 Conclusions

In this work, we have compared the description of DFT functionals for investigating PT reaction related to the metalation of nucleic acids. To this end, a chemical model that accounts for the main interactions inside DNA, namely hydration, interbase bonds and base stacking, has been used. Our calculations hints that both the B3LYP and BP86 functionals are probably not adequate to describe the PT transition state in DNA bases when applied in a ONIOM scheme. Since the barrier for the H1 PT is crucial for the stability of rare tautomeric form and hence for the possible biological impact of Mg^{2+}, alternative approaches offer a better compromise. As the M06-2X and ωB97X-D yield consistent results and have been shown previously to adequately model biochemical systems, this work corroborates the improved performance of refined hybrid functionals in the description of PT reactions. With respect to stacking effect, the pure dispersion-corrected B97-D functional provides a description for the border base pairs similar to the one of M06-2X. However, B97-D underestimates the energetic barrier and therefore it should be used cautiously for the description of the region where PT reactions takes place. Our calculations also revealed the decisive influence of the water molecules position in the PT energetic profile of the PT transition state as well as the relative small impact of bulk solvent effects on relative energies. Indeed, very similar results are obtained with or without PCM corrections once full geometry optimizations are carried out. We conclude that the predicted biological role of the magnesium strongly depends on the selected level of theory. This investigation is an illustration that the developments of new efficient functionals allows to tackle key biological problems.

Acknowledgments J.P.C.C. acknowledges the *Région des Pays de la Loire* and the fellowship provided by the Fundación Séneca, Agencia de Ciencia y Tecnología de la Región de Murcia, within its Postdoctoral Research Staff Training Program. D.J. is indebted to the *Région des Pays de la Loire* for financial support in the framework of a *recrutement sur poste stratégique*. The work was partially supported by the Fundación Séneca del Centro de Coordinación de la Investigación de la Región de Murcia under Project 08735/PI/08, and by the Ministerio de Educación y Ciencia of Spain under Projects CTQ2007-66528 and CONSOLIDER CSD2009-00038. This research used resources of (1) the CCIPL (*Centre de Calcul Intensif des Pays de Loire*) and (2) the GENCI-CINES/IDRIS (Grant C201108547).

References

1. Santoro F, Barone V, Improta R (2007) Proc Natl Acad Sci USA 104:9931–9936
2. Improta R, Barone V (2008) Theor Chem Acc 120:491–497
3. Improta R, Santoro F, Barone V, Lami A (2009) J Phys Chem A 113:15346–15354
4. Santoro F, Barone V, Improta R (2009) J Am Chem Soc 131:15232–15245
5. Santoro F, Barone V, Lami A, Improta R (2010) Phys Chem Chem Phys 12:4934–4948
6. Improta R, Scalmani G, Barone V (2000) Int J Mass Spectrom 201:321–336
7. Xerri B, Morell C, Grand A, Gadet J, Cimino P, Barone V (2006) Org Biomol Chem 4:3986–3992
8. Labet V, Morell C, Gran A, Cadet J, Cimino P, Barone V (2008) Org Biomol Chem 6:3300–3305
9. Müller J (2010) Metallomics 2:318–327
10. Hanna R, Doudna JA (2000) Curr Opin Chem Biol 4:166–170
11. Serebrov V, Clarke RJ, Gross HJ, Kisselev L (2001) Biochemistry 40:6688–6698
12. Anastassopoulou J (2003) J Mol Struct 651(653):19–26
13. Aich P, Dasgupta D (1995) Biochemistry 34:1376–1385
14. Huang HW, Li D, Cowan JA (1995) Biochimie 77:729–738
15. Keniry MA, Owen EA, Shafer RH (2000) Biopolymers 54:104–114
16. Sreedhara A, Cowan JA (2002) Biometals 15:221–223
17. Chen SY, Lin TH (2005) J Phys Chem B 109:9764–9772
18. Barceló F, Ortiz-Lombardía M, Martorell M, Oliver M, Méndez C, Salas JA, Portugal J (2010) Biochemistry 49:10543–10552
19. Rude RK, Singer FR (1981) Annu Rev Med 32:245–259
20. Chiu TK, Dickerson RE (2000) J Mol Biol 301:915–945
21. Egli M (2002) Chem Biol 9:277–286
22. Anastassopoulou J, Theophanides T (2002) Crit Rev Oncol Hematol 42:79–91
23. Bregadze VG, Gelagutashvili ES, Tsakadze KJ, Melikishvili SZ (2008) Chem Biodiv 5:1980–1989
24. Buckin VA, Tran H, Morozov V, Marky LA (1996) J Am Chem Soc 118:7033–7039
25. Kankia BI, Marky LA (1999) J Phys Chem B 103:8759–8767
26. Serganov A, Polonskaia A, Phan AT, Breaker RR, Patel DJ (2006) Nature 441:1167–1171
27. Every AE, Russu IM (2008) J Phys Chem B 112:7689–7695
28. Ruan C, Rodgers MT (2009) J Am Chem Soc 131:10918–10928
29. Russo N, Toscano M, Grand A (2001) J Am Chem Soc 123:10272–10279
30. Russo N, Toscano M, Grand A (2001) J Phys Chem B 105:4735–4741
31. Petrov AS, Lamm G, Pack GR (2002) J Phys Chem B 106:3294–3300
32. Burda JV, Šponer J, Hrabáková J, Zeizinger M, Leszczynski J (2003) J Phys Chem B 107:5349–5356
33. Burda J, Shukla MK, Leszczynski J (2005) J Mol Model 11:362–369
34. Šponer J, Lankas F (eds) (2006) . Computational Studies of RNA and DNA. Springer, Berlin, pp 411–432
35. Michalkova A, Kosenkov D, Gorb L, Leszczynski J (2008) J Phys Chem B 112:8624–8633
36. Noguera M, Bertrán J, Sodupe M (2008) J Phys Chem B 112:4817–4825
37. Noy A, Soteras I, Luque FJ, Orozco M (2009) Phys Chem Chem Phys 11:10596–10607
38. Šponer J, Sabat M, Burda JV, Leszczynski J, Hobza P (1999) J Phys Chem B 103:2528–2534
39. Muñoz J, Šponer J, Hobza P, Orozco M, Luque FJ (2001) J Phys Chem B 105:6051–6060
40. Florián J, Leszczyński J (1996) J Am Chem Soc 118:3010–3017
41. Guallar V, Douhal A, Moreno M, Lluch JM (1999) J Phys Chem A 103:6251–6256
42. Dannenberg JJ, Tomasz M (2000) J Am Chem Soc 122:2062–2068
43. Golo VL, Volkov YS (2003) Int J Mod Phys C 14:133–156
44. Gorb L, Podolyan Y, Dziekonski P, Sokalski WA, Leszczyński J (2004) J Am Chem Soc 126:10119–10129
45. Gervasio FL, Laio A, Iannuzzi M, Parrinello M (2004) Chem Eur J 10:4846–4852
46. Perun S, Sobolewski AL, Domcke W (2006) J Phys Chem A 110:9031–9038
47. Herrera B, Toro-Labbe A (2007) J Phys Chem A 111:5921–5926
48. Villani G (2010) J Phys Chem B 114:9653–9662
49. Pérez A, Tuckerman ME, Hjalmarson HP, von Lilienfeld OA (2010) J Am Chem Soc 132:11510–11515
50. Dedíková P, Neogrády P, Urban M (2011) J Phys Chem A 115:2350–2358
51. Basch H, Krauss M, Stevens WJ (1985) J Am Chem Soc 107:7267–7271
52. Schröder G, Lippert B, Sabat M, Lock CJL, Faggiani R, Song B, Sigel H (1995) J Chem Soc Dalton Trans 23:3767–3775
53. Song B, Zhao J, Griesser R, Meiser C, Sigel H, Lippert B (1999) Chem Eur J 5:2374–2387
54. Šponer J, Šponer JE, Gorb L, Leszczynski J, Lippert B (1999) J Phys Chem A 103:11406–11413

55. Noguera M, Bertran J, Sodupe M (2004) J Phys Chem A 108:333–341
56. Matsui T, Shigeta Y, Hirao K (2006) Chem Phys Lett 423:331–334
57. Matsui T, Shigeta Y, Hirao K (2007) J Phys Chem B 111:1176–1181
58. Šponer J, Sabat M, Gorb L, Leszczynski J, Lippert B, Hobza P (2000) J Phys Chem B 104:7535–7544
59. Cerón-Carrasco JP, Zúñiga J, Requena A, Perpète EA, Michaux C, Jacquemin D (2011) Phys Chem Chem Phys 13:14585–14589
60. Gil A, Branchadell V, Bertran J, Oliva A (2009) J Phys Chem B 113:4907–4914
61. Cauët E, Liévin J (2009) J Phys Chem A 113:9881–9890
62. Cauët E, Valiev M, Weare JH (2010) J Phys Chem B 114:5886–5894
63. Acosta-Silva C, Branchadell V, Bertran J, Oliva A (2010) J Phys Chem B 114:10217–10227
64. Chen HY, Kao CL, Hsu SCN (2009) J Am Chem Soc 131:15930–15938
65. Chen HY, Yeh SW, Hsu SCN, Kao CL, Dong TY (2011) Phys Chem Chem Phys 13:2674–2681
66. Šponer J, Burda JV, Sabat M, Leszczyński J, Hobza P (1998) J Phys Chem A 102:5951–5957
67. Cerón-Carrasco JP, Requena A, Zúñiga J, Michaux C, Perpète EA, Jacquemin D (2009) J Phys Chem A 113:10549–10556
68. Solt I, Simon I, Császár AG, Fuxreiter M (2007) J Phys Chem B 111:6272–6279
69. Schneider B, Berman HM (1995) Biophys J 69:2661–2669
70. Auffinger P, Westhof E (2000) J Mol Biol 300:1113–1131
71. Makarov V, Pettitt B, Feig M (2002) Acc Chem Res 35:376–384
72. Dapprich S, Komáromi I, Byun KS, Morokuma K, Frisch MJ (1999) J Mol Struct (Theochem) 462:1–21
73. Gaussian 09, Revision B01, Frisch MJ, Trucks GW, Schlegel HB, Scuseria GE, Robb MA, Cheeseman JR, Scalmani G, Barone V, Mennucci B, Petersson GA, Nakatsuji H, Caricato M, Li X, Hratchian HP, Izmaylov AF, Bloino J, Zheng G, Sonnenberg JL, Hada M, Ehara M, Toyota K, Fukuda R, Hasegawa J, Ishida M, Nakajima T, Honda Y, Kitao O, Nakai H, Vreven T, Montgomery JA Jr, Peralta JE, Ogliaro F, Bearpark M, Heyd JJ, Brothers E, Kudin KN, Staroverov VN, Kobayashi R, Normand J, Raghavachari K, Rendell A, Burant JC, Iyengar SS, Tomasi J, Cossi M, Rega N, Millam JM, Klene M, Knox JE, Cross JB, Bakken V, Adamo C, Jaramillo J, Gomperts R, Stratmann RE, Yazyev O, Austin AJ, Cammi R, Pomelli C, Ochterski JW, Martin RL, Morokuma K, Zakrzewski VG, Voth GA, Salvador P, Dannenberg JJ, Dapprich S, Daniels AD, Farkas Ö, Foresman JB, Ortiz JV, Cioslowski J, Fox DJ, Gaussian, Inc, Wallingford CT, 2010
74. Šponer J, Riley KE, Hobza P (2008) Phys Chem Chem Phys 10:2595–2610
75. Hobza P, Šponer J (2002) J Am Chem Soc 124:11802–11808
76. Zhao Y, Truhlar DG (2005) Phys Chem Chem Phys 7:2701–2705
77. van der Wijst T, Fonseca-Guerra C, Swart M, Bickelhaupt F (2006) Chem Phys Lett 426:415–421
78. Grimme S (2004) J Comput Chem 25:1463–1473
79. Riley KE, Vondrášek J, Hobza P (2007) Phys Chem Chem Phys 9:5555–5560
80. Noguera M, Ríos-Font R, Rodríguez-Santiago L, Solans-Monfort X, Oliva A, Bertran J, Sodupe M (2009) Theor Chem Acc 123:105–111
81. Antony J, Brüske B, Grimme S (2009) Phys Chem Chem Phys 11:8440–8447
82. Fonseca-Guerra C, van der Wijst T, Poater J, Swart M, Bickelhaupt FM (2010) Theor Chem Acc 125:245–252
83. Zhao Y, Schultz NE, Truhlar DG (2006) J Chem Theory Comput 2:364–382
84. Zhao Y, Schultz NE, Truhlar DG (2008) Acc Chem Res 41:157–167
85. Zhao Y, Truhlar DG (2008) Theor Chem Acc 120:215–241
86. Becke A (1986) J Chem Phys 84:4524–4529
87. Perdew J (1986) Phys Rev B 33:8822–8824
88. Grimme S (2006) J Comput Chem 27:1787–1799
89. Lee C, Yang W, Parr R (1988) Phys Rev B 37:785–789
90. Becke A (1993) J Chem Phys 98:1372–1377
91. Zhao Y, Truhlar DG (2011) Chem Phys Lett 502:1–13
92. Chai JD, Head-Gordon M (2008) Phys Chem Chem Phys 10:6615–6620
93. Valero R, Gomes JRB, Thrular DG, Illas F (2008) J Chem Phys 129:124710
94. Hobza P, Šponer J, Polasek M (1995) J Am Chem Soc 117:792–798
95. Šponer J, Leszczyński J, Hobza P (1996) J Phys Chem 100:5590–5596
96. Jurečka P, Šponer J, Hobza P (2004) J Phys Chem B 108:5466–5471
97. Šponer J, Jurečka P, Marchan I, Luque FJ, Orozco M, Hobza P (2006) Chem Eur J 12:2854–2865
98. Alemán C (1999) Chem Phys 244:151–162
99. Alemán C (2000) Chem Phys 253:13–19
100. Oliva R, Cavallo L (2009) J Phys Chem B 113:15670–15678
101. Tomasi J, Bonaccorsi R, Cammi R, del Valle FO (1991) J Mol Struct (THEOCHEM) 234:401–424
102. Cammi R, Tomasi J (1995) J Comput Chem 16:1449–1458
103. Tomasi J, Mennucci B, Cammi R (2005) Chem Rev 105:2999–3093
104. Hohenstein EG, Chill ST, Sherrill CD (2008) J Chem Theory Comput 4:1996–2000
105. Shukla MK, Leszczyński J (2010) Mol Phys 108:3131–3146
106. Paukku Y, Hill G (2011) J Phys Chem A 115:4804–4810
107. Kumar A, Sevilla MD (2011) J Phys Chem B 115:4990–5000
108. Zhao Y, Truhlar DG (2011) J Chem Theory Comput 7:669–676
109. Dabkowska I, Jurečka P, Hobza P (2005) J Chem Phys 122:204322
110. Jurečka P, Šponer J, Hobza P (2006) Phys Chem Chem Phys 8:1985–1993
111. Barone V, Orlandini L, Adamo C (1994) Chem Phys Lett 231:295–300
112. Barone V, Orlandini L, Adamo C (1995) Int J Quantum Chem 56:697–705
113. Fonseca-Guerra C, Bickelhaupt F, Snijders J, Baerends E (2000) J Am Chem Soc 122:4117–4128
114. Fonseca-Guerra C, Bickelhaupt FM (2002) Angew Chem Int Ed 41:2092–2095
115. Cerón-Carrasco JP, Requena A, Michaux C, Perpète EA, Jacquemin D (2009) J Phys Chem A 113:7892–7898
116. Cerón-Carrasco JP, Requena A, Michaux C, Perpète EA, Jacquemin D (2009) Chem Phys Lett 484:64–68
117. Gu J, Wang J, Leszczynski J, Xie Y, Schaefer-III HF (2008) Chem Phys Lett 459:164–166
118. Řezáč J, Fanfrlík J, Salahub D, Hobza P (2009) J Chem Theory Comput 5:1749–1760
119. Wigner EZ (1941) Phys Chem 19:203–216
120. Jolibois F, Grand A, Cadet J, Adamo C, Barone V (1999) Chem Phys Lett 301:255–262
121. Di Labio GA, Johnson ER (2007) J Am Chem Soc 129:6199–6203
122. Topal MD, Fresco JR (1976) Nature 263:285–289
123. Skodje RT, Truhlar DG (1981) J Phys Chem 85:624–628

Theor Chem Acc (2012) 131:1189
DOI 10.1007/s00214-012-1189-8

REGULAR ARTICLE

Adsorption of successive layers of H_2 molecules on a model copper surface: performances of second- to fifth-rung exchange-correlation functionals

Géraldine Cilpa · Jonathan Colin · Frédéric Labat · Carlo Adamo · Gilberte Chambaud

Received: 29 November 2011 / Accepted: 3 February 2012 / Published online: 9 March 2012

Abstract The interaction of H_2 molecules with a Cu(100) metallic surface has been investigated by DFT approaches using a $(H_2)_nCu_{13}$ cluster model. Nine exchange-correlation functionals, belonging to the generalized gradient approximations (GGA), meta GGA (mGGA), hybrid Kohn–Sham/Hartree–Fock models, either based on GGAs or mGGAs, range-separated hybrids, and double-hybrid families, have been tested on the chemisorption and physisorption processes involving one or two H_2 layers. The addition of an empirical correction for dispersion has also been tested for some of these functionals. The calculated energies and structural parameters were compared to sophisticated multi reference configuration interaction including Davidson's correction for quadruple excitations (MRCI + Q). Our results show that among the nine considered exchange-correlation functionals, none can accurately reproduce all processes involved in the successive layers adsorption. Although an hybrid based on a mGGA such as M06-2X can quantitatively describe both the physisorption step and dissociation barriers involved in the adsorption of the first H_2, it fails to reproduce its chemisorption. On the other hand, significant discrepancies with the reference post-HF data are obtained for the description of the second layer interaction, no matter which functional is considered, outlining the need of improvement and/or development of exchange-correlation functionals suitable for complex systems such as H_2/H_2Cu.

Keywords DFT · Physisorption · H_2 · Copper · Cluster

Dedicated to Professor Vincenzo Barone and published as part of the special collection of articles celebrating his 60th birthday.

G. Cilpa · G. Chambaud
Laboratoire Modélisation et Simulation Multi Echelle,
UMR 8208-CNRS, Université Paris-Est, 5 boulevard Descartes,
Marne-la-Vallée 77454, France

J. Colin · F. Labat (✉) · C. Adamo
Laboratoire d'Electrochimie,
Chimie des Interfaces et Modélisation pour l'Energie,
UMR CNRS 7575, ENSCP—Chimie Paristech,
11 rue P. et M. Curie, Paris Cedex 05 75231, France
e-mail: frederic-labat@chimie-paristech.fr

C. Adamo
Institut Universitaire de France,
103 Boulevard Saint Michel, Paris 75005, France

1 Introduction

Molecule-surface interactions have been widely studied, particularly for their interest in catalytic processes. Among them, the H_2-surface reaction has participated to the understanding of various mechanisms and it is one of the best known activated reaction yielding dissociative chemisorption process. Theoretical approaches have provided deep insights on the chemical phenomenon [1]. Many studies have focused on the adsorption of H_2 on clean copper surfaces and most of them were concerned with the dissociative chemisorption and the subtle effects governing the dissociation on surfaces. Using multi-dimensional potential energy surfaces (2, 4, and 6D PES) to describe the interaction potential, the dynamics of the chemisorption have been studied, taking into account the effects of the molecular ro-vibrational excitations on the dissociation, the scattering, the sticking probabilities, the anisotropy of the energy, the corrugative effects, and the reactivity at surfaces [2–9].

In most of these studies, the surface is described by a periodic 2D-representation, namely a slab representation, calculations being performed using density functional theory (DFT) methods with H_2 molecules either bound to the surface or very close to it (distances are smaller than 4 bohr). Alternative studies have also been performed, modeling the surface by a cluster. In such cases, the size of the cluster plays a relevant role since it influences the reactivity of the metal toward H_2, as demonstrated by Raghavan et al. [10] in the case of D_2/Ni, and also the quality of the results [11]. Moreover, it has been shown that, in the case of Cu, 3D metallic clusters [7] with more than 6 atoms can satisfactorily reproduce the experimental interatomic distances [12]. Similar conclusions were obtained by Siegbahn et al. [13] for Ni clusters with even (14) and odd (13) numbers of metallic atoms and one H_2 molecule. On the other hand, the convergence of the adsorption energy with the cluster size [14] is more problematic. Despite the large amount of theoretical works, some questions still remain, especially concerning the energetics of H_2 molecules interacting with a metallic surface already covered with a first layer of hydrogens. In a recent theoretical study on the H_2/Cu system [15] using a cluster model consisting of 13 Cu atoms, it has been shown that the chemisorption and physisorption of the first H_2 layer and the van der Waals (vdW) interactions responsible for the physisorption of additional H_2 layers can be satisfactorily reproduced by highly correlated electronic wavefunctions. Unfortunately, such refined approaches cannot be easily used in large or extended (2D or 3D) systems and other possibilities must be explored.

Density functional theory (DFT) techniques have been shown to be robust and reliable for the accurate evaluation of a large number of chemical and physical properties in different phases ranging from single molecules to solid states [16, 17]. Therefore, it appeared interesting to investigate the behavior of different DFT models for such an intriguing system, $(H_2)_nCu_{13}$. Several DFT studies have already been devoted to the specific problem of the interaction of one H_2 molecule with metallic surfaces, but more questionable is indeed the DFT description of the interaction of a second layer of H_2 molecules with the surface. This weak interaction is ruled by vdW forces and it is well known that standard exchange-correlation functionals provide a poor description of such dispersion interactions. Recent studies have shown, however, that sophisticated DFT approaches can give promising outcomes for such difficult cases. In this sense, meta-GGA models (see below) give interesting results for quantitative reproduction of weak interactions in small model systems like rare-gas dimers [18, 19] or in biological molecules [20]. More recently, the introduction of a perturbative second-order contribution in the DFT framework, through a semiempirical mixing, seems to be another promising route for vdW systems, even if, as for its predecessors it has been tested only on model systems [21]. More easily, addition of an empirical correction to a reparametrized functional [21] is another possibility, which has already been successfully applied to numerous weakly interacting systems [22–24].

The goal of the present work is thus to extend this study on weak interactions by considering a more complex (and chemically interesting) system, $(H_2)_nCu_{13}$, in which interactions of different nature (physisorption and chemisorption) occur and to assess the performances of different exchange-correlation functionals for the description of such a system.

2 Exchange-correlation functionals

In this section, we give a brief survey of the different classes of functionals selected for the present study. They belong to the second, third, fourth, and fifth levels of the so-called five level Jacob's ladder [25], which goes from local to fully non-local functionals, suggesting a complexity and quality improvement when climbing the different levels. Extensive literature is available on this subject (see Ref. 26 for instance).

In the second level, generalized gradient approximation (GGA) functionals include dependence on the electronic density ρ and its gradient $\nabla\rho$. DFT with GGAs includes self-exchange and self-correlation, both being unphysical [26]. Self-correlation can be eliminated in the next level, by including an additional dependence on the Laplacian of the density $\Delta\rho$ or on the kinetic energy density in meta-GGAs. Self-exchange on the other hand can be removed by adding a fraction of non-local Hartree–Fock (HF) exchange in the expression of the energy of a GGA or a mGGA, constituting the so-called hybrid functionals in the fourth level. Finally, in the fifth level, double-hybrid functionals such as B2PLYP [21] replace a fraction of the semi-local correlation energy by a non-local correlation energy expression that employs the Kohn–Sham orbitals in second-order many-body perturbation theory.

Global GGA hybrids (GH, such as B3LYP [27] or PBE0 [28, 29]), which include a *constant* percentage of HF exchange that is mixed with density functional exchange, are probably the most popular functionals since they tend to improve GGA or mGGA computed properties, both for molecular and solid-state systems. In some cases, however, range-separated hybrids (RSH), in which the Coulomb operator is usually separated in a short- and a long-range part with varying amounts of HF exchange depending on the point in space, perform better.

In this paper, nine functionals have been considered: (1) PW91PW91 [30] (referred to as PW91 in the following), a GGA which is well known to significantly overestimate van der Waals interactions [31, 32], (2) a dispersion-corrected GGA (B97-D [33]), based on a reparametrized B97 functional [34] and including damped atom-pairwise dispersion corrections, commonly used as a reference in the modeling of weakly interacting systems, (3) two popular hybrid GGAs, B3LYP [27] and PBE0 [28, 29], the first typically predicting unbound systems for this kind of interactions [35] while the latter has already been shown to give better results [18, 21], (4) M06-L [36], a mGGA parametrized on a diverse set of data including non-covalent interactions, (5) an hybrid mGGA, M06-2X [37], aimed at an accurate description of both thermochemistry and non-bonded interactions with 54% of HF exchange, (6) two RSH, ωB97X [37] and ωB97XD [38], including 15.77 and 100% of HF exchange at short- and long-range, respectively, with an additional empirical dispersion correction for the latter, which has already been successfully applied to the treatment of non-covalent interactions [39], (7) B2PLYP [21], a double hybrid belonging to the fifth level of the Jacob's ladder, which has already been shown to poorly describe the π stacking of the benzene dimer for instance [21].

For the sake of clarity, we note that the empirical dispersion contribution in the B-97D and ωB97XD functionals differ in two ways: a scaled contribution is used in the former, while it is not in the latter, and different dispersion parameters are also considered.

3 Computational details

The metallic surface corresponding to a (100) crystal face is described by a cluster containing 13 atoms of Cu : 9 on the first layer interacting with the H_2 molecule, and 4 below. A rigid unrelaxed cluster model, in which distances between Cu atoms are kept fixed at their solid-state values, has been chosen, since relaxation of such a system could induce spurious effects. This model is similar to the one used in a recent theoretical study [15] whose results are compared to the present ones.

All calculations have been carried out with the latest available commercial version of the Gaussian code [40], using an effective core potential (ECP) and associated basis set developed by the Stuttgart group [41] for Cu [19 electrons are treated explicitly ($3s^2$, $3p^6$, $3d^{10}$, $4s^1$)], while a large diffuse basis set 6-311++G(3df, 3pd) [42–45] is chosen for the hydrogen atoms in order to insure a proper description of the negative H^{δ} on the metallic surface.

For symmetry reasons, the position of the center of mass of H_2 has been constrained along the z axis, perpendicular to the surface at the center of the central Cu atom. The dissociation path investigated for the first H_2 molecule proceeds along the bridge-top-bridge (*btb*) orientation, as described in Fig. 1, where the center of mass of H_2 is on a top site and H atoms are dissociating on bridge sites.

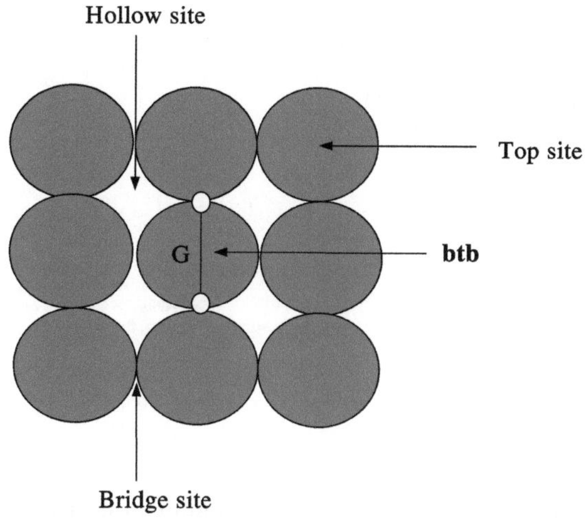

Fig. 1 Definition of the *btb* dissociation pathway, H atoms (*white circles*) on the (100) surface of Cu (*gray circles*) are localized on the bridge sites. *btb* corresponds to an approach where *G* center of mass of H_2 is on a *top site* and the H atoms are dissociating on *bridge sites*

According to recent calculations by Kokh et al. [46] employing the embedded cluster model and highly correlated wavefunctions, even though hollow sites are energetically favored for most of the molecular chemisorbed species, the difference in the binding energies between hollow, top, and bridge sites is rather small if the bond has a large fraction of ionic character, which is the case in the present situation. Such results confirmed the DFT calculations of Eichler et al. [47] who found an energy difference smaller than 0.1 eV between the binding energies in hollow or bridge sites. Since the different sites of chemisorption have roughly the same energy and correspond to stable or metastable minima, we have chosen to work here with the *btb* orientation of the H_2 molecule because it allows additional symmetry simplifications within our cluster model. Two degrees of freedom have been investigated, *Z* the distance of the center of mass of H_2 to the center of the central Cu atom, and *d* the intramolecular H–H bond distance. In the present work, we neglect the rotations of the first H_2 molecule, in particular we have fixed the molecular axis parallel to the surface, which is the most favorable orientation for the dissociation of the H_2 molecule. For the second layer of H_2 molecules, two orientations are selected: one with the molecular axis of H_2 parallel to the surface and to the first molecule, and the other one with the molecular axis perpendicular to the surface. These systems are depicted in Fig. 2, respectively. For each value of the distance *Z*, the H_2 bond length *d* has been optimized. On this 2D-Potential Energy Surface, the corresponding stationary points give the minimum energy pathway of H_2 along the z axis perpendicular to the cluster surface. The reference zero energy is defined as the sum of the energy of the free H_2 molecule and the energy of the bare Cu_{13} cluster.

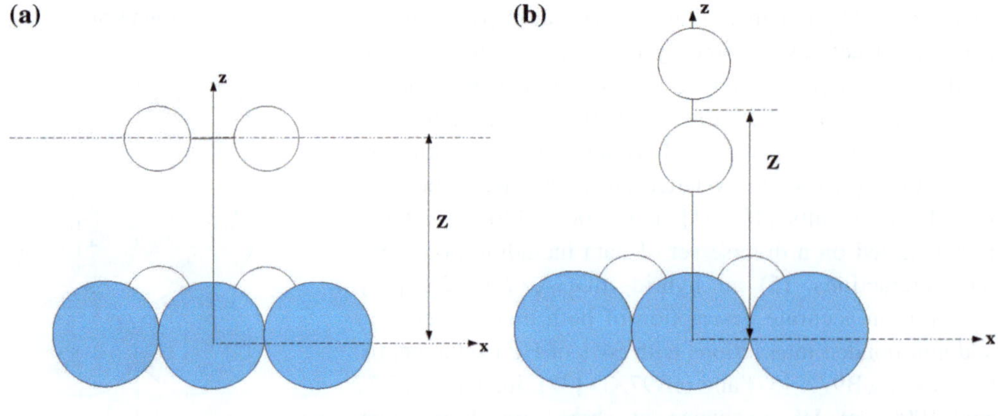

Fig. 2 Sketches of the second layer interaction for **a** *parallel* and **b** *perpendicular* approaches. The convention for the atoms is the same as in Fig. 1

In the DFT calculations, the contamination of the electronic state has been monitored through the expectation value of the S^2 operator, $<S^2>$ whose value was always close to 0.75 for all the physisorption processes (with one or two H_2 molecules). More involved is the situation for the chemisorption since two electronic states, namely $^2A_1$ and $^2B_2$ are very close in energy. For instance, their energy difference at the equilibrium geometry is about 0.1 eV at the B3LYP level. This is in line with a previous work carried out by some of us on $(H_2)_n/Ag_{13}$ [48], in which it was shown that several electronic states were interacting in the vicinity of the activation barrier, leading to complex electronic structures. Following the previous MRCI study, only the $^2A_1$ state has been considered in this work. Moreover, we want to notice that it was not possible to compute energies with the same basis used for the MRCI calculations due to problems related to the errors of the numerical integration for the exchange-correlation contribution to the total energy. In addition, the basis set superposition error (BSSE) has not been evaluated in order to make our DFT calculations directly comparable to the MRCI calculations, carried out under similar conditions [15]. In the MRCI + Q calculations, it is rather small in the present H_2/Cu system and estimated to 4×10^{-4} eV close to the minimum of physisorption, namely smaller than the expected accuracy of the MRCI calculation, which is rather in the range of 10^{-3} eV. With all considered DFT methods on the other hand, it is found below 0.1 eV.

4 Results and discussion

4.1 First layer interaction

In the study of the interaction of H_2 with the copper surface, three different steps are easily identified, namely the physisorption, the formation of the transition state, and the chemisorption. These steps are analyzed and discussed separately because they involve electronic effects of different nature. Interaction energies (E), dihydrogen bond lengths [d(H–H)], and distance from the cluster surface (Z) are displayed in Table 1 and compared to available data [7, 15].

4.1.1 *Physisorption*

An experimental physisorption energy of approximately 30 meV has been deduced from scattering experiments and selective adsorption measurements by Andersson et al. [49, 50]. From the data given in Table 1 one can see that the MRCI calculations [15] including the Davidson correction (MRCI + Q) are in excellent agreement with experimental results, and they concluded to the existence of a shallow physisorption minimum located far from the surface, at about 7.0 bohr. Typical shape of the potential energy curve of interacting fragments is obtained in Fig. 3 for all methods: repulsive component at short-range, followed by an attractive component at larger distances, with more or less deep wells.

We can first note that since the physisorption interaction does not change dramatically the internal structure of the H_2 molecule, it is not surprising that the d (H–H) data are weakly sensitive to the choice of the theoretical model, all methods providing very close results. On the other hand, vertical distances (Z) and interaction energies are much more functional-dependent. In particular, computed interaction energies range between 1.1 and 38 meV, the ωB97X, ωB97XD, and M06-2X functionals giving an error below 3 meV with respect to the MRCI + Q data, outperforming the original B97-D functional often used as a reference in the modeling of weakly interacting systems. The largest errors are obtained at the GH level for the interaction energies, B3LYP predicting almost unbound H_2 and PBE0 slightly improving the description. Interestingly, we note that a standard GGA functional such as PW91 underestimates the MRCI + Q value of about the same amount as the B97-D functional overestimates it.

Table 1 First layer interaction energy for the three processes. Data previously obtained with periodic boundary conditions (PBC) are also reported

	Physisorption			Barrier		Chemisorption		
	E/meV	Z/bohr	d(H–H)/bohr	E/eV	Z/bohr	E/eV	Z/bohr	d(H–H)/bohr
PW91	22.4	7.50	1.42	0.88	2.70	0.64	1.93	4.84
PBE0	6.5	8.30	1.41	1.02	2.65	0.53	1.96	4.72
B3LYP	1.1	9.70	1.40	1.26	2.65	0.54	1.94	4.77
B97-D	35.5	6.31	1.41	1.12	2.66	0.60	1.90	4.82
ωB97X	27.2	6.67	1.41	1.19	2.63	0.55	2.02	4.57
ωB97XD	30.5	6.43	1.41	1.16	2.63	0.57	1.98	4.63
M06-L	38.0	6.18	1.41	0.81	2.74	0.60	2.02	4.85
M06-2X	25.6	7.00	1.40	1.56	2.62	0.22	2.15	4.64
Other works								
MRCI + Q[a]	28.2	6.99	1.40	1.54	2.46	0.60	1.9	4.82
BP88–PBC	–	–	–	0.70[b]	2.62[b]	~0.5[b]	2.0[b]	4.82[b]
	–	–	–	0.63[c]	2.60[c]	–	–	–
PW91–PBC	–	–	–	1.21[d]	2.65[d]			
	–	–	–	–	–	0.47[e]	–	–
Experiment[f]	29.9–31.4	–	–	–	–	–	–	–

DFT calculations have been performed using SDD basis sets for the copper atoms and the large and diffuse 6-311++G(3df, 3dp) basis for the hydrogen atoms. Data previously obtained with periodic boundary conditions (PBC) are also reported

[a] See Ref. [15]

[b] See Ref. [7]

[c] See Ref. [4]

[d] See Ref. [8]

[e] See Ref. [53]

[f] See Refs. [49, 50]

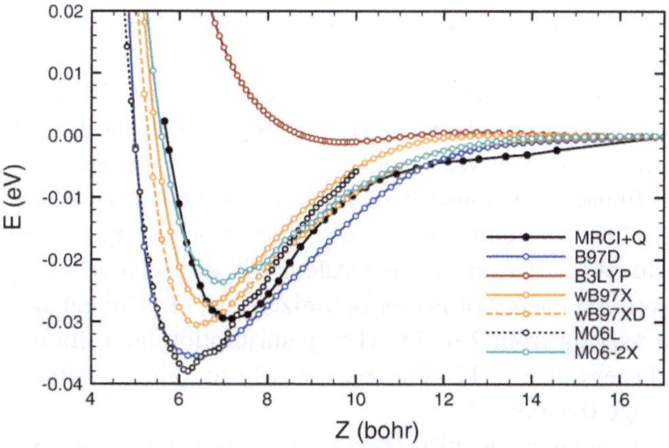

Fig. 3 Physisorption *curves* for the first H_2 layer interaction on the model Cu_{13} cluster (MRCI + Q data from Ref. [15])

The MRCI + Q vertical distance (Z) is only quantitatively reproduced at the M06-2X level, a GmH parametrized on a diverse set of data including non-covalent interactions, all other functionals predicting either too small (B97-D, ωB97X, ωB97XD, M06-L) or too large (PW91, PBE0, B3LYP) values. This is especially true for these two latter functionals for which computed errors with respect to the MRCI + Q data are larger than 1 bohr.

Surprisingly, we note that a mGGA such as M06-L poorly describes such an interaction, with significant errors of about 10 meV on the interaction energy and 0.6 bohr on the Z parameter. This might be related to the sawtooth shape of the M06-L curve around the minimum in Fig. 3, due to the default integration grid used. Finally, we note that significant differences are obtained in the shape of the energy profiles in the region between 10 and 12 bohr. In particular, the only functional reproducing nicely the MRCI + Q profile is ω B97X, and all other functionals leading to too steep or too flat curves. At larger distances ($Z > 12$ bohr), the interaction is also poorly described with an underestimation of the interaction with all functionals.

The B2PLYP approach deserves a separate comment since it leads to unphysical results, characterized by large (>1 Hartree) interaction energies. A deep analysis of the numerical data suggests that correct energies are obtained for the hybrid part of the functional, which is close to those provided by similar functionals, like B3LYP, while the erroneous values arise from the perturbative correlation corrections. This error could confirm that, at least in our case, the Kohn–Sham Hamiltonian is not the optimal choice for the perturbation expansion [51]. The same behavior has been found for the other systems studied in the present paper so that the results obtained with the B2PLYP functionals are not discussed or reported in the tables.

To sum up, physisorption of the first H_2 molecule on the Cu_{13} cluster represents a challenging system for DFT. Although a nice agreement with the MRCI + Q data can be obtained with some functionals, only one (M06-2X) simultaneously quantitatively reproduces the three chosen parameters (interaction energy, vertical H_2/Cu_{13} distance, and H–H distance) characterizing this process. At large cluster/H_2 distances, however, it fails to reproduce the MRCI + Q curve, predicting too steep curve in the 10–12 bohr region, and underestimating the interaction energy above.

4.1.2 Transition barrier

When the H_2 molecule approaches the surface and is close enough to experience the attractive potential of the substrate, its bond weakens and the two atoms are pulled apart before being chemisorbed, due to an electron donation of the Cu_{13} cluster in the first antibonding orbital of H_2. As previously shown in several studies on H_2/Cu(100) [7, 8], the molecule has to get over a barrier for the dissociation. This barrier is very sensitive to the models, and depends on: (1) the surface coverage, (2) the number of integration points in the **k**-space for periodic calculations, (3) the atomic basis set, (4) the method used to represent the core electrons (pseudopotential or frozen core). Investigations on both systems [H_2/Cu(100) and H_2/Cu(111)] already suggested that GGA methods with a periodic slab approach overestimate barriers for H_2 + Cu systems by about 0.1–0.2 eV [52]. These latter results mean that, according to Wiesenekker estimates (0.7 eV) [7], the barrier height would be rather about 0.5–0.6 eV and the difference between GGA/slab and MRCI + Q calculations would thus be large. Dynamical calculations carried out by Somers et al. [4] using a potential energy surface fitted on periodic GGA results are in line with the work of Wiesenekker et al. [7] (see Table 1).

In the present work, the same cluster as already considered in the previous MRCI calculations [15] has been used to model the copper surface, making a comparison with these reference post-HF data easier than with the DFT periodic works. The energy range covered by the chosen functionals is [0.81, 1.56] eV, and all functionals underestimate the barrier height if we except the M06-2X data. The lowest barriers and largest errors are computed at the M06-L and PW91 levels, the value obtained with this latter functional being in line with previously published periodic calculations [8]. All functionals including exact HF exchange improve the agreement with the reference post-HF data, an almost perfect agreement being obtained at the M06-2X level.

The positions of the barrier along the z axis are all slightly overestimated with a difference of about 0.28 bohr on Z (for the highest with M06-L) and about 0.16 bohr (for the lowest with M06-2X) with respect to the MRCI + Q calculations, all functionals performing nicely for the determination of the dissociation barrier.

Summarizing, we can say that DFT and MRCI + Q calculations are comparable for this process, with an overall better agreement when using a GmH functional such as M06-2X.

4.1.3 Chemisorption

For the dissociative chemisorption of H_2 on Cu, all functionals give similar equilibrium distances Z, slightly below 2 bohr. This distance is in excellent agreement with that obtained at the MRCI level or using periodic calculations within a GGA approach [7] (1.9 and 2.0 bohr, respectively). The agreement on the optimized H–H bond length is also quite satisfactory, the d value ranging between 4.57 (ωB97X) and 4.85 bohr (M06-L), with a reference post-HF value of 4.82 bohr.

The difference in the chemisorption energy between the MRCI and periodic approaches is typically of about 0.1–0.2 eV, which can be assigned to different reasons. In particular, in periodic calculations, due to interatomic interaction between H_2 molecules in the adlayer [7], the mean energy of one H_2 interacting with the unit cell of the surface is not equal to the energy of a single H_2 with the surface [6]. The results of Guvelioglu et al. [53] for the chemisorption energy of H_2 interacting with the Cu(100) surface confirm Wiesenekker et al. [7] energy value which is found to be around 0.5 eV. Overestimated energies when dealing with clusters have also been reported by Guvelioglu et al. [53] who have calculated dissociative chemisorption energy of H_2 for optimized Cu_n of different sizes, n varying from 2 to 15. The chemisorption interaction for clusters of $n > 12$ converges slowly to values within the range 0.6–0.9 eV.

Computed chemisorption energies are very similar with all methods, except for the M06-2X functional which significantly underestimates the MRCI + Q value. From Table 1, it is clear that all computed values are lower than the post-HF reference data, if we except the PW91 value for which an excellent agreement with previous periodic works is nevertheless obtained. All functionals perform very well, except M06-2X for which a significant underestimation of the chemisorption energy is obtained. Since functionals including 20–25% of HF exchange already reproduce nicely the MRCI + Q value, we can probably relate this discrepancy to the larger percentage included in M06-2X (54%) compared to other hybrids. On the other hand, the comparison with the two considered RSH (ωB97X and ω B97XD) is not straightforward, since their amount of HF exchange is not constant. A quantitative agreement with the MRCI + Q value is obtained at the M06-L and B97-D levels, i.e., with functionals which only include approximated exchange. In addition, the post-HF vertical distance (Z) is well-reproduced by all DFT methods, if we except once again the M06-2X functional.

In summary, among the nine functionals considered, none could be found to quantitatively reproduce the MRCI + Q data for the adsorption of one H_2 molecule on a model copper surface. Although M06-2X nicely performs both for the physisorption and energy barrier data, it surprisingly fails to reproduce the chemisorption. This might be related to its large amount of exact HF exchange, which is well known to provide large errors on geometrical parameters, thus affecting, as in the present case, the energetical features.

4.2 Second layer interaction

The second layer interaction is described as the interaction of an incoming H_2 with the chemisorbed system at its equilibrium geometry H + H + Cu_{13} ($Z = 1.9$ bohr and d about 4.80 bohr depending on the considered functional). Two approaches of the second H_2 molecule have been investigated: *parallel* or *perpendicular* to the first adsorbed H_2 see Figs. 2a and b, respectively.

The previous MRCI + Q calculations [15] are the only existing reference calculations for this process : they have shown that the incoming hydrogen molecule in this second layer is not dissociated and only slightly activated, with a negligible stretching of the H–H bond. Table 2 presents data computed for both approaches of the second H_2 molecule.

4.2.1 Parallel approach

The results for the physisorption along this orientation can be directly compared to those of the first layer since the H_2 molecule approaches the surface in the same way. The comparison of the two sets of data given in Tables 1 and 2 shows that, for all functionals, the physisorption interaction of the second layer is smaller than that of the first layer, giving longer Z distances and smaller interaction energies. This is not in agreement with the MRCI + Q data, for which the opposite behavior is obtained.

From Fig. 4, it is clear that all functionals considered fail to accurately reproduce the post-HF curve. In particular, computed DFT interaction energies range between 0.3 (B3LYP) and 19.8 meV (B97-D), corresponding to vertical distances Z of 10.11 and 7.99 bohr, that is significantly different from the 77.5 meV and 6.42 bohr values computed at the MRCI + Q level. In fact, all functionals underbind the second H_2 molecule of at least 60 meV and predict too large Z parameters of at least 1.5 bohr. This is especially true for B3LYP, for which the conclusions drawn for the adsorption of the first H_2 layer still hold, with a very shallow profile in this case. It is interesting to notice that very similar interaction energies are computed at the ωB97X and ωB97XD levels, outlining that, around the physisorption minimum obtained with these functionals, dispersion only plays a small role, the two profiles being notably different only at larger Z distances.

On the other hand, the H_2 bond length at the minimum of the physisorption well shows that H_2 is not dissociated but only slightly stretched, in agreement with the MRCI + Q results.

4.2.2 Perpendicular approach

In complete agreement with previous MRCI + Q results, for all functionals, calculated interaction energies upon

Table 2 Second layer physisorption minimum (H_2/H_2Cu_{13} system), considering both parallel and perpendicular approaches of the second H_2 molecule on the H_2Cu_{13} system

	Parallel			Perpendicular		
	E/meV	Z/bohr	d(H–H)/bohr	E/meV	Z/bohr	d(H–H)/bohr
PW91	19.1	8.56	1.42	41.5	7.80	1.42
PBE0	5.2	9.24	1.41	22.2	8.03	1.41
B3LYP	0.3	10.11	1.40	13.4	8.65	1.40
B97-D	19.8	7.99	1.41	15.6	8.02	1.41
ωB97X	15.8	8.00	1.41	68.5	6.96	1.41
ωB97XD	16.1	8.25	1.40	45.9	7.11	1.41
M06-L	10.7	8.00	1.41	45.4	6.66	1.41
M06-2X	11.2	8.45	1.39	37.5	6.67	1.40
Other works						
MRCI[a]	57.0	6.42	1.40	111.7	6.41	1.40
MRCI + Q[a]	77.5	6.42	1.40	131.9	6.24	1.40

[a] From ref. [15]

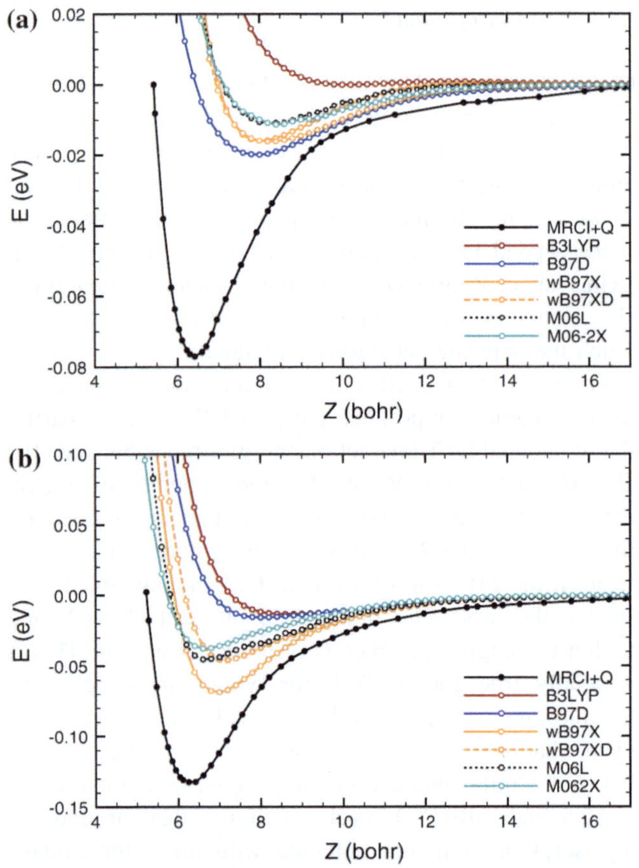

Fig. 4 Second H_2 layer interaction curves, with both **a** *parallel* and **b** *perpendicular* approaches with respect to the first adsorbed layer (MRCI + Q data from Ref. [15])

perpendicular approach are larger than those obtained when considering a parallel approach, except for B97-D for which a reverse trend is surprisingly obtained.

As noticed for the parallel approach, significant discrepancies are obtained at the DFT and MRCI + Q levels, with smaller interaction energies and larger equilibrium distances Z computed with the former techniques. These trends clearly appear in Fig. 4.

4.3 Comments

The analysis of the above discussed results is not straightforward since no trends are clearly evident in going from first to second layer interactions and from chemi- to physisorption. Nevertheless, some points can be underlined.

Although the physisorption and dissociation processes involved in the adsorption of the first H_2 molecule on the copper surface can be accurately described by a GmH functional such as M06-2X, this functional fails to reproduce the MRCI + Q data for the chemisorption. A more balanced description on such problems can be obtained with ab initio multiconfigurational methods, but these techniques are rapidly limited by the number of atoms and electrons to be considered. Description of the physisorption of the second H_2 layer is, however, much more challenging. In contrast to the MRCI + Q results, computed DFT interaction energies for this layer are smaller than those calculated for the first one. This stems probably from the fact that the DFT cannot reproduce the polarization of the surface due to the first chemisorbed layer as shown by the very small dipole moment calculated at the equilibrium geometry of the H + H + Cu_{13} chemisorbed system. Consequently, there is no enhancement of the second layer interaction. This suggests that the second layer H_2 molecule does not experience any significant distortion of its electronic cloud from the underlying system H + H + Cu_{13}.

From a more general viewpoint, it should also be noticed that functionals providing accurate results on small model systems show their limits when applied to larger and more complex systems, such as the one considered in the present study.

4.4 Conclusions

In this paper, the interaction of H_2 with a metallic Cu cluster has been studied at DFT level. Nine exchange-correlation functionals, belonging to the GGAs, meta GGA (mGGA), hybrid Kohn–Sham/HF models, either based on GGAs or mGGAs, range-separated hybrids (RSH), and double-hybrid families, have been tested on the chemisorption and physisorption processes involving one or two H_2 layers. The addition of an empirical correction for dispersion has also been tested for some of these functionals. Comparison has been made with previously reported accurate multi reference configuration interaction data.

Although the physisorption and dissociation processes involved in the adsorption of the first H_2 molecule can be accurately described by a GmH functional such as M06-2X, this functional fails to reproduce the post-HF data for the chemisorption step. Thus, among the nine tested DFT methods, none reproduced quantitatively the MRCI + Q values.

More critical is the description of the second H_2 layer, for which significant discrepancies on both physisorption energies and positions have been found with all DFT methods, no matter if a parallel or a perpendicular approach of the second H_2 molecule with respect to the first one is considered. In particular, all these methods predict interaction energies smaller than the computed physisorption energies of the first layer, in contrast to the post-HF data for which a reverse trend is obtained. This has been attributed to a poor description of the polarization of the surface due to the first chemisorbed layer, as outlined by the very small dipole moments calculated at the equilibrium geometry of the $H + H + Cu_{13}$ chemisorbed system.

From these results therefore, although recently developed DFT methods can be successfully applied to small- or medium-sized weakly interacting systems, their application to larger or more complex systems such as H_2/H_2Cu_{13} is still challenging, opening the route for improvement and/or development of exchange-correlation functionals suitable for such systems.

References

1. Groß A (1998) Surf Sci 32:291
2. Darling GR (1994) J Chem Phys 101:3268
3. Watts E, Sitz GO (2001) J Chem Phys 114:4171
4. Somers MF, Olsen RA, Busnengo HF, Kroes GJ (2004) J Chem Phys 121:11379
5. Diño WA, Kaisai H, Okiji A (2000) Prog Surf Sci 63:63
6. Wiesenekker G, Kroes GJ, Baerends EJ, Mowrey RC (1995) J Chem Phys 102:3873
7. Wiesenekker G, Kroes GJ, Baerends EJ (1996) J Chem Phys 104:7344
8. White JA, Bird DM (1994) Phys Rev Lett 73:1404
9. Kratzer P, Hammer B, Nϕrskov JK (1996) Surf Sci 359:45
10. Raghavan K, Stave MS, DePristo AE (1999) J Chem Phys 91:1904
11. Dominguez-Ariza D, Sousa C, Harrison NM, Ganduglia-Pirovano MV, Illas F (2003) Surf Sci 522:185
12. Jaque P, Toro-Labbé A (2002) J Chem Phys 117:3208
13. Siegbahn PEM, Blomberg MRA, Bauschlicher CW Jr (1984) J Chem Phys 81:2103
14. Whitten JL, Yang H (1996) Surf Sci Rep 24:55
15. Cilpa G, Chambaud G (2007) Surf Sci 601:320
16. Jones RO, Gunnarsson O (1989) Rev Mod Phys 61:689
17. Kurth S, Perdew JP, Blaha P (1999) Int J Quant Chem 75:889
18. Tao J, Perdew JP (2005) J Chem Phys 122:114102
19. Adamo C, Barone V (1998) J Chem Phys 108:664
20. Zhao Y, Truhlar DG (2007) J Chem Theor Comput 3:289
21. Grimme S (2006) J Chem Phys 124:34108
22. Grimme S (2004) J Comput Chem 25:1463
23. Huenerbein R, Schirmer B, Moellmann J, Grimme S (2010) Phys Chem Chem Phys 12:6940
24. Krishtal A, Vanommeslaeghe K, Olasz A, Veszprémi T, Van Alsenoy C, Geerlings P (2009) J Chem Phys 130:174101
25. Perdew JP, Ruzsinszky A, Tao J, Staroverov VN, Scuseria GE, Csonka GI (2005) J Chem Phys 123:62201
26. Cramer CJ, Truhlar DG (2009) Phys Chem Chem Phys 11:10757
27. Stephens PJ, Devlin FJ, Chabalowski CF, Frisch MJ (1994) J Phys Chem 98:11623
28. Adamo C, Barone V (1999) J Chem Phys 110:6158
29. Ernzerhof M, Scuseria GE (1999) J Chem Phys 110:5029
30. Perdew JP, Chevary JA, Vosko SH, Jackson KA, Pederson MR, Singh DJ, Fiolhais C (1992) Phys Rev B 46:6671
31. Patton DC, Pederson MR (1997) Phys Rev A 56:2495
32. Johnson ER, Dilabio GA (2006) Chem Phys Lett 419:333
33. Grimme S (2006) J Comput Chem 27:1787
34. Becke AD (1997) J Chem Phys 107:8554
35. Tsuzuki S, Lüthi HP (2001) J Chem Phys 114:3949–3957
36. Zhao Y, Truhlar DG (2006) J Chem Phys 125:194101
37. Zhao Y, Truhlar DG (2008) Theor Chem Acta 120:215
38. Chai J-D, Head-Gordon M (2008) J Chem Phys 128:084106
39. Chai J-D, Head-Gordon M (2008) Phys Chem Chem Phys 10:6615
40. Frisch MJ, Trucks GW, Schlegel HB, Scuseria GE, Robb MA, Cheeseman JR, Scalmani G, Barone V, Mennucci B, Petersson GA, Nakatsuji H, Caricato M, Li X, Hratchian HP, Izmaylov AF, Bloino J, Zheng G, Sonnenberg JL, Hada M, Ehara M, Toyota K, Fukuda R, Hasegawa J, Ishida M, Nakajima T, Honda Y, Kitao O, Nakai H, Vreven T, Montgomery JA Jr, Peralta JE, Ogliaro F, Bearpark M, Heyd JJ, Brothers E, Kudin KN, Staroverov VN, Kobayashi R, Normand J, Raghavachari K, Rendell A, Burant JC, Iyengar SS, Tomasi J, Cossi M, Rega N, Millam JM, Klene M, Knox JE, Cross JB, Bakken V, Adamo C, Jaramillo J, Gomperts R, Stratmann RE, Yazyev O, Austin AJ, Cammi R, Pomelli C, Ochterski JW, Martin RL, Morokuma K, Zakrzewski VG, Voth GA, Salvador P, Dannenberg JJ, Dapprich S, Daniels AD, Farkas O, Foresman JB, Ortiz JV, Cioslowski J, Fox DJ (2009) Gaussian 09 revision A.02. Gaussian, Wallingford
41. Kaupp M, Schleyer PvR, Stoll H, Preuss H (1991) J Chem Phys 94:1360
42. Krishnan R, Binkley JS, Seeger R, Pople JA (1980) J Chem Phys 72:650
43. Frisch MJ, Pople JA, Binkley JS (1984) J Chem Phys 80:3265
44. Clark T, Chandrasekhar J, Schleyer PVR (1983) J Comput Chem 4:294
45. Gill PMW, Johnson BG, Pople JA, Frisch MJ (1992) Chem Phys Lett 197:499
46. Kokh DB, Buenker RJ, Whitten JL (2006) Surf Sci 600:5104
47. Eichler A, Kresse G, Hafner J (1998) Surf Sci 397:116
48. Cilpa G, Guitou M, Chambaud G (2008) Surf Sci 602:2894
49. Andersson S, Wilzén L, Persson M (1988) Phys Rev B 38:2967
50. Andersson S, Persson M, Harris J (1996) Surf Sci Lett 360:L499
51. Bartlett R, Grabowski I, Hirata S, Ivanov S (2005) J Chem Phys 122:34104
52. Kroes GJ, Baerends EJ, Mowrey RC (1997) Phys Rev Lett 78:3583
53. Guvelioglu GH, Ma P, He X (2005) Phys Rev Lett 94:026103

Theor Chem Acc (2012) 131:1191
DOI 10.1007/s00214-012-1191-1

REGULAR ARTICLE

Heuristic approaches to the optimization of acceptor systems in bulk heterojunction cells: a computational study

Silvia Carlotto · Laura Orian · Antonino Polimeno

Received: 8 July 2011 / Accepted: 2 September 2011 / Published online: 10 March 2012

Abstract A computational protocol combining a heuristic search based on genetic algorithms (GAs) and quantum chemistry methods is implemented and applied to a family of acceptor compounds based on the 9,9′-bifluorenylidene backbone, to be coupled with the poly-3(hexylthiophene) polymer (donor) in a bulk heterojunction solar cell. Highly performing candidates are generated via GA from an initial generation, after a number of iterations (i.e., new generations), under the selective pressure of electronic constrains calculated at density functional theory level. The combination of heuristic search techniques and advanced electronic structure methodologies for characterization seems to be amenable to further applications in the field of molecular design.

Keywords Organic photovoltaics · Solar cells · Density functional theory · Genetic algorithms

1 Introduction

Computational chemistry can be a useful tool in helping to design a new molecular structure with defined electronic or structural properties. Usually, however, the search space is non-linearly dependent upon the problem dimensions. Standard optimization methods must make way to alternative approaches. For this reason there is a considerable interest in heuristic techniques that can usually find a near-optimal, or even just a reasonable solution in an acceptable computational time. A large class of techniques is based on genetic algorithms (GAs) [1], which are stochastic global optimization methods [2] inspired by the biological mechanisms of evolution and heredity, implementing processes such as mutation, selection, crossover, etc. The versatility of GAs lies in their flexibility and potential applicability to a wide range of problems [3, 4]. Recently, the application of GAs to computational chemistry and biology has spread considerably. Examples of computer-aided molecular design are reported by Clark [5] and Parrill [6]. Several GAs-based studies are related to the investigation of biomolecules [7–9]. Other examples are focused on the conformational analysis of small molecules [10], inclusion complexes [11], structure of interfaces [12], evolution of catalysts [13], and parameterization of NDDO wavefunctions [14].

We present here a computational study based on the combination of GA search and direct evaluation of molecular electronic properties via quantum mechanics (QM) methods. Previous combined applications of GAs and QM and/or molecular mechanics (MM) calculations are mostly concerned with the optimization of force field parameters [15–19] or energy minimization [20–23]. Arora [24] investigates via GA/QM methods novel ternary copolymers of donor–acceptor polymers [3] and polypeptide chains [24], and Cartensen [25] employs a GA approach to tune two excitation wavelengths of a molecular switch backbone.

In this work, we use a GA approach combined with QM calculations to optimize a family of acceptor compounds based on the 9,9′-bifluorenylidene backbone, to be coupled with the poly-3(hexylthiophene) (P3HT) polymer (donor) in a bulk heterojunction (BHJ) solar cell. We consider a

Dedicated to Professor Vincenzo Barone and published as part of the special collection of articles celebrating his 60th birthday.

S. Carlotto · L. Orian · A. Polimeno (✉)
Dipartimento di Scienze Chimiche, Università degli Studi di Padova, Via Marzolo 1, 35131 Padua, Italy
e-mail: antonino.polimeno@unipd.it

fixed backbone and explore the combinations of different substituents. Exploration of this "virtual chemistry space" of combination is rather large, over a million of possible molecules, and the problem can be suitably tackled with GAs methodologies. Interest in designing organic components of solar cells has spurred a considerable amount of theoretical and experimental studies in the last decade. Several strategies have been set up to improve the performance of these devices, but the screening of candidate molecular systems and of specific molecular properties can still be considered a bottleneck [26].

The paper is organized as follows: in Sect. 2 the molecular system is described; Sect. 3 summarizes the computational approach. In Sect. 4, results are presented and discussed.

2 Molecular system

We explore the design of acceptor compounds based on the 9,9′-bifluorenylidene (99′BF) backbone (Fig. 1a), which can be coupled with the P3HT polymer acting as donor in an ideal BHJ solar cell device. These systems have been proposed by Brunetti [27]. Since the idea of bulk heterojunction (BHJ) organic solar cell emerged over one decade ago [28], there has been an intense activity in this field [29]. Organic solar cells based on BHJ are particularly attractive due to their mechanical flexibility and potential low cost [30]. They are based on photoinduced charge transfer from an electron-donating material, such as a hole-conducting and light-absorbing polymer, to an electron-accepting component, typically fullerene (c60) [27]. The typical structure of these devices consists of an interpenetrating network of electron acceptor and donor material, sandwiched between two electrodes having different work functions [31].

The efficiency of a BHJ is controlled by several factors [32]. The donor must have a small band gap to efficiently absorb light in the visible region of the solar spectrum, i.e., an optimal $|E_{\mathrm{HOMO}-D} - E_{\mathrm{LUMO}-D}|$ separation is typically 2 eV. This value was estimated by Koster [30], Scharber [33], and Lenes [34]. Another factor is the necessity of improving the open-circuit voltage (V_{OC}), i.e., the electronic match between the donor and acceptor component must be optimized in order to minimize unnecessary internal loss of open-circuit voltage. Although the recent literature [35–38] suggests that there is no general consistent understanding of the origin of open-circuit voltage of the BHJ solar cells, experimental evidence reveals a linear relationship between the $E_{\mathrm{HOMO}-D}$ and $E_{\mathrm{LUMO}-A}$ energy difference and the V_{OC} of the bulk heterojunction device [31, 39]. Thus, a promising strategy to improve the BHJ cell efficiency is raising the $E_{\mathrm{LUMO}-A}$ [32]. However, one must take also into account that since after photoexcitation the electron is transferred from the LUMO of the donor to the LUMO of the acceptor, in the extreme case, if the two LUMO levels become too close in energy, the driving force would be lost [34]. An optimal condition turns out to be $\Delta E_{\mathrm{LUMO}} = E_{\mathrm{LUMO}-D} - E_{\mathrm{LUMO}-A}$ of 0.3 eV [33, 40].

On the basis of these considerations, different methods have been employed experimentally to improve the BHJ cell performance by acting on the molecular orbital energies (Fig. 1b): (1) lowering $E_{\mathrm{LUMO}-D}$ results in small-band gap donors that have the capability to absorb a larger part of the solar spectrum; (2) lowering both $E_{\mathrm{LUMO}-D}$ and $E_{\mathrm{HOMO}-D}$. In this case, the bandgap of the donor is constant and the device gains in efficiency due to an enhanced V_{OC}; (3) raising the $E_{\mathrm{LUMO}-A}$, also resulting in an increase of open-circuit voltage [32, 41]. Following these hints, several acceptor molecules have been synthesized and investigated by QM calculations [42].

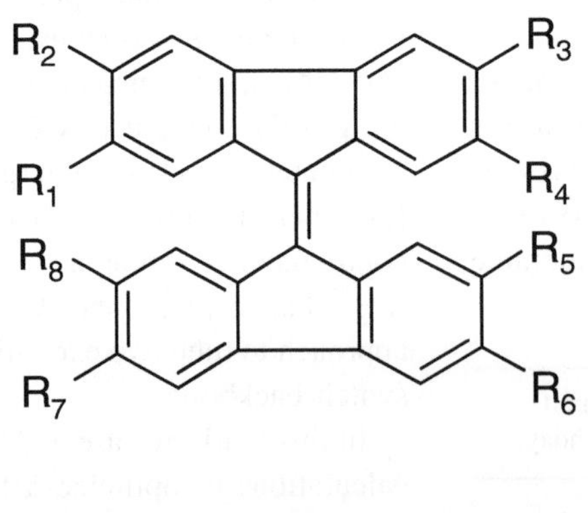

(a)

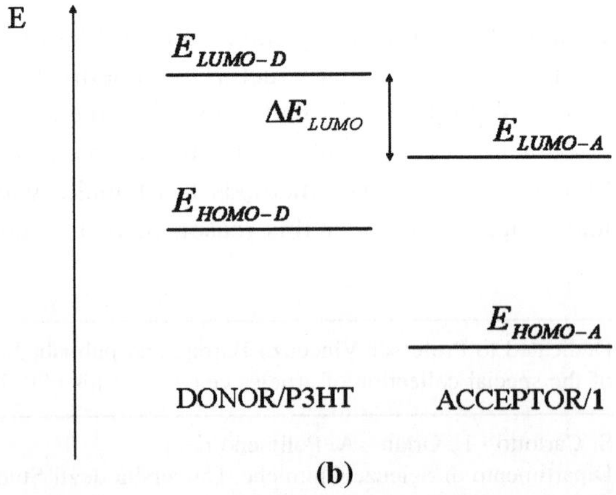

(b)

Fig. 1 **a** 9,9′-bifluorenylidene (99′BF) (**1**) and **b** energy level diagram for a donor–acceptor heterojunction

3 Methods

3.1 The genetic algorithm

GAs are global search heuristic techniques, employed currently in optimization problems of large dimensions [24]. The principle of evolution "Survival of the fittest" first described by Charles Darwin is exploited by mimicking the evolutionary principles and chromosomal processing in natural genetics [43], in which the basic idea is to maintain a population candidate solution, which evolves under selective pressure. A fitness value is assigned to every solution (i.e., individual of a given population), which is directly related to the objective of the optimization problem. Each individual (*chromosome*) represents a possible solution to the optimization problem. Chromosomes are commonly encoded as string of integer or binary numbers. An initial population (first generation) is created and the fitness of each individual is evaluated. The next generation is built by selecting individuals from the previous generation, based on their fitness values, and by forming new individuals via suitable *crossover* and *mutation* operators. New generations are created until a stop criterion (e.g., a certain number of generations reached) is fulfilled. In practice, the steps are (1) encoding a representation of the individuals and (2) defining a fitness function to evaluate the individuals. A number of control parameters need to be specified and optimized for each GA implementation; the most relevant are population size, mutation probability (p_m), crossover probability (p_c), the maximum number of generations, the type of selection. In our case, a chromosome corresponds to a combination of substituents and is represented by a string of integer numbers (see below). The roulette wheel parent selection [44] is chosen as a selection operation. Finally, the best performing individuals of the previous generation are always retained (*elitism*) [43]. The GA procedure was implemented in the form of a home-made code (Fortran90). The program works iteratively by (1) generating the input files and submitting them automatically to the Gaussian09 package, (2) analyzing the output Gaussian09 files, and (3) creating a new population (set of chromosomes) and starting again from 1.

3.2 Quantum–mechanical calculations

Molecular systems are characterized in silico as follows, using the Gaussian 09 package [45]. First the geometry of **1** is optimized by an hybrid DFT approach with Becke's three-parameter exchange energy density functional combined with the Lee–Yang–Parr correlation energy density functional, B3LYP and the Pople's 6-31G basis set [46, 47]. In this first accurate optimization process, the eight sites are linked to hydrogen atoms. During the GA search, the optimized geometry is taken as input, with random substituents in the eight available sites. Each individual is optimized by the semi-empirical method Austin Model (AM1). Tests performed randomly of different individuals show that AM1 geometries compare satisfactorily to full B3LYP/6-31G outcomes, but have a considerably smaller computational cost (cfr. Table 1). The energy $E_{\mathrm{LUMO}-A}$ and the total electric dipole moment of each individual are next estimated for the optimized geometry at B3LYP/3-21G level [29].

To examine the impact of the basis set on the accuracy of LUMO energy levels, we tested different basis sets, including the minimum basis sets STO-3G, 3-21G, 6-31G, 6-311G(d,p), 6-311++G(d,p), and a different level of theory, i.e., HF/6-31G. For each model, the optimized geometry is the same. As shown in Table 1, all methods fail to predict $E_{\mathrm{LUMO}-A}$, for the three test molecules T1 (fully hydrogenated parent compound), T2 with sites 1,4,5,8 = OCH_3, and T3 with 2,3,6,7 = OCH_3 (cfr. Fig. 1). $E_{\mathrm{LUMO}-A}$ is overestimated, as expected [48], but the differences between calculated and measured values [27] are roughly constant, within the same basis set. Therefore, we assume as a compromise required by the need of keeping the overall computational time per individual as low as possible that a constant systematic error affects $E_{\mathrm{LUMO}-A}$, as evaluated at B3LYP/3-21G level.

4 Results and discussion

Generations of $M = 24$ chromosomes each are employed. In this preliminary study, runs of $N = 100$ generations are investigated, with no other stop criteria than the completion

Table 1 Calculated and experimental LUMO energy values for the three test molecules T1 (fully hydrogenate parent compound), T2 with sites 1,4,5,8 = OCH_3, and T3 with sites 2,3,6,7 = OCH_3

Level of theory	T1	T2	T3
HF/6-31G[a]	0.77	0.96	0.68
B3LYP/STO-3G[a]	−0.34	0.29	−0.01
B3LYP/3-21G[a]	−2.46	−2.02	−2.31
B3LYP/6-31G[a]	−2.36	−2.02	−2.29
B3LYP/6-311G(d,p)	−2.65	−2.23	−2.54
B3LYP/6-311++G(d,p)	−2.75	−2.34	−2.64
B3LYP/6-31G[b]	−2.36	−2.02	−2.27
Experimental [19]	*−3.37*	*−3.11*	*−3.39*

All the energy values are expressed in eV

[a] Geometry optimized at AM1 level

[b] Geometry optimized at B3LYP/6-31G level

of the allowed number of generations. A chromosome is represented as an array of $m = 8$ possible replaceable positions. Each of the chosen $n = 7$ possible substituent is labeled by an integer number between 1 and n. The chosen substituents are H, Br, Cl, OCH_3, $C(CO)OCH_3$, $N(CH_3)_2$, and NO_2. Identical individuals, with all hydrogens (11111111), made up the first generation. New generations are obtained via a combination of crossover and mutation operations. In particular, the crossover-1-point operator is employed which selects a random cut point [49] and combines the first portion of one parent with the second portion of the other parent to produce two children. Genetic diversity is introduced by random variations or mutations [50]. Different fitness functions are investigated. The total fitness function for the generic α individual $(1 \leq \alpha \leq M)$ is defined as the sum of partial fitness functions: $f^{(\alpha)} = \sum_i c_i f_i\left[x_i^{(\alpha)}\right]$ where $f_i\left[x_i^{(\alpha)}\right]$ is the i-th partial fitness (normalized for simplicity in the range $[0, 1]$) defined with respect to the calculated property $x_i^{(\alpha)}$, c_i is the weight for the i-th fitness. The condition $\sum_i c_i = 1$ is imposed, which implies $0 \leq f^{(\alpha)} \leq 1$.

Different selection criteria based are tested, some of them chosen to clarify methodological aspects. The first main criterion is related to ΔE_{LUMO} and it has a direct bearing on the specific problematic of BHJ systems. Next, we consider an additional physical molecular property to be optimized together with ΔE_{LUMO}: for sake of simplicity, we choose the permanent overall electric dipole moment (see below). Finally, we consider a relevant requirement for real-life molecular design, i.e., possible synthetic difficulties. Therefore, we consider an additional constraint which imposes that the substituents must be placed symmetrically with respect to the in-plane molecular axes, which makes the resulting structure considerably easier to synthesize. All together, three cases are considered in this preliminary study:

4.1 Case I

The fitness function depends only on the ΔE_{LUMO} energy gap; the empirical fitness is defined as $f(\Delta E_{\mathrm{LUMO}}) = (1.225 - 0.75\Delta E_{\mathrm{LUMO}})/\mathrm{eV}$ for $0.3 \leq \Delta E_{\mathrm{LUMO}}/\mathrm{eV} \leq 1.63$ and $f(\Delta E_{\mathrm{LUMO}}) = 0$ otherwise. This choice is suggested by experimental data [29]. Below 0.3 eV the transfer from donor to acceptor is no more competitive with respect to other possible processes. The energy of $E_{\mathrm{LUMO}-D}$ for the P3HT polymer is set to −2.7 eV [29]. We have tested four different couples of crossover/mutation probability p_c, p_m equal to (0.8, 0.05) [44], (0.95, 0.01), (0.95, 0.05), and (0.95, 0.025), with and without the presence of elitism.

4.2 Case II

Same as Case I, with symmetry conditions imposed, $R_1 = R_8$, $R_2 = R_7$, $R_3 = R_6$, and $R_4 = R_5$ that are addressed to counter possible actual problems in the synthesis of non-symmetric compounds.

4.3 Case III

The fitness function is taken as $f = [f_1(\Delta E_{\mathrm{LUMO}}) + f_2(\mu)]/2$ where $f_1(\Delta E_{\mathrm{LUMO}})$ is the same as in Case I, and $f_2(\mu) = 0.1\mu/\mathrm{D}$ for $0 \leq \mu/\mathrm{D} \leq 10$ and 0 otherwise, where μ is the total electric dipole moment. Symmetry is imposed. In practice, we look in this case for symmetric molecules with

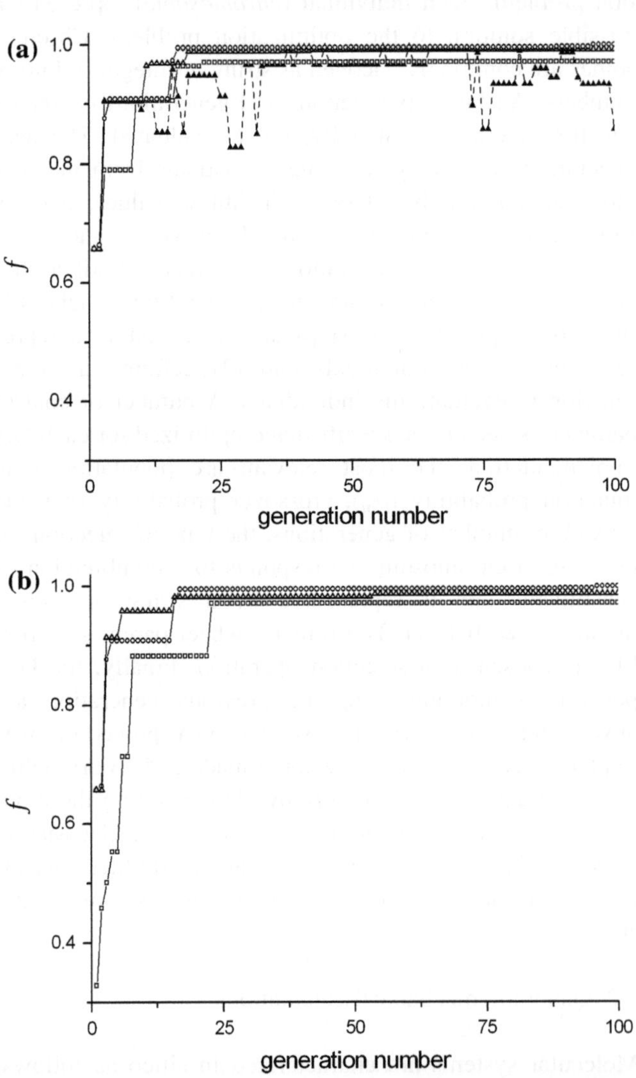

Fig. 2 **a** Evolution of maximum fitness (see text) in Case I for different choices of crossover and mutation probabilities: with elitism and p_c, p_mequal to 0.95, 0.025 (*triangles*), 0.95, 0.01 (*squares*), 0.95, 0.05 (*circles*); without elitism andp_c, p_mequal to 0.95, 0.025 (*filled triangles*) and **b** evolution of maximum fitness with optimized p_c, p_mequal to 0.95, 0.05, with elitism, for Cases I (*circles*), II (*triangles*), and III (*squares*)

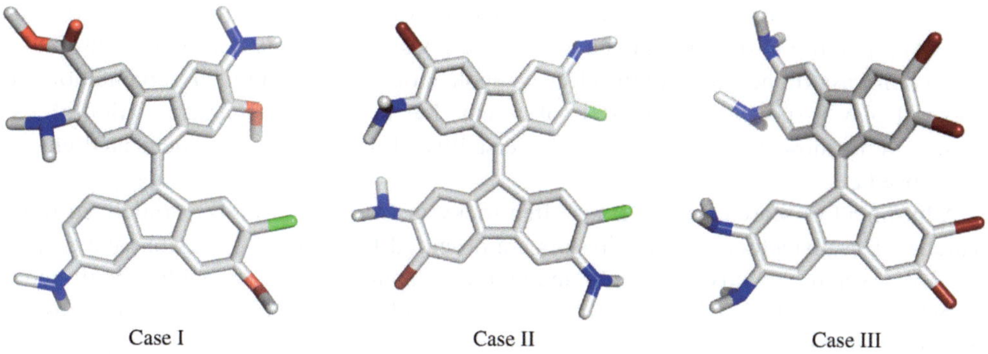

Fig. 3 Best performing selected molecular structures for the studied cases. (*Color code*: *gray* = C, *light red* = O, *dark red* = Br, *green* = Cl, and *blue* = N). H atoms are omitted for clarity

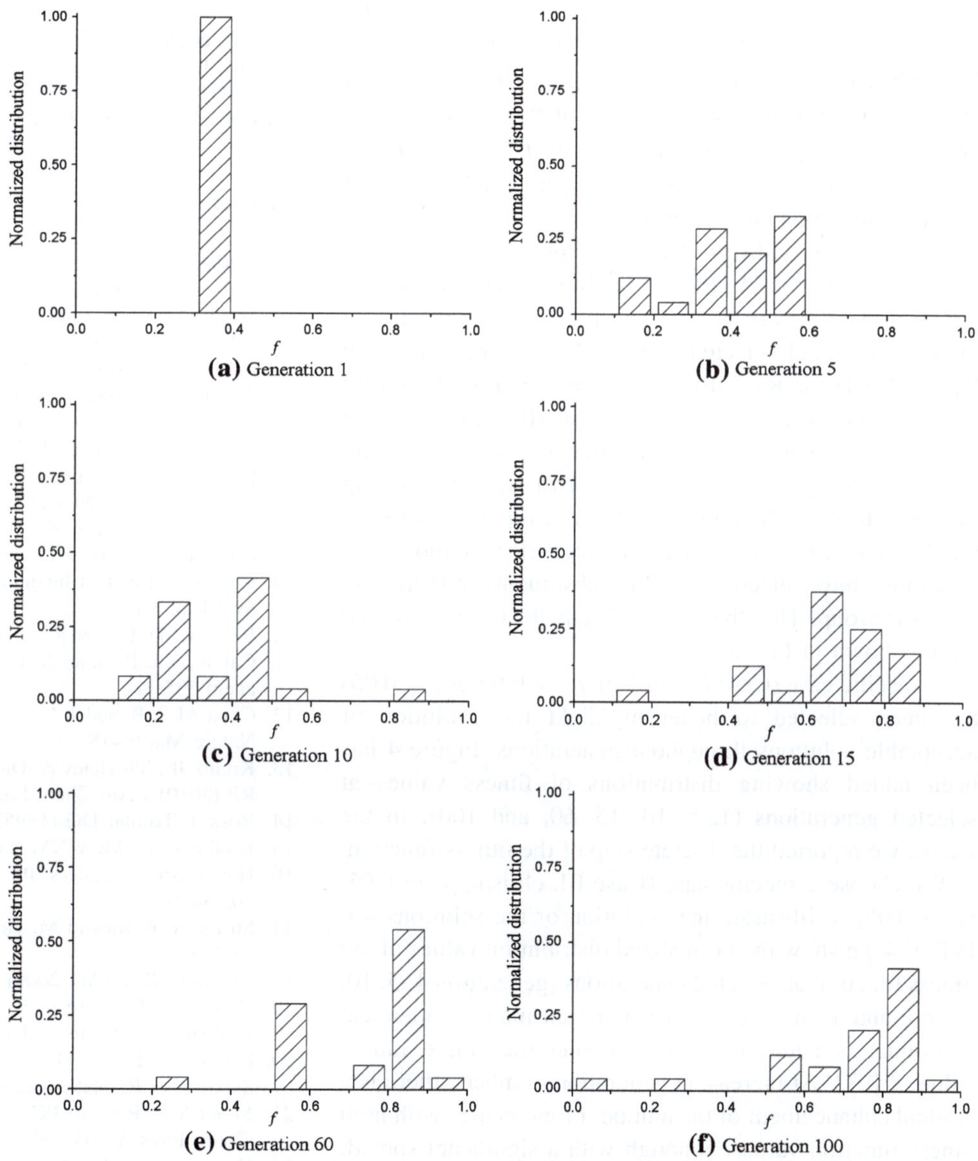

Fig. 4 Normalized distribution of the individuals at different generations for Case III with elitism and p_c, p_m equal to 0.95, 0.05, respectively. **a** Generation 1. **b** Generation 5. **c** Generation 10. **d** Generation 15. **e** Generation 60. **f** Generation 100

a dipole moment close to 10 D (an arbitrarily chosen value) and ΔE_{LUMO} close to 0.3 eV. In contrast with the constrain adopted in Case II, the choice of the electric dipole moment does not play a specific role in BHJ solar cells, but it is illustrative to show the capability and the versatility of the GA procedure.

Adjustment of search parameters is often tricky in GAs searches. Unsatisfactory searches' results are observed for low crossover probability ($p_c = 0.8$) and/or low mutation probability ($p_c = 0.95$ and $p_m = 0.01$). In addition, a high crossover probability ($p_c = 0.95$) is favored. The best results are obtained for $p_c = 0.95$ and $p_m = 0.05$, leading to a fitness function value of 0.9994. Elitism is shown to improve significantly the performance.

The "best" molecular structure in Case I is given by $R_{1,3,7} = N(CH_3)_2$, $R_2 = C(CO)OCH_3$, $R_{4,6} = OCH_3$, $R_5 = Cl$, and $R_8 = H$ with ΔE_{LUMO} equal to 0.3 eV. The GA optimization performance is illustrated in Fig. 2a showing the largest fitness f per generation, $f = \max\left[f^{(1)}, f^{(2)}, \ldots f^{(M)}\right]$, against the generation number for different combinations of crossover/mutation probabilities.

Next, Case II and Case III are explored, and results are summarized in Fig. 2b. Elitism is always included and the set of probabilities $p_c = 0.95$ and $p_m = 0.05$ is used. The "best" individual selected in Case II is characterized by $R_{1,3} = N(CH_3)_2$, $R_2 = Br$, and $R_4 = Cl$. The optimized ΔE_{LUMO} energy gap is 0.32 eV. Case III converges to a molecular structure with substituents $R_{1,2} = Br$ and $R_{3,4} = N(CH_3)_2$. In this case, the ΔE_{LUMO} energy gap increases to 0.38 eV, and the optimized dipole moment is 9.2 D. It is worth to notice that the selected molecular structures have electron donor substituents rather than acceptor group. The "best" individuals for Cases I, II, and III are shown in Fig. 3.

A specific case (Case III, elitism, $p_c = 0.95$, $p_m = 0.05$) has been selected to better highlight the evolution of acceptable solutions throughout generations. Figure 4 has been added showing distributions of fitness values at selected generations (1, 5, 10, 15, 60, and 100). In the x-axis, we reported the discrete step of the fitness function.

We choose a specific case (Case III, elitism, $p_c = 0.95$, $p_m = 0.05$) to illustrate the evolution of the solutions set. In Fig. 4 we show the normalized distribution values of the fitness function at selected generations (generation 1, 5, 10, 15, 60, and 100). The initial generation made of identical individual is naturally peaked around the single fitness value of 0.33. At increasing generation numbers, there is a gradual enhancement of the number of molecules with high fitness function values, although with a significant spread.

In conclusion, the present preliminary study confirms that the combination of heuristic search and in silico characterization may prove of help in selecting classes of molecules with desired structural electronic features, possibly contributing useful suggestions to synthetic efforts, at a minimal computational cost. Notice that in our case the full characterization of a single generation takes one CPU hour on a dedicated cluster; $M = 24$ nodes (four CPU per node) are employed, since the whole procedure is embarrassingly parallelized (one node per individual). The protocol is flexible and easily extendable to diverse molecular systems with specific requirements.

Further explorations are certainly required. The synthesis of the best performing individual obtained in Case II (cfr. Fig. 2a) has already been completed. Inclusion of the selected optimized acceptor molecule in BHJ devices is planned and will be thoroughly tested.

Acknowledgments This work is supported by Ministero dell'Istruzione, Università e Ricerca (MIUR), grant 2008J9RNB3 PRIN 2008 TIME and by the University of Padova, grant STPD08RCX5 "Progetto Strategico" HELIOS.

References

1. Holland J (1975) Adaptation in natural and artificial systems. University of Michigan Press, Ann Arbor
2. Liu J, Zhong W, Jiao L (2010) IEEE trans on system, man, and cybernetics-part B. Cybernetics 40:229
3. Arora V, Bakhshi AK (2010) Chem Phys 373:307
4. Assion A, Baumert T, Bergt M, Brixner T, Kiefer B, Seyfried V, Strehler M (1998) Gerber 282:919–922
5. Clark DE, Westhead R (1996) J Comput-Aided Mol Des 10:337
6. Parrill AL (1996) Drug Discov Today 1:516
7. Chen JH, Le SJ, Maizel JV (2000) Nucleic Acids Res 28:991
8. Fuhrmann J, Rurainski A, Lenhof HP, Neumann D (2010) J Comput Chem 31:1911
9. Chen CM, Lee C, Chuang CL, Wang CC, Shieh GS (2010) BMC Syst Biol 4:16
10. Bungay SD, Poirier RA, Charron RJ (2000) J Math Chem 28:389
11. Cai W, Xia B, Shao X, Guo Q, Maigret B, Pan Z (2001) Theochem 535:115
12. Chua ALS, Benedek NA, Chen L, Finnis MW, Sutton AP (2010) Nature Mat 9:418
13. Kreutz JE, Shukhaev A, Du W, Druskin S, Daugulis O, Ismagilov RF (2010) J Am Chem Soc 132:3128
14. Rossi I, Truhlar DG (1995) Chem Phys Lett 233:231
15. Brothers EN, Merz KM (2002) J Phys Chem B 106:2779
16. Hutter MC, Reimers JR, Hush MS (1998) J Phys Chem B 102:8080
17. Strassner T, Busold M, Herrmann WA (2002) J Comput Chem 23:282
18. Cundari TR, Fu W (2000) Inorg Chim Acta 300:113
19. Hunger J, Beyreuther S, Huttner G, Allinger K, Radelof U, Zsolnai L (1998) Eur J Inorg Chem 6:693
20. Pavlov T, Todov M, Stoyanova G, Schmieder P, Aladjov H, Serafimova R, Mekenyan O (2007) J Chem Inf Model 47:851
21. Moret MA, Pascutti PG, Bisch PM, Mundim MSP, Mundim KC (2006) Phys A 363:260
22. Hermann F, Suhai S (1995) J Comput Chem 16:1434
23. Alexandrova AN, Boldyrev AI (2005) J Chem Theory Comput 1:566
24. Arora V, Bakhshi AK (2010) Ind J Chem 49A:18

25. Carstensen NO, Dieterich JM, Hartke B (2011) Phys Chem Chem Phys 13:2903
26. Zeng G, Chua SJ, Huang W (2002) Thin Solid Films 417:194
27. Brunetti FG, Gong X, Tong M, Heeger AJ, Wudl F (2010) Angew Chem Int Ed 49:532
28. Yu G, Gao J, Hummelen JC, Wudl F, Heeger AJ (1995) Science 270:1789
29. Xiao S, Stuart AC, Liu S, Zhou H, You W (2010) Adv Funct Mater 20:635
30. Koster LJA, Mihailetchi VD, Blom PWM (2006) Appl Phys Lett 88:093511
31. Morvillo P, Bobeico E (2008) Phys Stat Sol (RRL) 6:260
32. Kooistra FB, Knol J, Kastenberg F, Popescu LM, Verhees WJH, Kroon JM, Hummelen JC (2007) Org Lett 9:551
33. Scharber MC, Mühlbacher D, Koppe M, Denk P, Waldauf C, Heeger AJ, Brabec CJ (2006) Adv Mater 18:789
34. Lenes M, Shelton SW, Sieval AB, Kronholm DF, Hummelen JC, Blom PWM (2009) Adv Funct Mater 19:3002
35. Brabec CJ, Cravino A, Meissner D, Sariciftci NS, Fromherz T, Rispens MT, Sanchez L, Hummelen JC (2001) Adv Funct Mater 11:374
36. Liu L, Shi Y, Yang Y (2001) Adv Funct Mater 11:420
37. Gadisa A, Svensson M, Andersson MR, Inganäs O (2004) Appl Phys Lett 84:1609
38. Frohne H, Shaheen SE, Brabec CJ, Müller DC, Sariciftci NS, Meerholz K (2002) ChemPhysChem 3:795
39. Morvillo P, Bobeico E (2008) Sol Energy Mater Sol Cells 92:1192
40. Dennler G, Scharber MC, Brabec CJ (2009) Solar Cells Adv Mater 21:1323
41. Kroon R, Lenes M, Hummelen JC, Blom PWM, De Boer B (2008) Polymer Rev 48:531
42. Cao D, Zhang X, Wang W (2009) Appl Surf Sci 255:5775
43. Majumdar J, Bhunia AK (2007) Eur J Oper Res 177:684
44. Chen BS, Chen PW (2010) Gene Reg Syst Biol 4:35
45. Gaussian 09, Revision B.01, Frisch MJ, Trucks GW, Schlegel HB, Scuseria GE, Robb MA, Cheeseman JR, Montgomery JA Jr, Vreven T, Kudin KN, Burant JC, Millam JM, Iyengar SS, Tomasi J, Barone V, Mennucci B, Cossi M, Scalmani G, Rega N, Petersson GA, Nakatsuji H, Hada M, Ehara M, Toyota K, Fukuda R, Hasegawa J, Ishida M, Nakajima T, Honda Y, Kitao O, Nakai H, Klene M, Li X, Knox JE, Hratchian HP, Cross JB, Bakken V, Adamo C, Jaramillo J, Gomperts R, Stratmann RE, Yazyev O, Austin AJ, Cammi R, Pomelli C, Ochterski JW, Ayala PY, Morokuma K, Voth GA, Salvador P, Dannenberg JJ, Zakrzewski VG, Dapprich S, Daniels AD, Strain MC, Farkas O, Malick DK, Rabuck AD, Raghavachari K, Foresman JB, Ortiz JV, Cui Q, Baboul AG, Clifford S, Cioslowski J, Stefanov BB, Liu G, Liashenko A, Piskorz P, Komaromi I, Martin RL, Fox DJ, Keith T, Al-Laham MA, Peng CY, Nanayakkara A, Challacombe M, Gill PAW, Johnson B, Chen W, Wong MW, Gonzalez C, Pople JA, Gaussian, Inc., Pittsburgh PA, 2002
46. Becke AD (1988) Phys Rev A 38:3098
47. Lee C, Yang W, Parr RG (1988) Phys Rev B 37:785
48. Zhang G, Musgrave CB (2007) J Phys Chem A 111:1554
49. Kellegöz T, Toklu B, Wilson J (2008) Appl Math Comput 199:590
50. Chakraborty B, Chaudhuri P (2003) Austrian J Stat 32:13

Theor Chem Acc (2012) 131:1195
DOI 10.1007/s00214-012-1195-x

REGULAR ARTICLE

The first solvation shell of Reichardt's dye in ionic liquids: a semiempirical study

Cinzia Chiappe · Christian Silvio Pomelli

Received: 15 July 2011 / Accepted: 16 February 2012 / Published online: 2 March 2012

Abstract The nature of solvation of Reichardt's dye in ionic liquids is investigated through semiempirical calculations. We build on the basis of electrostatic considerations a cluster of three ionic pairs that represent the first solvation shell of the dye. The spatial organization of this shell is a balance between sterical, electrostatic and, in the case of functionalized ionic liquids, specific interactions. This model is not sufficient to obtain values of ET(30) quantitatively comparable to the experimental ones, but some qualitative features can be rationalized. The resulting scenario of solute–solvent interactions is different with respect to molecular solvents. Thus, we suggest caution in comparing ionic and molecular solvents through solvatochromic scales.

Keywords Ionic liquids · Solvatochromic probes · PM6 · ZINDO · Reichardt's dye · CIS

1 Introduction

Recent years have seen a growing experimental focus on a novel class of solvents known as room-temperature ionic liquids (ILs). Generally constituted by an organic cation and a polyatomic anion, these salts liquid at or near room temperature have been found capable of solvating a wide range of organic, inorganic and polymeric compounds. The dissolution ability, however, strongly depends on cation and anion structure and the nature of the solute–solvent interactions responsible for their solvation properties remain one of the central themes in the ionic liquid research [1, 2]. Normally, the most important quantity determining the choice of a solvent is its polarity. In molecular liquids, polarity may be qualitatively understood by examination of the electrostatic dipole moment and hydrogen-bonding properties of individual solvent molecules. On the basis of these elements, it is possible to predict the solubility of a substrate in a selected solvent and the effects of the solvent on some reaction parameters, such as reaction rate. But, by definition, ions carry a nonzero electrostatic monopole moment (charge) that should dominate dipolar and higher-order terms. Furthermore, cation–anion interactions lead to considerable charge ordering in ILs, producing in particular in ILs bearing sufficiently long alkyl chains on cation or anion, an inhomogeneous environment characterized by polar and not polar regions. The unique nature of this environment makes often the application of models and concepts developed for molecular solvents suspect or misleading. Understanding polarity of ILs and their solvent power, therefore, may require a careful rethinking of the fundamental nature of solvation.

An experimental approach to evaluate such interactions and define polarity is the examination of the behavior of a chromophore in the medium of interest. Spectrophotometric studies of chromophores have been carried out both in molecular solvents and ILs showing that the solvatochromic properties of these latter media depend on the structure of cation and anion and their combination [3–8]. The effects due to a specific cation

Dedicated to Professor Vincenzo Barone and published as part of the special collection of articles celebrating his 60th birthday.

Electronic supplementary material The online version of this article (doi:10.1007/s00214-012-1195-x) contains supplementary material, which is available to authorized users.

C. Chiappe · C. S. Pomelli (✉)
Dipartimento di Chimica e Chimica Industriale,
Via Risorgimento 35, 56126 Pisa, Italy
e-mail: cris@dcci.unipi.it

are significantly affected by the counterion. Creating cation–anion interactions that compete with specific cation–solute or anion–solute interactions can control the polarity of the resulting IL.

To date, there have been relatively few simulation studies of IL systems, and most of these have centered on the structural, transport and thermodynamic properties of ILs. However, in 2004, a molecular dynamics simulation study was carried out to investigate the nature of solvent polarity in an IL system calculating the absorption spectrum of the Reichardt's dye 30 in 1-butyl-3-methylimidazolium hexafluorophosphate, [bmim][PF_6], and determining the solute–solvent interactions responsible for the solvatochromic shift [9]. This spectroscopic shift is a well-known indicator for polarity that has been widely applied to define IL polarity.

Here, we report a computational study using semiempirical methods to characterize the first solvation shell of some ILs. In a recent paper [10], we have reported a large set (about 60) of measures of ET(30) [11] and Kamlet Taft solvatochromic parameters [12] for different kinds of ILs. From this study emerges that:

1. as expected, ET(30) and α values are linearly correlated;
2. the range in which π^* values spreads is very small;
3. the values of β depend principally by the anion where ET(30) and α values depend on the cation;
4. the ET(30) and α values are strongly anion dependent. More coordinating anions, such as dicyanamide, are able to reduce the hydrogen bond ability of a specific cation.

Therefore, we decided to study computationally a subset of ILs that contain this anion and cations where the charge is carried out by a quaternary nitrogen enclosed in an aliphatic heterocyclic ring, for which experimental measurements suggested the presence of peculiar structural effects. Since the alkyl chain length on nitrogen and the heterocyclic ring size appeared to affect significantly the ET(30) and α values, we have included in this computation study different kind of rings and sidechains. Moreover, some sidechains presents one or more alcoholic moieties; the presence of this group significantly increases IL polarity increasing ET(30) and α values. The dicyanamide ILs shows a very narrow range of β values; thus, ET(30) (or α) is the only index that can discriminate between these ILs. The interactions between the IL components and the Reichadts dye, which determine the ET(30) and α values, are, therefore, fundamental for the rationalization of the solvation ability of these ILs.

Results arising from this investigation show that the solvation shell of Reichardt dye in ILs is the result of interplay between chemical, sterical and electrostatic factors.

2 The method and the model system

The remarkable size of the Reichardt's dye 30 is a challenge to computational methods. If the system do not consist of the dye only but include also some solvent molecules, the computational costs become prohibitive. Also, a RISM-KH description of ILs [13] is too computationally expensive. The MD-based study of Kobrak and Znamenskiy [9] give a nice picture of these kinds of systems both at local and global level. We refer to some results of this study when discuss our results. The use of semiempirical methods to study specific and bulk effect of solvent on Reichardt's dye is reported in a paper by Caricato et al. [14]. In this paper, the geometry of ET(30)-solvent clusters has been optimized using AM1 [15] semiempirical method followed by a ZINDO [16] calculation of the excited state. The aim of this paper is to describe specific effect of the solvent ions on the ET(30) solute, and thus, we prefer the semiempirical protocol rather than the MD one.

We decided to use the PM6 method, introduced by Steward in 2007 [17] and documented by the same author in a series of applicative papers [18, 19]. PM6 is one of the most recent semiempirical parameterizations of general use and in fact replace AM1 and PM3 [20]. PM6 is parameterized for a large number of chemical elements. This can be useful for future extensions to a larger set of ILs. We considered also DFTB [21] in its DFTB+ [22] implementation, but this method, at the present, is parameterized for fundamental organic elements and few metals. Furthermore, DFTB+ package does not allow performing excited states calculations. PM6 has been well received in the computational chemistry community (more than 200 papers reporting PM6 results in 2010 only). For this study, we used the implementation of PM6 in the Gaussian 09 package [23]. This implementation provides an interface with the CIS method for excited states [24]. We use both CIS/PM6 and ZINDO at PM6 geometries for the calculation of ET(30). We try also to perform some TDDFT calculations on the PM6 optimized geometries, but the hardware and CPU time requirement are beyond our computational resources. Finally, we considered also the use of IEFPCM [25] to estimate the bulk effect, but the values of dielectric constants of ILs are still a debated question [26, 27]. Some test calculations with an arbitrary dielectric constant of 20 lead to very chaotic behaviors, significantly worse than those in vacuo. Thus, we decided to avoid performing exhaustive IEFPCM calculations.

We adopt the supermolecular approach that has been already used to describe ILs features as reactivity [28, 29], solubility of small molecules [30] and ionic pairing [31]. The construction of the first shell is limited to the zones of the space that are immediately adjacent to the N and O atoms of the dye. For reasons of simplicity, we assume that they lie on the horizontal plane with an upper and lower side. The spatial arrangement of phenyl groups around the N atom leads to two narrow and equivalent holes one on the upper and the other on the lower side of the dye. The spatial accessibility of this moiety is limited. Since the N atom carries a positive charge, we place one anion (in this context dicyanamide only) near each of the holes. Two cations are placed nearby the oxygen atom. Another anion is inserted between the two cations. A third cation is inserted, to reach the electroneutrality near the third anion. This arrangement, which presents a strictly alternating of positive and negative centers, is schematized in Fig. 1.

When we tried to optimize the geometries using the standard Berny [32] algorithm of Gaussian 09, we had some troubles. This may be attributed to the fact that the system is very complex: there are seven not bounded groups of atoms and thus 36 intermolecular degrees of freedom. In fact, a consistent number of Berny optimizations did not converge showing a chaotic behavior. This is typical of complex system that reaches an energy plateau in the phase space. We decided to fix this issue by restarting from the point with the smallest gradient RMS using the GDIIS [33] algorithm. After this treatment, the structure converged in few steps. This does not introduce artifacts in the final obtained geometry since the structure acceptance criterion is the same. All the structures have been characterized as local minimums by frequency calculations. However, in these complex systems, there is not an absolute certain to have reached the global minimum. The final structures obtained for the different systems studied show, looking to geometrical parameters and by visual inspection, a degree of similarity and reasonability sufficient to ensure to have obtained, at least, snapshots representative of the real systems. On the optimized structures have been performed CIS/PM6 and ZINDO calculations of excited states. The number of states being optimized is 10.

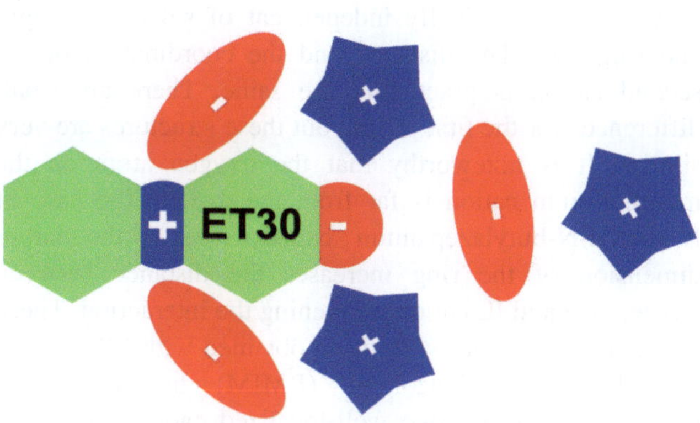

Fig. 1 Schematic representation of the first solvation shell of Reichardt's dye in a ionic liquid. The phenyl wings surround the zwitterionic part of the dye. The positive center (nitrogen, in *blue*) is sterically hindered. Nearby the nitrogen we found two anions on the *top* and *bottom* side. Two cations coordinate the oxygen. A further ionic pair is introduced in this region: the anion shields the two coordinating cations, and the cation determines the electroneutrality

3 Results and discussion

The results are arranged in Tables 1 and 2. In Table 1 are reported some geometrical quantities that define the geometry of the ionic clusters here considered. In Table 2 are reported the ET(30) values and the variation of HOMO and LUMO energies with respect the isolated dye. Some explicative molecular graphics are reported in Figs. 2, 3 and 4.

First, we analyze the geometrical quantities reported in Table 1. In particular, we start analyzing IL cation–dye interactions.

In this investigation, we have considered four aliphatic cations (pyrrolidinium, piperidinium, morpholinium and azepanium) characterized by localized quaternary nitrogen charged center (Scheme 1) and bearing a methyl group and an alkyl chain on nitrogen or a methyl group and a hydroxylethyl or a glyceryl group associated with the dicyanamide anion. In these systems, the center of the positive charge cannot interact directly with the dye oxygen; the interaction occurs through some hydrogen atoms. In Table 1, we report all the hydrogen atoms of cations that are <2.5 Å apart from the dye oxygen. Depending on cation structure, these hydrogens can be those of the ring in α positions with respect to the charged nitrogen (labeled as R), on the aliphatic chains (labeled by the number corresponding to the chain length 1, 2, 4, 8 and with the position α, β) or on the alcoholic moiety (OH) for 1,e and 1,g cations (in this latter case, we label as "e" the OH group nearest to the ring. However, we did not observe coordination by the further OH).

If the cation is functionalized with one or two hydroxyl groups on the alkyl chain, only the OH group closer to the ring is coordinated to the oxygen atom of Reichardt's dye. Furthermore, in this coordination is also involved one the four hydrogens of the ring located in the α positions. The coordination of the two cations is very similar. In the case of N-methyl-N-glycerylpiperidinium, $Pip_{1,g}$, reported in Fig. 2, there is no direct interaction between the terminal OH group and the dye oxygen. Probably, a chelation of the dye oxygen with two OH groups of the same cation takes away the second cation weakening the electrostatic interaction of this last one with the dye positive center. In this case, the electrostatic interaction appears to prevail on chelating effect. The terminal OH is thus free to coordinate other cations.

The pattern of interactions of non-functionalized cations is more complex. In this case, there is a neat distinction

Table 1 Geometrical quantities

Cation	N-A1	N-A2	O-C1	O-C2
$Pyr_{1,2}$	3.093(T)	3.207(T)	1.999(1), 2.099(R), 2.375(2β)	2.018(R)
$Mor_{1,e}$	3.179(T)	3.220(T)	1.751(e), 2344(R)	1.774(e), 1.988(R)
$Mor_{1,2}$	3.097(T)	3.283(T)	1.884(R), 2.093(1), 2.102(2β)	1.987(R)
$Mor_{1,4}$	3.117(T)	3.282(T)	1.934(R), 2.101(1)	2.034(R)
$Mor_{1,8}$	3.094(T)	3.309(T)	1.905(R), 2.162(1), 2.287(8α)	2.020(R)
$Pip_{1,e}$	3.174(T)	3.175(T)	1.776(e), 2168(R)	1.723(e), 2.312(R)
$Pip_{1,g}$	3.149(T)	3.234(T)	1.704(e)	1.735(e), 1.987(R)
$Aze_{1,4}$	3.234(T)	3.296(T)	2.207(R), 2.239(4β)	2.596(R)
$Aze_{1,e}$	3.131(T)	3.155(T)	1.788(e), 2.245(R)	1.724(e), 2.469(R)

All the data are expressed in Å

N, O Reichardt's dye heteroatoms, *A1, A2* nearest and next nearest anion, *C1, C2* nearest and next nearest cation, *T* terminal coordination of the anion, *R* coordination at hydrogen on the ring, *1, 2, 8* coordination at hydrogen on the sidechain (with the relative position), *e* coordination with alcoholic hydrogen

Table 2 Experimental and theoretical results for ET(30) parameters for the ionic liquids here considered

Cation	ET(30) exp[a]	ET(30) ZINDO	ET(30) CIS/PM6	ΔHOMO	ΔLUMO
In vacuo	27.10	26.75	38.77	0.00	0.00
$Pyr_{1,2}$	48.70	59.97	62.41	−17.24	17.21
$Mor_{1,e}$	55.80	61.61	68.53	−27.98	11.70
$Mor_{1,2}$	50.30	59.93	65.25	−21.87	15.32
$Mor_{1,4}$	49.50	55.33	66.97	−17.65	15.82
$Mor_{1,8}$	49.10	58.20	67.71	−17.50	16.91
$Pip_{1,e}$	55.65	60.61	68.04	−21.22	16.06
$Pip_{1,g}$	57.00	60.62	67.68	−19.76	15.72
$Aze_{1,4}$	54.50	53.01	64.17	−4.13	23.40
$Aze_{1,e}$	57.44	60.50	68.32	−20.39	17.09

The cation structures are defined in Scheme 1. All quantities are expressed in kcal/mol

[a] Values from Chiappe et. al [10] except from in vacuo value extrapolated reported by Reichardt [34, 35]

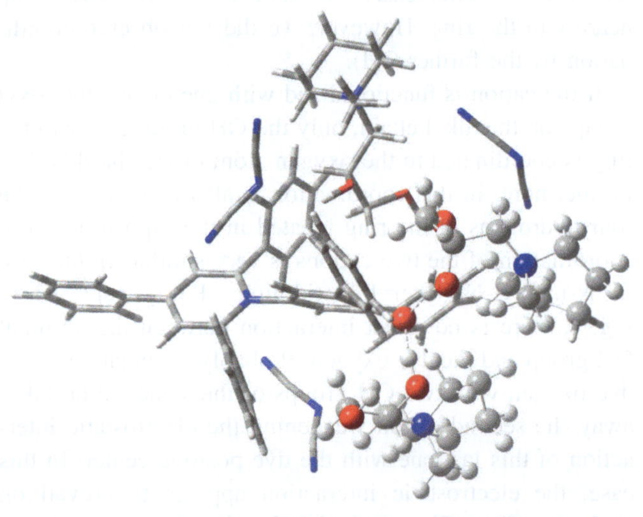

Fig. 2 Coordination of the Pip_{1e} cations and dicyanamide anions with Reichardt's dye. The use of two different representations is for visual purposes only

between a first cation closer to the dye and the second one. The behavior of pyrrolidinium and morpholiniun cations is very similar, practically independent of sidechain length and ring size. The distance and the coordination of the second cation is practically the same. There are small differences for the first cation, but these structures are very similar. It is noteworthy that the oxygen atom on the morpholinium cation is far from the dye. In the case of N-methyl-N-butylazepanium cation, $Aze_{1,4}$, the larger dimension of the ring increases the distance between oxygen dye and IL cation weakening the interaction. These results are consistent with those obtained with MD for the 1-butyl-3-methylimidazolium (BMIM) hexafluorophosphate [9]: there are two well-localized cations about the ET(30) oxygen differing only by orientation. The cations here considered have a different shape and some are functionalized. Furthermore, the methodology used is static and not dynamic. But the coordination number of the ET(30) oxygen is still two.

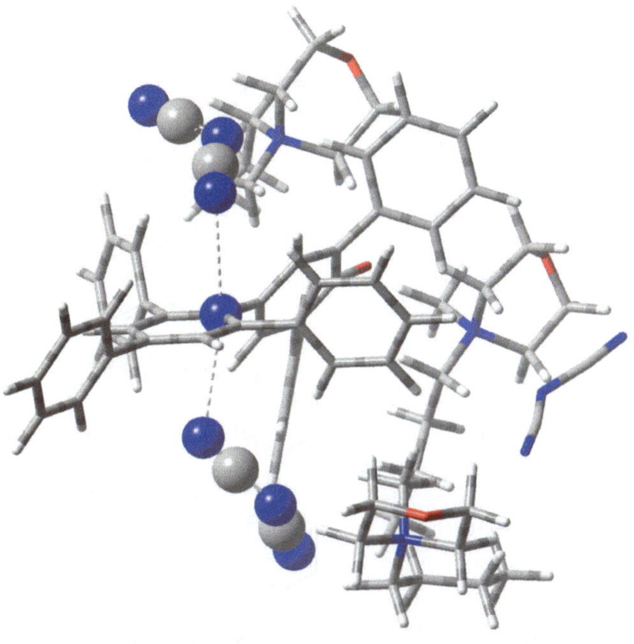

Fig. 3 Coordination of A1 and A2 with ET(30) in the presence of $Mor_{1,4}$ cations. The use of two different representations is for visual purposes only

The anion coordination is very similar in all examined situations. A pictorial example for N-methyl-N-ethylmorpholinium cation, $Mor_{1,2}$, is reported in Fig. 3. The dicyanamide ion is coordinated to the positive nitrogen center of the dye with one of its terminal nitrile nitrogens. This spatial arrangement leaves to the anion two degrees of freedom: it is able to rotate and vary its inclination (see Fig. 3). These two degrees of freedom are, however, partially limited by the surrounding phenyl groups. The anion exploits this feature to pair with the nearest cation that is one of those coordinated at the oxygen center. Consequently, there is an interaction between the species coordinated to the positive and negative sites. We can have a hint of this looking to the difference between the N-A1 and N-A2 distances in relation to the cations coordination patterns. When the two cations are similarly coordinated, as in the case of hydroxyl functionalized cations, N-A1 and N-A2 are very similar. The distances N-A1 and N-A2 increase as the size of the cation ring increases. The coordinating cations are subjected to two electrostatic forces: the ion pairing that reduces the cation–anion distance and the interaction with the dye oxygen that, for sterical reasons, tends to take away the cation from the anion. A small ring has more possibilities to rotate and move than a larger one. Thus, larger cations like $Aze_{1,4}$ are less able to optimize both the interactions. The PF6 anion studied by Kobrak et al. [9] is a spherical top and it is not possible for a direct comparison of the results as in the cation case.

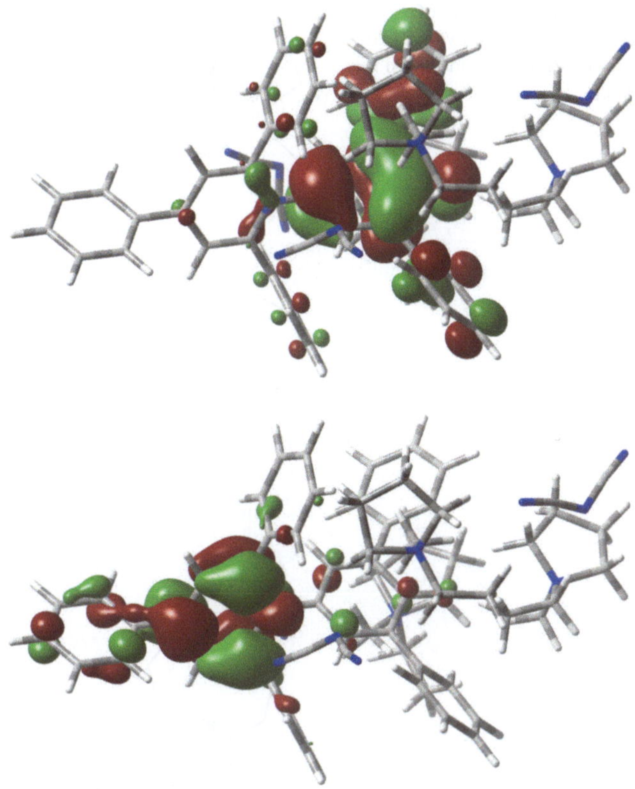

Fig. 4 HOMO (*top*) and LUMO (*bottom*) orbitals for Reichardt's dyes with $Pyr_{1,2}$ dicyanamide first shell of solvation

We perform calculation of the ET(30) solvatochromic parameters with CIS/PM6 and ZINDO methods at the PM6 optimized geometries. We discuss the result separately and compare later. These results are reported in Table 2 and, in graphical form in Fig. 5.

Values of the ET(30) parameter calculated at the CIS/PM6 level are largely overestimated with respect the experimental one. There are several motivations to rationalize this behavior, apart the level of calculation. First, we neglect completely the bulk effect. The model system here considered is a small but, in our intention, balanced network of a zwitterion interacting with six ions. These six ions in a real system are influenced by the bulk: other ions can move away or rotate the ions of the first solvation shell. Dynamics phenomena as sidechain conformational motions, ion exchange between first shell and the bulk are totally disregarded. However, some features can be observed. Experimental and calculated values of ET(30) for ILs with the OH moiety lie in a range of 1.3 kcal/mol regardless to size or the presence of heteroatoms on the ring. In the case of $Pyr_{1,2}$, experimental and calculated values of ET(30) are both the lower ones. The low experimental value for $Aze_{1,4}$ is in agreement with the low calculated value. This reduction is probably due to sterical hindering. The series of $Mor_{1,n}$ is different: ET(30)

Scheme 1 Structures of Reichardt's dye, dicyanamide and of cations considered in this paper

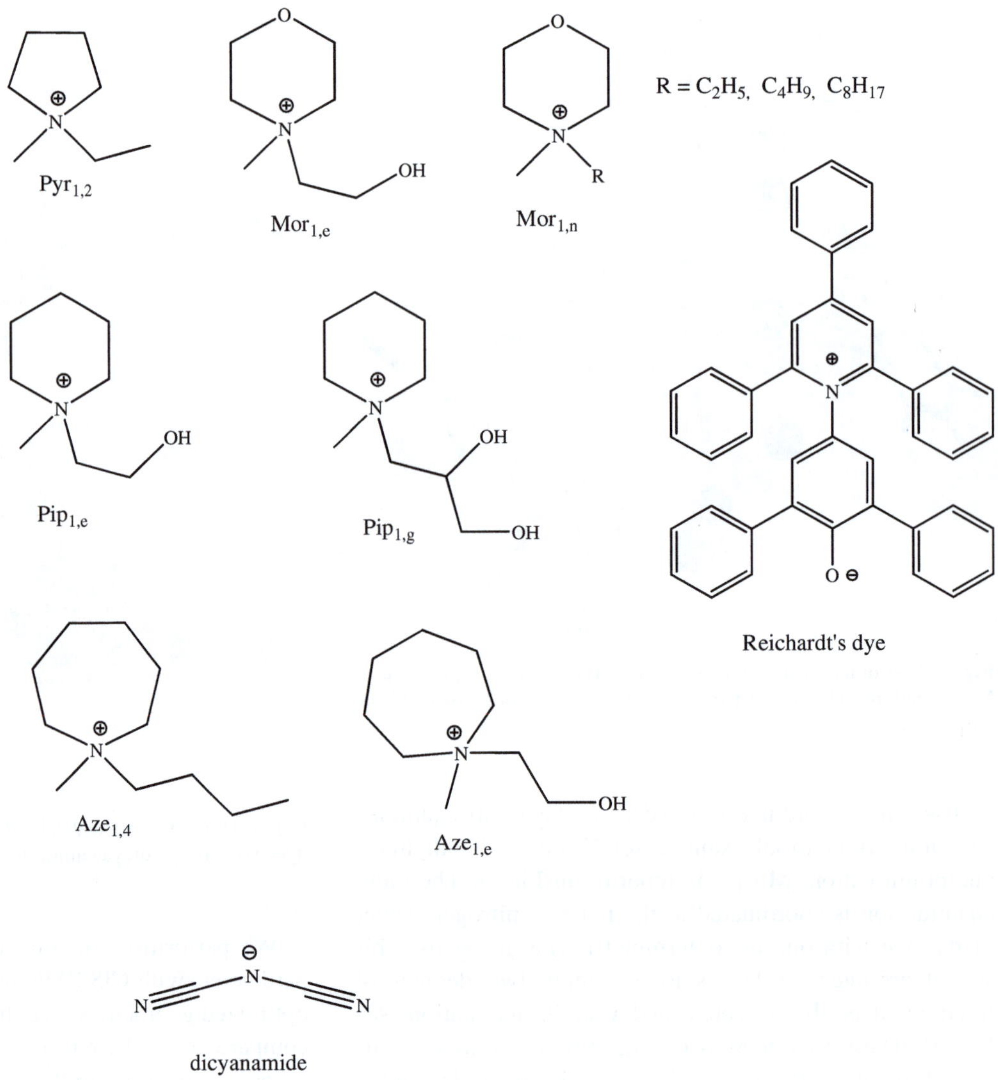

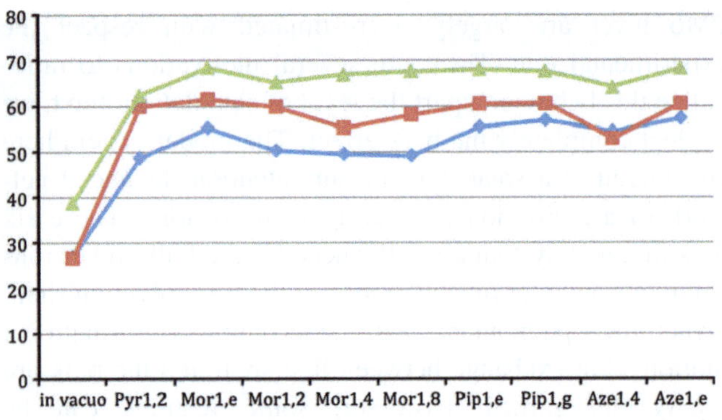

Fig. 5 Graphical representation of ET(30), in Kcal/mol, values obtained by experiment (*blue/diamonds*), CIS/PM6 (*green/triangles*) and ZINDO (*red/squares*)

experimental values decrease when the length of the side-chain increases, but the trend of calculated ones is opposite. Probably, in this case, dynamical and conformational factors cannot be disregarded.

ZINDO results are less overestimated and, in some cases, in good agreement with the experimental results. The value in vacuo is very similar to extrapolated value [34, 35], and thus, we can suppose that the not solvated system is better described by ZINDO than CIS/PM6. The value of $Mor_{1,4}$, obtained with ZINDO, is very different from the $Mor_{1,2}$ and $Mor_{1,8}$ in disagreement with the other two series of data. We have no explanation for this feature. In general, functionalized cations have a ZINDO calculated values that differ by 5 kcal/mol or less where non-functionalized ones differ of about 10 kcal/mol. An exception is $Aze_{1,4}$ value that lies in non-functionalized range. Kobrak [9] stated that in ILs, only the first and second shells are polarized by the Reichardt's dye. A possible interpretation is that in the case of functionalized or sterical hindering cations, the contribution of the second shell is less important, and our small supermolecular model is more similar to the real system. In general, CIS/PM6 values have a better trend but are less accurate than the ZINDO ones.

However, looking at Fig. 5, the presence of a well-defined trend (with the exception, again, of $Aze_{1,4}$) is evident. We decided to avoid a statistical analysis of ET(30) data due to the limited number of values and the heterogeneity of the ILs under investigation.

Table 2 contains also the differences of HOMO and LUMO values with respect the isolated dye. In Fig. 4, a graphical representation of these orbitals is reported. They are very similar to those reported by Caricato et al. [12]; HOMO/LUMO orbitals are stabilized/destabilized by the nearby positive/negative charges of cations/anions, as expected. Since the anion is always the same, the HOMO values are more differentiated than the LUMO ones. The stabilization of functionalized cations is larger, as expected, than the not functionalized ones. The anomaly represented by$Aze_{1,4}$ is reflected also in the orbital energies that show a consistent drift.

4 Conclusions

Solvation of Reichardt's dye in ILs is very different than in molecular solvents. The dimension of the ions, the fact that we added a 1:1 mixture of two very different chemical species, changes completely the scenario. The two extremes of the zwitterion *de facto* interact with different species, the anion and the cation. The balancing between sterical and electrostatic factors seems to be the key factor of ionic liquid–Reichardt's dye interaction. The presence of at least one hydroxyl group on the cation sidechain can overwield these factors. Given these considerations, ET(30) or similar solvatochromic parameters are useful to obtain information about the solvation ability of ILs. However, the agreement of the values here found for ILs with the experimental ones is at a qualitative level only, far from the today standards of computational spectroscopy. This is due to the low level of calculation and to the reduced size of the model system here used.

However, this rough scheme gives some information about the organization of the ionic liquid about the Reichardt's dye. This organization is very different from those of molecular solvents, and thus, a direct comparison can be misleading.

We plan to further extend both experimental and theoretical investigations to reach a better knowledge of the phenomena. Furthermore, a protocol to obtain reliable computational results for excited states of ET(30) needs to be established.

References

1. Wasserscheid P, Welton T (eds) (2007) Ionic liquids in synthesis. Wiley-VCH Weinheim, Germany
2. Chiappe C, Malvaldi M, Pomelli CS (2009) Pure Appl Chem 81(4):767–776
3. Carmichael AJ, Seddon KR (2000) J Phys Org Chem 13:591
4. Aki SNVK, Brennecke JF, Samanta A (2001) Chem Commun 413
5. Reichard C (2004) Pure Appl Chem 76:1903
6. Reichard C (2005) Green Chem 7:339
7. Crowhurst L, Mawdsley PR, Parez-Arlandis JM, Salter PA, Welton T (2001) Phys Chem Chem Phys 3:5192
8. Lee JM, Ruckes S, Prausnitz JM (2008) J Phys Chem B 112:1473
9. Znamenskiy V, Kobrak MN (2004) J Phys Chem B 108: 1072–1079
10. Chiappe C, Pomelli CS, Rajamani S (2011) J Phys Chem B 115(31):9653–9661
11. Dimroth K, Reichardt C, Siepmann T, Bohlmann F (1963) Liebigs Ann Chem 727:93
12. Kamlet MJ, Abboud JML, Taft RW (1981) Progr Phys Org Chem 13:485–630
13. Chiappe C, Malvaldi M, Pomelli CS (2010) J Chem Theory Comp 6:179–183
14. Caricato M, Mennucci B, Tomasi J (2006) Mol Phys 104: 875–887
15. Dewar MJS, Zoebisch EG, Healy EF (1985) J Am Chem Soc 107: 3902
16. Zerne MC (1991) Rev Comp Chem 2:313–366
17. Stewart JJP (2007) J Mol Model 13:1173–1213
18. Stewart JJP (2009) J Mol Model 15:765–805
19. Stewart JJP (2008) J Mol Model 14:499–535
20. Stewart JPP (1989) J Comp Chem 10:209–220
21. Frauenheim T, Seifert G, Elstner M, Hajnal Z, Jungnickel G, Porezag D, Suhai S, Scholz R (2000) Phys Stat Sol B 217:41–62
22. Aradi B, Hourahine B, Frauenheim Th (2007) J Phys Chem B 111(5678):2007
23. Frisch MJ, Trucks GW, Schlegel HB, Scuseria GE, Robb MA, Cheeseman JR, Scalmani G, Barone V, Mennucci B, Petersson GA, Nakatsuji H, Caricato M, Li X, Hratchian HP, Izmaylov AF, Bloino J, Zheng G, Sonnenberg JL, Hada M, Ehara M, Toyota K, Fukuda R, Hasegawa J, Ishida M, Nakajima T, Honda Y, Kitao O, Nakai H, Vreven T, Montgomery Jr. JA, Peralta JE, Ogliaro F, Bearpark M, Heyd JJ, Brothers E, Kudin KN, Staroverov VN, Kobayashi R, Normand J, Raghavachari K, Rendell A, Burant JC, Iyengar SS, Tomasi J, Cossi M, Rega N, Millam NJ, Klene M, Knox JE, Cross JB, Bakken V, Adamo C, Jaramillo J, Gomperts R, Stratmann RE, Yazyev O, Austin AJ, Cammi R, Pomelli C, Ochterski JW, Martin RL, Morokuma K, Zakrzewski VG, Voth GA, Salvador P, Dannenberg JJ, Dapprich S, Daniels AD, Farkas Ö, Foresman JB, Ortiz JV, Cioslowski J, Fox DJ (2009) Gaussian 09, revision A.1. Gaussian Inc., Wallingford
24. Foresman JB, Head-Gordon M, Pople JA, Frisch MJ (1992) J Phys Chem 96:135–149
25. Mennucci B, Cances E (1997) Tomasi. J Phys Chem B 101: 10506–10517
26. Mizoshiri M, Nagao T, Mizoguchi Y, Yao M (2010) J Chem Phys 132(164510):1–7
27. Huang MM, Jiang Y, Sasisanker P, Driver GW (2011) J Chem Eng Data 56:1494–1499
28. Bini R, Chiappe C, Pomelli CS, Parisi B (2009) J Org Chem 74:8522–8530
29. Bini R, Chiappe C, Mestre VL, Pomelli CS, Welton T (2009) Theor Chem Acc 123:347–352
30. Pomelli CS, Chiappe C, Vidis A, Laurenczy G, Dyson PJ (2007) J Phys Chem B 111:13014–13019
31. Hunt P, Kirchner B, Welton T (2010) Chem Eur J 12:6762–6775
32. Peng C, Ayala PY, Schlegel HB, Frisch MJ (1996) J Comp Chem 17:49–56
33. Császár P, Pulay P (1984) J Mol Struct 114:31–34
34. Reichardt C (2004) Solvent effects in organic chemistry, 3rd updated and enlarged edition. Wiley-VCH Weinheim, Germany
35. Lindner B (1992) J Phys Chem 96:10708–10712

Theor Chem Acc (2012) 131:1187
DOI 10.1007/s00214-012-1187-x

REGULAR ARTICLE

Comparison of the mechanism of deamination of 5,6-dihydro-5-methylcytosine with other cytosine derivatives

André Grand · Jean Cadet · Leif A. Eriksson · Vanessa Labet · Nelly L. Jorge · Maria L. Schreiber · Thierry Douki · Christophe Morell

Received: 8 July 2011 / Accepted: 8 December 2011 / Published online: 4 April 2012

Abstract The mechanism of deamination of 5,6-dihydro-5-methylcytosine has been investigated theoretically and compared to those of other cytosine derivatives. The main goal is to understand the effect of C5-methylation and C5–C6 saturation upon the deamination rate. It is found that C5–C6 saturation tends to increase the local electrophilicity of the cytosine derivative on carbon C4. It is also concluded that C5-methylation displays an opposite effect on saturated versus unsaturated systems: on unsaturated systems, C5-methylation tends to increase the local electrophilicity on C4, while it reduces the local electrophilicity on C4 for saturated ones.

Dedicated to Professor Vincenzo Barone and published as part of the special collection of articles celebrating his 60th birthday.

A. Grand · J. Cadet · T. Douki · C. Morell (✉)
Laboratoire Lésions des Acides Nucléiques,
INAC/SCIB-UMR-E no. 3 CEA-UJF, CEA Grenoble,
17 rue des Martyrs, 38054 Grenoble cedex 9, France
e-mail: christophe.morell@ujf-grenoble.fr

L. A. Eriksson
Department of Chemistry and Molecular Biology,
University of Gothenburg, 412 96 Gothenburg, Sweden

V. Labet
UPMC Univ Paris 06, UMR 7075, Laboratoire de Dynamique,
Interactions et Réactivité (LADIR), 75005 Paris, France

V. Labet
CNRS, UMR 7075, Laboratoire de Dynamique,
Interactions et Réactivité (LADIR), 75005 Paris, France

N. L. Jorge · M. L. Schreiber
Laboratorio de Investigaciones en Tecnología del Medio
Ambiente, FACENA-UNNE, Av. Libertad 5640,
3400 Corrientes, Argentina

Keywords Deamination · 5,6-Dihydro-5-methylcytosine · DNA damage · Mutation · DFT · Reactivity indices · Dual descriptor

1 Introduction

Damage to DNA is a major deleterious process in cells that can lead to lethality or modification of the genetic information. The latter process is of particular importance because it may represent through mutation the initial step toward malignant tumor formation. DNA damage arises as a result of exposure to a variety of exogenous agents such as ionizing and UV radiations, chemicals and alkylating agents. In addition, endogenous processes like mitochondrial respiration and inflammation lead to the release of genotoxic compounds including oxygen and nitrogen reactive species. This withstanding, water remains the most efficient endogenous damaging compound; spontaneous hydrolytic reactions of the *N*-glycosidic bonds are at the origin of the formation of abasic sites upon depurination and of the conversion of cytosine (**Cyt**) into uracil. In the latter reaction, water substitutes the exocyclic amino group of cytosine, giving rise to a tautomer of uracil. The biological relevance of deamination in genome stability is indirectly demonstrated by the very high uracil DNA *N*-glycosylase activity in cells.

5-Methylcytosine (**5mCyt**) also undergoes deamination. This minor base represents up to 5% of cytosine bases in human cells and plays a major role in gene regulation and defense against viruses [1, 2]. It is produced enzymatically from cytosine at CpG sites by DNA methyltransferases, with *S*-adenosyl-methionine as the methyl donor. 5-Methylcytosine is more sensitive than cytosine toward deamination. The rate constants for deamination in double-stranded DNA have been reported to be 5.8×10^{-13} s^{-1} (**5mCyt**) and 2.6×10^{-13} s^{-1} (**Cyt**) at 37 °C and physiological pH, respectively [3]. The deamination product of 5mCyt is thymine, which is excised from DNA by a specific repair glycosylase when thymine is paired with guanine [4]. However, mutational events at 5-methylated cytosines remain among the most frequent ones [5].

Saturation of the C5–C6 double bond of cytosine is another chemical modification that drastically affects the deamination rate. Values in the range of 10^{-6} s^{-1} were reported for 5,6-dihydrocytosine under various experimental conditions [6, 7]. Accelerated deamination is also observed in 5,6-saturated cytosine derivatives including oxidation products [8–10], photohydrates [11, 12], and cyclobutane pyrimidic dimeric photoproducts [13–17]. Interestingly, methylation at position 5 in such compounds decreases the deamination rate [18–21] in contrast to what is observed for the 5,6-unsaturated parent cytosine molecule [4].

In recent works, we have applied quantum chemistry tools to better describe the elementary steps involved in the deamination of cytosine [22, 23], 5-methylcytosine [24] and 5,6-dihydrocytosine (**5dhCyt**) [25]. The main conclusion of these studies is that more than one water molecule is needed to fully explain the observed rate constants. In each case, the assisted nucleophilic addition of one water molecule to C4 (see Fig. 1) of the cytosine derivative with the assistance of a second water molecule is the rate-determining step. It appears that the imino tautomer of cytosine can be viewed as the active form of the nucleobase for this reaction. For cytosine and 5-methylcytosine, the imino tautomer is formed in a concerted way during the nucleophilic addition, whereas for 5,6-dihydrocytosine it has to be generated prior to the addition step. We herein extend these studies to 5,6-dihydro-5-methylcytosine (**5mdhCyt**) with the purpose of exploring the mechanistic reasons for the opposite effect of methylation on deamination of cytosine and its 5,6-saturated derivative. The numbering of the atoms within all compounds is given in Fig. 1.

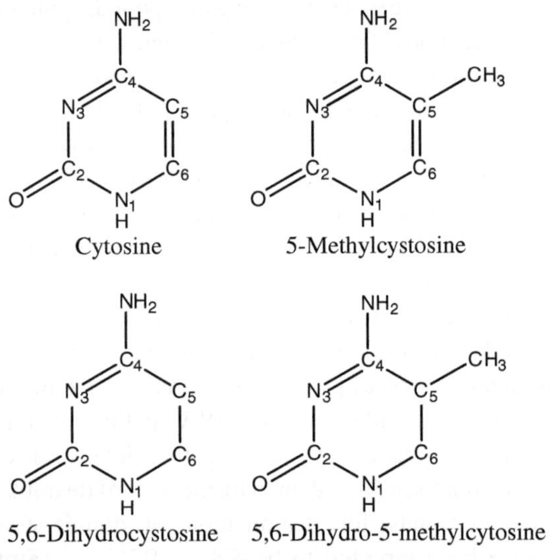

Fig. 1 Atom numbering of cytosine and cytosine derivatives

2 Computational details

2.1 Quantum chemical calculations

In order to enable a straightforward comparison with data previously reported for hydrolytic deamination of cytosine, 5-methylcytosine, and 5,6-dihydrocytosine, the same quantum methodology was adopted as in those studies [22–25]. All calculations were thus performed using the Gaussian 03 package [26] at the B3LYP/6-311G(d,p) level of theory [27–31]. As a first step, molecular systems of interest were fully optimized in vacuum. The stationary points were then characterized as minima or transition states by computing the vibrational frequencies within the harmonic approximation at the same level of theory, and thermal data were extracted to obtain thermodynamic functions of reaction and corresponding activation parameters at 298.15 K and 1 atm. The B3LYP functional is known to predict with moderate accuracy the barrier heights of H-transfers. Previous work [22–25] has however shown that, although B3LYP underestimates barrier heights in comparison with single-point MP2 calculations using the same basis set, the general energy trends are not affected.

Single-point energy calculations were subsequently performed on the optimized geometries, including the integral equation formalism polarized continuum model (IEF-PCM, hereafter referred to as PCM [32–34]), with dielectric constant $\varepsilon = 78.39$ to simulate the bulk effects of an aqueous environment, and to enable straightforward comparison with earlier work. The cavity was built using the united atom topologic model applied to the atomic radii of the UFF force field, with an average area of 0.2 Å^2 for the tesserae generated on each sphere. The default cavity was modified by adding individual spheres to all hydrogen atoms linked to nitrogen and oxygen atoms, using the keyword SPHEREONH.

Reaction paths in vacuum were characterized at the same level of theory through intrinsic reaction coordinate (IRC) calculations from the gas-phase-optimized transition structures to ensure that they connected the appropriate reactants and products.

2.2 Reactivity indices

In order to rationalize the difference in reactivity between the investigated systems, a range of global and local reactivity indices as derived from conceptual DFT [35–40] were computed for the imine tautomers of cytosine, 5-methylcytosine, and their 5,6-dihydrogenated counterparts. Calculations were performed on the optimized structures in *vacuum* and in bulk solvent. The different indices evaluated are as follows (full details and mathematical expressions can found in for instance Ref. [25]).

2.2.1 Global reactivity indices

These parameters, which give information on the reactivity of the molecular system as a whole, include the *chemical potential* [41] μ, giving a measure of the escaping tendency of the electrons; the *absolute hardness* [42, 43] η, which may be viewed as the resistance toward charge transfer; and the *global electrophilicity index* [44, 45] ω, which can be considered as a measure of the electrophilicity of the molecular system. From a computational perspective, these are evaluated from frontier orbital energies (HOMO, LUMO) together with expressions based on finite differences approximations and Koopmans' theorem [46, 47].

2.2.2 Local reactivity indices

In order to address differences in reactivity within a particular system, we instead need to explore local features. The first is the *net atomic charges*, in particular of carbon C4 at which the deamination reaction takes place and which gives a measure of the reactivity under electrostatic control. Three other local indices evaluated provide information about reactivity under electron-transfer control. These include the *local electrophilicity index* [48] $\omega^+(\mathbf{r})$ which is the product of the global electrophilicity index ω and the Fukui function [49–51] $f^+(\mathbf{r})$; the *excess electrophilicity* [52] $\Delta\omega(\mathbf{r})$ defined as the product of the *dual descriptor* [53–56] $\Delta f(\mathbf{r})$ and the global electrophilicity index ω and finally another local reactivity index $s^{(2)}(\mathbf{r}) = \frac{\Delta f(\mathbf{r})}{\eta^2}$ derived from the *dual descriptor within the grand canonical ensemble* [57] *(GCDD)*. The latter descriptor is size extensive and thus more suitable to compare molecules with different number of electrons. In its reduced form used in this paper, it can be calculated as the product of the dual descriptor and the inverse of the square of the global hardness. The more positive the *GCDD* is the more willing the site is to undergo nucleophilic attack, whereas a large negative number indicates a strong tendency to undergo electrophilic attack. The partial charges were evaluated using the ChelpG method [58], and the dual descriptors are based on comparisons of the electronic populations on carbon C4 upon adding or subtracting an electron to or from the normal system. For full details, we refer to Ref. [25].

3 Results and discussion

3.1 Geometries of the starting materials

5,6-Dihydro-5-methylcytosine possesses an asymmetric carbon, namely the carbon bearing the methyl group. As a consequence, two enantiomers can exist. It is obvious that the results observed are the same for both enantiomers.

We furthermore note that saturation of the C5=C6 bond produces a puckering of the ring system, and thus two orientations of the C5 methyl group, axial and equatorial, are possible, which will influence the energetics and interactions of the incoming water/OH addition at C4. Both cases are considered herein.

Cytosine and cytosine derivatives can furthermore exist under two tautomeric forms. The classical amino form that possess an exocyclic amino group and an imino form characterized by a double bond between the carbon C4 and the exocyclic nitrogen. Quite interestingly, for the 5,6-dihydrogenated forms, the axial conformer is the most stable one for the amino form, while it is the equatorial for the imino tautomer. All conformers do, however, lie very close in energy; the range is about 1.5 kJ/mol.

3.2 Reaction pathways—geometries and energies

3.2.1 Description of the mechanism studied

The reaction mechanism for spontaneous deamination of 5,6-dihydro-5-methylcytosine in protic medium is outlined schematically in Fig. 2. Following hydrogenation across the C5=C6 double bond (not shown), the reaction starts with a water-assisted tautomerization of the common amine form (**5mdhR**) to the corresponding imine (**5mdhIR1**). This generates a reactive C4=N4 double bond (atomic numbering given in Fig. 2), across which water is added to yield **5mdhIR2**. The exocyclic amine group is protonated from the medium (**5mdhIR3**), followed by NH_3 release from C4 and abstraction of the OH proton from the just added C4–OH group to give the **5mdhP** product (uracil tautomer) and an ammonium cation. Of the two water molecules included in the cluster calculations, one is involved in the proton shuttle mechanisms thus stabilizing hydrogen bonding, whereas the other actively adds across the C4–N4 bond as OH and H. The above labels for the different stationary structures will be used throughout, with the addition **eq** or **ax** for the conformers with C5 methyl group in equatorial or axial orientation, respectively.

3.2.2 Formation of the active form of the nucleobase

In Fig. 3a, the main structures for the stationary points along the initial tautomerization reaction for the equatorial conformer are shown, and in Fig. 3b the corresponding structures for the axial form are displayed. In the reactant complex **5mdhReq**, "water 7" is positioned with hydrogen bonding between O7 and H4a (2.096 Å) and between H7a and N3 (1.815 Å) to readily facilitate the proton shift from N4 to N3. The two protons transfer concertedly, with the

Fig. 2 Reaction pathways involved in the spontaneous deamination of 5-methyl-5,6-dihydrocytosine in protic medium. *Insert* atomic labeling used throughout the study

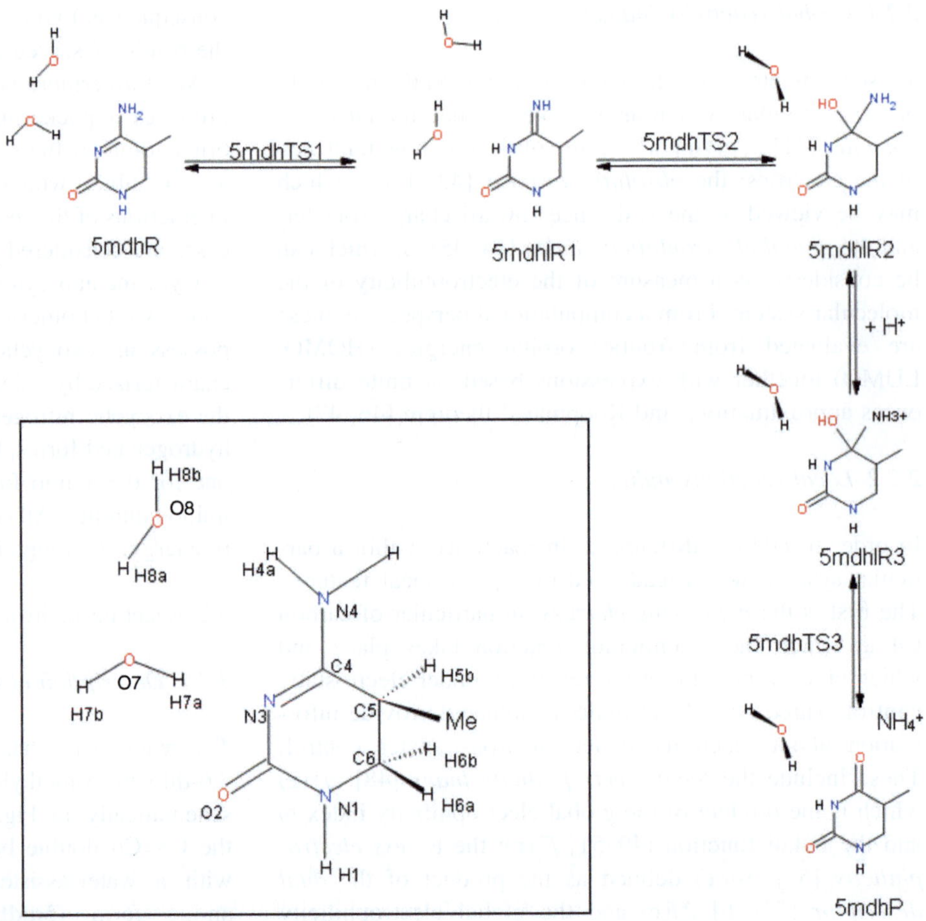

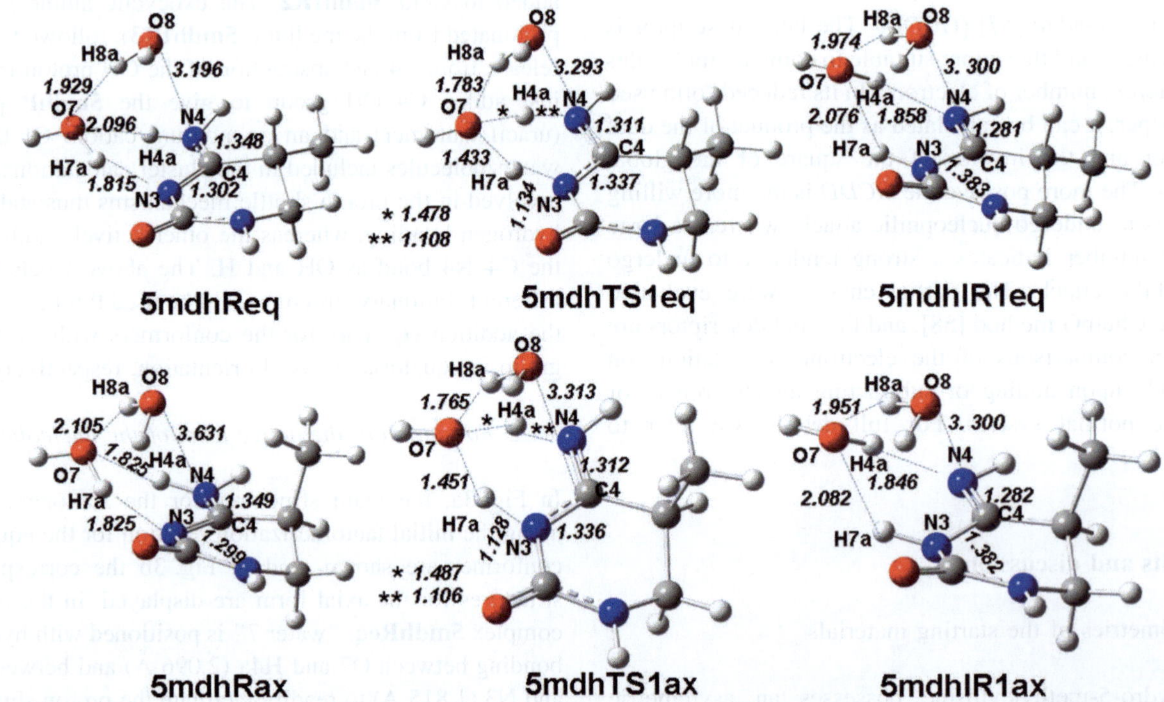

Fig. 3 **a** Optimized geometries of **5mdhReq**, **5mdhTS1eq**, and **5mdhIR1eq**; **b** **5mdhRax**, **5mdhTS1ax**, and **5mdhIR1ax** Distances in Å

Table 1 Relative energies in vacuum, free energies in vacuum, and relative free energies in aqueous solvent for the stationary points of the deamination pathway of 5-methyl-5,6-dihydrocytosine

System	ΔE_{vac}	ΔG_{vac}	ΔG_{aq}
a) Equatorial position			
5mdhR + H^+	0.0	0.0	0.0
5mdhTS1 + H^+	44.0	48.5	50.1
5mdhIR1 + H^+	−11.2	−10.8	−9.2
5mdhIR1 constrained + H^+	−11.1	−9.8	−7.9
5mdhTS2 + H^+	105.3	115.5	118.5
5mdhIR2 + H^+	15.9	24.6	16.7
5mdhIR3	−928.9	−918.6	−18.4
5mdhTS3	−916.3	−907.4	12.9
5mdhP	−1,035.8	−1,039.8	−150.5
b) Axial position			
5mdhR + H^+	0.0	0.0	0.0
5mdhTS1 + H^+	48.2	50.4	57.7
5mdhIR1 + H^+	−21.9	−22.4	−17.0
5mdhIR1 constrained + H^+	−6.2	−4.4	−1.5
5mdhTS2 + H^+	115.8	124.3	127.0
5mdhIR2 + H^+	25.8	31.4	26.0
5mdhIR3	−925.1	−922.3	−36.8
5mdhTS3	−914.8	−909.1	3.2
5mdhP	−1,056.1	−1,058.3	−152.8

All values in kJ/mo

H4a transfer representing an early TS and the H7a transfer a late one. The reaction free energy is 50.1 kJ/mol in aqueous solution, and the resulting **5mdhIR1eq** lies 9.2 kJ/mol below the reactant (Table 1a). For the axial form, the overall geometric features of the tautomerization reaction are highly similar to those of the equatorial conformer, except that the two water molecules are slightly (~0.02 Å) further out. Energetically, the reaction free energy is higher by 7.6 kJ/mol than for the equatorial system, and the **5mdhIR1ax** tautomer is 17 kJ/mol more stable than the reactant (Table 1b).

During the reaction, the C4–N4 bond shortens from 1.35 to 1.28 Å, and the N3–C4 bond lengthens from 1.30 to 1.38 Å. We also note that due to the hydrogenation of the C5–C6 bond, the conjugation of the ring system is broken and the molecule becomes highly non-planar. It is worth noticing that for 5,6-dihydro-5-methylcytosine the imino form is more stable than the amino form. The same trend is observed for the 5,6-dihydrocytosine [25]. One can hypothesize that it is a general tendency that when the C5–C6 bond is saturated, the imino tautomer is more stable than the amino tautomer. The overstabilization of the imino tautomer versus the amino tautomer is likely due to the presence of a urea moiety in the imino form. Urea fragments are very stable entities since the lone pair of each nitrogen interacts with the vicinal π^* (C=O) bond. A NBO [59] analysis confirms this hypothesis with a quite strong stabilizing energy between nitrogen N3 and the carbonyl π^* (C=O); about 48 kJ/mol.

The effects of free energy corrections relative to electronic energies in vacuum, as well as the effect of inclusion of the solvent, can also be seen from the data reported in Table 1. Introducing thermal corrections in the equatorial system increases the barrier by 4.5 kJ/mol, and adding the solvent effects raises this by another 1.6 kJ/mol. For the axial conformer, the corresponding increases are 2.2 and 7.3 kJ/mol, respectively. Similarly, the relative energy of the products becomes slightly more positive, following the same trends as for the barriers, with the exception of the thermal free energy correction to **5mhdIR1ax**, which lowers the energy by 0.5 kJ/mol.

3.2.3 *Preparation for the nucleophilic addition of water*

In order for the subsequent water addition to occur, the attacking oxygen (O8) must approach C4 in a perpendicular trajectory with respect to the molecular plane. In the current systems, the **5mCyt** ring is already distorted due to the hydrogenation at C5 and C6, whereas in the case of non-hydrogenated cytosine and 5-methylcytosine, the ring system remains planar and the two water molecules align with stable hydrogen bonds in that same plane. To overcome the unphysical situation of an in-plane water required to attack carbon C4 perpendicularly to the plane, the concept of reaction forces was employed herein as well as in our previous studies on deamination reactions.

The reaction force, introduced by Toro-Labbé [60], is defined as the negative derivative of the energy with respect to the intrinsic reaction coordinate. It represents the force acting on the system to bring the reactants into the products. The reaction force vanishes at stationary points, and it has been shown (see, e.g., Ref. [25]) that the reaction force becomes very close to zero in a region between reactant and TS. The coordinates at this point are then used in a constrained geometry optimization of a more physically realistic reactant complex. In the current context, the structural differences between the tautomerization product **5mdhIR1eq** and the reactive complex (Fig. 4) turned out to be relatively small due to the non-planarity of the cytosine derivative. This is also reflected in the close-to-zero electronic energy difference in vacuum between the two, and just over 1 kJ/mol difference in free energy in aqueous solution (Table 1a). The main difference involves a shortening of the C4–O8 distance from 3.300 to 3.181 Å. The situation is quite different for the axial system. Due to the repulsion between the axially oriented methyl group and the incoming water, the energy of the constrained system is raised by 16–18 kJ/mol. The energetics of the

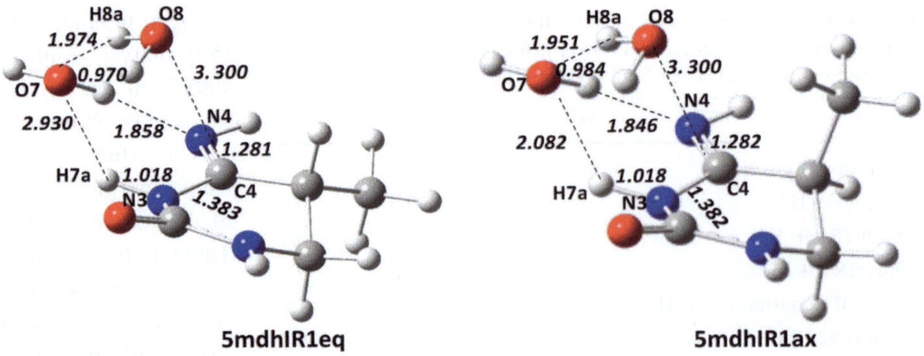

Fig. 4 Optimized constrained geometries of the **5mdhIR1** reactant complexes, in vacuum. Distances in Å

subsequent points on the reaction energy surfaces (Fig. 5a (**eq**), b (**ax**)) are shifted to be relative to the energy of the corresponding reactive complexes.

3.2.4 Nucleophilic addition of water to C4

Water addition has been shown in all our previous deamination reaction studies to be the rate-determining step. This is the case also in the current system, where the TS **5mdhTS2eq** lies 126.4 kJ/mol above the **5mdhIR1eq** reactant complex (Fig. 6). A free energy barrier of 128.5 kJ/mol is observed for the corresponding **5ndhTS2ax** system (Fig. 6). The geometries of both the TS's and the resulting adducts (**5mdhTS2, 5mdhIR2**; Fig. 6) are very similar to those of the corresponding 5,6-dihydrocytosine (**dhTS2, dhIR2**; Fig. 6)—the differences in bond distances throughout being within 0.01 Å between the corresponding stationary points for the equatorial conformer, and slightly larger differences (0.02–0.03 Å) for the distances to the two waters in the axial system. The TS forms a 6-membered ring (C4–N4–H4a–O7–H8a–O8) and thus involves the back-transfer of H4a that was originally transferred to O7 of water in the tautomerization reaction, concomitantly with the transfer of H8a to O7, as O8 binds to C4. The resulting adduct forms three hydrogen bonds to the O7 water molecule (O7–H7a, H8a–O8, and O7–H4a), all in the range 2.0–2.3 Å. In the case of the equatorial conformer, the inclusion of thermal corrections raises the barriers by 7–10 kJ/mol, whereas solvent effects are less pronounced, by 2–3 kJ/mol. For the axial system, inclusion of thermal effects raises the energy of the products by 6–9 kJ/mol, whereas inclusion of solvent effects provides product stabilization by a similar amount.

3.2.5 Protonation of the amino group and deamination

Water addition is followed by protonation of the N4 amine by the medium, which stabilizes the complex by 33–35 kJ/mol compared to the energy of **5mdhIR2** + H^+ [61]. Deamination of the protonated **5mdhIR3** leads to simultaneous

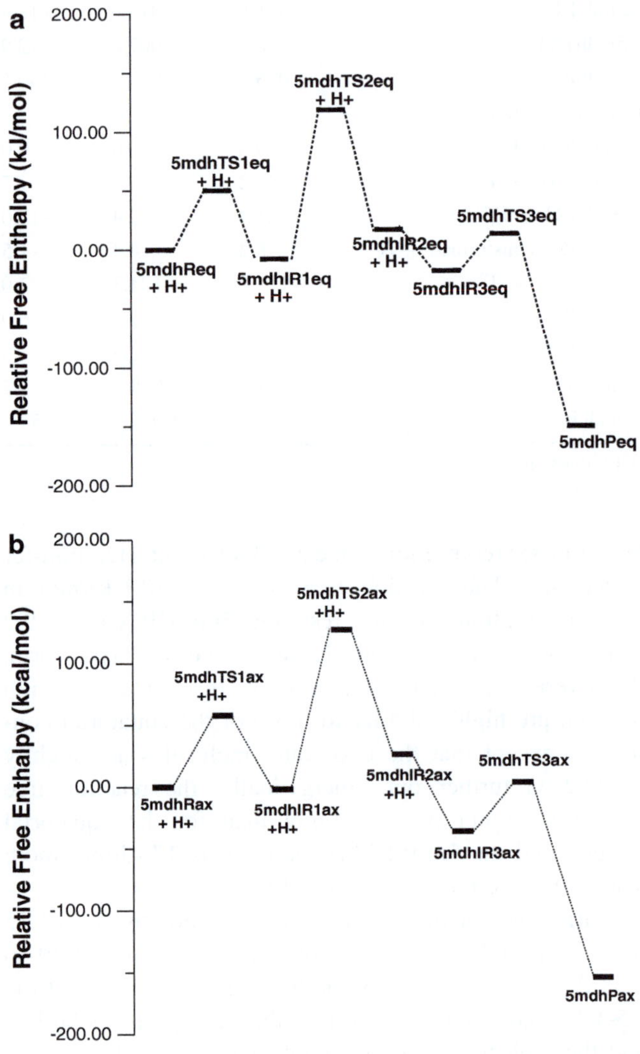

Fig. 5 **a** Relative free energies (kJ/mol) along the studied deamination pathway in aqueous solution for equatorial isomer; **b** As (**a**) for axial isomer

deprotonation by the leaving ammonia, of the just added O8 (Fig. 7). The barrier to deamination is 31 kJ/mol for the equatorial conformer, and the final product **5mdhPeq** is located 132 kJ/mol below the protonated reactant **5mdhIR3eq**. In the case of the axial conformer, the barrier is

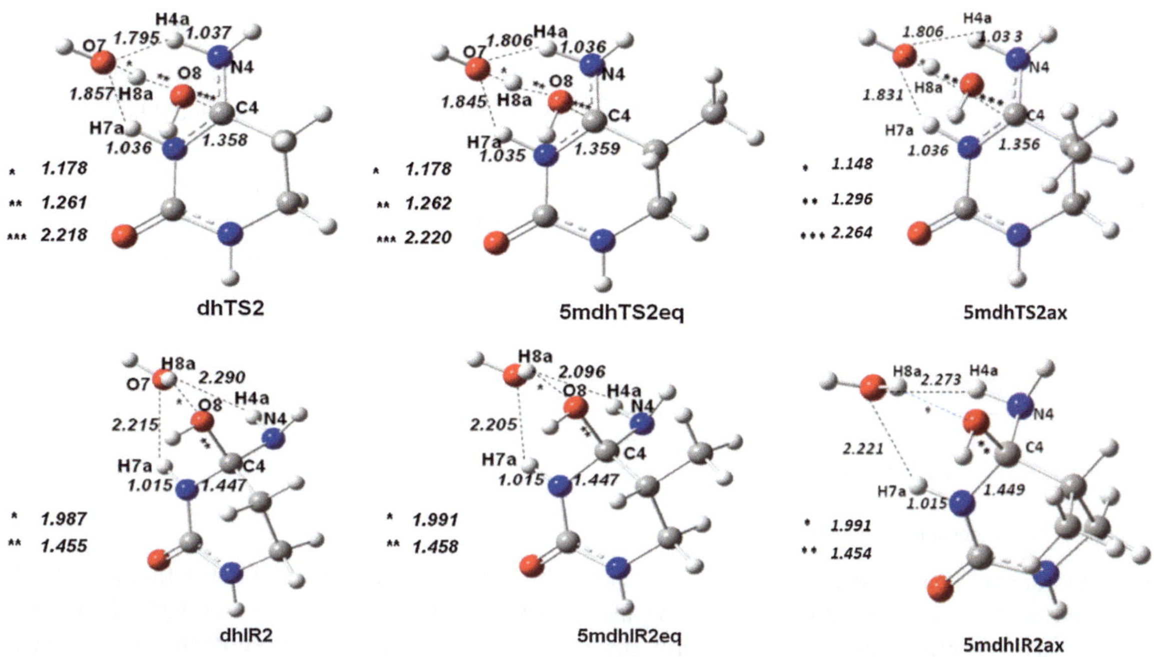

Fig. 6 Optimized geometries of IR2 and TS2 for **dhCyt**, **5mdhCyteq,** and **5mdhCytax**. Distances in Å

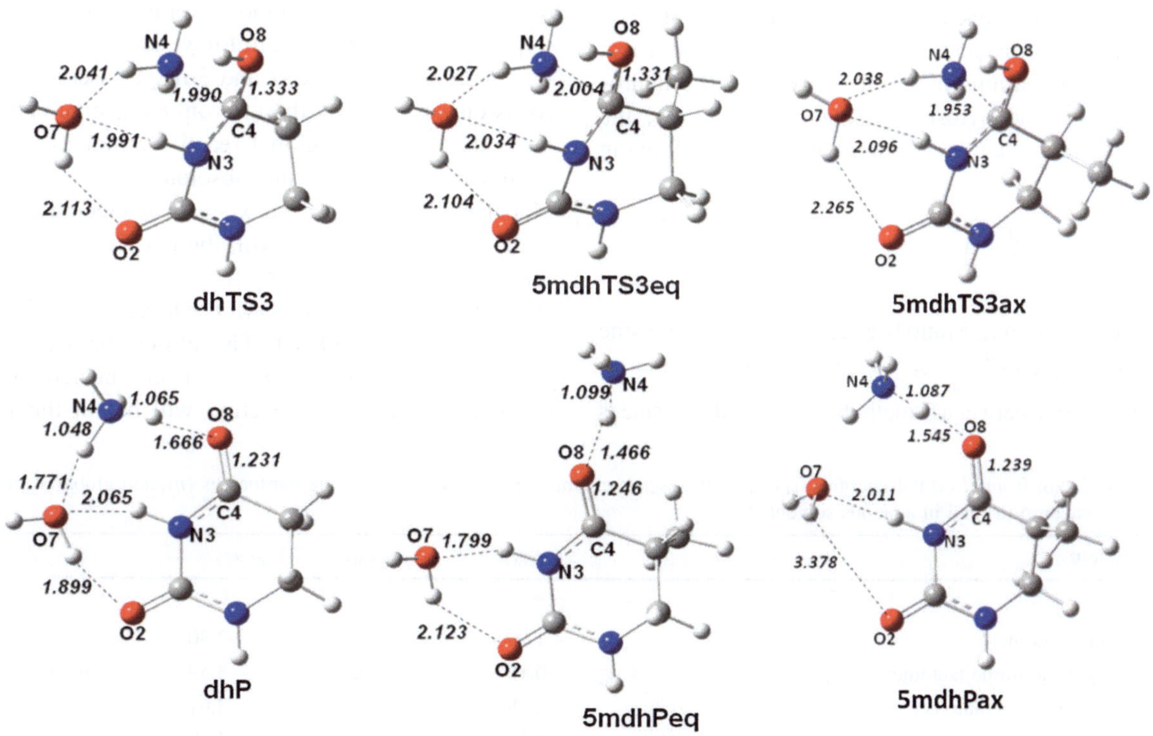

Fig. 7 Optimized geometries TS3 and product for **dhCyt**, **5mdhCyteq,** and **5mdhCytax**. Distances in Å

considerably higher, 40 kJ/mol, and the reaction less exergonic ($\Delta\Delta G_{aq}$ = 116.0 kJ/mol). The difference between the two conformers is largely due to the higher relative energy of the **5mdhIR3ax** complex over the corresponding equatorial one. The transition states for deamination are again very similar to the 5,6-dihydrocytosine counterpart, see Fig. 7.

3.3 Energetics

The computed energetics of the rate-determining steps for **Cyt**, **5mCyt**, **dhCyt,** and **5mdhCyt**, following similar pathways as outlined herein, are 138, 134, 110, and 128 kJ/mol, respectively. As seen from the energetics above,

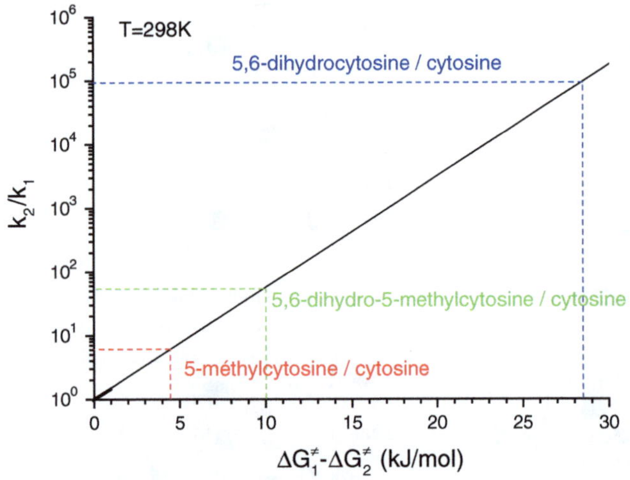

Fig. 8 Evolution of the deamination rate ratio

saturation of the double bond in cytosine gives a significantly increased rate of deamination, whereas for the **5mCyt** counterparts the effects are smaller but still significant. The trend agrees with that duly experimentally noted [16–19] and, interestingly, the effect of the methylation is opposite in cytosine and 5,6-dihydrocytosine.

From the data below and using the Eyring's law, one can estimate a deamination rate ratio between 5,6-dihydro-5methylcytosine and cytosine. Using Eyring's Law, the reaction rate ratio can be assessed using the following relation:

$$\frac{k_2}{k_1} \approx \exp\left(\frac{\Delta G_1^{\neq} - \Delta G_2^{\neq}}{\mathrm{RT}}\right)$$

As the deamination rate ratio between 5,6-dihydrocytosine and cytosine is about $\frac{k_{56\mathrm{dhcyt}}}{k_{\mathrm{cyt}}} \approx 10^5$ ($\Delta\Delta G = 138\text{–}110 = 28$ kJ/mol) and the one between 5-methylcytosine and cytosine is about $\frac{k_{5\mathrm{mcyt}}}{k_{\mathrm{cyt}}} \approx 5$ ($\Delta\Delta$G = 138–134 = 4 kJ/mol), one can assess that the deamination rate ratio between 5,6-dihydro-5-methylcytosine and cytosine would be about $\frac{k_{56\mathrm{dh5mcyt}}}{k_{\mathrm{cyt}}} \approx 50$ ($\Delta\Delta G = 138\text{–}128 = 10$ kJ/mol); cf. Fig. 8. Experimental value obtained at 25 °C for the deamination rate of cytosine is 2.1 10^{-10} s^{-1}. The estimated value for 5,6-dihydro-5-methylcytosine, based on the current analysis, is therefore 1.2 10^{-8} s^{-1}. Unfortunately, to our best knowledge, no accurate experimental determination of such value has been carried out so far.

3.4 Reactivity indices

3.4.1 Global reactivity indexes

In Table 2a and b, we report the frontier orbital energies (HOMO and LUMO) for **Cyt**, **5mCyt**, **dhCyt,** and the axial and equatorial forms of **5mdhCyt**. Since for each of these cytosine derivatives the imine tautomer has been shown to be the actual active form of the nucleobases for the rate-determining step, be it formed prior to the rate-determining step as in the **dhCyt** and **5mdhCyt** cases, or in a concerted way with it as in the **Cyt** and **5mCyt** cases, we therefore focus our analyses on these. From the orbital energies, we have also computed global reactivity indices such as the chemical potential μ, the absolute hardness η, and the global electrophilicity index ω. In nucleophilic reactions under frontier orbital control, the reactivity of the electrophile is, according to frontier molecular orbital theory (FMOT), driven by its LUMO, whereas that of the nucleophile is by its HOMO. The closer the LUMO of the electrophile and the HOMO of the nucleophile lie in energy, the easier the reaction will be. In the following

Table 2 Kohn-Sham frontier orbital energies ($\varepsilon_{\mathrm{LUMO}}$ and $\varepsilon_{\mathrm{HOMO}}$), chemical potentials ($\mu$), absolute hardnesses ($\eta$), and global electrophilicity indices (ω), in vacuum (a) and in aqueous solvent (b)

Molecular system	$\varepsilon_{\mathrm{LUMO}}$	$\varepsilon_{\mathrm{HOMO}}$	μ	η	ω
a) Vacuum					
Cytosine imine tautomer	−1.09	−6.50	−3.80	5.41	1.33
5,6-Dihydrocytosine imine tautomer	−0.16	−7.02	−3.59	6.86	0.94
5-Methylcytosine imine tautomer	−0.98	−6.36	−3.67	5.38	1.25
5-Methyl-5,6-dihydrocytosine imine tautomer(equatorial)	−0.19	−7.00	−3.59	6.82	0.95
5-Methylcytosine-5,6-dihydrocytosine imine tautomer(axial)	−0.20	−7.00	−3.60	6.81	0.95
b) Aqueous solvent					
Cytosine imine tautomer	−0.87	−6.37	−3.62	5.50	1.19
5,6-Dihydrocytosine imine tautomer	−0.22	−7.10	−3.66	6.88	0.97
5-Methylcytosine imine tautomer	−0.74	−6.16	−3.45	5.42	1.10
5-Methyl-5,6-dihydrocytosine imine tautomer(equatorial)	−0.11	−6.96	−3.53	6.85	0.91
5-Methylcytosine-5,6-dihydrocytosine imine tautomer(axial)	−0.22	−7.03	−3.63	6.80	0.97

All values are in eV

discussion, the effects of methylation and saturation of cytosine on the frontier orbital energies are discussed and analyzed.

3.4.1.1 Frontier orbital energies Methylation of cytosine has a mild effect on the frontier orbital energies. The energies of both frontier orbitals increase by 0.12–0.14 eV. For 5,6-dihydrocytosine, methylation has a weaker effect on the frontier orbital energies than for cytosine. Just like for cytosine, the frontier orbital energies increase, but only by 0.02 eV. The conformation does not significantly influence the frontier energies. The equatorial and axial conformers display identical values within a range of 0.02 eV. The same trends are found when solvation is added. Two comments can be drawn from these observations. First, the effect of methylation is much higher when the methyl group is directly connected to an atom of a double bond than when the bond is saturated. Second, the general tendency is that methylation increases the frontier orbital energies. As a consequence, following the FMOT, methylation renders both systems saturated and unsaturated less electrophilic. The deamination rates should therefore be higher for cytosine or 5,6-dihydrocytosine compared to their methylated counterparts. This is in agreement with experimental results for 5,6-dihydro-5-methylcytosine versus 5,6-dihydrocytosine but does not apply for 5-methylcytosine versus cytosine.

Saturation of the C5–C6 bond, on the other hand, has a dramatic effect on the LUMO energies which are raised (destabilized) by up to 0.9 eV relative to their unsaturated counterparts. The HOMO levels, on the other hand, become somewhat stabilized through saturation. As a result, unsaturated forms appear softer than their saturated counterparts. Solvation has a similar effect for the saturated systems and the unsaturated ones, in that it leaves the HOMO–LUMO gap essentially intact. According to FMOT, the deamination rate of the saturated system should be lower than for their unsaturated counterpart, which is in total contradiction with experiments. The same contradiction arises from the global conceptual DFT descriptors.

3.4.1.2 Conceptual DFT indices Turning to the global conceptual DFT indices of chemical potential, absolute hardness, and electrophilicity, we note that although the saturated compounds overall display less negative values for the chemical potentials *in vacuum*, the data when including bulk solvation become more similar. The saturated compounds have significantly larger hardness than the unsaturated ones. This is a mathematical consequence of the larger HOMO–LUMO gaps seen for these systems relative to the unsaturated ones and can be rationalized by that the system lacks one π-bond, which implies a reduction in the polarizability of the system. The larger the absolute hardness is, the bigger the resistance to charge transfer should be, which is also reflected in the smaller electrophilicity indices, since the chemical potentials are very similar. C5-methylation appears to have a relatively modest effect on these. As noted above, the two conformers behave identically in gas phase, whereas in aqueous solution, differences between the two conformers can be noted [62]. Comparing again with the saturated 5,6-dihydrocytosine system, the equatorial conformer of 5,6-dihydro-5-methylcytosine imine deviates more in terms of chemical potential, whereas for the absolute hardness the axial form deviates more from the values seen for **dhCyt**. The global electrophilicities, however, are affected to a much less degree.

Since the next reaction step is water addition across the C4–N4 bond, it is of interest to compare the data with the corresponding values for H_2O. At the same level of theory, the chemical potential, absolute hardness, and global electrophilicity index in aqueous solution are −3.58, 9.44, and 0.68 eV, respectively [25]. We note that the data for the 5,6-saturated derivatives of the imine tautomers are considerably closer to these values than ones for the unsaturated systems are.

From the global reactivity indices, we conclude that methylation at C5 has essentially the same effect on the saturated and unsaturated cytosine derivatives. The explanation as to why C5-methylation of the saturated forms of cytosine reduces the deamination rate while that of the unsaturated forms raises it, thus has to be found from other considerations.

3.4.2 Local reactivity indexes

In Table 3, we focus on the local reactivity indices at C4 for the same systems, *in vacuum* and in aqueous solution, the rationale being that addition of water oxygen to C4 is the rate-determining step in all these deamination reactions. First of all, we note that the electrostatic charges on C4 are higher, by 0.07–0.16 *e*, in aqueous solution than *in vacuum*. The same applies to the different local reactivity indices derived from conceptual DFT, in all cases.

3.4.2.1 Partial charges C5–C6 saturation of both **Cyt** and **5mCyt** leads to a reduced local charge on C4. This indicates that electrostatic effects cannot be at the root of the observed phenomenon. On the other hand, saturation of the C5–C6 increases the values of local DFT-based reactivity indices, such as the grand canonical dual descriptor. This indicates that C4 becomes more electrophilic in the saturated derivatives. This rationalizes the much reduced barrier for this reaction in **dhCyt** and **5mdhCyt** over **Cyt** and **5mCyt,** respectively, and in turn explains the much increased reaction rate as seen experimentally for the saturated derivatives of cytosine.

Table 3 Electrostatic charges on carbon C4 q_{C4}, local electrophilicity (ω^{+}_{C4}, in eV), excess electrophilicity ($\Delta\omega_{C4}$, in eV), and grand canonical dual descriptor ($\frac{\Delta f_{C4}}{\eta^2}$, in 10^{-2} eV^{-2}) condensed to carbon C4, in vacuum (a) and in aqueous solvent (b)

Molecular system	q_{C4}	ω^{+}_{C4}	$\Delta\omega_{C4}$	$\frac{\Delta f_{C4}}{\eta^2}$
a) Vacuum				
Cytosine imine tautomer	0.73	0.11	0.13	0.33
5,6-Dihydrocytosine imine tautomer	0.65	0.26	0.30	0.67
5-Methylcytosine imine tautomer	0.63	0.10	0.15	0.40
5-Methyl-5,6-dihydrocytosine imine tautomer (equatorial)	0.56	0.20	0.24	0.55
5-Methyl-5,6-dihydrocytosine imine tautomer (axial)	0.57	0.21	0.25	0.58
b) Aqueous solvent				
Cytosine imine tautomer	0.81	0.21	0.18	0.49
5,6-Dihydrocytosine imine tautomer	0.73	0.49	0.49	1.05
5-Methylcytosine imine tautomer	0.70	0.22	0.22	0.68
5-Methyl-5,6-dihydrocytosine imine tautomer (equatorial)	0.63	0.42	0.43	1.01
5-Methyl-5,6-dihydrocytosine imine tautomer (axial)	0.65	0.44	0.45	1.00

3.4.2.2 Fukui functions-based indices For the 5-methylated nucleobases, the local reactivity indices increase upon saturation of the C5–C6 bond. The trends are again more pronounced in bulk solvent. Interestingly, C5-methylation of cytosine induces an increase in $\Delta\omega_{C4}$ and $(\Delta f_{C4})/(\eta^2)$ values, both *in vacuum* and in aqueous solvent, while methylation of 5,6-dihydrocytosine results in a decrease in $\Delta\omega_{C4}$ and $(\Delta f_{C4})/(\eta^2)$ values. For the unsaturated system, the increase in $(\Delta f_{C4})/(\eta^2)$ results from a lowering of the conjugation between C5=C6 double bond and C4=N3 double bond. Indeed, the substitution of C5 by a methyl group induces a polarization of the C5=C6 electron cloud toward C6. Thus, in the case of a reaction under charge transfer control, C5-methlyation is expected to have opposite effect on the local reactivity of carbon C4 of saturated and unsaturated forms of cytosine. This is in agreement with the fact that experimentally methylation at position 5 in 5,6-saturated cytosine derivatives decreases the deamination rate in contrast to what is observed with the 5,6-unsaturated cytosine derivatives. However, the range of the deamination rate decrease cannot totally be explained by a relatively small decrease in the reactivity index values. It is therefore likely that besides these electronic effects, the steric hindrance of the methyl group affects the approach of the water molecules. The methyl group probably acts as a shield for at least one side of the ring system, especially when it is in axial position. As both axial and equatorial positions lie very close in energy, the odds that the methyl group is in an axial orientation during the process are considerable.

4 Conclusions

We have in the current study investigated the deamination of 5,6-dihydro-5-methylcytosine. The investigation has been carried out in close comparison with deamination of cytosine, 5-methylcytosine, and 5,6-dihydrocytosine. It appears that the deamination mechanism is very similar to that of cytosine and cytosine derivatives. Just like for cytosine and cytosine derivatives, the imine tautomer is the reactive form for the deamination process. But in contrast to cytosine and 5-methylcytosine, the imine tautomer is produced before the deamination and constitutes the more stable form for the saturated systems. Experimentally, the deamination rates is lower for 5,6-dihydro-5-methylcytosine than for 5,6-dihydrocytosine. This fact has been rationalized herein through quantum chemical calculations and conceptual DFT-based descriptors. It appears that besides the electronic effects of the methyl group on the carbon C4 of cytosine, steric hindrance caused by the methyl group, especially when it is in axial position, shields the approach of the reacting water molecules.

Acknowledgments The Faculty of Science at the University of Gothenburg is gratefully acknowledged for financial support (LAE). This research has beneficiated of the support of Aviesan ITMO cancer within the "Plan Cancer 2009–2013" and the application of the action 3.3.

References

1. Attwood JT, Yung RL, Richardson BC (2002) Cell Molr Life Sci 59:241
2. Ramsahoye BH, Davies CS, Mills KI (1996) Blood Rev 10:249
3. Shen J-C, Rideout WM III, Jones PA (1994) Nucleics Acids Res 22:972
4. Cortazar D, Kunz C, Saito Y, Steinacher R, Schar P (2007) DNA Repair(Amst) 6:489
5. Pfeifer GP (2006) Current topics in microbiology and immunology. In: DNA methylation: basic mechanisms, vol 301, Part IV. Springer, pp 259–281
6. Slae S, Shapiro R (1978) J Org Chem 43:1721
7. Green M, Cohen SS (1957) J Biol Chem 228:601
8. Decarroz C, Wagner JR, van Lier JE, Murali KC, Riesz P, Cadet J (1986) Int J Radiat Biol 50:491
9. Tremblay S, Douki T, Cadet J, Wagner R (1999) J Biol Chem 274:20833

10. Wagner JR, Hu CC, Ames BN (1992) Proc Natl Acad Sci USA 89:3380–3384
11. Boorstein RJ, Hilbert TP, Cunningham RP, Teebor GW (1990) Biochemistry 29:10455–10460
12. Douki T, Vadesne-Bauer G, Cadet J (2002) Photochem Photobiol Sci 1:565
13. Freeman KB, Hariharan PV, Johns HE (1965) J Mol Biol 13:833–848
14. Lemaire DGE, Ruzsicska BP (1993) Photochem Photobiol 57:755
15. Lemaire DGE, Ruzsicska BP (1993) Biochemistry 32:2525
16. Douki T, Cadet J (1992) J Photochem Photobiol B Biol 15:199–213
17. Liu F-T, Yang NC (1978) Biochemistry 17:4865
18. Bienvenu C, Cadet J (1996) J Org Chem 61:2632
19. Cannistraro VJ, Taylor JS (2009) J Mol Biol 392:1145
20. Celewicz L, Mayer M, Shetlar MD (2005) Photochem Photobiol 81:404
21. Douki T, Cadet J (1994) Biochemistry 33:11942
22. Labet V, Grand A, Morell C, Cadet J, Eriksson LA (2008) Theor Chem Acc 120:429
23. Labet V, Morell C, Grand A, Toro-Labbe A (2008) J Phys Chem A 112:11487
24. Labet V, Morell C, Cadet J, Eriksson LA, Grand A (2009) J Phys Chem A 113:2524
25. Labet V, Morell C, Douki T, Cadet J, Eriksson LA, Grand A (2010) J Phys Chem A 114:1826
26. Frisch MJ, Trucks GW, Schlegel HB, Scuseria GE, Robb MA, Cheeseman JR, Scalmani G, Barone V, Mennucci B, Petersson GA, Nakatsuji H, Caricato M, Li X, Hratchian HP, Izmaylov AF, Bloino J, Zheng G, Sonnenberg JL, Hada M, Ehara M, Toyota K, Fukuda R, Hasegawa J, Ishida M, Nakajima T, Honda Y, Kitao O, Nakai H, Vreven T, Montgomery JA Jr, Peralta JE, Ogliaro F, Bearpark M, Heyd JJ, Brothers E, Kudin KN, Staroverov VN, Kobayashi R, Normand J, Raghavachari K, Rendell A, Burant JC, Iyengar SS, Tomasi J, Cossi M, Rega N, Millam JM, Klene M, Knox JE, Cross JB, Bakken V, Adamo C, Jaramillo J, Gomperts R, Stratmann RE, Yazyev O, Austin AJ, Cammi R, Pomelli C, Ochterski JW, Martin RL, Morokuma K, Zakrzewski VG, Voth GA, Salvador P, Dannenberg JJ, Dapprich S, Daniels AD, Farkas O, Foresman JB, Ortiz JV, Cioslowski J, Fox DJ (2009) *Gaussian 09*, Revision A.02. Gaussian, Inc, Wallingford, CT
27. Becke AD (1993) J Chem Phys 98:5648–5652
28. Lee C, Yang W, Parr RG (1988) Phys Rev B 37:785–789
29. McLean AD, Chandler GS (1980) J Chem Phys 72:5639
30. Krishnan R, Binkley JS, Seeger R, Pople JA (1980) J Chem Phys 72:650
31. Frisch MJ, Pople JA, Binkley JS (1984) J Chem Phys 80:3265
32. Cancès MT, Mennucci B, Tomasi J (1997) J Chem Phys 107:3032
33. Mennucci B, Tomasi J (1997) J Chem Phys 106:5151–5158
34. Cossi M, Scalmani G, Rega N, Barone V (2002) J Chem Phys 117:43–54
35. Parr RG, Yang W (1989) Density functional theory of atoms and molecules. Oxford University Press, New York
36. Geerlings P, De Proft F, Langenaeker W (2003) Chem Rev 103:1793–1874
37. Chattaraj PK, Sarkar U, Roy DR (2006) Chem Rev 106:2065
38. Ayers PW, Anderson JSM, Bartolotti LJ (2005) Int J Quantum Chem 101:524
39. Chermette H (1999) J Comp Chem 20:129
40. Parr RG, Yang W (1995) Annu Rev Phys Chem 46:701
41. Parr RG, Donnelly RA, Levy M, Palke WE (1978) J Chem Phys 68:3801
42. Parr RG, Pearson RG (1983) J Am Chem Soc 105:7512
43. Mineva T, Sicilia E, Russo N (1998) J Am Chem Soc 120:9053
44. Parr RG, Szentpaly Lv, Liu SB (1922) J Am Chem Soc 1999:121
45. Morell C, Labet V, Grand A, Chermette H (2009) Phys Chem Chem Phys 11:3417
46. Koopmans TA (1934) Physica 1:104
47. Parr RG, Zhou Z (1993) Acc Chem Res 26:256
48. Chattaraj PK, Maiti B, Sarkar U (2003) J Phys Chem A 107:4973
49. Parr RG, Yang W (1984) J Am Chem Soc 106:4049
50. Ayers PW, Levy M (2000) Theor Chem Acc 103:353
51. Mineva T, Parvanov V, Petrov I, Neshev N, Russo N (1959) J Phys Chem A 2001:105
52. Padmanabhan J, Parthasarathi R, Elango M, Subramanian V, Krishnamoorthy BS, Gutierrez-Oliva S, Toro-Labbé A, Roy DR, Chattaraj PK (2007) J Phys Chem A 111:9130
53. Morell C, Grand A, Toro-Labbé A (2005) J Phys Chem A 109:205
54. Morell C, Grand A, Toro-Labbé A (2006) Chem Phys Lett 425:342
55. Morell C, Ayers PW, Grand A, Chermette H (2011) Phys Chem Chem Phys 13:9601
56. Cardenas C, Rabi N, Ayers P, Morell C, Jaramillo P, Fuentealba P (2009) J Chem Phys A 113:8860
57. Ayers PW, Morell C, De Proft F, Geerlings P (2007) Chem Eur J 13:8240
58. Breneman CM, Wiberg KB (1990) J Comp Chem 11:361
59. Reed AE, Curtiss LA, Weinhold F (1988) Chem Rev 88:899
60. Toro-Labbé A (1999) J Phys Chem A 103:4398
61. Llano J, Eriksson LA (2002) J Chem Phys 117:10193
62. De Luca G, Sicilia E, Russo N, Mineva T (2002) J Am Chem Soc 124:1494

Theor Chem Acc (2012) 131:1201
DOI 10.1007/s00214-012-1201-3

REGULAR ARTICLE

Integrated computational approaches for spectroscopic studies of molecular systems in the gas phase and in solution: pyrimidine as a test case

Malgorzata Biczysko · Julien Bloino · Giuseppe Brancato · Ivo Cacelli · Chiara Cappelli · Alessandro Ferretti · Alessandro Lami · Susanna Monti · Alfonso Pedone · Giacomo Prampolini · Cristina Puzzarini · Fabrizio Santoro · Fabio Trani · Giovanni Villani

Received: 24 July 2011 / Accepted: 3 March 2012 / Published online: 8 April 2012

Abstract An integrated computational approach built on quantum mechanical (QM) methods, purposely tailored inter- and intra-molecular force fields and continuum solvent models combined with time-independent and time-dependent schemes to account for nuclear motion effects is applied to the spectroscopic investigation of pyrimidine in the gas phase as well as in aqueous and CCl_4 solutions. Accurate post-Hartree–Fock methodologies are employed to compute molecular structure, harmonic vibrational frequencies, energies and oscillator strengths for electronic transitions in order to validate the accuracy of approaches rooted into density functional theory with emphasis also on hybrid QM/QM′ models. Within the time-independent approaches, IR spectra are computed including anharmonicities through perturbative corrections while UV–vis line-shapes are simulated accounting for the vibrational structure; in both cases, the environmental effects are described by continuum models. The effects of conformational flexibility, including solvent dynamics, are described through time-dependent models based on purposely DFT-tailored force fields applied to molecular dynamics simulations and on QM computations of spectroscopic properties. Such procedures are exploited to

Dedicated to Professor Vincenzo Barone and published as part of the special collection of articles celebrating his 60th birthday.

Electronic supplementary material The online version of this article (doi:10.1007/s00214-012-1201-3) contains supplementary material, which is available to authorized users.

M. Biczysko (✉) · J. Bloino · G. Brancato · G. Prampolini
Scuola Normale Superiore di Pisa Piazza dei Cavalieri 7, 56126 Pisa, Italy
e-mail: malgorzata.biczysko@sns.it

Present Address:
M. Biczysko
Center for Nanotechnology Innovation @NEST, Istituto Italiano di Tecnologia, Piazza San Silvestro 12, 56127 Pisa, Italy

Present Address:
J. Bloino
CNR–Consiglio Nazionale delle Ricerche, Istituto di Chimica dei Composti Organo Metallici (ICCOM-CNR), UOS di Pisa, Area della Ricerca, via G. Moruzzi 1, 56124 Pisa, Italy

I. Cacelli · C. Cappelli
Dipartimento di Chimica e Chimica Industriale, Università degli Studi di Pisa, Via Risorgimento 35, 56126 Pisa, Italy

A. Ferretti · A. Lami · S. Monti · F. Santoro · G. Villani
CNR, Consiglio Nazionale delle Ricerche, Istituto di Chimica dei Composti Organo Metallici (ICCOM-CNR), UOS di Pisa, Area della Ricerca, via G. Moruzzi 1, 56124 Pisa, Italy

A. Pedone
Dipartimento di Chimica, Unversità di Modena e Reggio Emilia, Via G. Campi 183, 41125 Modena, Italy

Present Address:
G. Prampolini
CNR, Consiglio Nazionale delle Ricerche, Istituto per i Processi Chimico-Fisici (IPCF-CNR), UOS di Pisa, Area della Ricerca, via G. Moruzzi 1, 56124 Pisa, Italy

C. Puzzarini
Dipartimento di Chimica "G. Ciamician", Università di Bologna, Via F. Selmi 2, 40126 Bologna, Italy

F. Trani
CECAM, Centre Européen de Calcul Atomique et Moléculaire, EPF Lausanne, Batochime (BCH), 1015 Lausanne, Switzerland

simulate IR and UV–vis spectra of pyrimidine in the gas phase and in solutions, leading in all cases to good agreement with experimental observations and allowing to dissect different effects underlying spectral phenomena.

Keywords Computational spectroscopy · Integrated approaches · QM/MM/PCM · DFT · Post-Hartree–Fock · CCSD(T) · Composite schemes · Hybrid models · Molecular dynamics · Force fields · Time-independent approaches · Time-dependent approaches · Solvent effects · Vertical excitation energies · Electronic spectra · Spectra line-shape · Vibronic transitions · UV–vis · Resonance Raman · Vibrational spectra · IR intensities · Anharmonicity

1 Introduction

Spectroscopy represents a powerful tool for the characterization of molecular systems [1–4]. However, the interpretation of most experimental spectra is difficult due to their inherent complexity caused by thermal or environmental effects and/or to intrinsic properties of the system itself. In this context, computational spectroscopy turned out to be a valuable tool to help unraveling the various contributions to the spectroscopic signal, allowing a better understanding of the underlying phenomena [1, 2, 5–9].

The predictive and interpretative capabilities of computational spectroscopy can be demonstrated by means of quantum mechanical (QM) approaches that nowadays are able to provide results comparable in accuracy to experimental measurements [1, 2, 5, 6, 10–13]. At the state-of-the-art, the most advanced theoretical models are unavoidably restricted to small-sized molecular systems [10–13], mainly because of the high computational cost required to sample the potential energy surface (PES) by means of highly correlated post-Hartree–Fock methods [14]. Currently, the main challenge of computational spectroscopy is the reliable prediction and interpretation of spectra for large molecular systems in complex environments [15–18]. A possible route to obtain accurate results, even for relatively large molecular systems (several dozens of atoms), is provided by hybrid QM/QM′ models [19], which combine less expensive yet reliable electronic structure approaches (e.g., rooted into the Density Functional Theory (DFT)) for the computation of molecular structure, vibrational properties and/or anharmonic force fields (in both ground and excited electronic states) with *a posteriori* refinement of selected properties (harmonic frequencies, relative energies, electronic transitions, etc.) at a higher level of theory. In this respect, we can mention the high accuracy of hybrid coupled cluster (CC)/DFT anharmonic frequencies [20], vibronic energy levels [21] or spin–spin coupling terms [22]. Additionally, even if the implementation of general and efficient computational approaches aimed at the spectroscopic investigation of macrosystems is a very challenging task, appropriate schemes can already be introduced for complex cases, like large flexible biomolecules. An example is provided by the development of *ad hoc* tailored force fields [23–26], which make possible an extensive sampling of the ground and/or excited electronic state PES, combined with *a posteriori* refinements or computation of selected properties on a set of representative configurations [9, 27, 28]. Furthermore, the role of the surrounding environment can be exploited in a time-independent fashion employing the continuum solvation models [29–33], which provide accurate results at a relatively low computational cost [34] and can be combined with several QM models, also including simulations of electronic spectra line-shapes [35–38]. To get a more reliable and quantitative description of the spectral features, considering both electronic and vibrational properties, the basic formulation of continuum models has to be extended beyond the so-called reaction field effects, that is, the direct solvent effect on the molecular wavefunction and the indirect effect on the molecular structure must be accounted for [39–44]. Furthermore, on the basis of the chemical nature of the solute–solvent couple, combined discrete-continuum models, in which a few solvent molecules strongly and specifically interacting with the solute are treated explicitly leading to a sort of supermolecule embedded in a polarizable continuum, can be successfully employed [45–49]. The librational motions of solvent molecules can also be included by means of molecular dynamics or Monte Carlo simulations with explicit solvent approaches [28, 50, 51]. Finally, the molecular system embedded in complex environments can be effectively described by integrated approaches combining advanced quantum mechanics models, density functional theory, molecular mechanics (MM) and the description of environment through polarizable continuum models (PCM) [29–31] leading to the effective QM/DFT/MM/PCM scheme.

In addition to problems related to a reliable description of molecular systems, a direct comparison with the experimental outcomes requires computational tools able to explicitly simulate the spectroscopic features [6]. To this aim, several computational tools, covering a large panel of spectroscopies, in particular those of vibrational and electronic origin, have been developed and coded (see Ref. [44] and [52, 53] for extensive reviews). However, for a broader use of these advanced theoretical models, two key challenges still remain: (1) development of user-friendly yet flexible problem-tailored packages and (2) validation of accurate yet effective computational models, which allow to extend computational spectroscopy studies to relatively large and complex molecular systems. This paper will present the example of an integrated computational approach for the simulation of IR and UV–vis spectra of a medium-sized dye in the gas and condensed phase. The research field touched by this paper is broad and many scientists have made relevant contributions, as well as several other integrated computational approaches have been worked out in a number of labs all over the world. Therefore, a balanced review of theoretical models and integrated computational approaches for spectroscopy is beyond the possibilities of this work, and we are obliged to refer the interested reader to the broad, existing recent literature, as, for example, Ref. [6, 27, 28, 50, 53–61] and references therein. Here, we make a partial excursus of the field, with the main idea of pointing out a number of key steps that have made possible the building of the specific integrated protocol here presented, following the general idea of developing and validating transferable and user-friendly computational tools. In this context, we will discuss general-purpose approaches aimed at providing computational support for interpreting IR and UV–vis experimental spectra of complex systems, including their environment [8]. These approaches represent a more accurate treatment with respect to the methods still commonly in use in such cases (e.g., harmonic approximation or vertical electronic transitions) and contribute to a better understanding of experimental spectra for the large molecular systems that are routinely studied nowadays. Complementary time-dependent and time-independent routes to simulate spectra will be applied. These allow the evaluation of the line-shapes for vibrational spectroscopy (e.g., infrared) at the anharmonic level and vibrationally resolved electronic spectroscopy (one-photon, electronic circular dichroism). It will be shown that the integration between different computational tools leads to a more complete picture of the overall spectroscopic properties for the system under study, allowing at the same time the validation of computationally affordable approaches and the setup of hybrid schemes that permit accuracy as well as feasibility. The development and validation of such tools should be carried out by comparing results with those obtained from more accurate models and/or with state-of-the-art experimental data. In the present paper, we have chosen, as a test case, pyrimidine, a medium-sized molecule for which investigation by means of highly accurate electronic QM approaches is still feasible. Furthermore, the large amount of theoretical and experimental data available in the literature allows a detailed comparison for various kinds of spectra along with a discussion on the accuracy of the presented computational models. Moreover, being the precursor of nucleobases, pyrimidine is a molecule of general interest, and the present study will provide some additional details to the understanding of its structural and spectroscopic properties.

The paper is organized as follows. First, the computational details and theoretical models of the various approaches used in this work are briefly summarized in the Sect. 2 (a more detailed account can be found in the Supplementary Information (SI)). Next, highly accurate single- and multi-reference post-Hartree–Fock methods are applied to evaluate molecular structure (Sect. 3.1), vertical excitation energies (Sect. 3.2) and harmonic vibrational frequencies (Sect. 3.3) of isolated pyrimidine. These computations, which are among the most accurate theoretical predictions thus far reported, are employed to assess the accuracy of less expensive QM schemes rooted into density functional theory (Sects. 3.1–3.3), which in turn are applied to the time-independent simulation of the vibrational and vibrationally resolved electronic spectra (Sects. 3.3–3.5). The continuum solvent models are discussed in Sect. 3.4 with respect to their applicability to computational spectroscopy and in particular to the specific case of infrared (IR) spectra simulations in solution. In Sect. 3.6, the geometry and Hessian obtained at the DFT level are used to establish system-specific MM force fields employed in classical molecular dynamics (MD) simulations, which allow to account for dynamical and explicit/implicit solvent effects in protic and aprotic media (H_2O and CCl_4, respectively) on the $S_1 \leftarrow S_0$ electronic transition within the time-dependent approach detailed in Sect. 3.7. Finally, the time-dependent (classical) and time-independent (quantum) approaches employed to simulate electronic spectra line-shapes are also compared and their complementarity is pointed out. The examples are chosen to demonstrate the overall good accuracy attainable by such spectroscopic models when used in conjunction with DFT and MM computations. Further possible improvements through adjustments based on more accurate QM computations are also pointed out.

2 Computational approaches

In view of investigating the balance of computational feasibility and accuracy of results as well as to validate effective approaches applicable to large systems, different computational methods have been adopted to explore step by step the accuracy of less demanding approaches with respect to the more accurate (and expensive) ones. Additionally, several time-dependent and time-independent methods that allow to simulate vibrational and/or vibrationally resolved electronic spectra of the isolated molecule as well as of the system embedded in more complex environments have been employed. The focus is here on user-friendly computational approaches implemented in widely available QM packages or stand-alone codes, which allow effective application for relatively large systems.

2.1 Electronic structure computations

Best estimates of the molecular structure and harmonic force field of pyrimidine in its electronic ground state have been evaluated at the coupled cluster (CC) level of theory by means of composite schemes to account for the incompleteness of basis set and core-correlation effects (see SI section 1.1 for details). The CC singles and doubles approximation augmented by a perturbative treatment of triple excitations [CCSD(T)] [62] has mainly been used in conjunction with the cc-pVnZ [63] basis sets. These calculations have been carried out with the quantum-chemical CFour program package [64].

Post-Hartree–Fock computations of vertical electronic excitation energies have been performed employing the multi-reference second-order diagrammatic perturbation theory (MRPT) in the barycentric Møller–Plesset scheme [65]. The CIPSI algorithm [66, 67] has been applied in order to determine the active space for the configuration interaction (CI) calculations (see SI section 1.2 for details).

The DFT/aug-N07D model [68] developed for spectroscopic studies of medium-to-large molecular systems, which provides an excellent compromise between reliability and computational effort [69, 20], has been used for geometry optimizations, energy, energy gradients and harmonic and anharmonic force field computations in the ground and excited electronic states. Within the DFT approach, the standard B3LYP [70] and its long-range corrected extension CAM-B3LYP [71] functionals and the double-hybrid functional B2PLYP [72] have been employed along with the aug-N07D [73] basis set built from the polarized double-ζ N07D [68] (see SI Section 1.3 for details). All DFT calculations have been performed with the Gaussian suite of quantum chemistry programs [74].

2.2 Force field (FF) development and molecular dynamics simulations

The intramolecular FF parameters for pyrimidine have been obtained with the Joyce [26] program. Details on the FF development and validation can be found in the supplementary information (SI Section 1.4). Classical molecular dynamics simulations of pyrimidine in water and CCl_4 have been performed using both standard periodic boundary conditions (PBC) and the recently proposed general liquid optimized boundary (GLOB) [75, 76] model, an effective hybrid discrete/continuum method for simulating complex systems in soft condensed matter. All simulations have been carried out with the GROMACS software [77] (PBC) and the JERSEY program [78] (GLOB). For both water and CCl_4 solvents, pre-equilibrated solvent boxes have been obtained through preliminary molecular dynamics runs in the NPT ensemble at 298 K and 1 atm, whereas geometry optimizations were performed according to the steepest-descent method [79] (see SI Section 1.5).

2.3 Vibrational spectra computations

Anharmonic vibrational spectra have been calculated by means of the vibrational second-order perturbation theory (VPT2) [80–86], exploiting the fully automated generalized second-order vibrational perturbation (GVPT2) approach [87–89]. The anharmonic force fields have been computed at the DFT/aug-N07D level and additionally, the hybrid CCSD(T)/DFT approach [20, 90, 91] has also been used to evaluate anharmonic frequencies. Concerning the calculation of anharmonic frequencies and intensities in solution, the extension of the GVPT2 model to the polarizable continuum model [92] has been used, and the reported data refer to both the fully equilibrium and the vibrational nonequilibrium solvation regimes [93, 94]. Vibrational harmonic and anharmonic intensities also account for cavity field contributions [43, 95]. For further details, see SI Section 1.6. All VPT2 computations have been performed using a locally modified version of the Gaussian suite of programs [74].

2.4 Electronic spectra simulations

Within the time-independent framework, vibrationally resolved electronic spectra have been simulated through an integrated procedure (described in detail in Ref. [21, 96, 97]) based on the computation of overlap integrals, also known as Franck–Condon (FC) integrals, between the vibrational wavefunctions of the electronic states involved in the transition. Both adiabatic and vertical approaches have been applied to simulate the one-photon absorption (OPA) and the vibrational resonance Raman (vRR) spectra, including the Duschinsky and Herzberg-Teller effects (see SI Section 1.7). Time-independent vibrationally resolved electronic spectra have been computed using locally modified versions of the Gaussian [74] and *FCclasses* [98] programs for OPA and vRR, respectively. They implement an effective prescreening technique of the relevant transitions developed by some of the authors [99]. Alternative routes have been proposed in literature, see for example Ref. [100–102]. In addition, the electronic spectra lineshapes have also been obtained within a time-dependent framework [9, 51] based on classical MD simulations and on-the-fly or *a posteriori* calculation of spectroscopic properties (see SI Section 1.8). Nonspecific solvent effects on the UV-vis spectrum of pyrimidine have been introduced by a conductor-like polarizable continuum model (CPCM) [103] with vertical excitation energies computed by the linear response LR-PCM/TD-DFT approach [41] within the non-equilibrium regime [104–106].

3 Results and discussion

3.1 The ground-state molecular structure of pyrimidine

The structure of pyrimidine is shown in Fig. 1 along with the atom labeling scheme, while Table 1 summarizes all the geometries obtained in the present work and compares them with experimental data available in the literature [107]. From the given results, it is first observed that frozen-core (fc) CCSD(T)/cc-pVQZ calculations already provide results close to the convergence with respect to the complete basis set (CBS) limit, as the corresponding differences are quite small. The bond lengths decrease by less than 0.002 Å, less than 0.0005 Å for those involving H, and the angles vary by about 0.1° or even less. On the other hand, the reliability of extrapolations carried out employing triple- and quadruple-zeta basis sets has been proven in Ref. [108]. Second, we note that the core-correlation effects are small. Indeed, for angles, the corrections are in all cases smaller than 0.1 degrees, and for bond distances, they amount to about 0.002 Å, about 0.001 Å for bonds involving H. Before proceeding, general considerations on the accuracy of CCSD(T) optimized geometries are warranted. On the basis of the literature available (see for instance Refs. [109, 110] and references therein), the accuracy of CBS+CV structures is expected to be of the order of 0.001–0.002 Å for distances and of about 0.05–0.1° for angles. These are conservative estimates and

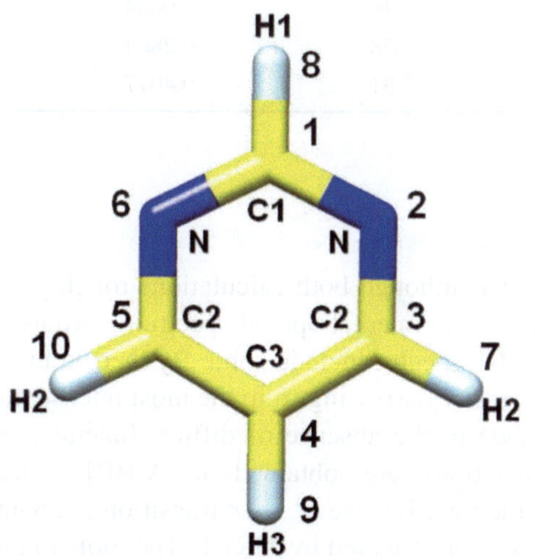

Fig. 1 Geometry and atom numbering scheme employed for pyrimidine

also account for the approximations made and for neglecting higher-order correlation effects. Concerning the accuracy of the DFT model, on the basis of the comparison reported in Table 1, the conclusion drawn is that both the B3LYP/aug-N07D and CAM-B3LYP/aug-N07D levels provide results of better quality than the CCSD(T)/cc-pVTZ ones. The only exception is given by the C–H bond lengths that result to be too long. The considerations are particularly encouraging as the DFT/aug-N07D approach can be easily applied to compute the structure of significantly large systems (e.g., more than 100 atoms) and then used to develop accurate intramolecular force fields, thus allowing molecular dynamics simulations (*vide infra*). As far as the comparison with experiment is concerned, we first note that the available experimental geometries are partial structures and none of them is an equilibrium structure. In most cases, CBS+CV and experimental values agree within the associated uncertainties. The largest discrepancies are observed for the $C_2{-}C_3$ distance for which experiment seems to overestimate the value by about 0.006 Å, and for a few angles in the case of the geometry denoted ED+LCNMR+MW, which is based on electron diffraction and liquid crystal NMR data combined with rotational constants of the parent isotopologue [107].

3.2 Post-Hartree–Fock and TD-DFT computation of vertical excitation energies in the gas phase

Energies of the ground and a number of singlet excited electronic states, as arising from excitations in the valence space, have been computed at the best estimated structure of the ground state (see Sect. 3.1 and Table 1). Two different methods have been employed for the vertical excitation (VE) spectrum: the post-Hartree–Fock MRPT method and the popular single-reference time-dependent density functional theory (TD-DFT) approach. For the latter, the CAM-B3LYP [71] functional in conjunction with the aug-N07D [73] and aug-cc-pVQZ [111, 112] basis sets has been used. The choice of functional has been related to its good performance over the large spectra range, as shown, for example, by TD-CAM-B3LYP results for Rydberg and charge transfer excitation energies or static electronic polarizabilities which are notably improved over the standard B3LYP functional [113]. Additionally, both basis sets used for TD-DFT computations include several diffuse functions, thus allowing the representation of a limited number of Rydberg states. In contrast, the MRPT computations have been performed with the 6-311G(d) basis set, and thus, the Rydberg states have not been considered.

In Table 2, the vertical excitation energies obtained from both approaches are compared to the experimental maxima from Ref. [114]. It is apparent that MRPT performs very well for the lowest excited states being

Table 1 Molecular structure of pyrimidine

	CCSD(T)/ cc-pVTZ	CCSD(T)/ cc-pVQZ	CBS	CBS+CV	B3LYP/ aug-N07D	CAM-B3LYP/ aug-N07D	Exp./r_s[a]	Exp./ ED+LCNMR+MW[a]
C_1–N	1.3402	1.3369	1.3351	1.3329	1.3376	1.3310	1.337(2)	1.328(7)
N–C_2	1.3411	1.3380	1.3363	1.3341	1.3386	1.3322	1.332(3)	1.350(7)
C_2–C_3	1.3932	1.3907	1.3891	1.3868	1.3924	1.3861	1.393(2)	1.393(4)
C_1–H_1	1.0837	1.0833	1.0831	1.0820	1.0887	1.0881	–	1.082
C_2–H_2	1.0845	1.0839	1.0837	1.0826	1.0893	1.0884	–	1.082
C_3–H_3	1.0813	1.0808	1.0806	1.0795	1.0856	1.0843	–	1.082
∠N–C_1–N	127.75	127.50	127.37	127.28	127.05	126.89	127.2(3)	128.3
∠C_1–N–C_2	115.36	115.58	115.69	115.75	115.94	116.06	115.8(3)	115.75
∠N–C_2–C_3	122.46	122.36	122.31	122.28	122.23	122.22	122.4(3)	121.2(3)
∠C_2–C_3–C_2	116.60	116.61	116.64	116.65	116.61	116.56	116.4(2)	117.8(2)
∠H_1–C_1–N	116.12	116.25	116.32	116.36	116.48	116.55	–	–
∠H_2–C_2–N	116.38	116.42	116.43	116.46	116.48	116.50	–	–
∠H_3–C_3–C_2	121.70	121.69	121.68	121.67	121.69	121.72	–	120.9(3)

Geometry parameters computed at the coupled cluster and DFT levels of theory and compared with available experimental data

Distances in Å and angles in degrees

[a] Ref. [107]

Table 2 Vertical excitation spectrum as computed at the MR-PT and TD-DFT levels

Assignment	Sym.	MRPT		TD-CAMB3LYP/aug-N07D		TD-CAMB3LYP/aug-cc-pVQZ		Exp.[a]
		E (eV)	f value	E (eV)	f value	E (eV)	f value	E (eV)
$\pi* \leftarrow n_N$	B_1	4.19	0.0042	4.58	0.0062	4.56	0.0053	4.18
$n_N \leftarrow n_N$	A_2	4.59	0.0000	4.94	0.0000	4.93	0.0000	4–5
$\pi* \leftarrow \pi$	B_2	5.16	0.0380	5.85	0.0453	5.84	0.0472	5.22
$\pi* \leftarrow n_N$	A_2	5.68	0.0000	6.03	0.0000	5.99	0.0000	
$\pi* \leftarrow \pi$	B_1	5.99	0.0034	6.33	0.0059	6.30	0.0056	~6
$\pi* \leftarrow \pi$	A_1	7.02	0.0715	6.71	0.0402	6.68	0.0394	6.69
$3s \leftarrow n_N$	B_2			6.86	0.0051	6.82	0.0053	6.42
$3p \leftarrow n_N$	B_1			7.37	0.0175	7.35	0.0160	7.44
$3p \leftarrow n_N$	A_1			7.5	0.2236	7.42	0.1629	6.97
$3p \leftarrow n_N$	B_2			7.56	0.0896	7.48	0.0684	7.05
$\pi* \leftarrow \pi$	A_1	7.79	0.4504	7.63	0.2292	7.58	0.2864	7.48
$\pi* \leftarrow \pi$	B_2	8.13	0.5118	7.84	0.4003	7.81	0.4077	

The oscillator strengths are in the mixed gauge

The experimental vertical transition energies are also reported

[a] From Ref. [114]

competitive with respect to previous calculations performed with the equation-of-motion (EOM)-CC method (see Ref. [114] for extensive review of experimental and theoretical results), whereas TD-CAM-B3LYP overestimates the excitation energies by 0.35–0.7 eV. It should be noted that the TD-CAM-B3LYP/aug-N07D level of theory provides results already converged with respect to the basis set limit, with energies higher than those computed at the TD-CAM-B3LYP/aug-cc-pVQZ level by only ~0.05 eV. The oscillator strengths appear to be systematically higher for TD-DFT although both calculations roughly reproduce the overall experimental optical spectrum. Around 6–7 eV, the TD-DFT spectra include some Rydberg transitions for which intensity borrowing from the most intense transitions occurs. Due to the absence of diffuse functions, no Rydberg transitions are obtained in MRPT calculations. Finally, the very intense $\pi^* \leftarrow \pi$ transition at about 7.5 eV is slightly overestimated by MRPT. The global intensity of these transitions is higher for MRPT because of the intensity borrowing of the TD-DFT transitions to Rydberg

Table 3 Harmonic vibrational frequencies (in cm^{-1}) and IR intensities (in km/mole) computed at the CC and DFT levels

Mode[a]	Sym.	CCSD(T)/ cc-pVTZ		CCSD(T)/ cc-pVQZ		CBS[b]	CBS+CV[c]	B3LYP/ aug-N07D		CAM-B3LYP/ aug-N07D		B2PLYP/ aug-cc-pVTZ	
		ω	Int	ω	Int	ω	ω	ω	Int	ω	Int	ω	Int
1	A1	3,219	7.3	3,217	6.6	3,217	3,223	3,203	9.3	3,230	5.3	3,222	7.9
2	A1	3,189	14.7	3,190	12.7	3,191	3,196	3,175	13.7	3,197	10.5	3,193	12.3
3	A1	3,173	9.4	3,174	9.5	3,175	3,180	3,158	12.7	3,182	12.9	3,178	11.4
4	A1	1,610	32.1	1,613	34.8	1,616	1,621	1,616	47.3	1,659	58.8	1,608	37.6
5	A1	1,425	55.7	1,430	55.6	1,432	1,437	1,437	55.8	1,461	56.3	1,436	57.2
6	A1	1,154	3.0	1,158	2.4	1,160	1,164	1,161	1.3	1,179	0.8	1,161	1.8
7	A1	1,077	1.4	1,079	1.3	1,080	1,084	1,079	1.0	1,095	1.0	1,083	1.0
8	A1	1,001	2.8	1,004	2.8	1,005	1,009	1,013	3.8	1,032	4.4	1,007	3.3
9	A1	684	3.5	686	3.2	687	689	692	3.0	700	3.2	692	2.9
12	B2	3,177	18.5	3,178	15.6	3,179	3,184	3,162	18.1	3,185	14.1	3,181	15.5
13	B2	1,614	72.1	1,616	74.5	1,618	1,623	1,616	84.9	1,659	94.7	1,609	78.0
14	B2	1,493	8.8	1,494	8.1	1,495	1,499	1,495	5.8	1,516	6.4	1,497	5.3
15	B2	1,390	0.3	1,392	0.4	1,392	1,395	1,390	0.3	1,401	0.3	1,400	0.3
16	B2	1,247	6.0	1,248	6.1	1,248	1,253	1,262	3.8	1,271	4.5	1,262	4.7
17	B2	1,193	4.4	1,195	5.4	1,196	1,199	1,228	8.2	1,230	8.4	1,226	7.9
18	B2	1,088	2.0	1,089	1.9	1,088	1,091	1,095	3.1	1,110	2.7	1,094	3.1
19	B2	625	11.5	626	11.4	626	628	634	11.3	641	11.6	631	11.5
20	B1	1,023	0.0	1,024	0.0	1,027	1,031	1,022	0.0	1,047	0.0	1,036	0.0
21	B1	978	0.0	977	0.0	976	980	980	0.1	1,000	0.3	991	0.0
22	B1	819	7.0	818	6.1	817	820	821	5.3	836	5.1	828	6.2
23	B1	732	39.9	732	39.7	731	734	738	41.0	747	42.9	742	39.7
24	B1	350	3.0	349	3.2	349	351	351	3.5	362	3.6	352	3.5
10	A2	1,001	0.0	1,001	0.0	1,001	1,004	998	0.0	1,019	0.0	1,010	0.0
11	A2	402	0.0	403	0.0	403	404	408	0.0	417	0.0	408	0.0

[a] Normal mode numbering as in Ref. [121, 120]

[b] Computations at the (fc)CCSD(T)/CBS level, see text for details

[c] Computations at the (fc)CCSD(T)/CBS+CV(MP2/CVTZ) level, see text for details

states. As far as the excitation spectrum is concerned, on the whole it seems that both methods are able to give reliable results. Concerning TD-B3LYP computations, it has been found that the excitation energies and oscillator strengths for the first two excited electronic states agree very well with MRPT as well as with previous theoretical and experimental data [114] (4.29 eV/0.0047 and 4.62 eV/0.0000, respectively), but, as expected, TD-B3LYP does not provide qualitatively correct results for the energy order and the nature of higher-lying excited electronic states. Thus, for the sake of consistency, the TD-CAM-B3LYP functional has been chosen to compute the electronic spectra line-shapes for all excited electronic states considered in this work; for the six lowest electronic transitions, a hybrid approach with energies corrected by means of the accurate MRPT computations has been used. These results will be discussed in Sect. 3.5, with emphasis on the comparison between experimental and computed band intensities in the 3.5–9.5 eV energy range.

3.3 Vibrational properties of isolated pyrimidine in the ground electronic state

Table 3 summarizes the harmonic frequencies and infrared (IR) intensities obtained at the CC and DFT levels. In regard to CCSD(T) calculations of harmonic frequencies, we note that the (fc)CCSD(T)/cc-pVQZ level of theory already provides converged results with respect to the CBS limit, the differences between CBS and cc-pVQZ being generally smaller than 1 cm^{-1} and being 4 cm^{-1} at most. Core-correlation corrections are in all cases positive (i.e., they enlarge the frequency value) and small (i.e., they range from 1 to 5 cm^{-1}). Moving to IR intensities, it can be noted that also in this case the (fc)CCSD(T)/cc-pVQZ computations are expected to yield results converged with respect to basis set completeness, as the differences between cc-pVTZ and cc-pVQZ are on average of 1 km/mol and not exceeding 3 km/mol. Taking into account that for IR intensities extrapolation schemes have not been formulated yet and that core-correlation corrections at the MP2/cc-pCVTZ level are smaller than 0.5 km/mol, the (fc)CCSD(T)/cc-pVQZ values can be considered as best estimates for the IR intensities.

The comparison of best theoretical estimates with DFT results confirms the overall good accuracy of the latter. With respect to CC reference values, the B3LYP functional shows a mean absolute error (MAE) of 8 cm^{-1} for frequencies and 2 km/mol for intensities. As already suggested (see Ref. [115, 116]), the accuracy of DFT results is further improved by using the double-hybrid functional B2PLYP and the aug-cc-pVTZ basis set, leading in the present case to a MAE of 6 cm^{-1} for frequencies and showing improvement also for the IR intensities (MAE of 1 km/mol). Thus, B2PLYP computations in conjunction with tiple-ζ quality basis sets can be considered as an alternative to coupled cluster computations in view of improving the accuracy of vibrational properties within hybrid models [20, 115] (*vide infra*). Finally, we also note that the CAM-B3LYP/aug-N07D results show a rather good agreement with the CC reference values (MAE of 16 cm^{-1} for frequencies and 3 km/mol for intensities); therefore, such a level of theory can be applied to cases where long-range effects are important, for example, computations in which both electronic ground and excited states need to be treated at the same time (see Sects. 3.5 and 3.7). Concerning computations beyond the harmonic approximation, it is well recognized that the GVPT2 model applied to anharmonic force fields evaluated at the DFT level (in particular with the Becke-family hybrid functionals, see Ref. [44, 20, 115] for general discussions) and coupled with appropriate criteria to define Fermi [117] and Darling–Dennison resonances is able to provide accurate results for semirigid systems [118]. For pyrimidine, it has been shown that such computations lead to frequencies with an average error with respect to experiment of about 10 cm^{-1} or less [119, 120]. In this work, we emphasize that such an (already good) agreement can be further improved by applying high-level corrections to the harmonic part of the force field.

The anharmonic frequencies computed at the DFT/aug-N07D level and those obtained by means of hybrid CC/DFT schemes are listed in Table 4 and compared to the most recent experimental data by Breda et al. [121], who reported the entire fully resolved Fourier-transform (FT)IR spectra recorded in solid argon. As expected, B3LYP/aug-N07D results show very good agreement with the experiment (MAE smaller than 10 cm^{-1}). Further improvements have been obtained by replacing the harmonic counterpart with the best theoretical estimate. Additionally, the small MAE in the case of hybrid CC/CAM-B3LYP computations points out that the larger discrepancies observed for the CAM-B3LYP itself can be entirely ascribed to the harmonic part of the potential. It should be noted that in the present case, resonant interactions are already well defined at the DFT level; thus, both hybrid approaches (see SI Section 1.6), the *a posteriori* one ($\omega_{CC}+\Delta\nu_{DFT}$) and the so-called *direct* model (ω_{CC}/K_{DFT}), lead to similar improvements over DFT results (for explanation of $\omega_{CC}+\Delta\nu_{DFT}$ and ω_{CC}/K_{DFT} models, see SI Section 1.6). In general terms, the *direct* model is necessary when harmonic DFT force field either gives an incorrect prediction of resonant terms or shows large inaccuracies. Computational results have also been used to simulate IR spectra line-shapes (convoluted with Gaussian functions with an half-width at half-maximum (HWHM) of 1 cm^{-1}) directly comparable with the experimental FTIR ones [121] (see Fig. 2). For such comparison, the theoretical spectrum beyond the harmonic approximation also includes overtones and combination bands, whose intensities have been evaluated from DFT GVPT2 computations, while intensities of fundamental bands have been evaluated through the hybrid CC/DFT *a posteriori* scheme. It is evident that anharmonic corrections to both vibrational frequencies and intensities are necessary to correctly reproduce the experimental outcomes; they are thus recommended for an unequivocal analysis and assignment of experimental spectra. The conclusion that can be drawn from this section is that the hybrid models presented in this work are cost-effective approaches to simulate vibrational spectra, which combine the feasibility of accurate harmonic frequency computations with the possibility of properly accounting for vibrational effects beyond the harmonic approximation. In this respect, the CC/B3LYP approach can be applied to systems with up to 10–15 atoms, while the B2PLYP/B3LYP one allows extension toward larger systems of biological and/or technological interest.

3.4 Vibrational anharmonic spectra of solvated pyrimidine: time-independent approach

Due to the availability of analytical expressions of the energy second derivatives, the harmonic oscillator/rigid rotor model can be nowadays routinely applied to solvated systems treated within continuum solvation models, giving reasonably accurate results and reproducing the main features of vibrational spectra of systems in solution [43, 44, 92, 93]. However, methods for evaluating anharmonic effects on vibrational frequencies of solvated systems are still in their infancy, with only few attempts reported in the literature so far [42, 92, 93]. In this view, recent papers by some of the authors [92, 93] have reported promising results, showing that both the accuracy and the scaling characteristics of the corresponding computations in vacuum can be retained for both IR frequencies and intensities. Before discussing the results of the application of the

Table 4 Anharmonic vibrational frequencies of pyrimidine obtained from GVPT2 computations with the semi-diagonal quartic force fields evaluated at the DFT and hybrid CC/DFT levels (see SI, Section 1.6 for details) and compared to the most recent experimental data (FTIR spectrum of pyrimidine isolated in solid argon form Ref. [121])

Mode[a]	Sym.	Assignment[b]	B3LYP/aug-N07D			CAM-B3LYP/aug-N07D			Exp[c]
			DFT	$\omega_{CC}+\Delta\nu_{DFT}$	ω_{CC}/K_{DFT}	DFT	$\omega_{CC}+\Delta\nu_{DFT}$	ω_{CC}/K_{DFT}	ν
1	A1	ν_{C3H3}	3,055	3,074	3,072	3,089	3,081	3,075	3,091.4
2	A1	ν_{C1H1}	3,053	3,074	3,073	3,050	3,049	3,074	3,057.4
3	A1	ν_{CH}sym	3,030	3,052	3,051	3,075	3,073	3,051	3,041
4	A1	ν_{CC}sym + ν_{CN}sym	1,573	1,578	1,578	1,616	1,579	1,576	1,567.3
5	A1	δ_{CH}sym + ν_{CN}sym	1,408	1,408	1,408	1,433	1,409	1,408	1,400.6
6	A1	ν_{CN}sym + δ ring	1,141	1,145	1,143	1,161	1,146	1,150	1,138.2
7	A1	ν_{CC}sym + δ ring	1,053	1,057	1,074	1,070	1,059	1,072	1,070.7
8	A1	δ ring + ν_{CN}sym	997	993	993	1,017	993	992	989.5
9	A1	δ ring	683	680	680	692	681	680	677.7
12	B2	ν_{CH}as	3,033	3,055	3,054	3,055	3,054	3,053	3,052.1
13	B2	ν_{CC}as + $\nu_{CN}as$	1,557	1,564	1,562	1,615	1,579	1,562	1,570.5
14	B2	δ_{CH}	1,463	1,467	1,467	1,486	1,468	1,467	1,464.8
15	B2	δ_{CH}	1,362	1,367	1,366	1,373	1,367	1,367	1,374
16	B2	ν_{CN} + δ_{CH}	1,228	1,220	1,220	1,245	1,227	1,224	1,223.2
17	B2	ν_{CN}as + $\nu_{CC}as$	1,197	1,167	1,170	1,197	1,166	1,163	1,156.8
18	B2	δ_{CH}	1,075	1,072	1,072	1,091	1,073	1,072	1,073.9
19	B2	σ ring	626	621	621	634	621	621	620.9
20	B1	γ_{CH}sym	1,002	1,010	1,011	1,024	1,008	1,008	
21	B1	γ_{CH}	958	958	959	979	958	957	
22	B1	τ ring + γ_{CH}	807	806	806	820	805	804	803.1
23	B1	τ ring + γ_{CH}	723	719	719	731	718	717	719.2
24	B1	τ ring	344	343	343	354	342	342	
10	A2	γ_{CH}as	979	985	985	998	983	982	
11	A2	τ ring	400	396	396	409	396	396	
MIN[d]			−36	−17	−20	−7	−12	−17	
MAX[d]			40	16	15	49	32	17	
MAE[d]			11	7	6	19	7	6	

[a] Normal mode numbering as in Ref. [120, 121]

[b] Assignment from Ref. [121] based on the potential energy distribution (PED) analysis, checked through visual inspection of atomic displacements along normal modes

[c] Ref. [121]

[d] Mean absolute error (MAE), and largest negative (MIN) and positive (MAX) deviations with respect to experiment

newly formulated PCM for anharmonic spectra of solvated systems (see Refs. [92, 93]), it is worth recalling some basic concepts. The formulation of a physically consistent continuum model to treat IR intensities in solution requires to consider that formally the "local field" acting on the solute does not coincide to the external field applied to the system, as in the case for the isolated molecule. However, the response of the molecule in the cavity within a dielectric medium depends on the field locally acting on the molecule in the cavity, so that, as already pointed out in the literature [43, 44, 93], in order to gain a reliable connection between the microscopic properties of the molecule and the macroscopic response, it is compulsory to account for the difference between the Maxwell and cavity fields. In addition, due to the fact that spectroscopic properties are by definition time-dependent phenomena (due to the oscillatory nature of the radiation field), a time dependence has also to be taken into account in the definition of the solvation model. In the field of IR spectroscopy, the time dependence of the radiation field is to be considered in the formulation of the cavity field, which is reflected in a proper definition of the dielectric permittivity. However, in the case of an oscillatory motion caused by an external electric field on a molecule surrounded by a dielectric

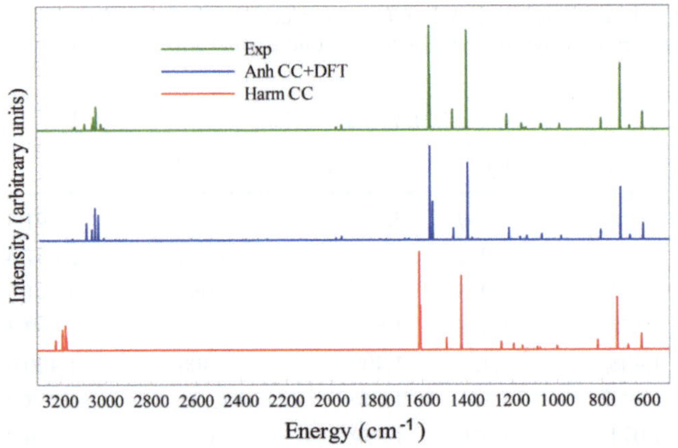

Fig. 2 Computed and experimental [121] IR spectra of isolated pyrimidine. IR spectra *line*-shapes have been convoluted with Gaussian functions with a HWHM of 1 cm^{-1}

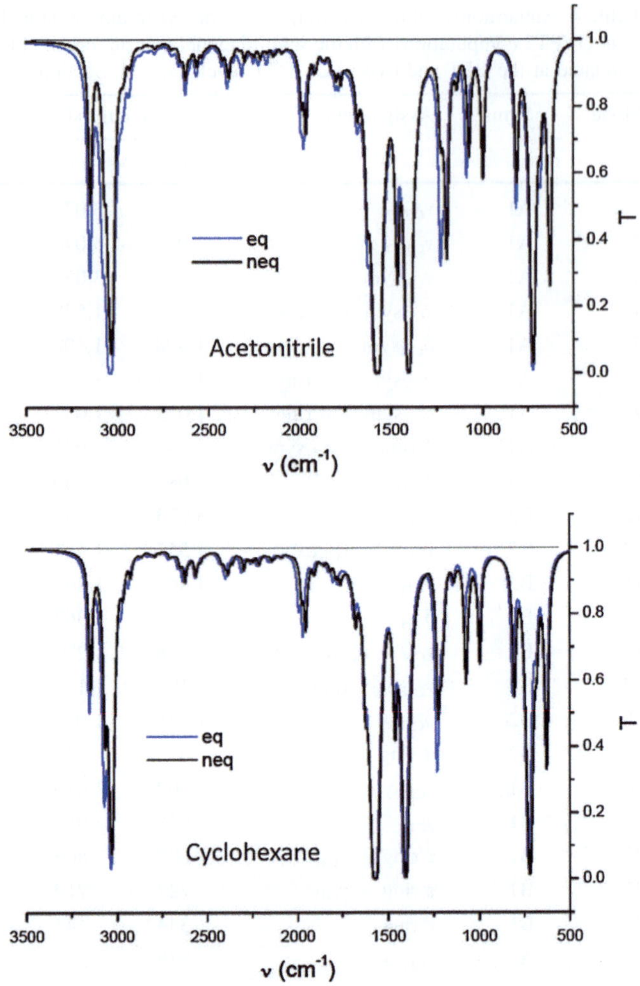

Fig. 3 Calculated IR spectra of pyrimidine in acetonitrile (*top*) and cyclohexane (*bottom*). IR spectra *line*-shapes have been convoluted with Gaussian functions with a HWHM of 2 cm^{-1}

medium, the definition of the solvent response is different whether the solvent nuclear and electronic distributions are assumed to instantaneously rearrange to follow the oscillating molecule (vibrational equilibrium solvation regime) or there exists some dephasing between the solute and the solvent motions (vibrational nonequilibrium solvation regime), so that the solvent distribution is only partly equilibrated to the oscillating molecule. Nonequilibrium effects are most usually addressed in the field of electronic phenomena (electronic excitations, emission or electronic response properties), for which there is a wide literature published in the field [39, 40]. Less standard and much less studied is the consideration of solvent nonequilibrium effects connected to the vibration of a solute surrounded by a continuum dielectric. In the latter case, in fact, to the best of our knowledge, only the contributions given to the groups of Rivail et al. [42] (limited to harmonic frequencies only) and Tomasi et al. [94] (addressing harmonic IR frequencies and intensities [94], Raman activities [122] and VCD rotational strengths [123]) were present in the literature before the recent extensions to treat IR frequencies and intensities beyond the harmonic approximation [93]. Back to pyrimidine, anharmonic GVPT2 vibrational frequencies and the deperturbed (DVPT2) IR intensities computed in the gas phase and CS_2 solution are compared in Table 5, clearly showing that the solvent effect cannot be simulated through simple scaling of gas-phase frequencies and intensities. For instance, the frequency of the ν_4 mode is only little affected by the solvent, while its intensity computed within the nonequilibrium regime is enhanced by more than 15 km/mol. Another example is provided by the ν_{13} mode, as its frequency increases by more than 10 cm^{-1} in solution (neq. regime), while its intensity remains almost unaffected.

More detailed insights on solvent effects and on the differences between the equilibrium and nonequilibrium regimes are provided by the simulated anharmonic spectra shown in Fig. 3 for pyrimidine in cyclohexane and acetonitrile solutions in the 500–3,500 cm^{-1} range. The usual equilibrium approach and the full vibrational nonequilibrium model are compared: in both cases, cavity field effects are included, so that the IR intensities obtained are directly comparable to experimental findings. In addition, not only fundamentals, but also overtones and combination bands are reported. In general, the frequencies of the various bands are quite similar, irrespective of the solvent and of the solvation regime assumed (although nonequilibrium values tend to be slightly higher in frequency). A notable difference is instead noticed for intensities (which are reported as transmittance values, as it is common in the IR spectroscopy): not only absolute values, but especially relative intensities differ remarkably by changing the solvating environment, and for a given solvent, moving from

Table 5 Anharmonic vibrational frequencies (cm^{-1}) and IR intensities (km/mol) of isolated pyrimidine and in CS_2 solution

Mode[a]	Sym.	gas phase		CS_2 eq.		CS_2 neq.	
		ν	Int	ν	Int	ν	Int
1	A1	3,055	14.7	3,073	16.0	3,066	7.1
2	A1	3,053	6.9	3,047	2.5	3,042	1.7
3	A1	3,030	22.3	3,034	20.1	3,028	15.9
4	A1	1,573	80.9	1,577	82.3	1,573	96.7
5	A1	1,408	55.5	1,412	69.1	1,402	77.1
6	A1	1,141	1.4	1,145	0.8	1,136	1.2
7	A1	1,053	1.1	1,059	0.6	1,052	0.0
8	A1	997	4.0	999	3.9	996	5.8
9	A1	683	2.9	684	3.1	680	2.1
12	B2	3,033	23.1	3,039	22.0	3032	12.4
13	B2	1,557	43.6	1,574	43.5	1,569	43.7
14	B2	1,463	5.6	1,468	6.2	1,461	9.1
15	B2	1,362	0.1	1,371	1.8	1,359	0.1
16	B2	1,228	6.5	1,232	14.2	1,223	8.3
17	B2	1,197	6.0	1,200	3.5	1,191	0.4
18	B2	1,075	3.6	1,082	3.1	1,071	6.8
19	B2	626	11.7	631	12.5	626	14.5
20	B1	1,002	0.0	1,010	0.0	1,000	0.0
21	B1	958	0.1	965	0.0	956	0.1
22	B1	807	5.5	816	7.5	805	7.2
23	B1	723	40.6	728	47.7	719	63.1
24	B1	344	3.5	347	3.7	340	5.4
10	A2	979	0.0	988	0.2	979	0.0
11	A2	400	0.0	402	0.0	397	0.0

For CS_2 solution, the PCM approach within equilibrium (eq.) and nonequilibrium (neq.) models has been applied. All computations at the B3LYP/aug-N07D level

[a] Normal mode numbering as in Ref. [121, 120]

equilibrium to nonequilibrium. Furthermore, the whole solvent effect and the nonequilibrium component do not act homogeneously on each band; therefore, the reported spectra cannot simply be obtained through a scaling procedure (see also Table 5) as bands intensities may be differently affected by changes in the solvating environment. Such a feature of the PCM model (i.e., the different effects on each vibrational mode) is strictly connected to the definition of the cavity, of molecular shape. Thus, the actual geometrical properties of the molecular system under study come into play, as it is physically consistent. Such a characteristics is already included in some continuum solvation models [31–33], but several others based on a spherical description of the molecular cavity [43, 44] are still used to analyze solvent effects on experimental vibrational spectra. In order to get a better insight into the reliability of the results computed with the PCM, calculated and experimental spectra are compared in Fig. 4. In this case, computed values refer to a CS_2 solution of pyrimidine, while experimental data [124] refer to a solution 10 % in CCl_4 for the 3,800–1,370 cm^{-1} range, and 10 % in CS_2 for the 1,370–450 cm^{-1} one. Due to the small difference in the dielectric properties of CS_2 and CCl_4, the calculations have only been performed in CS_2. On the whole, the experimental spectrum is well reproduced. A small shift in frequency absolute values is observed, which is of the same order as those associated with the B3LYP/aug-N07D GVPT2 computations in the gas phase. As far as intensities are concerned, the relative values are reliably reproduced, and notably, a good accuracy is reported for the 2,000–3,000 cm^{-1} spectra range, where anharmonicity plays a substantial role.

3.5 Vibrationally resolved electronic absorption and resonance Raman spectra of isolated pyrimidine, adiabatic and vertical time-independent models

3.5.1 *One-photon absorption spectra*

Pyrimidine S_0 equilibrium structure belongs to the C_{2v} symmetry point group. The first excited state S_1 belongs to the B_1 irreducible representation (irreps) in C_{2v}, and its PES presents a saddle point, with an imaginary frequency

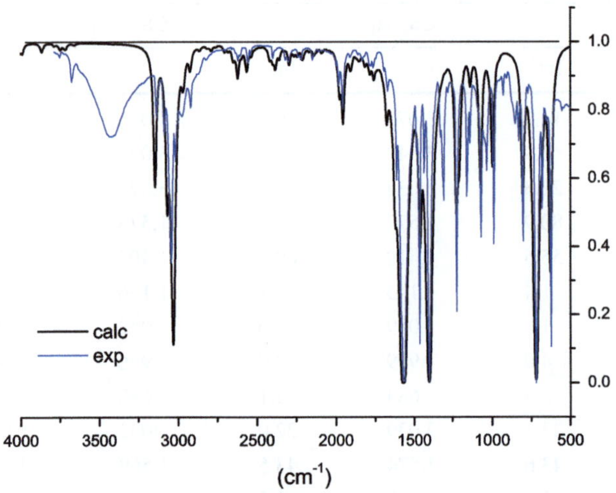

Fig. 4 Calculated versus experimental IR spectrum of pyrimidine in CS_2 solution. IR spectra *line*-shapes have been convoluted with Gaussian functions with a HWHM of 2 cm^{-1}. The experimental spectrum is taken from NIST [124] and refers to a solution 10 % in CCl_4 for 3,800–1,370 cm^{-1}, and 10 % in CS_2 for 1,370–450 cm^{-1} and has a resolution of 2 cm^{-1}

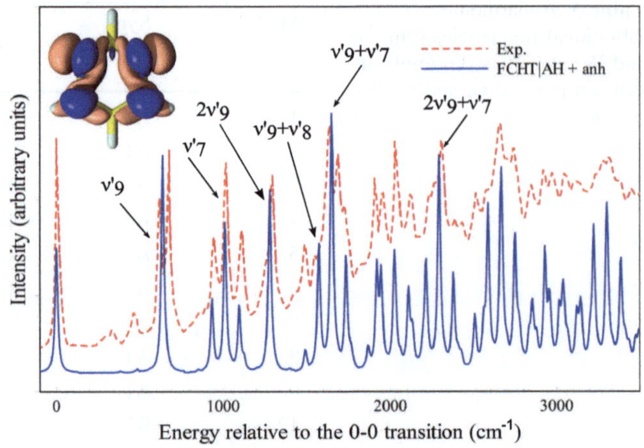

Fig. 5 Computed and experimental [114] spectra for the $S_1 \leftarrow S_0$ electronic transition of isolated pyrimidine, along with the assignment of the most intense vibronic bands. Both spectra are reported with the corresponding 0–0 transition energy set to 0. The theoretical spectrum has been simulated with the Gaussian function with HWHM of 10 cm^{-1} and corrected for anharmonicity though effective scaling procedure using the best CC/B3LYP estimates for the electronic ground-state frequencies. Electron density difference is also plotted (with isovalue of 0.002); regions that gain electron density upon $S_1 \leftarrow S_0$ transition are shown as dark blue, while regions that loose electron density are presented in light coral

of $i440$ cm^{-1}, along a b_2 mode ($Q_1^{(C_{2v})}$), roughly corresponding to the ground-state normal mode of ν_{17}). Distorting the geometry along such a mode, two degenerate global minima arise in C_s symmetry, with S_1 belonging to the A'' irreps. These features point out the existence of an inherently anharmonic double-well energy profile. Therefore, an accurate calculation of the $S_1 \leftarrow S_0$ absorption spectrum should require a full anharmonic treatment of both eigenenergies and eigenfunctions of S_1 vibronic levels [125], but such a challenging treatment is beyond the illustrative scopes of the present contribution. It has been postulated that, also in the presence of such a double-well profile, a harmonic description of the spectrum obtained building the harmonic PES from the Hessian computed at the saddle-point structure, and assigning a convenient real frequency to the imaginary-frequency mode, may deliver useful information and provide spectra in nice agreement with the experiment, provided that the main focus is on low-resolution spectra and/or high-frequency progressions [126]. The reliability of such a model is even improved when the energy barrier of the double-well is so small that no vibrational states are actually confined in the wells, since in that case anharmonic wavefunctions closely resemble the harmonic ones, once associated with a suitable "effective" frequency. In the specific example of the S_1 state of pyrimidine, the saddle point is only 71 cm^{-1} above the global minima. Furthermore, the $Q_1^{(C_{2v})}$ mode is projected on about ten normal modes of the C_s structure, with the largest projection being about 0.5 on a mode $Q_3^{(C_s)}$ with a frequency of 460 cm^{-1}. It can be, therefore, guessed that no vibrational states are confined in the C_s wells, since even the ground state along mode $Q_1^{(C_{2v})}$ has enough energy to overcome the barrier. This case should, therefore, fall in the category mentioned above, and a harmonic PES model with a minimum in the C_{2v} saddle-point structure should deliver rather reliable spectra. The nice agreement between simulated and experimental spectra shown in Fig. 5 confirms *a posteriori* this assumption and allows a more detailed analysis of experimental outcomes. The analysis of individual vibronic transitions points out that the most intense absorptions correspond to the in-plane ring deformations of A_1 symmetry which, due to a small Duschinsky mixing, are well described by their ground-state counterparts (ν'_9, ν'_8 and ν'_7). These results only partially agree with the experimentally proposed assignment, for example, suggesting that the band at about 3.98 eV is related to the ν'_7 fundamental rather than to some overtone transition. The detailed assignment analysis is beyond the scope of present, illustrative work, and thus has been postponed to future investigations. Concerning the nature of above mentioned energy barrier it can be observed that moving from C_{2v} to C_s stationary points, the x component of the S_1 transition dipole moment, perpendicular to the molecular plane, decreases from 0.29 to 0.16 Debye, while for S_2 (A_2 in C_{2v} and A'' in C_s), it increases from 0.0 to 0.15 Debye. This behavior strongly suggests that the S_1 and S_2 states couple when the geometry is distorted to C_s (where both belong to the same irreps), thus exchanging transition intensity with x-polarization. Symmetry arguments indicate that the B_1 and A_2 states can be coupled by movements along b_2 modes ($A_2 \otimes b_2 = B_1$). It is, therefore, reasonable that the S_1 double-well along $Q_1^{(C_{2v})}$ b_2 mode actually arises from the B_1/A_2 nonadiabatic coupling. However, the energy gap between these two states is about 0.4 eV at the Franck–Condon point, thus suggesting that at least the low-energy wing of the spectrum can reasonably be computed through adiabatic approaches like those adopted here [52].

According to our calculations, frequency changes and Duschinsky mixings blue-shift the first-moment (i.e., the center of gravity (CoG)) of the spectrum with respect to the vertical transition energy (VE) by about 0.03 eV, while the shift increases to about 0.04 eV if the Herzberg-Teller (HT) effect is also considered. Since CoG is strictly related to the maximum of the low-resolution absorption spectrum, being identical to the latter for symmetric line-shapes, and somewhat blue-shifted for asymmetric line-shapes, the present analysis allows a number of interesting considerations. First, the small deviation from CoG and VE confirms that very good agreement between MRPT VE and experimental band maximum is not biased by neglecting vibronic contributions. Second, the fact that HT effects blue-shift the CoG by only 0.01 eV supports the reliability of classical simulations of the spectrum (reported in following sections), even for the $S_1 \leftarrow S_0$ transition, for which the existence of HT effects would pose an additional challenge to such a classical treatment [127].

The application of the simplified vertical gradient (FC|VG) model to simulate the electronic spectra of pyrimidine in a broader energy range, covering several electronic transitions, has been validated by the comparison with simulated FCHT adiabatic Hessian (FCHT|AH) and experimental spectra [114] for the $S_1 \leftarrow S_0$ electronic transition (see SI, Figure 1). It is clear that such a simple single-state approach does not allow to take into account all possible effects influencing the spectral phenomenon (such as the nonadiabatic and anharmonic ones). Nevertheless, it should be noted that the FC|VG model usually allows to simulate qualitatively correct spectrum line-shapes even in broad energy range, encompassing several electronic transitions, as in the present case, for which TD-CAM-B3LYP inaccuracies in prediction of absolute VE have been overcome by applying the hybrid MRPT/DFT approach. Figure 6 documents in fact a remarkable agreement of the computed and experimental spectra in the whole energy range, especially if one considers that the most significant discrepancy, namely the underestimation of the relative intensity of the red-shoulder at about 6.7–7 eV with respect to the major band at ~7.5 eV, is probably connected to the fact that vibronic borrowing mechanisms, like HT and nonadiabatic effects, have not been included in the present simulation.

3.5.2 Resonance Raman spectra

To complete our analysis, we also report the theoretical vibrational resonance Raman (vRR) spectrum of pyrimidine in water solution for an incident frequency in resonance with the $S_1 \leftarrow S_0$ transition. Figure 7 shows the vRR spectrum for an incident frequency equal to the VE

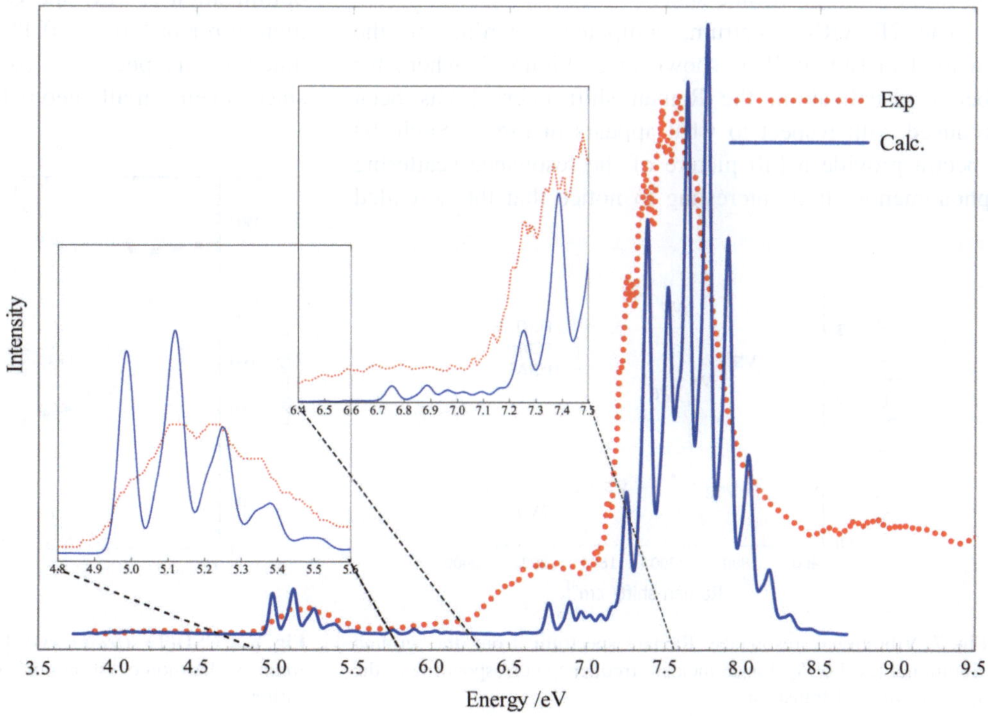

Fig. 6 Theoretical and experimental spectra [114] of isolated pyrimidine in the 3.5–9.5 eV energy range. Absorption spectrum has been computed at the FC|VG level, applying the hybrid MRPT/DFT approach (see text for the details). The theoretical spectrum has been simulated with the Gaussian function with HWHM of 0.025 eV

according to different models, namely Franck–Condon adiabatic shift (FC|AS), FC|AH and FCHT|AH. We assumed a linewidth for the S_1 excited state of 0.05 eV, and we broadened the peaks along the Raman shift coordinates with a Gaussian with HWHM = 25 cm^{-1}. Unfortunately, to the best of our knowledge, no experimental data for such a spectrum have been reported in literature till now, and it is not, therefore, possible to check the reliability of our predictions. Nonetheless, interesting features can be highlighted. In fact, the comparison of FC|AS and FC|AH spectra indicates that the effect of Duschinsky mixings and frequency changes in S_0 and S_1 introduces only modest changes in the spectrum, where all the dominant lines arise from fundamentals of total symmetric a_1 modes, and a few overtones (see the $2\nu_7'$ transition). As expected from the weakness of the resonant transition (nπ*), inclusion of HT effects has a strong impact on the spectrum. Besides a general enhancement of most of the peaks (observed also for OPA), new bands, absent in the FC spectra, arise. The most intense new bands are the fundamentals of b_1 (ν_{24}' and ν_{22}') and a_2 (ν_{10}', though hidden below the $1\nu_8'$ line) modes, while interestingly the overtone of the latter is also seen in the FC spectrum, due to the very strong frequency changes between the two states (~450 cm^{-1}). Symmetry considerations show that the S_1 (B_1) state can couple with A_1 states through distortions along the b_1 modes, borrowing z-polarized component of the transition dipole moment (TDM). Distortions along a_2 modes lead to S_1/B_2 coupling, with the latter gaining a y-polarized component of the TDM. Inspection of Table 2 shows that the first A_1 and B_2 symmetry states are actually the strongest OPA absorbers up to the sixth electronic state.

The 2D vRR spectrum, computed according to the method in Ref. [97], is shown in SI Figure 2, where for better visualization, the Raman shift interval has been reduced with respect to what appears in Fig. 7. Such 2D spectra provide a full picture of the resonance scattering phenomenon. It is interesting to notice that the so-called Raman excitation profiles (provided by 1D cuts of the 2D spectrum at fixed Raman shifts) show a clear dependence on the specific vibrational transition under analysis.

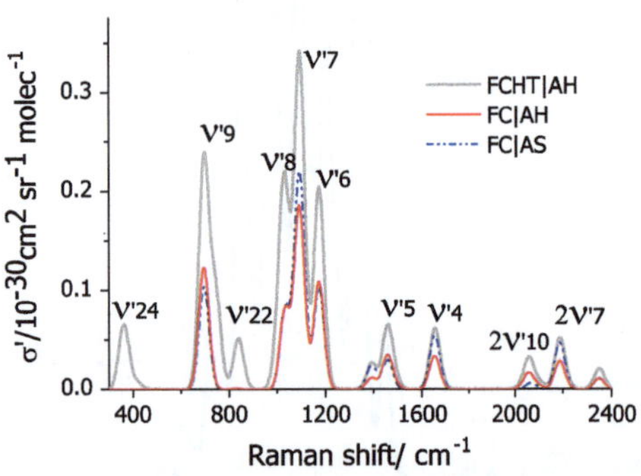

Fig. 7 Vibrational resonance Raman spectrum from the ground vibrational level of S_0, for an incident frequency corresponding to the $S_1 \leftarrow S_0$ vertical transition

3.6 Molecular dynamics simulations

Two separate sets of MD runs have been performed for pyrimidine in water and CCl_4 solvents, using the DFT-tailored FF's. For each solvent, after a 2.5 ns equilibration, a 3 ns production run has been carried out, saving the trajectories every 0.5 ps. It is worth noting that, as expected, no significant differences in the microsolvation of pyrimidine in aqueous solution have been observed between the GLOB and PBC MD simulations, thus validating the physical consistency of both approaches in the present case. Therefore, in the following, we report only the results obtained with the latter methodology. Before running MD simulations in solvent, a preliminary validation of the parameterized FF has been performed by vibrational harmonic frequency computation and MM geometry optimization, combined with *a posteriori* evaluation of UV-vis absorption spectra *in vacuo* at the TD-CAM-B3LYP/aug-N07D level of theory. The quality of the developed FF is illustrated by comparing the FF vibrational harmonic frequencies with those obtained from CAM-B3LYP/aug-N07D calculations (the reference for the FF development) as well as with the best theoretical estimates computed at the CC level. As shown in Fig. 8, the overall agreement is rather good, in particular, in view of the good match with the CC data. Moreover, all pyrimidine internal coordinates (IC's), obtained through geometry optimization at MM and QM levels, agree within a maximum error of 0.03 Å, 0.4° and 0.6° for bonds, angles and dihedrals, respectively (see SI, Table 2). Additionally, since even small geometrical differences can induce

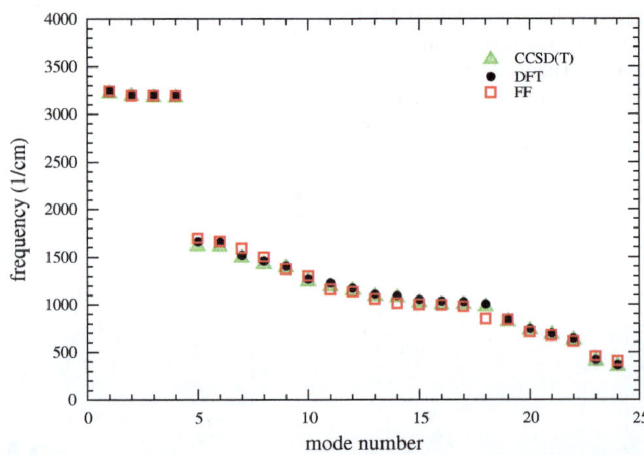

Fig. 8 CCSD(T) (*triangles*), DFT (*full circles*) and FF (*empty squares*) harmonic vibrational frequencies after JOYCE parameterization

nonnegligible shifts on the spectroscopic properties, the overall good agreement between energy and oscillator strength computed for the first 10 excited states at both DFT and MM optimized geometries is noteworthy (see SI, Table 3). Further details on FF validation are gathered in SI Section 1.4.

3.6.1 Water

Structural characterization of the solvation shells around pyrimidine has been performed by the analysis of the N–O_W, N–H_W and $CM_{pyrimidine}$–CM_{H2O} radial distribution functions (RDF's), which are reported in Fig. 9 and complemented by spatial distribution functions (SDF). As it appears from the examination of the SDF contours in water (red and cyan surfaces represent oxygen and hydrogen atoms, respectively), a strong clustering of the solvent molecules takes place near pyrimidine nitrogen atoms, which are engaged in quite stable and persistent intermolecular hydrogen bonds with water molecules. The percentage of occupancy of these specific interactions is about 98 %, which means that pyrimidine nitrogens are in contact with water for the whole simulation time. Also, the average number of hydrogen bonds (HB) between pyrimidine and water is about 3, where the largest amount comes from the pyrimidine nitrogens acting as hydrogen bond acceptors (about 1.3 HB for each nitrogen) plus a very small contribution from all the other hydrogen bonding forms. Note that hydrogen bonds were determined based on cutoffs for the donor–hydrogen–acceptor angle (60°, 0° corresponding to the linear donor–hydrogen–acceptor arrangement) and the hydrogen-acceptor distance (2.5Å). Such an hydrogen bonding pattern is also represented by the radial distribution functions where two sharp peaks are observed at 1.8 and 2.8Å in the $g(N \cdots H_W)$ and $g(N \cdots O_W)$ profiles, respectively (see Fig. 9).

The number of solvent molecules within 2.5 and 5.5 Å of at least one pyrimidine atom and in the shells between 2.5 and 3.5 (*sh*2), 3.5 and 4.5 (*sh*3), 4.5 and 5.5 Å (*sh*4) has also been monitored. The distributions of this solvation number are displayed in SI, Figure 4 and further support the conclusion drawn from RDF analysis. The most probable situation is to have 5 waters in the first solvation layer, where at least two of them are hydrogen bonded to the two pyrimidine nitrogens. Moreover, *sh*2 and *sh*3 are equally populated (14 waters), *sh*4 is filled with 20 waters on average, and 52 water molecules are found within 5.5 Å from at least one atom of the solute.

These data can be compared with those reported in Ref. [128], where a combined Monte Carlo-QM approach was carried out to study the solvatochromism of pyrimidine in water and carbon tetrachloride. Note that, apart small changes in the intermolecular FF parameters, the major

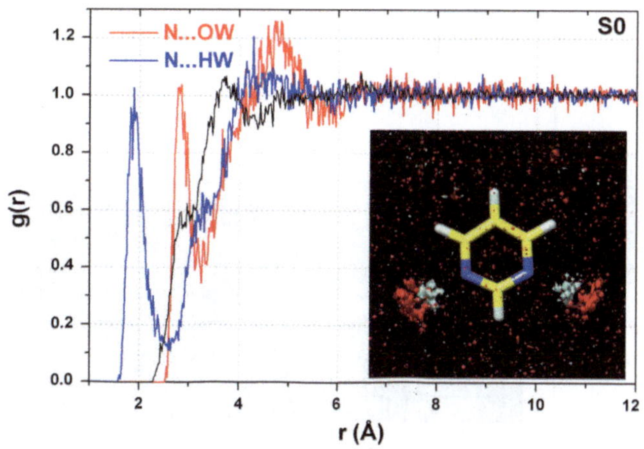

Fig. 9 Radial distribution functions for N–O_W (water oxygen, *red line*), N–H_W (water hydrogen, *blue line*) and $CM_{pyrimidine}$–CM_{H2O} (center of mass of pyrimidine–center of mass of water, *black line*) and three-dimensional contour plots of oxygen (*red*) and hydrogen (*cyan*) densities around pyrimidine

difference with Ref. [128] is the (intra) molecular flexibility included through the Joyce procedure in this work, while in Ref. [128], both solvent and solute molecules were kept rigid. It should be stressed that this approximation holds for fairly rigid molecules (like pyrimidine) but may cause large errors when applied to larger flexible molecules (see for instance Refs. [129, 130]). Despite the aforementioned differences, in the current case, both approaches provide similar results, for example, the first solvation shell, identified through the examination of the radial distribution function between pyrimidine center of mass and water center of mass, corresponds well to the first layer described in Ref. [128] and has the same number of coordinated waters (about 21). Nevertheless, the approach employed in this work should be considered generally more accurate, and thus, it is recommended whenever feasible.

3.6.2 CCl_4

The distribution of CCl_4 solvent molecules around pyrimidine is quite different from water organization. Indeed, CCl_4 molecules have been found at larger distances and their interactions with the solute are weaker. This behavior is clearly depicted in Fig. 10, where both the radial and spatial distribution functions are displayed. The RDF trend shows the presence of three shells, which can be identified with the small peaks centered at about 3.6 Å, 4.8 Å and the broader dominant peak centered at about 6.1 Å. The first two peaks are not well separated and together can be used to define the first solvent layer that extends to about 5.4 Å. This shell contains about 17 solvent molecules, whereas the third peak that stretches up to about 7.3 Å is filled with

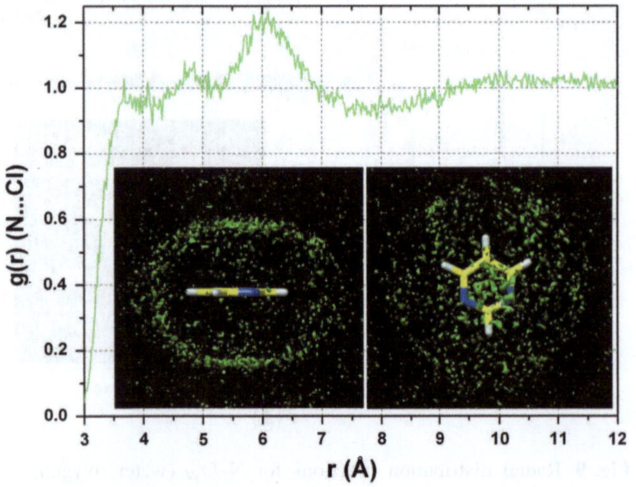

Fig. 10 Radial distribution function in CCl_4 and three-dimensional contour plot of chlorine density around pyrimidine

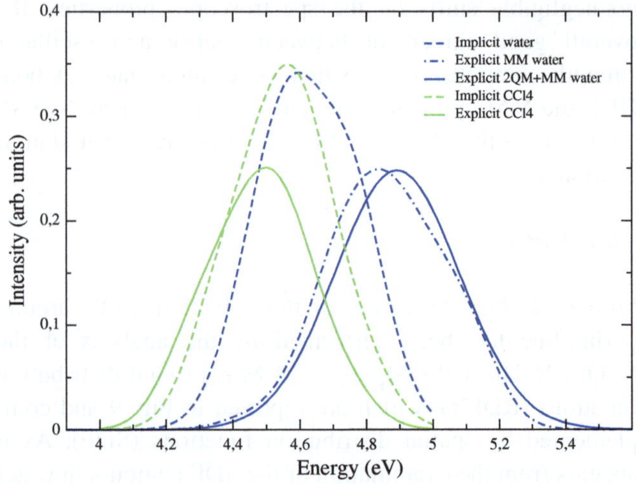

Fig. 11 UV-vis absorption spectra of pyrimidine in water (*blue lines*) and carbon tetrachloride (*green lines*), computed within a time-dependent approach, with implicit (*dashed lines*) and explicit (*solid* and *dash-dot lines*) solvent representations

about 40 CCl_4 molecules, and the shell contains 47 molecules, on average.

The probability distribution of the solvation number (SI, Figure 5) corresponding to the first and second layer is in line with the RDF data. Moreover, the SDF contour plot clearly supports a picture where pyrimidine seems more confined in the solvent cage and adopts a randomly distributed orientation with its motion rather limited. Similarly to what seen for water as solvent, these results match closely the more approximate models previously applied [128].

3.7 UV-vis spectra from a time-dependent approach

To estimate the improvement that can be achieved with the present multi-level time-dependent protocol, the UV-vis absorption $S_1 \leftarrow S_0$ spectrum has first been computed within a static approach, that is, by considering only the minimum energy geometry, optimized at the CAM-B3LYP level in both water and CCl_4, each represented with an implicit description by means of the nonequilibrium LR-CPCM [41]. The vertical excitation energies computed for water (4.69 eV) and carbon tetrachloride (4.61 eV) show only a small difference: in particular, when compared to the gas-phase VE of 4.55 eV, both values show a solvent blue shift of $\sim$1,100 and $\sim$500 cm^{-1}, for H_2O and CCl_4, respectively. As concerns water, the agreement with experimental data [131] is rather poor, the latter being 2,700 cm^{-1}. The blue shift found for CCl_4 can be compared to that computed by Canuto and coworkers [128], obtained with INDO/CIS calculations on supramolecular clusters extracted from Monte Carlo (MC) simulations ($\sim$100 cm^{-1}), the discrepancy is small and in opposite direction with respect to the results obtained for water.

In the time-dependent approach, dynamical averaging effects are taken into account by computing vertical transition energies and oscillator strengths on one hundred frames extracted from the classical MD simulations in each solvent. Moreover, the solvent description has been improved by explicitly considering all solvent molecules within a selected distance from pyrimidine geometrical center. The resulting spectra, obtained by broadening the vertical energies computed at the geometries extracted from the MD simulations with Gaussian functions with HWHM of 0.1 eV, are reported in Fig. 11.

The energies corresponding to the maximum absorption are 4.84 and 4.50 eV for spectra in H_2O and CCl_4, respectively, yielding to a solvent shift (with respect to the vertical transition energy *in vacuo* of 4.52 eV) of $\sim$2,600 and ~ -160 cm^{-1} for the former and latter solvent, respectively, thus improving the agreement with both the above mentioned experimental [131] and computed [128] values. Note that by treating the two closest hydrogen-bonded water molecules to pyrimidine at the QM level, along with the solute, and the remaining solvent molecules as point charges, the computed maximum absorption peak is found at 4.88 eV (see Fig. 11). The good agreement between the latter result and the previous fully electrostatic embedding (solvent) calculations suggests that the interaction between pyrimidine and water is basically electrostatic in nature. It is worth noting that the agreement with the experimental result (2,700 cm^{-1}) for water solvent is also improved with respect to previous calculations reported in the literature [128, 132] (2,223 cm^{-1} and 2,275 cm^{-1}, in Refs. [128] and [132], respectively).

As far as the absorption in CCl_4 is concerned, because of the weak nonpolar interactions between pyrimidine and the solvent molecules, it is not surprising that the vertical transition energy is only slightly shifted with respect to the value computed *in vacuo*. Indeed, these considerations, together with the less structured framework arising from the RDF analysis as compared to water, suggest that the explicit solvent model could be abandoned for the more computationally convenient PCM. Conversely, the strong local and specific interactions between pyrimidine and water make such a choice less advisable in this case. To verify this hypothesis, only pyrimidine geometries were extracted from MD frames, thus sampling only solute conformational dynamics (see SI, Figure 3). UV-vis absorption spectra have thereafter been computed on such geometries taking into account implicit solvent effects only, for both solvents. As expected, Fig. 11 shows small difference ($\sim$300 cm^{-1}) between implicit and implicit/explicit solvent models for the CCl_4, while larger difference arises in the case of water (blue shift of $\sim$700 cm^{-1}).

In view of the negligible specific solute–solvent interactions for the CCl_4 solution, time-dependent results computed with solvent modeled by PCM can be directly compared with their time-independent counterparts (SI, Figure 6), with both approaches agreeing on maxima positions and the width of the low-resolution spectra. It must be recalled that the presented approaches allow to study in detail different effects influencing solvent shift and electronic band shape: the former permits to take into account specific and bulk solute–solvent interactions, while the latter allows to analyze individual vibronic contributions. The good overall agreement on the final results confirms that integrated studies, which take advantage of both time-dependent and time-independent approaches, pave the route toward a better understanding of experimentally observed spectra in condensed phases.

4 Conclusions

Some of the most recent theoretical approaches, with particular focus on those either developed or extended and validated by Vincenzo Barone and co-workers, have been presented taking as an example a medium-sized molecule, pyrimidine. We first determined the structure and vibrational properties of the isolated molecule at the CCSD(T) level, while the energies and oscillator strengths of electronic excited states have been characterized at the MRPT level. It has been shown that the DFT method provides accurate results at a relatively low computational cost compared to the wavefunction-based methods that take into account electron correlation (CCSD(T) and MRPT). This allowed us to use hybrid methods, which combine the harmonic force field obtained at the CCSD(T) with the anharmonic corrections at the DFT level, to simulate the vibrational spectrum in good agreement with experiment. Moreover, DFT-based approaches have also been coupled with the PCM method to study the solvent effect on the calculation of anharmonic frequencies in an implicit way. While the $S_1 \leftarrow S_0$ electronic transition of pyridine is not a simple case for the time-independent approaches, which have been employed to simulate vibrationally resolved one-photon absorption and resonance Raman spectra, we have shown how a careful study allows to obtain accurate spectra and to interpret experimental data. Furthermore, the UV-vis electronic spectrum encompassing the first twelve excited electronic states has been computed, showing feasibility of spectra line-shape simulations even in the broad energy range.

Finally, a purposely DFT-tailored force field allowed us to study the conformational flexibility, solvation properties and their influence on the absorption spectrum of pyrimidine. The absorption spectra in water and CCl_4, obtained from several QM/MM/PCM calculations on sampled molecular configurations, have shown a solvatochromic shift in good agreement with experimental data, provided that the solvent is treated explicitly, thus validating the accuracy of the proposed method. In conclusion, we can point out that, although further developments are clearly needed, the accuracy of the computed IR and UV–vis spectra already allows a direct comparison with measurements in solution, and the QM/MM simulations represent a valuable complement to experimental results. This also represents a critical first step to study larger and more complex systems for which high-level theoretical data are unavailable and experimental data may be scarce.

Acknowledgments All authors are grateful to Professor Vincenzo Barone for fruitful collaborations, scientific advices and/or tutoring over the years. This work was supported by Italian MIUR (PRIN2008, PRIN2009). The large scale computer facilities of the VILLAGE network (http://m3village.sns.it) are acknowledged for providing computer resources. The COST-CMTS Action CM1002 "COnvergent Distributed Environment for Computational Spectroscopy (CODECS)" is also acknowledged.

References

1. Laane J (ed) (2009) Frontiers of molecular spectroscopy. Elsevier, Amsterdam
2. Quack M, Merkt F (eds) (2011) Handbook of high-resolution spectroscopy. Wiley, New Jersey
3. Siebert F, Hildebrandt P (eds) (2008) Vibrational spectroscopy in life science. Wiley-VCH Verlag GmbH and Co., KGaA, New Jersey
4. Astrid G, Rudolf R, Jerker W (eds) (2010) Single molecule spectroscopy in chemistry, physics and biology: nobel symposium. Springer,
5. Jensen P, Bunker PR (eds) (2000) Computational molecular spectroscopy. Wiley, New Jersey

6. Barone V (ed) (2011) Computational strategies for spectroscopy, from small molecules to nano systems. Wiley, New Jersey
7. Grunenberg J (ed) (2010) Computational spectroscopy. Wiley-VCH Verlag GmbH & Co., KGaA, New Jersey
8. Pedone A, Biczysko M, Barone V (2010) Chem Phys Chem 11:1812
9. Barone V, Biczysko M, Brancato G (2010) Adv Quantum Chem 59:17
10. Csaszar AG, Fabri C, Szidarovszky T, Matyus E, Furtenbacher T, Czako G (2012) Phys Chem Chem Phys 14:1085
11. Tennyson J (2011) WIREs Comput Mol Sci. doi:10.1002/wcms.94
12. Puzzarini C, Stanton J, Gauss J (2010) Int Rev Phys Chem 29:273
13. Carrington T, Wang XG (2011) WIREs Comput Mol Sci 1:952
14. Csaszar AG (2011) WIREs Comput Mol Sci. doi:10.1002/wcms.75
15. Clary D (2006) Science 314:5797
16. Truhlar DG (2008) J Am Chem Soc 130:16824
17. Visscher L, Bolhuis P, Bickelhaupt FM (2011) Phys Chem Chem Phys 13:10399
18. Pedone A, Prampolini G, Monti S, Barone V (2011) Chem Mater 23:5016
19. Riley KE, Pitonak M, Jurecka P, Hobza P (2010) Chem Rev 110:5023
20. Puzzarini C, Biczysko M, Barone V (2010) J Chem Theory Comput 6:828
21. Barone V, Bloino J, Biczysko M, Santoro F (2009) J Chem Theory Comput 5:540
22. Barone V, Cacelli I, Ferretti A, Prampolini G (2009) J Chem Phys 131:224103
23. Gordon MS, Fedorov DG, Pruitt SR, Slipchenko LV (2012) Chem Rev 112:632
24. Maple J, Dinur U, Hagler A (1988) Proc Natl Acad Sci USA 85:5350
25. Dasgupta S, Yamasaki T, Goddard W III (1996) J Chem Phys 104:2898
26. Cacelli I, Prampolini G (2007) J Chem Theory Comput 3:1803
27. Gao J (1996) Acc Chem Res 29:298
28. Canuto S, Coutinho K, Trzesniak D (2002) New developments in Monte Carlo/quantum mechanics methodology. The solvatochromism of carotene in different solvents. In: Sabin JR, Brandas E (eds) Advances in quantum chemistry, vol 41. Academic Press, pp 161–183
29. Miertus S, Scrocco E, Tomasi J (1981) Chem Phys 55:117
30. Cramer CJ, Truhlar DG (1999) Chem Rev 99:2161
31. Tomasi J, Mennucci B, Cammi R (2005) Chem Rev 105:2999
32. Tomasi J (2011) WIREs Comput Mol Sci 1:855
33. Klamt A (2011) WIREs Comput Mol Sci 1:699
34. Caricato M, Mennucci B, Scalmani G, Trucks GW, Frisch MJ (2010) J Chem Phys 132:084102
35. Improta R, Barone V, Santoro F (2007) Angew Chem Int Ed Engl 46:405
36. Wang YH, Halik M, Wang CK, Marder SR, Luo Y (2005) J Chem Phys 123:194311
37. Mennucci B, Cappelli C, Guido CA, Cammi R, Tomasi J (2009) J Phys Chem A 113:3009
38. Avila Ferrer F, Improta R, Santoro F, Barone V (2011) Phys Chem Chem Phys 13:17007
39. Cammi R, Tomasi J (1995) Int J Quantum Chem: Quantum Chem Symp 29:465
40. Mennucci B, Cammi R, Tomasi J (1998) J Chem Phys 109:2798
41. Marenich AV, Cramer CJ, Truhlar DG, Guido CA, Mennucci B, Scalmani G, Frisch MJ (2011) Chem Sci 2:2143
42. Rivail JL, Rinaldi D, Dillet V (1996) Mol Phys 89:1521
43. Cappelli C (2007) Continuum solvation approaches to vibrational properties. In: Mennucci B, Cammi R (eds) Continuum solvation models in chemical physics: theory and applications. Wiley, Chichester, p 167
44. Cappelli C, Biczysko M (2011) Time-independent approach to vibrational spectroscopies. In: Barone V (ed) Computational strategies for spectroscopy. Wiley, pp 309–360
45. Bandyopadhyay P, Gordon MS (2000) J Chem Phys 113:1104
46. Mennucci B, Martinez JM, Tomasi J (2001) J Phys Chem A 105:7287
47. Barone V, Improta R, Rega N (2008) Acc Chem Res 41:605
48. Cappelli C, Mennucci B, da Silva CO, Tomasi J (2000) J Chem Phys 112:5382
49. Santoro F, Barone V, Gustavsson T, Improta R (2006) J Am Chem Soc 128:16312
50. Gao J, Truhlar DG (2002) Annu Rev Phys Chem 53:467
51. Brancato G, Rega N (2011) Computational spectroscopy by classical time-dependent approaches. In: Barone V (ed) Computational strategies for spectroscopy. Wiley, pp 517–547
52. Biczysko M, Bloino J, Santoro F, Barone V (2011) Time-independent approaches to simulate electronic spectra lineshapes: from small molecules to macrosystems. In: Barone V (ed) Computational strategies for spectroscopy. Wiley, pp 361–443
53. Grimme S (2004) Calculation of the electronic spectra of large molecules. In: Reviews in computational chemistry. Wiley, pp 153–218
54. Cui Q, Karplus M (2000) J Chem Phys 112:1133
55. Aidas K, Mikkelsen KV, Kongsted J (2007) J Comp Methods Sci Eng 7:135
56. Barone V, Polimeno A (2007) Chem Soc Rev 36:1724
57. Ramos MJ (ed) (2008) Computational proteomics. Transw Res Netw
58. Grotendorst J, Attig N, Blügel S, Marx D (eds) (2009) Multiscale simulation methods in molecular sciences. Lecture Notes, NIC Series:42
59. Senn HM, Thiel W (2009) Angew Chem Int Ed Engl 48:1198
60. Sabin, JR, Brandas, E (eds) (2010) Advances in quantum chemistry, vol 59. Academic Press, Massachusetts
61. Arul Murugan N, Kongsted J, Rinkevicius Z, Aidas K, Ågren H (2010) J Phys Chem B 114:13349
62. Raghavachari K, Trucks GW, Pople JA, Head-Gordon M (1989) Chem Phys Lett 157:479
63. Helgaker T, Klopper W, Koch H, Noga J (1997) J Chem Phys 106:9639
64. CFOUR, a quantum chemical program package written by Stanton JF, Gauss J, Harding ME, Szalay PG with contributions from A.A. Auer, R.J. Bartlett, U. Benedikt, C. Berger, D.E. Bernholdt, Y.J. Bomble, L. Cheng, O. Christiansen, M. Heckert, O. Heun, C. Huber, T.-C. Jagau, D. Jonsson, J. Jusélius, K. Klein, W.J. Lauderdale, D.A. Matthews, T. Metzroth, L.A. Mück, D.P. O'Neill, D.R. Price, E. Prochnow, C. Puzzarini, K. Ruud, F. Schiffmann, W. Schwalbach, S. Stopkowicz, A. Tajti, J. Vázquez, F. Wang, J.D. Watts and the integral packages MOLECULE (J. Almlöf and P.R. Taylor) PROPS (P.R. Taylor) ABACUS (T. Helgaker, H.J. Aa. Jensen, P. Jørgensen, and J. Olsen) and ECP routines by A. V. Mitin and C. van Wüllen. For the current version, see http://www.cfour.de
65. Huron B, Malrieu JP, Rancurel P (1973) J Chem Phys 58:5745
66. Angeli C, Cimiraglia R, Persico M, Toniolo A (1997) Theor Chem Acc 93:57
67. Angeli C, Persico M (1997) Theor Chem Acc 98:117
68. Barone V, Cimino P, Stendardo E (2008) J Chem Theory Comput 4:751
69. Barone V, Bloino J, Biczysko M (2010) Phys Chem Chem Phys 12:1092
70. Becke AD (1993) J Chem Phys 98:5648
71. Yanai T, Tew DP, Handy NC (2004) Chem Phys Lett 393:51
72. Grimme S (2006) J Chem Phys 124:034108

73. Double and triple-ζ basis sets of n07 family, are available for download. Visit http://idea.sns.it (Accessed June 30, 2011)
74. Frisch MJ, Trucks GW, Schlegel HB, Scuseria GE, Robb MA, Cheeseman JR, Scalmani G, Barone V, Mennucci B, Petersson GA, Nakatsuji H, Caricato M, Li X, Hratchian HR, Izmaylov AF, Bloino J, Zheng G, Sonnenberg JL, Hada M, Ehara M, Toyota K, Fukuda R, Hasegawa J, Ishida M, Nakajima T, Honda Y, Kitao O, Nakai H, Vreven T, Montgomery Jr. JRJ. A. Peralta, Ogliaro F, Bearpark M, Heyd JJ, Brothers E, Kudin KN, Staroverov VN, Kobayashi R, Normand J, Raghavachari K, Rendell A, Burant JC, Iyengar SS, Tomasi J, Cossi M, Rega N, Millam JM, Klene M, Knox JE, Cross JB, Bakken V, Adamo C, Jaramillo J, Gomperts R, Stratmann RE, Yazyev O, Austin AJ, Cammi R, Pomelli C, Ochterski JW, Martin RL, Morokuma K, Zakrzewski VG, Voth GA, Salvador P, Dannenberg JJ, Dapprich S, Daniels AD, Farkas O, Foresman JB, Ortiz JV, Cioslowski J, Fox DJ, Gaussian 09 Revision B.01 (2009) Gaussian Inc. Wallingford CT 2009
75. Brancato G, Rega N, Barone V (2008) J Chem Phys 128:144501
76. Brancato G, Rega N, Barone V (2009) Chem Phys Lett 483:177
77. van der Spoel D, Lindahl E, Hess B, Groenhof G, Mark A, Barendsen H (2005) J Comp Chem 26:1701
78. Brancato G (2011) Jersey—a multi-purpose molecular simulation program
79. van der Spoel D, Lindahl E, Hess B, Buuren Avan, Apol E, Meulenhoff P, Tieleman D, Sijbers A, Feenstra K, Drunen Rvan, Berendsen H (2010) Gromacs user manual version, 4.5.4 http://www.gromacs.org
80. Mills IM (1972) Molecular spectroscopy: modern research. Academic Press, New York
81. Truhlar DG, Olson RW, Jeannotte AC II, Overend J (1976) J Am Chem Soc 98:2373
82. Isaacson AD, Truhlar DG, Scanlon K, Overend J (1981) J Chem Phys 75:3017
83. Amos RD, Handy NC, Green WH, Jayatilaka D, Willets A, Palmieri P (1991) J Chem Phys 95:8323
84. Gaw F, Willetts A, Handy N, Green W (1991) SPECTRO—a program for derivation of spectroscopic constants from provided quartic force fields and cubic dipole fields. In: Bowman JM (ed) Advances in molecular vibrations and collision dynamics, vol 1B. JAI Press, pp 169–185
85. Zhang Q, Day PN, Truhlar DG (1993) J Chem Phys 98:4948
86. Vazquez J, Stanton JF (2006) Mol. Phys. 104:377
87. Barone V (2005) J Chem Phys 122:014108
88. Barone V (2004) J Chem Phys 120:3059
89. Bloino J, Guido C, Lipparini F, Barone V (2010) Chem Phys Lett 496:157
90. Carbonniere P, Lucca T, Pouchan C, Rega N, Barone V (2005) J Comp Chem 26:384
91. Begue D, Benidar A, Pouchan C (2006) Chem Phys Lett 430:215
92. Cappelli C, Monti S, Scalmani G, Barone V (2010) J Chem Theory Comput 6:1660
93. Cappelli C, Lipparini F, Bloino J, Barone V (2011) J Chem Phys 135:104505. Erratum ibidem 2011, 135, 149901
94. Cappelli C, Corni S, Cammi R, Mennucci B, Tomasi J (2000) J Chem Phys 113:11270
95. Cammi R, Cappelli C, Corni S, Tomasi J (2000) J Phys Chem A 104:9874
96. Bloino J, Biczysko M, Santoro F, Barone V (2010) J Chem Theory Comput 6:1256
97. Santoro F, Cappelli C, Barone V (2011) J Chem Theory Comput 7:1824
98. Santoro F *FCclasses*, a fortran 77 code (2008) Available via the Internet at http://village.pi.iccom.cnr.it; last accessed July 14, 2011
99. Santoro F, Improta R, Lami A, Bloino J, Barone V (2007) J Chem Phys 126:084509
100. Dierksen M, Grimme S (2005) J Chem Phys 122:244101
101. Jankowiak HC, Stuber JL, Berger R (2007) J Chem Phys 127:234101
102. Hazra A, Nooijen M (2003) Int J Quantum Chem 95:643
103. Cossi M, Scalmani G, Rega N, Barone V (2003) J Comp Chem 24:669
104. Cammi R, Corni S, Menucci B, Tomasi J (2005) J Chem Phys 122:104513
105. Corni S, Cammi R, Menucci B, Tomasi J (2005) J Chem Phys 123:134512
106. Scalmani G, Frisch MJ, Menucci B, Tomasi J, Cammi R, Barone V (2006) J Chem Phys 124:094107
107. Kisiel Z, Pszczółkowski L, López J, Alonso J, Maris A, Caminati W (1999) J Mol Spectrosc 195:332
108. Puzzarini C (2009) J Phys Chem A 113:14530
109. Puzzarini C (2010) Int J Quant Chem 110:2483
110. Puzzarini C, Barone V (2011) Phys Chem Chem Phys 13:7158
111. Dunning T Jr (1989) J Chem Phys 90:1007
112. Kendall A, Dunning TH Jr, Harrison RJ (1992) J Chem Phys 96:6796
113. Peach MJG, Helgaker T, Salek P, Keal TW, Lutnaes OB, Tozer DJ, Handy NC (2006) Phys Chem Chem Phys 8:558
114. Silva F Ferreira da, Almeida D, Martins G, Milosavljevic AR, Marinkovic BP, Hoffmann SV, Mason NJ, Nunes NJ, Garcia G, Limao-Vleira G (2010) Phys Chem Chem Phys 12:6717
115. Biczysko M, Panek P, Scalmani G, Bloino J, Barone V (2010) J Chem Theory Comput 6:2115
116. Kozuch S, Gruzman D, Martin JML (2010) 114:20801
117. Martin JML, Lee TJ, Taylor PR, Francois JP (1995) J Chem Phys 103:2589
118. Barone V, Festa G, Grandi A, Rega N, Sanna N (2004) Chem Phys Lett 388:279
119. Barone V (2004) J Phys Chem A 108:4146
120. Boese AD, Martin J (2004) J Phys Chem A 108:3085
121. Breda S, Reva ID, Lapinski L, Nowak MJ, Fausto R (2006) J Mol Struct 786:193
122. Cappelli C, Corni S, Tomasi J (2001) J Chem Phys 115:5531
123. Cappelli C, Corni S, Mennucci B, Cammi R, Tomasi J (2002) J Phys Chem A 106:12331
124. See http://webbook.nist.gov/cgi/cbook.cgi?id=c289952. Last accessed june 23th 2011
125. Hazra A, Nooijen M (2005) Phys Chem Chem Phys 7:1759
126. Lin N, Santoro F, Zhao X, Rizzo A, Barone V (2008) J Phys Chem A 112:12401
127. Lami A, Santoro F (2011) Time-dependent approaches to calculation of steady-state vibronic spectra: from fully quantum to classical approaches. In: Barone V (ed) Computational strategies for spectroscopy. Wiley, pp 475–516
128. de Almeida K, Coutinho K, de Almeida W, Rocha W, Canuto S (2001) Phys Chem Chem Phys 3:1583
129. Barone V, Bloino J, Monti S, Pedone A, Prampolini G (2011) Phys Chem Chem Phys 12:10550
130. Barone V, Bloino J, Monti S, Pedone A, Prampolini G (2011) Phys Chem Chem Phys 13:2160
131. Baba H, Goodman L, Valenti P (1996) J Am Chem Soc 88:5410
132. Gao J, Byun K (1997) Theor Chem Acc 96:151

Theor Chem Acc (2012) 131:1211
DOI 10.1007/s00214-012-1211-1

REGULAR ARTICLE

Effects of molecular dynamics and solvation on the electronic structure of molecular probes

Pasquale Caruso · Mauro Causà · Paola Cimino · Orlando Crescenzi · Maddalena D'Amore · Roberto Improta · Michele Pavone · Nadia Rega

Received: 28 December 2011 / Accepted: 21 March 2012 / Published online: 10 April 2012

Abstract Most spectroscopic parameters are influenced by nuclear dynamics and by the chemical environment. However, proper inclusion of these effects still represents a challenge in computational spectroscopy studies. In many cases, a route coupling satisfactory accuracy with reasonable computational costs consists in the integration of DFT-based methods to compute spectroscopic parameters, with ab initio molecular dynamics simulations to sample from the classical phase space of the system. Here, we discuss the application of this approach in two case studies of remarkable practical interest, namely the simulation of the absorption spectrum of 9-methyladenine, an adenine nucleoside model; and the prediction of electron spin resonance parameters for nitroxyl radicals, the prototypical spin probes. In both cases, the accuracy of the results increases significantly when the subtle interplay of intramolecular dynamics and solvent effects is introduced.

Keywords Quantum mechanical calculations · Solution · Dynamics · Absorption spectra · EPR spectra · Adenine · Nitroxides · Electron density analysis

Dedicated to Professor Vincenzo Barone and published as part of the special collection of articles celebrating his 60th birthday.

P. Caruso
IIT@CRIB Center for Advanced Biomaterials for Healthcare, Italian Institute of Technology, p.le Tecchio 80, 80125 Napoli, Italy

M. Causà (✉) · O. Crescenzi (✉) · M. D'Amore · M. Pavone · N. Rega
Dipartimento di Chimica "Paolo Corradini", Università Federico II di Napoli, Complesso Monte S. Angelo, via Cintia, 80126 Napoli, Italy
e-mail: mauro.causa@unina.it

O. Crescenzi
e-mail: orlando.crescenzi@unina.it

P. Cimino
Dipartimento di Scienze Farmaceutiche, Università di Salerno, via Ponte don Melillo, 84084 Fisciano, SA, Italy

R. Improta
Istituto di Biostrutture e Bioimmagini, CNR, via Mezzocannone 16, 80134 Napoli, Italy

1 Introduction

Experimental studies on molecular spectroscopy and photophysics have greatly benefited from the recent advances in electronic structure methods [1]. The theoretical development of accurate yet feasible computational approaches and the availability of high-performance computing facilities have provided the chemical community with powerful tools to characterize molecular systems of higher and higher complexity [1]. On the one hand, the successful implementation of new models within the framework of density functional theory (DFT) [2] allows to study molecules of realistic dimensions, up to hundreds of atoms. On the other hand, computational spectroscopy techniques allow to investigate in detail the complex interaction of molecules with external electromagnetic fields (see Ref. [1] and references therein).

However, there is still a significant gap between numerical simulations and laboratory experiments. The most important issues are the inclusion in the electronic structure calculations of the nuclear dynamics (i.e., temperature) and of the chemical environment (e.g., the solvent). In the last decades, many theoretical efforts have been devoted to these topics, and recent extensive reviews cover the main results obtained by time-dependent and time-independent approaches [1].

From a general perspective, the influence of nuclear motion should be taken into account in the computation of essentially all kinds of spectroscopic parameters, even those that do not gauge vibrational structure directly (like IR and Raman). In the case of electronic transitions, for example, molecular vibrations are responsible for the fine structure of UV–vis absorption, fluorescence emission, or electronic circular dichroism spectra via the Franck-Condon selection rules (direct effect) [3]. By contrast, in magnetic resonance spectroscopy, the spectral line shapes are determined by long-time dynamics (e.g., diffusion and rotation of the magnetic molecular probe) rather than molecular motions on short-time scales. However, also in this case, temperature does affect the spectroscopic observables by averaging out the corresponding molecular parameters over the structural fluctuations due to molecular vibrations (indirect effect) [4]. A reliable route to account for this indirect effect is the calculation of the spectroscopic property via vibrational averaging with respect to normal modes, with temperature tuning the population of each mode. The method of choice for computing molecular vibrations is the solution of the nuclear Schrödinger equation. Reliable vibrational frequencies can be obtained in this way, but only with a proper correlation consistent description of the electronic potential energy surface and, possibly, by taking into account anharmonicity [5]. Such time-independent approach is very accurate for semirigid systems because it accounts for the quantum-mechanical nature of nuclear motions; however, the computational burden limits its feasibility.

An alternative approach is represented by the sampling of nuclear motions in the classical phase space via molecular dynamics (MD) techniques [6]. In this case, the spectroscopic property of interest is obtained by extracting a consistent set of configurations from the MD trajectories, and by averaging the molecular parameters computed on each of these frames. This time-dependent approach is quite effective, but once again, its accuracy depends on the description of the potential energy surface governing the MD. In most cases, ab initio MD (AIMD) simulations can be exploited to this purposes: thanks to the Car-Parrinello scheme [7, 8], the forces governing nuclear dynamics can be efficiently computed from first principles, within the framework of the density functional theory. From a practical point of view, it is also remarkable that time-dependent approaches allow the use of different levels of theory for the generation of trajectories and for the computation of spectroscopic observables on a subset of the sampled structures. For example, much better statistics can be obtained by MD simulations using reliable semi-empirical force fields, when these are available [6].

In the present study, we explore the performances of dynamical approaches to include nuclear and environmental effects by discussing two case studies: the simulation of the absorption spectrum of an adenine nucleoside model (9-methyladenine, henceforth 9-Me-Ade) in the gas phase and in aqueous solution, and the calculation of electron spin resonance (ESR) parameters for important spin probes such as nitroxyl radicals. As shown below, in both cases, the accuracy of the results considerably increases when the subtle interplay of intra-molecular dynamics and solvent effects is properly accounted for in the electronic structure calculations.

Besides their methodological interest, our results will provide additional insights into two topics of general chemical and biological interest, namely the photophysics of DNA components and the spectroscopic features of some prototypical 'spin labels'/'spin probes'.

The interaction between UV–visible radiation and nucleic acids is a topic of great biological relevance, since it can trigger a cascade a photochemical events leading to mutagenesis. This has motivated, in the last decade, a large number of theoretical and computational studies on the excited state properties of DNA and of its constituent nucleobases [9–13].

Concerning isolated nucleobases, time-resolved experiments evidenced very short (sub-ps) lifetimes of the lowest energy bright singlet excited states that have been rationalized by quantum mechanical calculations [9, 13]. For all the nucleobases, barrier-less paths exist for the spectroscopic states, connecting the Franck-Condon (FC) region with a conical intersection (CI) with the ground electronic state (S_0). However, several interesting features have not been assessed yet, as for example the involvement of dark excited states in the decay process [10, 13].

While for uracil derivatives, experiments and calculations both showed that a part of the population of the lowest energy bright excited state (with $\pi \rightarrow \pi^*$ character) can decay to a close lying $n\pi^*$ state [13, 14], the participation of dark excited states to the photophysics of adenine is still a matter of debate [15, 16].

As a first step toward the complete understanding of the factors ruling the interplay between bright and dark excited states in adenine derivatives, here we have computed the absorption spectra of 9-Me-Ade in the gas phase and in aqueous solution by exploiting a dynamical approach (vide infra). We focus our attention on the effects of intramolecular vibrational motion on the electronic spectrum and on the coupling among the different excited states. For the adenine derivative in aqueous solution, we mainly address here bulk solvent effects, while a detailed treatment of hydrogen bonding interactions will be tackled in a forthcoming paper.

The second case study concerns an ESR observable, namely the nitrogen hyperfine coupling constant (a_N), for nitroxide free radicals. Nitroxides are characterized by a long-living spin-unpaired electronic ground state and by molecular properties strongly dependent on the chemical environment embedding the NO moiety. These features led to such a widespread application of nitroxide derivatives as spin labels, spin probes, redox probes, or contrast agents that it is quite difficult to overemphasize the relevance of this class of compounds in many fields of modern chemistry [17]. In particular, we will focus on two of the nitroxide radicals most exploited in spin-probe/label ESR experiments, namely (2,2,6,6-tetramethylpiperidin-1-yl)oxidanyl (TEMPO) and (2,2,5,5-tetramethylpyrrolidin-1-yl)oxidanyl (PROXYL).

The molecular properties of these nitroxides strongly depend on the chemical environment, and recent theoretical simulations have achieved a detailed characterization of such dependencies for isotropic solutions in non-polar/polar and protic/aprotic solvents [18–22]. Besides the effects of complex chemical environment, ESR parameters are also affected by intra-molecular dynamics.

Here, we discuss the effects of intra-molecular dynamics on the a_N value of two different nitroxides that have a five-atom- (PROXYL) and six-atom- (TEMPO) ring backbone. In particular, we focus our analysis on the flexibility of the nitroxide moiety and its role in tuning the ESR observable. To this aim, we performed a characterization of minimum-energy structures, a rigid scan of the PES, and ab initio molecular dynamics simulations for both PROXYL and TEMPO in the gas phase. It has been recently demonstrated for these radicals that the AIMD-based time-dependent approach is in full agreement with the aforementioned time-independent approach [23]. In order to gain further chemical information, we also report a topological analysis of the nitroxide electronic structure by means of the maximum probability domains (MPD) approach [24, 25], which has allowed us to identify and discuss the subtle variations of the electron density with respect to structural fluctuations.

2 Methods

2.1 Time-dependent approach to the computation of optical and EPR spectra

Countours of optical spectroscopic bands can be obtained from a classical treatment of nuclear motion. In the classical limit, the absorption cross section can be expressed as

$$c\sigma(\omega) = \frac{4\pi\omega}{3}\int \rho_g(Q)|\mu_{ge}(Q)|^2\delta(V_g + \omega - V_e)dQ \qquad (1)$$

where $\rho_g(Q)$ is the classical density of normal coordinates Q in the ground state, μ_{ge} is the transition dipole between the ground and the excited state, δ is the Dirac delta, and $V_{g(e)}$ is the energy potential in the ground (excited) state. Therefore, absorption band shapes can be retrieved by a configurational sampling in the classical phase space of the electronic ground state, along with the corresponding transition dipole and vertical transition energy. While in this classical approach, the details of vibrational progression are lost, finite temperature effects in complex environment (solvent) can be easily taken into account by analysis of molecular dynamics trajectories. In this work, absorption spectra are constructed by means of a posteriori time-dependent DFT analysis of ab initio dynamics.

Classical sampling can also be exploited to introduce finite temperature effects on spectroscopic parameters. In the present work, we focus on the nitrogen EPR hyperfine coupling constant (a_α), which has been computed according to the following expression:

$$a_\alpha = \frac{8}{3}\pi g_\alpha g_e \mu_\alpha \mu_e \rho_s(r_\alpha) \qquad (2)$$

where μ_α and g_α are, respectively the magneton and the g factor of nucleus α, μ_e is the Bohr magneton, g_e is the g value for the electron, and $\rho_s(r_\alpha)$ is the electron spin density at the α nuclear position [4].

2.2 Maximum probability domains analysis on nitroxide probes

To exhaustively analyze the chemical bonding in open-shell nitroxyl systems, we resorted both to electron localization function [26] (ELF) and the maximum probability domains (MPDs) [24, 25]. In MPDs analysis, our first objective has been to identify spatial subsystems $\Omega \subset R^3$ in the nitroxide electronic system by specifying a given electron number v; we chose $v = 2$, 1 when trying to find by our calculations an electron pair and an unpaired electron, respectively. The next choice is the arbitrary definition of a volume Ω where we are going to find a given number of electrons v with a certain probability p; in essence, this definition reflects the quantum-mechanical view that, for a given state Ψ, the occurrence of a certain number of electrons v in Ω can only be formulated in terms of probability. Thus, we compute the probability of finding exactly v electrons in Ω based on Hartree-Fock wavefunctions and define the regions of space for which the probability to find v electrons becomes a maximum. Finally, for each given value of v, the shape of Ω (now more appropriately indicated as Ω_v) was optimized to find out local maxima of a certain probability by making use of shape derivatives [27]. As in all distributions, information about the probability distribution $p_{(0)}(\Omega), p_{(1)}(\Omega), p_{(2)}(\Omega), \ldots p_{(N)}(\Omega)$ is obtained by means of some parameters: among others, we focused on the mean $\langle N \rangle$. For an N-electrons system, the average number of electrons in Ω is given by:

$$\langle N\rangle(\Omega) = \sum_{\nu=0}^{N} \nu p_\nu(\Omega) \quad (3)$$

This has been obtained by integrating the electron density (ρ) over the domain, as indicated in the following equation:

$$\int_\Omega \rho(r)d^3r = \int_\Omega \langle\Psi|\hat{\rho}(r)|\Psi\rangle d^3r = \langle\Psi|\int_\Omega \hat{\rho}(r)d^3r|\Psi\rangle. \quad (4)$$

2.3 Computational details

Most of the calculations discussed in the present study were performed at the PBE0 level of theory. PBE0 [28] is an hybrid functional in which the spurious self-interaction is partly cured by mixing PBE and Hartree-Fock exchange with amounts determined in order to fulfill a number of physical conditions, without resorting to any fitting procedure [28]. In particular, PBE0 obeys both the Levy condition [29] and the Lieb-Oxford bound [30], providing a fairly accurate description of the low-density/high-gradient regions.

Regarding 9-Me-Ade, geometries were provided by PBE0/6-31G(d) optimizations in both the gas phase and the aqueous solvent, by means of the polarizable continuum model (PCM) [31–33]. Since the presence of explicit solute–solvent hydrogen bonds is particularly important for the excited states with $n \rightarrow \pi^*$ character, we have also performed test calculations by studying the model depicted in Fig. 1, which includes 4 water molecules of the first solvation shell. The excited state vertical excitation energies (VEE) have been computed at the TD-PBE0 level, by using standard 6-31G(d), 6-31+G(d,p), and 6-311+G(2d,2p) basis sets.

AIMD simulations were performed in both the gas phase and aqueous solvent within the atom-centered density matrix propagation (ADMP) MD scheme [34–37], which allows to use the most accurate hybrid HF-DFT levels of theory [38]. Bulk solvent effects were accounted for by the PCM, and explicit solute–solvent interactions were not included in ADMP dynamics runs nor in spectra calculations, since such more detailed treatment was deemed to fall outside the explorative purposes of the present study.

The configurational space accessible in the ground state was sampled by ADMP trajectories obtained at the B3LYP/6-31G(d,p) and B3LYP/6-31G(d,p)/CPCM levels of theory for the gas phase and the aqueous solution, respectively. After 5 ps of equilibration, trajectories have been collected for about 10 ps, with a time step of 0.2 fs. The fictitious value of the electronic mass was set to 0.5 amu bohr^2, and a velocity rescaling procedure maintained the average value of the total kinetic energy corresponding to a temperature of 300 K.

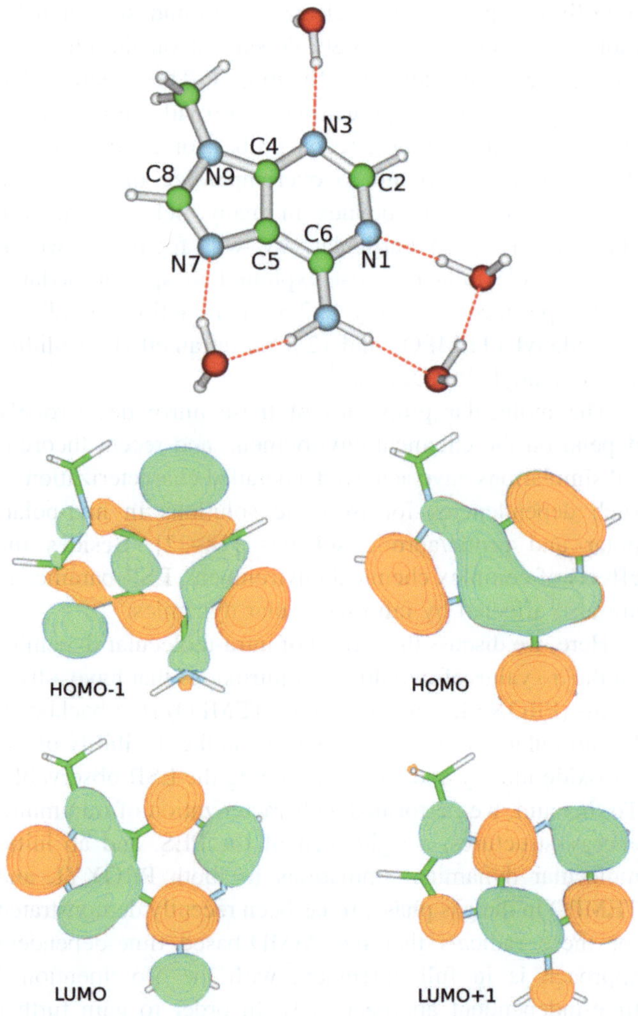

Fig. 1 Schematic drawings of 9-methyladenine and of its frontier orbitals

A number of configurations (approximatively 2000) uniformly extracted from the ADMP trajectories have been considered to compute excited state VEEs at the TD-PBE0/6-31+G(d,p) and TD-PBE0/6-31+G(d,p)/CPCM levels of theory. In order to evaluate direct and indirect bulk solvent effects on the electronic transitions, we considered gas phase VEE calculations on samplings obtained in both the gas phase and aqueous solution. To produce graphs of the absorption spectra, each transition was convoluted by a gaussian with FWHM = 0.05 eV, by a procedure similar to that adopted by Thiel and coll [39].

All the calculations carried out on the TEMPO and PROXYL nitroxide radicals were performed using the PBE0 density functional in combination with the polarized double-ζ basis set N07D [40], purposely tailored for DFT-based calculations of magnetic parameters [41, 42]. Gas phase minimum-energy structures were optimized using tight criteria of convergence. From the minima, we performed a rigid scan of the NO out-of-plane dihedral angle (vide infra). ADMP simulations were performed in the gas phase with a time step of 0.2 fs, for a trajectory of ~ 15 and ~ 18 ps in the case of PROXYL and TEMPO, respectively. The electron fictitious mass parameter was set to 0.5 amu bohr^2. The translational motion and the total angular momentum were projected out at each time step, while an average temperature of 298 K was pursued via rescaling of nuclei velocity modulus. For both PROXYL and TEMPO, the hyperfine constant a_N has been computed on minima, along the rigid scan and along the AIMD trajectories.

All the calculations were carried out by using a development version of the Gaussian package [43]. A 3D extension of MPDs [24, 25] analysis implemented in a development version of CRYSTAL code [44], able to deal both with molecules and crystalline systems in the Hartree-Fock and DFT approximation, has been adopted. The code has been modified by considering spins separately, and for the first time, we apply the method to wave functions obtained from unrestricted Hartree-Fock calculations.

3 Results and discussion

3.1 Solvent effect on the absorption spectra of 9-methyladenine

As anticipated in the introduction, for adenine several studies interpreted the very short excited state lifetime, in aqueous solution [45, 46] and the gas phase [47, 48], as an ultrafast radiation-less decay of the spectroscopic bright state (L_a, see below) to S_0 ($L_a \rightarrow S_0$) [15, 49]. On the other hand, recent studies suggested that L_a is strongly mixed in the FC region with a close lying dark $n\pi^*$ state and that ground state recovery follows a different pathway, passing through the dark $n\pi^*$ state ($L_a \rightarrow n\pi^* \rightarrow S_0$) [15, 49]. Interestingly, the same mechanism has been proposed to be active also for Ade bases within DNA [16].

It is clear that a key factor modulating the relative importance of the two different pathways is the mixing between the different excited states in the FC region: since a barrier-less path on the L_a surface connects the FC region with a CI with S_0, only an ultrafast $L_a \rightarrow n\pi^*$ decay would be competitive with the direct $L_a \rightarrow S_0$ decay. Intramolecular vibrational motion, distorting the planar geometry of the purine ring and potentially coupling L_a and $n\pi^*$, could play a crucial role. As a consequence, we have decided to investigate the effect of nuclear motion on the absorption spectrum of 9-Me-Ade.

Solvent effects are also expected to be important because they can influence the $L_a \rightarrow n\pi^*$ interplay directly, by affecting the energy of the different electronic transitions, or indirectly, by changing the equilibrium geometry and vibrational properties of adenine.

The vertical excitation energies (ν_A) of 9-methyladenine in the gas phase and in aqueous solution are reported in Table 1. Confirming the results of the previous computational studies [15, 39], three electronic transitions contribute to the lowest energy absorption band: two with

Table 1 Vertical excitation energies (in eV) computed for 9-Me-Ade and the 9-Me-Ade(H_2O)$_4$ water cluster in the gas phase (TD-PBE0) and in aqueous solution (TD-PBE0/PCM)

	TD/PBE0			TD-PBE0/PCM	
	9-Me-Ade[a]	9-Me-Ade[b]	9-Me-Ade(H_2O)$_4^b$	9-Me-Ade[a]	9-Me-Ade(H_2O)$_4^b$
6-31G(d)					
$Sn\pi$	5.18 (0.007)	5.14 (0.003)	5.10 (0.000)	5.27 (0.020)	5.63 (0.001)
L_a	5.26 (0.194)	5.28 (0.185)	5.26 (0.210)	5.21 (0.223)	5.10 (0.245)
L_b	5.49 (0.050)	5.45 (0.060)	5.48 (0.040)	5.42 (0.072)	5.37 (0.120)
6-31+G(d,p)					
$Sn\pi$	5.15 (0.013)	5.10 (0.096)	5.07 (0.05)	5.27 (0.002)	5.60 (0.000)
L_a	5.09 (0.222)	5.13 (0.140)	5.08 (0.237)	5.08 (0.295)	4.98 (0.305)
L_b	5.39 (0.028)	5.36 (0.028)	5.39 (0.022)	5.32 (0.045)	5.26 (0.087)
6-311+G(2d,2p)					
$Sn\pi$	5.13 (0.006)	5.10 (0.037)	5.04 (0.06)	5.26 (0.002)	5.57 (0.000)
L_a	5.03 (0.225)	5.05 (0.195)	5.02 (0.229)	5.02 (0.292)	4.92 (0.300)
L_b	5.33 (0.024)	5.30 (0.021)	5.32 (0.010)	5.26 (0.038)	5.20 (0.078)

Geometries were obtained by PBE0/6-31G(d) and PBE0/6-31G(d)/PCM optimizations. Oscillator strengths are given in parentheses

Experimental band maximum: dA 4.77 eV [60]; adenine in the gas phase at the TD-PBE0/6-31+G(d,p)//PBE0/6-31G(d) level of theory: $Sn\pi$ 5.13(0.018), L_a 5.19(0.219), L_b 5.42(0.033)

[a] Gas phase optimization

[b] Aqueous solvent optimization

$\pi \rightarrow \pi^*$ character (usually labeled π^* L_a and π^* L_b, respectively, hereafter simply L_a and L_b) and one with $n \rightarrow \pi^*$ character (hereafter $Sn\pi$). All the methods agree in predicting that $S_0 \rightarrow L_a$ is the most intense transition: it can be described as a HOMO → LUMO excitation. The L_b state corresponds mainly to a HOMO → LUMO+1 $\pi \rightarrow \pi^*$ transition. $Sn\pi$ can be described as HOMO-1 → LUMO transition, involving the transfer of an electron from a non bonding Kohn-Sham (KS) orbital (HOMO-1 in Fig. 1) to the π^* orbital.

The prediction of TD/PBE0 calculations [50, 51] is consistent with those of the most accurate ab initio methods [15, 39]. As reported in the Notes of Table 1, for adenine in the gas phase (the system most thoroughly studied), we predict that $Sn\pi$ corresponds to the lowest energy excited states at the FC point, being ~0.1 eV more stable than L_a. In the literature, there is more disagreement concerning the relative energy of the L_a and L_b transitions: while according to some studies, L_b is slightly more stable than L_a [15, 52, 53], our calculations and other computational studies [39] predict that L_b is ~0.2 eV less stable than L_a. In any case, L_b should not be significantly involved in the excited state decay of adenine derivatives [49, 52–54].

As shown in Table 1, the 9-methyl substituent and, especially, the inclusion of solvent effects strongly favor L_a over $Sn\pi$. When only bulk solvent effects are included by PCM, for 9-Me-Ade L_a corresponds to S_1, being more stable than $Sn\pi$ by 0.05–0.2 eV (an increase in the size of the basis set favors L_a). The most significant consequence of the presence of solute/solvent hydrogen bonds (9-Me-Ade 4 H_2O model) is the destabilization of the $n \rightarrow \pi^*$ transition by ~0.3 eV. In analogy with what found in uracil-like molecules [55], this latter kind of transitions is blue-shifted in the presence of hydrogen bonds with the solvent molecules. The transfer of an electron from the nitrogen lone pair (acting as donor in the hydrogen bonds with water molecules) to a π^* orbital delocalized on the ring decreases the strength of the solute–solvent interactions.

From the quantitative point of view, our results are in very good agreement with experiments. For the 9-Me-Ade · 4 H_2O model, the computed ν_a is blue-shifted by only 0.2 eV with respect to the maximum of the absorption band of Adenosine in solution. Furthermore, we predict a solvent red-shift in the range 0.10–0.16 eV (depending on the basis set), the corresponding experimental estimate being 0.15 eV (see Ref. [39] and references therein). For what concerns the main purpose of this study, the coupling between L_a and $Sn\pi$ is always rather small. The contributions of the HOMO → LUMO excitation to $Sn\pi$ and, conversely, that of the HOMO-1 → LUMO excitation to L_a are always smaller than 0.2. Furthermore, also when L_a and $Sn\pi$ ν_A are within 0.05 eV, the L_a oscillator strength is always at least ten times that of $Sn\pi$: a strong coupling between the two states would have instead produced a much larger intensity borrowing between the two states.

Before analyzing the spectra computed from our ADMP/PBE0//TD-PBE0 procedure, it is important to remind that, especially in the presence of a strong vibronic coupling between L_a and $Sn\pi$, our approach is not expected to provide an accurate description of the absorption line-shape, which would require a quantum-mechanical treatment [56, 57] of nuclear dynamics. However, we can provide qualitative insights on the role of intramolecular vibrations in promoting the L_a and $Sn\pi$ coupling. In this spirit, without resorting to a detailed diabatization procedure, we shall use the ratio (r) between the oscillator strengths of the 'dark' and the 'bright' adiabatic excited states as a qualitative measure of the coupling between L_a and $Sn\pi$ diabatic states. For 9-Me-Ade geometries close to planarity (as in the PBE0 equilibrium minimum) L_a bears indeed all the oscillator strength (the L_b contribution is small) while $Sn\pi$ is almost dark. On the contrary, a coupling between L_a and $Sn\pi$ will be mirrored by a transfer of oscillator strength between the two lowest adiabatic excited states.

Inspection of Fig. 2 shows that in the gas phase, the curve associated withS_1 and S_2 has comparable intensities, the former being only 30 % less intense than the latter. This picture is very different with respect to that reported in Table 1: at the same level of theory, S_1 (in this case almost coinciding with the $Sn\pi$ diabatic state) is a hundred times less intense than S_2 (corresponding to L_a). Actually, the similar intensities of the S_1 and S_2 absorption bands are not only due to the L_a/$Sn\pi$ vibronic coupling. For several geometries, it depends simply on the interchange between L_a and $Sn\pi$ energy, that is L_a corresponds to S_1 and $Sn\pi$ to S_2, without a significant mixing between the two adiabatic states. A preliminary analysis indicates that only in ~50 % of the geometries considered r is ≥ 0.1 (moderate coupling), in ~30 % r is ≥ 0.3, and only in ~15 % r is ≥ 0.5 (strong coupling).

When bulk solvent effects are included in our analysis (middle panel of Fig. 2), the picture is more similar to that provided by the spectrum computed in the S_0 minimum (Table 1). Most of the oscillator strength is carried by S_1, the intensity of the S_2 band being ~40 % of the S_1 one. However, our analysis shows a distribution of the computed r values similar to that found in the gas phase. In other words, inclusion of bulk solvent effects does not seem to dramatically affect the L_a/$Sn\pi$ coupling. In order to further check this conclusion, we have computed the spectra in the gas phase by using the geometries provided by PCM/ADMP calculations (dashed lines in the top panel of Fig. 2). The computed spectra are very similar to those analyzed above. Interestingly, the use of the PCM sampling leads to a slightly larger coupling, in line with results

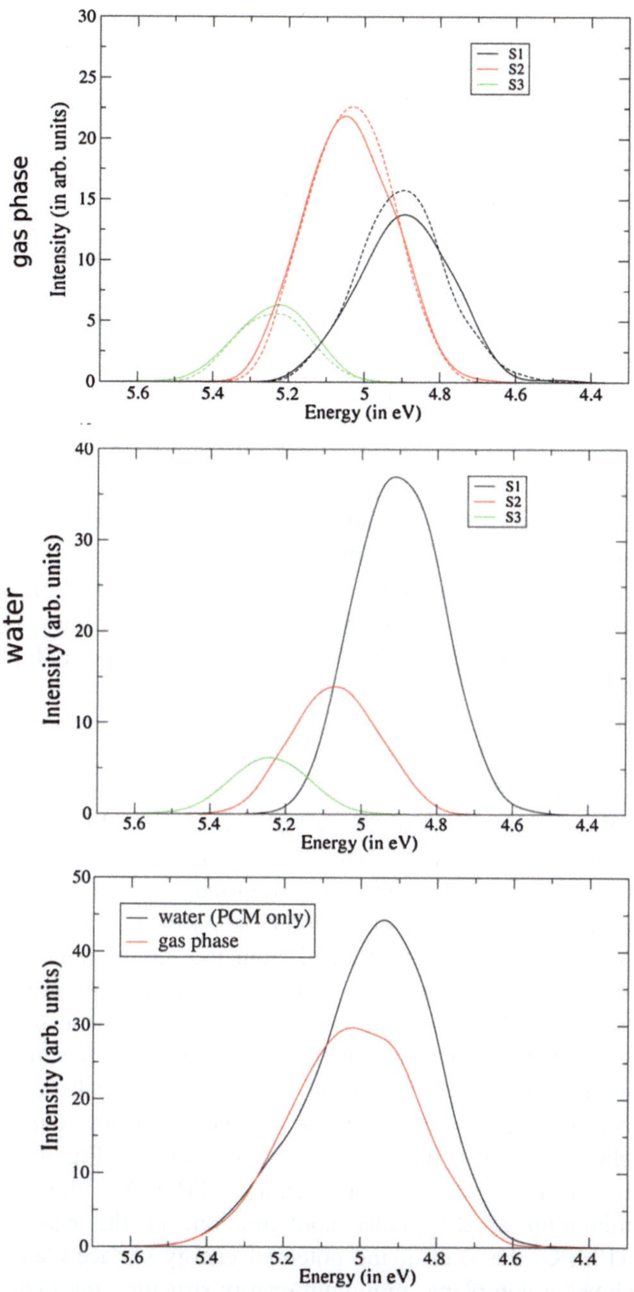

Fig. 2 Absorption bands associated with the three lowest adiabatic excited states for 9-Me-Ade in the gas phase (*upper panel*; the *dashed curves* display the results obtained by using the geometries issuing from ADMP simulations performed in water) and in aqueous solution (*middle panel*), computed by using the TD-PBEO/6-31+G(d,p)//ADMP/PBE0/6-31G(d) procedure described in the text. The resulting absorption spectra are shown in the *bottom panel*

obtained by computing the TD-PBE0 v_A in the gas phase on the geometries optimized in aqueous solution (see the first 2 columns of Table 1).

A detailed characterization of vibronic effects on 9-Me-Ade absorption spectrum in water is outside the scope of the present paper, and, in our case, the absence of explicit solute–solvent interactions could bias the quantitative reliability of our conclusion. A preliminary analysis indicates that, not surprisingly, the out-of-plane motion of the purine nitrogen atoms (whose lone pairs are involved in the $Sn\pi$ excitation) can act as very effective coupling modes. In order to get additional insights, we have extracted and examined more carefully the 10 structures exhibiting the largest r values ($r \geq 0.9$). Our analysis indicates that small departures from planarity (<5°) are sufficient to induce a significant coupling between L_a and $Sn\pi$ states and that the combination of different collective low-frequency torsional mode, not necessarily involving nitrogen atoms, can affect the vibronic coupling. For example, in several structures, the largest deviation from the molecular plane occurs at the C6 position (i.e., the site bearing the amino-substituent, see Fig. 1). In other cases, the coupling is mainly due to the 'bending' of the angle between the 5- and the 6-membered rings (the two rings remaining essentially close to planarity). Finally, pyramidalization of the C2 and C8 carbon atoms can also increase the $L_a/Sn\pi$ coupling.

It is worth noting that our calculations (bottom panel of Fig. 2) predict a noticeable solvent red-shift (~0.1 eV). The computed value is slightly underestimated with respect to the experimental one, but we recall that at this stage of the analysis, we are not considering explicit solute–solvent interactions; as shown in Table 1, these lead to an additional red-shift of the L_a band in water with respect to the gas phase.

Our picture is globally rather similar to that obtained by Thiel and coll. [39], which points out to a very strong mixing between $Sn\pi$ and L_a states in the FC region; However, a more detailed comparison is not easy, since only 10 optimized snapshot geometries were reported [39]. The simulated absorption spectrum in solution [39] confirms that both the S_1 and S_2 adiabatic states contribute to the lowest absorption band, carrying similar intensities. The ratio between the bands associated with S_1 and S_2 is rather similar to what we found in the gas phase, despite the fact that solute–solvent hydrogen bond interactions were explicitly included in the aforementioned computations [39]. In this respect, it is important to remind that the results reported in Ref. [39] concern adenine, while we have shown the methyl substitution affects the relative energies of L_a and $Sn\pi$ states, favoring the former by ~0.1 eV. Therefore, we can expect that for 9-Me-Ade in a larger fraction of the sampled geometries, L_a would correspond to S_1 and $Sn\pi$ to S_2.

3.2 Dynamical effects on EPR parameters and maximum probability domains of nitroxyl radicals

The molecular structures of TEMPO and PROXYL radicals are shown in Fig. 3. The isotropic hyperfine coupling

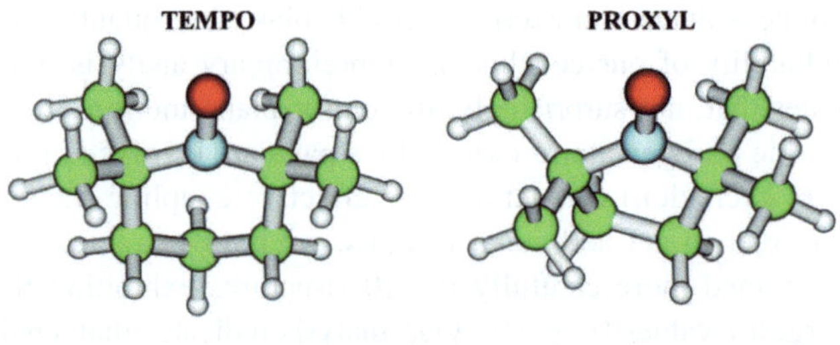

Fig. 3 Molecular structure of the TEMPO (**a**) and PROXYL (**b**) nitroxide radicals; the *white*, *green*, *light-blue* and *red spheres* represent hydrogen, carbon, nitrogen and oxygen atoms, respectively

is a direct measure of the coupling between the spin-unpaired electron and the magnetic moment of a specific nucleus. For nitroxide spin-label or spin-probe experiments, the quantity of interest is the nitrogen hyperfine coupling constant (a_N). For this ESR observable, the most relevant structural parameter of the nitroxide moiety is the NO out-of-plane angle (i.e., the C–N–O…C improper dihedral angle) [18–20]. Due to their cyclic backbone, TEMPO and PROXYL display different minimum-energy conformations in the gas phase, as listed in Table 2. In the case of PROXYL, the five-atom ring leads to a planar conformation of the NO moiety (the oxygen atom lies in the plane defined by the C–N–C atoms). The six-atom ring of TEMPO leads to a nonplanar nitroxide moiety, the oxygen center lying 20° off the C–N–C plane. As a direct consequence of that, the NO bond length in TEMPO is slightly longer than in PROXYL. As a matter of facts, in the planar NO moiety of PROXYL, both the N and O atoms show formal sp^2 hybridization, whereas in the non-planar case of TEMPO, the N hybridization is close to sp^3, leading to an elongation of the NO equilibrium bond length. Table 2 also lists the a_N values computed on the minimum-energy configurations, which are different for the two nitroxides (11.8 and 14.8 Gauss for PROXYL and TEMPO, respectively). Such values can be directly compared to the experimental a_N measured in solution of a non-polar and aprotic solvent with very low dielectric constant, like, for example, dodecane or cyclohexane. The experimental a_N for PROXYL and TEMPO are 14.02 and 15.28 Gauss, respectively [58, 59]. The predicted a_N for the TEMPO minimum-energy structure is more accurate than the one of PROXYL, which is off by 2 Gauss. Part of this discrepancy is due to vibrational averaging. Therefore, a_N was computed on frames extracted along the AIMD trajectories at regular 25-fs intervals for a total of ~500 frames and ~600 frames in the case of PROXYL and TEMPO, respectively. In both cases, the first 3 ps of the AIMD simulations were taken as an equilibration phase and were discarded: this sampling scheme ensures well-converged average values of the magnetic parameters [23]. Table 2 lists structural and a_N values averaged over the AIMD trajectories for both PROXYL and TEMPO.

Clearly, the most significant discrepancy between 'static' and 'dynamic' pictures concerns the out-of-plane bending of the NO moiety, which, judging by the standard deviations, gives rise to large-amplitude motions. The average value corresponds to a planar configuration for both PROXYL and TEMPO. However, by taking the absolute values of the C–N–O–C improper dihedral angle, the averages are now pointing to a pyramidal configuration for both nitroxides. In other words, the NO moieties of both PROXYL and TEMPO oscillate between positive and negative values that are almost equivalent with respect to the pseudo-symmetric C–N–C plane.

Despite the large standard deviation of 2 Gauss, the a_N values averaged over the AIMD trajectories (14 and 15 Gauss for PROXYL and TEMPO, respectively) agree remarkably well with the experimental counterparts. Moreover, in the case of TEMPO, the AIMD average value is very close to that obtained from the minimum-energy structure, while for the PROXYL radical, the proper treatment of temperature and vibrational motions leads to an a_N value that is substantially different from the minimum-energy one. This behavior can be rationalized by considering how the a_N value changes along the out-of-plane bending of the NO moiety, as shown in Fig. 4. As found from geometry optimization, PROXYL shows a minimum at the planar configuration. In the case of TEMPO, the scan of the potential energy surface clearly shows a non-planar minimum-energy structure, the planar configuration being a transition state, even though the barrier is quite low (~2 kcal/mol). Despite this difference in energy profiles, the a_N values of both nitroxides display a

Table 2 Structural and magnetic parameters of TEMPO and PROXYL radicals

	$TEMPO_{min}$	$TEMPO_{\langle ADMP \rangle}$	$PROXYL_{min}$	$PROXYL_{\langle ADMP \rangle}$
N–O	1.269	1.27 (0.02)	1.260	1.27 (0.02)
C–N–O–C (α)	21.4	0 (20)	0.0	0 (16)
$\langle \lvert\alpha\rvert \rangle$		16 (10)		12 (10)
a_N	14.8	15 (3)	11.8	14 (2)

Bond lengths in Angstroms, improper dihedral angles in degrees and nitrogen hyperfine coupling constants (a_N) in Gauss. Standard deviations in parentheses

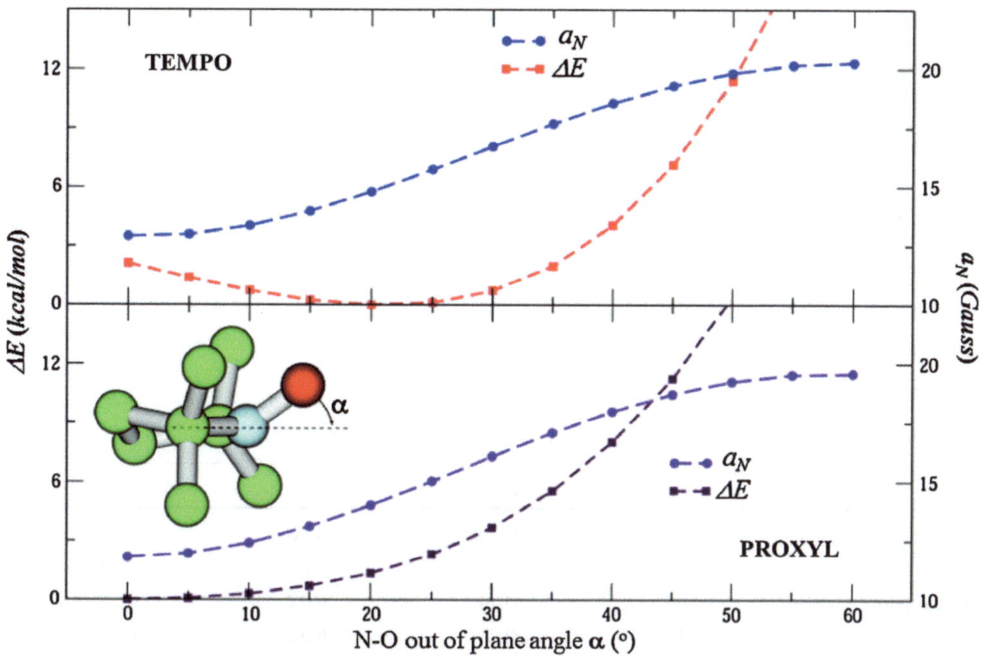

Fig. 4 Energy and a_N changes along the out-of-plane bending of the NO moiety for the TEMPO (*upper panel*) and PROXYL (*lower panel*) nitroxide radicals

very similar trend, from both the quantitative and qualitative points of view. These results highlight the importance of taking properly into account this motion for predicting reliable isotropic hyperfine couplings of nitroxide spin probes.

To better analyze the electronic structure of the nitroxide moiety, we performed a topological study of the PROXYL electron density by means of the ELF and the MPD schemes. Figure 5 depicts the results of both approaches on the nitroxide moiety of the PROXYL minimum-energy planar structure. The ELF core basins can be easily located close to the N and C centers. The oxygen core basin is hidden by lone pairs. The C–N bond is polarized toward the more electronegative nitrogen atom; the very small N–O basin suggests a lower localization of electrons, consistently with the expected multiple nature of the N–O bond. However, the ELF results also show two additional lone pairs around the N center, a quite unexpected location for electrons. In other words, the ELF analysis of the planar PROXYL minimum provides a chemically unsound picture, in which the N center has three σ bonds and two lone pairs. On the contrary, the MPD analysis agrees nicely with chemical common sense. The $\langle N \rangle$ values computed according to Eq. (2) are listed in Table 3. The picture that emerges is in agreement with the chemical imagine of a planar nitroxide structure, with a N=O double bond, two lone pairs on oxygen, and the unpaired electron delocalized on the NO bond. The sum of $\langle N \rangle$ on all bonds, lone pairs and unpaired electrons amounts to 13.38 e, in agreement with the electronegative character of the NO moiety. As indicated in Fig. 5, the C–N σ MPDs enhance the polarization toward the N center. The N–O σ MPD is smaller. The oxygen lone pair MPD dominates the O side of the molecule. The MPD relative to a pair of electrons can be associated with a π bond between N and O. This MPD is very mobile: similar and close optimal domains can be found with very similar probability values. This is a situation of 'multiple solution' probability domain already identified by Savin [24]; the physical multiplicity may reflect a situation in which several solutions can be obtained in the optimization of Ω for a given $\nu(\Omega_\nu)$ and will be the object of future analysis on MPDs method. Furthermore, the MPDs relative to a single electron for three representative PROXYL structures with different piramidalization degrees at the nitrogen atom have been computed and are reported in Fig. 6. The unpaired electron MPD reproduces the spatial description of spin density found in the case of a flat NO moiety. The single electron MPD has very small overlap with other domains: it inserts in the space left by the oxygen lone pairs and the N–O π domain. As the degree of pyramidalization increases, the population of this domain shifts clearly toward the oxygen center, while the residual probability at the nitrogen center still plays a crucial role in tuning the ESR observable.

The trends observed for a_N can be now put into a more general perspective by recalling that both direct and spin-polarization contributions to a_N are roughly proportional to the spin population in the singly occupied highest energy molecular orbital (SOMO) [4]. In the planar conformation (which we have seen does not necessarily correspond to a minimum on the PES), the only contribution to a_N is related to spin polarization, because the SOMO has a π nature. However, pyramidalization at the nitrogen center

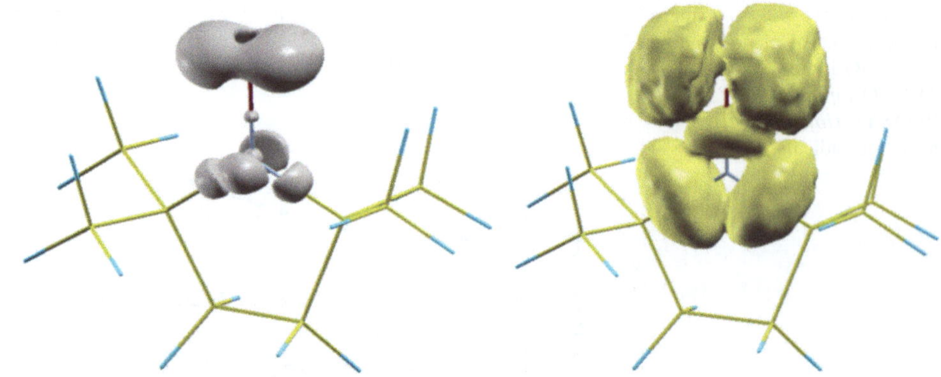

Fig. 5 *Left* ELF iso-surface of the NO moiety in the equilibrium structure; a ELF value of 0.83 has been chosen for the electron function. *Right* maximum probability domains relative to electron pairs associated with C–N σ bonds, N–O σ bond and oxygen lone pairs

Table 3 Results of MPD analysis for the equilibrium structure of PROXYL

Domain name	p_0	p_1	p_2	p_3	$\langle N \rangle$	Volume
C–N	0.04	0.23	**0.38**	0.24	2.12	13.3 (1.47)
N–O σ	0.07	0.27	**0.37**	0.21	1.99	7.6 (1.22)
O lone pair	0.04	0.24	**0.42**	0.23	2.04	22.2 (1.74)
N–O π	0.10	0.27	**0.32**	0.20	1.96	19.1 (1.66)
N–O unpaired	0.30	**0.39**	0.22	0.07	1.11	1.5 (0.72)

The probability p_ν value for which the domain is optimized is highlighted in bold; the average number of electrons in Ω, $\langle N \rangle$ is reported for each domain. The volume of the domain is expressed in bohr^3; in brackets the radius (in bohr) of the equivalent sphere is also reported

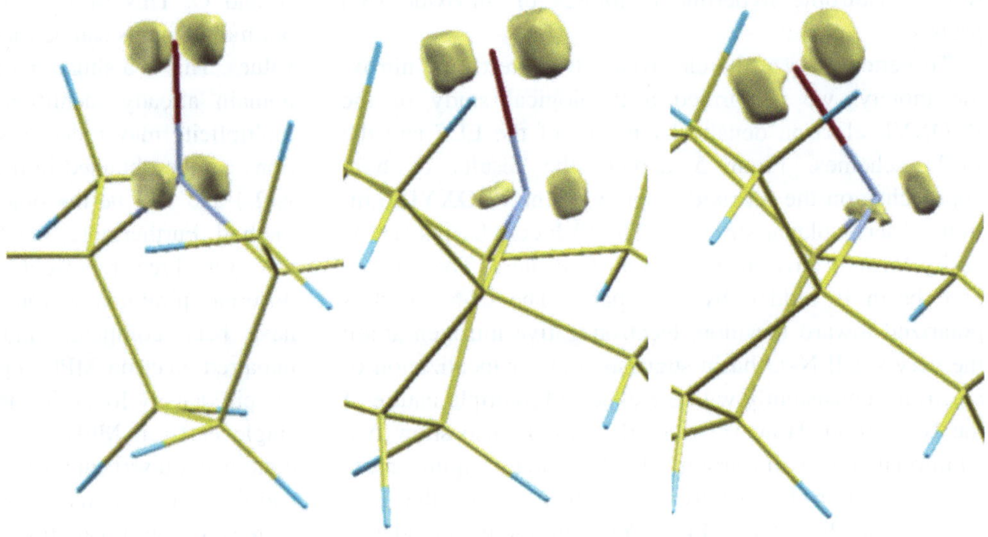

Fig. 6 Maximum probability domain relative to a single electron for three representative structures. *Left* equilibrium structure with an improper dihedral angle C–N–O–C equal to 180°. *Center* dihedral angle C–N–O–C equal to 136°. *Right* dihedral angle C–N–O–C equal to 118°

allows for the direct involvement of the nitrogen *s*-type orbital in the SOMO, leading to a large direct contribution to the Fermi contact term, which overcomes the spin-polarization term and determines the observed increase in a_N.

4 Concluding remarks

In the present account, we discussed two case studies by focusing our attention on the interplay of nuclear motions and solvation in tuning the electronic structure of molecules that are relevant for their spectroscopic properties and photophysical behavior.

For what concerns the 9-Me-Ade excited state decay, our results cannot be considered definitive since a fully reliable description of the vibronic coupling between L_a and $Sn\pi$ states would require a complete treatment of the explicit solute–solvent interactions and an accurate diabatization procedure. However, our analysis shows that the intramolecular vibrational motion at finite temperature significantly increases the mixing between L_a and $Sn\pi$, which is larger than that found in the absolute ground state minimum at 0 K. On the other hand, we have shown that for 9-Me-Ade in solution in 50 % of the sampled geometries, the ratio between the less intense ($Sn\pi$ like) and most intense (L_a like) transition is smaller the 0.1, indicating a very weak coupling between the states. Since inclusion of solute–solvent hydrogen bonds is expected to increase the energy gap between the two diabatic states, our analysis casts some doubts on a significant involvement of the $Sn\pi$ state in the deactivation mechanism.

Regarding the results on ESR parameters of nitroxide spin probes, we have shown how considering only minimum-energy structures in the computation of a_N leads to predictions that are qualitatively unreliable and quantitatively far from the experimental counterparts, for example in the case of PROXYL. In the case of TEMPO, the agreement between static minimum-energy values and those obtained when dynamics is properly taken into account through AIMD average is caused by a fortuitous cancelation of effects. Evidently, this does not mean that neglecting molecular dynamics may represent in general a safe and reliable computational strategy.

In conclusion, the above examples highlight the relevance of a proper description, within a QM computational protocol, of two key factors, namely thermal averaging effects and the role of the chemical surroundings. Theoretical developments of multi-scale methods along these directions represent a significant step toward the challenging task of a direct comparison between experiments and simulations and promise to become a valuable tool for supporting the interpretation of ambiguous laboratory data and for driving the design and synthesis of innovative chemical systems.

Acknowledgments Financial support was provided by MIUR (PRIN 2008 and FIRB Futuro in Ricerca). The authors gratefully acknowledge the inspiration, guidance and support provided over the years by Professor Vincenzo Barone, and wish him a happy 60th birthday.

References

1. Barone V (ed) (2012) Computational strategies for spectroscopy: from small molecules to nano systems. Wiley, Hoboken
2. Parr RG, Yang W (1989) Density-functional theory of atoms and molecules. Oxford University Press, New York
3. Santoro F, Barone V (2010) Int J Quantum Chem 110:476
4. Improta R, Barone V (2004) Chem Rev 104:1231
5. Barone V (2005) J Chem Phys 122:014108
6. Leach AR (2001) Molecular modeling: principles and applications. Pearson Education, Harlow
7. Carr R, Parrinello M (1985) Phys Rev Lett 55:2471
8. Marx D, Hutter J (2000) Modern methods and algorithms of quantum chemistry, John von Neumann Institute for Computing, Julich, vol 1, chap. Ab initio molecular dynamics: theory and implementation. p 301
9. Crespo-Hernandez CE, Cohen B, Kohler B (2004) Chem Rev 104:1977
10. Middleton CT, de La Harpe K, Su C, Law YK, Crespo-Hernandez CE, Kohler B (2009) Annu Rev Phys Chem 60:13
11. Markovitsi D, Gustavsson T, Talbot F (2007) Photochem Photobiol Sci 6:717
12. Markovitsi D, Gustavsson T, Vayá I (2010) J Phys Chem Lett 1:3271
13. Gustavsson T, Improta R, Markovitsi D (2010) J Phys Chem Lett 1:2025
14. Hare PM, Crespo-Hernandez CE, Kohler B (2007) Proc Natl Acad Sci USA 104:435
15. Perun S, Sobolewski AL, Domcke W (2005) J Am Chem Soc 127:6257
16. Lu Y, Lan ZG, Thiel W (2011) Angew Chem Int Edit 50:6864
17. Berliner LJ (ed) (1998) Biological magnetic resonance, spin labeling in the next millennium, vol 14. Plenum Press, New York
18. Pavone M, Cimino P, Crescenzi O, Sillanpaa A, Barone V (2007) J Phys Chem B 111:8928
19. Pavone M, Sillanpaa A, Cimino P, Crescenzi O, Barone V (2006) J Phys Chem B 110:16189
20. Cimino P, Pavone M, Barone V (2007) J Phys Chem A 111:8482
21. Rinkevicius Z, Murugan NA, Kongsted J, Frecu B, Steindal AH, Agren H (2011) J Chem Theory Comput 7:3261
22. Houriez C, Ferre N, Masella M, Siri D (2008) J Chem Phys 128:244504
23. Pavone M, Biczysko M, Rega N, Barone V (2010) J Phys Chem B 114:11509
24. Causà M, Savin A (2011) J Phys Chem A 115:13139
25. Savin A (2002) In: Reviews of modern quantum chemistry: a celebration of the contributions of Robert G. Parr. World Scientific, Singapore, p 43
26. Becke A (1990) J Chem Phys 92:5397
27. Cancès E, Keriven R, Lodier F, Savin A (2004) Theor Chem Acc 111:373
28. Adamo C, Barone V (1999) J Chem Phys 110:6158
29. Levy M, Perdew JP (1993) Phys Rev B 48:11638
30. Lieb EH, Oxford S (1981) Int J Quantum Chem 19:427
31. Tomasi J, Persico M (1994) Chem Rev 94:2027
32. Tomasi J, Mennucci B, Cammi R (2005) Chem Rev 105:2999
33. Cossi M, Barone V (2001) J Chem Phys 115:4708
34. Schlegel HB, Millam JM, Iyengar SS, Voth GA, Daniels AD, Scuseria GE, Frisch MJ (2001) J Chem Phys 114:9758
35. Iyengar SS, Schlegel HB, Millam JM, Voth GA, Scuseria GE, Frisch MJ (2001) J Chem Phys 115:10291
36. Schlegel HB, Iyengar SS, Li X, Millam JM, Voth GA, Scuseria GE, Frisch MJ (2002) J Chem Phys 117:8694
37. Rega N, Iyengar SS, Voth GA, Schlegel HB, Vreven T, Frisch MJ (2004) J Phys Chem B 108:4210
38. Prendergast D, Grossman JC, Galli G (2005) J Chem Phys 123:014501
39. Lan ZG, Lu Y, Fabiano E, Thiel W (2011) Chem Phys Chem 12:1989
40. Barone V, Cimino P, Stendardo E (2008) J Chem Theor Acc 4:751
41. Barone V, Cimino P (2008) Chem Phys Lett 454:139
42. Barone V, Cimino P (2009) J Chem Theory Comput 5:192
43. Frisch MJ, Trucks GW, Schlegel HB, Scuseria GE, Robb MA, Cheeseman JR, Scalmani G, Barone V, Mennucci B, Petersson GA, Nakatsuji H, Caricato M, Li X, Hratchian HP, Izmaylov AF, Bloino J, Zheng G, Sonnenberg JL, Hada M, Ehara M, Toyota K, Fukuda R, Hasegawa J, Ishida M, Nakajima T, Honda Y, Kitao O, Nakai H, Vreven T, Montgomery JA Jr, Peralta JE, Ogliaro F, Bearpark M, Heyd JJ, Brothers E, Kudin KN, Staroverov VN, Kobayashi R, Normand J, Raghavachari K, Rendell A, Burant JC, Iyengar SS, Tomasi J, Cossi M, Rega N, Millam JM, Klene M,

Knox JE, Cross JB, Bakken V, Adamo C, Jaramillo J, Gomperts R, Stratmann RE, Yazyev O, Austin AJ, Cammi R, Pomelli C, Ochterski JW, Martin RL, Morokuma K, Zakrzewski VG, Voth GA, Salvador P, Dannenberg JJ, Dapprich S, Daniels AD, Farkas, Foresman JB, Ortiz JV, Cioslowski J, Fox DJ. Gaussian 09 Revision A.2. Gaussian Inc. Wallingford CT 2009
44. Dovesi R, Saunders VR, Orlando R, Zicovich-Wilson CM, Pascale F, Civalleri B, Doll K, Bush IJ, DArco P, Lunell M. Crystal 2009 user manual. Turin University 2009
45. Pecourt JML, Peon J, Kohler B (2001) J Am Chem Soc 123:10370
46. Gustavsson T, Sharonov A, Onidas D, Markovitsi D (2002) Chem Phys Lett 356:49
47. Ullrich S, Schultz T, Zgierski MZ, Stolow A (2004) J Am Chem Soc 126:2262
48. Bisgaard CZ, Satzger H, Ullrich S, Stolow A (2009) Chem Phys Chem 10:101
49. Conti I, Garavelli M, Orlandi G (2009) J Am Chem Soc 131:16108
50. Santoro F, Barone V, Improta R (2007) Proc Natl Acad Sci USA 104:9931
51. Improta R, Barone V (2008) Theor Cheom Acc 120:491
52. Serrano-Andres L, Merchan M, Borin AC (2006) Proc Natl Acad Sci USA 103:8691
53. Serrano-Andres L, Merchan M, Borin AC (2006) Chem Eur J 12:6559
54. Fabiano E, Thiel W (2008) J Phys Chem A 112:6859
55. Santoro F, Barone V, Gustavsson T, Improta R (2006) J Am Chem Soc 128:16312
56. Improta R, Barone V, Santoro F (2007) Angew Chem Int Edit 46:405
57. Santoro F, Lami A, Improta R, Barone V (2007) J Chem Phys 126:184102
58. Keana JFW, Lee TD, Bernard EM (1976) J Am Chem Soc 98:3025
59. Knauer BR, Napier JJ (1976) J Am Chem Soc 91:4395
60. Onidas D, Markovitsi D, Marguet S, Sharonov A, Gustavsson T (2002) J Phys Chem B 106:11367

Theor Chem Acc (2012) 131:1225
DOI 10.1007/s00214-012-1225-8

REGULAR ARTICLE

Theoretical and experimental investigation on the near-infrared and UV–vis spectral regions of a newly synthesized triarylamine electrochromic system

Bruna Clara De Simone · Angelo Domenico Quartarolo · Sante Cospito · Lucia Veltri · Giuseppe Chidichimo · Nino Russo

Received: 17 January 2012 / Accepted: 18 April 2012 / Published online: 8 May 2012

Abstract A new triarylamine derivative, *N,N′*-Bis (4-heptanoylamidophenyl)-*N,N′*-di(4-methoxyphenyl)-1,4-phenylenediamine, with stable electrochromism in near-infrared and visible light regions, has been synthesized and characterized at theoretical and experimental level. The detected and simulated spectra, with and without the presence of an external potential at different values, clearly show that this mixed-valence system undergoes ionization at a low value of the applied potential, and the formed radical cation absorbs in the near-infrared region with an intense peak located at 1,040 nm. Density functional computations give the geometrical structure and absorption properties in very good agreement with experiment, allowing assigning the electronic transition and contributing to an understanding of the electron-transfer process between the two redox centers.

Keywords Electrochromism · Density functional theory · Electron transfer · Mixed-valence compound

Dedicated to Professor Vincenzo Barone and published as part of the special collection of articles celebrating his 60th birthday.

Electronic supplementary material The online version of this article (doi:10.1007/s00214-012-1225-8) contains supplementary material, which is available to authorized users.

B. C. De Simone (✉) · A. D. Quartarolo · S. Cospito · L. Veltri · G. Chidichimo · N. Russo (✉)
Dipartimento di Chimica, Università della Calabria, 87036 Rende, CS, Italy
e-mail: desimone@unical.it

N. Russo
e-mail: nrusso@unical.it

1 Introduction

Electrochromism is the property of a material to reversibly change color in response to an externally applied potential: this effect is connected with the use of electrochromic (EC) molecules, whose UV–visible spectrum exhibits new optical absorption bands when an electron is gained or lost in a redox process [1, 2]. Recently the "electrochromic" definition has been extended to molecules that show modulation of radiation in the near-infrared (NIR) or microwave regions, and "color" can mean response of detectors to these wavelengths, not just of the human eye. In the last years, different molecular systems that show interesting electrochromic behavior have been synthesized, characterized, and proposed for a series of applications [3–6]. Special interest [7] has been devoted to the mixed-valence (MV) systems containing more than one redox state in the same molecule or molecular unit. These systems play an important role in the study of the intramolecular electron transfer called intervalence transfer (IT) by Hush [8]. In MV molecules, IT bands are typically formed in the NIR region from 1,000 to 2,000 nm.

Generally, the used organic or inorganic MV compounds contain at least two redox states linked throughout by bridged molecules and assume a mono-dimensional topology. The properties of MV species essentially depend on the extend of electronic interaction between the redox centers and have been classified by Robin and Day [9] on the basis of this characteristic as negligible (Class I), moderate (Class II), and strong (Class III). Among the proposed organic MV systems that contain *N,N,N′,N′*-tetraphenyl-*p*-phenylenediamine units are particularly interesting for NIR applications because of their intramolecular electron transfer in the oxidized states [10]. The major part of the studies available in the literature on these systems concerns the *N,N,N′,N′*-tetraphenyl-*p*-phenylenediamine (TPPA), for which X-ray diffraction, electrochemical, spectroscopic, and theoretical data are available [11–15]. Furthermore, experimental and theoretical studies on a series of substituted triarylamine redox systems including fluorine-based conjugated polymers containing propeller-shape di-triarylamine with electron-donating propylphenyl groups and other chemical substituents have been recently published [10, 16–22]. Despite these studies, we are far from an exhaustive knowledge of the elementary mechanism that explains the electron-transfer reactions that occur in the mixed-valence systems. More experimental and theoretical works are required to fully explain the phenomenon as well as to propose efficient new molecules with electrochromic properties. In this context and with the aim to contribute to the clarification of the electron-transfer process and to design new systems active in the near-infrared band region, a novel highly stable molecule based on the electroactive tetraphenyl-*p*-phenylenediamine unit, *N,N′*-Bis(4-heptanoylamidophenyl)-*N,N′*-di(4-methoxyphenyl)-1,4-phenylenediamine (TPPAHM) (Scheme 1), has been synthesized and investigated through UV–vis–NIR spectroscopy and density functional theory (DFT) calculations. Our strategy is to incorporate electron-donating substituents at the *para* position of the phenyl groups in order to greatly prevent the coupling reactions by affording stable cation radical [10]. Moreover, the methoxy and amide groups para-substituted make this molecule readily soluble in polar aprotic solvents. The main characteristic of this new triarylamine is the presence of a strong peak in the NIR region when an external electric field is applied. In addition to the applications as smart windows, car mirrors, and display panels [23–25], this NIR EC molecule could be exploitable for heating control in buildings. In this paper, we present the experimental (UV–vis—NIR spectra upon increasing electric field) and theoretical characterization (time-dependent–density functional theory, TD-DFT, spectral simulation) of TPPAHM. In addition, for the purpose of comparison, we report also new spectra and their theoretical analysis for the TPPA system.

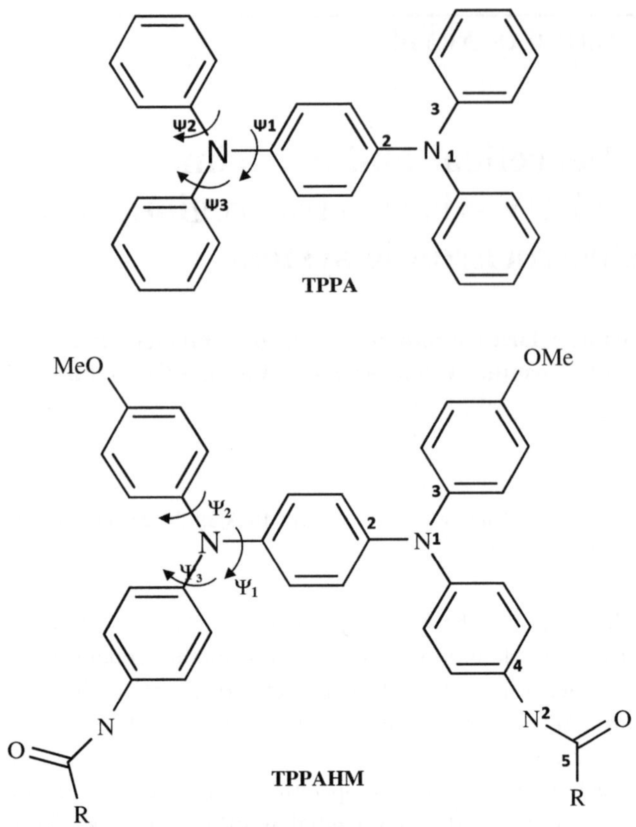

Scheme 1 Molecular structures for TPPA and TPPAHM with main bond length and torsional angle (Ψ_1–Ψ_3) labeling. R = $-CH_3$ and R = $-(CH_2)_5CH_3$ are the substituent groups, respectively, for the model and real systems

2 Experimental and theoretical details

2.1 Synthesis

N,N′-Bis(4-nitrophenyl)-*N,N′*-di(4-methoxyphenyl)-1,4-phenylenediamine was prepared by Ullmann coupling between 1,4-di-iodobenzene and 4-methoxy-4′-nitrodiphenylamine according to previously published procedures [10]. *N,N′*-Bis(4-heptanoylamidophenyl)-*N,N′*-di(4-methoxyphenyl)-1,4-phenylenediamine was prepared in a two-step procedure as described below. All other materials were commercially available and were used without further purification. In a 250-mL three-neck round-bottomed flask equipped with a stirring bar under nitrogen atmosphere, *N,N′*-Bis(4-nitrophenyl)-*N,N′*-di(4-methoxyphenyl)-1,4-phenylenediamine (1.01 g, 1.79 mmol) was dissolved in 10 mL of ethanol and 60 mL of tetrahydrofurane (THF). To the stirred solution was slowly added hydrazine monohydrate (1 g, 20.0 mmol), and the mixture was refluxed. Just when it began to reflux, Raney Ni (about 100 mg) was added. The solution frothed, and as the reaction proceeded, the color changed from yellow to almost colorless. After further 6 h of heating, the hot solution was filtered under nitrogen to remove Raney Ni, and the filtrate was washed with THF. The THF was removed by distillation at 65–70 °C under nitrogen atmosphere. The remaining ethanol solution was cooled to 0 °C under nitrogen in order to have the precipitation of crude *N,N′*-Bis(4-aminophenyl)-*N,N′*-di(4-methoxyphenyl)-1,4-phenylenediamine that was recovered by filtration, washed with cold ethanol, and dried in vacuum. The *N,N′*-Bis(4-aminophenyl)-*N,N′*-di(4-methoxyphenyl)-1,4-phenylenediamine was unstable and was immediately used in the next step without further purification. To a stirred solution of crude *N,N'*-Bis(4-aminophenyl)-*N,N'*-di(4-methoxyphenyl)-1,4-phenylenediamine, prepared as described above, in 16 mL of dry THF, 0.9 mL of NEt_3 was added under nitrogen. To the resulting mixture, a solution of heptanoyl chloride (0.46 mL, 2.98 mmol) in 11 mL of dry THF was dropped, and the mixture was stirred at room temperature. After 20 h, a colorless precipitate was formed. The precipitate was collected by filtration, washed with cold ethanol, and dried in vacuum to give 0.58 g (45 % based on starting *N,N'*-Bis(4-nitrophenyl)-*N,N'*-di (4-methoxyphenyl)-1,4-phenylenediamine) of crude *N,N'*-Bis(4-heptanoylamidophenyl)-*N,N'*-di(4-methoxyphenyl)-1, 4-phenylenediamine, TPPAHM, that was sufficiently pure by TLC (silica gel 60 F_{254}; 1:1 Hexane-AcOEt) and NMR analysis and was used without further purification. Mp 218–221 °C; IR (KBr): $v = 3{,}303$ (m), 2,976 (m), 2,938 (m), 2,756 (m), 2,758 (m), 2,677 (m), 2,491 (m), 1,657 (s), 1,505 (s), 1,398 (m), 1,171 (m), 1,036 (m), 805 (m) cm^{-1}; ^1H NMR (300 MHz, DMSO-d_6): $\delta = 0.86$ (d, $J = 6.4$, 6 H), 1.24–1.40 (m, 8 H), 1.48–1.62 (m, 4 H), 2.21–2.34 (m, 8 H), 3.73 (s, 6 H), 6.80 (s, 4 H), 6.88 (d, $J = 9.1$, 4 H), 6.89 (d, $J = 9.1$, 4 H), 6.98 (d, $J = 9.1$, 4 H), 7.46 (d, $J = 9.1$, 4 H), 9.77 (s br, 2 H); MS (ESI^+, direct infusion): $m/z = 749$ $[(M + Na)^+]$; Anal. Calcd for $C_{46}H_{54}N_4O_4$ (726.95): C 76.00, H 7.49, N 7.71; found C 76.19, H, 7.47, N 7.73.

Melting point analyses were performed on a Linkam (LTS350 stage, TP94 System Controller) at a scan rate of 10 °C/min. ^1H NMR spectra were recorded at 25 °C on a Bruker DPX Avance 300 Spectrometer in DMSO-d_6 solutions at 300 MHz with Me_4Si as internal standard. IR spectra were taken with a JASCO FT-IR 4200 spectrometer, and microanalyses were carried out with a Carlo Erba Elemental Analyzed Mod. 1106. The electrospray ionization mass spectra were acquired by direct infusion on an ABSciex API 2000 mass spectrometer equipped with a turbo ion spray ionization source. The spectra in the positive ion mode were obtained under the following conditions: ion spray voltage (IS) 4,500 V; curtain gas 10 psi; temperature 25 °C; ion source gas (1) 20 psi; declustering and focusing potentials 50 and 400 V, respectively.

2.2 Electrochromic response

The electrochromic behavior of the two EC molecules was investigated by UV–vis–NIR spectroscopy.

Two solutions were prepared dissolving each molecule in aprotic organic solvents. The first solution was prepared dissolving TPPA in *N*-methyl-2-pyrrolidinone (NMP), while the second one is composed of TPPAHM dissolved in *N,N*-dimethylformamide (DMF). A small quantity of these solutions were introduced by capillarity in homemade cells, whose thickness was set to be about 30 μm by means of glass spheres. Cell walls had an indium tin oxide (ITO) conductive substrate in order to perform electro-optical characterization.

UV–vis–NIR spectra were recorded with a Thermo Scientific GENESYS 10S UV–Vis spectrophotometer. In our experimental setup, the cells have been powered by two electrodes fixed to the opposite extremes of the cell. The potential difference was supplied by means of an Amel 2049 model potentiostat. Measurements were performed at 25 °C.

In order to verify the oxidation reversibility, we applied to the sample a series of square potential pulses monitoring the absorbance at fixed wavelength ($\lambda = 846$ nm for TPPA and $\lambda = 1{,}040$ nm for TPPAHM) as a function of time. The absorbance increased during the pulse application and rapidly returned to its initial state once the pulse was removed.

2.3 Computational details

All calculations were performed with the TURBOMOLE V-6.0 software package [26]. Due to the medium–high molecular size of TPPAHM that requires high computational costs, the computation has been performed in a model system in which the $-(CH_2)_5CH_3$ moiety is substituted with the $-CH_3$ one. This substitution does not affect the spectral properties that are essentially due to the aromatic core. Geometry optimizations, without imposing symmetry constraints, as well as vibrational frequency analysis were carried out at the density functional level of theory in conjunction with the non-empirical PBE0 hybrid functional [27, 28] that adds up a fixed amount of Hartree–Fock exchange energy (25 %) to the gradient-corrected PBE exchange–correlation functional. The SV(P) double-ζ quality basis set containing polarization functions on C, N, and O atoms ([3s2p1d]/[2s]) of Ahlrichs et al. [29, 30] was used for all the computations. Vibrational frequency analysis confirmed each optimized structure as absolute energy minimum on the relative potential energy surfaces. The closed-shell systems (neutral species) were treated with the restricted Kohn–Sham (RKS) formalism, while for the open-shell systems (cation and di-cation species), unrestricted Kohn–Sham (UKS) calculations were performed. For the open-shell compounds, the $\langle S^2 \rangle$ values of the ground-state Kohn–Sham determinant computed at the PBE0/SV(P) level of theory were found in the range 0.76–0.77 (for the cation) and 2.02–2.05 (for the di-cation), indicating a small effect of spin contamination. The lowest twenty vertical excitation energies were calculated by time-dependent density-functional linear response theory [31, 32] (TD-DFT) on the PBE0/SV(P) optimized geometries. The reliability of the excitation energies obtained from the combination of SV(P) basis set and hybrid functional (PBE0) has been previously tested for a series of molecular

Table 1 Main geometrical parameters for TPPA and TPPAHM neutral, radical cation, and di-cation species

Parameter	TPPA	$TPPA^+$	$TPPA^{2+}$	TPPAHM	$TPPAHM^+$	$TPPAHM^{2+}$
N1–C2	1.412 (1.433)	1.374 (1.363)	1.419	1.410	1.380	1.418
N1–C3	1.410 (1.404)	1.422 (1.444)	1.399	1.408	1.415	1.400
C4–N2	/	/	/	1.403	1.390	1.371
N2–C5	/	/	/	1.506	1.510	1.409
C5–O	/	/	/	1.371	1.385	1.204
C2–N1–C3	119.8 (116.9)	121.1 (120.9)	119.7	119.5	120.7	119.5
Ψ_1	41.8 (53.8)	24.7 (16.0)	43.0	40.7	26.9	43.6
Ψ_2	40.1 (55.7)	48.5 (49.1)	36.4	44.5	48.6	38.2
Ψ_3	39.4 (23.0)	48.4 (64.8)	36.5	37.6	44.5	34.2

Distances are in Å, and valence and dihedral angles in degree. The experimental results (from ref. [11]) are given in parenthesis

systems, giving a mean absolute deviation from experiment within 0.2–0.3 eV [33–37]. Solvent effects on geometries and excitation energies, due to electrostatic interaction in polar solvent, were taken into account with the conductor-like screening model [38, 39] assuming the dielectric constant value of 37.2 and 32.2 for DMF and NMP, respectively, and default parameters for the cavity generation. The simulation of the electronic spectra band profile was obtained by a sum of Gaussian shape functions centred at each excitation energy and with a constant full width at half maximum of 0.3 eV.

Default threshold for the geometry optimizations and for XC grid integrations has been used. In short, the used protocol is very similar to that suggested by Renz et al. [40] for the study of mixed-valence radical cations.

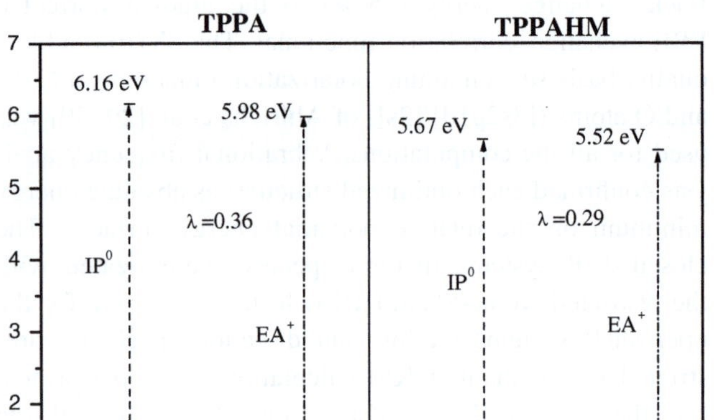

Fig. 1 Vertical and adiabatic relative energies (eV) at neutral and cation geometries for TPPA and TPPAHM molecules. For each compound, the neutral geometry is taken as energy reference. IP^0, EA^+, and λ are the ionization potentials (at neutral geometry), electron affinity (at cation geometry), and the reorganization energy

3 Results and discussion

The two studied molecules are shown in Scheme 1. Both show N–N distance of 5.654 Å, and a number of bonds linking the redox center equal to 5. Following the indication of Nelsen et al. [41], even though the distance between the two redox centers is short, these species can have a

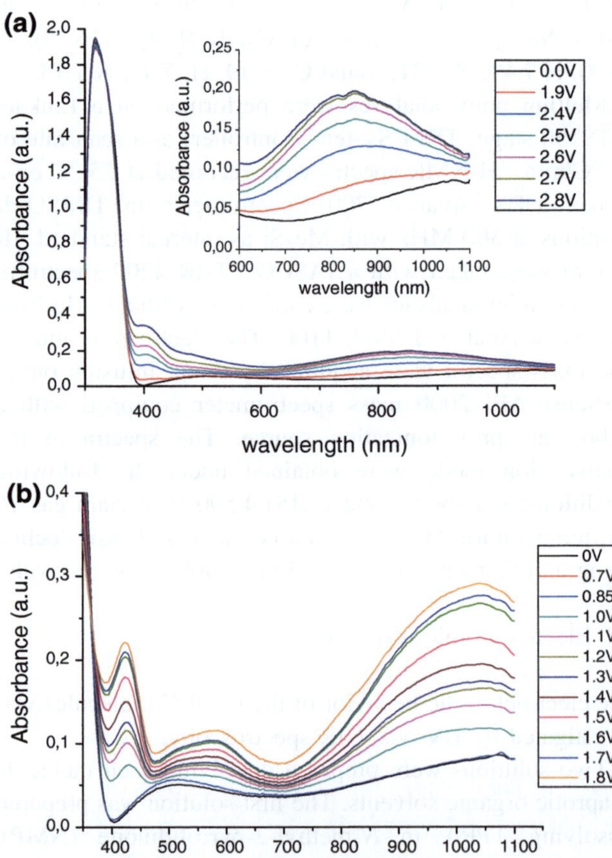

Fig. 2 **a** vis–NIR absorption spectra of 30 mM TPPA in NMP solution under different applied potentials; **b** vis–NIR absorption spectra of 1.5 mM TPPAHM in DMF solution under different applied potentials

Table 2 Computed vertical excitation energies, main configuration, and oscillator strengths f for molecule TPPA in *N,N*-dimethylformamide ($\varepsilon = 37.2$) and molecule TPPAHM in *N*-methyl-2-pyrrolidinone ($\varepsilon = 32.2$)

TPPA				TPPAHM			*Exptl.*
State	ΔE (eV, nm)	Main configuration (%)	f	ΔE (eV, nm)	Main configuration (%)	f	
1	3.65, 338	H → L (94.7)	0.0102	3.40, 365	H → L (95.6)	0.0046	318 (1.9257)
2	3.70, 335	H → L + 1 (97.6)	0.6169	3.48, 356	H → L + 1 (93.7)	0.7344	
3	3.97, 312	H → L + 2 (93.7)	0.1988	3.71, 335	H → L + 2 (84.0)	0.0192	
4	3.99,311	H → L + 3 (56.5)	0.0385	3.78, 328	H → L + 2 (94.0)	0.3432	
5	4.04, 307	H → L + 4 (50.9)	0.1608	3.88, 319	H → L + 3 (90.4)	0.3094	
Cation							
1	1.47, 842	H-1 → L β (96.8)	0.4247	1.21, 1025	H-1 → L β (97.9)	0.4812	848 (0.1983)
2	2.42, 513	H-6 → L β (69.6)	0.0038	2.09, 593	H-3 → L β (97.5)	0.0520	
3	2.46, 504	H-3 → L β (97.6)	0.0000	2.14, 580	H-4 → L β (97.8)	0.1668	
4	2.66, 466	H-9 → L β (97.7)	0.0176	2.63,471	H-17 → L β (63.2)	0.0004	
5	3.37, 368	H → L + 1 α (74.3)	0.3800	3.14, 395	H → L + 1 α (43.7)	0.2058	
Di-cation							
1	1.69, 733	H-3 → L β (83.6)	0.0272	1.53, 810	H-2 → L β (95.0)	0.1916	
2	1.72, 721	H-4 → L + 1 β (81.8)	0.0000	1.58, 786	H-3 → L β (93.9)	0.2547	
3	1.77, 699	H-2 → L β (66.7)	0.0262	2.03, 611	H-2 → L + 1 β (92.9)	0.0927	
4	1.88, 658	H-10 → L β (59.3)	0.1911	2.07, 599	H-3 → L + 1 β (94.3)	0.0823	
5	2.02, 615	H-8 → L β (58.8)	0.1022	2.11, 588	H-6 → L β (87.1)	0.2745	
6	2.05, 605	H-5 → L β (64.8)	0.1276	2.23, 556	H-11 → L β (71.3)	0.0003	
7	2.29, 541	H-14 → L β (86.1)	0.2404	2.24, 553	H-12 → L β (67.3)	0.0333	

good electronic coupling due to the presence of the five interlink bonds.

The computed and measured properties will be reported and discussed in specific sections.

3.1 Structural properties

The main geometrical parameters for the ground state of the neutral, cationic, and bi-cationic species of both the considered systems are reported in Table 1. The molecular structures computed with and without symmetry constraints are almost identical. In all cases, the nitrogen centers, surrounded by the three phenyl groups, are trigonal planar, and the CNC valence angle values are about 120°. This means that the central nitrogen atoms are sp^2 hybridized, and the lone electron pairs are involved in conjugation with the phenyl moieties.

In TPPA, the N1–C2 (see Scheme 1 for the definitions) bond lengths assume values of 1.410, 1.374, and 1.419 Å for the neutral, radical cation, and bi-radical species, respectively. The same bond distances in TPPAHM assume very similar values (1.410, 1.380, and 1.418 Å for neutral, radical cation, and bi-cation, respectively). For the neutral species, this distance is slightly larger than in aromatic amines and shorter than in aliphatic ones. We underline the sensible decrease of this bond distance, in going from the neutral to the radical cationic system that occurs in both molecules. In the second ionization process, the N1–C2 bond lengths assume almost the same values with that in the neutral species for both TPPA and TPPAHM systems. Other sensible variations that the ionization process induces in the geometrical parameters are found in the torsional angles. As indicated in the Table 1, the major variation concerns the Ψ1 angle that in the neutral form assumes a value of 41.8° (TPPA) and 40.7° (TPPAHM) that becomes 24.7° and 26.9° for TPPA and TPPAHM radical cations, respectively. Concerning the N1–C3 bond, connecting the nitrogen atom with the peripheral phenyl moiety, our results show that during the first ionization process, it slightly increases, while decreases in the second ionization. The C4–N2 distance in TPPAHM decreases during the formation of the radical cation and continues to decrease upon the second ionization. From these data, it seems that the N1 center is more involved in the first ionization, while the N2 atom is more involved in the second ionization process. Looking at the C5–O bond, we note that both ionization processes cause a decrease in the length. The C–C distances of the phenyl rings correspond to that expected for delocalized aromatic moieties.

Comparison between the computed and crystallographic structures [11] for neutral and radical TPPA species is satisfactory for the bond distances and valence angles, while discrepancies are found for the dihedral angles (see Table 1). The same trend has been noted in a previous DFT

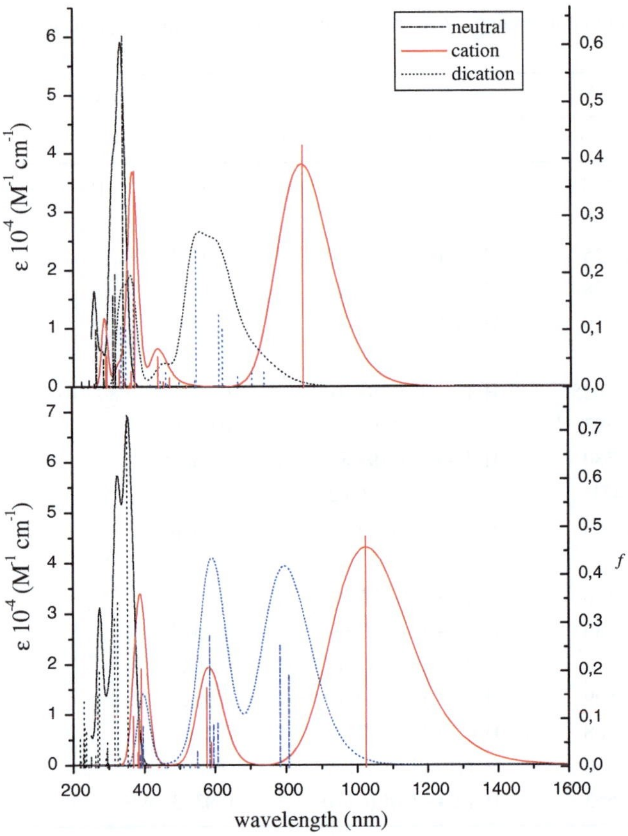

Fig. 3 TDDFT simulated spectra of TPPA (*top*) and TPPAHM (*bottom*)

study employing a different exchange–correlation functional (BPW91) [11]. If we consider that in solid state, there is a different environment (intermolecular interactions) and that around the considered dihedral angles the potential energy profile is flat, these discrepancies can be justified.

The theoretical predictions for the geometrical parameters of TPPAHM, for which experimental data are absent, follow the trends of the parent TPPA and give indications regarding the geometry of the substituted moiety as well as a useful guideline for future crystallographic measurements.

The hole transport process can be approximated as an electron transfer from a neutral molecule N to a neighboring radical cation N^{+}:$N(a) + N(b)^{+} \rightarrow N(a)^{+} + N(b)$.

As it is evident the reactants and the products in this reaction are the same, and the reaction Gibbs free energy is zero. It was pointed out [42, 43] that, neglecting the contributions arising from the polarization of the medium and from molecular vibrations, the activation energy (ΔG^*) for this intermolecular charge transfer should be related with the re-organization energy (λ) through the relation $\Delta G^* = \lambda/4$. The re-organization energy is directly associated with the geometric rearrangement in going from neutral to charged systems and from charged to neutral ones. From the computations of the vertical and adiabatic ionization potentials and electron affinities (see Fig. 1), we obtain a reorganization energy values of 0.36 and 0.29 eV for TPPA and TPPAHM, respectively. These values account for a sensible geometrical relaxation process occurring in both systems as previously discussed. In a recent study [42], the λ values have been associated with the electron–hole mobility, observing that a decrease in the reorganization energy corresponds to an increase in the electron–hole mobility. On the basis of our results, we can predict a higher electron–hole mobility for the newly synthesized TPPAHM molecule compared to its TPPA counterpart.

3.2 UV–vis–NIR spectra

Spectroelectrochemical experiments have been performed in order to evaluate the optical properties of TPPA and TPPAHM electrochromic systems. The obtained spectra are shown in Fig. 2a (TPPA) and b (TPPAHM).

Without the applied electric field, the optical spectrum of TPPA is characterized by only one strong absorption peak in the B-band region. Increasing the applied voltage from 0.0 to 2.4 V, a new broad band centered around 820 nm gradually increases in intensity, and the maximum absorption peak is located at 840 nm when the applied voltage reaches its higher value (2.8 V). This band is due to an intervalence charge-transfer (IV-CT) excitation with a photoexcited electron transfer from a neutral triphenylamine center to the corresponding radical cation ones. Upon oxidation, another less intense transition appears at 402 nm. The computed TD-DFT spectrum (see Table 2) of the neutral TPPA gives also strong intensity transitions in the B-region of the spectrum (the maximum peak at 335 nm has a value of oscillator strength of 0.62). The simulated spectra show that the main peak is due to a HOMO → LUMO + 1 transition (see Table 2; Figs. 3, 4). In the simulated spectrum of the TPPA radical cation, the B-band is shifted to the lower energy value (368 nm) and the IV-CT peak is located at 842 nm with an oscillator strength value of 0.42. The analysis of the orbital transition (Fig. 4) clearly indicates its HOMO-1 → LUMO nature (96.8 %). The peak at 466 nm is assigned to a transition from an internal orbital (HOMO-9) to the LUMO one.

The newly synthesized TPPAHM molecule differs from TPPA due to the presence in *para* position of the external phenyl rings of electron-donating groups (O–CH3 and –NCOR). It was previously shown that the presence of these meta O–CH3 groups in triarylamine-like electrochromic systems prevents coupling reactions and lowers the oxidation potential [19, 44]. Thus, we will expect a pronounced NIR response in this newly synthesized

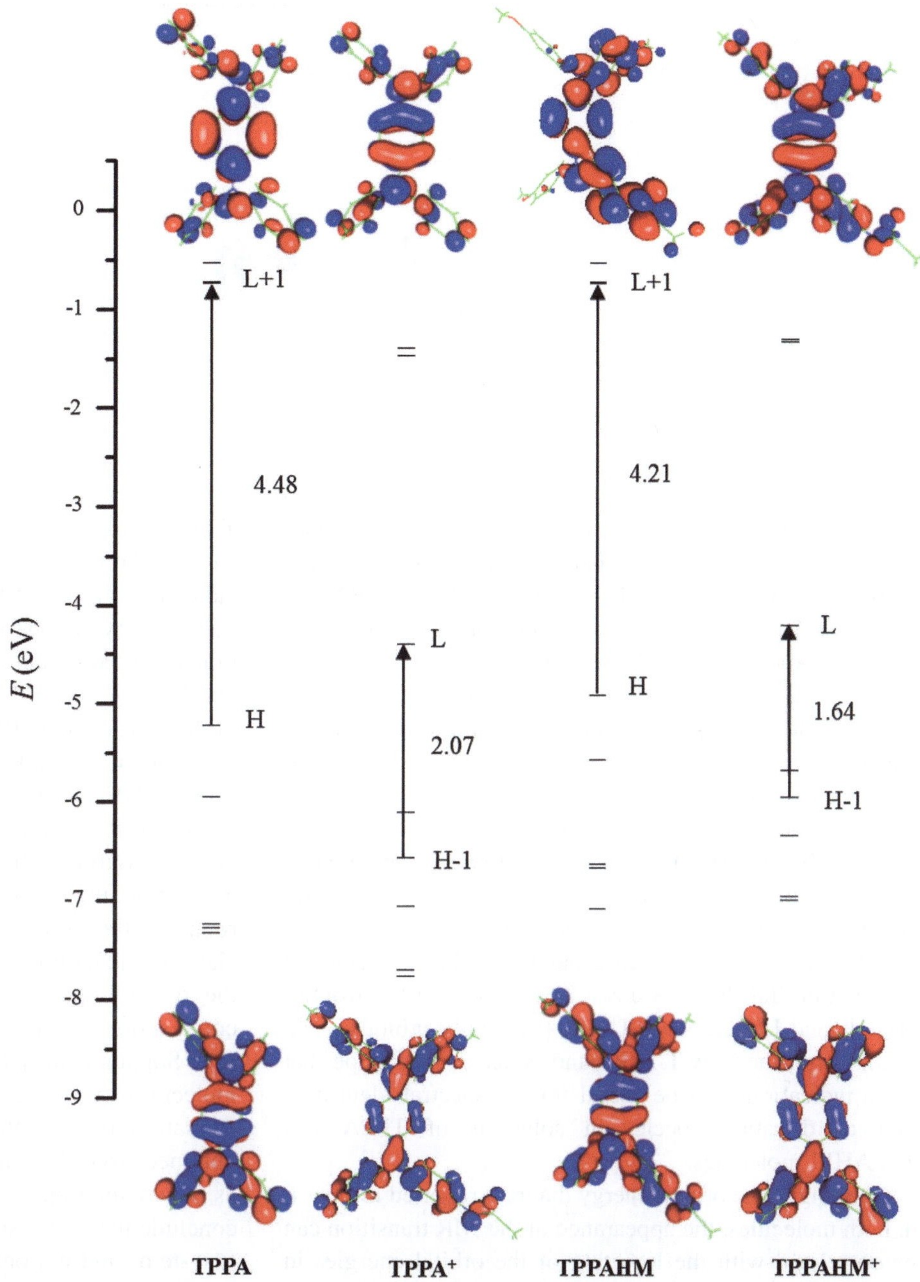

Fig. 4 Partial energetic diagram (eV) for the highest occupied and lowest virtual unoccupied molecular orbitals of molecules TPPA and TPPAHM (neutral and cationic forms). The orbital compositions are reported for the most intense transitions

system. The results of the spectrochemical experiments are reported in Fig. 2. The spectrum feature, in the neutral species, essentially shows the presence of an intense B-band at about 360 nm. The computed counterpart (see Table 2) gives almost the same spectral features with one strong peak due to the HOMO → LUMO + 1 π–π* orbital transition at 356 nm (see Fig. 4). When the applied electric field reaches the value of 0.7 V, two new bands appear in the visible region. Upon oxidation (increasing the applied voltage up to 1.8 V), the intensity of these new absorption transitions increases, having the maximum absorption wavelength at 424 and 580 nm. The broad absorption NIR band centered at 1,040 nm was the characteristic result due to IV-CT excitation associated with the electron transfer during the oxidation of TPPAHM. The simulated TD-DFT spectra agree well with the experimental evidence as shown in Table 2 and Fig. 3. In particular, the DFT NIR peak reaches the maximum intensity (0.48) at 1,025 nm, and the transition is mainly (97.9 %) HOMO-1 → LUMO in nature (Fig. 4). The transition at 580 nm and with an oscillator strength value of 0.17 is essentially due to the HOMO-4 → LUMO transition.

Although in our experimental conditions we do not note the presence of a second oxidation process until an applied voltage of about 5 V, we have simulated the spectra for both TPPA and TPPAHM di-cations as reported in Table 2

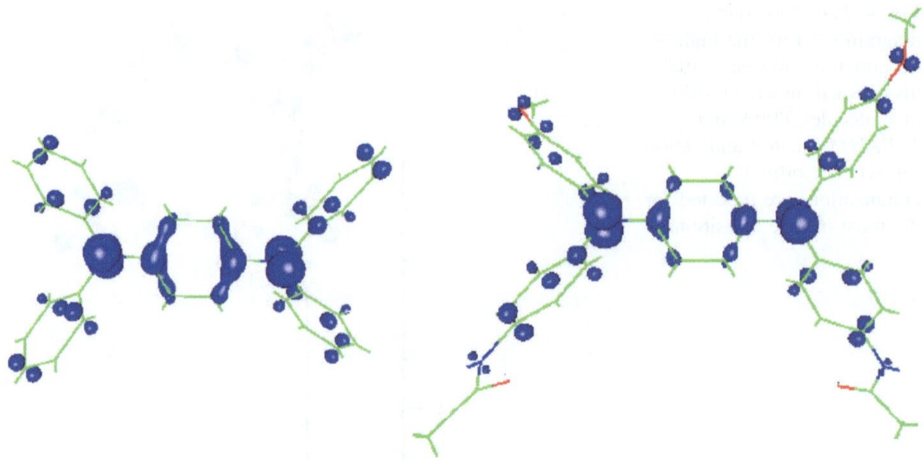

Fig. 5 Spin density contour plot for $TPPA^+$ (*left*) and $TPPAHM^+$ (*right*) (isodensity value of 0.005 a.u.)

and Fig. 3. The TPPA di-cation shows essentially a broad peak in the visible region that contains different transitions between the internal orbital and the LUMO and LUMO + 1 orbital with the maximum intensity located at 541 nm. Another less intense band is present around 280 nm. As in previous experiments, on other tetraphenylamine derivatives [10], also in our computed spectrum, the NIR band disappears when the second oxidation process occurs. Contrary to that of TPPA, the spectrum of TPPHM di-cation has a different behavior showing three different bands. The first and less intense is centered at 400 nm, the second reaches its maximum intensity (0.88) at 599 nm, and the last has oscillator strength of 0.26 at 786 nm. The last two intense bands are due to transition between orbital that lies down in energy than the frontier orbitals and LUMO and LUMO + 1 empty orbitals. Also in this case, the IV-CT NIR band is absent. We hope that these indications can be useful for the spectral identification of di-cation species in solutions of TPPA and TPPAHM molecules.

Looking at the orbital energy diagram reported in Fig. 4, in both molecules, the appearance of the NIR transition can be correlated with the lowering of the orbital energies in going from the neutral to the cation radical species.

The oxidation mechanism of triarylamine-like electrochromic systems has been the subject of many previous publications [20, 45–47]. In the major part of these works, it is supposed that the first oxidation is due to the removal of an electron from the nitrogen atom with the larger electron density, while the second one derives from the removal of a second electron from the other nitrogen atom. Recently, Wu et al. [16], in order to explain the spectrochemical behavior of a series of fluorine-based conjugated polymers containing propeller-shape di-triarylamine, have proposed an other mechanism for which the first electron must be removed by the HOMO orbital instead from the nitrogen lone pair and the second from the SOMO orbital. This proposal is based on coupling the spectrochemical data by the net charges of neutral and oxidized states calculated at the B3LYP/6-31G* level of theory. Following this strategy, we have computed, by using our computational protocol, the natural bond charges and the spin density (Fig. 5) for the neutral, cation radical, and di-cation of the TPPA and TPPAHM studied systems. From our data, it is evident that in both systems, although the N atom loss 0.11|e| (TPPA) and 0.10|e| (TPPAHM) in going from the neutral to the cation radical, these quantities are not sufficient to assert that the oxidation is mainly generated by this atom. Looking at the other net charges of the atoms surrounding the nitrogen, we note other loss of charge especially in the carbon located in the phenyl moiety that links the two nitrogen centers. Furthermore, from the orbital composition of the LUMO orbital in the neutral species (See Supplementary Informations Fig. 6), we note, in both molecules, a strong involvement of the nitrogen in the formation of the phenyl π system. Similar results are obtained from the analysis of the spin density distribution as shown in Fig. 5. On the basis of these data, we can conclude that the first oxidation involves the entire LUMO π systems and not only the nitrogen atom. Concerning the second oxidation, the analysis of Table 2, as well as that of Figs. 4 and 5, suggests that the electron must be removed from the SOMO orbital created in the first oxidation process.

4 Summary

In this study, we have employed spectroelectrochemical experiments and DFT first principles computations in order to characterize a newly synthesized triarylamine electrochromic system (TPPAHM). In addition, also the TPPA has been considered. The UV–vis–NIR spectra have been registered with and without an applied electric field at different voltages. On the basis of our results, we can draw the following conclusions:

- Both radical cations have a delocalized IV state in which the unpaired electron is fully delocalized over the whole molecules;
- The two oxidation processes occur with the involvement of the delocalized HOMO orbital that releases an electron in the first ionization, while in the second, the electron is given by the SOMO orbital;
- The presence of electron-rich moieties on the TPPAHM shifts the NIR resonance at higher values with respect to the TPPA;
- Although the used hybrid exchange–correlation functional can give an overestimation of the charge delocalization, the agreement between the calculated and experimental UV–vis–NIR spectral behavior is very satisfactory;
- The presence of a low applied voltage is able to ionize reversibly both molecules.

Acknowledgments The University of Calabria and MIUR (PRIN 2008ALLB79) are gratefully acknowledged for financial support.

References

1. Platt JR (1961) J Chem Phys 34:862
2. Monk PMS, Mortimer RJ, Rosseinsky DR (1995) Electrochromism: fundamentals and applications. VCH Weinheim, Germany
3. Monk PMS, Mortimer RJ, Rosseinsky DR (2007) Electrochromism and electrochromic devices. Cambridge University Press, Cambridge
4. Beaujuge PM, Reynolds JR (2010) Chem Rev 110:268
5. Beaupre S, Breton AC, Dumas J, Leclerc M (2009) Chem Mater 21:1504
6. Zheng J, Qiao W, Wan X, Gao JP, Wang ZY (2008) Chem Mater 20:6163
7. Barbara PF, Meyer TJ, Ratner MA (1996) J Phys Chem 100: 13148
8. Hush NS (1961) Trans Faraday Soc 57:557
9. Robin M, Day P (1967) Adv Inorg Radiochem 10:247
10. Yen HJ, Liou GS (2009) Chem Mater 21:4062
11. Szeghalmi AV, Erdmann M, Engel V, Schmitt M, Amthor S, Kriegich V, Noll G, Stahl R, Lambert C, Leusser D, Stalke D, Zabel M, Popp J (2004) J Am Chem Soc 126:7834
12. Coropceanu V, Gruhn NE, Barlow S, Lambert C, Durivage JC, Bill TG, Noll G, Marder SR, Brédas JL (2004) J Am Chem Soc 126:2727
13. Ramos AM, Meskers SCJ, Van Hal PA, Knol J, Hummelen JC, Janssen RAJ (2003) J Phys Chem A 107:9269
14. Zhang H-G, Yu W-T, Wang L, Liu Z, Tao X-T, Jiang M-H (2005) Z Kristallogr 220:101
15. Lambert C, Noll G (1999) J Am Chem Soc 121:8434
16. Wu H-Y, Wang K-L, Liaw D-J, Lee K-R, Lai J-Y, Chen C-L (2010) J Polym Sci Part A: Polym Chem 48:3913
17. Chiu KY, Su T-H, Huang CW, Liou G-S, Cheng S-H (2005) J Electroan Chem 578:283
18. Yeh SJ, Tsai CY, Huang C-Y, Liou G-S, Cheng S-H (2003) Electrochem Commun 5:373
19. Hirao Y, Ito A, Tanaka K (2007) J Phys Chem A 111:2951
20. Selby TD, Kim K-Y, Blackstock SC (2002) Chem Mater 14:1685
21. Barlow S, Risko C, Coropceanu V, Tucker NM, Jones SC, Levi Z, Khrstalev VN, Antipin MY, Kinnibrugh TL, Timofeeva T, Marder SR (2005) Chem Commun 6:764
22. Low PJ, Paterson MAJ, Puschmann H, Goeta AE, Howard JAK, Lambert C, Cherryman JC, Tackley DR, Leeming S, Brown B (2004) Chem Eur J 10:83
23. Tsutsumi H, Nakagawa Y, Tamura K (1995) Sol Energy Mater Sol Cells 39:341
24. Leventis N, Chung Y (1998) U.S. Patent 5818636
25. Nishikitani Y, Uchida S, Asano T, Minami M, Oshima S, Ikai K, Kubo T (2008) J Phys Chem C 112:4372
26. Ahlrichs R, Bär M, Häser M, Horn H, Kölmel C (1989) Chem Phys Lett 162:165
27. Ernzerhof M, Scuseria GE (1999) J Chem Phys 110:5029
28. Adamo C, Barone V (1999) J Chem Phys 110:6158
29. Schaefer A, Horn H, Ahlrichs R (1992) J Chem Phys 97:2571
30. Eichkorn K, Weigend F, Treutler O, Ahlrichs R (1997) Theor Chem Acc 97:119
31. Bauersnschmitt R, Ahlrichs R (1996) Chem Phys Lett 256:454
32. Casida ME (1995) In: Chong DP (ed) Recent advances in density functional methods Part I. Word Scientific, Singapore
33. Quartarolo AD, Russo N, Sicilia E (2006) Chem Eur J 12:6797
34. Quartarolo AD, Sicilia E, Russo N (2009) J Chem Theory Comput 5:1849
35. Quartarolo AD, Lanzo I, Sicilia E, Russo N (2009) Phys Chem Chem Phys 11:4586
36. Lanzo I, Quartarolo AD, Russo N, Sicilia E (2009) Photochem Photobiol Sci 8:386
37. Jacquemin D, Mennucci B, Adamo C (2011) Phys Chem Chem Phys 13:16987
38. Klamt A, Jonas V (1996) J Chem Phys 105:9972
39. Klamt A (1996) J Phys Chem 100:3349
40. Renz M, Thellacker K, Lambert C, Kaupp M (2009) J Am Chem Soc 131:16292
41. Nelsen SF, Ismagilov RF, Trieber DA (1997) Science 278:846
42. Sakarnane K, Motoda M, Sugimoto M, Sakaki F (1999) J Phys Chem A 103:5551
43. Malagoli M, Brédas JL (2000) Chem Phys Lett 327:13
44. Yen H-J, Lin H-Y, Liou G-S (2011) Chem Mater 23:1874
45. Chen WH, Wang KL, Liaw DJ, Lee KR, Lai JY (2010) Macromolecules 43:2236
46. Grazulevicius JV, Strohriegl P, Pielichowski J, Pielichowski K (2003) Prog Polym Sci 28:1297
47. Ito A, Ino H, Tanaka K, Kanemoto K, Kato T (2002) Org Chem 67:491

MIX
Papier aus verantwortungsvollen Quellen
Paper from responsible sources
FSC® C105338

If you have any concerns about our products,
you can contact us on
ProductSafety@springernature.com

In case Publisher is established outside the EU,
the EU authorized representative is:
Springer Nature Customer Service Center GmbH
Europaplatz 3, 69115 Heidelberg, Germany

Printed by Libri Plureos GmbH
in Hamburg, Germany